AF576053

Developing Temperature-Resilient Plants: Responses and Mitigation Strategies

Developing Temperature-Resilient Plants: Responses and Mitigation Strategies

Editors

Channapatna S. Prakash
Ali Raza
Xiling Zou
Daojie Wang

MDPI • Basel • Beijing • Wuhan • Barcelona • Belgrade • Manchester • Tokyo • Cluj • Tianjin

Editors

Channapatna S. Prakash
Tuskegee University
Tuskegee
AL
USA

Ali Raza
Chinese Academy of
Agricultural Sciences (CAAS)
Wuhan
China

Xiling Zou
Chinese Academy of
Agricultural Sciences (CAAS)
Wuhan
China

Daojie Wang
Henan University
Henan
China

Editorial Office
MDPI
St. Alban-Anlage 66
4052 Basel, Switzerland

This is a reprint of articles from the Special Issue published online in the open access journal *Agronomy* (ISSN 2073-4395) (available at: https://www.mdpi.com/journal/agronomy/special_issues/temperature_resilient).

For citation purposes, cite each article independently as indicated on the article page online and as indicated below:

LastName, A.A.; LastName, B.B.; LastName, C.C. Article Title. *Journal Name* **Year**, *Volume Number*, Page Range.

ISBN 978-3-0365-8256-6 (Hbk)
ISBN 978-3-0365-8257-3 (PDF)

Contents

Preface to "Developing Temperature-Resilient Plants: Responses and Mitigation Strategies"

Climate-change-induced extreme temperatures significantly impact crop production worldwide. Therefore, this Special Issue is focused on "Developing Temperature-Resilient Plants: Responses and Mitigation Strategies". Several authors have contributed novel research and review articles to this Special Issue, covering various directions of genetics, genomics, agronomic and physiological interventions in developing temperature-resilient plants.

Channapatna S. Prakash, Ali Raza, Xiling Zou, and Daojie Wang
Editors

Editorial

Developing Temperature-Resilient Plants: A Matter of Present and Future Concern for Sustainable Agriculture

Ali Raza [1,*], Daojie Wang [2,*], Xiling Zou [3,*] and Channapatna S. Prakash [4,*]

1 College of Agriculture, Fujian Agriculture and Forestry University (FAFU), Fuzhou 350002, China
2 College of Agriculture, Henan University, Jin Ming Avenue, Kaifeng 475004, China
3 Oil Crops Research Institute, Chinese Academy of Agricultural Sciences (CAAS), Wuhan 430062, China
4 College of Arts & Sciences, Tuskegee University, Tuskegee, AL 36088, USA
* Correspondence: alirazamughal143@gmail.com (A.R.); drdaojiewang@163.com (D.W.); zouxiling01@caas.cn (X.Z.); cprakash@tuskegee.edu (C.S.P.)

1. Introduction

Plants are decisive for nurturing life on Earth, but climate change threatens global food security, poverty decrease, and sustainable agriculture [1,2]. Climate change events, such as altered rainfall patterns, mega-fires, droughts, soil salinity, floods, extreme temperatures, and spreading pests and diseases, are becoming more frequent and severe. These events directly and indirectly influence sustainable agriculture, food security, and people's livelihoods (FAO, Climate Change; https://www.fao.org/climate-change/en/, accessed on 8 March 2023). According to current climate change predictions, extreme temperatures exert a significant risk to the sustainability of major crops globally. These extreme temperatures hinder plant growth and development, trigger damage, and eventually cause yield shortfalls, making it difficult to reach the "ZERO HUNGER (Sustainable Development Goal 2)" (FAO-UN, https://www.fao.org/sustainable-development-goals/goals/goal-2/en/, accessed on 8 March 2023).

Plants cannot avoid temperature stress by relocating as they are immobile [3,4]. Consequently, plants have developed diverse mechanisms to acclimatize to stressful environments by changing their developmental, physio-biochemical, and molecular activities [1,4–8]. To fast-track the development of stress-resilient crops for sustainable agriculture, discovering approaches to improve plant stress tolerance is a vital mission for plant biologists worldwide. Therefore, this Special Issue entitled "Developing Temperature-Resilient Plants: Responses and Mitigation Strategies" (https://www.mdpi.com/journal/agronomy/special_issues/temperature_resilient, accessed on 8 March 2023) was designed at the right time to collect the recent scientific advances on different mitigation strategies as well as stress adaptation and tolerance mechanisms to support the rising population. Twenty papers are published in this Special Issue, including 15 research and 5 review articles authored by a diverse group of scientists worldwide. Based on the published articles, this editorial presented the scientific advances in two sections, i.e., (1) genetics and genomics interventions and (2) agronomic and physiological interventions in developing temperature-resilient plants for the sustainable future.

Citation: Raza, A.; Wang, D.; Zou, X.; Prakash, C.S. Developing Temperature-Resilient Plants: A Matter of Present and Future Concern for Sustainable Agriculture. *Agronomy* **2023**, *13*, 1006. https://doi.org/10.3390/agronomy13041006

Received: 22 March 2023
Accepted: 26 March 2023
Published: 29 March 2023

2. Genetics and Genomics Interventions in Developing Temperature-Resilient Plants

Heat stress at different developmental stages substantially impacts the production of cotton (*Gossypium hirsutum* L.). Recent advances in genetics and genomics tools have revolutionized crop improvement against temperature stress by identifying stress-responsive regions, genes, and pathways. In this context, Rani et al. [9] aimed to discover and map the quantitative trait loci (QTLs) associated with heat tolerance in cotton using a microsatellite marker approach. They used an F2 population originating from a cross of MNH-886 (heat-tolerant) and MNH-814 (heat-sensitive) varieties and mapped different heat tolerance-related QTLs during diverse morphological stages in cotton. Authors have identified a total of 17 QTLs which are located on different chromosomes and are associated with other traits, including heat stress. They also discovered that some QTLs were highly significant and accounted for a considerable fraction of the phenotypic variation (7.76%–36.62%). In short, these findings may facilitate marker-assisted breeding on cotton heat tolerance and contribute to developing heat-tolerant cotton varieties [9].

Improving quantitative attributes associated with crop yield, quality, and stress tolerance is the primary objective of breeding programs. Hence, Arslan et al. [10] examine the genetic diversity of grass pea (*Lathyrus sativus* L.) germplasm (94 accessions), a legume crop is grown under low- and high-land conditions in Turkey, to identify genetic markers associated with stress tolerance and yield-related traits. Authors have observed high genetic diversity among the grass pea accessions, and there were significant variations between genotypes for all agronomic attributes in low-land locations. The information on differences in agronomic, quality, and forage attributes identified in this study presented valuable genetic resources. The parental genotypes with preferred traits can be utilized in grass pea improvement programs for developing stress-resilient new cultivars [10].

It is vital to predict approved green super rice (GSR) to uphold the high production of rice (*Oryza sativa* L.) in Pakistan [11]. Zaid et al. [11] analyzed the genetic diversity, heritability, and stability of yield-connected traits of GSR grown under multi-environmental conditions (eight regions) in Pakistan. This research revealed three stable GSR lines (GSR 305, GSR 252, and GSR 112) with the lowest stability values in univariate stability data. It is observed that GSR 48 demonstrated the highest stability than all other lines in the univariate model across the two years for grain yield and associated attributes data. It is also determined that multivariate parametric stability models are suitable for choosing the most appropriate and stable GSR lines for particular and diverse environments. AMMI and GGE biplot analysis classified GSR 305 and GSR 252 as the most stable genotypes across eight examined locations. Furthermore, Swat, Narowal, and Muzaffargarh are the best locations to commercialize GSR lines in Pakistan [11].

Genome-wide identification and characterization of various gene families help to uncover their putative roles in plant growth/development and stress tolerance [12–15]. For instance, Kilwake et al. [12] identified 16 members of S-adenosyl-L-methionine synthetase (SAMS) gene family in upland cotton, named *GhSAMS*. Different bioinformatics tools were used to characterize their properties, and structures and predict their putative functions. Gene expression analysis showed that the *GhSAMS2* gene was highly induced by salinity and drought stress. The *ghSAMS2* gene was knockdown using the virus-induced gene silencing method, and the resultant knock-down plants demonstrated sensitivity to both salinity and drought stresses. The knock-down plants revealed *GhSAMS2* was involved in cotton's growth and physiological performances. These discoveries deliver perceptions into *SAMS* gene structure, classification, and roles in abiotic stress responses in upland cotton [12]. Maize (*Zea mays* L.) is a major cash crop grown globally; however, its growth and yield are impacted by numerous stresses. To get insight into the mechanisms of stress responses by maize, Haider et al. [13] performed a comprehensive genome-wide analysis. They identified 25 Heat shock transcription factors (HSFs) in the maize (*ZmHSFs*) genome. A diverse set of *in silico* tools were used to characterize the *ZmHSFs*, including chromosomal location, gene structure, phylogenetic analysis, motif analysis, localization, protein–protein interaction, and gene ontology. RNA-seq-based expression analysis showed that different *ZmHSFs* are highly expressed in various organs (seed, vegetative, and reproductive development) and upregulated against other abiotic stresses such as temperature, salinity, UV, and drought. This study offers novel visions for functional dissection of the *ZmHSFs* in maize and will benefit breeding programs [13].

Wheat (*Triticum aestivum* L.) is an essential supplier of starch, protein, and minerals in the diet of >35% of the world's population [16,17]. To stabilize wheat production under different conditions, it is crucial to predict novel genes associated with growth and stress tolerance. Altaf et al. [14] identified 40 *TUBBY* genes (*TaTLPs*), and Rasool et al. [15] identified 37 phenylalanine ammonia-lyase genes (*TaPALs*) in the wheat genome. Both studies have used different in silico tools to comprehensively characterize their physio-biochemical properties, gene structures, *cis*-elements, duplication types, etc. qRT-PCR-based expression analysis of *TaTLPs* confirmed that most of the genes were upregulated in response to various hormones, and temperature stress, indicating their vital roles in wheat improvement under extreme temperature [14]. On the other hand, *TaPAL* genes showed higher expression in roots of drought tolerant than in those of drought-sensitive varieties. RNA-seq-based expression analysis showed that all *TaPAL* genes were highly expressed in shoots and roots under abiotic stress conditions [15]. These studies lay the grounds for functional studies of *TaPAL* and *TaTLPs* genes to uncover their roles in wheat growth/development and stress tolerance.

Pervaiz et al. [18] explore cDNA-microarray and miRNAs to understand how plants respond to abiotic stresses and survive in challenging environments. MicroRNA notably impacts the response of plants to environmental stressors, plant growth, and development and controls diverse biological and metabolic functions. Despite the availability of relevant miRNAs, there is still a limited way to identify them and the application of cDNA-microarray. Advanced sequencing and bioinformatics techniques are necessary for miRNA identification and target gene network prediction. This article recommends the application of miRNAs for detecting and organizing new practical genes conferring a significant practical role in stress tolerance [18].

3. Agronomic and Physiological Interventions in Developing Temperature-Resilient Plants

There is rising importance in using chemical approaches to enhance plant growth and productivity in stressful conditions. These approaches involve the exogenous application or seed priming of natural or synthetic substances, such as phytohormones, osmolytes, neurotransmitters, gaseous molecules, amino acids, etc. [6,7,19,20]. A study by Hmmam et al. [21] reported the protective role of salicylic acid (SA, 0, 0.5, 1, and 1.5 mM L^{-1}) in alleviating the harmful consequences of chilling stress (4 ± 1 °C) on "Seddik" mango transplants. Results showed that the application of SA (mainly 1.5 mM L^{-1}) helped to cope with the chilling stress by sustaining the integrity of the cell membrane in the leaves, lessening the electrolyte leakage, enhancing the photosynthetic pigment contents, antioxidant enzyme activities, total sugar contents, 2, 2-diphenyl-1-picrylhydrazyl radical scavenging activity and by decreasing proline and total phenolic contents in the "Seddik" mango transplants' leaves [21].

In another study, Waraich et al. [22] showed that seed priming with thiourea (500 ppm) improves the efficiency of Camelina (*Camelina sativa* L.) plants under heat stress (32 °C) by regulating several physiological and yield-related traits. Thiourea seed priming increases shoot and root length, biomass, gas exchange rate, water relations, and seed yield, and helps Camelina plants to mitigate the adverse influences of heat stress [22]. A combination of heat, salinity, and waterlogging reduces wheat's growth and yield-associated attributes, whereas Altaf et al. [23] reported that applying sulfur-coated urea (SCU, 130 kg ha^{-1}) helps boost the grain yield under all three stresses. They found a strong association between soil nitrogen content and growth rate, spike length, yield, and physiological limitations in wheat plants. While the effectiveness of SCU fertilizer was limited under heat stress, it was observed to cause better tolerance of wheat to salinity [23].

Zulfiqar et al. [20] reviewed glycine betaine's beneficial role in plants' heat stress tolerance. The exogenous application of glycine betaine has displayed promise in advancing heat stress tolerance by increasing osmolyte contents, protein modifications, photosynthetic mechanisms, stress-responsive gene expression, and oxidative defense. Under heat stress, glycine betaine accumulation in plants varies; consequently, engineering genes for glycine betaine accumulation in non-accumulating plants is an integral approach for increasing heat stress tolerance.

The production of cotton, a fiber crop, is being hindered by the unpredictable increase in temperature caused by rapidly changing climate conditions, as reviewed by Majeed et al. [24]. In this review, the authors comprehensively presented the impacts of heat stress on various traits such as seed germination, development of seedlings at early and lateral vegetative growth phases, yield and quality of fiber, floral components, and physiological aspects. Various breeding and mitigation strategies for heat stress tolerance have been discussed, including conventional and genomics-assisted breeding (QTL mapping, GWAS, and genomic selection), transgenic breeding, and CRISPR/Cas-mediated genome editing. Omics approaches have greatly improved our knowledge of how cotton responds to heat stress by identifying genes, proteins, and metabolites that appear differentially. These novel markers can be utilized for genetic engineering to produce cotton cultivars resilient to heat stress [24].

Rana et al. [25] examined thirteen upland cotton genotypes for variations in physiological and morphological attributes associated with heat stress during the vegetative and reproductive phases. The authors hypothesize that different parts of a single cotton plant may display variable responses to stress, which was tested by observing two flowering positions [25]. They collected data on various traits from different genotypes' top and bottom branches. The bottom branches performed better for most traits except boll weight. AA-933 genotype had the best pollen germination and boll retention, while CYTO-608 had the highest pollen viability. MNH-1016 and CIM-602 had better cell membrane thermostability and chlorophyll content, respectively. This variability within genotypes can be helpful in breeding programs to develop stress-tolerant varieties [25].

In tomato (*Solanum lycopersicum* L.) plants, heat stress causes changes in diverse responses during all vegetative and reproductive growth stages, resulting in poor fruit quality and low yield, as reviewed by Lee et al. [26]. The authors have reviewed past and present research efforts to identify heat-resistant tomato varieties through screening procedures conducted under varying heat stress conditions and temperature thresholds. They have also presented information on the correlation between heat tolerance and physiological and biochemical characteristics at different vegetative and reproductive growth stages. This article has explored the numerous parameters utilized to assess the heat tolerance of tomatoes, which include factors related to both vegetative and reproductive growth, such as leaf growth parameters, plant height, stem size, number of flowers, fruit set, yield, and the development of pollen and ovules, thus suggesting techniques for developing tomato cultivars that are more tolerant to heat stress [26].

When the temperature is too high, plants cannot make or use energy properly through respiration and photosynthesis, which are vital metabolic events for plants to endure and grow [27]. In this context, Sharma et al. [27] explore the mechanistic underpinnings of how heat stress triggers mitochondrial dysfunction, ultimately controlling dark respiration in plants. They also analyze the influence of hormones on the intricate network of processes involved in retrograde signaling. This review suggested various approaches for mitigating carbon loss under heat stress, such as choosing genotypes with lower respiration levels or employing gene editing techniques to modify carbon pathways by relocating, switching, or reorganizing metabolic events.

At the flowering stage, Zafar et al. [28] evaluated the consequences of heat stress on agronomic and physiological attributes of green super rice, such as plant height (PH), tillers per plant (TPP), grain yield per plant (GY), straw yield per plant (SY), harvest index (HI), 1000-grain weight (GW), grain length (GL), cell membrane stability (CMS), normalized difference vegetative index (NDVI), and pollen fertility percentage (PFP). The results explained that GY, TPP, SY, HI, and CMS were substantially impacted by heat stress, while other traits like PH, GW, GL, PFP, and NDVI were altered in only a few genotypes. These findings can aid in preventing and managing heat stress in rice. NGSR-16 and NGSR-18 genotypes performed well under heat stress and can be employed to create heat-tolerant rice.

Thoroughly examining the adaptability mechanisms of short- and long-duration japonica rice cultivars under different temperature circumstances is beneficial for probing improved adaptation and management ways. Given this background, Farooq et al. [29] experimented with two locations in China to examine the adaptability mechanisms of four japonica rice cultivars in response to varying temperature conditions. They concluded that earlier transplantation and anthesis were helpful in stress tolerance, and high temperature at the start of the day aided plants in escaping from high temperatures later in the day. The temperature had a considerable impact on rice, while the precipitation of the growing season did not. The influence of daily sunshine was substantial but less spatially consistent. These discoveries are contributed to sustainable and rewarding rice cultivation in the Northeast China region.

Pea (*Pisum sativum* L.) is frequently cultivated in semi-arid temperate zones. Tafesse et al. [30] examined the role of leaf pigments and surface wax in heat avoidance in pea canopies and their correlation with spectral vegetation indices. Field trials were conducted on 24 pea cultivars with varying leaf traits across six environments in western Canada. Heat stress reduced leaf pigment concentrations but increased chlorophyll a/b ratio, anthocyanin, and wax concentrations. Higher pigment and wax concentrations were linked with cooler canopy temperatures and increased heat tolerance. Spectral vegetation indices, including photochemical reflectance index, green normalized vegetation index, normalized pigment chlorophyll ratio index, and water band index, were revealed to have strong correlations with heat avoidance attributes and heat tolerance index. This report emphasized heat avoidance attributes in crop canopies and spectral lengths for choosing heat-tolerant genotypes.

Creating new varieties that can survive abiotic stresses is vital for meeting the challenge for natural fibers. Thus, Dey et al. [31] aimed to detect cold-tolerant varieties and understand the mechanisms that improve cold tolerance in various jute varieties. Findings revealed that Y49 and M33 varieties had high levels of chlorophyll and carotenoid contents and enzymatic and non-enzymatic antioxidants and phenolics, which helped lessen oxidative damage triggered by cold stress. Furthermore, osmolytes such as soluble sugars and proline were found to play roles in lowering impairment caused by cold stress. This study verified the cold tolerance ability of some chosen varieties, suggesting their perspective as an efficient adaptation strategy and as candidates for cold-resilient breeding programs.

The above-overviewed articles present the latest scientific advancements and developments in the mechanisms of crops' resilience to temperature stress, aimed at ensuring their sustainable production in the future. Consequently, our Special Issue will be a valuable one-stop source for scientists developing temperature-resilient plants to feed the growing population.

Funding: This research was funded by the National Natural Science Foundation of China (NSFC: grant number 32161143021), and by the Key R&D program of Henan province (221111110200).

Acknowledgments: We thank all the authors who have contributed to this Special Issue and the reviewers who have provided valuable feedback to enhance the quality of the articles. We would also like to thank the editorial team for their helpful assistance in compiling this Special Issue.

Conflicts of Interest: The authors declare no conflict of interest.

References

1. Zandalinas, S.I.; Fritschi, F.B.; Mittler, R. Global warming, climate change, and environmental pollution: Recipe for a multifactorial stress combination disaster. *Trends Plant Sci.* **2021**, *26*, 588–599. [CrossRef] [PubMed]
2. Fujimori, S.; Wu, W.; Doelman, J.; Frank, S.; Hristov, J.; Kyle, P.; Sands, R.; Van Zeist, W.-J.; Havlik, P.; Domínguez, I.P. Land-based climate change mitigation measures can affect agricultural markets and food security. *Nat. Food* **2022**, *3*, 110–121. [CrossRef]
3. Zandalinas, S.I.; Mittler, R. Plant responses to multifactorial stress combination. *New Phytol.* **2022**, *234*, 1161–1167. [CrossRef] [PubMed]
4. Rivero, R.M.; Mittler, R.; Blumwald, E.; Zandalinas, S.I. Developing climate-resilient crops: Improving plant tolerance to stress combination. *Plant J.* **2022**, *109*, 373–389. [CrossRef] [PubMed]

5. Raza, A.; Tabassum, J.; Kudapa, H.; Varshney, R.K. Can omics deliver temperature resilient ready-to-grow crops? *Crit. Rev. Biotechnol.* **2021**, *41*, 1209–1232. [CrossRef]
6. Raza, A.; Charagh, S.; García-Caparrós, P.; Rahman, M.A.; Ogwugwa, V.H.; Saeed, F.; Jin, W. Melatonin-mediated temperature stress tolerance in plants. *GM Crops Food* **2022**, *13*, 196–217. [CrossRef]
7. Raza, A.; Charagh, S.; Abbas, S.; Hassan, M.U.; Saeed, F.; Haider, S.; Sharif, R.; Anand, A.; Corpas, F.J.; Jin, W. Assessment of proline function in higher plants under extreme temperatures. *Plant Biol.* **2023**, *25*, 379–395. [CrossRef]
8. Saeed, F.; Chaudhry, U.K.; Raza, A.; Charagh, S.; Bakhsh, A.; Bohra, A.; Ali, S.; Chitikineni, A.; Saeed, Y.; Visser, R.G. Developing future heat-resilient vegetable crops. *Funct. Integr. Genom.* **2023**, *23*, 47. [CrossRef]
9. Rani, S.; Baber, M.; Naqqash, T.; Malik, S.A. Identification and Genetic Mapping of Potential QTLs Conferring Heat Tolerance in Cotton (*Gossypium hirsutum* L.) by Using Micro Satellite Marker's Approach. *Agronomy* **2022**, *12*, 1381. [CrossRef]
10. Arslan, M.; Yol, E.; Türk, M. Disentangling the Genetic Diversity of Grass Pea Germplasm Grown under Lowland and Highland Conditions. *Agronomy* **2022**, *12*, 2426. [CrossRef]
11. Zaid, I.U.; Zahra, N.; Habib, M.; Naeem, M.K.; Asghar, U.; Uzair, M.; Latif, A.; Rehman, A.; Ali, G.M.; Khan, M.R. Estimation of genetic variances and stability components of yield-related traits of green super rice at multi-environmental conditions in Pakistan. *Agronomy* **2022**, *12*, 1157. [CrossRef]
12. Kilwake, J.W.; Umer, M.J.; Wei, Y.; Mehari, T.G.; Magwanga, R.O.; Xu, Y.; Hou, Y.; Wang, Y.; Shiraku, M.L.; Kirungu, J.N. Genome-Wide Characterization of the *SAMS* Gene Family in Cotton Unveils the Putative Role of *GhSAMS2* in Enhancing Abiotic Stress Tolerance. *Agronomy* **2023**, *13*, 612. [CrossRef]
13. Haider, S.; Rehman, S.; Ahmad, Y.; Raza, A.; Tabassum, J.; Javed, T.; Osman, H.S.; Mahmood, T. *In silico* characterization and expression profiles of heat shock transcription factors (HSFs) in maize (*Zea mays* L.). *Agronomy* **2021**, *11*, 2335. [CrossRef]
14. Altaf, A.; Zada, A.; Hussain, S.; Gull, S.; Ding, Y.; Tao, R.; Zhu, M.; Zhu, X. Genome-wide identification, characterization, and expression analysis of Tubby gene family in wheat (*Triticum aestivum* L.) under biotic and abiotic stresses. *Agronomy* **2022**, *12*, 1121. [CrossRef]
15. Rasool, F.; Uzair, M.; Naeem, M.K.; Rehman, N.; Afroz, A.; Shah, H.; Khan, M.R. Phenylalanine ammonia-lyase (PAL) genes family in wheat (*Triticum aestivum* L.): Genome-wide characterization and expression profiling. *Agronomy* **2021**, *11*, 2511. [CrossRef]
16. Miransari, M.; Smith, D. Sustainable wheat (*Triticum aestivum* L.) production in saline fields: A review. *Crit. Rev. Biotechnol.* **2019**, *39*, 999–1014. [CrossRef]
17. Singh, J.; Chhabra, B.; Raza, A.; Yang, S.H.; Sandhu, K.S. Important wheat diseases in the US and their management in the 21st century. *Front. Plant Sci.* **2022**, *13*, 1010191. [CrossRef]
18. Pervaiz, T.; Amjid, M.W.; El-kereamy, A.; Niu, S.-H.; Wu, H.X. MicroRNA and cDNA-microarray as potential targets against abiotic stress response in plants: Advances and prospects. *Agronomy* **2021**, *12*, 11. [CrossRef]
19. Raza, A.; Salehi, H.; Rahman, M.A.; Zahid, Z.; Haghjou, M.M.; Najafi-Kakavand, S.; Charagh, S.; Osman, H.S.; Albaqami, M.; Zhuang, Y.; et al. Plant hormones and neurotransmitter interactions mediate antioxidant defenses under induced oxidative stress in plants. *Front. Plant Sci.* **2022**, *13*, 961872. [CrossRef]
20. Zulfiqar, F.; Ashraf, M.; Siddique, K.H. Role of glycine betaine in the thermotolerance of plants. *Agronomy* **2022**, *12*, 276. [CrossRef]
21. Hmmam, I.; Ali, A.E.; Saleh, S.M.; Khedr, N.; Abdellatif, A. The Role of Salicylic Acid in Mitigating the Adverse Effects of Chilling Stress on "Seddik" Mango Transplants. *Agronomy* **2022**, *12*, 1369. [CrossRef]
22. Waraich, E.A.; Ahmad, M.; Soufan, W.; Manzoor, M.T.; Ahmad, Z.; Habib-Ur-Rahman, M.; Sabagh, A.E. Seed priming with sulfhydral thiourea enhances the performance of *Camelina sativa* L. under heat stress conditions. *Agronomy* **2021**, *11*, 1875. [CrossRef]
23. Altaf, A.; Zhu, X.; Zhu, M.; Quan, M.; Irshad, S.; Xu, D.; Aleem, M.; Zhang, X.; Gull, S.; Li, F. Effects of environmental stresses (heat, salt, waterlogging) on grain yield and associated traits of wheat under application of sulfur-coated urea. *Agronomy* **2021**, *11*, 2340. [CrossRef]
24. Majeed, S.; Rana, I.A.; Mubarik, M.S.; Atif, R.M.; Yang, S.-H.; Chung, G.; Jia, Y.; Du, X.; Hinze, L.; Azhar, M.T. Heat stress in cotton: A review on predicted and unpredicted growth-yield anomalies and mitigating breeding strategies. *Agronomy* **2021**, *11*, 1825. [CrossRef]
25. Rana, I.A.; Majeed, S.; Chaudhary, M.T.; Zulfiqar, M.; Yang, S.-H.; Chung, G.; Jia, Y.; Du, X.; Hinze, L.; Azhar, M.T. Intra-Plant Variability for Heat Tolerance Related Attributes in Upland Cotton. *Agronomy* **2021**, *11*, 2375.
26. Lee, K.; Rajametov, S.N.; Jeong, H.-B.; Cho, M.-C.; Lee, O.-J.; Kim, S.-G.; Yang, E.-Y.; Chae, W.-B. Comprehensive understanding of selecting traits for heat tolerance during vegetative and reproductive growth stages in tomato. *Agronomy* **2022**, *12*, 834. [CrossRef]
27. Sharma, N.; Thakur, M.; Suryakumar, P.; Mukherjee, P.; Raza, A.; Prakash, C.S.; Anand, A. 'Breathing Out' under Heat Stress—Respiratory Control of Crop Yield under High Temperature. *Agronomy* **2022**, *12*, 806. [CrossRef]
28. Zafar, S.A.; Arif, M.H.; Uzair, M.; Rashid, U.; Naeem, M.K.; Rehman, O.U.; Rehman, N.; Zaid, I.U.; Farooq, M.S.; Zahra, N. Agronomic and physiological indices for reproductive stage heat stress tolerance in green super rice. *Agronomy* **2022**, *12*, 1907. [CrossRef]
29. Farooq, M.S.; Gyilbag, A.; Virk, A.L.; Xu, Y. Adaptability mechanisms of japonica rice based on the comparative temperature conditions of harbin and qiqihar, heilongjiang province of Northeast China. *Agronomy* **2021**, *11*, 2367. [CrossRef]

30. Tafesse, E.G.; Warkentin, T.D.; Shirtliffe, S.; Noble, S.; Bueckert, R. Leaf Pigments, Surface Wax and Spectral Vegetation Indices for Heat Stress Resistance in Pea. *Agronomy* **2022**, *12*, 739. [CrossRef]
31. Dey, S.; Biswas, A.; Huang, S.; Li, D.; Liu, L.; Deng, Y.; Xiao, A.; Birhanie, Z.M.; Zhang, J.; Li, J. Low temperature effect on different varieties of *Corchorus capsularis* and *Corchorus olitorius* at seedling stage. *Agronomy* **2021**, *11*, 2547. [CrossRef]

Article

Identification and Genetic Mapping of Potential QTLs Conferring Heat Tolerance in Cotton (*Gossypium hirsutum* L.) by Using Micro Satellite Marker's Approach

Shazia Rani [1], Muhammad Baber [2,*], Tahir Naqqash [2] and Saeed Ahmad Malik [1]

[1] Institute of Pure and Applied Biology, Bahauddin Zakariya University, Multan 60800, Pakistan; shaziarani19@yahoo.com (S.R.); saeedbotany@yahoo.com (S.A.M.)

[2] Institute of Molecular Biology & Biotechnology, Bahauddin Zakariya University, Multan 60800, Pakistan; tahirnaqqash@gmail.com

* Correspondence: muhammadbabar@bzu.edu.pk

Abstract: High-temperature stress can cause serious abiotic damage that limits the yield and quality of cotton plants. Heat Tolerance (HT) during the different developmental stages of cotton can guarantee a high yield under heat stress. HT is a complex trait that is regulated by multiple quantitative trait loci (QTLs). In this study, the F_2 population derived from a cross between MNH-886, a heat-tolerant cultivar, and MNH-814, a heat-sensitive variety, was used to map HT QTLs during different morphological stages in cotton. A genetic map covering 4402.7 cm, with 175 marker loci and 26 linkage groups, was constructed by using this F_2 population (94 individuals). This population was evaluated for different 23 morpho-physiological HT contributing traits QTL analysis via composite interval mapping detected 17 QTLs: three QTLs each for Total Number of Sympodes (TNS), Length of Bract (LOB), and Length of Staminal-column (LOS); two QTLs for First Sympodial Node Height (FSH), and one QTL each for Sympodial Node Height (SNH), Percent Boll set on second position along Sympodia (PBS), Total Number of Nodes (TNN), Number of Bolls (NOB), Total Number of Buds (TNB), and Length of Petal (LOP). Individually, the QTLs accounted for 7.76%–36.62% of phenotypic variation. QTLs identified linked with heat tolerance traits can facilitate marker-assisted breeding for heat tolerance in cotton.

Keywords: *G. hirsutum*; molecular markers; morpho physiological characteristics; quantitative trait loci; heat tolerance

Citation: Rani, S.; Baber, M.; Naqqash, T.; Malik, S.A. Identification and Genetic Mapping of Potential QTLs Conferring Heat Tolerance in Cotton (*Gossypium hirsutum* L.) by Using Micro Satellite Marker's Approach. *Agronomy* **2022**, *12*, 1381. https://doi.org/10.3390/agronomy12061381

Academic Editors: Channapatna S. Prakash, Ali Raza, Xiling Zou and Daojie Wang

Received: 8 April 2022
Accepted: 26 May 2022
Published: 8 June 2022

1. Introduction

Cotton is a miracle of the plant realm as it fulfills most of the vital needs and provides more than 90% of the world's total production of fiber for the textile industry and edible oil for almost half of the world's population [1]. It has been observed that more than 50% of cotton around the globe is affected by abiotic stress such as salinity, drought, and heat stress that lead to deficient production of this field crop, especially when affected at the seedling stage [2]. Cotton growth requires sufficient fresh water for better fiber quality, but if it faces drought or heat stress the fiber production is reduced [3]. Many new drought tolerant cultivars of cotton have been introduced with improved plant growth, and even other genetically engineered genotypes of cotton by breeding techniques are being cultivated that can tolerate many abiotic stresses [4]. However, the genetic basics and amendments behind these stresses need to be evaluated more to combat these problems from the genetic roots. Cotton is divided into eight genomes (groups) from A to G and K including 45 diploids and the basic seven tetraploid [5,6]. Evolutionary data based on DNA sequencing suggested that about six to seven million years, ago due to trans-oceanic dispersal, D genome divergence gave rise to the A genome and in America (primarily Mexico), it became a separate lineage [7,8]. An incredible diversification occurred over this time that resulted in the worldwide spread of the Gossypium species. Domestication of wild varieties of cotton by human beings resulted in lot of change in all phenotypic and genotypic characteristics.

In terms of production, Pakistan is at the fourth position among the cotton growers of the world; raw cotton exported from Pakistan holds third position in the world as per records of 2012–2013 [9]. Pakistan is more prone to climate changes due to its geographical location [10]. Heat stress is a combination of different intricate functions of intensity duration of temperature. Because of its geographical position, in Pakistan during the summer in some locations, the temperature reaches up to 50 °C and the scorching heat adversely affects cotton plants. Cotton is cultivated in hot areas in Pakistan [11]. High temperature affects growth and development of the plant as well as fiber quality traits [12,13]. Episodes of periodic heat stress and increase in average temperature for the full season enhances the detrimental effects on almost all the factors of plant growth, and that is the reason there is great reduction in the seed number, fiber quality, and content [14]. Cotton yield is suppressed when the plant faces heat and drought stress due to decreased plant transpiration and reduced biomass accumulation, resulting in an inadequate yield [15]; these stresses adversely affect cell elongation, differentiation, and division and also suppress stomatal conductance [16].

The cotton plant has a wide range of adaptability [17], but high temperature is one of the major constraints in cotton productivity and greatly reduces seed cotton yield and quality, which can be addressed by breeding methods. Marker-assisted selection fastens the breeding technology with an accurate approach towards the desired phenotypic traits among the breeding population [18], and it requires detection and analysis of genetic variations using advanced genetic approaches, leading to phenotypic traits of quantitative and agro-economic importance [19]. Genomic selection (GS) and MAS developed by molecular markers techniques has made it possible to map quantitative trait loci (QTL) and identifying QTLs for high-temperature stress and breeding heat-tolerant varieties is an effective way to address this issue. MAS methodology has been used globally to acquire ordered and swift ways for cotton improvement on large scales internationally, with both highly demanded attributes like high seed production and excellent quality of fiber [20]. For dissection of QTLs related to traits with complex genetic patterns of inheritance, molecular marker use has been an efficient tool and these markers have also facilitated MAS breeding [21].

Both agronomic and economically important traits are approached by researchers for obtaining the aim of better yield of cotton [22]. The main challenging goal for current cotton breeders is to further enhance cotton production. However, this aim is hindered by the use of locally available germplasm and extreme environmental fluctuations that influence yield attributing traits [23,24]. Certain different genes cause different expressions of characters regarding tolerance of heat stress at vegetative and reproductive growth stages [25]. Genes attributing to relative water content, stomatal conductance, especially along with Percent Boll set on the First Position along Sympodia (PBF), Percent Boll set on the second Position along Sympodia (PBS), Cell Injury (CIY), Boll Number (BON), Total number of Buds (TNB), Size of Petiole (SOP), Total number of Flowers (TNF), Length of Bract (LOB), Length of Petal (cm) (LOP), Length of Staminal Column (LOS), Length of Pistil (LPI), and Proline Con. (μg mL^{-1}) (PCO) have been reported as crucial for heat stress determination [26,27]. Therefore, during the selection of heat tolerant varieties, both vegetative and reproductive traits should be considered equally.

Molecular genetic methods, especially molecular markers, have been applied widely in cotton in recent couple of decades. Recently, the development of molecular markers was accelerated with the release of assembled genome sequences of *G. hirsutum* [28,29]. Numerous genetic linkage maps including the intraspecific map of *G. hirsutum* have been constructed using restriction fragment length polymorphisms (RFLPs), simple sequence repeats (SSRs), and single nucleotide polymorphisms (SNPs). Thousands of quantitative trait loci (QTLs) for yield and fiber quality in cotton have been documented in Cotton

QTLdb, Release 2.3 [30,31]. However, there are few studies about the simultaneous dissection of the genetic basis underlying complex traits and their genetic correlations in multiple upland cotton populations by QTL mapping. In the situation of changing weather and elevating temperature around the globe, it is of the utmost importance to recognize QTLs for morphological, architectural, and physiological traits that are directly or indirectly affected by high heat stress at some stages of cotton plant development. This study was conducted to identify and map quantitative trait loci (QTLs) conferring heat tolerance in an Intraspecific cross and used microsatellite markers to identify polymorphism between two upland cotton cultivars in the scorching heat of Multan (Pakistan) during summer. QTL identified in this project could be helpful for future cotton growers of high-temperature regions in the world.

In this study, F_2 populations were used, which were derived from hybridization of two *G. hirsutum* normal lines (MNH-886 and MNH-814). The corresponding genetic linkage map was constructed using 175 polymorphic SSR markers. QTL mapping was implemented with the integration of the genotypic and phenotypic data of twenty-three agronomic and economic traits contributing towards heat tolerance; the aim of this study was to (a) screen cotton cultivars for heat tolerance, (b) select diverse cultivars as parental lines and then their assessment by SSRs for parental survey, (c) develop the segregating/mapping population (F_2) of selected parents and collect phenotypic trait data at different time intervals, (d) survey the F_2 population by polymorphic markers obtained from the parental survey, (e) evaluate phenotypic traits with the association of genotypic markers (SSR) data, (f) identify QTLs directing heat tolerance by QTL cartographer software, and (g) construct a genetic linkage map of *Gossypium* from the obtained information. The outcomes of this study will help plant breeders to produce heat-resistant varieties that will help farmers and countries with agriculture-dependent economies, especially in high-temperature areas around the globe.

2. Materials and Methods

This study was conducted to identify and map QTLs conferring heat tolerance in an Intraspecific cross and used microsatellite markers to identify polymorphism between two upland cotton cultivars in the scorching heat of Multan (Pakistan) during summer. QTL identified in this project could be helpful for future cotton growers of high temperature regions in the world. The research was arranged at Cotton Research Station (CRS) Multan to coincide the reproductive phase with higher temperature. The field work encompassed 14 cultivars sown in Randomized Complete Block Design (RCBD) replicated three times during the year 2012. All fourteen cultivars were tagged randomly altogether to evaluate 23 morphological and physiological parameters contributing to heat tolerance for identifying the genomic regions under plant breeding techniques; F2 generation was observed for screening purposes. The cultivars named as CIM-557, CIM-573, NN-3, Cyto-108, NIAB-852, CIM-588, BH-172, GH-102, NIAB-2008, MNH 886, CIM-554, Shahbaz-12, MNH-2007, and MNH 814 were chosen for screening of heat tolerance based on different agronomic traits related to heat, and their genomic basics were screened out. Different morpho-physiological characters included plant height (PH), fully dehiscent anther (FDA %), Total number of sympods (TNS), Total Number of Nodes (TNN), Pollen Viability (%) (POV), First Sympodial Node Number (FSN), First Sympodial Node Height (cm) (FSH), Sympodial Node Number bearing first effective boll (SNF), Sympodial Node Height (cm) bearing first effective boll (SNH), Sympodial Node Number bearing Last effective boll (SNL), Sympodial Node Height (cm) bearing last effective boll (SNB), Percent Boll set on First Position along Sympodia (PBF), Percent Boll set on second Position along Sympodia (PBS), Cell Injury (CIY), Boll Number (BON), Total number of Buds (TNB), Size of Petiole (SOP), Total number of Flowers (TNF), Length of Bract (LOB), Length of Petal (cm) (LOP), Length of Staminal Column (LOS), Length of Pistil (LPI), and Proline Con. (μg mL^{-1}) (PCO).

2.1. Heat Stress Estimation

Heat stress was measured in plants that were sown late in the month of April and traits were compared to plants sown earlier in May because temperature in the latter was higher than 46–48 °C during the research time period and the heat stress-related 23 morphophysiological traits were observed to be affected by temperature in late-sown irrigated conditions. The heat was estimated by a weather forecast taken from the automated metrological station of cotton research station, Multan, as given in the Table 1:

Table 1. Comparative Monthly Meteorological Data Recorded at CCRI, Multan.

Month	Air Temperature (°C)		Relative Humidity		Rainfall (mm)	Evapotranspiration (cm Day)	Soil Temperature (°C)	
	Max	Min	Max	Min			5 cm	10 cm
January	5.3	19.1	63	92	1.5	0.24	9.4	10.5
February	6.9	20.5	52	76	0.0	0.39	12.3	12.7
March	13.9	27.4	45	65	0.0	0.67	19.2	19.7
April	20.6	32.8	55	72	24.7	0.86	26.5	27.0
May	25.7	39.4	54	57	1.10	1.22	31.7	32.0
June	28.6	39.4	58	64	0.0	1.26	35.4	35.4
July	28.8	38.1	61	73	16.9	1.11	35.8	36.0
August	28.0	35.6	72	76	16.1	0.84	34.9	35.1
September	25.7	33.1	80	87	167.0	0.59	29.8	30.2
October	18.9	31.7	62	83	3.2	0.48	24.3	25.1
November	13.1	26.8	81	87	0.0	0.28	17.7	18.6
December	7.8	21.9	80	87	4.0	0.19	12.8	13.8

Heat-susceptible and -resistant varieties (MNH-814 and MNH-886 respectively) were selected on the basis of data for relative water content, osmotic potential, cell injury, and proline concentration. Relative water content was measured by the following [32] formula:

$$RWC = \frac{fresh\ weight - Dry\ weight}{Turgid\ weight - Dry\ weight} \times 100$$

Cell Injury (CIY) was measured when the crop was 55–60 days old, and a sufficient number of leaves was taken from the upper portion and stored in a paper bag. By the use of a punching machine, 15–25 discs of 1 cm diameter were cut. With distilled water, leaf discs were washed three times, were put in test tubes, and then the test tubes were filled up to 40 mL with distilled water. Three sets were made each of 14 test tubes containing leaf discs of 14 cultivars. The first set of test tubes was kept at room temperature as control and the electrical conductivity of the water was noted. The second set was heated at 48 °C for 45 min in a water bath. When water was cooled after 6 h, its electrical conductivity was recorded, while the third set of test tubes was autoclaved at 15 lbs (pressure) for 15 min and electrical conductivity was noted when water was cooled.

The greater the EC, greater the damage caused to plant cells due to heat stress as the maximum number of electrolytes came out of the cell due to cell injury. Consequently, cell injury was also greater. Cell injury was expressed in percentage. Proline is an organic compound synthesized from glutamine. It is located in cytoplasm under stressed conditions as nontoxic compatible organic solute to compensate for the dehydrating effects of high osmotic pressure in the vacuole and in the external media. The proline concentration at 700 mol m^3 was not inhibitory to enzymes and develops in consequences of poor plant growth under toxic effects. Therefore, its exogenous application should promote tolerance [33]. Different workers stated that upon heat stress, when starch and protein synthesis are inhibited, proline might be used by the plant for growth [34,35]. Proline from different tissues was measured by Spectrophotometry based on the method of ref. [36].

2.2. Parental Lines Screening

Fourteen tetraploid cotton cultivars were chosen, named CIM-557, CIM-573, NN-3, Cyto-108, NIAB-852, CIM-588, BH-172, GH-102, NIAB-2008, MNH 886, CIM-554, Shahbaz-12, MNH-2007, and MNH 814 for altogether 23 morphological and physiological characteristics, viz Total Plant Height (TPH), Fully Dehiscent Anther (%) (FDA), Total Number of Sympodes (TNS), Total Number of Nodes (TNN), Pollen Viability (%) (POV), 1st Sympodial Node number (FSN), 1st Sympodial node Height (cm) (FSH), Sympodial Node number having 1st effective boll (SNF), Sympodial Node Height (cm) having 1st effective boll (SNH), Sympodial Node Number having Last effective boll (SNN), Sympodial Node Height (cm) having last effective boll (SNH), Percent boll set on 1st position with sympodia (PBF), Percent boll set on 2nd position with sympodia (PBS), Cell Injury (CIY), Number of Bolls (NOB), Total number of buds (TNB), Size of Petiole (SOP), Total Number of Flowers (TNF), Length of Bract (LOB), Length of Petal (cm) (LOP), Length of Staminal column (LOS), Length of Pistil (LPI), and Proline Con. ($\mu g\ mL^{-1}$) (PCO), contributing to heat tolerance to identify the genomic regions. Arithmetic means of three replicates were calculated for fourteen cultivars for each characteristic. The data were compared. The variance and standard deviation were also calculated. The computation of trait correlation was carried out using Minitab Inc., University Park, PA, USA and the following shortlisted traits had considerably varying phenotypes among two genotypes, i.e., MNH 886 and MNH 814 (Figure 1).

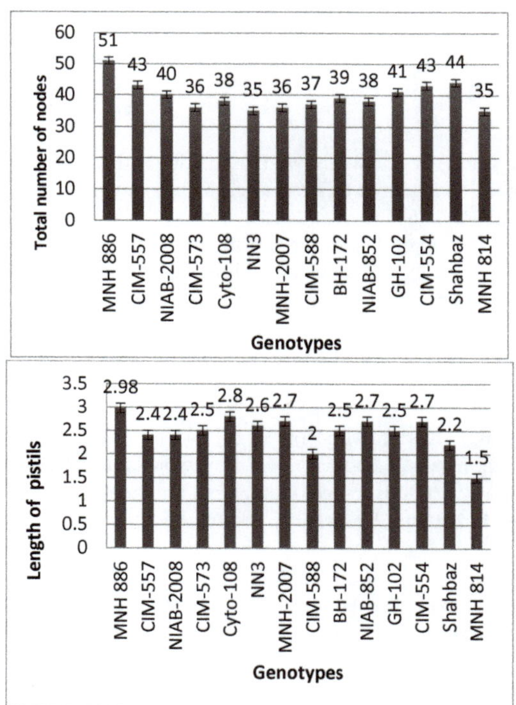

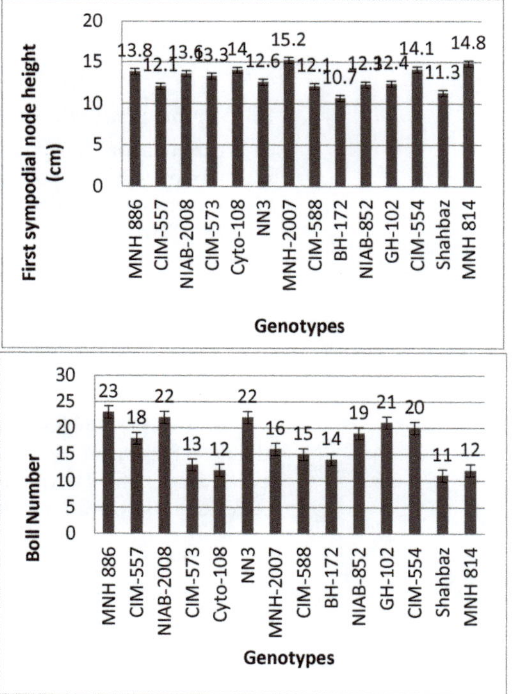

Figure 1. *Cont.*

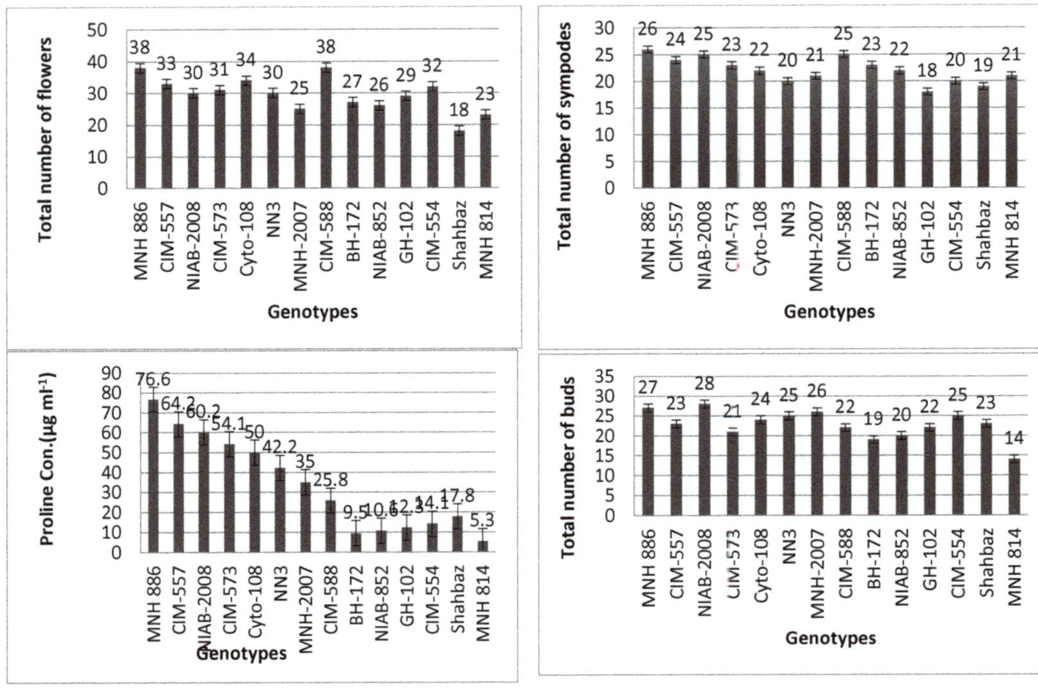

Figure 1. Mean values of phenotypic variations of morpho-physiological traits related to the heat stress of 14 cotton genotypes.

2.3. Mapping Population

Based on highly significant differences between two parental lines, the F_2 population was developed by self-pollinating F_1 plants from a cross between upland cotton line MNH-886 (a heat-tolerant cultivar), and MNH-814 (a heat-sensitive cultivar), and was used to map HT-QTLs during different morphological stages in cotton. Five plants were tagged at random in each line for recording physiological traits data. Ninety-four plants from the F_2 population were selected to derive phenotypic and molecular data along with two parents. The experimental field area of Cotton Research Station Multan under natural conditions was selected for experiment to coincide the reproductive phase with higher temperature.

2.4. Phenotypic Data Collection Statistical Analysis

Selected parental lines and 94 F_2 individuals' phenotypic data were collected from fields at different time intervals. Arithmetic means of 3 replicates were calculated for each parent for each characteristic. The data for heat characteristics were compared. The computation of trait correlation was carried out using Minitab Inc., University Park, PA, USA.

2.5. Microsatellite Analysis

Laboratory techniques for DNA extraction were performed as described by Peterson. Amplification reactions were carried out in 15 uL reaction volumes containing 30 mg genomic DNA, 1.0 µM each of SSR primers sequences, which were drawn from the following sources: BNL primers from the Research Genetics Co. (Huntsville, AL, USA, http://www.resgen.com, accessed on 7 April 2022); JESPR primers [37]; CIR primers [37]; and NAU primers [38,39], 100 uM each of dATP, dCTP, dGTP, and dTTP, 1 unit of Taq DNA Polymerase (Fermentas), 1x*Taq* Polymerase Buffer, and 2.5 mM $MgCl_2$. PCR amplifications were performed as described [40] using a Peltier Thermal Cycler (MJ Research, Waltham,

MA, USA) programmed as follows: an initial denaturation of 5 min at 94°; 35 cycles of 94° for 1 min (denaturation), 55° for 1 min (annealing), and 72° for 2 min (extension). One additional cycle of 10 min at 72° was used for final extension. The amplified products were electrophoresed on a 10% non-denatured polyacrylamide gel using a DYCZ-30 electrophoresis apparatus (Beijing WoDeLife sciences instrument company, Beijing, China).

2.6. QTL Mapping

Genetic mapping and QTL analysis were performed on each population separately and combined across populations. Linkage maps were constructed using MAPMAKER/Exp Version 3.0b software [41]. QTLs were identified by composite interval mapping [42] using Windows QTLs Cartographer 2.5 [43]. A LOD threshold of 3.0 was used [44]. Marker's order was confirmed with the "ripple" command. Recombination frequencies were converted into map distances (cm) using the Kosambi mapping function [45].

Tests for independence of QTLs were also conducted when 2 or more QTLs of a trait were located on the same chromosome [46]. QTLs were declared significant if the corresponding LR score were greater than 11.5 (equal to a LOD score of 2.5). The proportion of the phenotypic variation explained by each QTL was calculated as R^2 (%) = Phenotypic variability explained by QTL/all of the variation in the population × 100. The total phenotypic variance explained together by all the putative QTLs for each trait was estimated by fitting a multiple-QTL model in the Mapmaker/QTL program.

3. Results

3.1. Average Performance of Cotton Varieties Based on Morpho-Physiological Traits

Based on statistically significant differences for various morpho-physiological characteristics, two cultivars, MNH-886 and MNH-814, were selected. Significant variations in heat tolerance characteristics were observed among both varieties. The mean value for fully dehiscent anthers (FDA) was 92 and 64 for MNH-886 and MNH-814 respectively. MNH-814 showed less pollen viability (66.4) than MNH886 (88.3). SNF was 34 for MNH-886 and for MNH-814 its average value was 27 (Figure 2a). The trait PBF average data of both parents were 51 and 38. MNH-886 showed less CIY while exposed to high temperatures, with an average value of 65, while MNH-814 was susceptible to extreme temperatures and the CIY was greater, with a value of 80.

Likewise, MNH 886 excelled in NOB with an average value of 23 while MNH-814 showed 12 TNN under heat-stress conditions. MNH 886 showed TNF even under heat stress with an average value of 35 while MNH 814 showed retention of a smaller number of flowers with an average value of 23. Maximum variation was observed in trait PCO; its value was 5.3 for MNH-886 and 76.6 for MNH 814. The average values of morpho-physiological traits showed that both varieties vary in most of the traits and showed that MNH-886 excelled in heat tolerance considering each trait compared with other cultivars, while MNH-814 was the most susceptible as compared with other varieties (Table 2).

Phenotypic distribution of F_2 population for morpho-physiological traits is shown in Figure 2a,b. The phenotypic values of morpho-physiological traits are presented in Table 3. Twenty-one morpho-physiological traits displayed a normal distribution (skewness did not exceed 1.0), while two traits, TNF and NOB, showed a non-normal distribution. These results indicated the trend of having major QTL involvement in this population and it was thus suitable for QTL analysis.

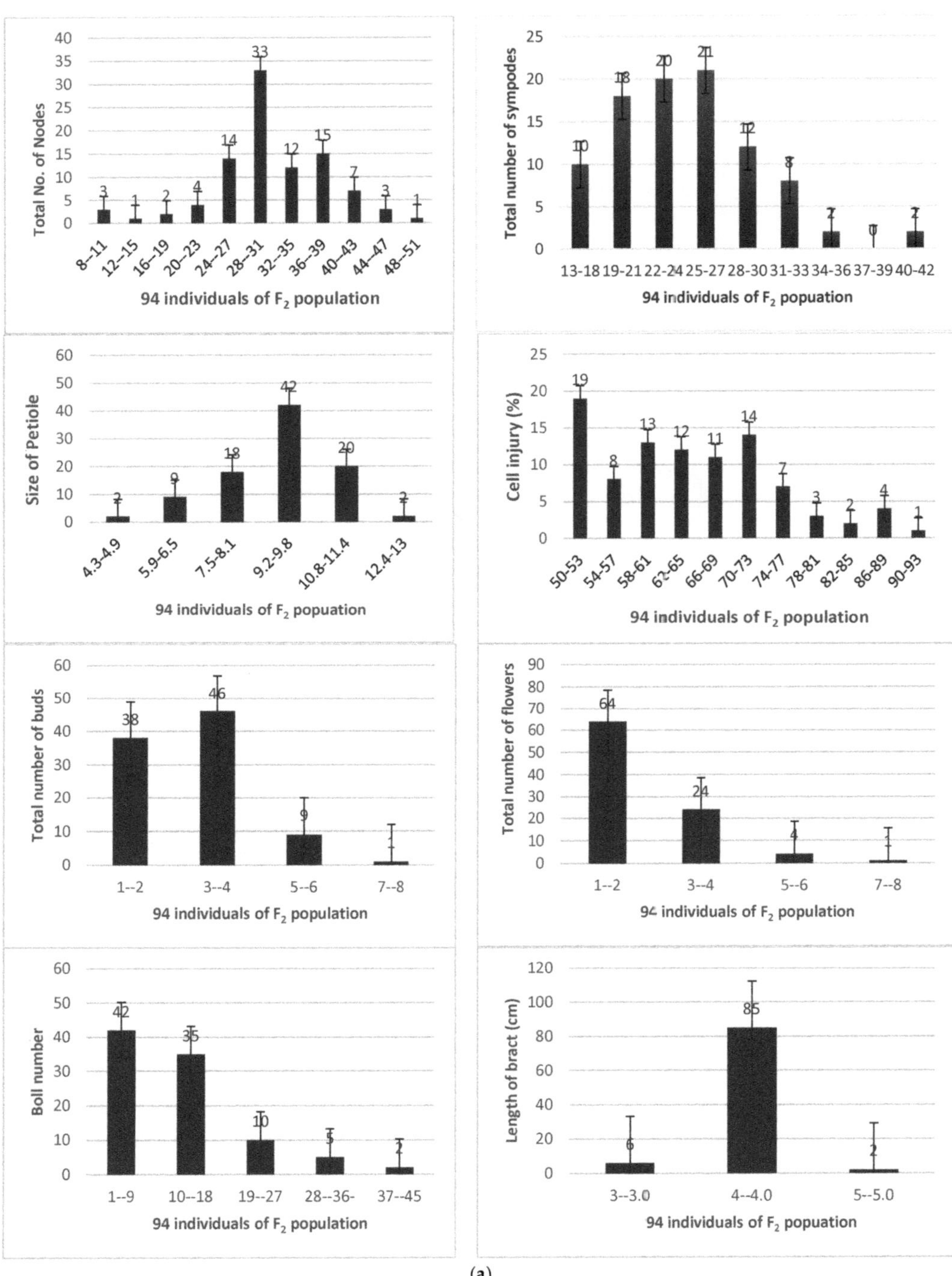

(a)

Figure 2. *Cont.*

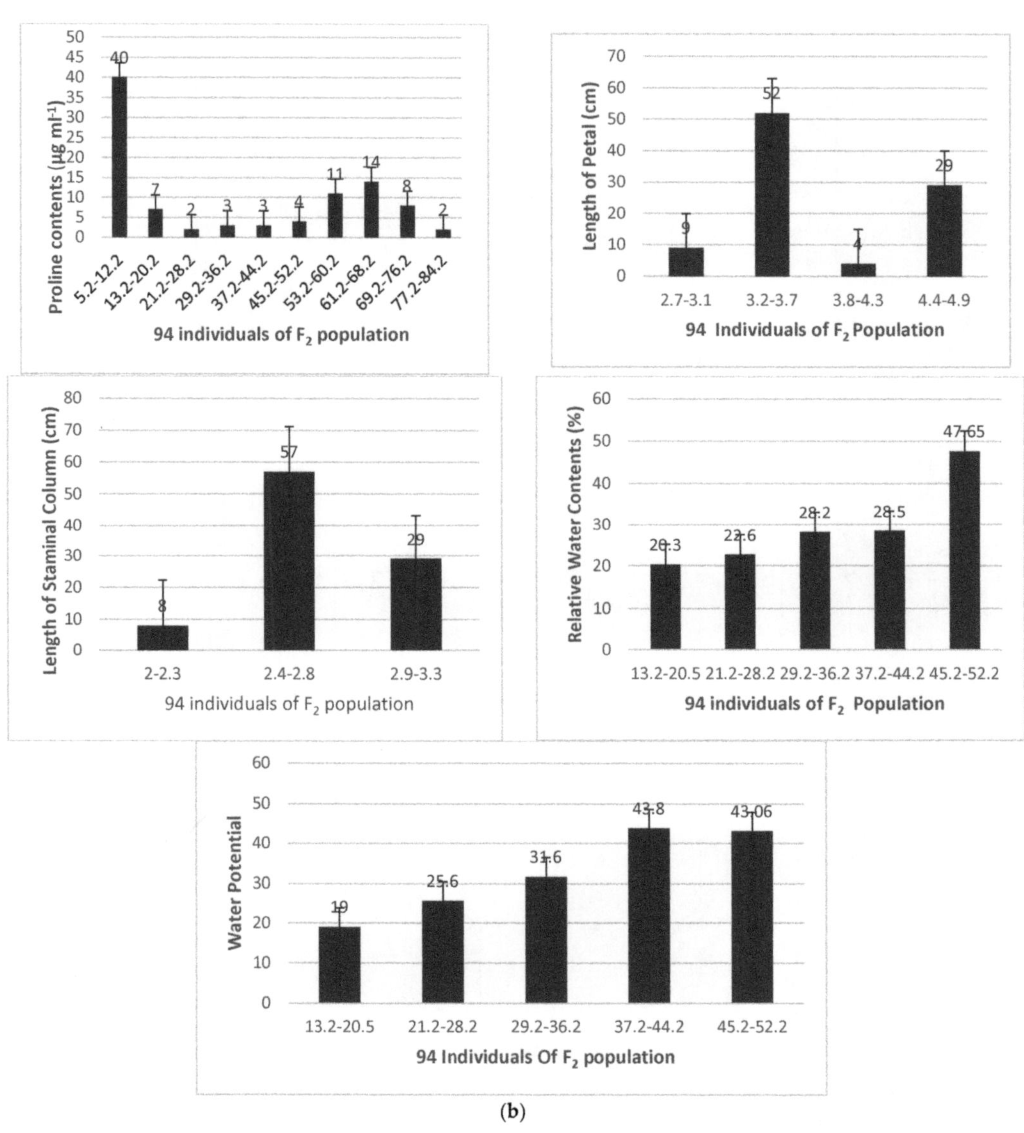

(b)

Figure 2. (**a**,**b**) Frequency distribution of morpho-physiological traits related to heat stress across F_2 population.

Table 2. Mean values data for 23 morpho-physiological traits of fourteen cotton cultivars.

Cultivars	Total Plant Height (cm)	Fully Dehiscent Anther (%)	Pollen Viability (%)	First Sympodial Node Number	First Sympodial Node Height (cm)		Sympodial Node Number Bearing First Effective Boll	Sympodial Node Height (cm) Bearing First Effective Boll	Sympodial Node Number Bearing Last Effective Boll	Sympodial Node Height (cm) Bearing Last Effective Boll	Percent Boll Set on First Position along Sympodia	Percent Boll Set on Second Position along Sympodia
MNH 886	90	92	88.3	7	13.8		8	15	34	114.2	51	32
CIM-557	67	91	87.4	7	12.1		8	14.8	32	93.4	49	31
NIAB-2008	55	89	86.1	7	13.6		8	15.2	33	82.5	50	32
CIM-573	54	88	85.8	8	13.3		9	15.9	31	112.3	48	31
Cyto-108	67	87	85.1	7	14		8	16.3	33	108.9	49	30
NN3	88	85	83.5	7	12.6		8	14.5	32	105.6	47	31
MNH-2007	67	83	82.2	8	15.2		9	17.9	31	87.3	47	30
CIM-588	77	82	80.5	7	12.1		8	15.6	33	92.7	45	29
BH-172	80	80	79.3	7	10.7		8	13.8	32	84.5	44	29
NIAB-852	85	79	77.5	7	12.3		8	14.9	33	113.8	45	28
GH-102	77	77	76.3	7	12.4		8	15.3	31	103.2	43	27
CIM-554	66	75	74.1	7	14.1		8	16.8	31	121.2	42	29
Shahbaz	88	73	71.2	7	11.3		8	13.7	29	81.7	41	26
MNH 814	54	64	66.4	8	14.8		9	17.3	27	106.5	38	23
Max	90	92	88	8	15		9	17	34	121	51	32
Min	54	64	66	7	10		8	13	27	81	38	23
Variance	219	60	42	181	1.70		181	1.52	3.34	176	14	6
Std. Dev.	±14.79	±7.80	±6.53	±42	±1.30		±42	±1.23	±1.82	±13.2	±3.75	±2.50
Cultivars	**Cell Injury (%)**	**Total Number of Sympodes**	**Total Number of Nodes**	**Size of Petiole**	**Total Number of Flowers**	**Number of Bolls**	**Total Number of Buds**	**Length of Bract (cm)**	**Length of Petal (cm)**	**Length of Staminal Column**	**Length of Pistil**	**Proline Con. ($\mu g\ mL^{-1}$)**
MNH-886	65	26	51	9.3	38	23	27	5	4	2.90	2.98	76.7
CIM-557	66	24	43	7.3	33	18	23	4.5	2.9	2.4	2.4	64.2
NIAB-2008	67	25	40	7.6	30	22	28	3.6	2.8	2.3	2.4	60.2
CIM-573	67	23	36	8.1	31	13	21	3.7	2.6	2.6	2.5	54.1
Cyto-108	68	22	38	8.3	34	12	24	3.9	2.7	2.7	2.8	50
NN3	68	20	35	7.5	30	22	25	3.4	2.6	2.1	2.6	42.2

Table 2. *Cont.*

Cultivars	Total Plant Height (cm)	Fully Dehiscent Anther (%)	Pollen Viability (%)	First Sympodial Node Number	First Sympodial Node Height (cm)		Sympodial Node Number Bearing First Effective Boll	Sympodial Node Height (cm) Bearing First Effective Boll	Sympodial Node Number Bearing Last Effective Boll	Sympodial Node Height (cm) Bearing Last Effective Boll	Percent Boll Set on First Position along Sympodia	Percent Boll Set on Second Position along Sympodia
MNH-2007	68	21	36	7.8	25	16	26	3.3	2.5	2.5	2.7	35
CIM-588	70	25	37	9.0	38	15	22	3.7	2.2	2	2	25.8
BH-172	71	23	39	8.0	27	14	19	3.1	2.8	2.2	2.5	17.8
NIAB-852	72	22	38	8.7	26	19	20	3.2	2.8	2.6	2.7	14.1
GH-102	73	18	41	8.5	29	21	22	3.5	2.7	2.7	2.5	12.3
CIM-554	74	20	43	8.3	32	20	25	3.8	2.4	2.5	2.7	10.6
Shahbaz	76	19	44	7.9	18	11	23	3.9	2.9	3	2.2	9.5
MNH-814	80	21	35	7	23	12	14	3	2	4.5	1.5	5.3
Max	80	26	51	9	38	23	28	5	4	4	2	76
Min	65	18	35	7	18	11	14	3	2	2	1.5	5
Variance	18	5	19	0.41	27	19	13	0.29	0.20	0.36	0.13	56
Std. Dev.	±4.25	±2.40	±4.41	±0.64	±5.2	±4.4	±3.6	±0.54	±0.45	±0.60	±0.37	±23

Table 3. Phenotypic values for heat tolerance traits of F_2 population and their parents.

Population Size	Traits	Parents		F₂ Population Statistical Data				
		MNH-886	MNH-814	Max	Min	Mean	SD	Skew
94	TPH	90	54	48	105	74.37	12.07	0.390
	FDA	92	64	60	92.	75.03	9.98	0.365
	POV	88.3	66.4	58	666	78.48	61.89	0.391
	FSN	7	8	7	9	7.82	0.824	0.328
	FSH	13.8	14.8	10	16.10	13.64	1.134	−0.224
	SNF	8	9	6	11	8.64	0.912	0.934
	SNH	15	17.3	11.10	17.30	14.30	1.568	0.211
	SNL	34	27	22	34	28.92	3.26	0.002
	SNB	114.2	106.5	80.5	115	100.19	9.63	−0.315
	PBF	51	38	31	51	43.44	5.30	−0.210
	PBS	32	23	23	3191	62.47	326.15	0.691
	CIY	65	80	50	90	64.60	10.36	0.477
	TNS	26	21	13	39	21.77	5.33	0.732
	TNN	45	51	6	48	30.93	7.12	−0.577
	SOP	9.3	7	4.30	12.30	8.69	1.51	−0.428
	TNF	35	23	1	8	2.13	1.25	1.62
	NOB	23	12	1	45	12.10	8.80	1.26
	TNB	27	14	1	7	2.92	1.32	0.796
	LOB	5	3	3	5	3.94	0.33	−0.770
	LOP	4	2	2.70	4	3.65	0.317	−0.760
	LOS	2.90	4.5	2.0	3.10	2.64	0.261	−0.434
	LPI	2.98	1.5	2.30	4.00	2.99	0.308	0.240
	PCO	5.3	76.6	5.20	76.7	33.43	27.05	0.274

3.2. Stress Determining Physiological Traits

Physiological traits measure the response of plants to different phenomena taking place internally, such as cell injury and production of certain proteins, such as proline, in response to heat stress. MNH-886 showed less CIY while exposed to high temperatures, with an average value of 65, while MNH-814 was susceptible to extreme temperatures and CIY was higher with a value of 80. CIM-557 showed 64.2, while MNH-886 showed a significant value of 76.6 for proline content in heat stress. MNH-814 was found as the most susceptible among fourteen experimental cultivars and showed a proline content value of 5.3 under stress (Table 4).

Table 4. Stress determining physiological traits (Relative Water Content = RWC, water potential = WP, Osmotic Potential = OP, CIY = Cell injury = CIY, proline Contents = PCO).

Population Size	Traits	Parents (Means)		F$_2$ Population Statistical Data				
		MNH-886	MNH-814	Max	Min	Mean	SD	Skew
94	RWC	47.65	43.06	54.91	40.28	47.65308	3.815905	0.321
	WP	20.30	19.00	27	15	19.65	2.511	0.283
	OP	860.76	805.92	975	727	833.34	65.07	0.382
	CIY	65	80	50	90	64.60	10.36	0.477
	PCO	5.3	7.66	7.66	5.3	6.48	2.05	0.274

3.3. Correlation

Correlation (Figure 3) was observed by OriginPro 8.5 software and it was observed that plant height showed a positive correlation with the number of fruiting branches per plant, total number of nodes, size of petiole and balls, length of bracts, length of petals, and length of pistil but it had no correlation with total number of flowers, whereas plant height was negatively correlated with total number of nodes, first sympodial node height, sympodial nose number bearing first effective boll, sympodia node height, bearing last effective boll, cell injury, total number of sympods, length of staminal column, and proline content. Fully dehiscent anther had a positive correlation with sympodial node number, percent boll set on first position, percent boll set on second position along sympodia, total number of sympods, total number of nodes, size of petiole, boll number, total number of bolls, and length of bract. Hence, the length of petiole, proline contents, sympodial node number, percent boll set on first and second position along sympodia, total number of nodes, size of petiole, branch number, total number of bolls, length of bract, and length of petiol were all positively correlated with each other and a significant effect was observed among the traits.

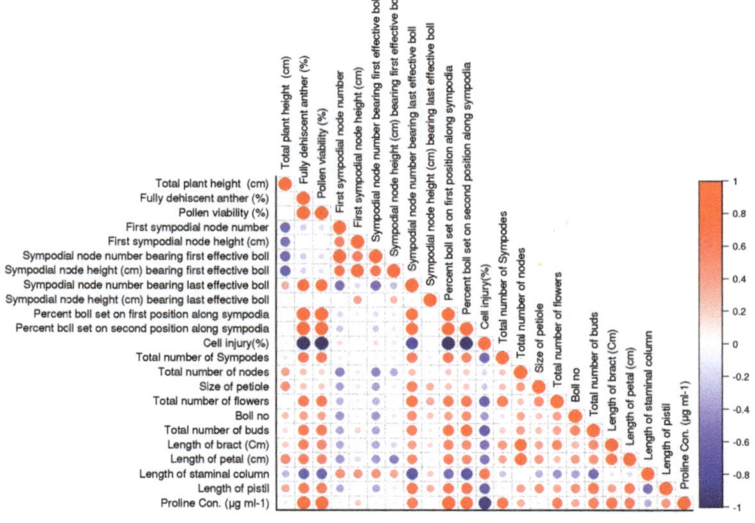

Figure 3. Pearson correlation among phenotypic traits of cotton under heat stress (= + ve = - ve).

3.4. Construction and Characterization of Intra Specific Linkage Map

Among the 1450 SSR primer pairs tested on parental lines, 175 markers were found to be polymorphic. These markers were applied on population. Using a LOD score > 3.0, these markers were assigned to 26 chromosomes for population based on the information on the cotton SSR map [47]. The linkage map was constructed for the F_2 population. Each linkage group was assigned to specific chromosome (Figure 4). The linkage maps covered approximately 4402.7 cm (Table 5) with an average distance of 20 cm within the markers which, according to the position of SSR markers, is common with the cotton map [48]. We estimate that we surveyed close to 70% of the cotton map, comparing the length of our map with that of the cotton map. The genetic map for the population was generated by MAPMAKER/version 3.0. Genotypic frequencies deviation from the expected segregation ratio of 1:2:1 for the co-dominant locus or 3:1 for the dominant locus was detected with the legitimacy of the additive-dominance model by means of the Chi square (χ^2) method [49].

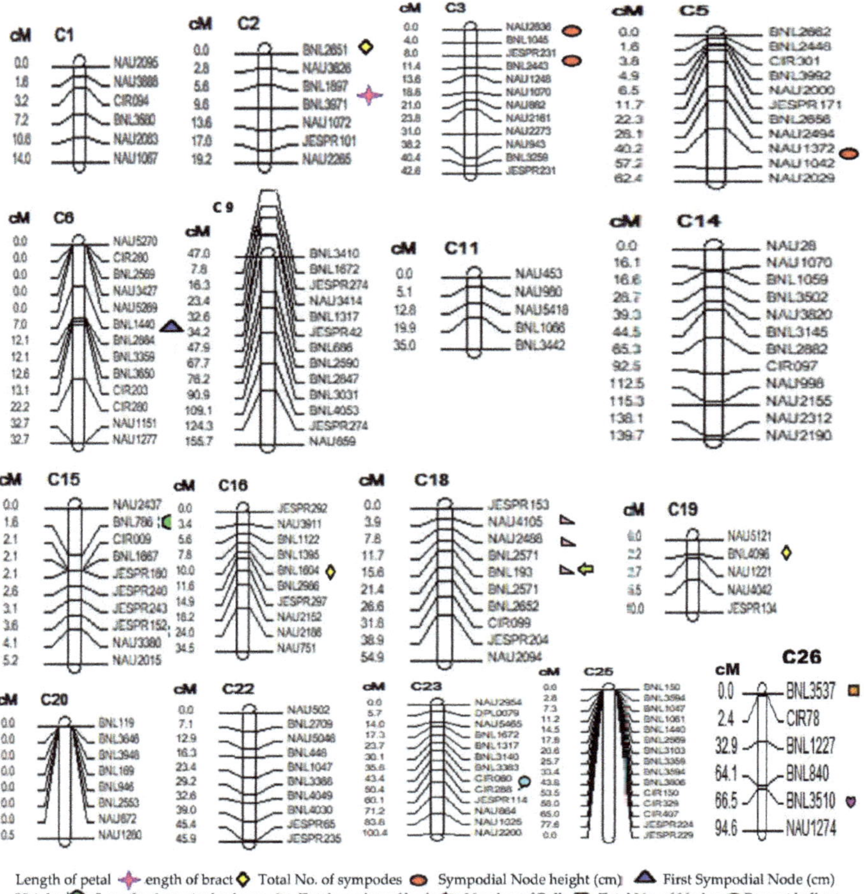

Figure 4. Linkage map and QTLs for heat stress tolerance determined in an F_2 population made from cross among Intraspecific MNH-886 and MNH-814 (*G. hirsutum*). The gap indicates that the linkage distance of the primer loci > 50 (cm) indicate significant QTLs (Kosambi).

Table 5. Basic characteristics of the genetic map.

Item	Field Exp. Pop
Total no. of SSR loci	175
No. of mapped loci	171
No. of individuals	94
No. of linkage groups	17
No. of unlinked loci	4
Length of map (cm)	4402.7
Total no. of skewed loci	24

3.5. QTLs Mapping for Traits Associated with Heat Tolerance in Cotton

A summary of statistically important QTLs is shown in Table 6. All QTLs for First Sympodial node height, Sympodial node height, percent boll set along sympodia at 2nd position along sympodia, Total number of Sympodes, Total no. of nodes, number of bolls, Total no of buds, Length of bract, Length of Staminal column, and and Length of petal are shown in Figure 3. A total of 17 regions were recognized that contain the QTLs with LOD value 3.0 and above. The most noteworthy QTLs are described in Table 6.

Table 6. QTLs related to heat tolerance in Intraspecific cross among MNH-886 and MNH-814. (LOD = Logarithm of odds, Additive = Additional effects, Dominance/additive = ratio between dominance and additive effects, PV% = Phenotypic variance).

QTLs	Chr. No.	SSR Markers	LOD Value	Additive	Dominance	Dominance/Additive	PV% Age
First Sympodial Node Height (cm)							
qFSHa1	15	BNL786-CIR009	6.10	0.59	−0.80	−1.36	36.62
qFSHa2	15	JESPR152-NAU3380	6.09	0.58	−0.81	−1.39	35.98
Sympodial Node Height (cm)							
qSNH1	6	BNL1440-BNL2884	3.42	0.77	−0.31	−0.40	17.59
Percent Boll Set on Second Position Along Sympodia							
qPBS1	26	BNL3510-NAU1274	18.19	0.69	0.35	0.50	14.56
Total No. of Sympodes							
qTNSa1	03	NAU2836-BNL1045	3.59	6.00	0.41	0.07	10.05
qTNSa2	03	JESPR231-BNL2443	3.71	6.27	0.38	0.06	10.12
qTNSa3	05	NAU1372-NAU1042	3.98	2.89	−0.50	−0.17	16.93
Total No. of Nodes							
qTNN1	23	CIR080-CIR288	4.05	0.18	0.03	0.17	12.91
Number of Bolls							
qNOB1	26	BNL3537-CIR078	3.80	4.25	−3.15	−0.74	21.52
Total Number of Buds							
qTNB1	18	BNL193-BNL2571	3.79	1.05	−0.74	−0.70	17.67
Length of Bract							
qLOBa1	02	BNL2651-NAU3626	3.24	0.18	0.04	0.20	8.59
qLOBa2	16	BNL1604-BNL2986	3.01	−0.13	−0.03	0.23	7.76
qLOBa3	19	NAU5121-BNL4096	4.05	0.18	0.03	0.17	12.91
Length of Staminal Column							

Table 6. *Cont.*

QTLs	Chr. No.	SSR Markers	LOD Value	Additive	Dominance	Dominance/Additive	PV% Age
qLOSa1	18	JESPR153-NAU4105	3.78	0.52	0.11	0.20	16.30
qLOSa2	18	NAU2488-BNL2571	3.76	0.52	0.11	0.20	15.84
qLOSa3	18	BNL193-BNL2571	3.07	0.30	0.11	0.36	14.57
			Length of Petal				
qLOP1	02	BNL1897-BNL3971	3.56	0.45	-0.02	-0.05	19.46

3.5.1. QTLs for First Sympodial Node Height (FSH)

Two QTLs, *qFSHa1* and *qFSHa2*, for first sympodial node height were detected on chromosome 15 with LOD $\geq$ 6.0, which explain 35% and 36% of phenotypic variance in F_2 respectively. These Loci detected 35%–36% of the PV value. When two QTLs were assembled together, they explained 71% of the PV value. The additive values for both QTLs were 0.59 and 0.58 respectively (Table 6). Position of the QTLs on the linkage map is shown in Figure 4.

3.5.2. QTLs for Sympodial Node Height Bearing First Effective Boll Set (SNH)

One QTL, *qSNH1*, influencing Last Effective Boll Set with a LOD score of 3.42 was detected in the F_2 population and it was located on chromosome 6. Putative QTL in this region accounted for 17% of phenotypic variance. So, this QTL explained 17% of the total phenotypic variance (Table 6 and Figure 4).

3.5.3. QTLs for Percent Boll Set on Second Position along Sympodia (PBS)

In the F_2 population of one QTL, *qPBS1*, the total influencing number of nodes was identified with a LOD score of 18.21 and it was located on chromosome 26. Phenotypic variance in this region was 14.56% (Table 6). The additive value for this QTL was 0.69.

3.5.4. QTLs for Total No of Sympodes (TNS)

Two QTLs, *qTNSa1* and *qTNSa2*, on chromosome 03 were detected for a total number of sympodes with LOD values 3.59 and 3.71 respectively. Phenotypic variance was observed between 10.05% and 10.12%, and the additive effect was 6.00 and 6.27 respectively. A total of 22% of phenotypic variance was seen when two QTL were fitted simultaneously. The third QTL*qTNSa3* was detected on chromosome 05 with LOD value 3.98. The additive effect was 2.89. Phenotypic variance observed was 16.93%. (Table 6).

3.5.5. QTL for Total No of Nodes (TNN)

On chromosome 23 single QTL *qTNN1*was detected for total no of nodes with LOD value 4.05. Positive additive effect was seen with value 0.18. Phenotypic variance seen was 12.91% (Table 6).

3.5.6. QTLs for Number of Bolls (NOB)

In experiment, one QTL, *q NOB1*, for Length of bract was identified on chromosome 26 with accumulative phenotypic variance of 21.52%. The LOD value was 3.80. So, this QTL showed phenotypic variance of 22%. Additive positive effect of *q NOB1* was 4.25 (Table 6).

3.5.7. QTLs for Total No of Buds (TNB)

In the F_2 population, one QTL, *qTNB1*, influencing the Total No. of buds was identified with a LOD score of 3.79 and it was located on chromosome 18. Phenotypic variance in this region was 17.67%. Additive effect was positive with value 1.05 (Table 6).

3.5.8. QTLs for Length of Bract (LOB)

Three QTLs, *qLOBa1*, *qLOBa2*, and *qLOBa3*, for length of bract were detected during analysis. The first QTL was on chromosome 2 with LOD ≥ 3.24 and with a positive additive effect of 0.18. Phenotypic variance observed was 8.59%. The second QTL was detected on chromosome 16 with LOD ≥ 3.01 and a negative additive effect of 0.13 and phenotypic variance 7.76%. The third QTL was detected on chromosome 19 with a positive additive effect of 0.18 and phenotypic variance of 12.91%. When three QTLs were fitted together simultaneously the phenotypic variance was 28% (Table 6).

3.5.9. QTL for Length of Staminal Column (LOS)

Three QTL, *sqLOSa1*, *qLOSa2*, and *qLOSa3*, were detected on chromosome number 18 for length of staminal column. The LOD values were 3.78, 3.76, and 3.07 respectively. Results showed positive additive effects of 0.52, 0.52, and 0.30 for the three QTLs, while 16.30, 15.84, and 14.57 were the values for phenotypic variance for all QTLs. When the three QTL's were fitted together simultaneously the phenotypic variance was 47% (Table 6).

3.5.10. QTLs for Length of Petal (LOP)

One QTL, *qLOP1*, was identified that influenced Length of petal trait. The QTL was located on chromosome number 2. The LOD value for QTL was 3.56. Phenotypic variance was 19.46. The QTL showed a positive additive effect of 0.45 (Table 6).

4. Discussion

The current study was carried out to identify the genetic basis responses of cotton plants under heat stress. The data collected were at the parental line and then after by F_2 generation from which heat-susceptible and heat-tolerant genotypes were selected for the screening process. Initially, the emergence of the first sympodial branch at lower nodes determined the early maturity of cotton plants. Theoretically, it is implicated for the 1st sympodial branch to appear on lower nodes as it is highly correlated with earliness and heat tolerance [50,51]. The strong relationship between early maturity and lower sympodial branch node number was reported in previous studies [52]. It was reported that there was a strong association of the 1st sympodial branch node number and heat tolerance. Highly significant differences were found in analysis of variance for the 1st sympodial node number (Table 4). The data of correlation (Figure 3) showed a positive correlation of the 1st sympodial node number expressed with all the traits except sympodial node number, present boll set on first position along sympodia, cell injury, and length of pistil. Node number to set the initial fruiting sympodia is a reliable and realistic morphological trait of heat tolerance [53]. Minimum and maximum temperature significantly affected the first sympodial branch with 1st boll [54]. All genotypes under study differed significantly for this trait (Table 4). Hussain et al. (2000) revealed similar results for plant height under heat stress, presenting a familiar correlation among traits that plant height has a positive correlation with the morphological traits under study [55]. Boll development was affected by the high temperature stress as compared with vegetative phase and a similar reduction in boll weight was observed when the temperature fluctuated [56]. Morris (1964) also reported a reduction in cotton boll maturity time at high temperature stress [57]. After screening the genotypes on morphological parameters, one genotypes was selected as tolerant against heat stress and another one was selected as heat susceptible, among others, on the basis of physiological characteristics, i.e., relative water contents, water potential, osmotic potential, cell injury, and proline contents. Highly significant differences were perceived by analysis of variance for all the physiological traits among the genotypes, except photosynthesis rate, which is significant (Table 3). The membrane structure of plant cells was distorted under severe temperature stress, which caused the increased permeability of membrane. As a result, electrolyte leakage increased and eventually led to cell death [58]. Azhar et al. (2009) measured the heat tolerance in term of relative cell injury percentage in cotton and found that thermal stress-tolerant genotypes were more stable for seed cotton yield and also maintained fiber quality as compared with heat-susceptible genotypes. A significant decrease was observed in leaf relative water content % (RWC) for heat-susceptible genotypes when exposed to heat stress, and similar findings were also obtained by Rahman et al. (2000), Siddique et al. (2000), and Parida et al. (2007) under stress conditions [59–61]. Higher leaf relative water content (RWC) could be a criterion for selection of a parent for hybridization to develop stress-tolerant genotypes [62,63]. On the basis of grand mean attained from normal and heat-stress situations, the protein contents was variable among genotypes and Raison et al. (1982) revealed that for temperature conditions above the optimum, significant reticence of photosynthesis takes place, resulting in substantial reduction in protein formation [64].

Finally, it was observed that high heat tolerance is a multigenic trait and its expression is controlled by many QTLs. Almost all the vegetative and floral characteristics of cotton plants were affected adversely because of this stress. The identification of QTLs activated to combat heat stress allowed the estimation of genetic architecture and improvement of heat-tolerance traits by molecular marker-assisted selection (MAS). A total of 1450 markers were applied, among which 175 SSR markers were observed to be polymorphic and were found to be significant; the observations were also in accordance with some other researchers [65]. In order to dissect the genetic basis of heat tolerance, two upland cotton cultivars (MNH-886 and MNH-884) were selected as parents and an F_2 population was developed. A high LOD (logarithm of odds) value provided strong evidence that the reported QTLs are actually associated with the respective traits. We only reported QTLs whose LOD score values were greater than three and which showed a significant additive or dominance genetic effect. A total of 17 QTLs with different effects on ten morphological and physiological traits such as First sympodial node height (FSH), sympodial node height (SNH), Percent boll set along sympodia on 2nd position (PBS), total no. of sympods (TNS), total no. of nodes (TNN), number of bolls (NOB), total no. of buds (TNB), length of bracts (LOB), length of staminal column (LOS), and length of petal (LOP) were detected in the present study. These QTLs were mapped on chromosome numbers 2, 3, 5, 6, 15, 16, 18, 19, 23, and 26. QTLs for length of petal and length of bracts were located on Chr. 2 while QTLs for total no. of buds and length of staminal column were located on Chr. 18 [66,67]. Likewise, QTLs for Boll no. and Percent boll set along sympodia on 2nd position were located on Chr. 26. Our findings are in accordance with work carried out by [68,69].

5. Conclusions

The purpose of cotton breeding is to boost and stabilize its yield in abiotic and biotic stress environments and to make cultivars with such physiological and architectural characteristics that can tolerate heat stress conditions. A low level of polymorphism is one of the major constraints for plant breeders and geneticists that can be attributed to the different processes like selection and domestication. It resulted in narrowing genetic shuffling in cotton. The use of an enormous number of SSRs can overcome the constraint of low polymorphism. In this study project, more than 1450 SSRs were assessed and the polymorphism rate was 12%, meaning the genetic diversity level was low owing to Intraspecific cross and segregation distortions. In spite of Intraspecific cross, 17 QTLs were detected by evaluating earlier-used and some novel traits. QTL detection can be attributed to a high rate of diversity in both parents. SSR markers were found best to deal with and easiest to assess polymorphism. The main goal of cotton breeding is to help increase and stabilize its productivity in stress environments and to develop cultivars with morphological traits which can withstand heat conditions. Our data suggest that favorable alleles for morphological traits can be combined to improve heat stress tolerance in cotton. Comparisons could be made to evaluate the consistency of QTL detection for the same trait in various backgrounds, which will help to determine the value of targeting these loci for selection in breeding programs.

6. Future Recommendation

Such coverage in the localization of QTLs controlling different quantitative traits suggested a close genotypic correlation among these traits or a pleiotropic effect of a single gene. It remains to be tested whether these common genomic regions have pleiotropic effects or there are clusters of tightly linked genes for some related traits in these regions. A more numerous mapping population and more closely spaced markers in the map are needed to determine whether the QTLs correspond to a gene with pleiotropic effects or to several separate but closely linked genes, each controlling a single character.

Author Contributions: Conceptualization, S.R., M.B. and S.A.M.; writing and draft preparation, S.R., M.B., T.N. and S.A.M.; writing, review, and editing, S.R., M.B., T.N. and S.A.M.; supervision, M.B. and S.A.M. All authors have read and agreed to the published version of the manuscript.

Funding: Authors are grateful to the Higher Education Commission (HEC) Pakistan, for providing funds under PIN 074-0747-Bm4-151, Indigenous 5000 Fellowship Program for this study.

Acknowledgments: The presented study is part of research proposal Mapping QTL's associated with heat tolerance in cotton using microsatellite markers. Author is grateful to Allah almighty, supervisors, Peng Chee (Associate University of Georgia, USA), Edward Lubbers (Research Associate at cotton molecular breeding lab, university of Georgia, Tifton campus, USA) Bahauddin Zakariya University Multan and Higher Education Commission (HEC) Pakistan.

Conflicts of Interest: The authors declare no conflict of interest. The funders had no role in the design of the study; in the collection, analyses, or interpretation of data; in the writing of the manuscript, and in the decision to publish the results.

References

1. IVAN Study Investigators; Chakravarthy, U.; Harding, S.P.; Rogers, C.A.; Downes, S.M.; Lotery, A.J.; Wordsworth, S.; Reeves, B.C. Ranibizumab versus bevacizumab to treat neovascular age-related macular degeneration: One-year findings from the IVAN randomized trial. *Ophthalmology* **2012**, *119*, 1399–1411. [CrossRef] [PubMed]
2. Dabbert, T.A.; Gore, M.A. Challenges and perspectives on improving heat and drought stress resilience in cotton. *J. Cotton Sci.* **2014**, *18*, 393–409.
3. Chapagain, A.K.; Hoekstra, A.Y.; Savenije, H.H.; Gautam, R. The water footprint of cotton consumption: An assessment of the impact of worldwide consumption of cotton products on the water resources in the cotton producing countries. *Ecol. Econ.* **2006**, *60*, 186–203. [CrossRef]
4. Ashraf, M. Inducing drought tolerance in plants: Recent advances. *Biotechnol. Adv.* **2010**, *28*, 169–183. [CrossRef] [PubMed]
5. Wendel, J.F.; Brubaker, C.; Alvarez, I.; Cronn, R.; Stewart, J.M. Evolution and natural history of the cotton genus. In *Genetics and Genomics of Cotton*; Springer: New York, NY, USA, 2009; pp. 3–22.
6. Grover, C.E.; Zhu, X.; Grupp, K.K.; Jareczek, J.J.; Gallagher, J.P.; Szadkowski, E.; Seijo, J.G.; Wendel, J.F. Molecular confirmation of species status for the allopolyploid cotton species, *Gossypium ekmanianum* Wittmack. *Genet. Resour. Crop Evolut.* **2015**, *62*, 103–114. [CrossRef]
7. Senchina, D.S.; Alvarez, I.; Cronn, R.C.; Liu, B.; Rong, J.; Noyes, R.D.; Wendel, J.F. Rate variation among nuclear genes and the age of polyploidy in *Gossypium*. *Mol. Biol. Evol.* **2003**, *20*, 633–643. [CrossRef] [PubMed]
8. Cronn, R.; Wendel, J.F. Cryptic trysts, genomic mergers, and plant speciation. *New Phytol.* **2004**, *161*, 133–142. [CrossRef]
9. Banuri, T. Pakistan: Environmental impact of cotton production and trade. *Int. Inst. Sustain. Dev.* **1998**, *161*.
10. Janjua, P.Z.; Samad, G.; Khan, N.U.; Nasir, M. Impact of climate change on wheat production: A case study of Pakistan [with comments]. *Pak. Dev. Rev.* **2010**, *49*, 799–822. [CrossRef]
11. Riaz, M.; Farooq, J.; Sakhawat, G.; Mahmood, A.; Sadiq, M.A.; Yaseen, M. Genotypic variability for root/shoot parameters under water stress in some advanced lines of cotton (*Gossypium hirsutum* L.). *Genet. Mol. Res.* **2013**, *12*, 552–561. [CrossRef]
12. Farooq, J.; Anwar, M.; Rizwan, M.; Riaz, M.; Mahmood, K.; Mahpara, S. Estimation of correlation and path analysis of various yield and related parameters in cotton (*Gossypium hirsutum* L.). *Cotton Genom. Genet.* **2015**, *6*, 1–6.
13. Khan, N.; Faqir, M.A.; Khan, A.A.; Rashid, A. Measurement of canopy temperature for heat tolerance in upland cotton: Variability and its genetic basis. *Pak. J. Agri. Sci.* **2014**, *51*, 359–365.
14. Reddy, A.R.; Reddy, K.R.; Padjung, R.; Hodges, H.F. Nitrogen nutrition and photosynthesis in leaves of Pima cotton. *J. Plant Nutr.* **1996**, *19*, 755–770. [CrossRef]
15. Rosolem, C.A.; Sarto, M.V.M.; Rocha, K.F.; Martins, J.D.L.; Alves, M.S. Does the introgression of BT gene affect physiological cotton response to water deficit? *Planta Daninha* **2019**, *37*, 1–7. [CrossRef]
16. Jarwar, A.H.; Wang, X.; Iqbal, M.S.; Sarfraz, Z.; Wang, L.; Ma, Q.; Shuli, F. Genetic divergence on the basis of principal component, correlation and cluster analysis of yield and quality traits in cotton cultivars. *Pak. J. Bot* **2019**, *51*, 1143–1148. [CrossRef]

17. Wang, M.; Tu, L.; Yuan, D.; Zhu, D.; Shen, C.; Li, J.; Liu, F.; Pei, L.; Wang, P.; Zhao, G.; et al. Reference genome sequences of two cultivated allotetraploid cottons, *Gossypium hirsutum* and *Gossypium barbadense*. *Nat. Genet.* **2019**, *51*, 224–229. [CrossRef]
18. Tester, M.; Langridge, P. Breeding technologies to increase crop production in a changing world. *Science* **2010**, *327*, 818–822. [CrossRef]
19. Swinnen, J.; Vandeplas, A. Rich consumers and poor producers: Quality and rent distribution in global value chains. *J. Glob. Dev.* **2012**, *2*, 1–30. [CrossRef]
20. Wang, K.; Song, X.; Han, Z.; Guo, W.; John, Z.Y.; Sun, J.; Pan, J.; Kohel, R.J.; Zhang, T. Complete assignment of the chromosomes of *Gossypium hirsutum* L. by translocation and fluorescence in situ hybridization mapping. *Theor. Appl. Genet.* **2006**, *113*, 73–80. [CrossRef]
21. Park, Y.-H.; Alabady, M.S.; Ulloa, M.; Sickler, B.; Wilkins, T.A.; Yu, J.; Stelly, D.; Kohel. R.J.; El-Shihy, O.M.; Cantrell, R.G. Genetic mapping of new cotton fiber loci using EST-derived microsatellites in an interspecific recombinant inbred line cotton population. *Mol. Genet. Genom.* **2005**, *274*, 428–441. [CrossRef]
22. Li, H.; Pan, Z.; He, S.; Jia, Y.; Geng, X.; Chen, B.; Wang, L.; Pang, B.; Du, X. QTL mapping of agronomic and economic traits for four F_2 populations of upland cotton. *J. Cotton Res* **2021**, *4*, 1–12. [CrossRef]
23. Tyagi, P.; Gore, M.A.; Bowman, D.T.; Campbell, B.T.; Udall, J.A.; Kuraparthy, V. Genetic diversity and population structure in the US Upland cotton (*Gossypium hirsutum* L.). *Theor. Appl. Genet.* **2014**, *127*, 283–295. [CrossRef] [PubMed]
24. Zhang, J.; Stewart, J.M. Economical and rapid method for extracting cotton genomic DNA. *J. Cotton Sci.* **2000**, *4*, 193–201.
25. Iqbal, M.; Ul-Allah, S.; Naeem, M.; Ijaz, M.; Sattar, A.; Sher, A. Response of cotton genotypes to water and heat stress: From field to genes. *Euphytica* **2017**, *213*, 1–11. [CrossRef]
26. Akhtar, K.P.; Ullah, R.; Khan, I.A.; Saeed, M.; Sarwar, N.; Mansoor, S. First symptomatic evidence of infection of *Gossypium arboreum* with Cotton leaf curl Burewala virus through grafting. *Int. J. Agric. Biol.* **2013**, *15*, 157–160.
27. Wahid, A.; Gelani, S.; Ashraf, M.; Foolad, M.R. Heat tolerance in plants: An overview. *Environ. Exp. Bot.* **2007**, *61*, 199–223. [CrossRef]
28. Li, F.; Fan, G.; Lu, C.; Xiao, G.; Zou, C.; Kohel, R.J. Genome sequence of cultivated Upland cotton (*Gossypium hirsutum* TM-1) provides insights into genome evolution. *Nat. Biotechnol.* **2015**, *33*, 524–530. [CrossRef]
29. Zhang, T.; Hu, Y.; Jiang, W.; Fang, L.; Guan, X.; Chen, J.; Zhang, J.; Saski, C.A.; Scheffler, B.E.; Stelly, D.M.; et al. Sequencing of allotetraploid cotton (*Gossypium hirsutum* L. acc. TM-1) provides a resource for fiber improvement. *Nat. Biotechnol.* **2015**, *33*, 531–537. [CrossRef]
30. Said, J.I.; Lin, Z.; Zhang, X.; Song, M.; Zhang, J. A comprehensive meta QTL analysis for fiber quality, yield, yield related and morphological traits, drought tolerance, and disease resistance in tetraploid cotton. *BMC Genom.* **2013**, *14*, 1–22. [CrossRef]
31. Said, J.I.; Knapka, J.A.; Song, M.; Zhang, J. Cotton QTLdb: A cotton QTL database for QTL analysis, visualization, and comparison between *Gossypium hirsutum* and *G. hirsutum* × *G. barbadense* populations. *Mol. Genet. Genom.* **2015**, *290*, 1615–1625. [CrossRef]
32. Burke, H.R.; Clark, W.E.; Cate, J.R.; Fryxell, P.A. Origin and dispersal of the boll weevil. *Bull. ESA* **1986**, *32*, 228–238. [CrossRef]
33. Stewart, J.M.; Hsu, C.L. In-ovulo embryo culture and seedling development of cotton (*Gossypium hirsutum* L.). *Planta* **1977**, *137*, 113–117. [CrossRef]
34. Barnett, N.M.; Naylor, A.W. Amino acid and protein metabolism in Bermuda grass during water stress. *Plant Physiol.* **1966**, *41*, 1222–1230. [CrossRef]
35. Bates, L.S.; Waldem, R.P.; Teare, I.D. Rapid determination of free proline for water-stress studies. *Plant Soil* **1973**, *39*, 205–207. [CrossRef]
36. Paterson, A.H.; Brubaker, C.L.; Wendel, J.F. A rapid method for extraction of cotton (*Gossypium* spp.) genomic DNA suitable for RFLP or PCR analysis. *Plant Mol. Biol. Report.* **1993**, *11*, 122–127. [CrossRef]
37. Reddy, O.U.K.; Pepper, A.E.; Abdurakhmonov, I.; Saha, S.; Jenkins, J.N.; Brooks, T.; El-Zik, K.M. New Dinucleotide and Trinucleotide Microsatellite Marker Resources for Cotton Genome Research. *J. Cotton Sci.* **2001**, *5*, 103–113.
38. Nguyen, T.-B.; Giband, M.; Brottier, P.; Risterucci, A.-M.; Lacape, J.-M. Wide coverage of the tetraploid cotton genome using newly developed microsatellite markers. *Theor. Appl. Genet.* **2004**, *109*, 167–175. [CrossRef]
39. Han, Z.G.; Guo, W.Z.; Song, X.L.; Zhang, T.Z. Genetic mapping of EST-derived microsatellites from the diploid *Gossypium arboreum* in allotetraploid cotton. *Mol. Genet. Genom.* **2004**, *272*, 308–327. [CrossRef]
40. Han, Z.; Wang, C.; Song, X.; Guo, W.; Gou, J.; Li, C. Characteristics, development and mapping of *Gossypium hirsutum* derived EST-SSRs in allotetraploid cotton. *Theor. Appl. Genet.* **2006**, *112*, 430–439. [CrossRef]
41. Zhang, J.; Guo, W.; Zhang, T. Molecular linkage map of allotetraploid cotton (*Gossypium hirsutum* L. × Gossypium barbadense L.) with a haploid population. *Theor. Appl. Genet.* **2002**, *105*, 1166–1174.
42. Tanksley, S.D.; Young, N.D.; Paterson, A.H.; Bonierbale, M.W. RFLP mapping in plant breeding: New tool for an old science. *Biotechnology* **1989**, *7*, 257–264. [CrossRef]
43. Soller, M.; Brody, T.; Genizi, A. On the power of experimental designs for the detection of linkage between marker loci and quantitative loci in crosses between inbred lines. *Theor. Appl. Genet.* **1976**, *47*, 35–39. [CrossRef] [PubMed]
44. Lander, E.S.; Botstein, D. Mapping mendelian factors underlying quantitative traits using RFLP linkage maps. *Genetics* **1989**, *121*, 185–199. [CrossRef]
45. Knapp, T.R. Treating ordinal scales as interval scales: An attempt to resolve the controversy. *Nurs. Res.* **1990**, *39*, 121–123. [CrossRef] [PubMed]

46. Haley, C.S.; Knott, S.A. A simple regression method for mapping quantitative trait loci in line crosses using flanking markers. *Heredity* **1992**, *69*, 315–324. [CrossRef]
47. Song, K.; Slocum, M.; Osborn, T. Molecular marker analysis of genes controlling morphological variation in *Brassica rapa* (syn. campestris). *Theor. Appl. Genet.* **1995**, *90*, 1–10. [CrossRef]
48. Kosambi, D.D. The estimation of a map distance from recombination values. *Ann. Eugen.* **1994**, *12*, 172–175. [CrossRef]
49. Larntz, K. Small-sample comparisons of exact levels for chi-squared goodness-of-fit statistics. *J. Am. Stat. Assoc.* **1978**, *73*, 253–263. [CrossRef]
50. Mutschler, M.A.; Doerge, R.W.; Liu, S.C.; Kuai, J.P.; Liedl, B.E.; Shapiro, J.A. QTL analysis of pest resistance in the wild tomato *Lycopersicon pennellii*: QTLs controlling acylsugar level and composition. *Theor. Appl. Genet.* **1996**, *92*, 709–718. [CrossRef]
51. Baloch, M.J.; Veesar, N.F. Identification of plant traits for characterization of early maturing upland cotton varieties. *Biol. Sci.-PJSIR* **2007**, *50*, 128–132.
52. Reddy, K.R.; Hodges, H.F.; Reddy, V.R. Temperature effects on cotton fruit retention. *Agron. J.* **1992**, *84*, 26–30. [CrossRef]
53. Azhar, F.M.; Ali, Z.; Akhtar, M.M.; Khan, A.A.; Trethowan, R. Genetic variability of heat tolerance, and its effect on yield and fibre quality traits in upland cotton (*Gossypium hirsutum* L.). *Plant Breed.* **2009**, *128*, 356–362. [CrossRef]
54. Hussain, M.; Azhar, F.M.; Khan, A.A. Genetics of inheritance and correlations of some morphological and yield contributing traits in upland cotton. *Pak. J. Bot.* **2009**, *41*, 2975–2986.
55. Wang, D. *RFLP Mapping, QTL Identification, and Cytogenetic Analysis in Sour Cherry*; Michigan State University: East Lansing, MI, USA, 1998.
56. Siddique, M.R.B.; Hamid, A.I.M.S.; Islam, M.S. Drought stress effects on water relations of wheat. *Bot. Bull. Acad. Sin.* **2000**, *41*, 35–39.
57. Baloch, A.; Rind, A.; Jamali, K. Genetic maps and marker assisted selection for major gene traits in rice. *Pak. J. Biotechnol.* **2004**, *1*, 33–46.
58. Parida, A.K.; Dagaonkar, V.S.; Phalak, M.S.; Umalkar, G.V.; Aurangabadkar, L.P. Alterations in photosynthetic pigments, protein and osmotic components in cotton genotypes subjected to short-term drought stress followed by recovery. *Plant Biotechnol. Rep.* **2007**, *1*, 37–48. [CrossRef]
59. Guo, W.; Cai, C.; Wang, C.; Han, Z.; Song, X.; Wang, K.; Niu, X.; Wang, C.; Lu, K.; Shi, B.; et al. A microsatellite-based, gene-rich linkage map reveals genome structure, function and evolution in *Gossypium*. *Genetics* **2007**, *176*, 527–541. [CrossRef]
60. Malik, R.S.; Dhankar, J.S.; Turner, N.C. Influence of soil water deficits on root growth of cotton seedlings. *Plant Soil* **1979**, *53*, 109–115. [CrossRef]
61. Baloch, A.W.; Solangi, A.M.; Baloch, M.; Baloch, G.M.; Abro, S. Estimation of heterosis and heterobeltiosis for yield and fiber traits in F1 hybrids of upland cotton (*Gossypium hirsutum* L.) genotypes. *Pak. J. Agri. Agril. Engg. Vet. Sci.* **2015**, *31*, 221–228.
62. Peng, M.W. *Global Strategy. Cengage Learning*; Cengage 200 Pier 4 Boulevard: Boston, MA, USA, 2021.
63. Raison, J.K.; Berry, J.A.; Armond, P.A.; Pike, C.S. Membrane properties in relation to the adaptation of plants to temperature stress. In *Adaptation of Plants to Water and High Temperature Stress*; Turner, N.C., Kramer, P.J., Eds.; John Wiley and Sons: New York, NY, USA, 1980; pp. 261–273.
64. Kushanov, F.N.; Turaev, O.S.; Ernazarova, D.K.; Gapparov, B.M.; Oripova, B.B.; Kudratova, M.K.; Abdurakhmonov, I.Y. Genetic diversity, QTL mapping and MAS technology in cotton (*Gossypium* spp.). *Front. Plant Sci.* **2021**, *12*, 29–71. [CrossRef]
65. Iqbal, M.; Naeem, M.; Rizwan, M.; Nazeer, W.; Shahid, M.Q.; Aziz, U.; Aslam, T.; Ijaz, M. Studies of genetic variation for yield related traits in upland cotton. *Am. Eurasian J. Agric. Environ. Sci.* **2013**, *13*, 611–618.
66. Gipson, J.R.; Joham, H.E. Influence of night temperature on growth and development of cotton (*Gossypium hirsutum* L.). III. Fiber elongation. *Crop Sci.* **1969**, *9*, 127–129. [CrossRef]
67. Odongo, I.; Ssemambo, R.; Kungu, J.M. Prevalence of Escherichia Coli and its antimicrobial susceptibility profiles among patients with UTI at Mulago Hospital, Kampala, Uganda. *Interdiscip. Perspect. Infect. Dis.* **2020**, *2020*, 8042540. [CrossRef] [PubMed]
68. Rahman, M.; Yasmin, T.; Tabassum, N.; Ullah, I.; Asif, M.; Zafar, Y. Studying the extent of genetic diversity among *Gossypium arboreum* L. genotypes/cultivars using DNA fingerprinting. *Genet. Resour. Crop Evol.* **2008**, *55*, 331–339. [CrossRef]
69. Shaheen, T.; Zafar, Y.; Rahman, M.-U. QTL mapping of some productivity and fibre traits in *Gossypium arboreum*. *Turk. J. Bot.* **2013**, *37*, 802–810. [CrossRef]

 agronomy

Article

Disentangling the Genetic Diversity of Grass Pea Germplasm Grown under Lowland and Highland Conditions

Mehmet Arslan [1,*], Engin Yol [1] and Mevlüt Türk [2]

1 Department of Field Crops, Faculty of Agriculture, Akdeniz University, 07058 Antalya, Turkey
2 Department of Field Crops, Faculty of Agriculture, Isparta University of Applied Science, 32200 Isparta, Turkey
* Correspondence: mehmetarsalan@akdeniz.edu.tr; Tel.: +90-242-227-44-00 (ext. 6501)

Abstract: Grass pea is recognized as one of the most resilient and versatile crops, thriving in extreme environments. It has also high protein content and suitable for forage production. These abilities make the crop a superior product for guaranteeing food security in changing climate conditions. To address this concern, a total of 94 accessions were assessed in relation to three qualitative and 19 quantitative traits in lowland (Antalya, Turkey) and highland (Isparta, Turkey) conditions. There were significant differences among genotypes for all agronomic traits in lowland location. The maximum biological yield was detected in GP104 and GP145 with values of 22.5 and 82.4 g in lowland and highland, respectively. The *t*-test of significance for mean values indicated that there were significant differences between the growing areas for all agronomic traits except for number of pods. Principal component analysis using the 11 agronomic traits including maturity, yield and yield related-traits showed that 76.4% and 72.2% variability were accounted for the first four principal components (PCs) with eigenvalues ≥ 1 in collection grown in highland and lowland, respectively. The data on variations in agronomic, quality and forage traits detected in this research provided useful genetic resources. The parental genotypes which have desired traits can be used in grass pea improvement programs to develop new cultivars.

Keywords: *Lathyrus sativus*; climate change; neglected legume; forage quality

Citation: Arslan, M.; Yol, E.; Türk, M. Disentangling the Genetic Diversity of Grass Pea Germplasm Grown under Lowland and Highland Conditions. *Agronomy* **2022**, *12*, 2426. https://doi.org/10.3390/agronomy12102426

Academic Editors: Channapatna S. Prakash, Ali Raza, Xiling Zou and Daojie Wang

Received: 23 September 2022
Accepted: 4 October 2022
Published: 6 October 2022

1. Introduction

The impact of climate change on agriculture varies depending on the region [1]. The environmental effects on plants can emerge as reduced water availability, rising temperatures, the pest and disease epidemics and an increase in the frequency of different extreme events [2]. Alternative varieties/types or new crops are required to provide a steady food supply under changing environmental conditions [3]. Most of agricultural systems in different countries, therefore, have re-designed their breeding approaches based on growing population and climate change. Increasing research focused on developing and improving plants that are underutilized or neglected should be one of the important parts of new breeding studies. *Lathyrus sativus* L. is a legume crop that belongs to the Lathyrus genus, which includes 187 different species and subspecies [4,5]. Grass pea is recognized as one of the most resilient and versatile crops, thriving in extreme environments and climatic conditions such as cold, heat, drought, salinity-affected soils, and flooding, and is resistant to insect attacks, when compared to other legume crops [4,6]. These characteristics make it a superior product for guaranteeing food security, particularly in the face of anticipated climate challenges [3]. Its seeds contain about 8.6–34.6% protein content, which is higher than chickpea [7], and can replace rapeseed and soybean meal in animal feed, moderately [8]. It is therefore used, not only for human food, but also for livestock feed, forage and green manure [9]. According to Hanbury et al. [10], the grass pea offers a cheap, high-protein, and currently under-utilized feed source due to the rising demand for animal products.

However, excessive consumption of grass pea seeds can cause lathyrism, occurred by the non-protein amino acid β-N-oxalyl-L-α, β-diaminopropionic acid (β-ODAP). It is thought that the presence of β-ODAP as a free amino acid in seeds and in significant concentrations in drought-tolerant grass pea is what causes this debilitating illness [11] affecting both animals and humans [12]. Indeed, when consumed in high quantities over an extended period of time (as is frequently the case during famine), "lathyrism" can result in permanent paralysis and brain damage [12]. According to a report by Abd El Moneim et al. [13], grass pea seeds should have an β-ODAP level of less than 0.22% for safe consumption to reduce the danger of lathyrism. Environmental conditions have a significant impact on the amount of β-ODAP content in grass pea seeds [4]. Its concentration varies commonly among both genotypes and environments [14]. Further domestication and improving of this crop for food (as low β-ODAP) and fodder (as high as biological and seed yield) have been made necessary [15]. Therefore, generating germplasm/cultivars with a low β-ODAP content should be the main goal of both traditional and contemporary breeding programs on Lathyrus [15]. Additionally, the majority of traditional breeding programs for grass peas have emphasized using the selection criterion to increase yield (number of branches per plant). The single node double blooms or pods, higher protein content, 100-seed weight, and forage traits are also important characteristics that can be used in grass pea breeding studies [15,16].

The improvement of quantitative traits related to yield and quality is the main target of breeding programs [17]. With the use of only few elite lines and/or cultivars make a limited contribution to the improvement studies because of their narrow genetic base [18]. Selection from a collection with high diversity makes it more possible to discover the desired traits [19]. Lots of genetic diversity studies have been conducted in grass pea for different regions [5,20–26] to find new traits and develop cultivars in the respective region. However, there is no study which was conducted in two different climatic conditions (lowland and highland) in grass pea. From this perspective, we evaluated a total of 94 grass pea accessions with agro-morphological traits, nutritional contents and β-ODAP contents in lowland and highland environmental areas. The evaluation of agronomic, food and forage traits for economic importance should be useful for choosing the appropriate genetic resources for crop improvement.

2. Materials and Methods

2.1. Genetic Material

The grass pea collection, 250 accessions from USDA and 24 accessions from ICARDA, were evaluated as an initial genetic material in this study (Table S1). This genetic resource was coded "GP" and ordered. It was grown in the experimental field of Akdeniz University in Antalya, Turkey (36°53′56.2″ N 30°38′30.3″ E) in 2018. However, 184 out of 274 accessions were discarded from the study because they did not produce sufficient seeds. Remaining genetic material (90 accessions) and four registered controls (Karadağ, İptaş, Gürbüz and Ceora) (Table S2) were used to conduct field trials in both locations, Antalya and Isparta, Turkey (Figure 1).

2.2. Field Trials

2.2.1. First Year in Field Trials (2019)

A total of 94 grass pea genotypes were grown in the experimental field of Akdeniz University at Antalya, Turkiye (36°53′56.2″ N 30°38′30.3″ E) (lowland). Before seeding, 15-15-15 (N-P-K) compound fertilizer was applied as 10 kg da^{-1} during the soil tillage stage. Grass pea seeds were sown on 5 December 2019 in a randomized complete blocks design with two replications. Each genotype was grown in two rows of 3 m length with a row-to-row distance of 50 cm and plant to plant within a row of 10 cm. The plants were not irrigated during the trial period. Standard agronomic practices were applied to all plots. The harvest was performed on 30 May 2020.

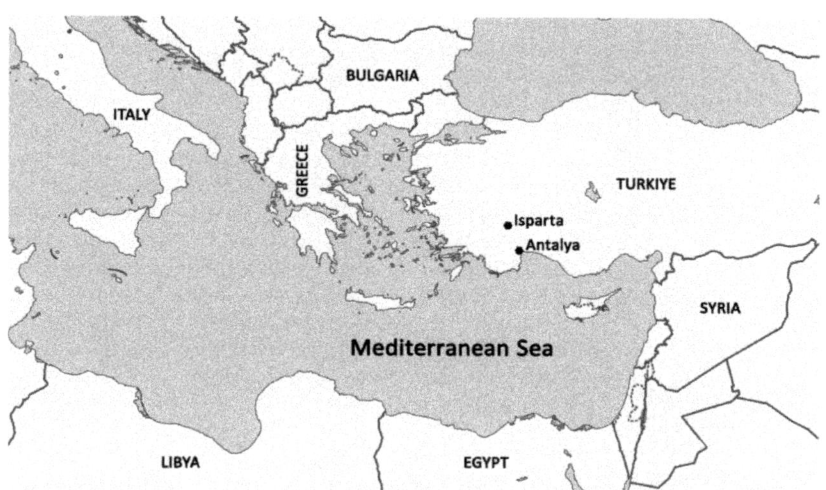

Figure 1. (•) The site of experimental areas (Isparta (37°50′12.3″ N 30°32′31.1″ E) and Antalya (36°53′56.2″ N 30°38′30.3″ E), Turkiye).

2.2.2. Second Year in Field Trials (2020)

The 2nd year studies were carried out with the same genetic material in both locations, Antalya (lowland) and Isparta (highland). The field trial in Antalya was established on 23 December 2020 in the same experimental area with a randomized complete blocks design with two replications. The seeds were also sown on 20th October 2020 in the field of Isparta Applied Sciences University (37°50′12.3″ N 30°32′31.1″ E) in Isparta with an augmented experimental design. The experimental field was divided into three blocks of equal size and each block had four checks replicated across the three blocks in augmented treatment. Soil type of each location was monitored in Table 1. It shows that the pH of the soils in both Antalya and Isparta fields are neutral, the lime content is high, slightly salty and sufficient in terms of organic matter. Each experimental field consisted of a 3 m row on a ridge. Spacing was 50 cm between rows and 10 cm between plants in both locations. Similarly, standard agronomic practices were applied in both fields. The seeds were harvested on 25 May 2021 and 15 June 2021 in Antalya and in Isparta respectively.

Table 1. The chemical and physical properties of soil in the experiment fields.

Characteristics	Values		Evaluation
	Antalya	Isparta	
pH	7.30	7.66	Slightly alkaline
Lime (%)	21.6	28.7	Too limy
EC micromhos/cm (25 °C)	485	375	Slightly salted
Texture	Clay-loam	Clay-loam	
Organic material, (%)	2.3	1.54	Enough
P ppm (Olsen)	5	23.5	20–25
K ppm	275	176.2	200–320

2.3. *Climatic Conditions*

The monthly temperature, precipitation, and humidity were recorded by the State Meteorology Station for Antalya (Table S3) and Isparta (Table S4). For 2018–2019 and 2019–2020, there were similar trends for temperatures, with the lowest temperatures occurring in January and February and the highest in June at Antalya. The highest rainfall was recorded in December. Additionally, there was less rainfall in 2018–2019 than in 2019–2020 in all growing seasons except for January, April and June. Similarly, the highest rainfall was observed in December and January in Isparta (highland), however at least 40% total rainfall was recorded compared to Antalya. The coldest month in highland is January, with 5.2 (°C), followed by February and March (Table S4)

2.4. Data Collection

Although all of the genotypes successfully passed the seedling period and bloomed, limited seed holding was occurred in 184 genotypes. The qualitative traits were recorded for the initial collection (274 accessions). However, 94 lines were used to get observations for quantitative and nutritional traits. Grass pea descriptors [4] were used for recording qualitative and quantitative characteristics. Plants were characterized by the following traits: days of first flowering, days of 50% flowering, plant height, pod number per plant, number of branches per plant, pod height, pod width, stem diameter, hundred seed weight, biological yield per plant and seed yield per plant. In addition, β-ODAP, raw protein, acid detergent fiber (ADF), and neutral detergent fiber (NDF) were determined from the seeds of grass pea.

2.5. Chemical Analysis

2.5.1. β-ODAP Content

In the first stage of analysis, a blender was used for homogenization. These samples were divided into 1-g portions in individual 50-mL sample tubes. Then, 25 mL of extraction solution, 0.1% (*v*/*v*) formic acid in water:methanol (50:50) (*v*/*v*), was added to the sample tube. For recovery studies, a standard was added to the tube at this stage. The mixture was extracted using Ultra-Turrax (IKA, Germany) for 2 min at 10,000 rpm. Centrifuging was done on the extracted samples for 10 min at 4 °C and 4000 rpm. The supernatant was passed through a 0.2-μm PTFE membrane filter. Filtered samples were diluted with a mobile phase and injected at 10 μL volumes to LC-MS/MS. β-ODAP was purchased from ChemFaces (Wuhan, China), with high purity (>98%). After extraction, β-ODAP content was identified with the UHPLC-MS/MS method, which has been detailed in the study by Arslan et al. [27].

2.5.2. Forage Traits

The Kjeldahl method was used for nitrogen content, and the crude protein ratio was calculated using a conversion factor of 6.25. ADF and NDF concentrations were determined according to standard laboratory procedures for forage quality analysis outlined by Ankom Technology. ANKOM F57 filter bags were used for ADF and NDF analysis in this study. Total digestible nutrients (TDN), dry matter intake (DMI), digestible dry matter (DDM), and relative feed value (RFV) were estimated [28] according to the following equations adapted from:

$$\text{TDN} = (-1.291 \times \text{ADF}) + 101.35$$

$$\text{DMI} = 120\%\ \text{NDF}\%\ \text{dry matter basis}$$

$$\text{DDM} = 88.9 - (0.779 \times \text{ADF}\%\ \text{dry matter basis})$$

$$\text{RFV} = \text{DDM}\% \times \text{DMI}\% \times 0.775$$

2.6. Data Analysis

Qualitative data were analyzed using percentage distribution. Analysis of variance was conducted using SAS 9.1 (Cary, NC, USA) [29]. Augmented randomized complete block design was performed using the "augmentedRCBD" package developed by Aravind et al. [30] in R-Studio (version 2022.02.0) (Boston, MA, USA) [31]. The least significant difference (LSD) test was used for mean comparison in the analysis of variance. Principal component analysis (PCA) was performed with the quantitative traits data using the Minitab 19.1 software (State College, PA, USA) [32]

3. Results

In this study, the large and diverse grass pea collection grown in highland and lowland conditions was evaluated with three qualitative, 11 quantitative, six forage and two quality traits. The frequency distribution of qualitative traits, flower colors, plant growth habits and seed color of the grass pea genotypes were shown in Figure 2. There was a large variation among grass pea genotypes in terms of flower colors. Nine different flower colors were observed in the grass pea collection as blue, white, blue-white, dark blue, purple, light blue, pink, blue-purple and red with percentages of 28.1, 24.45, 21.53, 13.5, 7.3, 2.55, 1.46, 0.73 and 0.36%. Furthermore, three plant growth habits were determined as erect (50%), semi-erect (40%) and spreading (10%). The seeds of the collection had a gray color with a value of 57%.

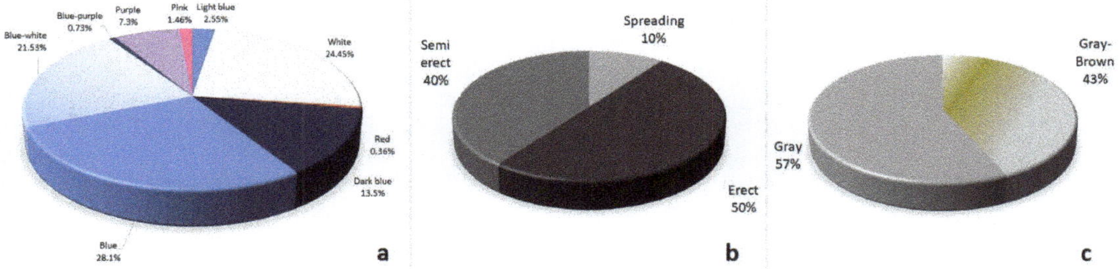

Figure 2. Frequency distribution of qualitative traits in the whole collection. (**a**) the color of flowers; (**b**) growing type; (**c**) seed color.

According to the analysis of variance, there were significant differences among genotypes for all traits in Antalya (lowland) location (Table S2). The mean number of days of the first flowering varied from 88.8 to 109, and the number of days of 50% flowering ranged from 99.5 to 120.5 in the collection. In the average of two years, the genotypes GP23, GP11, and GP114 had the earliest flowering date. The highest plant height was recorded in GP213 with a value of 90.8, while the shortest plant was GP10 had a plant height of 41.5 cm. The grand mean of number of pods was 23.77 and check cultivars, Corea and Gürbüz showed a higher number of pods than the mean of the germplasm. The maximum and minimum biological yield was detected in GP104 and GP23 with values of 22.5 and 3.3 g, respectively. The seed yield ranged from 1.2 to 7.3 g. The highest seed yield was observed in genotype GP105, followed by GP 104 and GP 249, the mean of collection was 3.01 g. Among the check cultivars, İptaş gave the highest seed yield, however 40 genotypes of the collection had higher means for seed yield than this check cultivar.

A total of 90 genotypes along with four check cultivars were evaluated at field condition in Isparta (highland) for quantitative agronomic traits using an augmented experimental design (Table S5). The analysis of variance showed that there was a significant (<0.01) difference among the genotypes for all traits except for days of the first flowering. The results revealed no significant differences were observed for blocks except for traits of number of pods, number of branches and stem diameter (Table S6). The number of days to first flowering ranged from 172.1 to 206.3 days, and the number of days to 50% flowering ranged from 181.3 to 213.3 days. The check cultivar, Corea had the earliest 50% flowering date among the check cultivars with a value of 178, only genotypes GP230 and GP247 had higher values compared to this cultivar (Table S5). GP107 and GP105 had the tallest plants, while GP246 was the shortest (15.6 cm). Genotype GP156 and GP145 gave the highest number of pods (62) as well as GP145 produced the highest biological yield being 82.4 g in the highland conditions. There was also a lot of variation in the collection for hundred-seed weight and seed yield traits, which ranged from 4.1 to 82.4 g and 0.9 to 33 g, respectively (Table S5). The genotype GP40 had the highest seed yield followed by GP161, GP18 and GP 19.

Significant differences were observed in the germplasm for quality and forage traits among genotypes grown in Antalya (lowland) (Table S7). The β-ODAP content (%) ranged from 0.25 to 0.49, the average being 0.38. The genotypes GP213, GP49, GP58, GP60 and GP110 had the lowest values for this trait while the genotypes GP248 and GP227 had the highest content. The lowest value was 0.38 among the check cultivars. The maximum amount of protein content was recorded for GP53 in the germplasm, with a notable higher value in the quantity of this trait observed in the genotypes of GP40, GP270 and GP197. Regarding ADF (%), genotypes GP251, GP243, GP248 and GP23 had the highest values over the 9.0 in the mean of two years, the check cultivar, Karadağ had the highest check cultivar with value of 8.82. There were four genotypes (GP34, GP156, GP225 and GP149) had an NDF > 17%, the mean of the collection was 13.98. The DMI and DDM ranges in the collection varied from 6.48 to 9.96 and 81.48 to 83.4 with mean values of 8.45 and 82.45, respectively.

The ANOVA analysis of the genotypes showed a highly significant variation in all quality and forage traits in highland conditions (Table S8). There is a non-significant difference among the blocks for these traits in augmented experiment design. Overall, the β-ODAP content (%) corresponding to the genotypes was 0.35 (Table S9). The lowest values were recorded for GP17, GP18 andGP49 with the means of 0.19, 0.21, 0.22, respectively. GP215 had the greatest mean value (0.51) for this trait. The maximum and minimum protein content (%) were detected in the genotypes GP242 and GP225. The genotype, GP248 also had >24% protein content which is higher than the mean of all check cultivars (Table S9). The mean of ADF and NDF values were 8.43 and 15.72, respectively. The highest mean was 10.22 for ADF and 24.23 for NDF. The genotype GP248 was superior for these traits whose means were 10.15 and 19.78, respectively, they were higher than general means and check cultivars. The highest DMI was recorded for GP242, followed by GP43 and GP207. The check cultivar, Corea was the fourth genotype for this trait. When all genotypes were combined, the total mean of total digestible nutrients was 90.47%, with GP213 being the top genotype for this trait. The highest relative feed value was observed in GP242 (668.63) while the lowest value was measured in GP199 (316.72) in the collection.

The grass pea collection examined in this study had 94 genotypes and was evaluated in lowland and highland conditions. The *t*-test of significance for mean values indicated that there were significant differences between the growing areas for all agronomic traits except for the number of branches (Table 2). These genotypes grown in two different environmental area were also compared with quality and forage traits. The mean values for β-ODAP content, raw protein ADF, NDF, DMI and RFV were significantly different between the genotypes grown in lowland and highland (Table 2).

The PCA using the 11 quantitative traits including maturity traits, yield and yield related-trait showed that more than 76.4% and 72.2% variability were accounted for the first four principal components (PCs) with eigenvalues $\geq$ 1 in the collection grown in Isparta (highland) and Antalya (lowland) (Table 3). The 1st principal component (PC1) had an eigenvalue of 3.89 and explained 35.4% of the total variation in highland. Seed yield and biological yield had the highest positive eigenvectors in PC1, while the pod height had the highest negative eigenvector. The second component (PC2) explained 17.1% of the total variance with an eigenvalue of 1.87 and was mainly correlated with flowering traits, negatively (Table 3). In lowland, the first principal component's (PC1) eigenvalue was 3.36, explaining 30.36% of the total variation, the highest positive eigenvector was biological yield (Table 3). The PC2 explained 16.3% of the total variance and was correlated with days to first flowering, days to 50% flowering, plant height, and the number of pods and stem diameter, positively. The traits of quality and forage were also evaluated with PCA for the grass pea collection grown in lowland and highland (Table 4). Results showed that in the analysis, three components had eigenvalues > 1 for highland andthey explained 93.7% of the variability among the 94 genotypes grown in highland. The PC1 explained 57.2% of the total variance and was positively correlated with all quality and forage traits except for ADF and NDF. The PC1 explained 52.79% of the total variance (75.9%) and was positively correlated with raw protein, TDN, DMI, DDM and RFV in the collection that was evaluated in lowland.

Table 2. Means and standard errors for 11 quantitative and eight quality traits in 94 genotypes produced in Antalya (lowland) and Isparta (highland).

	Antalya (Lowland)		**Isparta (Highland)**		
Agronomic Traits	**Mean**	**S.E. †**	**Mean**	**S.E. †**	**Differences #**
First flowering (day)	99.15	0.51	187.8	0.76	★★
50% flowering (day)	107.06	0.54	192.9	0.76	★★
Plant height (cm)	64.40	0.73	39.7	1.30	★★
Number of pods	23.77	0.55	22.0	1.34	ns
Number of branches	9.65	0.21	3.9	0.15	★★
Pod height (cm)	3.53	0.22	3.19	0.05	★★
Pod width (cm)	1.29	0.01	1.03	0.01	★★
Stem diameter (mm)	1.81	0.01	4.1	0.07	★★
Hundred seed weight (g)	13.15	0.18	14.3	0.48	★
Biological yield (g)	10.30	0.21	20.8	1.68	★★
Seed yield (g)	3.01	0.09	6.0	0.56	★★
Forage and quality traits	Mean	S.E. †	Mean	S.E. †	Differences #
Beta-ODAP	0.38	0.01	0.35	0.01	★
Raw protein	24.00	0.12	21.06	0.18	★★
Acid detergent fiber	8.20	0.03	8.43	0.08	★
Neutral detergent fiber	13.98	0.13	15.72	0.27	★★
Total digestible nutrients	90.66	0.04	90.47	0.10	ns
Dry matter intake	8.45	0.07	7.83	0.12	★★
Digestible dry matter	82.45	0.02	82.33	0.06	ns
Relative feed values	539.95	4.7	499.86	8.0	★★

† S.E.: standard error of the mean. # Differences between means of entire and core collection were tested by t test; ns is non-significant, $p = 0.05$ and $p = 0.001$, and ★ and ★★, respectively.

Table 3. Eigenvectors for the four principal components (PCs) of traits associated with agronomic performance of 94 grass pea genotypes produced in two different regions.

	Isparta (Highland)						**Antalya (Lowland)**	
				PC Axis				
	PC1	**PC2**	**PC3**	**PC4**	**PC1**	**PC2**	**PC3**	**PC4**
Eigenvalues	3.89	1.87	1.53	1.10	3.36	1.79	1.53	1.24
Explained proportion of variation, %	35.4	17.1	13.9	10.0	30.6	16.3	14.0	11.3
Cumulative proportion of variation, %	35.4	52.5	66.4	76.4	30.6	46.9	60.9	72.2
Traits				Eigenvectors				
First flowering (day)	−0.000	−0.640	0.233	−0.013	0.293	0.562	−0.155	−0.108
50% flowering (day)	−0.005	−0.669	0.148	−0.103	0.291	0.563	−0.154	−0.061
Plant height (cm)	0.169	0.032	0.182	−0.700	0.265	0.249	0.424	0.034
Number of pods	0.413	0.114	−0.036	−0.273	0.114	0.091	0.644	0.345
Number of branches	0.364	0.035	−0.034	−0.016	0.278	−0.163	0.406	0.037
Pod height (cm)	−0.104	−0.152	−0.690	0.109	0.307	−0.182	0.181	−0.502
Pod width (cm)	−0.035	−0.225	−0.624	−0.396	0.374	−0.231	−0.023	−0.457
Stem diameter (mm)	0.329	−0.069	−0.096	0.396	0.291	0.092	−0.246	0.298
Hundred seed weight (g)	0.341	−0.222	0.016	0.304	0.351	−0.257	−0.169	−0.084
Biological yield (g)	0.465	0.001	−0.109	0.054	0.382	−0.216	−0.231	0.375
Seed yield (g)	0.463	0.012	−0.042	−0.055	0.284	−0.253	−0.127	0.407

Table 4. Eigenvectors for principal components (PCs) of traits associated with forage and quality value of 94 grass pea genotypes produced in two different regions.

	Isparta (Highland)			Antalya (Lowland)	
	PC Axis				
	PC1	PC2	PC3	PC1	PC2
Eigenvalues	4.57	1.89	1.02	4.22	1.83
Explained proportion of variation, %	57.2	23.7	12.9	52.9	23.0
Cumulative proportion of variation, %	57.2	80.9	93.7	52.9	75.9
Traits	Eigenvectors				
Beta-ODAP	0.007	−0.086	−0.963	−0.151	−0.007
Raw protein	0.311	0.253	−0.252	0.307	−0.001
Acid detergent fiber	−0.372	0.440	−0.040	−0.393	0.369
Neutral detergent fiber	−0.398	−0.341	−0.055	−0.368	−0.422
Total digestible nutrients	0.372	−0.440	0.040	0.382	−0.424
Dry matter intake	0.402	0.353	0.022	0.382	0.418
Digestible dry matter	0.371	−0.440	0.040	0.382	−0.424
Relative feed values	0.411	0.328	0.023	0.394	0.389

4. Discussion

This study was carried out in two different locations; lowland and highland, showed that this special grass pea collection has great variation with respect to seed yield, yield components, quality and forage traits. Obtaining a higher seed yield for different environmental conditions is one of the most important challenges in plant breeding [19]. However, improving the traits related to yield characteristics such as double podding, more seeds per pod, plant height or branches are also highly critical to obtain desired grass pea lines [4]. Our collection was characterized by three qualitative and 19 quantitative traits to develop desired cultivars. Similarly, different germplasm resources have been characterized with different agro-quality traits in grass pea [20,22,25,33,34].

Based on comparison between their altitudes, the mean number of days to the first flowering was found to be shorter in lowland environmental conditions. Altitude levels and sun exposure times are thought to be responsible for these differences [35]. The results obtained in our study were found to be higher when compared to the studies conducted in European countries and India [35–37]. Furthermore, the highest day of 50% flowering was determined as 213.3 days (GP107) in highland (Table S5) and the lowest mean was 120.5 days (GP237) in lowland (Table S6). While our findings regarding the flowering period are in agreement with the results of Çakmakçı and Çeçen [38], Şeydoşoğlu et al. [39] and Öten et al. [40], they are higher than the findings of Kumari [37]. Grela et al. [41] stated that variation of these traits depends on the environmental factors, especially on soil type and precipitation amount during vegetation period. Plant height effects both seed and biological yield in grass pea. In the present study, plant height ranged from 41.5 cm to 90.8 cm in lowland conditions and 15.6–76.8 cm in highland conditions. In the research of grass pea, it was reported that the plant height was determined between 24.5 cm and 172.0 cm by different studies [35,42,43]. These results clearly revealed that environmental differences highly affect the plant height trait of grass pea. We also had large variation for the number of branches and comparable results were also observed as 6.10–13.00 [44], 3.73–6.00 [45], 4.6–8.6 [46]. The hundred-grain weight is considerable in terms of giving an idea about the grain's size, fullness, thinness and flour yield. We obtained the maximum value as 23.8 in highland and 29.39 in lowland. These are higher values than those obtained by Aksu et al. [43] and Başaran et al. [44]. However, Grela et al. [35], Ribinski et al. [47] monitored higher weights in this trait compared to our maximum results. Biological yield frequently used as yield selection criteria, especially in studies to increase grain yield in cereal and legume plants [48]. The maximum biological yield was detected in GP104 and GP145 in lowland and highland, respectively. These genotypes, therefore, should be used in cultivar development program and also as parents in crossing programs to obtain superior lines. Looking at the seed yield, we observed increased mean values compared to the check cultivar average values. Moreover, in comparison with Antalya and Isparta for seed yields, we observed significant differences between genotypes (Table S8). According to Pandey et al. [34], seed yield per plant in grass pea ranged from 0.5 to 19.7 g, whereas Ribinski et al. [47] claimed that it ranged from 7.20 to 21.19 g. Especially, we obtained lower values in lowland compared to highland and these previous studies. According to Das et al. [15] one of the important factors influencing seed yield is ecological difference. In addition, PCA analysis demonstrated that seed yield and biological yield had high and positive values in PC1 both environmental areas. Similar these traits positively contributed to PC was obtained by Polignano et al. [49] when they characterized of genetic diversity of grass pea entries.

Genotype x environment effects play an important role in the phenotypic expression of ODAP content, which is polygenically inherited [50]. Low ODAP is frequently linked to undesired features such as late flowering and low seed and biological yields [51]. Therefore, new breeding programs have been successful in achieving both low ODAP, high yield and protein in recent years [5]. The β-ODAP content of the genotypes grown in lowland (Antalya) and highland (Isparta) regions has significantly different (Tables S7 and S9). We obtained large variation for β-ODAP content among the genotypes and lands and it is showed that this trait is affected by both genetic and environmental factors [52,53]. The genotypes GP2 and GP49 genotypes in collection are notable for having low β-ODAP concentrations in both Antalya and Isparta. These accessions therefore provide better opportunities for developing high seed yielding and low B-ODAP cultivars suitable for studied regions. Previous studies indicated that there was no genotype of grass pea that was β-ODAP free, although in several genotypes the β-ODAP content was low [6]. Futhermore, many researchers [14,54,55] found that these low-toxin cultivars did not have stable -ODAP levels in grass pea seeds when grown under different environmental conditions. The β-ODAP content in grass pea seed had high variation depends on genotype and environmental conditions and it ranged from 0.02 to 7.2% [56]. Onar et al. [57] reported that they found the amount of β-ODAP in local grass pea varieties grown in Turkey, ranged from 0.10% to 0.87% (*w*/*w*). Arslan et al. [27] investigated the β-ODAP contents of 173 local grass pea genotypes in lowland conditions and obtained values ranging from 1.55 mg/g to 20.8 mg/g, showing the genotype effect on this trait.

Crude protein content is a significant indicator of feed quality [58]. The minimum crude protein content in ruminant diet should be around 6.0 to 8.0% of dry matter for adequate activity of rumen microorganisms [59], suggesting that hay crude protein content in investigated grass peas is more than twice or thrice the needed ratios. The highest crude protein ratios were found in GP242 (24.64%) and GP248 (24.08%) genotypes in highland (Table S9), GP53 (27.09%), GP40 (26.88%) and GP270 (26.55%) in lowland (Table S7). There were significant differences among genotypes grown in two different locations (Table 2). Differences in crude protein ratios of the genotypes were mainly resulted from plant genetics, but leaf, spike and stem ratios, ripening periods, temperature and fertilization might also have significant effects on crude protein contents [60]. Present crude protein contents of some genotypes were similar to the findings of Basaran et al. [61]. In fact, Basaran et al. [61] stated that due to ecological differences in the regions where grass pea genotypes are collected, variation in seed crude protein concentrations can be linked to ecological factors rather than genetic variation.

Increasing ADF ratios reduces digestibility of the feeds and increasing NDF ratios reduces feed intake and make the animals feel full, thus limit feed intake and feed availability. Since high ADF and NDF ratios have negative effects on feed intake and digestibility, feeds with ideal ADF and NDF values are usually preferred [62]. ADF content of grass pea varied between 6.87 and 9.74% in lowland, while they varied between 7.19 and 10.22% in highland. Lower ADF values are preferred for animal production due to the negative correlation between ADF values and ruminant digestion [63]. Therefore, genotypes having the lowing values should be taken into consideration for forage breeding. NDF content of grass pea varied between 11.15 and 17.58% in lowland, while it varied between 11.52 and 24.23% in highland. Grela et al. [35] found NDF between 11.25 and 18.92%, while Karadag and Yavuz [63] found it between 10.18 and 13.55%. The monogastric and ruminants should have lower NDF content in their feed. Furthermore, ruminant animals may require a certain amount of NDF values, but higher NDF values may reduce animal intake [63]. The TDN refers to the nutrients that are available for livestock and are related to the ADF concentration of the forage [64]. As ADF increases, there is a decline in TDN which means that animals are not able to utilize the nutrients that are present in the forage [65]. In Antalya ecological conditions, line GP270 had the highest TDN and DDM values, while line GP248 had the lowest. When the averages of both regions in terms of TDN, DMI, DDM and RFV were compared, it was determined that the values in Antalya were higher. According to the Hay Market Task Force of American Forage and Grassland Council standards, the genotype is classed as premium quality when it has protein content > 19, ADF < 31%, NDF < 40%, and RFV > 151. The scale showed that lots of genotypes in the collection should be classed as premium with regard to forage quality. Considering all the results, GP60 in lowland and GP40 in highland were considered the most promising lines for grass pea breeding with their high crude protein content, low ADF, NDF and β-ODAP ratios, as well as biological yield.

Supplementary Materials: The following supporting information can be downloaded at: https://www.mdpi.com/article/10.3390/agronomy12102426/s1, Table S1: The initial genetic material of the study; Table S2: Means of 11 agronomic traits for grass pea collection produced in the 2020 and 2021 growing seasons, Antalya (lowland), Turkey; Table S3: Monthly temperature, precipitation and humidity values of experimental area in Antalya (lowland); Table S4: Climatic data for experimental area in Isparta (highland); Table S5: Means of 11 agronomic traits for grass pea collection produced 2021 growing season, Isparta (highland), Turkey; Table S6 Analysis of variance of augmented block design for 11 agronomic traits in grass pea collection produced 2021 growing season in Isparta (highland), Turkey; Table S7: Means of quality and forage traits for grass pea collection produced in the 2020 and 2021 growing seasons, Antalya (lowland), Turkey; Table S8: Analysis of variance of augmented block design for forage and quality traits in grass pea collection produced 2021 growing season in Isparta (highland), Turkey; Table S9. Means of eight quality and forage traits for grass pea collection produced in the 2021 growing season, Isparta (lowland), Turkey

Author Contributions: Conceptualization, M.A. and E.Y.; methodology, E.Y.; validation, M.A. and M.T.; formal analysis, E.Y.; investigation, M.A.; resources, M.A.; data curation, M.A and M.T.; writing—original draft preparation, M.A. and E.Y.; writing—review and editing, E.Y.; visualization, E.Y.; supervision, E.Y.; project administration, M.A.; funding acquisition, M.A. All authors have read and agreed to the published version of the manuscript.

Funding: This study was funded by the Scientific Research Projects Coordination Unit of Akdeniz University, Turkey with the project code: FBA-2020-5294.

Institutional Review Board Statement: Not applicable.

Informed Consent Statement: Not applicable.

Data Availability Statement: Not applicable.

Acknowledgments: We are grateful to USDA, ARS Plant Genetic Resources Conservation Unit and International Center for Agricultural Research in the Dry Areas (ICARDA) for supplying genetic material several times.

Conflicts of Interest: The authors declare no conflict of interest. The funders had no role in the design of the study; in the collection, analyses, or interpretation of data; in the writing of the manuscript; or in the decision to publish the results.

References

1. Araujo, S.S.; Beebe, S.; Crespi, M.; Delbreil, B.; González, E.M.; Gruber, V.; Lejeune-Henaut, I.; Link, W.; Monteros, M.J.; Prats, E.; et al. Abiotic stress responses in legumes: Strategies used to cope with environmental challenges. *Crit. Rev. Plant Sci.* **2015**, *34*, 237–280. [CrossRef]
2. Hanjra, M.A.; Qureshi, M.E. Global water crisis and future food security in an era of climate change. *Food Policy* **2010**, *35*, 365–377. [CrossRef]
3. Gonçalves, L.; Rubiales, D.; Bronze, M.R.; Vaz Patto, M.C. Grass pea (*Lathyrus sativus* L.)—A sustainable and resilient answer to climate challenges. *Agronomy* **2022**, *12*, 1324. [CrossRef]
4. Campbell, C.G. *Grass Pea, Lathyrus sativus L.*; Promoting the Conservation and Use of Underutilized and Neglected Crops. Nr 18; Institute of Plant Genetics and Crop Plant Research: Rome, Italy; International Plant Genetic Resources Institute: Gatersleben, Germany, 1997; p. 92.
5. Hanbury, C.D.; Siddique, K.H.M.; Galwey, N.W.; Cocks, P.S. Genotype-environment interaction for seed yield and ODAP concentration of *Lathyrus sativus* L. and *L. cicera* L. in Mediterranean-type environments. *Euphytica* **1999**, *110*, 45–60. [CrossRef]
6. Vaz Patto, M.C.; Skiba, B.; Pang, E.C.K.; Ochatt, S.J.; Lambein, F.; Rubiales, D. Lathyrus improvement for resistance against biotic and abiotic stresses: From classical breeding to marker assisted selection. *Euphytica* **2006**, *147*, 133–147. [CrossRef]
7. Longvah, T.; Ananthan, R.; Bhaskarachary, K.; Venkaiah, K. *Indian Food Composition Table*; National Institute of Nutrition: Hyderabad, India, 2017; pp. 1–578.
8. Castell, A.G.; Cliplef, R.L.; Briggs, C.J.; Campbell, C.G.; Bruni, J.E. Evaluation of lathyrus (*Lathyrus sativus* L.) as an ingredient in pig starter and grower diets. *Can. J. Anim. Sci.* **1994**, *74*, 529–539. [CrossRef]
9. Fikre, A.; Korbu, L.; Kuo, Y.-H.; Lambein, F. The contents of the neuro-excitatory amino acid β-ODAP (β-N-oxalyl-L-α,β-diaminopropionic acid), and other free and protein amino acids in the seeds of different genotypes of grass pea (*Lathyrus sativus* L.). *Food Chem.* **2008**, *110*, 422–427. [CrossRef]
10. Hanbury, C.D.; White, C.L.; Mullan, B.P.; Siddique, K.H.M. A review of the use and potential of *Lathyrus sativus* L. and *L. cicera* L. grain for animal feed. *Anim. Feed. Sci. Technol.* **2000**, *87*, 1–27. [CrossRef]
11. Lambein, F.; Travella, S.; Kuo, Y.-H.; Van Montagu, M.; Heijde, M. Grass pea (*Lathyrus sativus* L.): Orphan crop, nutraceutical or just plain food? *Planta* **2019**, *250*, 821–838. [CrossRef]
12. Hillocks, R.J.; Maruthi, M.N. Grass pea (*Lathyrus sativus*): Is there a case for further crop improvement? *Euphytica* **2012**, *186*, 647–654. [CrossRef]
13. Abd El Moneim, A.M.; Van Dorrestein, B.; Baum, M.; Mulugeta, W. *Role of ICARDA in Improving the nutritional Quality and Yield Potential of Grass Pea (Lathyrus sativus) for Subsistence Farmers in Developing Countries: CGIAR-Wide Conference on Agriculture Nutrition*; International Food Policy Research Institute: Washington, DC, USA, 1999; pp. 5–6.
14. Dahiya, B.S.; Jeswani, L.M. Genotype and environment interactions for neurotoxic principle (BOAA) in grass pea. *Indian J. Agric. Sci.* **1975**, *45*, 437–439.
15. Das, A.; Parihar, A.K.; Barpete, S.; Kumar, S.; Gupta, S. Current perspectives on reducing the β-ODAP content and improving potential agronomic traits in grass pea (*Lathyrus sativus* L.). *Front. Plant Sci.* **2021**, *12*, 703275. [CrossRef] [PubMed]
16. Basaran, U.; Mut, H.; Onal-Asci, O.; Acar, Z.; Ayan, I. Variability in forage quality of Turkish grass pea (*Lathyrus sativus* L.) landraces. *Turk. J. Field Crops* **2011**, *16*, 9–14.
17. Upadhyaya, H.D.; Swamy, B.P.M.; Goudar, P.V.K.; Kullaiswamy, B.Y.; Singh, S. Identification of diverse groundnut germplasm through multienvironment evaluation of a core collection for Asia. *Field Crops Res.* **2005**, *93*, 293–299. [CrossRef]
18. Gupta, S.K.; Baek, J.; Carrasquilla-Garcia, N.; Penmetsa, R.V. Genome-wide polymorphism detection in peanut using next-generation restriction-site-associated DNA (RAD) sequencing. *Mol. Breed.* **2015**, *35*, 145. [CrossRef]
19. Yol, E.; Furat, S.; Upadhyaya, H.D.; Uzun, B. Characterization of groundnut (*Arachis hypogaea* L.) collection using quantitative and qualitative traits in the Mediterranean basin. *J. Integr. Agric.* **2018**, *17*, 63–75. [CrossRef]
20. Tay, J.; Valenzuela, A.; Venegas, F. Collecting and evaluating Chilean germplasm of grasspea (*Lathyrus sativus* L.). *Lathyrus Lathyrism Newsl.* **2000**, *1*, 21.
21. Abd El-Moneim, A.M.; Dorrestein, B.V.; Baum, M.; Ryan, J.; Bejiga, G. Role of ICARDA in improving the nutritional quality and yield potential of grasspea (*Lathyrus sativus* L.) for subsistence farmers in dry areas. *Lathyrus Lathyrism Newsl.* **2001**, *2*, 55–58.
22. Kumari, V. Stable genotypes of grasspea for mid hill conditionsof Himachal Pradesh. *Indian J. Genet.* **2000**, *60*, 399–402.
23. Tadesse, W.; Bekele, E. Variation and association of morphological and biochemical characters in grass pea (*Lathyrus sativus* L.). *Euphytica* **2003**, *130*, 315–324. [CrossRef]
24. Tadesse, W.; Bekele, E. Phenotypic diversity of Ethiopian grass pea (*Lathyrus sativus* L.) in relation to geographical regions and altitudinal range. *Genet. Resour. Crop Evol.* **2003**, *50*, 497–505. [CrossRef]
25. Tavoletti, S.; Iommarini, L.; Crinò, P.; Granati, E. Collection and evaluation of grasspea (*Lathyrus sativus* L.) germplasm of central Italy. *Plant Breed.* **2005**, *124*, 388–391. [CrossRef]
26. Sammour, R.H. Genetic diversity in Lathyrus sativus L. germplasm. *Res. Rev. BioSci.* **2014**, *8*, 325–336.
27. Arslan, M.; Oten, M.; Erkaymaz, T.; Tongur, T.; Kilic, M.; Elmasulu, S.; Cinar, A. β-N-oxalyl-L-2,3-diaminopropionic acid, L-homoarginine and asparagine contents in the seeds of different genotypes *Lathyrus sativus* L. as determined by UHPLC-MS/M. *Int. J. Food Prop.* **2017**, *20*, S108–S118. [CrossRef]
28. Horrocks, R.D.; Vallentine, J.F. *Harvested Forages*; Academic Press: London, UK, 1999; p. 426.

29. SAS Institute. *SAS/STAT Software 9.1*; SAS Institute Inc.: Cary, NC, USA, 2003.
30. Aravind, J.; Mukesh Sankar, S.; Wankhede, D.P.; Kaur, V. augmentedRCBD: Analysis of Augmented Randomised Complete Block Designs, R Package Version 0.1.5. 2021. Available online: https://aravind-j.github.io/augmentedRCBD/index.html (accessed on 20 August 2022).
31. R Core Team. *R: A Language and Environment for Statistical Computing*; R Foundation for Statistical Computing: Vienna, Austria, 2022. Available online: https://www.R-project.org/. (accessed on 24 August 2022).
32. MINITAB. Minitab Package Program. 2019.
33. Jain, H.K.; Somayajulu, N.; Barat, G.K. *Final Technical Report on Investigation in Lathyrus sativus*; Indian Agricultural Research Institute: New Delhi, India, 1994.
34. Pandey, R.L.; Chitale, M.W.; Sharma, R.N.; Geda, A.K. Evaluation and characterization of germplasm of grass pea (*Lathyrus sativus*). *J. Med. Aromat. Plants* **1997**, *19*, 14–16.
35. Grela, E.R.; Rybiński, W.; Klebaniuk, R.; Matras, J. Morphological characteristics of some accessions of grass pea (*Lathyrus sativus* L.) grown in Europe and nutritional traits of their seeds. *Genet. Resour. Crop Evol.* **2010**, *57*, 693–701. [CrossRef]
36. De la Rosa, L.; Martin, I. Morphological characterization of Spanish genetic resources of *Lathyrus sativus* L. *Lathyrus Lathyrism Newsl.* **2001**, *2*, 31–34.
37. Kumari, V. Field evoluation of grasspea (Lathyrus sativus L.) germplasm for its toxicty in the Norhwestern Hills of India. *Lathyrus Lathyrism Newsl.* **2001**, *2*, 82–84.
38. Çakmakçı, S.; Çeçen, S. The possibilities at entering crop rotation system of certain annual legume plants in Antalya. *Turk. J. Agric. For.* **1999**, *23*, 119–123.
39. Seydoşoğlu, S.; Saruhan, V.; Kökten, K.; Karadağ, Y. Determination of yield and yield components of some grasspea (*Lathyrus sativus* L.) genotypes in ecological conditions of Diyarbakır. *J. Agric. Fac. Gaziosmanpasa Univ.* **2015**, *32*, 98–109. [CrossRef]
40. Öten, M.; Kiremitçi, S.; Erdurmuş, C. The determination of yield characteristics of some grass pea (*Lathyrus sativus* L.) lines collected from Antalya natural flora. *Ege J. Agric. Res.* **2017**, *54*, 17–26.
41. Grela, E.R.; Rybinski, W.; Matras, J.; Sobolewska, S. Variability in phenotypic and morphological characteristics of some Lathyrus sativus L. and Lathyrus cicera L. accessions and nutritional traits of their seeds. *Genet. Resour. Crop Evol.* **2012**, *59*, 1687–1703. [CrossRef]
42. Ahmadi, J.; Vaezi, B.; Pour-Aboughadareh, A. Assessment of heritability and relationships among agronomic characters in grass pea (Lathyrus sativus L.) under rainfed conditions. *Biharean Biol.* **2015**, *9*, 29–34.
43. Aksu, E.; Dogan, E.; Arslan, M. Agro-morpholoogical performance of grass pea (*Lathyrus sativus* L.) genotypes with low B-ODAP content grown under Mediterranean environmental conditions. *Fresenius Environ. Bull.* **2021**, *30*, 638–644.
44. Basaran, U.; Acar, Z.; Karacan, M.; Onar, N. Variation and correlation of morpho-agronomic traits and biochemical contents (protein and β-Odap) in Turkish grass pea (*Lathyrus sativus* L.) landraces. *Turk. J. Field Crops* **2013**, *18*, 166–173.
45. Kosev, V.I.; Vasileva, V.M. Morphological characterization of grass pea (*Lathyrus sativus* L.) varieties. *J. Agric. Sci.-Sri Lanka* **2019**, *14*, 67–76. [CrossRef]
46. Kumar, S.; Dubey, D.K. Genetic diversity among induced mutants of grasspea (*Lathyrus sativus* L.). *Lathyrus Lathyrism Newsl.* **2003**, *3*, 15–17.
47. Rybinski, W.; Szot, B.; Rusinek, R. Estimation of morphological traits and mechanical properties of grass pea seeds (*Lathyrus sativus* L.) originating from EU countries. *Int. Agrophys.* **2008**, *22*, 261–275.
48. Enneking, D. The nutritive value of grass pea (*Lathyrus sativus* L.) and allied species, their toxicity to animals and the role of malnutrition in nerulathyrism. *Food. Chem. Toxicol.* **2011**, *49*, 694–709. [CrossRef]
49. Polignano, G.B.; Uggenti, P.; Olita, G.; Bisignano, V.; Alba, V.; Perrino, P. Characterization of grass pea (*Lathyrus sativus* L.) entries by means of agronomically useful traits. *Lathyrus Lathyrism Newsl.* **2005**, *4*, 9–14.
50. Sharma, R.N.; Kashyap, O.P.; Chitale, M.W.; Pandey, R.L. Genetic analysis for seed attributes over the years in grass pea (*Lathyrus sativus* L.). *Ind. J. Gen. Plant Breed.* **1997**, *57*, 154–157.
51. Kumar, S.; Bejiga, G.; Ahmed, S.; Nakkoul, H.; Sarker, A. Genetic improvement of grass pea for low neurotoxin (β-ODAP) content. *Food Chem. Toxicol.* **2011**, *49*, 589–600. [CrossRef] [PubMed]
52. Lambein, F.; Khan, J.K.; Kuo, Y.-H.; Campbell, C.G.; Briggs, C.J. Toxins in the seedlings of some varieties of grass pea (*Lathyrus sativus*). *Nat. Toxins* **1993**, *1*, 246–249. [CrossRef]
53. Abd El-Moneim, A.M.; Cocks, P.S. Adaptation and yield stability of selected lines of *Lathyrus* spp. under rainfed conditions in West Asia. *Euphytica* **1992**, *66*, 89–97. [CrossRef]
54. Siddique, K.H.M.; Loss, S.P.; Herwig, S.P.; Wilson, J.M. Growth, yield and neurotoxin (ODAP) concentration of three Lathyrus species in Mediterranean type environments of Western Australia. *Aust. J. Exp. Agric.* **1996**, *36*, 209–218. [CrossRef]
55. Dixit, G.P.; Parihar, A.K.; Bohra, A.; Singh, N.P. Achievements and prospects of grass pea (*Lathyrus sativus* L.) improvement for sustainable food production. *Crop J.* **2016**, *4*, 407–416. [CrossRef]
56. Despande, S.S.; Campbell, C.G. Genotype variation in BOOA, condensed tannins, phenolics and enzyme inhibitors in grass pea (*Latyhrus sativus*). *Can. J. Plant Sci.* **1992**, *72*, 1037–1047. [CrossRef]
57. Onar, A.N.; Erdoğan, B.Y.; Ayan, I.; Acar, Z. Homoarginine, β-ODAP, and asparagine contents of grass pea landraces cultivated in Turkey. *Food Chem.* **2014**, *143*, 277–281. [CrossRef]

58. Assefa, G.; Ledin, I. Effect of variety, soil type and fertilizer on the establishment, growth, forage yield, quality and voluntary intake by cattle of oats and vetches cultivated in pure stand and mixtures. *Anim. Feed. Sci. Technol.* **2001**, *92*, 95–111. [CrossRef]
59. Van Soest, P.J. *Nutritional Ecology of the Ruminant*, 2nd ed.; Comstock Publishing Associates: Ithaca, NY, USA, 1994.
60. Ball, D.M.; Collins, M.; Lacefield, G.D.; Martin, N.P.; Mertens, D.A.; Olson, K.E.; Putnam, D.H.; Undersander, D.J.; Wolf, M.W. *Understanding Forage Quality*; American Farm Bureau Federation Publication: Park Ridge, IL, USA, 2001.
61. Basaran, U.; Asci, O.O.; Mut, H.; Acar, Z.; Ayan, I. Some quality traits and neurotoxin β-N-oxalyl-L-α,β- diaminopropionic acid (β-ODAP) contents of Lathyrus sp. cultivated in Turkey. *Afr. J. Biotechnol.* **2011**, *10*, 4072–4080.
62. Kiraz, A.B. Determination of relative feed value of some legume hays harvested at flowering stage. *Asian J. Anim. Vet. Adv.* **2011**, *6*, 525–530. [CrossRef]
63. Karadag, Y.; Yavuz, M. Seed yields and biochemical compounds of grasspea (*Lathyrus sativus* L.) lines grown in semi-arid regions of Turkey. *Afr. J. Biotechnol.* **2010**, *9*, 8343–8348.
64. Surmen, M.; Yavuz, T.; Cankaya, N. Effects of phosphorus fertilization and harvesting stages on forage yield and quality of common vetch. *J. Food Agric. Environ.* **2011**, *9*, 353–355.
65. Aydin, N.; Mut, Z.; Mut, H.; Ayan, I. Effect of autumn and spring sowing dates on hay yield and quality of oat (*Avena sativa* L.) genotypes. *J. Anim. Vet. Adv.* **2010**, *9*, 1539–1545. [CrossRef]

Article

Estimation of Genetic Variances and Stability Components of Yield-Related Traits of Green Super Rice at Multi-Environmental Conditions in Pakistan

Imdad Ullah Zaid †, Nageen Zahra †, Madiha Habib, Muhammad Kashif Naeem, Umair Asghar, Muhammad Uzair, Anila Latif, Anum Rehman, Ghulam Muhammad Ali and Muhammad Ramzan Khan *

National Institute for Genomics and Advanced Biotechnology (NIGAB), National Agricultural Research Centre, Islamabad 45500, Pakistan; imdadcas@gmail.com (I.U.Z.); nageenzahra@hotmail.com (N.Z.); madihahabib217@gmail.com (M.H.); kashifuaar@gmail.com (M.K.N.); umairasghar308@gmail.com (U.A.); uzairbreeder@gmail.com (M.U.); anilalatif87@gmail.com (A.L.); anumrehman92@hotmail.com (A.R.); chairman@parc.gov.pk (G.M.A.)

* Correspondence: mrkhan@parc.gov.pk; Tel.: +92-5190733808

† These authors contributed equally to this work.

Abstract: Identifying adopted Green Super Rice (GSR) under different agro-ecological locations in Pakistan is crucial to sustaining the high productivity of rice. For this purpose, the multi-location trials of GSR were conducted to evaluate the magnitude of genetic variability, heritability, and stability in eight different locations in Pakistan. The experimental trial was laid out in a randomized complete block (RCB) design with three replications at each location. The combined analysis of variance (ANOVA) manifested significant variations for tested genotypes (g), locations (L), years (Y), genotype × year (GY), and genotype × location (GL) interactions revealing the influence of environmental factors (L and Y) on yield traits. High broad-sense heritability estimates were observed for all the studied traits representing low environmental influence over the expression of traits. Noticeably, GSR 48 showed maximum stability than all other lines in the univariate model across the two years for grain yield and related traits data. Multivariate stability analysis characterized GSR 305 and GSR 252 as the highest yielding with optimum stability across the eight tested locations. Overall, Narowal, Muzaffargarh, and Swat were the most stable locations for GSR cultivation in Pakistan. In conclusion, this study revealed that G×E interactions were an important source of rice yield variation, and its AMMI and biplots analysis are efficient tools for visualizing the response of genotypes to different locations.

Keywords: variability; heritability; univariate stability analysis; AMMI; GGE biplot analysis

Citation: Zaid, I.U.; Zahra, N.; Habib, M.; Naeem, M.K.; Asghar, U.; Uzair, M.; Latif, A.; Rehman, A.; Ali, G.M.; Khan, M.R. Estimation of Genetic Variances and Stability Components of Yield-Related Traits of Green Super Rice at Multi-Environmental Conditions in Pakistan. *Agronomy* **2022**, *12*, 1157. https://doi.org/10.3390/agronomy12051157

Academic Editor: Federica Zanetti

Received: 24 March 2022
Accepted: 28 April 2022
Published: 11 May 2022

1. Introduction

Rice is considered an important staple food crop across the globe, including in Pakistan. In 2020, Pakistan produced ten percent of the world's rice and ranked among the top ten rice producers worldwide (FAO, 2020) [1]. Rice is the sixth-largest export commodity of Pakistan. Pakistan exports more than 4.59 million t (making up 8% of the world's total rice trade), equivalent to 2.3 billion USD, which accounted for 10% of the world's total exports, ranking Pakistan third-largest rice exporter in terms of volume and value (International Trading Center (ITC) 2020). Rice cultivars grown in Pakistan are mainly divided into IIRI type, Basmati, and non-Basmati type. Basmati rice, an exclusive trademark of Pakistan with elongated and slender grains, soft and fluffy texture when cooked, and an aromatic taste, is one of the most appealing high-end rice in the international market. From September to December 2020, Basmati rice increased its footprint in the European market, retaining the minimum level of pesticide contamination per the European Union's standard. Moreover, rice exports rose in the country during November 2020, with 78,160 t valuing USD 76 m from 43,032 t to fetching USD 41 m in October 2020. According to the Rice Exporters Association of Pakistan (REAP), exports of coarse rice also expanded sharply to 379,944 t, with earnings of USD 154 m in November compared to 220,674 t fetching USD 98 m in October 2020.

Unfortunately, in Pakistan, the unavailability of certified seeds, diseases, and insect pests attack, uneven and limited distribution of water for paddy irrigation, fertilizer management, and post-harvest losses are critical factors in rice production. Moreover, occurrences of floods, temperature rises, droughts, and unusual rainfalls subsequently increase the skirmishes between rice production and environmental resources. Under these consequences, the fundamental breeding objective is to develop rice cultivars that reveal green traits, i.e., tolerance against multiple stresses, high nutrients-yield potential, and fertilizer–water-use efficiencies.

In the term "Green Super Rice (GSR)", the word "Green" means environmentally friendly as it grows more grains under fewer inputs while "Super" means more stress-tolerant. In the light of growing fluctuating resources, the development and adaptation of GSR also represent resources-saving, high-yielding, efficient, and ecologically stable rice [2]. Recently, 552 GSR advanced lines were introduced at National Institute for Genomics and Advanced Biotechnology (NIGAB) National Agriculture Research Council (NARC), Islamabad (Pakistan), to develop rice cultivars that retain sustainable yield even under unfavorable environmental conditions.

Before releasing a new variety for commercial purposes, plant breeders usually evaluate the set of genotypes across multi-environments. A stable genotype produces the expected yield in a particular environment [3]. The stronger a genotype–environment interaction is, the more unpredictable it is to assess the performance of a genotype in multi-environments [4]. Selection of a particular genotype becomes difficult due to genotype $\times$ environment interaction [5]. Hence, it is significant to assess the adaptation and stability of a group of genotypes before commercial release. Various statistical methods that have been developed for this purpose are divided into parametric and non-parametric stability statistics. Parametric stability statistics is further divided into univariate and multivariate methods. The univariate methods include Wricke's ecovalence (W_i^2) [3], Shukla's stability variance (σ^2) [6], coefficient of variance (CV) [7], Environmental variance (S^2) [8], Mean-variance component (θ) [9], GE variance component (θ') [10], Regression coefficient (b_i) [11], and many others. The multivariate methods imply the additive main effects and multiplicative interaction (AMMI) model [12] and the GGE biplot method [13]. Multivariate methods can effectively predict the genotype $\times$ environment interactions by using the approaches such as the 'which-won-where' pattern, identifying mega environments, ideal genotypes across different testing environments, and ranking environments [14]. Non-parametric methods include Nassa and Huhn's and Huhn's statistics (S) [15], Kang's rank-sum (KR) [16], TOP-Fox (TOP) [17], Thennarasu's non-parametric statistics (NP) [18], and Genotype stability index (GSI) [19].

The present study aims to identify superior rice genotypes with stable yield performance over eight different locations for two consecutive years by evaluating the efficacy of various univariate and multivariate stability parameters.

2. Materials and Methods

2.1. Plant Material

Five experimental genotypes and two commercial check cultivars were evaluated at eight different locations using RCBD with three replications in three provinces of Pakistan (Table 1). The experimental rice genotypes were: GSR-48, GSR-82, GSR-112, GSR-252, and GSR-305. The check cultivars evaluated were IRRI-6 and Kissan Basmati. The GSR lines were selected based on the two-year agro-morphological performance for yield and yield-related traits at the National Institute for Genomics and Advanced Biotechnology (NIGAB) National Agriculture Research Council (NARC) Islamabad, Pakistan.

Table 1. Code for genotypes name and locations evaluated during the two years.

Genotypes with Codes	Locations with Environment Codes	Years
GSR-48 = G1	Pindi Bhattian = E1	2020, 2021
GSR-82 = G2	Kala Shah Kaku = E2	
GSR-112 = G3	Narowal = E3	
GSR-252 = G4	Swat = E4	
GSR-305 = G5	Islamabad = E5	
IRRI6 = G6	Dera Ismail Khan = E6	
Kissan Basmati = G7	Muzaffargarh = E7	
	Dokri = E8	

2.2. Experimental Location

All genotypes were evaluated at eight different locations: Soil Salinity Research Institute Pindi Bhattian; Rice Research Institute Kala Shah Kaku, Narowal, and Muzaffargarh in Punjab; Agriculture Research Institute Swat and Dera Ismail Khan in Khyber Pakhtunkhwa; Dokri in Sindh; and National Agriculture Research Centre (NARC) Islamabad for two years cropping season of 2020–2021. Climatic characteristics (average rainfall and temperature) of test locations for 2020–2021 from transplantation to harvesting are given in Table 2.

2.3. Experimental Procedures and Cultural Practices

In the first week of each year June 2020 and 2021, 1000 cleaned seeds of GSR lines with two check cultivars were sown on nursery trays with 98 holes, where each hole was seeded with two healthy seeds. The plastic trays were filled with a mix of 70% sandy clay loam soil and 25% peat moss. The trays were labeled with genotype code and name, respectively. The 30-day-old rice seedlings were shifted to paddy fields at eight different locations and transplanted manually. Transplantations of all rice genotypes were performed on the third of July (2020 and 2021) in a straight-rows method in three replications at each location. Each plot was set with a net size of 2.1 m $\times$ 0.90 m containing five rows with eight seedlings per row. There was a 17 cm row-to-row and 20 cm plant-to-plant spacing within the plot. All the yield and yield-related traits were measured at the physiological maturity stage. Data were collected from five randomly selected plants from each plot in each replication. The plant height (PH) of each genotype was measured with the help of a meter rod in centimeters (cm). Tillers per plant (TPP) was determined by counting all productive tillers' numbers. Straw yield per plant (SYPP) and grain yield per plant (GYPP) were recorded with 14% moisture content. Nitrogenous fertilizers were applied in three splits (after seven days, 37 days, and 60 days of transplantation); phosphorus and potash were used in full doses after the two weeks of transplantation. During the rice growth stages, weeds were removed by two times hand-weeding. However, neither herbicides nor insecticides were applied in the experimental trials.

2.4. Statistical Analysis

2.4.1. Analysis of Variance

The obtained morphological data of five GSR lines along with check cultivars at eight different locations over two years were subjected to the combined ANOVA, using R statistical software version 4.1.1. Furthermore, ANOVA results were used to determine the effect of genotypes (G), locations (L), replications (R), and years (Y) effect and the magnitude of the G $\times$ L, G $\times$ Y, and G $\times$ L $\times$ Y interactions.

Table 2. Mean temperature, rainfall, and soil texture of each experimental site during 2020 and 2021.

Locations	Month	July		August		September		October		November		Soil Type
	Year	Temp (°C)	Rain (mm)	Temp (°C)	Rain (mm)	Temp (°C)	Rain (mm)	Temp (°C)	Rain (mm)	Temp (°C)	Rain (mm)	
NARC	2020	35	162.5	32	165.1	31	68.5	29	22.8	22	12.7	loam
	2021	28	174	26	162	25	73	21	31	15	39	
Swat	2020	33	89.8	35	55.8	28	91.8	23	36.8	19	64.4	sandy
	2021	28	55.8	27	55.8	25	27.9	19	20.3	13	15.2	
Kala Shah Kaku	2020	36	50.6	33	57.2	32	39.2	32	3.2	23.5	2.8	silty clay
	2021	31	134.6	31	124.4	30	55.8	25	12.7	20	5	
Pindi Bhattian	2020	34	71.1	33	67.5	32	45.3	32	6.4	24	4.4	sandy loam
	2021	35	19.2	36	91.4	34	40.6	30	7.6	24	5	
Narowal	2020	36	160.5	33	232.6	35	139.9	32	14.3	23.5	12.4	silty, loamy
	2021	33	89	31	59	28.5	43	23	11	19	4	
Muzaffargarh	2020	40	37.5	37	43.7	36	20.8	36	2.21	28.5	1.8	salinity
	2021	39.2	52	38.1	40	37.2	19	34.4	2	28.3	3	
Dokri	2020	41	118.8	38.5	106.8	37	50.1	33	12.23	26.5	5.8	sandy clay loam
	2021	41.1	41	39	24	38.3	9	35.7	2	30.1	2	
Dera Ismail Khan	2020	33	61	32	58	30	18	25	5	19	2	sandy/loamy sand
	2021	41.4	60	38.4	57	37.3	18	36.6	5	33.2	2	

2.4.2. Genetic Parameters

Genetic and environmental effects among the genotypes for traits were measured by using their mean sum of squares [20]. The heritability estimate was categorized as low (0–30%), medium (30–60%), and high (>60%).

(a) Genotypic variance

$$\sigma^2 g = \frac{GMS - EMS}{r}$$

Here, GMS is the genotype mean square and EMS denotes the error mean square, and r is the number of replications of genotypes.

(b) Phenotypic variance

$$\sigma^2 p = \sigma^2 g + \sigma^2 e$$

Here, σ^2 p is the phenotypic variance, $\sigma^2 g$ is the genotypic variance, and $\sigma^2 e$ is the environmental variance.

(c) Environmental variance

$$\sigma^2 e = \frac{EMS}{r}$$

Here, $\sigma^2 e$ is the environmental variance, EMS is the error mean square, and r is the number of replications of genotypes.

(d) H^2

$$h_B^2 = \frac{\sigma^2 g}{\sigma^2 p}$$

where h_B^2 is the broad-sense heritability, which is equal to the ratio of σ^2 g (genotypic variance) and $\sigma^2 p$ (phenotypic variance).

2.4.3. Estimation of Stability Parameters

The univariate and multivariate parametric stability analyses were performed to assess genotype yield and yield-related traits across multiple environments and predict stable genotypes. Both univariate and multivariate stability analyses were performed year-wise due to the presence of significant variation between the year effect.

2.4.4. Univariate Stability Analysis

Univariate stability of the 7 genotypes for plant height, number of tillers per plant, grain yield per plant, and straw yield per plant was calculated by using AMMI Stability Value (ASV) [21] and AMMI Stability Index (ASI) [22], Shukla's stability variance (σ^2) [6] and Wricke's ecovalence (Wi^2) [3].

1. AMMI Stability Value (ASV)

As suggested by Purchase et al. [21], AMMI Stability Value (ASV) parameter for stability assessment is calculated by the following equation

$$ASV = \sqrt{\left(\frac{SS_{IPCA1}}{SS_{IPCA2}} \quad (IPCA1)\right)^2 + (IPCA2)^2}$$

SS_{IPCA1} and SS_{IPCA2} are the sum of squares in the first two principal component interactions. IPCA1 and IPCA2 are the scores of genotypes in the first and second principal components interactions.

2. AMMI Stability Index (ASI)

Jambhulkar et al. [22] suggested the AMMI-model based AMMI Stability Index (ASI), which is calculated by using the following equation:

$$ASI = \sqrt{\left[\left(IPCA1 \times \theta_1^2\right)^2 + \left(IPCA2 \times \theta_2^2\right)^2\right]}$$

IPCA1 and IPCA2 are the values of the first two principal component interactions and θ_1^2 and θ_2^2 are the values of the percentage sum of square explained by these two components.

3. Wricke's Ecovalence

Wricke [3] introduced the idea of ecovalence parameter to calculate the share of each genotype to the sum of squares of genotype × environment interaction by using the following equation:

$$W_i^2 = \sum (X_{ij} - \overline{X}_{i.} - \overline{X}_{.j} + \overline{X}_{..})^2$$

Here, X_{ij} represents the mean of ith genotype in the jth environment, $\overline{X}_{i.}$ is the mean of the yield of ith genotype, $\overline{X}_{.j}$ is the mean of the yield of the jth environment, and $\overline{X}_{..}$ is the grand mean.

4. Shukla's Stability Variance

Shukla [6] proposed the Shukla's stability variance of genotypes across different environments based on the following equation:

$$\sigma^2 = \left[\frac{\mathrm{p}}{(\mathrm{p}-2)(\mathrm{q}-1)}\right] \mathrm{W}_\mathrm{i}^2 - \frac{\sum W_i^2}{(\mathrm{p}-1)(\mathrm{p}-2)(\mathrm{q}-1)}$$

Here, p and q represent the genotypes and environments number while ${W_i}^2$ is the Wricke's ecovalence of the ith genotype.

2.4.5. Multivariate Stability Analysis

Multivariate stability analysis; AMMI [23] and GGE biplot [13] were performed to identify the ideal genotype across each testing environment with high performance and stability, mega-environments, and understanding of the genotype × environment interactions.

2.4.6. Additive Main Effect and Multiplicative Interaction (AMMI) Model

In the present study, multivariate stability based on the AMMI model was assessed for G×E interaction and stability analysis to predict the stability of GSR lines. The AMMI model combines the application of pooled ANOVA to evaluate the additive main effects; then factorization of a complex matrix (SVD) is applied to the total error for computing interaction principal components (IPCs). We estimated the additive main effect and AMMI model in R using the metan library [24]. As suggested by Zobel et al. [23] the base of the additive main effect and multiplicative interaction (AMMI) model was computed as follows:

$$Y_{ij} = \mu + \alpha_i + \beta_j + \sum_{k=1}^{n} \lambda_k \gamma_{ik} \delta_{jk} + \varepsilon_{ij}$$

where Y_{ij} is the mean performance of ith genotype in the jth individual environment, μ is the overall mean, α_i is the *fixed effect of the GSR line*, β_j is the *environmental effect*, n is the number of IPCA kept in the AMMI model, λ_k is the singular value for IPC axis k, γ_{ik} is the ith genotype eigenvector value for IPC axis k, δ_{jk} is the jth environment eigenvector value for IPC axis k, and ε_{ij} is the average residual.

2.4.7. Biplot Analysis

After ranking the most adoptable GSR lines with the AMMI model, a study of the sustainable phenotypic reliability of the multi-locations analysis of the biplot graphic was designed. Biplots are performance and stability-related graphs where factors of both genotypes and locations are plotted on the same axis so the inter-relationships can be visualized.

In our constructed biplots, the abscissa represents the variables that affect the values of a genotype, and the ordinate is the first interaction axis (IPCA1). The GSR line with IPCA1 close to zero will be considered a stable and "ideal" GSR line while low stability will be associated with low productivity [25,26].

3. Results

3.1. Combined Analysis of Variance

Table 3 and Supplementary Table S1 represent the combined ANOVA of rice genotypes for plant height, tillers per plant, straw yield per plant, and grain yield per plant across eight different locations (Supplementary Table S1). The mean square of genotypes showed significant differences ($p \leq 0.01$) for traits. The mean square of locations, years, and genotypes by locations, genotypes by years, and genotypes by locations by years (G × L interaction, G × Y interaction, and G × L × Y interaction) showed significant differences ($p \leq 0.01$) for traits. In our study, the significant G × L interaction effects revealed that rice genotypes responded differently against fluctuation in the environment, which indicated the necessity of testing rice genotypes at multiple locations. Moreover, interaction among genotypes, locations, and years also revealed a highly significant ($p \leq 0.01$) difference for studied traits. Therefore, further general adaptability and stability analysis across genotypes should be followed before their selection.

Table 3. Estimation of significant levels for yield and related traits of seven rice genotypes revealed by combined ANOVA.

Source of Variation	df	Plant Height	Tillers Per Plant	Grain Yield per plant	Straw Yield per plant
Genotype	6	1.07×10^{-52}	7.07×10^{-4}	3.14×10^{-16}	1.54×10^{-15}
location	7	2.76×10^{-12}	2.12×10^{-52}	1.05×10^{-86}	2.74×10^{-60}
Year	1	1.78×10^{-3}	2.43×10^{-2}	1.52×10^{-72}	7.03×10^{-36}
Replication	2	7.59×10^{-6}	6.57×10^{-1}	2.30×10^{-1}	9.54×10^{-3}
Genotype: Location	42	3.78×10^{-5}	4.77×10^{-3}	1.97×10^{-10}	3.23×10^{-7}
Genotype: Year	6	1.76×10^{-6}	6.72×10^{-2}	1.88×10^{-3}	6.03×10^{-3}
Genotype:Location:Year	42	2.84×10^{-7}	2.22×10^{-3}	4.11×10^{-6}	3.97×10^{-15}

df = degree of freedom.

3.2. Analysis of Genetic and Phenotypic Variances

In our study, the phenotypic variance for plant height, tillers per plant, grain yield per plant, and straw yield per plant were distributed into genotypic and environmental variances (Table 4). A low magnitude of genotypic coefficient of variation was found in the corresponding phenotypic coefficient of variation for all traits studied. The broad-sense heritability for four yield traits ranged from 75.3% (tillers per plant) to 98.7% (plant height), respectively. Accordingly, all the yield-related traits considered in our study showed high heritability (>60%), constituting a high breeding value with more additive genetic effects, which is important for rice grain yield enhancement.

Table 4. Estimation of genetic parameters in rice genotypes for yield and yield-related traits.

Genetics Parameters	Plant Height	Tillers Per Plant	Grain Yield Per Plant	Straw Yield per Plant
Vg	1104.9	9.9	466	2468
Ve	14	3.2	28.6	161.2
Vp	1118	13.1	490	2629
h^2_B (%)	98.7	75.3	94.2	94

Vg; Genotypic variance, Ve; Environmental variance, Vp; Phenotypic variance, h^2_B; Broad sense heritability.

3.3. Univariate Models

3.3.1. Univariate Parametric Stability Statistics (First-Year 2020)

The results of different univariate parametric stability statistics are given in Table 5. The stability parameter designed by Shukla (σ^2), Wricke's ecovalence (Wi^2), AMMI stability value (ASV), and AMMI stability index (ASI) are based on the concept that genotypes with the smallest stability value are the most stable ones. The stability values were worked out for rice genotypes over eight locations and are presented in Table 5. Based on σ^2, Wi^2, ASV, and ASI genotype, G1 (GSR-48) was found as the most stable genotype for plant height. Genotype G6 (IRRI-6) was found the most stable genotype for tillers per plant. Genotype G7 (Kissan basmati) was found most stable for grain yield per plant, and genotype G3 (GSR-112) was found as most stable for straw yield per plant. These genotypes are stable because their values are relatively close to zero.

3.3.2. Univariate Parametric Stability Statistics (Second-Year 2021)

The univariate parametric stability statistics for 2021 found that a different trend for the stability of the same genotypes had changed from 2020. Using σ^2, Wi^2, ASV, and ASI stability indicators, genotype G4 (GSR-252) was identified as the most stable genotype for plant height. G3 (GSR-112) was also a stable genotype for plant height based on ASV and ASI values. Using σ^2 and Wi^2 values, genotype G3 (GSR-112) was found as the most stable genotype, and using ASV and ASI values, genotype G1 (GSR-48) was found to be the most stable genotype for tillers per plant. Genotype G4 (GSR-252) was found stable for grain yield per plant, as indicated by its lowest values for all studied stability statistics. Genotype G3 (GSR-112) was the most stable genotype based on σ^2 and Wi^2, while genotype G4 (GSR-252) was also identified as the stable genotype due to its lowest values for ASV and ASI.

3.4. Multivariate Models

3.4.1. AMMI Analysis of Variance (First-Year 2020)

The AMMI model for yield and yield-related traits revealed significant variations ($p < 0.05$) for both the main (genotypes and locations) and interaction effects revealing the presence of considerable variability among the studied genotypes, locations, and their interactions (Supplementary Table S1). The maximum part of the total variance in the AMMI analysis was attributed to the locations factor, followed by genotypes and genotype by location interaction. In our study, locations explained the maximum (53%) of the total sum of squares for all traits, indicating that varied environmental conditions could cause most variations among genotype traits. Genotypes explained only 25% of the total sum of squares on average for traits, whereas the G $\times$ L interaction accounted for 20% of total variations.

The AMMI analysis generated two significant PCs from the G $\times$ L interaction. The PC1 and PC2 accounted for 80% of the variation for plant height, 73% for tillers per plant, 75% for grain yield per plant, and 84.5% for straw yield per plant, respectively (Supplementary Table S1). The extracted PCs are informative by elucidating information on the interaction effect; although, their degree decreases gradually from the first to the last PC.

Table 5. Parametric stability statistics for Plant height, Tillers per plant, Grain yield per plant, and Straw yield per plant of seven rice genotypes grown in eight different locations in Pakistan.

Year 2020	**Plant Height**				**Tillers per Plant**				**Grain Yield per Plant**				**Straw Yield per Plant**			
Lines	σ^2	Wi^2	ASV	ASI	σ^2	Wi^2	ASV	ASI	σ^2	Wi^2	ASV	ASI	σ^2	Wi^2	ASV	ASI
G1	−0.6	63.5	1.5	0.3	7.3	126.1	2.7	0.8	59.9	1277.4	2.3	0.5	399.5	8503.5	4.3	1.7
G2	32.9	568.4	3.5	0.8	7.9	135.0	2.0	0.6	84.0	1637.9	6.5	1.4	672.7	12601.5	6.0	2.4
G3	8.7	204.6	2.9	0.6	12.0	196.1	3.5	1.0	280.5	4586.2	11.4	2.6	269.2	6548.4	2.4	1.0
G4	43.6	727.7	7.8	1.7	3.5	68.4	1.2	0.3	230.7	3839.2	10.6	2.4	352.8	7803.6	4.6	1.8
G5	23.3	423.4	2.2	0.5	3.9	74.4	1.0	0.3	118.8	2159.7	4.6	1.0	1598.0	26481.4	8.4	3.4
G6	33.0	569.8	6.6	1.5	0.6	25.5	0.5	0.1	74.5	1495.7	3.9	0.8	931.2	16478.9	6.2	2.5
G7	30.6	533.2	6.3	1.4	1.2	34.2	1.1	0.3	32.7	869.0	0.5	0.1	1633.6	27014.9	8.6	3.5
Year 2021	**Plant height**				**Tillers Per Plant**				**Grain Yield per Plant**				**Straw Yield per Plant**			
Lines	σ^2	Wi^2	ASV	ASI	σ^2	Wi^2	ASV	ASI	σ^2	Wi^2	ASV	ASI	σ^2	Wi^2	ASV	ASI
G1	22.5	484.5	4.1	0.6	2.1	52.3	0.6	0.1	74.1	1272.1	3.5	0.9	133.1	3218.3	2.5	0.7
G2	95.7	1582.1	15.6	2.5	2.8	61.7	2.2	0.4	40.4	766.7	3.4	0.8	397.0	7176.7	7.7	2.1
G3	29.3	585.5	3.0	0.4	0.6	29	1.7	0.3	70.0	1210.7	5.1	1.3	−10.6	1061.5	1.8	0.5
G4	8.3	270.1	3.0	0.4	7.9	138	3.4	0.7	−1.2	141.2	0.9	0.2	87.3	2531.9	0.8	0.2
G5	136.5	2194.3	18.3	2.9	5.2	97.8	2.3	0.4	55.2	988.2	2.7	0.7	765.7	12707.5	9.0	2.5
G6	25.6	530.2	4.8	0.7	14.4	235	6.1	1.2	62.5	1098.0	5.4	1.4	1012.1	16403.0	12.4	3.4
G7	21.7	472.5	5.8	0.9	13	215	5.1	1.0	70.6	1218.6	6.3	1.6	465.7	8208.4	4.7	1.3

σ^2: Shukla's stability variance; Wi^2: Wricke's Ecovalence for stability; ASV: AMMI Stability Value; ASI: AMMI Stability Index.

3.4.2. AMMI Analysis of Variance (Second-Year 2021)

We also conducted the AMMI analysis for the second year of the multi-location trials to reveal the effect of tested genotypes, locations, and their interaction with traits. Here, the AMMI model showed significant differences among tested genotypes, locations, and their interaction at ($p < 0.05$) probability for all the studied traits as analyzed in Supplementary Table S1. The greater contribution for the total sum of squares in AMMI analysis was caused by locations (66%), followed by genotype by location interaction effect (20%) and genotypes (11.8%). The maximum variation due to the interaction effect confirmed that tested genotypes responded significantly to the fluctuation in environmental conditions at locations. The proportion of PC1 and PC2 from the interaction effect explained 83% of the variation for plant height, 74.3% for tillers per plant, 72% for grain yield, and 75.5% for straw yield, respectively.

3.4.3. GGE Biplot Analysis (First-Year 2020)

'Mean vs. Stability' Analysis of GGE Biplot

The GSR lines' stability pattern across different locations was analyzed using the mean vs. stability pattern of the GGE biplot. It facilitates genotype evaluation based on mean performance and stability across various environments. The biplot graph is formed by the intersection of a vertical AEC abscissa and a horizontal AEC ordinate line. Each line has a single arrowhead that points towards a higher mean performance for the studied trait. In our investigation, the mean vs. stability analysis revealed 95.9% for plant height (Figure 1A), 75.66% for tillers per plant (Figure 1B), 75.63% for grain yield per plant (Figure 1C), and 88.34% variation for straw yield (Figure 1D), of G + G × E variation. Here, G5 (GSR-305) revealed maximum plant height in E2 (Kala Shah Kaku), E3 (Narowal), and E8 (Dokri); followed by two check cultivars, G6 (IRRI-6) and G7 (Kissan basmati) that showed maximum plant height in E1 (Pindi Bhattian), E4 (Swat), E5 (Islamabad), E6 (Dera Ismail Khan), and E7 (Muzaffargarh). G1 (GSR-48) and G3 (GSR-112) were the most stable GSR lines tested across different locations these lines recorded lower heights in all locations.

The maximum numbers of tillers per plant were recorded by a check cultivar G7 (Kissan basmati) followed by G4 (GSR-252) in E1 (Pindi Bhattian), E2 (Kala Shah Kaku), E3 (Narowal), E4 (Swat), and E5 (Islamabad). Only G3 (GSR-112) showed performance in E7 (Muzaffargarh). G2 (GSR-82), G6 (IRRI-6) and G5 (GSR-305) were the stable genotypes even though they produced fewer tillers and their performance is limited to the E8 (Dokri) location only (Figure 1B). G5 (GSR-305) was the most stable and high-performing genotype for grain yield per plant trait. The second-best high-performing genotype was G3 (GSR-112) in E8 (Dokri) and E7 (Muzaffargarh), even though it was not stable and the only genotype performing in E7 (Muzaffargarh). G1 (GSR-48) was the stable genotype after G5 (GSR-305) and, together with G5 (GSR-305), performed well in E1 (Pindi Bhattian), E2 (Kala Shah Kaku), E3 (Narowal), E4 (Swat), E5 (Islamabad), E6 (Dera Ismail Khan), and E8 (Dokri). Both check varieties and G4 (GSR-252) were not stable, and neither did they perform in any tested location for grain yield (Figure 1C). G5 (GSR-305) was also the best performing and most stable genotype for straw yield, followed by G3 (GSR-112) in E4 (Swat), E5 (Islamabad), and E8 (Dokri). G6 (IRRI-6) performed in E2 (Kala Shah Kaku), E3 (Narowal), E6 (Dera Ismail Khan), and E7 (Muzaffargarh). G1 (GSR-48), G4 (GSR-252), G7 (Kissan basmati), and G2 (GSR-82) showed some performance in E1 (Pindi Bhattian), but these were not stable genotypes for any tested location (Figure 1D).

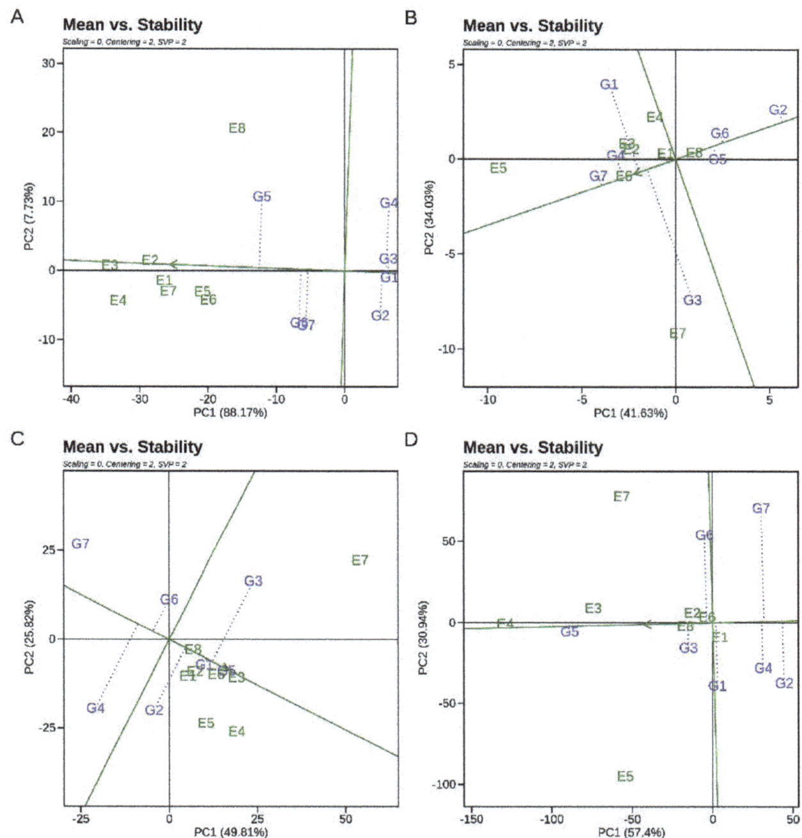

Figure 1. The GGE biplot 'Mean vs. stability' pattern of genotype × environment interaction of 5 GSR lines and 2 control lines grown under eight environments in the year 2020 for (**A**) plant height, (**B**) the number of tillers, (**C**) grain yield, and (**D**) straw yield. The biplots were created with cent–ering 0, SVP = 2, and scaling = 0 parameters.

'Which-Won-Where' GGE Biplot

Figure 2 represents the polygon view of the GGE biplot and it revealed the best performing genotypes for traits in a single group of locations. The G + G × E biplot ('which-won-where' pattern) explained 95.9%, 75.66%, 75.63%, and 88.34% variation for plant height, tillers per plant, grain yield per plant, and straw yield per plant, respectively (Figure 2). As explained by Oladosu et al. [27], the genotypes lying on the vertex of a polygon with no environmental indicator nearby are poorly performed genotypes, and the genotypes that are present on the vertex of a polygon where one or more environmental indicators are present are the best performing genotypes in the relevant environments. The genotypes lying inside a polygon are less responsive to any testing environment. All environmental indicators formed a single sector for plant height, and G5 (GSR-305) was the winning genotype in all testing environments. No other genotype won in any testing environment and thus defined poorly performing genotypes for plant height trait (Figure 2A). Eight environments were divided into four sectors for tillers per plant, with different genotypes winning in each sector. Sector one has environment E1 (Pindi Bhattian), E2 (Kala Shah Kaku), E3 (Narowal), and E4 (Swat); sector two has environment E5 (Islamabad) and E6 (Dera Ismail Khan); sector three has environment E7 (Muzaffargarh), and sector four has environment E8 (Dokri). G1 (GSR-48) was the winning genotype in sector one, G7 (Kissan basmati) was the winning genotype in sector two, G3 (GSR-112) was the winning genotype in sector three, and G2 (GSR-82) was the winning genotype in sector four for tillers per plant (Figure 2B). The which-won-where GGE biplot of grain yield divided the eight locations into three sectors. Sector one has only environment E7 (Muzaffargarh) with G3 (GSR-112) the winning genotype; sector two has environment E1 (Pindi Bhattian), E2 (Kala Shah Kaku), E3 (Narowal), E4 (Swat), E6 (Islamabad), and E8 (Dokri) with G5 (GSR-305) the winning genotype in these environmental indicators; and sector three has environment E5 (Islamabad) with no winning genotype in it. G2 (GSR-82), G3 (GSR-112), and G7 (Kissan basmati) were poor-performing genotypes for grain yield (Figure 2C). For straw yield, which-won-where GGE biplot divided eight testing environments into three sectors. Sector one has environment E7 (Muzaffargarh) with no genotype winning in it; sector two has environment E2 (Kala Shah Kaku), E3 (Narowal), E4 (Swat), E5 (Islamabad), E6 (Dera Ismail Khan), and E8 (Dokri) with G5 (GSR-305) winning in all these testing environments; and sector three has environment E1 (Pindi Bhattian) with G2 (GSR-82) the winning genotype. Both check varieties were poorly performing genotypes for straw yield (Figure 2D).

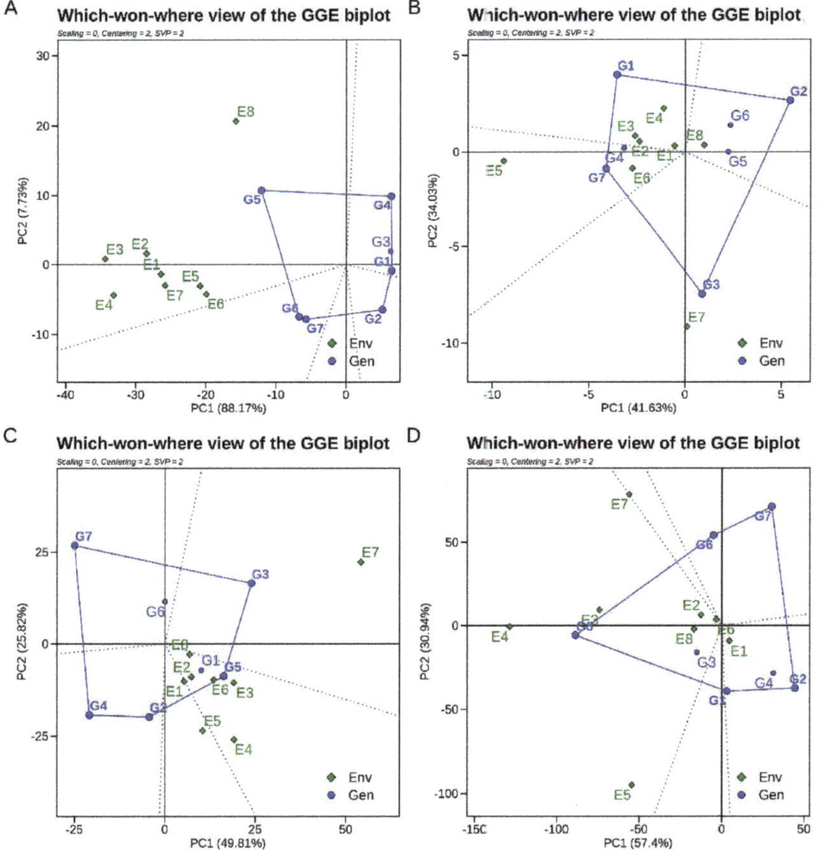

Figure 2. The GGE biplot polygon of the 'Which−won−where' pattern to identify the best cultivar in each location of 5 GSR lines and 2 control lines grown under eight environments in the year 2020 for (**A**) plant height, (**B**) the number of tillers, (**C**) grain yield, and (**D**) straw yield. The biplots were created with centering = 0, SVP = 2, and scaling = 0 parameters.

Locations and Genotypes Ranking: Best and Stable Location/Genotypes Evaluation

Figure 3 shows the GGE biplot 'Ranking environments' pattern to rank locations with respect to ideal environment or tester for genotypes. The genotypes are treated random and focus is placed on testing environments. E3 (Narowal) appeared to be as best locations for plant height (Figure 3A); E5 (Islamabad) for number of tillers (Figure 3B); and E4 (Swat) for both grain yield and straw yield (Figure 3C or Figure 3D). Testers were ranked based on their closeness to the concentric center. The rank of environments based on pattern of ranking environments GGE biplot for plant height is E3 > E2 ≈ E4 > E1 > E7 > E5 > E6 > E8; ranking for tillers per plant is E5 > E6 > E3 ≈ E2 > E1 > E4 > E8 > E7; ranking for grain yield is E4 > E3 > E5 ≈ E6 > E2 > E1 > E7 > E8; and ranking for straw yield is E4 > E3 > E7 ≈ E8 > E2 ≈ E5 > E6 > E1.

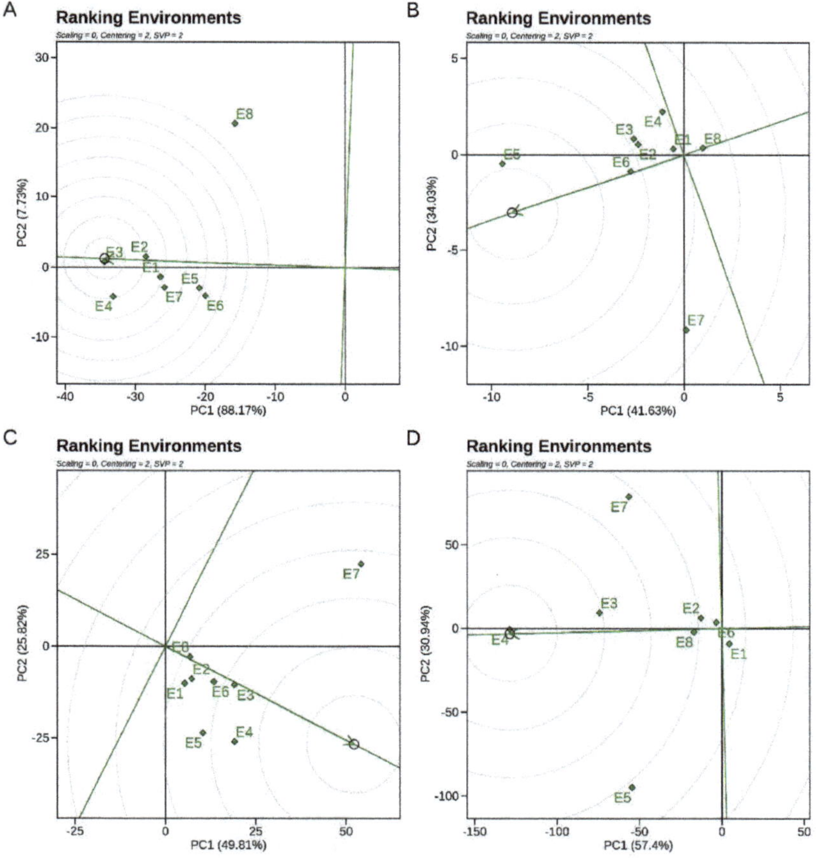

Figure 3. The GGE biplot 'Ranking environments' pattern to rank environments for the ideal environment of 5 GSR lines and 2 control lines grown under eight environments in the year 2020 for (**A**) plant height, (**B**) the number of tillers, (**C**) grain yield, and (**D**) straw yield. The biplots were created with centering = 0, SVP = 2, and scaling = 0 parameters.

The GGE biplot of ranking genotypes concerning the ideal genotype revealed the unique genotype compared to the others evaluated (Figure 4). The blue arrowhead points toward the ideal genotype that performs best in all testing environments. Ideal entry is placed in the center of the concentric circle, followed by other circles. If no entry is located in the center, then the most closely located entry to the concentric circle is ideal. Environments are treated as random samples of testing environments, and the concentration points are genotypes. G6 (IRRI-6) was the best genotype for plant height (Figure 4A) based on its nearness to the innermost circle. G7 (Kissan basmati) was the best genotype among others for tillers per plant (Figure 4B); for grain yield per plant, G5 (GSR-305) was the ideal genotype that was present within the innermost circle (Figure 4C). G5 (GSR-305) was also the best genotype for straw yield (Figure 4D). The genotypes ranking for plant height was G6 > G7 > G5 > G3 ≈ G1 ≈ G2 > G4; for tillers per plant G7 > G4 > G1 > G5 > G6 > G3 > G2; for grain yield G5 > G1 > G3 > G2 > G6 > G4 > G7; and for straw yield the ranks were G5 > G3 > G1 ≈ G6 > G4 > G2 ≈ G7.

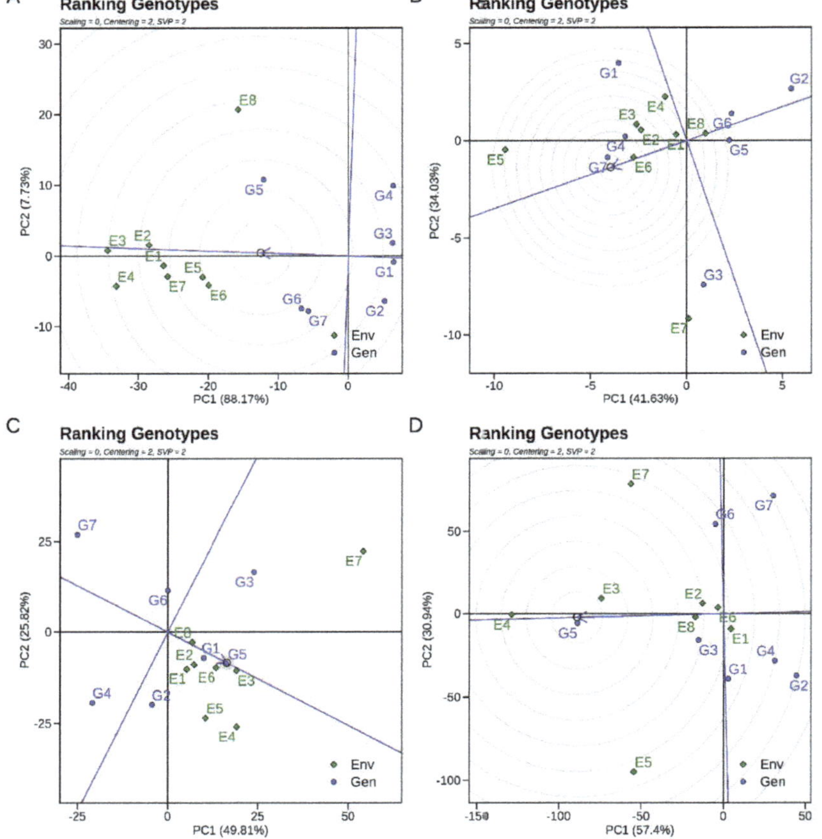

Figure 4. The GGE biplot 'Ranking genotypes' pattern to rank genotypes for the ideal genotype of 5 GSR lines and 2 control lines grown under eight environments in the year 2020 for (**A**) plant height, (**B**) the number of tillers, (**C**) grain yield, and (**D**) straw yield. The biplots were created with centering = 0, SVP = 2, and scaling = 0 parameters.

3.4.4. GGE Biplot Analysis (Second-Year 2021)

Mean vs. Stability' Analysis of GGE Biplot (First-Year 2021)

The genotype evaluation is based on the average performance and stability in various environments. The mean versus stability pattern of GGE biplots explained 90.23% for plant height, 69.74% for tillers per plant, 65.08% for grain yield, and 79.06% of the total variation for straw yield (Figure 5). Check variety G6 (IRRI-6) showed maximum plant height in E1 (Pindi Bhattian), and E3 (Narowal), followed by G5 (GSR-305) was high performing genotype in E2 (Kala Shah Kaku), E4 (Swat), E5 (Islamabad), E6 (Dera Ismail Khan), E7 (Muzaffargarh), and E8 (Dokri). G7 (Kissan basmati) performed better in E1 (Pindi Bhattian) and E3 (Narowal). G3 (GSR-112) and G4 (GSR-252) were the stable genotypes even though with less performance (Figure 5A). G4 (GSR-252) produced the maximum number of tillers per plant in E1 (Pindi Bhattian), E2 (Kala Shah Kaku), and E3 (Narowal). After that, G5 (GSR-305) and check variety G6 (IRRI-6) showed good performance in E4 (Swat), E5 (Islamabad), E6 (Dera Ismail Khan), E7 (Muzaffargarh), and E8 (Dokri). G1 (GSR-48), G2 (GSR-82), and G3 (GSR-112) were the stable genotypes with fewer tillers. G5 (GSR-305) was also a stable genotype with good performance (Figure 5B). G3 (GSR-112) followed by G1 (GSR-48) and G5 (GSR-305) were high grain yielding genotypes in G7 (Muzaffargarh). G7 (Kissan basmati) was the stable genotype for grain yield in E2 (Kala Shah Kaku), E3 (Narowal), and E6 (Dera Ismail Khan). G3 (GSR-112), G2 (GSR-82), and G4 (GSR-252) showed performance in E1 (Pindi Bhattian), E4 (Swat), and E5 (Islamabad). At the same time, G5 (GSR-305) and G1 (GSR-48) were performing genotypes in E7 (Muzaffargarh) (Figure 5C). G5 (GSR-305) was high performing genotype for straw yield in E2 (Kala Shah Kaku), E3 (Narowal), and E7 (Muzaffargarh). In environments E1 (Pindi Bhattian), E5 (Islamabad), E6 (Dera Ismail Khan), and E8 (Dokri), the only performing genotype was check cultivar G6 (IRRI-6). G1 (GSR-48), G3 (GSR-112), and G4 (GSR-252) were the stable genotypes. In E4 (Swat) some performance was shown by G1 (GSR-48), G2 (GSR-82), G3 (GSR-112), G4 (GSR-252), and G7 (Kissan basmati) (Figure 5D).

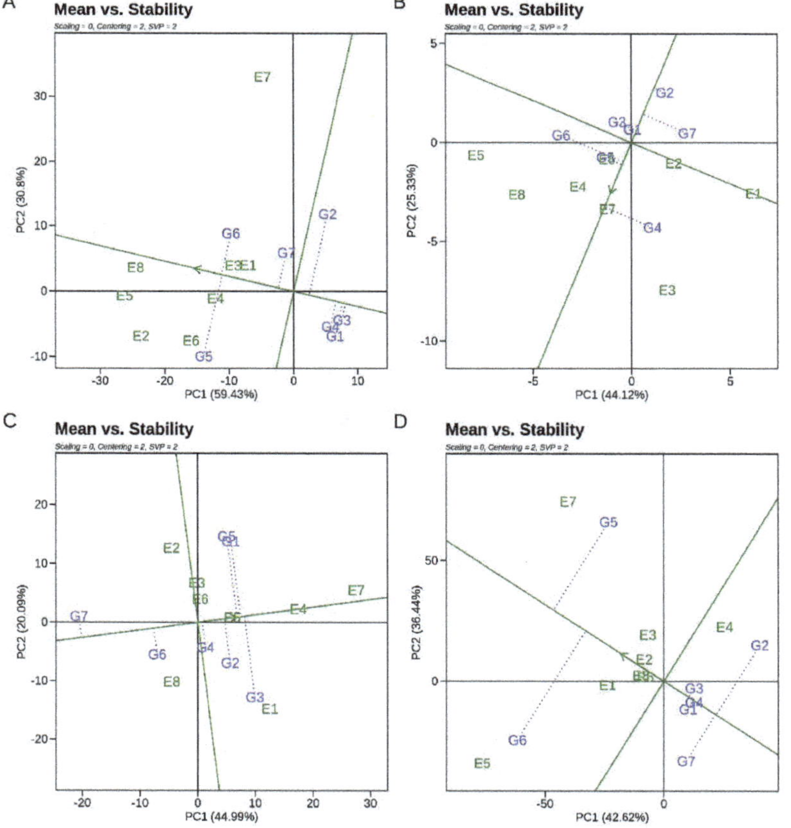

Figure 5. The GGE biplot 'Mean versus stability' pattern of genotype × environment interaction of 5 GSR lines and 2 control lines grown under eight environments in the year 2021 for (**A**) plant height, (**B**) the number of tillers, (**C**) grain yield, and (**D**) straw yield. The biplots were created with centering = 0, SVP = 2, and scaling = 0 parameters.

'Which-Won-Where' GGE Biplot

The PC1 and PC2 scores of the constructed GGE biplot of 'which-won-where' for the year 2021 explained 90.23%, 69.74%, 65.08%, and 79.06% of total variations for plant height, tillers per plant, grain yield per plant, and straw yield per plant, respectively (Figure 6). The genotypes positioned at the corners of the polygons for the studied traits were considered elite in that location. The genotypes placed at vertexes with no tester are regarded as poor genotypes. This GGE biplot divided eight testing environments into three sectors for plant height traits. Sector one has one environment E7 (Muzaffargarh) with G2 (GSR-82) the winning genotype in it; sector two has E1 (Pindi Bhattian) and E3 (Narowal) with check variety G6 (IRRI-6) winning in it; and sector three has E2 (Kala Shah Kaku), E4 (Swat), E5 (Islamabad), E6 (Dera Ismail Khan), and E8 (Dokri) with G5 (GSR-305) as the winning genotype (Figure 6A). For tillers per plant, testing environments formed two sectors. Sector one has E4 (Swat), E5 (Islamabad), E6 (Dera Ismail Khan), and E8 (Dokri) with check variety G6 (IRRI-6) as the winning genotype in these testing environments; and sector two has E1 (Pindi Bhattian), E2 (Kala Shah Kaku), E3 (Narowal), and E7 (Muzaffargarh) with G4 (GSR-252) as a winning genotype for these environmental indicators (Figure 6B). The which-won-where GGE biplot for grain yield divided the eight testers into four sectors. G1 (GSR-48) and G5 (GSR-305) were winning genotypes in sector one that has environment E2 (Kala Shah Kaku), E3 (Narowal), and E6 (Dera Ismail Khan). Sector two has environment E7 (Muzaffargarh) with no winning genotype. G3 (GSR-112) was the winning genotype in sector three that has environment E1 (Pindi Bhattian), E4 (Swat), and E5 (Islamabad). Sector four has an environment with no genotype winning in it. Both check varieties were poorly performing for grain yield (Figure 6C). The which-won-where pattern of straw yield separated eight testers into three sectors. Sector one has environment E2 (Kala Shah Kaku), E3 (Narowal), and E7 (Muzaffargarh) with G5 (GSR-305) the winning genotype in these environments; sector two has environment E4 (Swat) with G2 (GSR-82) as the winning genotype; and sector three has environment E1 (Pindi Bhattian), E5 (Islamabad), E6 (Dera Ismail Khan), and E8 (Dokri) with check variety G6 (IRRI-6) as the winning genotype in these testers. G7 (Kissan Basmati) was regarded poorly performing genotype for straw yield (Figure 6D).

Locations and Genotypes Ranking: Best and Stable Location/Genotypes Evaluation

Ranking location pattern of GGE biplots reveals ideal testing environments for all entries. The green arrow points towards the ideal environment, which is placed in the inner most circle. Genotypes are treated as random samples of entries and focus is placed on testers. In our investigation for the second year E8 (Dokri) was found ideal location for genotypes plant height (Figure 7A) and E7 (Muzaffargarh) for tillers per plant, grain yield, and straw yield (Figure 7B–D). The rank of environments based on pattern of ranking environments GGE biplot for plant height is E8 > E5 > E2 > E4 ≈ E6 > E3 > E1 > E7; ranking for tillers per plant is E7 > E3 ≈ E4 > E8 > E6 > E2 ≈ E5 > E1; ranking for grain yield is E7 > E4 > E5 > E1 > E6 > E3 > E2 > E8; and ranking for straw yield is E7 > E1 ≈ E3 > E2 > E8 > E6 ≈ E5 > E4.

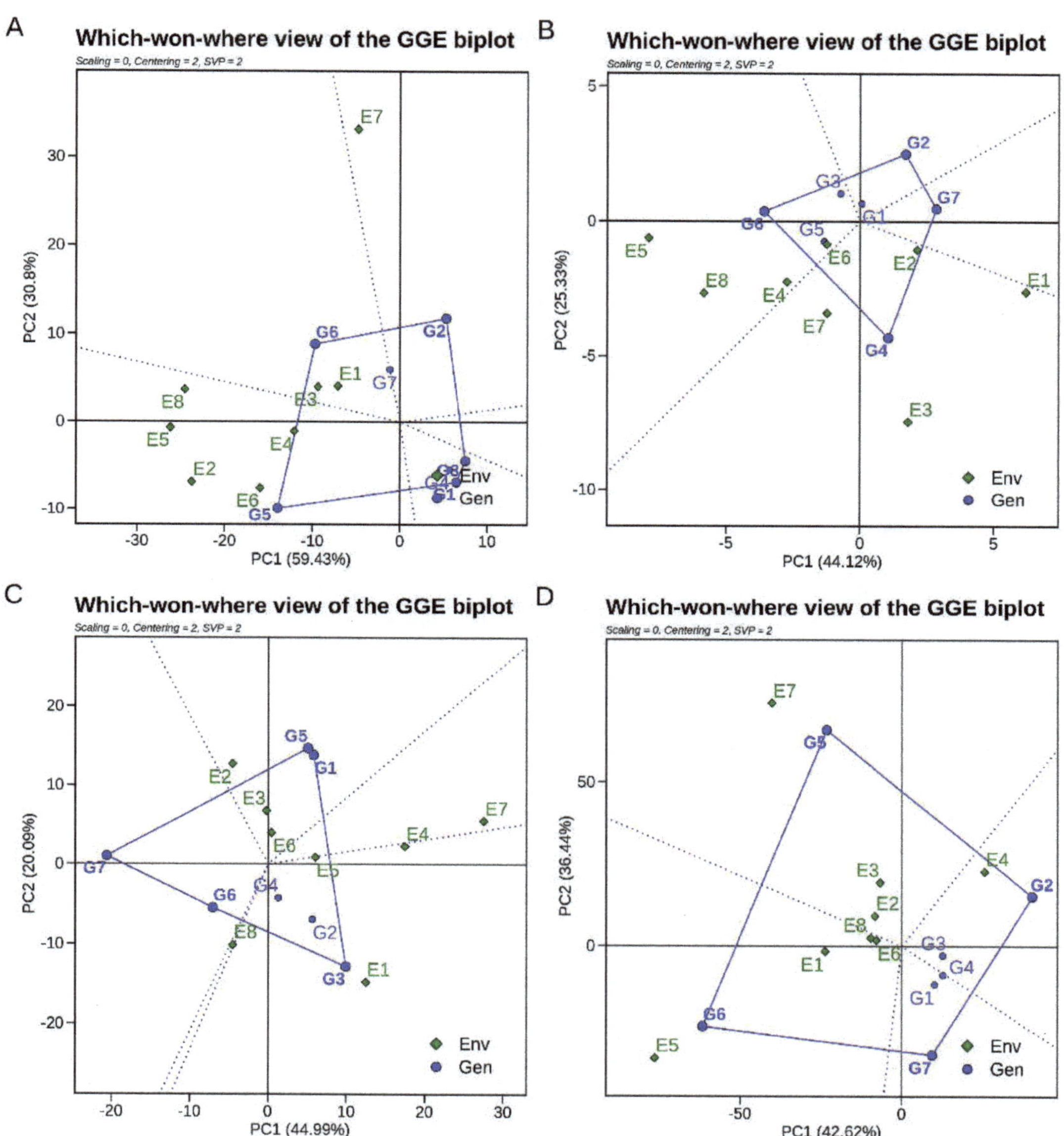

Figure 6. The GGE biplot polygon of the 'Which-won-where' pattern to identify the best cultivar in each location of 5 GSR lines and 2 control lines grown under eight environments year in 2021 for (**A**) plant height, (**B**) the number of tillers, (**C**) grain yield, and (**D**) straw yield. The biplots were created with centering = 0, SVP = 2, and scaling = 0 parameters.

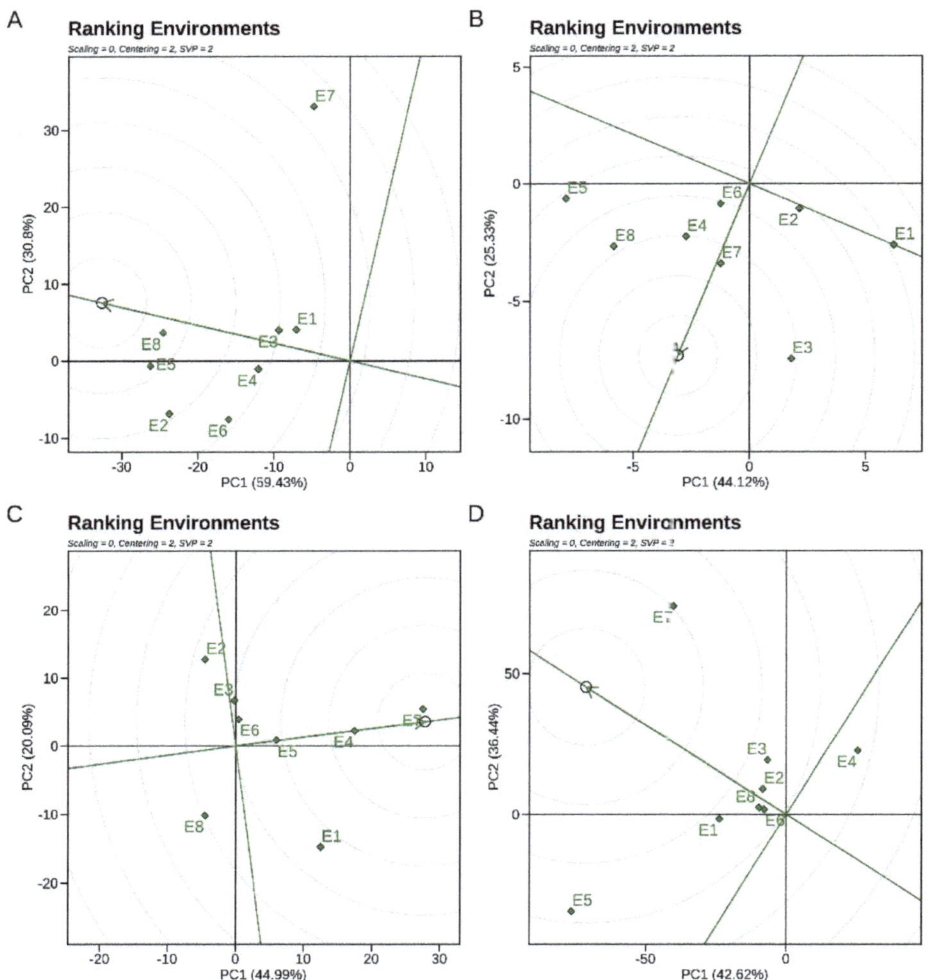

Figure 7. The GGE biplot 'Ranking environments' pattern to rank environments for an ideal environment of 5 GSR lines and 2 control lines grown under eight environments in the year 2021 for (**A**) plant height, (**B**) the number of tillers, (**C**) grain yield, and (**D**) straw yield. The biplots were created with centering = 0, SVP = 2, and scaling = 0 parameters.

Using the genotype ranking GGE biplot (Figure 8) we can identify the best entry in comparison to other entries tested in all testers. GGE biplot noted G6 (IRRI-6) as high performing genotypes for plant height (Figure 8A); G5 (GSR-305) for tillers per plant (Figure 8B); G5 (GSR-305) for grain yield per plant (Figure 8C); and again G5 (GSR-305) for straw yield per plant (Figure 8D). The genotypes ranking for plant height was G6 > G7 > G5 > G4 > G1 > G2 ≈ G3; for tillers per plant G5 > G4 > G1 > G6 ≈ G3 > G7 > G2; for grain yield G5 > G1 > G3 > G2 > G6 > G4 > G7; and for straw yield the ranks were G5 > G3 > G1 ≈ G6 > G4 > G2 ≈ G7.

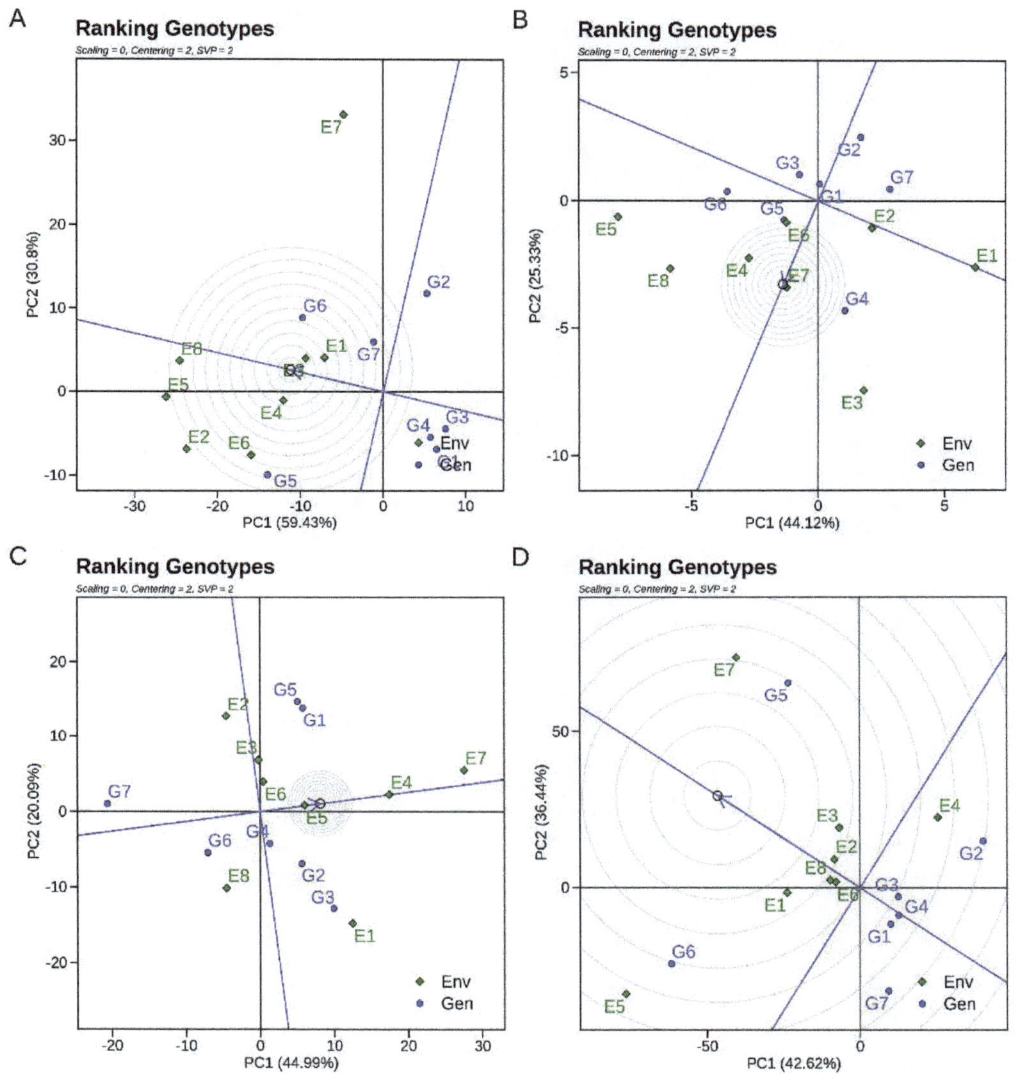

Figure 8. The GGE biplot 'Ranking genotypes' pattern to rank genotypes for the ideal genotype of 5 GSR lines and 2 control lines grown under eight environments in the year 2021 for (**A**) plant height, (**B**) the number of tillers, (**C**) grain yield, and (**D**) straw yield. The biplots were created with centering = 0, SVP = 2, and scaling = 0 parameters.

4. Discussion

The Green Super Rice in Pakistan (GSRP) project is one of the research components of megaprojects on "Productivity Enhancement of Rice" in Pakistan, where the task is to rapidly increase rice grain yield from 10 to 20 t/ha. The pedigree of the newly introduced GSR advanced lines is the mixture of more than 250 different potential rice varieties and hybrids adapted to challenging growing conditions. The GSR breeding project has been successful in attaining the most satisfactory traits, including strong and erect plant architecture, early maturity, maximum tillering, long and dense panicle, and disease/insect pest resistance (Figure 9).

Figure 9. Salient features of GSR traits over BASMATI rice: (**A**) Architecture, left-side plant is GSR and right-side plant is BASMATI, scale = 10 cm; (**B**) Stature, short BASMATI plant vs. long GSR plant, scale = 10 cm; (**C**) Maturity, early maturing GSR plant vs. late maturing BASMATI plant, scale = 10 cm; (**D**) Tillering, left-side plant is BASMATI and right-side plant is GSR, scale = 10 cm; (**E**) Panicle density, left-side plant is BASMATI, center and right-side plant are GSR, scale = 1 cm; (**F**) Bacterial blight disease, left-side plant is BASMATI, center and right-side plant are GSR, scale = 1 cm.

Univariate stability parameters: AMMI stability value (ASV), AMMI stability index (ASI), Shukla (σ^2), Wricke's ecovalence (W_i^2), multivariate stability parameters; AMMI-model and GGE biplots, were determined to find out the stable GSR line. GSR 48 was identified as the most stable genotype as a result of univariate stability analysis, while multivariate analyses have identified GSR 305 and GSR 252 as the most stable genotypes. Haider et al. [28] evaluated 18 rice varieties for yield and stability in Pakistan using the data from Rice Research Institute, Kala Shah Kaku for two years over nine different environments [28]. Our results demonstrated that the tested genotypes across different locations for two consecutive years are highly vulnerable to climatic zones and environmental factors [29,30]. Such variances could be due to the difference in topography and climatic conditions across different locations and years where the experimentations were conducted [31]. The breeding protocol must quantify genotype, environment, years, and their interaction factors to obtain successful breeding results of yield and related traits in rice [32]. The present findings of significant sources of variation have been previously noted in rice [33,34] and other cereal crops [35,36].

Stability analysis for multi-location data has been evaluated in both univariate and multivariate statistics [37]. Among the multivariate methods, the additive main effects and AMMI analysis are widely used for G × E interactions. The AMMI model combines ANOVA and G × E interactions to identify the genotypes and environmental variables [23,26]. The relative contributions of the total sum of squares of location, genotype, and GL interactions in the AMMI model of two-year data for grain yield per plant showed a similar pattern in the previous rice stability analysis [31,38]. Significant interactions between locations and tested genotypes in plant height and tillers per plant, as a high portion of the first two interaction principle components (IPCA1 and IPCA2), have been reported [32].

In our study, the univariate stability analysis screened out highly stable (GSR 112 and GSR 252) GSR lines for most of the studied traits. The GGE biplot analysis showed that IIRI-6 was the most stable genotype for plant height. GSR-305 and Kissan basmati were the most stable genotypes for tillers per plant. GSR 305 was closed to the biplot origin, depicting less response than the vertex genotypes. Moreover, it also reveals low environmental interaction in terms of grain and straw yield per plant. On the contrary, the other genotypes were farther from the biplot origin and demonstrated higher vulnerability towards environmental factors that affect their stability. Based on the adaptation pattern, Narowal and Dokri were found to be the most dynamic locations for genotypes plant height, Muzaffargarh and the NARC for tillers per plant, and Swat and Muzaffargarh for grain and straw yield per plant in 2020 and 2021, respectively. However, the tested genotypes showed different yields concerning their locations for the yield traits. Similar observations of the biplot model for multi-location studies using rice genotypes were also concluded earlier [39,40]. However, high-performing GSR lines for yield traits with less stability across locations can be stabilized following the backcross approach [41] with the most stable GSR line.

5. Conclusions

In our study, multi-location adoptability trials were aimed to predict the most promising rice genotypes across multi-environmental conditions in Pakistan. In this regard, several univariate and multivariate parametric stability models were analyzed to determine the stability performances of genotypes across environments. This study revealed three consistently stable GSR lines with minimum stability values in univariate stability statistics: GSR 305, GSR 252, and GSR 112. It is noted that GSR 48 showed the maximum stability when compared to all other lines in the univariate model across the two years for grain yield and related traits data. Furthermore, it is also concluded that multivariate parametric stability models (AMMI analysis of variance and GGE biplot) are great components to select the most suitable and stable GSR lines for specific as well as diverse environments. In this study, the combined ANOVA of the AMMI model showed that genotypes, locations, G × L interactions, and AMMI components (PCs one and two) were found significant. Therefore, yield and significant PCs were taken into account simultaneously to define the effect of GL interactions and, then, to predict the most stable GSR line. Resultantly, AMMI and GGE biplot analysis classified GSR 305 and GSR 252 as the most stable genotypes across eight tested locations. Moreover, Swat, Narowal, and Muzaffargarh tend to be the best locations to commercialize GSR lines in Pakistan.

Supplementary Materials: The following supporting information can be downloaded at: https://www.mdpi.com/article/10.3390/agronomy12051157/s1, Supplementary Table S1: Combined analysis of variances and AMMI stability model.

Author Contributions: M.R.K. and I.U.Z. initiated the concept. M.H., M.K.N., U.A., M.U., A.L. and A.R. contributed in conducted research and data collection. M.R.K. and G.M.A. acquired the funding and other resources. I.U.Z. and N.Z. wrote the paper. M.R.K. and M.U. helped with manuscript revision. M.R.K. supervised the experiment. All authors have read and agreed to the published version of the manuscript.

Funding: This research was funded by the PSDP project "Productivity Enhancement of rice sub-component Green Super Rice in Pakistan" at the National Institute for Genomics and Advanced Biotechnology, National Agriculture Research Centre Islamabad.

Data Availability Statement: All the data are presented in the main text and supplementary file.

Acknowledgments: We thank Zhikang Li, Jianlong Xu, and Shahzad Amir Naveed from the Institute of Crop Sciences, Chinese Academy of Agricultural Sciences (CAAS), who provided germplasm and valuable suggestions about Green Super Rice breeding in Pakistan.

Conflicts of Interest: The authors declare no conflict of interest.

References

1. FAO. *World Food and Agriculture—Statistical Yearbook 2020*; FAO Statistical Yearbook—World Food and Agriculture; FAO: Rome, Italy, 2020; ISBN 978-92-5-133394-5.
2. Yu, S.; Ali, J.; Zhang, C.; Li, Z.; Zhang, Q. Genomic Breeding of Green Super Rice Varieties and Their Deployment in Asia and Africa. *Theor. Appl. Genet.* **2020**, *133*, 1427–1442. [CrossRef] [PubMed]
3. Wricke, G. Uber eine Methode zur Erfassung der okologischen Streubreite in Feldverzuchen. *Z. Pflanzenzuchtg* **1962**, *47*, 92–96.
4. Piepho, H.P. A Comparison of the Ecovalence and the Variance of Relative Yield as Measures of Stability. *J. Agron. Crop Sci.* **1994**, *173*, 1–4. [CrossRef]
5. Farshadfar, E.; Mohammadi, R.; Aghaee, M.; Vaisi, Z. GGE biplot analysis of genotype x environment interaction in wheat-barley disomic addition lines. *Aust. J. Crop Sci.* **2012**, *6*, 1074–1079. [CrossRef]
6. Shukla, G. Some statistical aspects of partitioning genotype environmental components of variability. *Heredity* **1972**, *29*, 237–245. [CrossRef]
7. Francis, T.R.; Kannenberg, L.W. Yield stability studies in short-season maize. i. a descriptive method for grouping genotypes. *Can. J. Plant Sci.* **1978**, *58*, 1029–1034. [CrossRef]
8. Roemer, J. Sinde die ertagdreichen Sorten ertagissicherer. *Mitt DLG* **1917**, *32*, 87–89.
9. Plaisted, R.L.; Peterson, L.C. A technique for evaluating the ability of selections to yield consistently in different locations or seasons. *Am. Potato J.* **1959**, *36*, 381–385. [CrossRef]
10. Plaisted, R.L. A shorter method for evaluating the ability of selections to yield consistently over locations. *Am. Potato J.* **1960**, *37*, 166–172. [CrossRef]
11. Finlay, K.W.; Wilkinson, G.N. The analysis of adaptation in a plant-breeding programme. *Aust. J. Agric. Res.* **1963**, *14*, 742–754. [CrossRef]
12. Gauch, H.G., Jr. Model selection and validation for yield trials with interaction. *Biometrics* **1988**, *44*, 705–715. [CrossRef]
13. Yan, W.; Cornelius, P.L.; Crossa, J.; Hunt, L.A. Two Types of GGE Biplots for Analyzing Multi-Environment Trial Data. *Crop Sci.* **2001**, *41*, 656–663. [CrossRef]
14. Gauch, H.G., Jr.; Piepho, H.-P.; Annicchiarico, P. Statistical Analysis of Yield Trials by AMMI and GGE: Further Considerations. *Crop Sci.* **2008**, *48*, 866–889. [CrossRef]
15. Nassar, R.; Hühn, M. Studies on Estimation of Phenotypic Stability: Tests of Significance for Nonparametric Measures of Phenotypic Stability. *Biometrics* **1987**, *43*, 45–53. [CrossRef]
16. Kang, M. A rank-sum method for selecting high-yielding, stable corn genotypes. *Cereal Res. Commun.* **1988**, *16*, 113–115.
17. Fox, P.N.; Skovmand, B.; Thompson, B.K.; Braun, H.-J.; Cormier, R. Yield and adaptation of hexaploid spring triticale. *Euphytica* **1990**, *47*, 57–64. [CrossRef]
18. Thennarasu, K. On Certain Non-Parametric Procedures for Studying Genotype-Environment Inertactions and Yield Stability. Ph.D. Thesis, IARI, Division of Agricultural Statistics, New Delhi, India, 1995.

19. Farshadfar, E. Incorporation of AMMI stability value and grain yield in a single non-parametric index (GSI) in bread wheat. *Pak. J. Biol. Sci. PJBS* **2008**, *11*, 1791–1796. [CrossRef]
20. Burton, G.W.; DeVane, E.H. Estimating Heritability in Tall Fescue (Festuca Arundinacea) from Replicated Clonal Material1. *Agron. J.* **1953**, *45*, 478–481. [CrossRef]
21. Purchase, J.L.; Hatting, H.; van Deventer, C.S. Genotype × environment interaction of winter wheat (*Triticum aestivum* L.) in South Africa: II. Stability analysis of yield performance. *South Afr. J. Plant Soil* **2000**, *17*, 101–107. [CrossRef]
22. Jambhulkar, N.; Rath, N.; Bose, L.; Subudhi, H.; Mondal, B.; Das, L.; Meher, J. Stability analysis for grain yield in rice in demonstrations conducted during *rabi* season in India. *ORYZA Int. J. Rice* **2017**, *54*, 234. [CrossRef]
23. Zobel, R.W.; Wright, M.J.; Gauch, H.G., Jr. Statistical Analysis of a Yield Trial. *Agron. J.* **1988**, *80*, 388–393. [CrossRef]
24. Olivoto, T.; Lúcio, A.D. metan: An R package for multi-environment trial analysis. *Methods Ecol. Evol.* **2020**, *11*, 783–789. [CrossRef]
25. Kempton, R.A. The use of biplots in interpreting variety by environment interactions. *J. Agric. Sci.* **1984**, *103*, 123–135. [CrossRef]
26. Gauch, H.G., Jr.; Zobel, R.W. Optimal Replication in Selection Experiments. *Crop Sci.* **1996**, *36*, 838–843. [CrossRef]
27. Oladosu, Y.; Rafii, M.Y.; Abdullah, N.; Magaji, U.; Miah, G.; Hussin, G.; Ramli, A. Genotype × Environment interaction and stability analyses of yield and yield components of established and mutant rice genotypes tested in multiple locations in Malaysia. *Acta Agric. Scand. Sect. B Soil Plant Sci.* **2017**, *67*, 590–606. [CrossRef]
28. Haider, Z.; Akhter, M.; Mahmood, A.; Khan, R.A.R. Comparison of GGE biplot and AMMI analysis of multi-environment trial (MET) data to assess adaptability and stability of rice genotypes. *Afr. J. Agric. Res.* **2017**, *12*, 3542–3548. [CrossRef]
29. Bose, L.; Jambhulkar, N.; Singh, O. Additive Main effects and Multiplicative Interaction (AMMI) analysis of grain yield stability in early duration rice. *JAPS J. Anim. Plant Sci.* **2014**, *24*, 1885–1897.
30. Zeigler, R.S. Rice research for poverty alleviation and environmental sustainability in Asia. *GeoJournal* **2006**, *35*, 286–298.
31. Kanfany, G.; Ayenan, M.A.T.; Zoclanclounon, Y.A.B.; Kane, T.; Ndiaye, M.; Diatta, C.; Diatta, J.; Gueye, T.; Fofana, A. Analysis of Genotype-Environment Interaction and Yield Stability of Introduced Upland Rice in the Groundnut Basin Agroclimatic Zone of Senegal. *Adv. Agric.* **2021**, *2021*, e4156167. [CrossRef]
32. Balakrishnan, D.; Subrahmanyam, D.; Badri, J.; Raju, A.K.; Rao, Y.V.; Beerelli, K.; Mesapogu, S.; Surapaneni, M.; Ponnuswamy, R.; Padmavathi, G.; et al. Genotype × Environment Interactions of Yield Traits in Backcross Introgression Lines Derived from Oryza sativa cv. Swarna/Oryza nivara. *Front. Plant Sci.* **2016**, *7*, 1530. [CrossRef]
33. Sharifi, P.; Aminpanah, H.; Erfani, R.; Mohaddesi, A.; Abbasian, A. Evaluation of Genotype × Environment Interaction in Rice Based on AMMI Model in Iran. *Rice Sci.* **2017**, *24*, 173–180. [CrossRef]
34. Shrestha, J.; Kushwaha, U.K.S.; Maharjan, B.; Kandel, M.; Gurung, S.B.; Poudel, A.P.; Karna, M.K.L.; Acharya, R. Grain Yield Stability of Rice Genotypes. *Indones. J. Agric. Res.* **2020**, *3*, 116–126. [CrossRef]
35. Shojaei, S.H.; Mostafavi, K.; Omrani, A.; Omrani, S.; Nasir Mousavi, S.M.; Illés, Á.; Bojtor, C.; Nagy, J. Yield Stability Analysis of Maize (*Zea mays* L.) Hybrids Using Parametric and AMMI Methods. *Scientifica* **2021**, *2021*, 5576691. [CrossRef] [PubMed]
36. Tremmel-Bede, K.; Szentmiklóssy, M.; Tömösközi, S.; Török, K.; Lovegrove, A.; Shewry, P.R.; Láng, L.; Bedő, Z.; Vida, G.; Rakszegi, M. Stability analysis of wheat lines with increased level of arabinoxylan. *PLoS ONE* **2020**, *15*, e0232892. [CrossRef] [PubMed]
37. Pour-Aboughadareh, A.; Khalili, M.; Poczai, P.; Olivoto, T. Stability Indices to Deciphering the Genotype-by-Environment Interaction (GEI) Effect: An Applicable Review for Use in Plant Breeding Programs. *Plants* **2022**, *11*, 414. [CrossRef] [PubMed]
38. Katsura, K.; Tsujimoto, Y.; Oda, M.; Matsushima, K.; Inusah, B.; Dogbe, W.; Sakagami, J.-I. Genotype-by-environment interaction analysis of rice (*Oryza* spp.) yield in a floodplain ecosystem in West Africa. *Eur. J. Agron.* **2016**, *73*, 152–159. [CrossRef]
39. Gauch, H.G.J. AMMI analysis on yield trials. *CIMMYT Wheat Spec. Rep. CIMMYT* **1992**, 9–12.
40. Wade, L.J.; McLaren, C.G.; Quintana, L.; Harnpichitvitaya, D.; Rajatasereekul, S.; Sarawgi, A.K.; Kumar, A.; Ahmed, H.U.; Sarwoto; Singh, A.K.; et al. Genotype by environment interactions across diverse rainfed lowland rice environments. *Field Crops Res.* **1999**, *64*, 35–50. [CrossRef]
41. Singh, R.; Huerta Espino, J. The use of single-backcross, selected-bulk breeding approach for transferring minor genes based rust resistance into adapted cultivars. *Cereals* **2004**, *54*, 48–51.

Article

Genome-Wide Characterization of the SAMS Gene Family in Cotton Unveils the Putative Role of *GhSAMS2* in Enhancing Abiotic Stress Tolerance

Joseph Wanjala Kilwake [1,†], Muhammad Jawad Umer [1,†], Yangyang Wei [2], Teame Gereziher Mehari [3], Richard Odongo Magwanga [1,4], Yanchao Xu [1], Yuqing Hou [1], Yuhong Wang [1], Margaret Linyerera Shiraku [1], Joy Nyangasi Kirungu [1], Xiaoyan Cai [1], Zhongli Zhou [1], Renhai Peng [2,*] and Fang Liu [1,5,*]

1 State Key Laboratory of Cotton Biology/Institute of Cotton Research, Chinese Academy of Agricultural Science, Anyang 455000, China
2 Biological and Food Engineering, Anyang Institute of Technology, Anyang 455000, China
3 School of Life Sciences, Nantong University, Nantong 226007, China
4 School of Biological, Physical, Mathematics and Actuarial Sciences (SBPMAS), Jaramogi Oginga Odinga University of Science and Technology (JOOUST), Bondo P.O. Box 210-40601, Kenya
5 School of Agricultural Sciences, Zhengzhou University, Zhengzhou 450001, China
* Correspondence: aydxprh@163.com (R.P.); liufcri@163.com (F.L.); Tel.: +86-139-4950-7902 (F.L.)
† These authors contributed equally to this work.

Abstract: The most devastating abiotic factors worldwide are drought and salinity, causing severe bottlenecks in the agricultural sector. To acclimatize to these harsh ecological conditions, plants have developed complex molecular mechanisms involving diverse gene families. Among them, S-adenosyl-L-methionine synthetase (SAMS) genes initiate the physiological, morphological, and molecular changes to enable plants to adapt appropriately. We identified and characterized 16 upland cotton SAMS genes (*GhSAMSs*). Phylogenetic analysis classified the *GhSAMSs* into three major groups closely related to their homologs in soybean. Gene expression analysis under drought and salt stress conditions revealed that *GhSAMS2*, which has shown the highest interaction with *GhCBL10* (a key salt responsive gene), was the one that was most induced. *GhSAMS2* expression knockdown via virus-induced gene silencing (VGIS) enhanced transgenic plants' susceptibility to drought and salt stress. The TRV2:*GhSAMS2* plants showed defects in terms of growth and physiological performances, including antioxidative processes, chlorophyll synthesis, and membrane permeability. Our findings provide insights into SAMS genes' structure, classification, and role in abiotic stress response in upland cotton. Moreover, they show the potential of *GhSAMS2* for the targeted improvement of cotton plants' tolerance to multiple abiotic stresses.

Keywords: S-adenosyl-L-methionine synthetase; virus-induced gene silencing; *SAMS2*; abiotic stress; upland cotton

Citation: Kilwake, J.W.; Umer, M.J.; Wei, Y.; Mehari, T.G.; Magwanga, R.O.; Xu, Y.; Hou, Y.; Wang, Y.; Shiraku, M.L.; Kirungu, J.N.; et al. Genome-Wide Characterization of the SAMS Gene Family in Cotton Unveils the Putative Role of *GhSAMS2* in Enhancing Abiotic Stress Tolerance. *Agronomy* **2023**, *13*, 612. https://doi.org/10.3390/agronomy13020612

Academic Editor: Tristan Edward Coram

Received: 28 April 2022
Revised: 24 May 2022
Accepted: 25 May 2022
Published: 20 February 2023

1. Introduction

Cotton is a valued economic crop worldwide. The long growth cycle of cotton coupled with its large genome size have rendered many available traditional methods complicated and labor-intensive in analyzing its gene function [1]. *Gossypium hirsutum*, commonly referred as upland cotton, is the most popular cotton germplasm due to its high yield. About 90% of all cotton cultivars being produced globally are derived from upland cotton. Due to climate change, crops are exposed to various abiotic stresses affecting plant growth, development, yield components, and productivity [2]. Among them, drought and salinity are the harshest environmental adversities, causing dramatic losses in cotton production [3]. Drought stress induces extensive crop loss, and predictions have revealed that it will intensify in the future [3]. It is estimated that no less than 6% of landmass globally is affected by salinity [4]. Sodium chloride is the primary salt responsible for soil salinity, and its continued accumulation poses a severe threat to farmers worldwide as agriculture productivity dwindles due to considerable defects in plant growth [5,6]. The presence of sodium chloride in high concentrations usually induces deficiency diseases (the unavailability of crucial nutrients for plants' healthy growth) and disrupts cellular ionic balance [7].

Plants have developed complex and dynamic mechanisms to adapt to these stressful environments, including various morphological, physiological, and molecular changes [8]. The common strategies employed by plants to tolerate drought and salt stresses are the reinforcement and maintenance of biological membranes' structure and properties and the escalated synthesis of antioxidant enzymes [3,9]. Many gene families, such as S-adenosyl-L-methionine synthase (SAMS), are involved in the dynamic complex regulatory networks of plants' stress responses to modulate continued development and enhance stress tolerance [10]. The SAMS genes contain a methionine binding site and an ATP binding motif in their N-terminal and C-terminal domain, respectively [10]. They catalyze the combination of methionine and ATP to produce SAM (S-Adenosyl-L-methionine), a critical molecule involved in essential biological processes in eukaryotic cells [11]. SAM provides methyl groups for DNA, RNA, lipids, and proteins methylation and participates in transsulfuration reactions and the biosynthesis of polyamine, nicotianamine, and lignin [11–14]. Moreover, SAM is the precursor for synthesizing ethylene and polyamines (PAs), which are essential for plant growth, development, and responses to environmental stresses [15–18].

Regarding the importance of SAMS, studies have focused on SAMS' function in regulating plants' stress response. The overexpression of the potato *SbSAMS* improved drought and salt stress tolerance in transgenic *Arabidopsis* plants [2]. In rice, the knockdown of *OsSAMS1*, 2, and 3 altered the histones and DNA methylation, leading to late flowering [19]. The overexpression of the Sugar Beet M14 *SAMS2* in transgenic *Arabidopsis* enhanced its tolerance to oxidative stress and salt [14]. The targeted reduction of PAs biosynthesis induced a decrease in pollen viability and plant length and promoted sensitivity to abiotic stress in rice [20]. The overexpression of *Medicago sativa subsp. falcata SAMS1* induced oxidation and polyamine synthesis in transgenic tobacco plants, improving their tolerance to chilling and freezing stress [21]. The overexpression of the cucumber *CsSAMS1* and its interacting protein *CsCDPK6* promoted ethylene and PAs biosynthesis, leading to the enhancement of salt stress tolerance in transgenic tobacco [22]. The SAMS gene family has been well studied in diverse monocotyledonous and dicotyledonous plants such as rice, sugar beet M14, *Arabidopsis*, barley, tomato, soybean, sunflower, sorghum, *Medicago truncatula*, eggplant, *Triticum urartu* [11], and *Chorispora bungeana* [23]. However, in upland cotton, no study has focused on SAMS genes and their potential to enhance stress tolerance.

Moreover, it was recently found that *GhCBL10* plays a central role in upland cotton's tolerance to salt stress [24]. Therefore, it is of particular interest to identify the *GhSAMS* with strong co-expression interaction with *GhCBL10* for the targeted improvement of cotton plants' tolerance to multiple abiotic stresses.

In the present study, SAMS genes were identified in upland cotton, and their structure, chromosomal distribution, subcellular localization, phylogeny, cis-acting elements, and conserved motifs were revealed through comprehensive bioinformatic analyses. We performed yeast two-hybrid experiments and detected the *GhSAMS* that exhibited the strongest co-expression relationship with *GhCBL10*. Furthermore, we explored the expression patterns of *GhSAMS* genes in response to salt and drought treatments, and the most promising *GhSAMS* for enhancing plant tolerance to multiple abiotic stresses was identified and functionally validated via transgenic experiments. Our data represent important resources for deciphering *GhSAMSs* in plant functions and insights into the complex molecular regulatory networks of abiotic stress response in cotton.

2. Materials and Methods

2.1. Protein Identification and Physiochemical Analysis of SAMS Genes in Gossypium hirsutum

SAMS proteins were retrieved from three Pfam domain accessions in the NAU assembly: PF00438 (1), PF02772 (2), and PF02773 (3). The three accession domains carry 16, 17, and 16 genes, respectively, though the gene names are similar. Pfam Scan was specifically used to query the genes (https://www.ebi.ac.uk/Tools/pfa/pfamscan/; accessed on 5 May 2020), and SMART search provided the identity of SAMS genes present in *Gossypium hirsutum* (http://smart.emblheidelberg.de/smart/; accessed on 20 May 2020). *GhSAMS* genes' identity was further confirmed via the official website of the Cotton genomic database (https://cottonfgd.org/; accessed on 29 May 2020), using PF02772. The physical and chemical properties of *GhSAMS* proteins (excluding the scaffolded gene), including the instability index, protein length, isoelectric point (pI), grand average of hydropathy (GRAVY), and molecular weight (MW), were predicted by ExPASy ProtParam software [25].

2.2. Chromosomal Location, Phylogenetic Analysis, Prediction of Subcellular Localization, Gene Structure, Cis-Acting Elements, and Conserved Motifs Analyses

For gene location visualization on the respective chromosomes, the retrieved gene ID information in gtf3 file format of all the *GhSAMSs* was used in Tbtools software to map the genes onto the chromosomes. The coding sequence of *GhSAMS* members was downloaded from the official website of Phytozome (https://Phytozome.jgi.doe.gov; accessed on 9 June 2020). Homolog genes from closely related plant species (Table S1) were also downloaded from the Phytozome website and later used to perform the phylogenetic analysis via the neighbor-joining method in the MEGA 7.0 program, with the specification of 1000 bootstrap replicates [26]. ClustalX software was used to align all the protein sequences before generating a phylogenetic tree diagram for evolutionary relationships analysis [27]. The Poisson correction was applied to estimate the distance between sequences. WoLF Psort online software was used to predict *GhSAMS* genes' subcellular localization (https://wolfpsort.hgc.jp; accessed on 15 June 2020) [28].

GhSAMS genes' structure analysis was conducted via the Gene Structure Display Server website (http://gsds.cbi.pku.edu.cn; accessed on 19 June 2020) [29]. In-depth prediction of the cis-regulatory DNA elements in the *GhSAMS* promoter region (2000 bp upstream nucleotide sequence) was achieved by the online PlantCARE server software (Bristol, England) [30]. MEME server (https://meme-suite.org, version 5.4.1; accessed on 13 Jully 2020), with the default setting, was used to predict conserved motifs within the gene structures ().

2.3. Plant Materials and Treatments

Marie Galante-85 and CRI-12 semi-wild accessions of *G. hirsutum* were used, as they are tolerant to drought and salt stresses. Seeds of the two accessions were provided by the Institute of Cotton Research, Chinese Academy of Agricultural Sciences, where the entire experiment was performed with a Complete Random design. In the preliminary steps, the cotton seeds were soaked in dd H_2O overnight to allow the seed coat to soften. The soaked seeds were then grown in folded absorbent papers vertically placed in mini rectangular plastic buckets, which had been filled with distilled water halfway and left for a period of four days [31]. Upon germination (the sixth day), the healthy seedlings were transplanted to a Hoagland nutrient solution medium in the greenhouse. In the greenhouse, transplanted seedlings were treated by a 16 h light-8 h dark photoperiod with specified temperatures of 28 °C during the day and 25 °C at night. The relative humidity in the experimental room was maintained at 60–70%, as previously described [32]. The entire Hoagland nutrient solution medium was replenished when the seedlings reached the three-true-leaf stage, and freshly prepared solutions of 17% of glycol PEG-6000 and 250 mM of sodium chloride compounds were immediately added to simulate drought and salt stresses, respectively [33]. The healthy tissues of the root and leaf were collected from nine plants of each category for RNA extractions after stress exposure at the following time intervals: 0 h, 3 h, 6 h, 9 h, 12 h, and 24 h. Three biological replications were considered in each case. Untreated plants were considered as the control. The harvested tissues were directly frozen in liquid nitrogen and transferred to the fridge at −80 °C for storage up to the total RNA extraction.

2.4. RNA Extraction and RT-qPCR Assays

The total RNA was extracted using the RNAprep Pure Plant kit (Tiangen, Beijing, China), and its quality and concentration were determined using a NanoDrop 2000 spectrophotometer. cDNA synthesis was conducted by treating 1 g of total RNA using RNase-free DNase I and a reverse transcriptase, strictly following the guidelines given by the manufacturer (Thermo Fisher Scientific, Shanghai, China). We investigated the expression patterns of *GhSAMSs* under drought and salt stress at different time intervals, using previously released RNA-Seq data (https://cottonfgd.org/analyze/ accessed on 5 May 2020). According to the genes' expression patterns, we selected 14 genes, including 5 genes that were induced under salt and drought stress, 5 genes that were downregulated under the stress conditions, and 4 genes that showed similar expression under normal and stress conditions for the RT-qPCR analysis. Using the Premier Premier5 software, the primers of all the selected genes were designed for the RT-qPCR assays (Table S2). *GhActin* was chosen to serve as a standard reference gene. The SYBR Green Real-Time PCR Master Mix (Thermo Scientific, Rockford, IL, USA) was used to perform qPCR assays following the procedure described previously [34]. The reactions comprised the following reagents: cDNA template (5 µL), forward primer (0.5 µL), reverse primer (0.5 µL), SYBR green master mix (10 µL), and dd H_2O (4 µL). The final mixture, whose concentration was at 10 mM, was centrifuged at 12,000 rpm for 1 min and placed into PCR thermal cycling conditions, as previously described [35]. The PCR procedure was performed according to the manufacturer's instructions. Pre-incubation, 1 cycle: 95 °C for 30 s; Amplification, 40 cycles: 95 °C for 10 s, 60 °C for 30 s; Melting curve, 1 cycle: 95 °C for 15 s, 60 °C for 60 s, 95 °C for 15 s; Cooling, one cycle: 40 °C for 30 s. The real-time analysis of each gene was performed with three independent biological replicates under the same conditions. The expression levels of the genes were analyzed using the $2^{-\Delta\Delta CT}$ method [36].

2.5. Identification of Pray Proteins

First, the CBL10 gene-specific protein sequence for the *Arabidopsis* plant was obtained from the official website of the *Arabidopsis* Information Resource database (ftp://ftp.arabidopsis.org; accessed on 21 February 2021). Then, the isolated protein sequence was used in a BLASTp analysis as a query against the proteomes of upland cotton, and the NAU assembly was used to identify the CBL10 homolog. The identified cotton CBL10 gene (*Gh_D05G0440.1*) was later used in the Y2H system experiment to screen for its interacting proteins from the AD library.

2.6. Construction of Yeast Two-Hybrid Library, Bait Cloning, and Auto Activation Analysis

The Yeast Two-Hybrid (Y2H) fusion library of *Gossypium hirsutum* Marie-Galante leaves, stems, and roots under drought and salt conditions (pGADT7-library) was prepared by Oebiotech (Shanghai, China). The BD-*GhCBL10* bait plasmid was constructed as previously described [37]. In summary, the full length of the *GhCBL10* CDS was amplified by PCR, using the primers F-TGCATATGGCCATGGAGGCCGAATTC and R-TGCGGCCGCTGCAGGTCGAC GGATCC, and cloned at the pGBKT7 vector sites NCO1 and BamH1. It was crucial to confirm the transcriptional activation of the bait in the Y2HGold competent cell in the absence of a prey protein. We independently transformed the plasmids of bait, the negative control, and the positive control into Y2H Gold competent cells. The constructs were grown on different growth media, as described by Chen et al. [37], for three days. Table 1 presents the annotation of the negative and positive controls and the empty vector.

Table 1. The bait auto-activation and toxicity test sampling.

Reaction	Plasmid 1	Plasmid 2
Positive Control	pGBKT7-53	pGADT7-T
Negative Control	pGBKT7-Lam	GADT7-T
BD (Target gene)	pGBKT7-GhCBL10	
Empty vector	pGADT7	

2.7. cDNA Libraries Screening and Yeast Two-Hybrid Interaction Assay

Yeast two-hybrid screening was conducted following the Oebiotech (Shanghai) mating protocol, as previously described. Briefly, we mated the bait strain (Y2HGold (pGBKT7-*GhCBL10*)) and the pGADT7-library plasmid, plated on the SD/–Ade/–His/–Leu/– Trp/X-α-gal/AbA (QDO/X/A) and SD/–Leu/–Trp/–His/X- α-gal/AbA (TDO/X/A) plates, and incubated the plates at 30 °C for five days [38]. We conducted colony PCR and sequencing using the T7 primer to determine the positive interaction and the duplicates. After sequencing, we used the BLASTn of the CottonFGD database to analyze the nucleotide sequence. We then co-transformed the potential positive prey identified with the pGBKT7-*GhCBL10* bait into Y2HGold competent cells. The CDS of CBL10 was cloned into the DNA-binding domain (BD) vector pGBKT7, while the CDSs of PRA1 B1, DSP8, and SAMS2 were cloned into the activation domain (AD) vector pGADT7, respectively, using the primers presented in Table S3. The generated transformants were grown on TDO/X/A and QDO/X/A plates and incubated at 30 °C until colonies appeared. PGBKT7-Lam and pGBKT7 (empty vector) denoted the negative control, while pGBKT7-53 denoted the positive control [38,39].

2.8. Virus-Induced Gene Silencing of GhSAMS2 in G. hirsutum and Stress Treatments

Tobacco rattle virus (pTRV) was used to elucidate *GhSAMS2* (*Gh_A08G1067*) gene function with the RNAi technique [40]. VIGS TRV2:PDS, TRV2:00, TRV2:*GhSAMS2*, and WT plants were investigated under both drought and salt stress conditions. The CDS fragment of *GhSAMS2* was 1182 bp in length. The *GhSAMS2* cDNA was amplified using the specific primers F-TGCATATGGCCATGGAGGCCGAATTC and R-TGCGGCCGCTGCAGGTCGACGGATCC. Next, the PCR products were cloned into the *Xba*1 and *Xho*1 sites of the pTRV to generate pTRV:*GhSAMS2* [41]. Subsequently, recombinant DNA transformation into the LBA4404 bacteria strain (*Agrobacterium tumefaciens*) was conducted as previously described [42]. The LBA4404 strain containing the pTRV2-PDS, pTRV1, pTRV2-*Gh_A08G1067*, and pTRV2 vectors was cultured in a shaking incubator at 28 °C in the Luria-Bertani (LB) liquid medium, with freshly prepared 10 mM 2-(N-morpholino)- ethane sulfonic acid (MES) added in. Kanamycin and rifampicin antibiotics were first added to the LB medium. Then, the cultures were put in the shaking incubator overnight, as previously prescribed [43]. This was followed by the centrifugation of the cultures for 10 min at 8000 rpm after the OD had been determined at 1.5, and the cells then were re-suspended into the infiltration buffer containing 200 μM of acetosyringone (As), 10 mM of magnesium chloride, and 10 mM of MES to a final OD600 = 1.5. To obtain the final infiltration medium, the pTRV1 re-suspension was mixed with pTRV2-PDS, pTRV2-*GhSAMS2*, and pTRV2, separately, at a ratio of 1:1 before the seedlings were infiltrated by the infusion, as previously described [44]. The functional analysis experiment via VIGS involved the inoculation of 60 plants with the TRV:*GhSAMS2* and TRV: PDS inoculum, respectively. The empty vector (TRV2:00) was inoculated into 60 other plants to represent the wild type. Then, 60 other plants were left to grow without any inoculum, serving as the control in this experiment. When the plants reached the three-leaf-stage, the Hoagland nutrient solution medium into which they had been transplanted was treated with freshly prepared solutions of 17% of glycol PEG-6000 and 250 mM of sodium chloride compounds to simulate drought and salt stress, respectively [45]. The duration of each stress was 48 h. After the stress exposure, the healthy tissues of the stem, root, and leaf were collected from ten plants of each category in triplicate for RNA extraction and physiological and biochemical analyses.

2.9. Measurement of the Physiological and Morphological Parameters

The morphological and physiological parameters were equally determined to help assess the extent of susceptibility between the silenced and non-silenced plants under drought and salt stress conditions. Plant height (PH), root length (RL), shoot fresh weight (SFW), root fresh weight (RFW), relative leaf water content (RLWC), cell membrane stability (CMS), chlorophyll content (SPAD/Chlo), and excised leaf water loss (ELWL) were measured. The relative leaf water loss, cell membrane stability through ion leakage, and chlorophyll content were determined, as described previously [46]. Excised leaf water loss was determined by first weighing the collected fresh leaf samples immediately after harvesting to note the initial leaf weight in grams. After the leaf sample had lasted for 24 h on the bench at room temperature, the second weight measurement was taken and recorded as wilted weight (WW). Finally, the third measurement was taken and recorded as dry weight (DW) after the leaf sample had stayed inside an oven (50 °C) for four days. To calculate the ELWL, the formula below was applied.

$$\text{ELWL} = \left\{\frac{\text{FW} - \text{WW}}{\text{DW}}\right\}$$

Regarding the relative leaf water content and fresh weight (FW), the leaf samples were placed into dd H_2O at room temperature for 24 h using tissue paper; they were then dried on both surfaces before being weighed again to obtain the saturated weight (SW). Finally, the dry weight (DW) was measured and recorded after the leaf samples had stayed inside an oven at 50 °C for four days. The formula applied in the calculation of RLWC was:

$$\text{RLWC} = \left\{\frac{\text{FW} - \text{DW}}{\text{SW} - \text{DW}}\right\} \times 100$$

Ion leakage in the plant tissues, which is also referred to as cell membrane stability (CMS), was assessed using the fresh leaf tissues. First, the plant electrolyte was quantified in the process of determining cell membrane stability, as previously described [47]. Then, plastic cylindrical tubes filled with 5 mL of dd H_2O and kept in the dark for 24 h were used to harbor leaf samples weighing 0.5 g each. Two electrical conductivities were measured per sample, the first one being measured after the 24 h dark period stage (T1), while the second one was conducted after the leaves had been boiled in a water bath at 99 °C for 30 min and cooled to room temperature (T2). The CMS was calculated using the following formula [48]:

$$\text{CMS} = \left[\left(1 - \frac{\text{TI}}{\text{T2}}\right) / \left(1 - \frac{\text{C1}}{\text{C2}}\right)\right] \times 100$$

where C is the electrical conductivity of dd H_2O.

2.10. Estimation of Oxidant and Antioxidant Enzyme Activities

The plant tissue samples were collected in three replicates, wrapped in aluminum foil, and kept at −80 °C until the biochemical analyses. Two oxidant (hydrogen peroxide, H_2O_2 and malondialdehyde, MDA) and two antioxidant (catalase, CAT, and peroxidase, POD) enzyme activities were evaluated. According to the manufacturer's protocols, the extraction and spectrometric analysis of H_2O_2 and the antioxidant enzymes were achieved using their respective assay kits supplied by Beijing Solarbio Science and Technology, China [46]. As for the malondialdehyde (MDA), which is a byproduct of liquid metabolism, its cellular concentration was measured following the method described previously [49]. The physiological parameters that were measured are very significant for water stress tolerance in plants, and they have been used extensively in evaluating various field crops [50].

2.11. Statistical Analysis

All the samples used in these experiments were in three bio-replicates. Data analysis and visualization were conducted with the aid of GraphPad Prism 8.4.3 (GraphPad Software Inc., La Jolla, CA, USA). The analysis of variance (ANOVA) determined the mean difference of the samples statistically. The significant difference was set at $p < 0.05$.

3. Results

3.1. Identification, Physiochemical Properties, Chromosomal Distribution, Phylogenetic Analysis, and Subcellular Localization

In total, seventeen (17) SAMS genes were identified in the *Gossypium hirsutum* species using the NAU assembly. They are distributed on all of the fourteen (14) chromosomes of the *G. hirsutum* genome, with the chromosomes At11 and Dt11 harboring, respectively, two *GhSAMSs* (Figure 1a). One of them is located on a scaffold. We also investigated the chromosomal distribution of *GhSAMSs* in the genomes of *G. arboretum* (A genome) and *G. raimondii* (D genome). We found that the *GhSAMS* genes are located on chromosomes At02, At04, At07, At08, At09, At11, and At12 of the A genome and on chromosomes Dt02, Dt04, Dt07, Dt08, Dt09, Dt11, and Dt12 of the D genome (Figure 1b,c).

The physicochemical properties of the *GhSAMS* genes are presented in Table 2. Their protein sequence length ranged from 256 (*Gh_A07G1193*) to 393 aa (*Gh_A08G1067*), with a molecular weight (MW) varying from 28.12 to 42.61 kDa. Both the proteins were stable and exhibited negative GRAVY values. The isoelectric points (*PI*) of the protein ranged from 5.579 to 8.974 (Table 2).

The phylogenetic analysis of *GhSAMS* proteins and SAMS proteins from other related species was performed to examine their relationships (Figure 1d). The results indicated that SAMS proteins could be classified into five major groups (I–V). The Group III genes were all from monocotyledonous plant species, while the genes from the dicotyledonous plant species were scattered across all four groups. *GhSAMS* genes were classified into groups I, II, and IV, where they showed a high degree of similarity with the SAMS genes from soybean, *T. cacao*, and *M. truncatula* (Figure 1d). The subcellular localization analysis showed that SAMS proteins are located mainly in the cytoplasm and cytoskeleton of *G. hirsutum* cells (Table 2; Figure 1e,f).

Table 2. The physicochemical properties of cotton SAMS genes.

Transcript ID	Length (aa)	MW (kDa)	Charge	PI	GRAVY	Instability Index	Subcellular Localization
Gh_A02G0578.1	393	43.061	−4	5.941	−0.308	Stable	Cytoplasm
Gh_A04G0603.1	390	42.855	3.5	6.983	−0.326	Stable	Cytoplasm
Gh_A07G1193.1	256	28.12	7	8.974	−0.326	Stable	Cytoplasm
Gh_A08G1067.1	393	43.091	−5	5.772	−0.325	Stable	Cytoplasm
Gh_A09G1368.1	390	42.61	4	7.118	−0.299	Stable	Cytoskeleton
Gh_A11G0966.1	393	43.026	−6.5	5.579	−0.36	Stable	Cytoplasm
Gh_A11G2886.1	390	42.682	2	6.786	−0.332	Stable	Cytoplasm
Gh_A12G1098.1	393	43.071	−6	5.594	−0.335	Stable	Cytoplasm
Gh_D02G0636.1	393	43.044	−4.5	5.909	−0.3	Stable	Cytoplasm
Gh_D04G1064.1	390	42.812	3.5	6.983	−0.306	Stable	Cytoplasm
Gh_D07G1294.1	393	43.039	−6	5.594	−0.328	Stable	Cytoplasm
Gh_D08G1348.1	393	43.04	−5.5	5.618	−0.338	Stable	Cytoskeleton
Gh_D09G1369.1	390	42.695	4	7.118	−0.294	Stable	Cytoskeleton
Gh_D11G1117.1	393	43.062	−6	5.596	−0.358	Stable	Cytoplasm
Gh_D11G3272.1	390	42.616	2	6.786	−0.32	Stable	Cytoskeleton
Gh_D12G1222.1	393	43.042	−6	5.594	−0.321	Stable	Cytoplasm

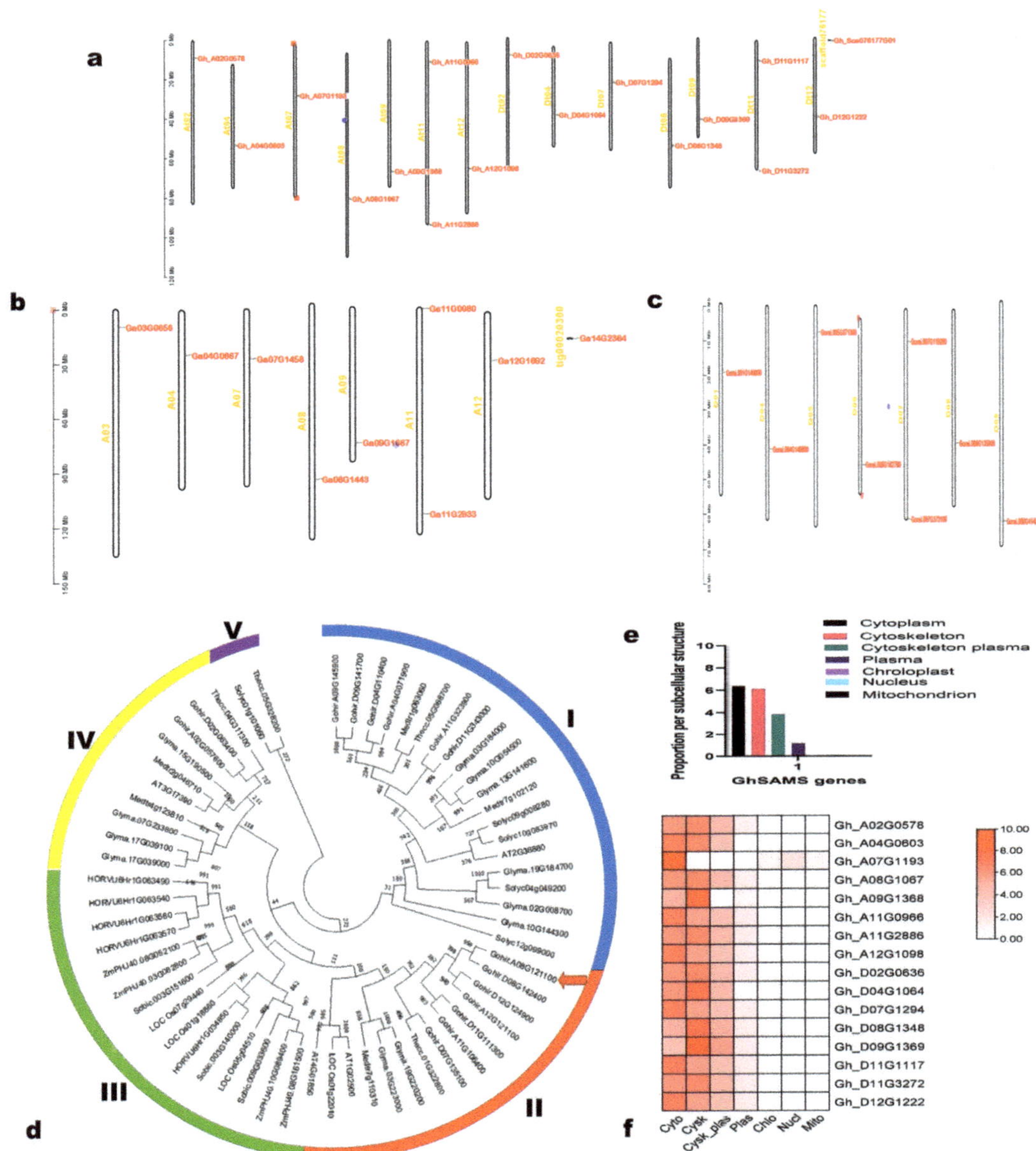

Figure 1. Chromosomal distribution, phylogenetic analysis, and subcellular localization of *GhSAMS* genes. (**a–c**) Distribution of *GhSAMS* genes on the chromosomes of *G. hirsutum*, *G. arboreum*, and *G. raimondii*, respectively. (**d**) Phylogenetic tree of *GhSAMS* genes and their homologs from *A. thaliana*, *T. cacao*, Soya bean, Rice, Tomato, *Medicago truncatula*, Scrghum, Maize, and Barley. (**e**,**f**) Subcellular localization of *GhSAMS* genes. The neighbor-joining method was used to construct the phylogenetic tree with replicates of 1000 bootstrap values in the MEGA 7.0 software.

3.2. Gene Structure, Conserved Motifs, and Cis-Acting Elements Analyses

The gene structure analysis revealed that all of the sixteen *GhSAMS* genes are intronless and contain only one exon (Figure 2a). In total, we identified five (5) conserved motifs in the sequence of the *GhSAMS* genes. Both *GhSAMS* genes contained the five motifs, except for *Gh_A07G1193*, which lacked motifs 3 and 5 (Figure 2b). Furthermore, the cis-acting regulatory elements analysis in the promoter region of the *GhSAMS* genes indicated that they might be primarily involved in the plant defense and stress responsiveness within the plant cells, considering the phytohormonal signals (Table S4).

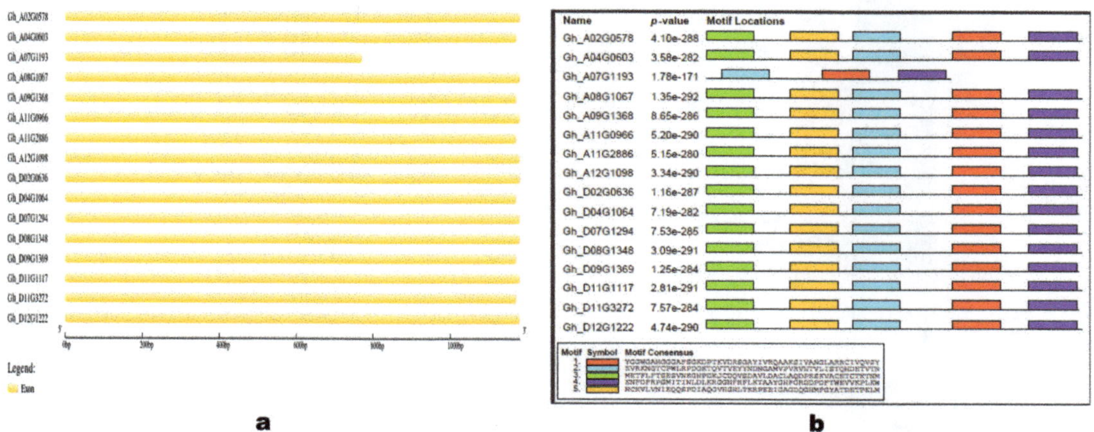

Figure 2. *GhSAMS* genes' structure (**a**), and conserved motifs in their promoter region (**b**).

3.3. GhSAMS Genes Expression under Drought and Salt Stress

Since the cis-acting elements analysis predicted that *GhSAMS* genes' function might be closely related to stress response, we examined their expression patterns under drought and salt stress in leaves and roots via RT-qPCR and using available RNA-Seq data (Figures 3 and S1). The *GhSAMS* genes' expression patterns showed significant variations, as some were down-regulated while others were highly up-regulated within the leaves and roots under the stress conditions. The RT-qPCR results were significantly correlated with the RNA-Seq in the leaf and root tissues. In general, most of the analyzed genes exhibited higher expression in the leaves (Figure 3a,c) than in the roots (Figure 3b,d). The expression of *Gh_A08G1067*, *Gh_A09G1368*, *Gh_A12G1098*, *Gh_D02G0636*, *Gh_D07G1294*, *Gh_D08G1348*, *Gh_D11G1117*, and *Gh_D12G1222* was significantly induced by both the drought and salt stress in the leaves. It is worth noting that *Gh_A08G1067* (*GhSAMS2*) expression was significantly up-regulated under the drought and salt stress in both the tissues (Figure 3), indicating that it might be critical for upland cotton's tolerance to abiotic stresses.

3.4. Identification of CBL10 Interacting Proteins from the Cotton AD Library under Drought and Salt Stress Using the Y2H System

To confirm the potential role of *GhSAMS2* in abiotic stress tolerance in upland cotton, we searched for prey proteins interacting with *GhCBL10*, the salt responsive gene, using the Yeast two-hybrid (Y2H) system. The summary of the experiments, including the self-auto-activation state, the toxicity test, the verification of the interactions, and the mating efficiency determination of the *GhCBL10* bait gene in the Y2H system, is shown in Figure S2. The zygotes that resembled a cloverleaf with a three-lobed structure (Figure S2e) confirmed the successful mating between *GhCL10* and the prey proteins contained in the cotton AD library. The *GhCBL10* bait protein did not auto-activate the reporter genes in the Y2HGold cells in the absence of a prey protein, confirming the suitability of the results. We identified 23 prey proteins that showed interaction with *GhCBL10* (Table 3). Of them, only four proteins, *PRA1 B1, DSP8, CAB-151,* and *SAMS2,* could activate the expression of the reporter genes in diploid yeast cells. The SAMS2 gene showed the highest interaction frequency with the CBL10 bait protein, supporting its importance for upland cotton's tolerance to abiotic stress.

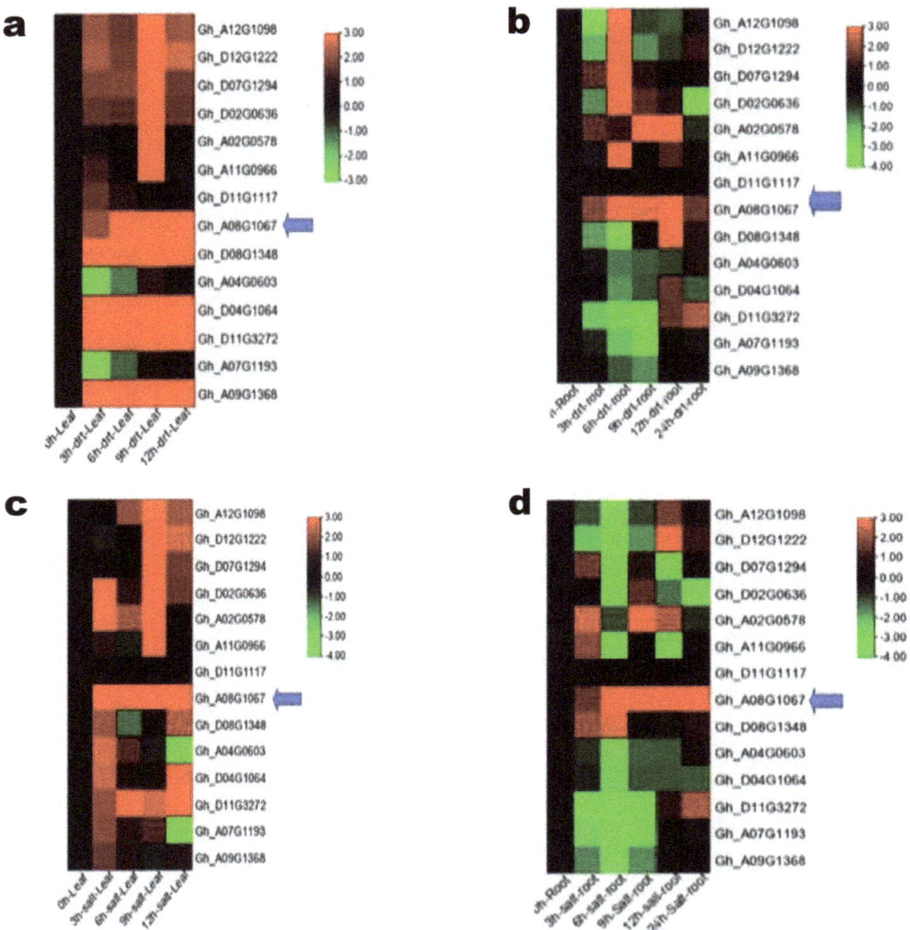

Figure 3. Heat maps showing *GhSAMS* genes' differential expression in *G. hirsutum* under drought and salt stress conditions. RT-qPCR analysis of the expression of *GhSAMS* genes under drought stress in the leaf (**a**) and roots (**b**). RT-qPCR analysis of the expression of *GhSAMS* genes under salt stress in the leaf (**c**) and roots (**d**). The higher expression level, the lower expression level, and no expression of the *GhSAMS* genes at a particular time are depicted by red, green, and black colors, respectively.

3.5. GhSAMS2 Gene Silencing Significantly Increased Sensitivity to Drought and Salt Stress

To verify the function of *GhSAMS2* in response to abiotic stresses, it was knocked down through VIGS in cotton seedlings, and the plants' morphological and physiological characteristics were analyzed under drought and salt stress conditions. The phenotypes of the cotton seedlings grown in hydroponics under various conditions are shown in Figure S3. The plants infiltrated with pTRV2: PDS exhibited photo-bleached leaves after 14 days of post-inoculation (Figure S3a). The WT- and TRV2:00-infected seedlings had rapid growth and, morphologically, looked much healthier after three weeks of inoculation (Figure S3b,c). To confirm that *GhSAMS2* was effectively silenced, we analyzed its expression in the different cotton seedlings via RT-qPCR assays (Figure S3e). The results confirmed that the expression of *GhSAMS2* in WT was significantly higher than that in TRV2:*GhSAMS2* VIGS plants. Figure S4 presents the phenotypes of the WT, TRV2:00, and TRV2:*GhSAMS2* plants under the drought and salt stresses.

Table 3. Isolated prey proteins from the Y2H system's AD library of cotton leaves.

Transcript ID	Name	Gene Description	Chr	Starting	Ending	Length
Gh_D06G1756.1	*PRA1B1*	PRA1 family protein B1	D06	57,193,276	57,193,932	657
Gh_A11G0688.1	*DSP8*	Putative dual-specificity protein DSP8 phosphatase	A11	6,717,943	6,719,957	945
Gh_A07G1725.1	*CAB-151*	Chlorophyll a-b binding protein 151, chloroplastic	A07	70,403,379	70,404,266	798
Gh_AO8G1067.1	*SAMS2*	S-adenosylmethionine synthase-2	A08	73,601,857	73,603,038	1182
Gh_D12G0158.1	*PYD3*	Beta-ureidopropionase	D12	2,003,668	2,006,254	1251
Gh_D04G1908.1	*RPL34*	60S ribosomal protein L34	D04	51,393,192	51,394,076	363
Gh_D02G0037.1	*UBC28*	Ubiquitin-conjugating enzyme E2 28	D02	190,259	192,000	447
Gh_D06G1538.1	*PSAF*	Photosystem I reaction center subunit III, chloroplastic	D06	51,265,731	51,266,405	675
Gh_D08G1752.1	*LON2*	Lon protease homolog 2, peroxisomal	D08	53,762,846	53,770,001	2670
Gh_D02G0914.1	*PAH2*	Phosphatidate phosphatase PAH2	D02	19,402,617	19,409,204	2934
Gh_A11G2956.1	*BEE3*	Transcription factor BEE 3	scaffold2723_A11	67,019	68,759	708
Gh_D12G0965.1	*Rnf25*	E3 ubiquitin-protein ligase RNF25	D12	35,117,734	35,120,355	1026
Gh_A13G2030.1	*RAX2*	Transcription factor RAX2	A13	79,732,246	79,733,388	903
Gh_A12G2413.1	*ALMT9*	Aluminum-activated malate transporter 9	A12	86,624,248	86,627,577	1839
Gh_D11G0245.1	*ARF9*	Auxin response factor 9	D11	2,017,754	2,033,373	3696
Gh_D11G2402.1	*NA*	NA	D11	47,820,689	47,823,871	1290
Gh_D09G1701.1	*NA*	NA	D09	44,755,344	44,757,734	2070
Gh_A05G3519.1	*At1g54200*	Protein BIG GRAIN 1-like B	A05	90,846,177	90,847,466	1290
Gh_D08G0705.1	*NA*	Ent-copalyl diphosphate synthase, chloroplastic	D08	9,782,732	9,788,296	2538
Gh_A04G1028.1	*At4g26680*	Pentatricopeptide repeat-containing, containing protein At4g26680	A04	60,318,577	60,320,187	1611
Gh_D05G3560.1	*RH32*	DEAD-box ATP-dependent RNA helicase 32	D05	58,950,106	58,954,430	2262
Gh_D13G0219.1	*AN11010*	Putative GTPase-activating protein	D13	2,155,213	2,163,523	2538
Gh_A12G0039.1	*NA*	NA	A12	598,420	600,787	1281

We investigated various morphological and physiological parameters under stress conditions. We found minimal differences in the plant heights and root lengths between the VIGS plants and the controls (Figure 4A,C). The control plants had slightly longer roots compared to the treated ones. The root fresh weight and shoot fresh weight of WT were significantly higher than those of the silenced *Gh_A08G1067* plants after stress treatment (Figure 4B,D). The TRV2:*GhSAMS2* plants showed a significant reduction in leaves' RLWC (relative water content) and chlorophyll content compared to the controls (Figure 4E,F). As expected, the *Gh_SAMS2*-infiltrated leaves exhibited a significantly increased ELWL (excised leaf water loss) and ion leakage compared to the WT and TRV2:00 plants (Figure 4G,H), indicating the deterioration of biological membranes.

We further analyzed biochemical parameters, including malondialdehyde (MDA) and H_2O_2 contents and the activity of antioxidant enzyme peroxidase (POD) and catalase (CAT). The contents of MDA and H_2O_2 in the TRV2:*GhSAMS2* plants were significantly higher than those in the WT under the drought and salt stress conditions (Figure 5c,d). Supportively, the antioxidative activities of POD and CAT were significantly lower in the VIGS plants compared to those in the WT under the conditions of drought and salt stress (Figure 5a,b).

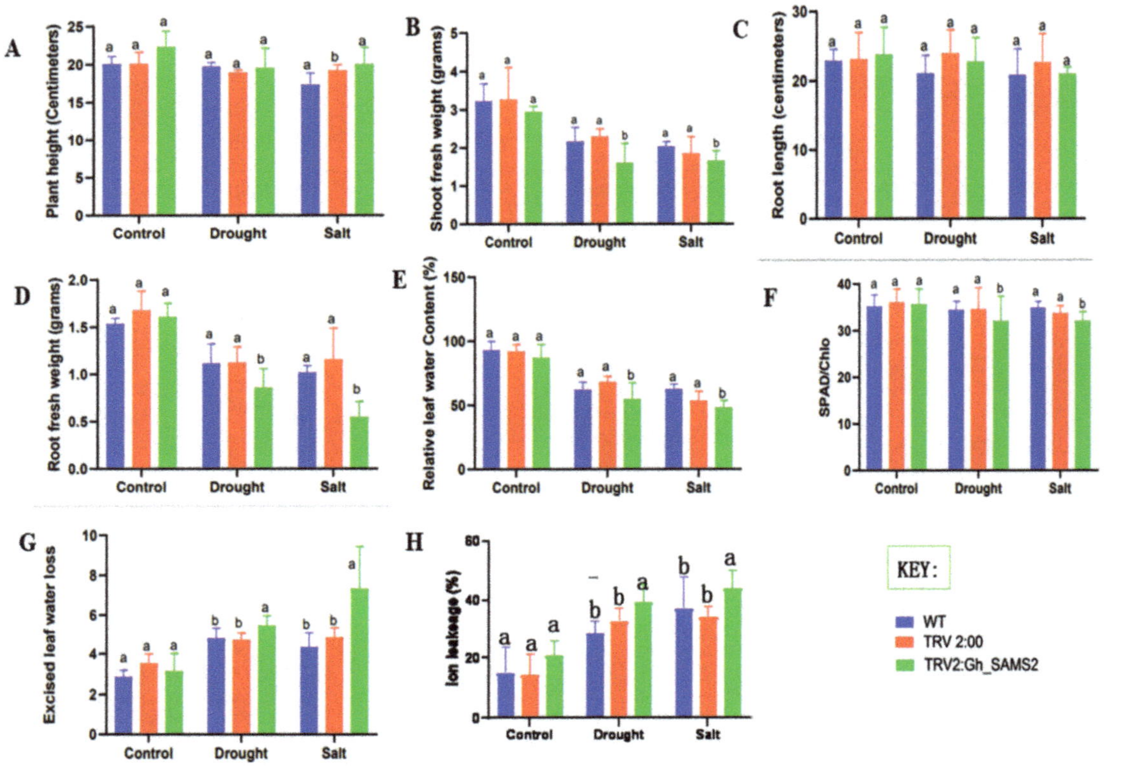

Figure 4. VIGS and WT plants' physiological and morphclogical traits analyzed under the conditions of drought and salt stress. (**A**) Plant height. (**B**) Shoot fresh weight. (**C**) Root length. (**D**) Root fresh weight. (**E**) Relative leaf water content. (**F**) Leaves' chlorophyll content. (**G**) Excised leaf water loss. (**H**) Ion leakage in the leaf. TRV2:00, Positive control; WT, Wild type; and TRV2:*GhSAMS2*, VIGS plants. Different letters above the bars indicate statistically significant differences at $p < 0.05$.

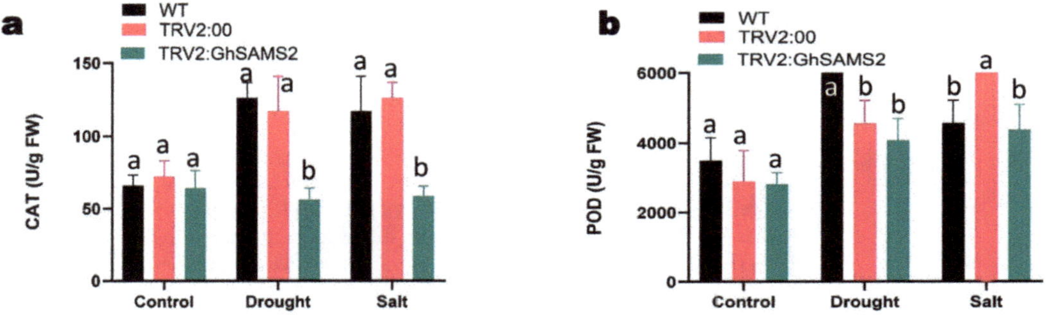

Figure 5. *Cont.*

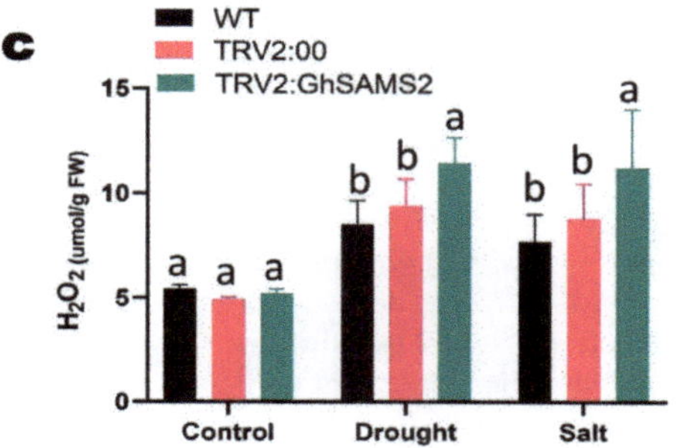

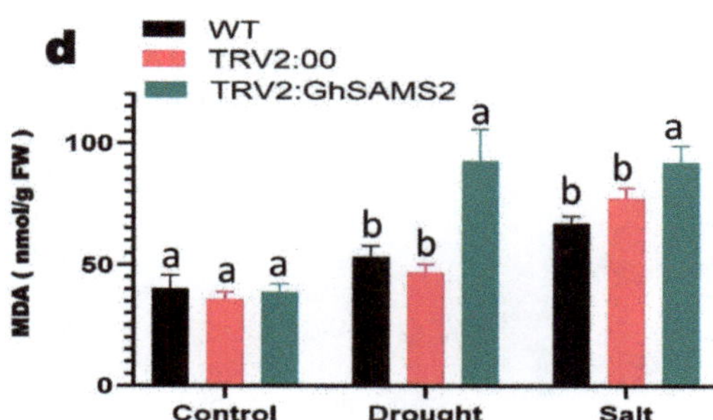

Figure 5. Oxidant and antioxidant enzyme biochemical assays in the leaves of WT and VIGs plants after 24 h post-stress-exposure. (**a**) Determination of catalase quantity. (**b**) Determination of peroxidase quantity. (**c**) Determination of hydrogen peroxide quantity. (**d**) Determination of Malondialdehyde quantity. TRV2:00, Positive control; WT, Wild type; and TRV2:*GhSAMS2*, VIGS plants. Different letters above the bars indicate statistically significant differences at $p < 0.05$.

4. Discussion

Crop production is negatively affected by salinity and alkalinity in semi-arid and arid regions. It is estimated that 831 million hectares of soils in the world are affected by excessive salinity and alkalinity, of which 397 million hectares are saline soils compared to 434 million hectares of alkaline soils [51]. Hence, propagating cultivars that are salt-tolerant to utilize saline soils is of absolute urgency [52]. Previous studies in cotton have pointed out that cotton may have thousands of putative functional genes, but since it is labor-intensive and ineffective to carry out the stable genetic transformation of cotton, the majority of these genes have not yet been characterized [1]. The *SAMS2* gene from previous studies has been found to play a crucial role in plant development regulation, metabolism, and abiotic and biotic stress tolerance mechanisms [12]. The SAMS gene family has been well studied in many different dicot and monocot plants such as tomato, *Arabidopsis*, sunflower, eggplant, soybean, *Medicago truncatula*, barley, sorghum, *Triticum urartu*, and rice [11]. This study identified and functionally characterized the cotton *SAMS2* gene for the targeted enhancement of multiple abiotic stresses tolerance in *G. hirsutum*.

According to the prediction of the subcellular localization, cytoplasm and cytoskeleton are the key sites where *GhSAMS* genes are localized. Besides the cytoplasm, cytosol and chloroplasts also contain substantial SAMS proteins, as previously reported [53]. Interestingly, all the sixteen cotton SAMS genes lacked introns in their gene structure. It is reported that, in eukaryotic organisms, many genes are intronless [54]. Additionally, intronless genes are enriched in plant species such as Populus, *Arabidopsis*, and rice [6]. Therefore, the datasets provided by intronless genes have great potential for comparative genomics and evolutionary studies in eukaryotic organisms. However, studies within a phylogenetic framework on intronless genes are limited to few species, based on the previous evolutionary studies that have been carried out [55]. The lack of intron in the *GhSAMS* genes suggests that they might play important roles in biotic and abiotic stress acclimation mechanisms [56]. The results of the cis-acting regulatory elements analysis support this statement. Key abiotic stress responsiveness cis-elements were detected within the promoter regions of the *GhSAMS* genes [57]. The specific function of each *GhSAMS* gene could be predicted through the phylogenetic relationships.

The *GhSAMS* proteins recorded negative GRAVY values, indicating that they are hydrophilic [33]. Hydrophilic proteins have been highly linked to plant protection through antioxidants and membrane stabilizers during water stress conditions [58]. Furthermore, they prevent the collapse of cells in deficient water conditions by acting as space fillers [59]. Additionally, the presence of hydrophilic proteins in certain plants, invertebrates, and microorganisms has been highly associated with their adaptations to water-scarce ecological conditions [60]. *GhSAMS* genes are stable proteins, as shown by the instability index values, a property that allows cellular biochemical reactions to proceed despite unfavorable environmental conditions. Enzymes' stability within cells is often shown by the instability index of various proteins involved in multiple reactions for a particular time [61]. Moreover, the proteins encoded by *GhSAMSs* have significantly higher thermal stability, as recorded in high aliphatic index values [62].

Gene expression analysis revealed that most of the *GhSAMS* genes are significantly induced by abiotic stresses. These findings are consistent with previous reports on SAMS genes in various crops such as tomato, *Arabidopsis*, rice, and soybean [63]. Among upland cotton SAMS genes, *GhSAMS2* exhibited the highest expression under both salt and drought stress conditions, suggesting its pivotal role in the plant's adaptation to unfavorable environmental conditions. Supportively, *GhSAMS2* exhibited the highest stability index and interaction frequency with the *GhCBL10* bait protein. The gene's function has been previously studied in many plant species via knocking down using the VIGS tool [64]. The down-regulation of *GhSAM2* via VIGS and the post-exposure of VIGS plants to drought and salt stress confirmed the key role of this gene in moderating abiotic stress tolerance. The TRV2:*GhSAMS2* plants showed growth and biomass accumulation defects compared to the controls under drought and salt stress conditions. Their leaves contained less chlorophyll and exhibited higher ion leakage, indicating the high sensitivity of VIGS-*GhSAM2* plants to abiotic stresses. In general, plants exhibit wilting behaviors when exposed to drought and salt stress [65]. The disruption of stress tolerance mechanisms by abiotic stress in plants often exacerbates the transpiration rate, biological membrane deterioration, and cells' function perturbation [66]. Damage to the phospholipid membrane structure due to oxidation is mainly induced by drought and salt stresses. Hydrogen peroxide and Malondialdehyde contents are the biochemical parameters usually used to determine the cellular damage within the organism's tissues [3]. Oxidative stress in living organisms is dictated by the level of MDA and ROS contents accumulated at a particular time [14]. ROS production is often promoted by reducing the usage of absorption light energy caused by Calvin cycle enzyme inhibition under abiotic stress conditions. [67]. The *GhSAMS2* knockdown in VIGS plants incapacitated the scavenging ability of excess ROS, resulting in acute oxidative stress and high H_2O_2 and MDA accumulation. The VIGS-*GhSAMS2* plants showed deficiency in terms of CAT and POD activities compared to the control plants, supporting the deterioration of enzymatic oxidation defense systems [68]. These results demonstrate that *GhSAMS2* (Gh_A08G1067) is a promising gene for enhancing upland cotton and other crops' tolerance to drought and salt stress through molecular breeding. Moreover, they confirm the successful gene knockdown and effectiveness of the tobacco virus rattle vector [3,32]. Further functional characterization of identified *GhSAMSs* via subsequent knockout and overexpression coupled with transcriptomic and metabolomic analyses is required to understand cotton plant stress response mechanisms better.

5. Conclusions

This study identified sixteen (16) SAMS genes in upland cotton and comprehensively explored their chromosomal locations, gene structure, phylogenetic relationships, cis-acting elements, conserved motifs, and expression under drought and salt stress conditions. We found that *GhSAMS* genes might be primarily involved in the network's regulation of various environmental stresses. Particularly, *GhSAMS2* was identified as a promising candidate gene for the targeted improvement of upland cotton's tolerance to multiple abiotic stresses. The downregulation of *GhSAMS2* expression via VIGS confirmed its pivotal role in mediating the plant's response to abiotic stress. Our findings provide reference information for the in-depth investigation of *GhSAMS* genes' functions and for dissecting the complex molecular networks associated with abiotic stress tolerance in cotton.

Supplementary Materials: The following supporting information can be downloaded at: https://www.mdpi.com/article/10.3390/agronomy13020612/s1, Additional file 1, Table S1: SAMS genes from other plant species analyzed in this study; Table S2: List of primers used for the RT-qPCR analysis; Table S3: List of primers used in Y2H experiments; Table S4: Cis-regulatory elements detected within the promoter region of *GhSAMS* genes. Additional file 2, Figure S1: Heat maps showing *GhSAMS* genes' differential expression in *G. hirsutum* under drought and salt stress conditions; Figure S2: Self-auto-activation state, toxicity test, verification of interactions, and mating efficiency determination of the *GhCBL10* bait gene in a Yeast Two-Hybrid system; Figure S3: Phenotypic observation of cotton seedlings under VIGS and *GhSAMS2* expression analysis via RT-qPCR; Figure S4: Morphology of VIGS and wild type plants under the conditions of drought and salt stress.

Author Contributions: Conceptualization, methodology, and validation, J.W.K., T.G.M., Y.W. (Yangyang Wei), Y.X. and F.L.; formal analysis, data curation, and writing—original draft preparation, M.J.U., R.O.M., M.L.S., J.N.K. and R.P.; writing—review and editing, Y.H., Y.W. (Yuhong Wang), X.C. and J.W.K.; supervision, F.L. and Z.Z.; project administration and funding acquisition, F.L. All authors have read and agreed to the published version of the manuscript.

Funding: This research was funded by the National Natural Science Foundation of China (32171994, 32072023) and the National Key R&D Program of China ((2021YFE0101200), PSF/CRP/18th Protocol (07).

Informed Consent Statement: Informed consent was obtained from all subjects involved in this study.

Data Availability Statement: The data presented in this study are available on request from the corresponding author.

Conflicts of Interest: The authors declare no conflict of interest. The funders had no role in the design of the study; in the collection, analyses, or interpretation of data; in the writing of the manuscript; or in the decision to publish the results.

References

1. Gu, Z.; Huang, C.; Li, F.; Zhou, X. A versatile system for functional analysis of genes and microRNAs in cotton. *Plant Biotechnol. J.* **2014**, *12*, 638–649. [CrossRef]
2. Kim, S.H.; Kim, S.H.; Palaniyandi, S.A.; Yang, S.H.; Suh, J.W. Expression of potato S-adenosyl-l-methionine synthase (*SbSAMS*) gene altered developmental characteristics and stress responses in transgenic *Arabidopsis* plants. *Plant Physiol. Biochem.* **2015**, *87*, 84–91. [CrossRef]
3. Mehari, T.G.; Xu, Y.; Umer, M.J.; Shiraku, M.L.; Hou, Y.; Wang, Y.; Yu, S.; Zhang, X.; Wang, K.; Cai, X.; et al. Multi-Omics-Based Identification and Functional Characterization of *Gh_A06G1257* Proves Its Potential Role in Drought Stress Tolerance in *Gossypium hirsutum*. *Front. Plant Sci.* **2021**, *12*, 2092. [CrossRef]
4. Turan, M.A.; Elkarim, A.H.A.; Taban, N.; Taban, S. Effect of salt stress on growth, stomatal resistance, proline and chlorophyll concentrations on maize plant. *Afr. J. Agric. Res.* **2009**, *4*, 893–897.
5. Shi, D.; Sheng, Y. Effect of various salt-alkaline mixed stress conditions on sunflower seedlings and analysis of their stress factors. *Environ. Exp. Bot.* **2005**, *54*, 8–21. [CrossRef]
6. Yang, C.W.; Xu, H.H.; Wang, L.L.; Liu, J.; Shi, D.C.; Wang, D.L. Comparative effects of salt-stress and alkali-stress on the growth, photosynthesis, solute accumulation, and ion balance of barley plants. *Photosynthetica* **2009**, *47*, 79–86. [CrossRef]
7. Ahmad, S.T.; Sima, N.A.K.K.; Mirzaei, H.H. Effects of sodium chloride on physiological aspects of *Salicornia persica* growth. *J. Plant Nutr.* **2013**, *36*, 401–414. [CrossRef]
8. Ahuja, I.; de Vos, R.C.H.; Bones, A.M.; Hall, R.D. Plant molecular stress responses face climate change. *Trends Plant Sci.* **2010**, *15*, 664–674. [CrossRef]
9. Yang, L.; Zhang, Y.; Zhu, N.; Koh, J.; Ma, C.; Pan, Y.; Yu, B.; Chen, S.; Li, H. Proteomic analysis of salt tolerance in sugar beet monosomic addition line M14. *J. Proteome Res.* **2013**, *12*, 4931–4950. [CrossRef]
10. He, M.W.; Wang, Y.; Wu, J.Q.; Shu, S.; Sun, J.; Guo, S.R. Isolation and characterization of S-Adenosylmethionine synthase gene from cucumber and responsive to abiotic stress. *Plant Physiol. Biochem.* **2019**, *141*, 431–445. [CrossRef]
11. Heidari, P.; Mazloomi, F.; Nussbaumer, T.; Barcaccia, G. Insights into the SAM Synthetase Gene Family and Its Roles in Tomato Seedlings under Abiotic Stresses and Hormone Treatments. *Plants* **2020**, *9*, 586. [CrossRef]
12. Roje, S. S-Adenosyl-l-methionine: Beyond the universal methyl group donor. *Phytochemistry* **2006**, *67*, 1686–1698. [CrossRef]
13. Bürstenbinder, K.; Rzewuski, G.; Wirtz, M.; Hell, R.; Sauter, M. The role of methionine recycling for ethylene synthesis in Arabidopsis. *Plant J.* **2007**, *49*, 238–249. [CrossRef]
14. Ma, C.; Wang, Y.; Gu, D.; Nan, J.; Chen, S.; Li, H. Overexpression of S-adenosyl-L-methionine synthetase 2 from sugar beet M14 increased arabidopsis tolerance to salt and oxidative stress. *Int. J. Mol. Sci.* **2017**, *18*, 847. [CrossRef]

15. Jang, S.J.; Wi, S.J.; Choi, Y.J.; An, G.; Park, K.Y. Increased polyamine biosynthesis enhances stress tolerance by preventing the accumulation of reactive oxygen species: T-DNA mutational analysis of *Oryza sativa* lysine decarboxylase-like protein 1. *Mol. Cells* **2012**, *34*, 251–262. [CrossRef]
16. Gupta, K.; Dey, A.; Gupta, B. Plant polyamines in abiotic stress responses. *Acta Physiol. Plant.* **2013**, *35*, 2015–2036. [CrossRef]
17. Wang, K.L.C.; Li, H.; Ecker, J.R. Ethylene biosynthesis and signaling networks. *Plant Cell* **2002**, *14*, 131–152. [CrossRef]
18. Vandenbussche, F.; Vaseva, I.; Vissenberg, K.; Van Der Straeten, D. Ethylene in vegetative development: A tale with a riddle. *New Phytol.* **2012**, *194*, 895–909. [CrossRef]
19. Li, W.; Han, Y.; Tao, F.; Chong, K. Knockdown of SAMS genes encoding S-adenosyl-l-methionine synthetases causes methylation alterations of DNAs and histones and leads to late flowering in rice. *J. Plant Physiol.* **2011**, *168*, 1837–1843. [CrossRef]
20. Chen, M.; Chen, J.; Fang, J.; Guo, Z.; Lu, S. Down-regulation of S-adenosylmethionine decarboxylase genes results in reduced plant length, pollen viability, and abiotic stress tolerance. *Plant Cell Tissue Organ Cult.* **2014**, *116*, 311–322. [CrossRef]
21. Guo, Z.; Tan, J.; Zhuo, C.; Wang, C.; Xiang, B.; Wang, Z. Abscisic acid, H2O2 and nitric oxide interactions mediated cold-induced S-adenosylmethionine synthetase in *Medicago sativa* subsp. *Falcata* that confers cold tolerance through up-regulating polyamine oxidation. *Plant Biotechnol. J.* **2014**, *12*, 601–612. [CrossRef] [PubMed]
22. Zhu, H.; He, M.; Jahan, M.S.; Wu, J.; Gu, Q.; Shu, S.; Sun, J.; Guo, S. Cscdpk6, a cssams1-interacting protein, affects polyamine/ethylene biosynthesis in cucumber and enhances salt tolerance by overexpression in tobacco. *Int. J. Mol. Sci.* **2021**, *22*, 11133. [CrossRef]
23. Ding, C.; Chen, T.; Yang, Y.; Liu, S.; Yan, K.; Yue, X.; Zhang, H.; Xiang, Y.; An, L.; Chen, S. Molecular cloning and characterization of an S-adenosylmethionine synthetase gene from *Chorispora bungeana*. *Gene* **2015**, *572*, 205–213. [CrossRef] [PubMed]
24. Xu, Y.; Magwanga, R.O.; Yang, X.; Jin, D.; Cai, X.; Hou, Y.; Wei, Y.; Zhou, Z.; Wang, K.; Liu, F. Genetic regulatory networks for salt-alkali stress in Gossypium hirsutum with differing morphological characteristics. *BMC Genom.* **2020**, *21*, 15. [CrossRef] [PubMed]
25. Gasteiger, E.; Hoogland, C.; Gattiker, A.; Duvaud, S.; Wilkins, M.R.; Appel, R.D.; Bairoch, A. *The Proteomics Protocols Handbook*; Humana Press: Totowa, NJ, USA, 2005; pp. 571–608. [CrossRef]
26. Tamura, K.; Peterson, D.; Peterson, N.; Stecher, G.; Nei, M.; Kumar, S. MEGA5: Molecular evolutionary genetics analysis using maximum likelihood, evolutionary distance, and maximum parsimony methods. *Mol. Biol. Evol.* **2011**, *28*, 2731–2739. [CrossRef] [PubMed]
27. Thompson, J.D.; Gibson, T.J.; Higgins, D.G. Multiple Sequence Alignment Using ClustalW and ClustalX. *Curr. Protoc. Bioinform.* **2003**, *1*, 2–3. [CrossRef]
28. Hortona, P.; Park, K.J.; Obayashi, T.; Nakai, K. Protein subcellular localization prediction with WoLF PSORT. *Ser. Adv. Bioinform. Comput. Biol.* **2006**, *3*, 39–48. [CrossRef]
29. Hu, B.; Jin, J.; Guo, A.Y.; Zhang, H.; Luo, J.; Gao, G. GSDS 2.0: An upgraded gene feature visualization server. *Bioinformatics* **2015**, *31*, 1296–1297. [CrossRef]
30. Mehari, T.G.; Xu, Y.; Magwanga, R.O.; Umer, M.J.; Kirungu, J.N.; Cai, X.; Hou, Y.; Wang, Y.; Yu, S.; Wang, K.; et al. Genome wide identification and characterization of light-harvesting Chloro a/b binding (LHC) genes reveals their potential role in enhancing drought tolerance in *Gossypium hirsutum*. *J. Cott. Res.* **2021**, *4*, 15. [CrossRef]
31. Hoagland, D.R.; Arnon, D.I. Preparing the nutrient solution. In *The Water-Culture Method for Growing Plants Without Soil*; Circular California Agricultural Experiment Station: Berkeley, CA, USA, 1950; Volume 347, pp. 29–31.
32. Mehari, T.G.; Xu, Y.; Magwanga, R.O.; Umer, M.J.; Shiraku, M.L.; Hou, Y.; Wang, Y.; Wang, K.; Cai, X.; Zhou, Z.; et al. Identification and functional characterization of Gh_D01G0514 (GhNAC072) transcription factor in response to drought stress tolerance in cotton. *Plant Physiol. Biochem.* **2021**, *166*, 361–375. [CrossRef]
33. Magwanga, R.O.; Lu, P.; Kirungu, J.N.; Lu, H.; Wang, X.; Cai, X.; Zhou, Z.; Zhang, Z.; Salih, H.; Wang, K.; et al. Characterization of the late embryogenesis abundant (LEA) proteins family and their role in drought stress tolerance in upland cotton. *BMC Genet.* **2018**, *19*, 1–31. [CrossRef] [PubMed]
34. Mustafa, R.; Shafiq, M.; Mansoor, S.; Briddon, R.W.; Scheffler, B.E.; Scheffler, J.; Amin, I. Virus-Induced Gene Silencing in Cultivated Cotton (*Gossypium* spp.) Using Tobacco Rattle Virus. *Mol. Biotechnol.* **2016**, *58*, 65–72. [CrossRef] [PubMed]
35. Kirungu, J.N.; Magwanga, R.O.; Pu, L.; Cai, X.; Xu, Y.; Hou, Y.; Zhou, Y.; Cai, Y.; Hao, F.; Zhou, Z.; et al. Knockdown of *Gh_A05G1554* (*GhDHN_03*) and *Gh_D05G1729* (*GhDHN_04*) Dehydrin genes, Reveals their potential role in enhancing osmotic and salt tolerance in cotton. *Genomics* **2020**, *112*, 1902–1915. [CrossRef]
36. Livak, K.J.; Schmittgen, T.D. Analysis of relative gene expression data using real-time quantitative PCR and the $2^{-\Delta\Delta CT}$ method. *Methods* **2001**, *25*, 402–408. [CrossRef]
37. Shiraku, M.L.; Magwanga, R.O.; Cai, X.; Kirungu, J.N.; Xu, Y.; Mehari, T.G.; Hou, Y.; Wang, Y.; Wang, K.; Peng, R.; et al. Knockdown of 60S ribosomal protein L14-2 reveals their potential regulatory roles to enhance drought and salt tolerance in cotton. *J. Cott. Res.* **2021**, *4*, 27. [CrossRef]
38. Chen, Y.; Li, C.; Zhang, B.; Yi, J.; Yang, Y.; Kong, C.; Lei, C.; Gong, M. The Role of the Late Embryogenesis-Abundant (LEA) Protein Family in Development and the Abiotic Stress Response: A Comprehensive Expression Analysis of. *Genes* **2019**, *10*, 148. [CrossRef]
39. Lei, K.; Liu, A.; Fan, S.; Peng, H.; Zou, X.; Zhen, Z. Identification of *TPX2* Gene Family in Upland Cotton and its Functional Analysis in Cotton. *Genes* **2019**, *10*, 508. [CrossRef]

40. Fu, D.Q.; Zhu, B.Z.; Zhu, H.L.; Jiang, W.B.; Luo, Y.B. Virus-induced gene silencing in tomato fruit. *Plant J.* **2005**, *43*, 299–308. [CrossRef]
41. Gao, X.; Britt, R.C.; Shan, L.; He, P. *Agrobacterium*-mediated virus-induced gene silencing assay in cotton. *J. Vis. Exp.* **2011**, *54*, e2938. [CrossRef]
42. Dupadahalli, K. A modified freeze-thaw method for the efficient transformation of *Agrobacterium tumefaciens*. *Curr. Sci.* **2007**, *93*, 770–772.
43. Tuttle, J.R.; Haigler, C.H.; Robertson, D. Method: Low-cost delivery of the cotton leaf crumple virus-induced gene silencing system. *Plant Methods* **2012**, *8*, 27. [CrossRef] [PubMed]
44. Kirungu, J.N.; Magwanga, R.O.; Lu, P.; Cai, X.; Zhou, Z.; Wang, X.; Peng, R.; Wang, K.; Liu, F. Functional characterization of Gh-A08G1120 (*GH3.5*) gene reveal their significant role in enhancing drought and salt stress tolerance in cotton. *BMC Genet.* **2019**, *20*, 62. [CrossRef] [PubMed]
45. Yang, X.; Jawdy, S.; Tschaplinski, T.J.; Tuskan, G.A. Genome-wide identification of lineage-specific genes in *Arabidopsis*, *Oryza* and *Populus*. *Genomics* **2009**, *93*, 473–480. [CrossRef] [PubMed]
46. Magwanga, R.O.; Kirungu, J.N.; Lu, P.; Yang, X.; Dong, Q.; Cai, X.; Xu, Y.; Wang, X.; Zhou, Z.; Hou, Y.; et al. Genome wide identification of the trihelix transcription factors and overexpression of *Gh_A05G2067* (*GT-2*), a novel gene contributing to increased drought and salt stresses tolerance in cotton. *Physiol. Plant.* **2019**, *167*, 447–464. [CrossRef] [PubMed]
47. Agarie, S.; Hanaoka, N.; Kubota, F.; Agata, W.; Kaufman, P.B. Measurement of Cell Membrane Stability Evaluated by Electrolyte Leakage as a Drought and Heat Tolerance Test in Rice (*Oryza sativa* L.). *J. Fac. Agric. Kyushu Univ.* **1995**, *40*, 233–240. [CrossRef]
48. Rehman, S.U.; Bilal, M.; Rana, R.M.; Tahir, M.N.; Shah, M.K.N.; Ayalew, H.; Yan, G. Cell membrane stability and chlorophyll content variation in wheat (*Triticum aestivum*) genotypes under conditions of heat and drought. *Crop Pasture Sci.* **2016**, *67*, 712–718. [CrossRef]
49. Magwanga, R.O.; Lu, P.; Kirungu, J.N.; Dong, Q.; Cai, X.; Zhou, Z.; Wang, X.; Hou, Y.; Xu, Y.; Peng, R.; et al. Knockdown of cytochrome *P450* genes *Gh_D07G1197* and *Gh_A13G2057* on chromosomes D07 and A13 reveals their putative role in enhancing drought and salt stress tolerance in *Gossypium hirsutum*. *Genes* **2019**, *10*, 226. [CrossRef]
50. Majidi, M.M.; Rashidi, F.; Sharafi, Y. Physiological traits related to drought tolerance in *Brassica*. *Int. J. Plant Prod.* **2015**, *9*, 541–560. [CrossRef]
51. Parihar, P.; Singh, S.; Singh, R.; Singh, V.P.; Prasad, S.M. Effect of salinity stress on plants and its tolerance strategies: A review. *Environ. Sci. Pollut. Res.* **2015**, *22*, 4056–4075. [CrossRef]
52. Flowers, T.J. Improving crop salt tolerance. *J. Exp. Bot.* **2004**, *55*, 307–319. [CrossRef]
53. Hazarika, P.; Rajam, M.V. Biotic and abiotic stress tolerance in transgenic tomatoes by constitutive expression of S-adenosylmethionine decarboxylase gene. *Physiol. Mol. Biol. Plants* **2011**, *17*, 115–128. [CrossRef] [PubMed]
54. Bhandari, B.; Roesler, W.J.; DeLisio, K.D.; Klemm, D.J.; Ross, N.S.; Miller, R.E. A functional promoter flanks an intronless glutamine synthetase gene. *J. Biol. Chem.* **1991**, *266*, 7784–7792. [CrossRef]
55. He, S.; Zou, M.; Guo, B. The roles and evolutionary patterns of intronless genes in deuterostomes. *Comp. Funct. Genom.* **2011**, *2011*, 680673. [CrossRef]
56. Shiraku, M.L.; Magwanga, R.O.; Cai, X.; Kirungu, J.N.; Xu, Y.; Mehari, T.G.; Hou, Y.; Wang, Y.; Agong, S.G.; Peng, R.; et al. Functional Characterization of *GhACX3* Gene Reveals Its Significant Role in Enhancing Drought and Salt Stress Tolerance in Cotton. *Front. Plant Sci.* **2021**, *12*, 1120. [CrossRef]
57. Zhou, M.L.; Ma, J.T.; Pang, J.F.; Zhang, Z.L.; Tang, Y.X.; Wu, Y.M. Regulation of plant stress response by dehydration responsive element binding (DREB) transcription factors. *Afr. J. Biotechnol.* **2010**, *9*, 9255–9269. [CrossRef]
58. Gao, J.; Lan, T. Functional characterization of the late embryogenesis abundant (LEA) protein gene family from *Pinus tabuliformis* (Pinaceae) in *Escherichia coli*. *Sci. Rep.* **2016**, *6*, 19467. [CrossRef]
59. Tunnacliffe, A.; Wise, M.J. The continuing conundrum of the LEA proteins. *Naturwissenschaften* **2007**, *94*, 791–812. [CrossRef]
60. Chakrabortee, S.; Boschetti, C.; Walton, L.J.; Sarkar, S.; Rubinsztein, D.C.; Tunnacliffe, A. Hydrophilic protein associated with desiccation tolerance exhibits broad protein stabilization function. *Proc. Natl. Acad. Sci. USA* **2007**, *104*, 18073–18078. [CrossRef]
61. Heidari, P.; Ahmadizadeh, M.; Izanlo, F.; Nussbaumer, T. In silico study of the CESA and CSL gene family in *Arabidopsis thaliana* and *Oryza sativa*: Focus on post-translation modifications. *Plant Gene* **2019**, *19*, 100189. [CrossRef]
62. Bachmair, A.; Finley, D.; Varshavsky, A. In vivo half-life of a protein is a function of its amino-terminal residue. *Science* **1986**, *234*, 179–186. [CrossRef]
63. Gong, B.; Li, X.; Vandenlangenberg, K.M.; Wen, D.; Sun, S.; Wei, M.; Li, Y.; Yang, F.; Shi, Q.; Wang, X. Overexpression of S-adenosyl-l-methionine synthetase increased tomato tolerance to alkali stress through polyamine metabolism. *Plant Biotechnol. J.* **2014**, *12*, 694–708. [CrossRef] [PubMed]
64. Singh, B.; Kukreja, S.; Salaria, N.; Thakur, K.; Gautam, S.; Taunk, J.; Goutam, U. VIGS: A flexible tool for the study of functional genomics of plants under abiotic stresses. *J. Crop Improv.* **2019**, *33*, 567–604. [CrossRef]
65. Ma, Y.; Dias, M.C.; Freitas, H. Drought and Salinity Stress Responses and Microbe-Induced Tolerance in Plants. *Front. Plant Sci.* **2020**, *11*, 1750. [CrossRef] [PubMed]
66. Suzuki, N.; Rivero, R.M.; Shulaev, V.; Blumwald, E.; Mittler, R. Abiotic and biotic stress combinations. *New Phytol.* **2014**, *203*, 32–43. [CrossRef]

67. Zago, E.; Morsa, S.; Dat, J.F.; Alard, P.; Ferrarini, A.; Inzé, D.; Delledonne, M.; Van Breusegem, F. Nitric oxide- and hydrogen peroxide-responsive gene regulation during cell death induction in tobacco. *Plant Physiol.* **2006**, *141*, 404–411. [CrossRef]
68. Asada, K. The water-water cycle in chloroplasts: Scavenging of active oxygens and dissipation of excess photons. *Annu. Rev. Plant Biol.* **1999**, *50*, 601–639. [CrossRef]

Article

In Silico Characterization and Expression Profiles of Heat Shock Transcription Factors (HSFs) in Maize (*Zea mays* L.)

Saqlain Haider [1,*], Shazia Rehman [2], Yumna Ahmad [1], Ali Raza [3,*], Javaria Tabassum [4], Talha Javed [5,6], Hany S. Osman [7] and Tariq Mahmood [1]

1 Plant Biochemistry and Molecular Biology Laboratory, Department of Plant Sciences, Quaid-i-Azam University, Islamabad 45320, Pakistan; yumna789ahmad@gmail.com (Y.A.); tmahmood.qau@gmail.com (T.M.)
2 University Institute of Biochemistry and Biotechnology, PMAS, Arid Agriculture University, Rawalpindi 46300, Pakistan; shaziarehman7@gmail.com
3 Fujian Provincial Key Laboratory of Crop Molecular and Cell Biology, Oil Crops Research Institute, Center of Legume Crop Genetics and Systems Biology/College of Agriculture, Fujian Agriculture and Forestry University (FAFU), Fuzhou 350002, China
4 State Key Laboratory of Rice Biology, China National Rice Research Institute, Chinese Academy of Agricultural Science (CAAS), Hangzhou 311400, China; javaria.tabassum@outlook.com
5 College of Agriculture, Fujian Agriculture and Forestry University, Fuzhou 350002, China; mtahaj@fafu.edu.cn or talhajaved54321@gmail.com
6 Department of Agronomy, University of Agriculture, Faisalabad 38040, Pakistan
7 Department of Agricultural Botany, Faculty of Agriculture, Ain Shams University, Hadayek Shubra, Cairo 11241, Egypt; hany_osman1@agr.asu.edu.eg
* Correspondence: shaider@bs.qau.edu.pk (S.H.); alirazamughal143@gmail.com or aliraza6@fafu.edu.cn (A.R.)

Citation: Haider, S.; Rehman, S.; Ahmad, Y.; Raza, A.; Tabassum, J.; Javed, T.; Osman, H.S.; Mahmood, T. In Silico Characterization and Expression Profiles of Heat Shock Transcription Factors (HSFs) in Maize (*Zea mays* L.). *Agronomy* **2021**, *11*, 2335. https://doi.org/10.3390/agronomy11112335

Academic Editor: Carlos Iglesias

Received: 6 October 2021
Accepted: 16 November 2021
Published: 18 November 2021

Abstract: Heat shock transcription factors (HSFs) regulate many environmental stress responses and biological processes in plants. Maize (*Zea mays* L.) is a major cash crop that is grown worldwide. However, the growth and yield of maize are affected by several adverse environmental stresses. Therefore, investigating the factors that regulate maize growth and development and resistance to abiotic stress is an essential task for developing stress-resilient maize varieties. Thus, a comprehensive genome-wide identification analysis was performed to identify *HSFs* genes in the maize genome. The current study identified 25 *ZmHSFs*, randomly distributed throughout the maize genome. Phylogenetic analysis revealed that *ZmHSFs* are divided into three classes and 13 sub-classes. Gene structure and protein motif analysis supported the results obtained through the phylogenetic analysis. Segmental duplication is shown to be responsible for the expansion of *ZmHSFs*. Most of the *ZmHSFs* are localized inside the nucleus, and the *ZmHSFs* which belong to the same group show similar physio-chemical properties. Previously reported and publicly available RNA-seq analysis revealed a major role of class A HSFs including *ZmHSFA-1a* and *ZmHSFA-2a* in all the maize growth stages, i.e., seed, vegetative, and reproductive development. Under abiotic stress conditions (heat, drought, cold, UV, and salinity), members of class A and B *ZmHSFs* are induced. Gene ontology and protein–protein interaction analysis indicated a major role of *ZmHSFs* in resistance to environmental stress and regulation of primary metabolism. To summarize, this study provides novel insights for functional studies on the *ZmHSFs* in maize breeding programs.

Keywords: abiotic stress; HSFs; genomics; gene ontology; maize breeding; protein 3D structures

1. Introduction

In recent years, there has been an increasing trend to focus on the responses of plants to abiotic stresses due to global climate change [1]. Particularly, research has been focused on plant heat stress (HS) tolerance mechanisms, as higher temperatures have a negative effect on plant growth and production [2,3]. Although plants are susceptible to HS throughout their lifespan, reproductive tissues are specifically characterized by vulnerability to HS [4,5]. Heatwaves are expected to become more frequent in the future, and atmospheric temperature is on course to rise ~4 °C by the end of this century [6,7]. HS negatively affects plant morphology, physiology, and growth in diverse ways [5,8]. HS alters plasma membrane fluidity, creates proteotoxic stress, causes the overproduction of reactive oxygen species (ROS), and dismantles cellular organization by inducing the collapse of the cytoskeleton apparatus [3,8–10]. Nonetheless, plants possess an efficient system that allows them to perceive the environmental stimulus, activate the signaling pathways, and alter their gene expression to ensure survival [9,11,12]. A major step in this process is the activation of stress-inducible genes, the expression of which is controlled by transcription factors (TFs). TFs represent a group of regulatory proteins that control the expression pattern of genes under various developmental and stressful conditions [13,14]. Several TFs families including heat shock transcription factors (HSFs), WRKY (named due to conserved WRKYGQK motif), v-myb avian myeloblastosis viral oncogene homolog (MYB), Petunia NAM, *Arabidopsis* ATAF1/2 and CUC2 (NAC), and dehydration responsive-element binding transcriptional activator (DREB), etc., have been shown to positively regulate HS-responsive gene expression and improve plant HS tolerance [14,15]. Among them, the most comprehensively analyzed family is HSFs, the members of which are considered the master regulators of plant HS-response (HSR) and are also involved in regulating plant responses to other abiotic and biotic stress conditions [12,16]. Since the first HSF gene was identified in yeast [17], the HSF gene family has been characterized in several plant species [18,19]. The pioneering study by Nover et al. [20] allowed the identification of the HSF gene family in various plant species, including essential crop plants [19]. The Heatster database (http://www.cibiv.at/services/hsf, accessed on 1 August 2021) currently holds 848 HSF sequences from 33 different plant species.

Generally, HSFs are divided into three classes, i.e., A, B, and C, based on phylogenetic analysis and structural properties [21]. All the HSFs contain two highly conserved domains, the DNA-binding domain (DBD), which binds with "heat-shock elements" (5′-nGAAnnTTCn-3′) present in regulatory sequences of target genes, through helix-turn-helix (HTH) motif, and oligomerization domain (OD), which has a bipartile heptad repeat pattern of the hydrophobic-associated region (HR-A/B) and is responsible for the trimerization of HSFs [20,22]. Based on the linker length between the HR-A/B region, HSFs are classified into different classes. The linker length comprises 21 amino acid residues in the case of class A and 7 for class C HSFs. On the other hand, HSFs of class B, lack any insertion. The classification of HSFs is also supported by the length of the linker between DBD and OD, 9–39 amino acids for class A, 50–78 for class B, and 14–49 for class C [20–22]. The HSFs of class A are transcriptional activators, and class B are repressors [20,21]. However, in tomato, the HSFB1 possesses both co-activator and repressor functions [23,24]. The class C HSFs are activators like class A [25]. However, they lack activator peptide motif (AHA), and thus cannot initiate transcription on their own [19,20]. The AHA motif is present towards the C-terminal of HSFs and is specific to class A HSFs [26]. In addition to these domains, HSFs also possess nuclear localization signals (NLS), and some contain nuclear export signals (NES) [27]. The NLS and NES are responsible for the nuclear import and export of HSFs, an essential step in cellular functioning [27].

The function of HSFs as the master regulators of plant HSR has been demonstrated mainly in *Arabidopsis* and tomatoes [28,29]. In tomato, the overexpression of *HSFA1a*, improved plant thermotolerance, while co-suppression mutants were susceptible to HS [28]. The mutant plants and their fruits were characterized by extreme sensitivity to HS. *HSFA1a*, *HSFA2*, and *HSFB1* control the fundamental HSF network in tomato [30]. The role of the master regulator is shared by *HSFA1a*, *HSFA1b*, and *HSFA1d* in *Arabidopsis* [29]. In *hsfa1a/b/d* triple mutants, the expression of TFs and chaperons was severely hampered under HS, while the expression of several genes was downregulated even under normal growth conditions [29]. *HSFA2* is responsible for the extension of acquired thermotolerance (AT) in *Arabidopsis* [31]. In *hsfa2* mutants, the duration of AT was compromised, and the expression of HS-inducible genes was downregulated. Lämke et al. [32] reported that HSFA2 promotes the sustained activation of several HS-memory genes through methylation of the target genomic region [32]. The transcripts of *HSFA2* are undetectable under normal conditions. However, after HS, the *HSFA2* becomes the most strongly induced HSF in plants [18,33]. Yoshida et al. reported that the overexpression of *HSFA3* improves plant thermotolerance, while the T-DNA mutants showed reduced thermotolerance [34]. Lin et al. reported that *HSFA2*, *HSFA4a*, and *HSFA7a* are essential for HSR and cytoplasmic protein response. HSFs have also been characterized in several crop plants [35]. It was reported that *OsHSFA4a* and its homolog in wheat *TaHSFA4a*, confers cadmium tolerance to plants [36]. The expression of *TaHSFA2-10* is induced in response to HS, oxidative stress, salicylic acid, and its overexpression improves plant thermotolerance [37]. In addition, HSFs are also involved in the regulation of growth and development in *Arabidopsis thaliana* [16]. For example, *HSFA9* is expressed specifically during embryogenesis and maturation in *Arabidopsis* seeds [38]. Albihlal et al. [39] reported that in *Arabidopsis*, at least 85 development-associated genes are controlled by *HSFA1b* [39]. The authors proposed that the *HSFA1b* allows plants to adjust growth and development under continuously varying environments by transducing external stimuli to stress-associated and development-related genes.

The HSF gene family has been characterized in several plant species, including *Arabidopsis thaliana* [20], *Oryza sativa* [40], *Zea mays* [41], *Glycine max* [42], *Populus trichocarpa* [43], *Solanum lycopersicum* [44], *Brachypodium distachyon* [45], and *Triticum aestivum* [46]. However, the role of HSFs in plant growth and development and in responses to stresses other than HS, is not very well understood in maize. Computational biology approaches provide a convenient and reliable platform upon which further wet-lab studies could be carried out. Here, we perform an extensive in silico analysis of maize HSFs to gain better insights into the genomic distribution, phylogeny, gene duplication history, gene structure and protein motif, physio-chemical properties, gene annotation, protein networks, and expression profiling of maize HSFs in growth and development and tolerance to multiple abiotic stresses.

2. Materials and Methods

2.1. Sequence Retrieval

The protein sequences of *Zea mays* HSFs were extracted from PLAZA 4.5 (https://bioinformatics.psb.ugent.be/plaza/versions/plaza_v4_5_monocots/, accessed on 1 August 2021) and Ensembl Plants (https://plants.ensembl.org/index.html, accessed on 1 August 2021) databases. For this, the BLASTP search was performed using the *Arabidopsis* (AT4G17750, AT2G26150, AT5G03720, AT4G36990, and AT3G24520) and *Zea mays* (GRMZM2G115456, GRMZM2G002131, GRMZM2G086880) HSFs against maize genome Zm-B73-REFERENCE-NAM-5.0, as a query to obtain a putative list of HSFs in the maize genome using default parameters (e-value $\leq 10^{-5}$ and identity % = 80%). These sequences were checked for the presence of DNA binding domain (DBD) and oligomerization domain (OD) through EMBL-EBI employing the hidden Markov model (HMM) (https://www.ebi.ac.uk/Tools/hmmer/search/phmmer, accessed on 1 August 2021) and SMART (http://smart.embl-heidelberg.de/, accessed on 1 August 2021) tools [47]. The coiled-coil structure, which is a property of OD was predicted using MARCOIL [48]. After carefully analyzing the sequence, the proteins that lacked DBD and/or OD (HR-A/B region) were removed. In addition, the redundant proteins that were the product of a single gene were also discarded from further analysis. Finally, a total of 25 maize HSFs genes were selected for further analysis.

2.2. Sequence Analysis

The coding sequence (CDS) and genomic DNA sequences were obtained from the maize genome database (MaizeGDB) (https://www.maizegdb.org/, accessed on 1 August 2021). The fundamental data such as gene chromosomal location, position, strand position, the total number of transcripts, and intron and exon number was also retrieved from MaizeGDB. The physio-chemical properties of HSFs were predicted using the Expasy website (https://www.expasy.org/, accessed on 1 August 2021). Subcellular localization was predicted using two online tools, i.e., WoLFPSORT [49] and CELLO server.2.5 [50]. The conserved protein domains (through protein sequences) were identified using the MEME suite (https://meme-suite.org/meme/tools/meme, accessed on 1 August 2021) [51]. Twenty conserved motifs were identified using default parameters. Using CDS, the intron–exon structure of maize HSF genes was analyzed through GSDS 2.0 (http://gsds.gao-lab.org/, accessed on 1 August 2021) [52].

2.3. Sequence Alignment and Phylogenetic Analysis

The protein sequences of *Arabidopsis thaliana*, *Oryza sativa*, *Brachypodium distachyon*, and *Sorghum bicolor* were aligned with *Zea mays* HSFs through Clustal W [53]. Phylogenetic analysis was performed with the neighbor-joining method implemented in MEGA7.0 and tests were carried out with 1000 bootstrap replicates [54].

2.4. Gene Duplication and Evolutionary Analysis

Gene duplication events were investigated by following two parameters: (1) the length of an alignable sequence covers > 80% of the longer gene; and (2) the similarity of the aligned regions > 70% [55,56]. To analyze the molecular evolutionary rates of duplicated gene pairs, the non-synonymous substitution (Ka) and synonymous substitution (Ks) ratio were calculated using Ka/Ks calculation tool (http://services.cbu.uib.no/tools/kaks, accessed on 1 August 2021). In principle, the value of Ka/Ks < 1 signifies the purifying selection (negative selection), Ka/Ks > 1 signifies positive selection, and Ka/Ks = 1 means neutral selection [57]. Based on a rate of 6.1×10^{-9} substitutions per site per year, the divergence time (T) was calculated as $T = Ks/(2 \times 6.1 \times 10^{-9}) \times 10^{-6}$ million years ago (Mya) for HSF genes [57]. The duplicated HSF genes were connected using Tbtools [58].

2.5. Chromosomal Distribution

Based on their initial positions on the maize genome, the HSF genes were named, and a chromosomal graph was constructed using Tbtools.

2.6. Expression Profiling of HSF Genes

The RNA-seq data utilized in the current study was retrieved from the maize MaizeGDB database (https://qteller.maizegdb.org/genes_by_name_B73v4.php, accessed on 1 August 2021). Previously, a comprehensive maize gene expression analysis was performed by Stelpflug et al. [59], used in the current study. We analyzed HSF profiles in 3 different growth stages (seed, vegetative, and reproductive) and across 20 different tissues (embryo, endosperm, whole seed, primary root, tap root, whole root, stem and shoot apical meristem, immature leaves, tip of stage 2 leaf, mature leaf tissue, pooled leaves, topmost leaves, vegetative meristem and surrounding tissues, immature tassel, meiotic tassel, anthers, mature pollen, mature female spikelet, pre-pollination cob, immature cob, and silks) (Table S4). Furthermore, the expression patterns were investigated at different timescales in a particular developmental stage to get an overview of the spatiotemporal expression of HSFs. In addition, the expression of *Zm*HSFs was also analyzed under abiotic (heat, drought, salinity, cold, UV) stress conditions. For the construction of the heatmap, FPKM values were used, which are already available on MaizeGDB. The heatmap was constructed using Tbtools.

2.7. Protein 3D Structure, Network Interaction, and Gene Ontology Analysis

The three-dimensional (3D) structure of maize HSFs was predicted through AlphaFold (https://www.alphafold.ebi.ac.uk/, accessed on 1 August 2021) [60]. For this, the protein IDs were entered into the search bar and the structures were obtained. The protein interaction network analysis was performed using the STRING database (https://string-db.org/, accessed on 1 August 2021) using default parameters, i.e., sequences showing more than 40% identity in the database were included for interaction networking [61]. The network interaction file was downloaded and visualized using Cytoscape V. 3.8.2 (https://cytoscape.org/, accessed on 1 August 2021) [62]. Gene ontology annotation analysis was performed by uploading the gene IDs of *Zm*HSFs to the GENE ONTOLOGY RESOURCE (http://geneontology.org/, accessed on 1 August 2021) [63].

3. Results

3.1. Identification and Chromosomal Distribution of Maize HSFs

With the availability of the genomic sequences of the number of plant species, including maize, it is now possible to obtain the protein sequences of all the HSF members. In the present study, a total of 25 HSFs were identified from the maize genome (Figure 1; Table 1). All the HSF proteins were surveyed for the presence of DBD and OD through EMBL-EBI, employing HMM. Furthermore, SMART was used to search the HSF-DBD to check the accuracy of the results. After discarding redundant sequences, 25 *Zm*HSFs were selected for analysis. These HSFs were named based on their chromosomal locations (*ZmHSF-01* to *ZmHSF-25*) (Table S1). The characteristics of maize HSF genes are presented in Table 1. All the HSFs were mapped on the chromosomes of maize (Figure 1). The maize genome was shown to possess HSF genes on all of its chromosomes, though the number of HSFs between different chromosomes varied considerably. Chromosome 1 had a maximum of 6 HSFs genes, whereas a single HSF gene copy was localized on each of chromosomes 4, 6, and 10. On the other hand, chromosomes 2, 3, 7, and 9 harbor two gene copies each, while three gene copies were recognized on each of chromosomes 5 and 8. Except for *ZmHSF01*, *ZmHSF02*, *ZmHSF12*, *ZmHSF16*, and *ZmHSF23*, all the other HSF genes were present on the lower arm of the chromosomes.

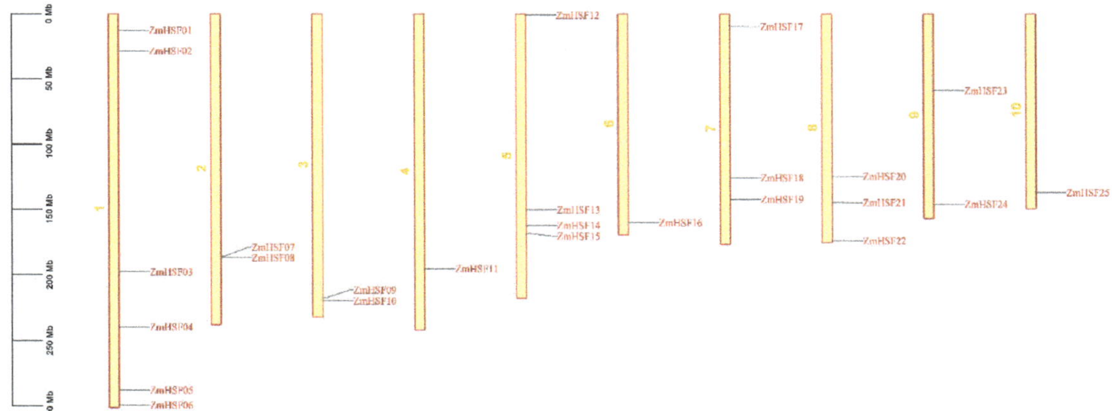

Figure 1. The chromosomal position of HSF genes in maize. The scale on the left side represents megabases (Mb). The chromosome number is indicated at the middle of each bar.

Table 1. Detailed sequence annotation of identified maize HSF genes.

Gene	Transcript ID	Gene ID	Chromosome	Location	Strand	Transcript Count	Genomic DNA	CDS Length	Exon	Intron
ZmHSF-01	GRMZM2G165972_T02	Zm00001d027757	1	12518410–12520734	+	4	2325	1590	2	1
ZmHSF-02	GRMZM2G118485_T02	Zm00001d028269	1	28522823–28529222	+	3	7416	1254	3	2
ZmHSF-03	GRMZM2G164909_T01	Zm00001d029270	1	197781323–197783002	−	1	1680	1245	2	1
ZmHSF-04	GRMZM2G010871_T01	Zm00001d032923	1	239675070–239678097	−	6	3628	1074	2	1
ZmHSF-05	GRMZM2G132971_T01	Zm00001d034433	1	287786016–287790222	+	3	4207	1080	2	1
ZmHSF-06	GRMZM2G115456_T01	Zm00001d034886	1	299272434–299277717	+	1	5284	1584	2	1
ZmHSF-07	GRMZM2G088242_T01	Zm00001d005843	2	185666362–185668432	+	1	2071	1185	2	1
ZmHSF-08	GRMZM2G002131_T01	Zm00001d005888	2	186829423–186833487	+	1	4065	1125	2	1
ZmHSF-09	GRMZM2G089525_T01	Zm00001d044168	3	217438918–217440728	−	1	1811	996	2	1
ZmHSF-10	GRMZM2G005815_T01	Zm00001d044259	3	219545854–219552097	+	2	6244	1389	2	1
ZmHSF-11	GRMZM2G098696_T01	Zm00001d052738	4	195298840–195300584	−	2	1745	1113	2	1
ZmHSF12	GRMZM2G384339_T01	Zm00001d016520	5	889476–893867	−	2	4392	1494	2	1
ZmHSF-13	GRMZM2G105348_T01	Zm00001d016674	5	150228697–150230232	−	1	1536	1185	2	1
ZmHSF-14	GRMZM2G179802_T02	Zm00001d012823	5	162047221–162053475	−	3	6261	1920	2	1
ZmHSF-15	GRMZM2G059851_T01	Zm00001d016255	5	168263116–168266561	+	2	3446	1527	2	1
ZmHSF-16	AC206165.3_FGT007	Zm00001d038746	6	159812447–159814105	−	1	1659	1410	2	1
ZmHSF-17	GRMZM2G125969_T01	Zm00001d020714	7	9712577–9716596	−	3	4020	1284	2	1
ZmHSF-18	GRMZM2G139535_T01	Zm00001d021263	7	125896425–125900126	+	2	3702	897	2	1
ZmHSF-19	GRMZM2G165272_T01	Zm00001d022295	7	142270318- 142272208	+	1	1891	1185	2	1
ZmHSF-20	AC205471.4_FGT003	Zm00001d010812	8	124710390–124712153	+	1	1764	1341	2	1
ZmHSF-21	GRMZM2G086880_T01	Zm00001d011406	8	144887947–144889308	+	1	1364	1248	2	1
ZmHSF-22	GRMZM2G118453_T01	Zm00001d012749	8	173979796–173982203	−	1	2407	1870	2	1
ZmHSF-23	GRMZM2G173090_T02	Zm00001d046204	9	59460696–59462007	+	2	1473	1357	2	1
ZmHSF-24	GRMZM2G026742_T01	Zm00001d048041	9	146029209–146032662	−	4	4461	1840	5	4
ZmHSF-25	GRMZM2G301485_T01	Zm00001d026094	10	137124859–137126235	−	2	1376	1300	2	1

Note: +: Positive strand −: Negative strand.

3.2. Phylogenetic Analysis and Classification of Maize HSFs

In present study, the evolutionary relationship among *At*HSFs, *Os*HSFs, *Sb*HSFs, *Bd*HSFs, and *Zm*HSFs was explored. A total of 118 HSFs were divided into three classes and 15 sub-classes based on the phylogenetic tree (Figure 2; Table S2). Variation in HSF gene number was observed among different plant species and between sub-groups (Table 2). For example, *Arabidopsis thaliana* contains 21 HSFs (15 HSFAs, 5 HSFBs, and 1 HSFC), the *Oryza sativa* possess 25 HSFs (13 HSFAs, 8 HSFBs, and 4 HSFCs), *Zea mays* harbors 25 HSFs in its genome (15 HSFAs, 7 HSFBs, and 3 HSFCs), *Brachypodium distachyon* 24 HSFs (14 HSFAs, 7 HSFBs, and 4 HSFCs), while *Sorghum bicolor* contains 23 HSFs (12 HSFAs, 7 HSFBs, and 4 HSFCs) (Table 2). Results indicated that maize HSFs were close to sorghum HSFs. Similarly, HSFs of rice were closer to *Brachypodium* HSFs, which is in line with the botanical classification among monocots. In contrast to *Arabidopsis*, sub-class A1 contains fewer HSFs in monocots (Table 2). On the other hand, sub-class A2 appears to have expanded in monocots. There was no ortholog of *Arabidopsis* HSFA9, HSFB3 in monocots that might reflect the loss of these sub-groups in family Poaceae (Figure 2, Table 2). Contrarily, this could also signify the gain of these sub-groups in dicots. A higher number of class C HSFs was observed in monocots.

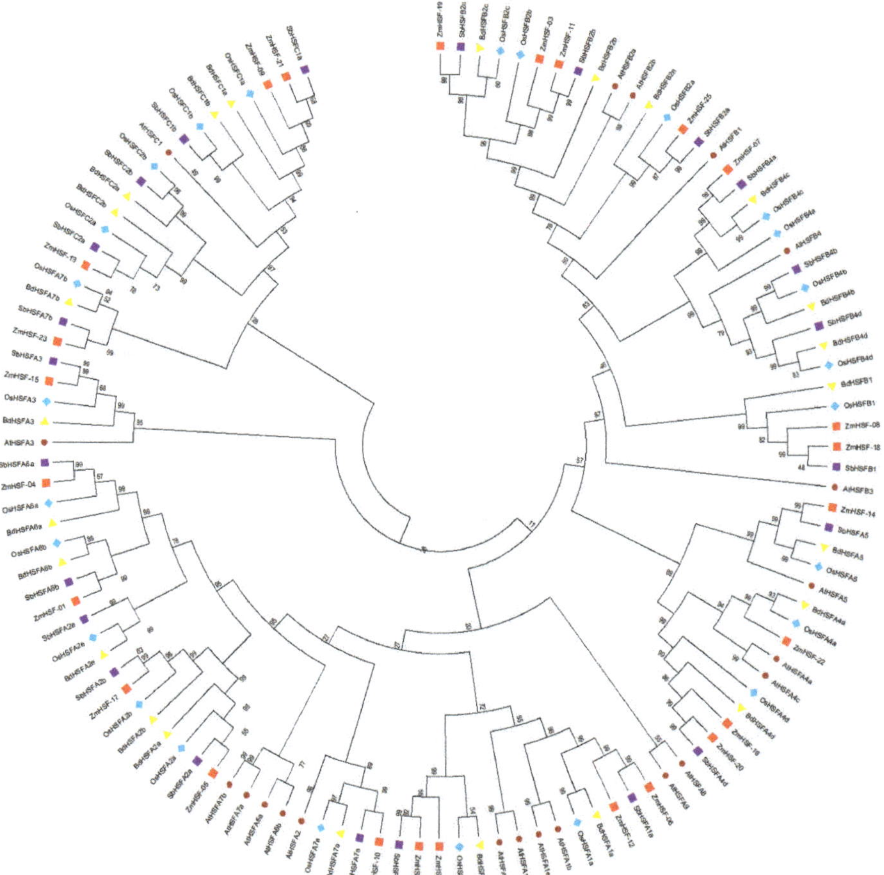

Figure 2. A neighbor-joining phylogenetic tree was constructed after aligning protein sequences of *Arabidopsis thaliana*, *Oryza sativa*, *Brachypodium distachyon*, *Sorghum bicolor*, and *Zea mays*. Overall, 21 AtHSFs (marron circle), 25 OsHSFs (turquoise rhombus), 23 SbHSFs (purple rectangular), 24 BdHSFs (yellow triangle), and 25 ZmHSFs (red square) were divided into three classes and further into 15 sub-classes.

Table 2. Distribution of HSF genes in different sub-classes in selected plant species.

Sub-Class	*Arabidopsis* *At*HSFs	*Zea mays* *Zm*HSFs	*Sorghum bicolor* *Sb*HSFs	*Oryza sativa* *Os*HSFs	*Brachypodium distachyon* *Bd*HSFs
A1	4	2	1	1	1
A2	1	2	3	3	3
A3	1	1	1	1	1
A4	2	3	1	2	2
A5	1	1	1	1	1
A6	2	2	2	2	2
A7	2	2	2	2	2
A8	1	2	1	1	1
A9	1	0	0	0	0
B1	1	2	1	1	1
B2	2	4	3	3	3
B3	1	0	0	0	0
B4	1	1	3	4	3
C1	1	2	2	2	2
C2	0	1	2	2	2
Total	21	25	23	25	24

3.3. Gene Duplication Analysis and Evolutionary Rate Calculation

In the present study, a total of 18 (18/25; 72%) maize HSF genes were shown to be duplicated (Table 3). Further, only one pair of a gene (*ZmHSF-01*/*Zm-HSF-04*) appeared to be tandemly duplicated, which was recognized on chromosome number 1 (Figure 3). The rest of duplicated genes were all segmentally duplicated, with eight different clusters containing 16 genes. These genes were recognized on chromosomes 1–9. Moreover, the molecular evolutionary rate of tandemly and segmentally duplicated HSF genes was calculated to gain insights into the selective constraints on the duplicated HSF genes. The ratio of Ka and Ks substitution rate is an effective method to investigate the selective constraint among duplicated gene pairs [64]. Hence, in the present study, Ka, Ks, and Ka/Ks values for each paralogous gene pair were calculated (Table 3). Here, 18 *Zm*HSF genes were shown to be duplicated. The Ka/Ks ratio for duplicated *Zm*HSF genes ranged from 0.3415 to 0.7572. All the HSF genes in the present study have Ka/Ks value < 1.

Table 3. Duplicated gene pairs, non-synonymous substitution rate (Ka), synonymous substitution rate (Ks), non-synonymous to synonymous substitution rate ratio (Ka/Ks), estimated time of duplication event in a million years ago (MYA), and mode of gene duplication.

Duplicated Genes	Ka	Ks	Ka/Ks	Estimated Time (MYA)	Mode of Duplication
ZmHSF-01/ZmHSF-04	0.1837	0.3733	0.4921	30.59	Tandem
ZmHSF-02/ZmHSF-24	0.0248	0.0727	0.3415	5.96	Segmental
ZmHSF-03/ZmHSF-11	0.2497	0.3726	0.6702	30.54	Segmental
ZmHSF-05/ZmHSF-17	0.1810	0.3250	0.5570	26.63	Segmental
ZmHSF-06/ZmHSF-12	0.2772	0.4600	0.6026	37.70	Segmental
ZmHSF-07/ZmHSF-19	0.1538	2.2641	0.5143	21.64	Segmental
ZmHSF-08/ZmHSF-18	0.1814	0.2395	0.7572	19.63	Segmental
ZmHSF-09/ZmHSF-21	0.0389	0.0839	0.4644	6.87	Segmental
ZmHSF-16/ZmHSF20	0.0733	0.1040	0.7049	8.52	Segmental

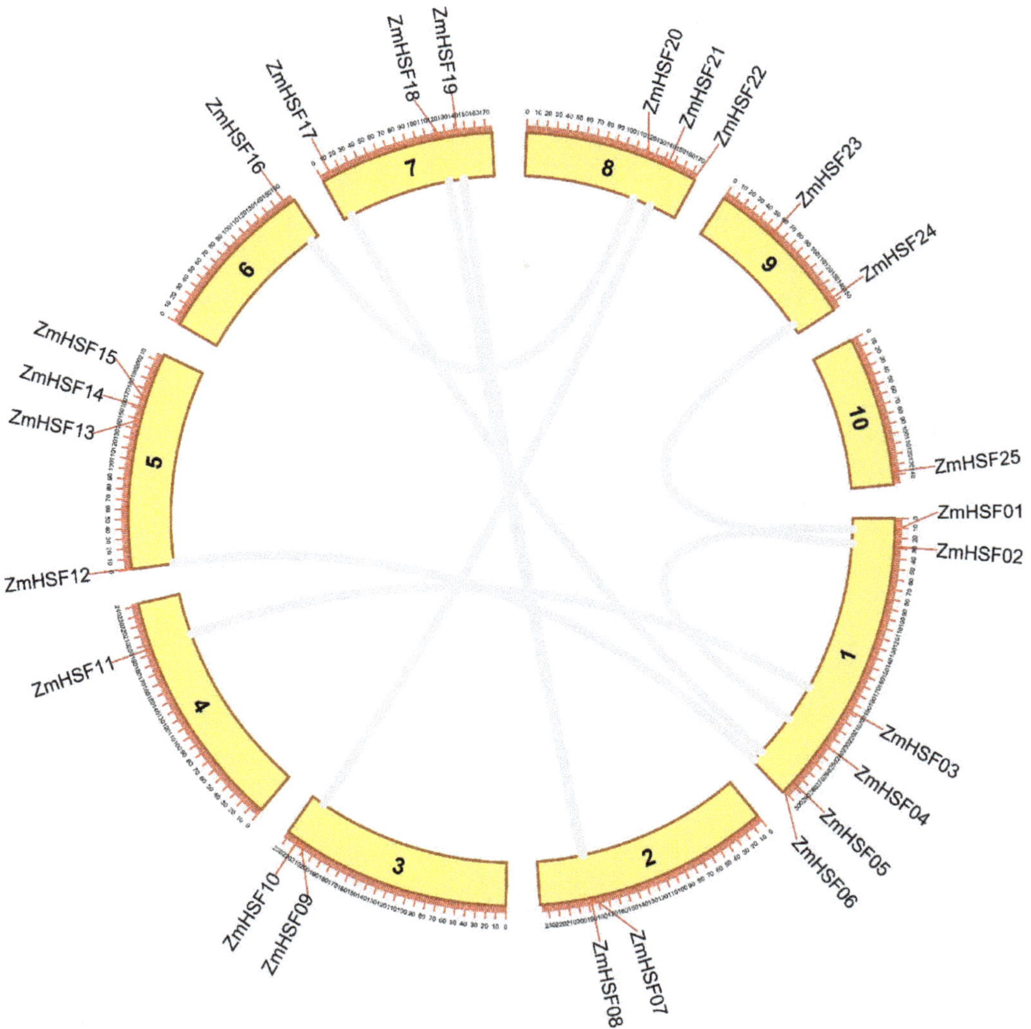

Figure 3. Circular illustration of the maize genome. The paralogous HSF genes are connected with grey lines. The red lines on top of the chromosomal bar represent the position of HSFs.

The outcome suggests that these genes were under strong purifying selection pressure by natural selection during the course of evolution. Further, the divergence periods of tandemly and segmentally duplicated *ZmHSF* genes were estimated to range from 5.96 to 38.04 with an average of 20.89 million years ago (MYA). Some paralogous gene pairs (*ZmHSF-02/ZmHSF-24*, *ZmHSF-09/ZmHSF21*, and *ZmHSF-16/ZmHSF20*) appeared to be recently duplicated (Table 3). The grass species are estimated to have diverged around 56–73 MYA [65,66]. In the present analysis, all the *HSF* genes in maize appeared to have duplicated after the divergence of grass species. Further, most of the HSF genes in maize are paralogs, and it can be concluded that duplication events (primarily segmental) played a significant role in the expansion of the *HSF* gene family in maize.

3.4. Gene Structure and Protein Motif Analysis

To investigate the structural relationship among the HSF genes, the intron–exon organization of all the targeted HSFs was analyzed using GSDS software. The intron–exon structure and number play a key role in the evolution of gene families [67,68]. The gene structure analysis was in line with the phylogenetic relationship among maize HSFs (Figure 4). In general, the intron and exon numbers were shown to be highly consistent. Particularly, 92% (23/25) HSFs contain only one intron except for HSF-02 and HSF-24 (Figure 4). Similarly, HSF-02 and HSF-24 were shown to contain three and five exons. In contrast, the rest of the HSF genes contained two exons. Further, 17 HSFs contained 5′ UTR and 3′ UTR. The HSF genes belonging to the same class and sub-class showed a similar intron–exon pattern in terms of intron number, exon length, intron phase, and overall gene length (Figure 4).

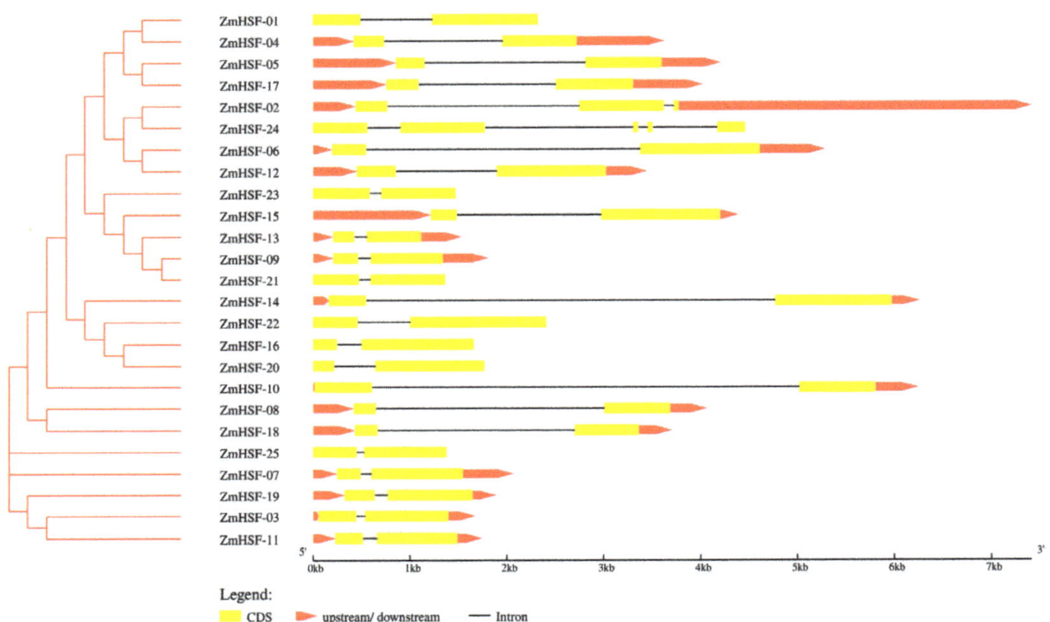

Figure 4. The intron–exon structure of maize HSF genes. A black line represents the introns. Exons are represented by a yellow rectangle and 5′ UTR and 3′ UTR by reddish-pink wedges. The gene structure of *Zea mays* HSFs was in alignment with the phylogenetic relationship.

MEME was used to identify the conserved motifs/regions responsible for DNA-binding, oligomerization, nuclear localization, nuclear export, and biological activation of HSFs (Figure 5). In total, 20 motifs designated as motifs 1–20 were identified among maize HSF proteins (Table 4). The highly conserved DBD is represented by motifs 1, 2, and 4. Motif 3 corresponds to OD of class A and C HSFs. The HR-A core region of OD is represented by motif 5 and is present in all HSFs. OD of class B HSFs is depicted by motif 7. Motif 6 constitutes the NLS of class A HSFs and is present in eight members. The AHA motif is shown by motif 8 and is present in 11 class A HSFs. The NES of class A HSFs is represented by motif 16. Motif 15 constitutes the NLS of Class B2 HSFs. Certain HSFs also possess class-specific conserved motifs i.e., *Zm*HSF-02 and *Zm*HSF-24 contain motifs 10 and 13, HSFs of subclass B1 harbor motif 14, sub-class C1 HSFs motif 19, the sub-class B2 HSFs motif 11, and sub-class A4 members (*Zm*HSF-16, *Zm*HSF-20) contain motif 18. Motifs 9 and 12 are present in the same members of class A HSFs. Four members of class A

HSFs contain motif 17. Motif 18 is found in two members of class A HSFs. Finally, motif 20 is found in members of sub-class A2 and A6 HSFs.

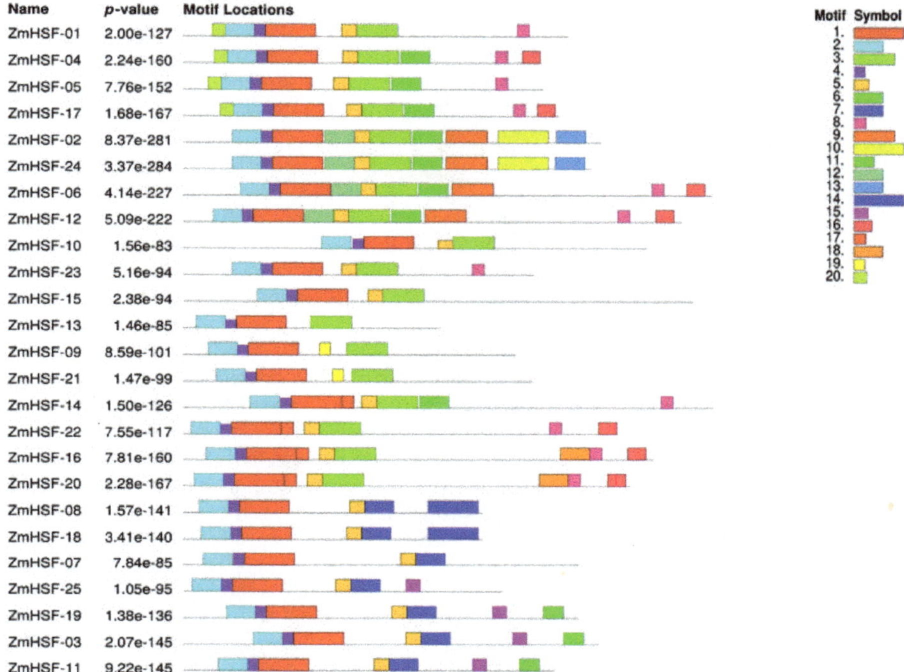

Figure 5. Conserved motifs in maize HSFs as identified by MEME. The motifs were identified by using full-length protein sequences of *Zm*HSFs. Name of HSFs and p-values are indicated on the left side. Each motif is represented by a unique color, as indicated at the bottom.

Table 4. Conserved motif sequence of *Zea mays* HSFs.

Motif	Consensus Sequence
1	LPKYFKHNNFSSFVRQLNTYGFRKVDPDRWEFANEGFLRGQKHLLKNIHR
2	PFLTKTYEMVDDPATDAVVSWGAAGNSFV
3	LLAELVRLRQZQQSTSEQLQALERRLQGMEQRQQQMMAFLA
4	VWBPAEFARDL
5	GLEEEIERLKRDKAL
6	MQNPDFLRQLVQQQEKSKELEDAINKKRR
7	LARELAQMRKLCNNILLLMSKYADTQQPD
8	VNDDFWEZFLTE
9	DGGPVDDSEAAGGGGQIIKYQPPIPEAAKQPLPKNLAFDSS
10	MPMDVEMASNNVGTFDSTGNDFTDTSALCEWDDMDIFGGELEHILQQPEQ
11	QSWPIYRPRPVYHPLRACNG
12	KPSQDGPSDPQQPPVKTAPGPENIEIGKY
13	TIEDYGYDRPWLEQDCQMEAQQNCKNPQY
14	KSVKLFGVLLKDAARKRGRCEEAAASERPIKMIRIGEPWIGVPSSGPGRC
15	RLFGVSIGRKRMRD
16	DNVBQLTEQMGYLSSANH
17	RKPIHSHSPQTQ
18	VDMCSDTTTGDTSQDETTSETGGSHGPAK
19	CCISMGGEDHR
20	PRPMEGLHDVGPP

3.5. Domain Analysis and Physio-Chemical Properties

The modular structure and the functional domains of HSFs have been studied and described extensively [22]. The HSF-type DBD was highly conserved and consisted of approximately 100 amino acids (Figure 6). The locations of DBD and OD were predicted using SMART and MARCOIL (Table 5). The DBD of most maize HSFs was located at the beginning of the N-terminal. Few exceptions were *Zm*HSF-03, *Zm*HSF-10, *Zm*HSF-14, and *Zm*HSF-15. As expected, the linker length between DBD and OD of HSFBs was larger than HSFAs and HSFCs.

The physio-chemical properties of HSFs such as amino acid length, Mw, and pI were investigated using Expasy (Table 5). In addition, the amino acid composition of each group was analyzed using the online tool CoPId (Table S3). The amino acid length of class A HSF ranged from 350 (*Zm*HSF-23) to 528 (*Zm*HSF-14), for class B, 298 (*Zm*HSF-08) to 414 (*Zm*HSF-03), and in class C the amino acid length ranged from 257 (*Zm*HSF-13) to 348 (*Zm*HSF-21). The pI of class A HSFs ranged from 4.70 (*Zm*HSF-17) to 8.87 (*Zm*HSF-10). For class B it varied from 5.00 (*Zm*HSF-19) to 9.53 (*Zm*HSF-18), and for class C, pI ranged from 5.85 (*Zm*HSF-13) to 8.09 (*Zm*HSF-21). For class A, the Mw varied from 38.154 (*Zm*HSF-23) to 58.138 (*Zm*HSF-14), for class B it ranged from 33.258 (*Zm*HSF-18) to 44.381 (*Zm*HSF-03). While for class C HSFs, Mw ranged from 27.836 (*Zm*HSF-13) to 37.409 (*Zm*HSF-21). In general, the pI of class A HSFs was in acidic ranges except for *Zm*HSF-10, which was shown to be in a slightly basic range. The pI of class B was in slightly acidic and basic ranges. Finally, for class C the pI was in similar ranges as class B HSFs. The average amino acid composition of *Zm*HSFs ranged from 1.1 (cysteine) to 10.0 (alanine) (Figure 7A). The average amino acid composition of class A HSFs ranged from 0.8 (cysteine) to 8.7 (alanine) (Figure 7B). In contrast, the average amino acid composition of class B ranged from 1.3 (tryptophan) to 12.6 (alanine) (Figure 7C). Finally, the class C amino acid composition ranged from 1.1 (tryptophan) to 11.2 (alanine) (Figure 7D).

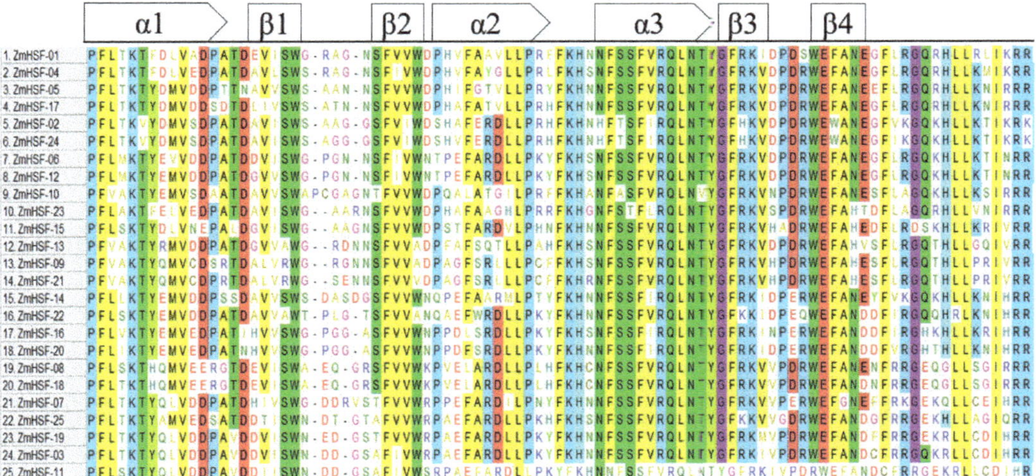

Figure 6. Multiple sequence alignment of maize HSFs. The highly conserved DBD consists of four antiparallel β-stranded sheets (β1, β2, β3, β4) and a bundle of three α-helices (α1, α2, α3) which form the secondary structure. Rectangular boxes represent α-helices while square boxes represent β-stranded sheets.

Table 5. The physio-chemical properties of maize HSFs proteins including group, protein ID, domains, amino acid physio-chemical properties, and sub-cellular localization.

Gene	Group	Protein ID	DBD	OD	Amino Acid	pI	Molecular Weight	Localization CELLO	Localization WoLFPSORT
*Zm*HSF-01	A-6b	GRMZM2G165972_P02	44–133	164–184	384	5.30	43,268.46	Nucleus	Nucleus
*Zm*HSF-02	A-8a	GRMZM2G118485_P02	51–140	170–190	417	5.09	46,816.21	Nucleus	Nucleus
*Zm*HSF-03	B-2d	GRMZM2G164909_P01	72–161	231–260	414	7.14	44,381.82	Nucleus	Nucleus
*Zm*HSF-04	A-6a	GRMZM2G010871_P01	46–135	165–206	357	5.03	40,550.58	Nucleus	Nucleus
*Zm*HSF-05	A-2a	GRMZM2G132971_P01	40–129	149–204	359	5.57	40,587.23	Nucleus	Nucleus
*Zm*HSF-06	A-1a	GRMZM2G115456_P01	59–148	176–203	527	5.11	56,724.50	Nucleus	Nucleus
*Zm*HSF-07	B-4a	GRMZM2G088242_P01	22–112	216–243	394	7.81	41,742.83	Nucleus	Nucleus
*Zm*HSF-08	B-1a	GRMZM2G002131_P01	18–107	165–192	298	9.13	32,270.31	Nucleus	Nucleus
*Zm*HSF-09	C-1c	GRMZM2G089525_P01	27–116	168–188	331	5.94	35,883.69	Nucleus	Nucleus
*Zm*HSF-10	A-7a	GRMZM2G005815_P01	140–231	260–280	462	8.87	50,970.67	Chloroplast	Nucleus
*Zm*HSF-11	B-2b	GRMZM2G098696_P01	37–126	196–216	370	5.89	39,562.03	Nucleus	Nucleus
*Zm*HSF-12	A-1b	GRMZM2G059851_P01	76–165	190–231	508	4.96	56,061.66	Nucleus	Nucleus
*Zm*HSF-13	C-2a	GRMZM2G105348_P01	15–104	135–171	257	5.85	27,836.98	Nucleus	Nucleus
*Zm*HSF-14	A-5	GRMZM2G179802_P02	69–159	177–225	528	5.57	58,138.65	Nucleus	Nucleus
*Zm*HSF-15	A-3	GRMZM2G384339_P01	32–121	149–176	497	5.06	54,104.58	Nucleus	Nucleus
*Zm*HSF-16	A-4b	AC206165.3_FGP007	25–114	135–155	469	5.41	51,623.79	Nucleus	Nucleus
*Zm*HSF-17	A-2b	GRMZM2G125969_P01	52–141	169–210	375	4.70	42,043.73	Nucleus	Nucleus
*Zm*HSF-18	B-1b	GRMZM2G139535_P01	20–109	169–189	298	9.53	32,258.35	Nucleus	Nucleus
*Zm*HSF-19	B-2c	GRMZM2G165272_P01	45–134	207–234	394	5.00	41,468.10	Nucleus	Nucleus
*Zm*HSF-20	A-4d	AC205471.4_FGP003	13–102	123–143	446	5.15	49,718.46	Nucleus	Nucleus
*Zm*HSF-21	C-1a	GRMZM2G086880_P01	35–124	173–193	348	8.09	37,409.50	Nucleus	Nucleus
*Zm*HSF-22	A-4a	GRMZM2G118453_P01	10–99	120–175	433	5.30	48,829.92	Nucleus	Nucleus
*Zm*HSF-23	A-7b	GRMZM2G173090_P02	51–140	164–205	350	4.95	38,154.66	Cytoplasm	Cytoplasm
*Zm*HSF-24	A-8b	GRMZM2G026742_P01	51–140	170–190	407	4.97	45,317.63	Nucleus	Nucleus
*Zm*HSF-25	B-2a	GRMZM2G301485_P01	51–140	170–190	318	5.89	39,562.03	Nucleus	Nucleus

Note: a, b, c, d represents the corresponding protein groups to which these genes belong.

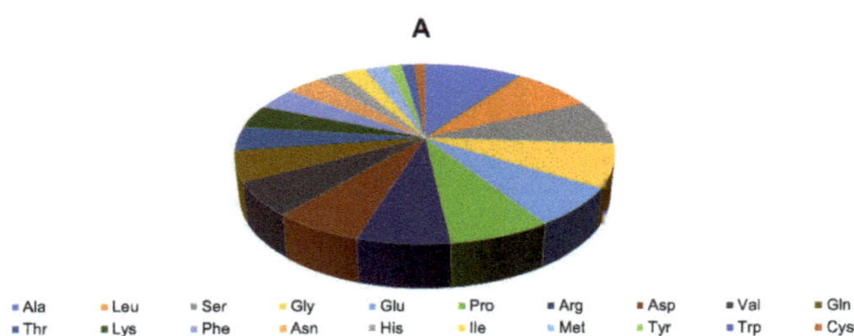

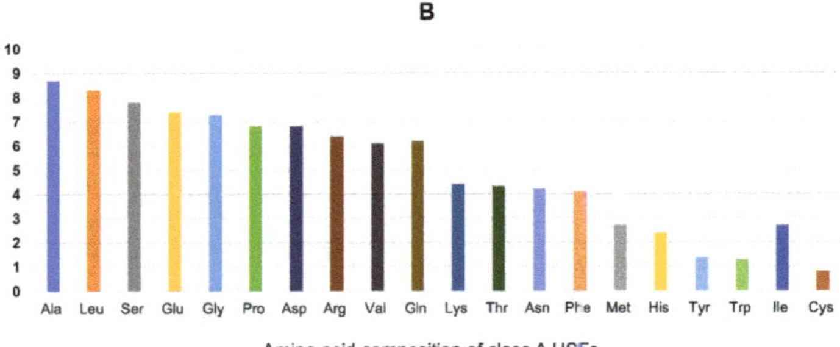

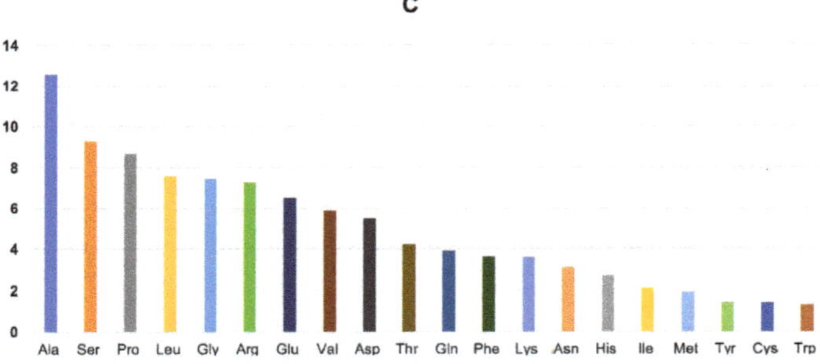

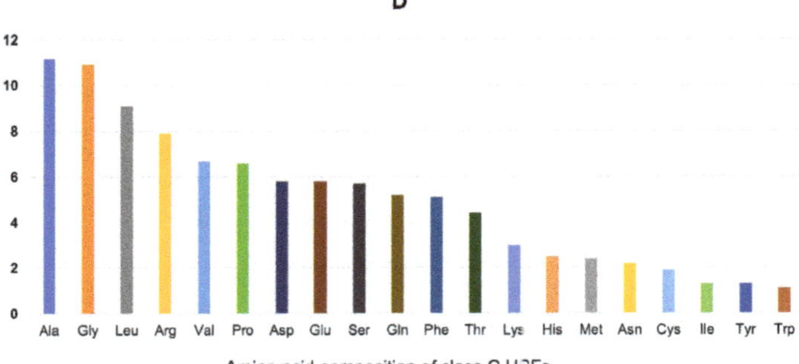

Figure 7. (**A**) The overall amino acid composition of maize HSFs, (**B**) class A HSFs, (**C**) class B HSFs, and (**D**) class C HSFs.

3.6. Proteins Structure and Sub-Cellular Localization of Maize HSFs

The protein structures of maize HSFs were predicted through Alphafold. The predicted models were downloaded to view their 3D structure (Figure 8). The highly conserved DBD is represented by α-helices and β-sheets. The OD can be seen to be linked with DBD through linker residues. Most maize HSFs were predicted to be localized inside the nucleus (Table 5). Exceptions were *Zm*HSF-10 and *Zm*HSF-23 (HSFA7a and HSFA7b). This indicates that class 7 HSFs might possess distinct properties.

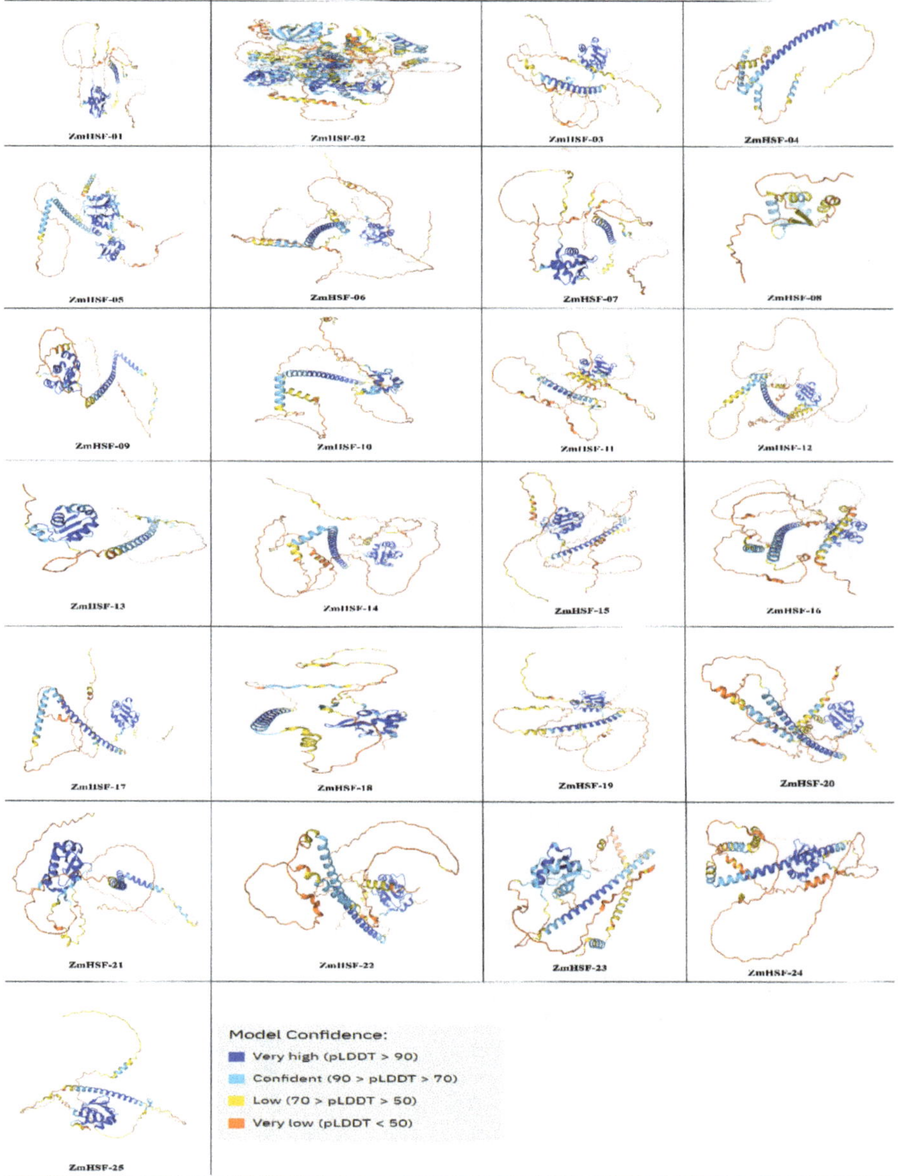

Figure 8. Protein structure of maize HSFs. The prediction confidence level is presented at the bottom. For detailed information about proteins, see Table 5.

3.7. Expression Profiles of Zea mays HSFs during Different Developmental Stages

The expression patterns of *Zea mays* HSF genes were investigated during different growth stages and across different time scales. Although gene expression pattern is not always directly related to protein abundance, the transcriptome profiles can still provide insights into the probable role of genes in particular biological processes [69]. The transcriptome data utilized in the present study was downloaded from the maize genome database [59].

The *ZmHSF-04*, *ZmHSF-05*, and *ZmHSF-06* are shown to be the most highly induced HSFs across all the tissues during different growth stages (Figure 9A–C). During seed developmental stages, *ZmHSF-02*, *ZmHSF-03*, *ZmHSF-04*, *ZmHSF-07*, *ZmHSF-08*, *ZmHSF-09*, *ZmHSF-13*, *ZmHSF-15*, *ZmHSF-19*, *ZmHSF-20*, *ZmHSF-23*, and *ZmHSF-25* are upregulated (Figure 9A). A total of 8 HSFs showed very little or no expression at all. Interestingly, the transcripts of all ZmHSFs completely disappear 12 days after pollination. During vegetative stage, the *ZmHSF-01*, *ZmHSF-02*, *ZmHSF-09*, *ZmHSF-13*, *ZmHSF-14*, *ZmHSF-15*, *ZmHSF-19*, *ZmHSF-23*, and *ZmHSF-25* are upregulated (Figure 9B). This indicates that these genes might play putative regulate role in these tissues and control vegetative growth. *ZmHSF-22* and *ZmHSF-24* showed no expression during vegetative stage. The highest expression of ZmHSFs was observed in root tissues. Similarly, at reproductive stage across different tissues, *ZmHSF-02*, *ZmHSF-04*, *ZmHSF-13*, *ZmHSF-15*, *ZmHSF-19*, and *ZmHSF-25* are upregulated (Figure 9C). Interestingly, the transcripts of most HSFs except for *ZmHSF-02*, *ZmHSF-04*, *ZmHSF-05*, *ZmHSF-06*, *ZmHSF-16*, and *ZmHSF-25* completely disappear in pollen tissues. ZmHSF-17 showed almost no expression in any tissue during reproductive stage. Most of the HSFs were highly expressed in silk tissues.

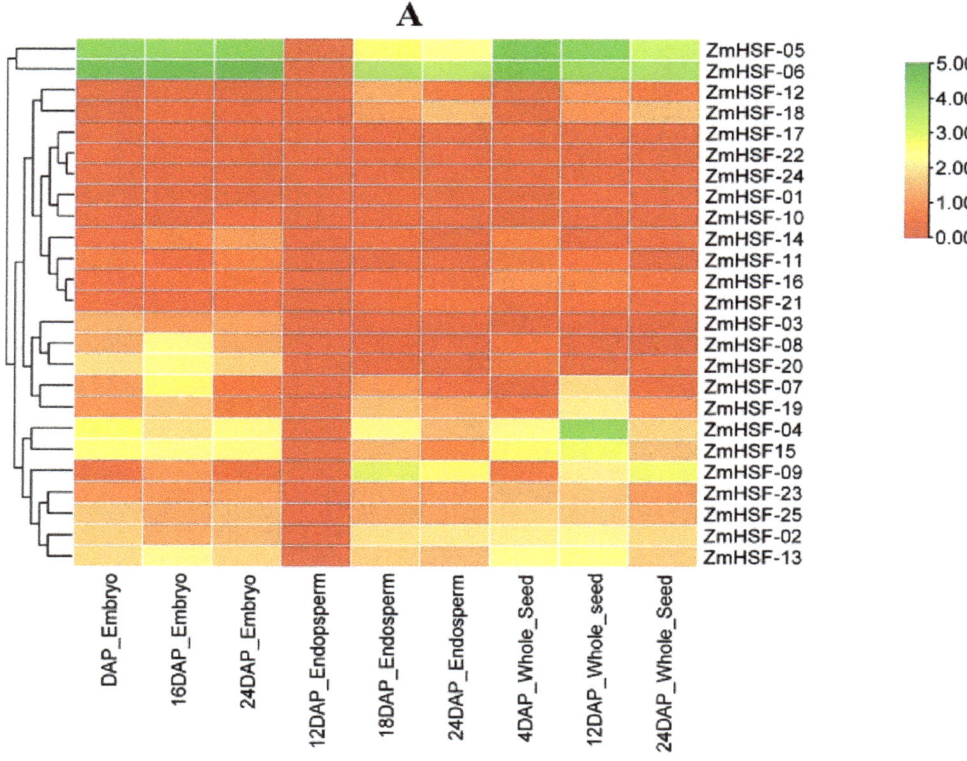

Figure 9. *Cont.*

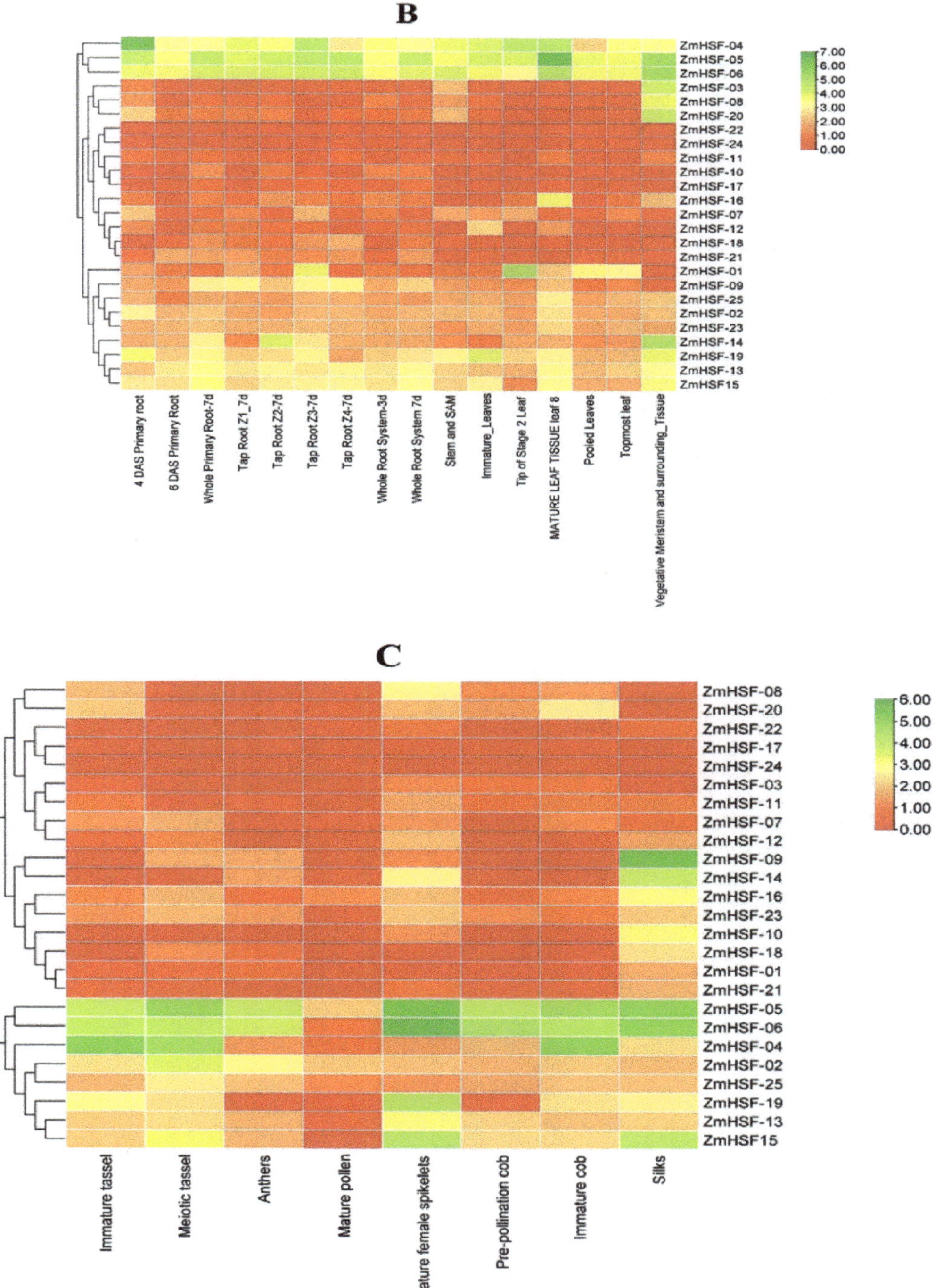

Figure 9. Expression profiles of maize HSFs in different seed (**A**), vegetative (**B**), and reproductive (**C**) tissues based on transcriptome data. The heat map was constructed using Tbtools. The color bar at the top right represents the log transformed FPKM values. Green represents higher, red lower, and yellow, medium transcript values.

3.8. Expression Pattern of *Zea mays* HSFs under Abiotic Stresses

Maize growth, development, and yield is adversely affected by several abiotic stresses [70]. Therefore, to examine maize HSFs expression under different abiotic stress events, RNA-seq data was analyzed and a heat map was constructed (Figure 10; Table S5). In response to HS, members of class A and B HSFs showed the highest expression. Interestingly, under the drought stress, only the transcript of *ZmHSF-05* was moderately overexpressed. Under cold stress, three members of class A HSFs (*ZmHSF-05*, *ZmHSF-06*, and *ZmHSF-16*) showed relatively higher expression. Moderate expression of *ZmHSF-05*, *ZmHSF-06*, *ZmHSF-04*, and *ZmHSF-19* was observed under UV stress. Under salinity stress, *ZmHSF-05* and *ZmHSF-14* showed the highest expression. Under abiotic stress conditions, *ZmHSF-01*, *ZmHSF-02*, *ZmHSF-05*, *ZmHSF-06*, *ZmHSF-07*, *ZmHSF-09*, *ZmHSF-12*, *ZmHSF-13*, *ZmHSF-14*, *ZmHSF-15*, *ZmHSF-19*, *ZmHSF-23*, and *ZmHSF-25* are upregulated (Figure 10). However, some HSFs were only induced by a particular stress. For example, higher transcripts of *ZmHSF-24* are only detected after HS treatment. Similarly, ZmHSF-16 is slightly overexpressed after cold treatment. While, some HSFs showed almost no expression under any stress condition. These include *ZmHSF-03*, *ZmHSF-08*, *ZmHSF-10*, *ZmHSF-11*, *ZmHSF-17*, *ZmHSF-18*, *ZmHSF-20*, *ZmHSF-21*, and *ZmHSF-22* (Figure 10).

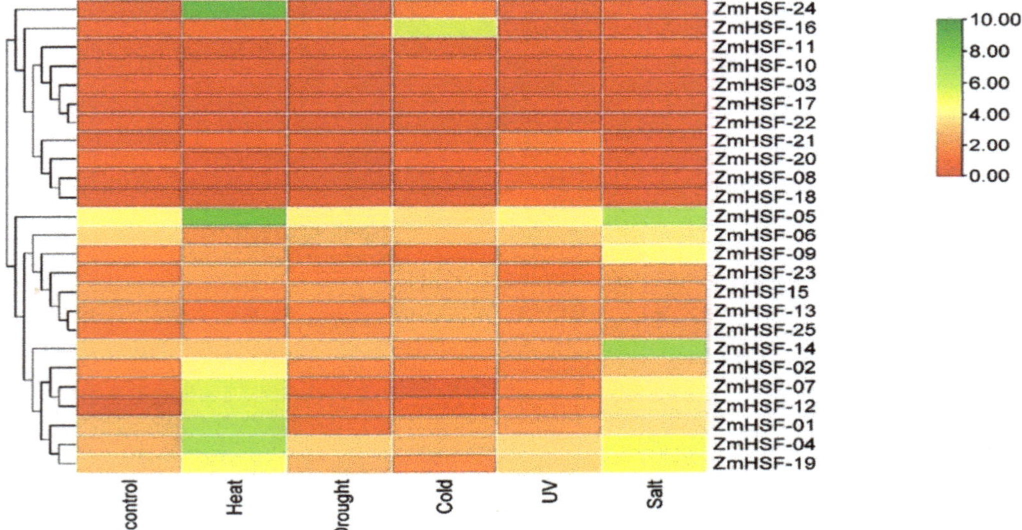

Figure 10. Expression profiles of maize HSFs under different abiotic stress conditions. Color bar at the top right represents the log-transformed FPKM values. Green represents higher, red lower, and yellow, medium transcript values.

3.9. Functional Annotation of Maize HSFs

HSFs have been reported to play a major role not only under stressful conditions but also in plant growth and development [16,69]. Therefore, the regulatory functions of maize HSFs were predicted through GO annotation investigation based on the biological process (BP), molecular function (MF), and cellular component (CC) classes (Figure 11; Table S6). The BP annotation analysis indicated that maize HSFs are mainly involved in cellular response to heat (GO:0034605), response to heat (GO:0009408), response to temperature stimulus (GO:0009266), regulation of transcription by RNA polymerase II (GO:0006357), response to abiotic stimulus (GO:0009628), cellular response to stress (GO:0033554), regulation of RNA biosynthetic process (GO:2001141), etc. (Figure 11). With regard to MF annotation analysis, it was revealed that maize HSFs are mostly involved in RNA polymerase II cis-regulatory region sequence-specific DNA binding (GO:0000978), cis-regulatory region sequence-specific DNA binding (GO:0000987), RNA polymerase II transcription regulatory region sequence-specific DNA binding (GO:0000977), transcription regulatory region nucleic acid binding (GO:0001067), transcription cis-regulatory region binding (GO:0000976), sequence-specific double-stranded DNA binding (GO:1990837), double-stranded DNA binding (GO:0003690), etc. (Figure 11). The CC annotation study showed that ZmHSFs are majorly involved in the nucleus (GO:0005634), intracellular membrane-bounded organelle (GO:0043231), membrane-bounded organelle (GO:0043227), intracellular organelle (GO:0043229), organelle (GO:0043226), etc. (Figure 11). To conclude, the GO annotation study confirms the role of maize HSFs in regulating abiotic stresses and plant metabolism.

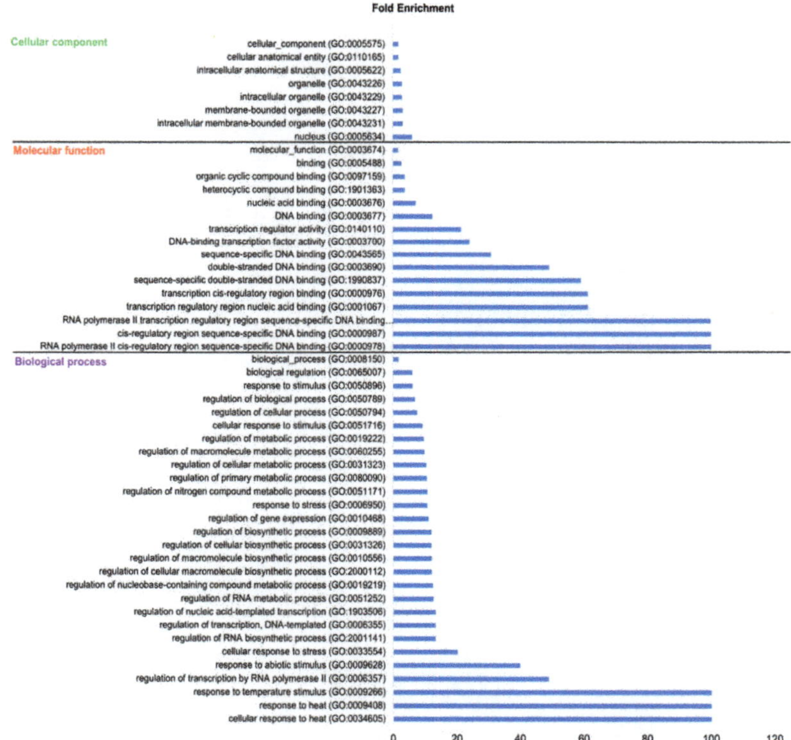

Figure 11. Gene ontology enrichment analysis. The figure depicts the predicted role of HSFs in biological processes, molecular functions, and cellular components.

3.10. Protein–Protein Interaction Network Analysis

The protein network interaction analysis can help understand protein biological function's and mechanisms [71]. Since both the RNA-seq data and GO annotation analysis suggested the role of HSFs in stress conditions and normal growth, we performed network analysis to predict the interacting partners of *Zm*HSFs (Table S7). The results showed that maize HSFs interact with themselves and a range of proteins with well-known functions in cellular growth and stress responses (Figure 12). For example, HSFs were shown to interact with molecular chaperons HSP101, HSP82 (belongs to HSP90 family), HSBP-2, and DnaJ-like protein (belongs to the HSP40 family). It was reported that HSP101 and HSA32 interact with each other and promote acquired thermotolerance in *Arabidopsis* [72]. The HSP82 was reported to be induced by higher temperatures. A higher concentration of HSP82 is required for normal cellular growth in yeast at higher temperatures [73]. Gu et al. reported that maize HSBP-2 and HSFA2 interact with each other and modulate raffinose biosynthesis [74]. HSFA2 was shown to bind to the promoter sequence of HSBP-2 and activate its expression. Higher raffinose synthesis improved HS tolerance of *Arabidopsis thaliana*. The DnaJ-like proteins are molecular co-chaperones that interact with HSP70s and control protein homeostasis [75]. DnaJ proteins have been reported to play a critical role in plant growth, development, and HS tolerance [75–77]. *Zm*HSFs also interact with two major proteins, i.e., multi-protein bridging factor 1c (MBF1c) and DREB2A. Both these proteins have been shown to accumulate under diverse abiotic stress conditions. DREB2A is a major protein, and its overexpression improves plant HS, drought stress, cold stress, etc., tolerance [78]. MBF1c is a transcriptional co-activator that modulates the expression of DREB2A, some HSFs, and phytohormones [3]. Interestingly, MBF1c is necessary for basal thermotolerance but not for acquired thermotolerance [79]. In addition, MBF1c is also shown to be required for plant developmental responses [80].

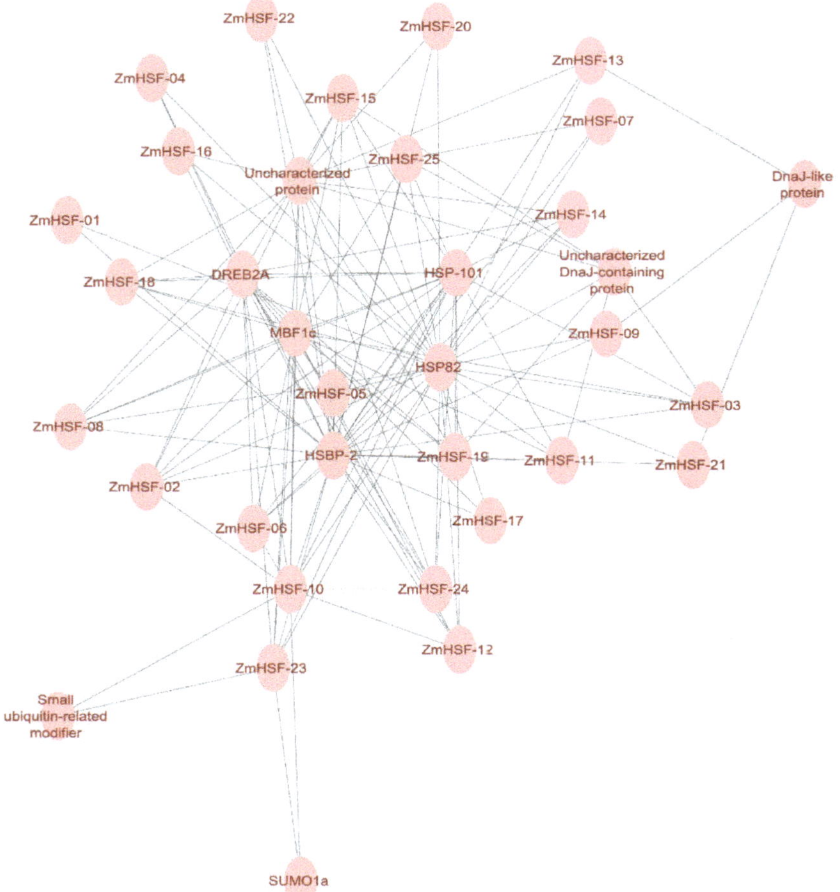

Figure 12. Protein–protein network of maize HSFs. The line connecting two proteins represents that an interaction exists between them.

Maize HSFs are also predicted to interact with SUMO proteins. SUMOylation is a post-translational phenomenon where SUMO proteins are covalently attached and detached to target proteins [81]. This process affects several biological processes inside the cell, including transcriptional regulation of gene expression, apoptosis, programmed cell death, cellular response to stress, stability of proteins, etc. [81]. Rytz et al. reported that SIZI, a SUMO protein, targets multiple TFs, chromatin remodelers, transcriptional co-activators/repressors connected to abiotic and biotic stress responses [82]. This suggests maize HSFs may also be SUMOylated under diverse biological conditions and stress responses. To conclude, PPI analysis aligned with the RNA-seq and GO annotation analysis, which indicated that HSFs of *Zea mays* play an important role in abiotic stress conditions and in maize growth and metabolism.

4. Discussion

Maize (*Zea mays*) is a major cereal crop that is widely cultivated worldwide for food, feed, fiber, and fuel. Maize is also considered a model plant for basic and applied research in plant science [83]. Unraveling the factors regulating the growth and stress resistance would contribute significantly to the development of climate-smart, stress-resilient maize cultivars with higher agricultural productivity. The sequencing of the maize genome (B73 inbred line) in 2009 opened a plethora of opportunities to identify, analyze, and characterize stress-associated genes in maize [84]. To provide food security in the scenario of climate change and ever-growing world population, it is imperative to understand the molecular mechanisms behind plant stress resistance and explore genetic resources associated with the higher crop yield [3,15]. HSFs have been identified in several plant species, including important crops. The *Arabidopsis thaliana, Oryza sativa, Zea mays, Glycine max, Populus trichocarpa, Solanum lycopersicum, Brachypodium distachyon, Sorghum bicolor,* and *Triticum aestivum* contain 21, 25, 25, 38, 28, 26, 24, 23, and 61 HSFs in their genomes, respectively [20,40–45,71,85]. Following the sequencing of several plants, it is found out that the number of HSFs may be independent of the genome size [71]. For example, *Arabidopsis thaliana* (135 Mb) contains 21 HSFs [20,86], while *Medicago truncatula* (375 Mb) harbors 15 HSFs [43,87]. Similarly, 25 HSFs are found in *Oryza sativa* (430 Mb) [40,88] and an equal number of HSFs are also present in *Zea mays* (2.4 Gb) [41,84]. Even though the HSF gene family was previously characterized by Lin et al. [41], our work differs from theirs in multiple aspects. Their research was mostly restricted to the identification and classification of *Zm*HSFs. On the other hand, this comprehensive study particularly focused on the evolutionary analysis, expression profiling, GO, and PPI networks to explore the probable regulatory role played by *Zm*HSFs under benign and stress conditions.

The distribution of the HSF gene family in maize was analyzed by constructing a chromosomal map (Figure 1). The fact that all the chromosomes harbor at least one HSF gene suggests that *Zea mays'* most recent common ancestor has HSF genes distributed widely in its genome. Phylogenetic analysis indicated the *At*HSFA2, *At*HSFC1 did not align with sub-class A2, C1 in the present study, which aligns with the results reported by Lin et al. [41]. Maize HSFs are divided into three classes and further into 13 sub-classes which is consistent with the HSF class number observed in other monocots. For example, the HSFs of *Oryza sativa, Brachypodium distachyon, Sorghum bicolor,* and *Triticum aestivum* are also divided into three classes and 13 sub-classes [40,45,46,85]. Despite that, differences among HSF numbers were observed in different sub-classes between monocots (Figure 2, Table 2). For example, compared to rice, *Brachypodium*, and *Sorghum*, the sub-families B4, A2, and C2 contain fewer HSF members in maize. On the other hand, the sub-classes A1, A4, A8, B1, and B2 are expanded in *Zea mays* (Figure 2, Table 2). Gene duplications generate new genes and provide novel possibilities for evolutionary success [89,90]. In fact, it has been proposed that tandem and segmental duplications have been the primary driving source of evolution as these events lead to expansion of gene families and generation of proteins with novel functions [91]. Tandem duplication involves the duplication of two or more genes located on the same chromosome, while segmental duplication refers to the phenomenon when genes belonging to the same clade but located on different chromosomes are duplicated [92]. In the present analysis, nine pairs of *Zm*HSFs were shown to be paralogs (Table 3, Figure 3). The results indicate that segmental duplication events have played a major role in the expansion of the HSF gene family in maize. An increase in gene regulatory repertoire such as transcriptional regulators, developmental regulators, signal transducers, etc., is a prerequisite for the evolution of complex systems in different organisms [22,93]. Since the gene duplication events result in the doubling of a single gene which cannot account for such large expansions, it has been suggested that whole-genome duplication (WGD) events have been instrumental in expanding the regulatory repertoire of plants [90]. It is assumed that the *Arabidopsis* genome experienced two rounds of WGD in the past 60–70 million years [94,95]. More than 90% increase in regulatory genes has been caused by duplication events *Arabidopsis* in the past 150 million years [94]. This suggests that an increase in the HSF gene members in plants accounts for WGDs. Additionally, segmental duplications occur in gene families, which evolve at a slower rate [91]. It is thought that the increase or decrease in exon number plays an important role in the evolution of a gene family [96]. Therefore, we investigated the number and distribution of introns and exons in *ZmHSF* genes. Our results showed that the *Zm*HSF genes contain 2 exons and 1 intron except for *ZmHSF-02* and *ZmHSF-24* (Figure 4; Table 1). Moreover, the length and position of exons and introns were well conserved in the same sub-classes but varied considerably between different sub-classes.

The previous investigations showed that HSFs play an important role in plant growth [39]. Therefore, we investigated the tissue-specific expression of *Zm*HSFs in 20 different developmental tissues using RNA-seq data (Figure 9A–C). Several genes showed an enhanced expression that reflects their role under various developmental stages. In particular, *ZmHSF-05* (A-2a) and *ZmHSF-06* (A-1a) were highly expressed almost across all the growth phases. The *hsfa1abde* quadruple mutants displayed abnormal phenotype and growth retardation, implying HSFA1s is also involved in developmental processes [29]. Interestingly, *HSFA2* could rescue the developmental defects of *hsfa1abde* quadruple mutants [97]. This further supports the result obtained from our analysis and provides a strong base for further wet-lab studies to characterize the function of *ZmHSF-05* and *ZmHSF-06* in plant growth and development. Similarly, HSFs have been reported to play a key role in plant acclimation to abiotic stress conditions. Kumar et al. reported that *TaHSFA6e* modulates tolerance of wheat to HS and drought stress during pollination and grain filling stages [98]. Yokotani et al. reported that *OsHSFA2e* improves Arabidopsis tolerance to HS and salinity stress by activating the expression of HSPs [99]. Thus, the expression patterns of *Zm*HSFs were evaluated under abiotic stress conditions. Most of *Zm*HSFs displayed stress-specific expression, with some HSFs showing upregulation only under particular stress events (Figure 10). Jiang et al. reported that *Zm*HSF-04 improves plant tolerance to HS, salinity stress and increases the sensitivity to abscisic acid [100]. Similarly, *ZmHSF-12* overexpression improves plant basal thermotolerance and AT [101]. These results are in line with our analysis which showed a higher transcript accumulation of these TFs under respective stress conditions (Figure 10).

The PPI analysis indicated that maize HSFs interact with molecular chaperones and stress-associated proteins (Figure 12). Molecular chaperones are present inside the cells and are constitutively expressed under normal conditions or are induced under specific developmental stages or stress conditions [102]. These chaperons perform various functions under physiological conditions inside the cells, such as signaling, folding, and stabilization, translocation, and degradation of proteins [11,102]. Under harsh environmental conditions, molecular chaperones act as powerful buffers to limit protein misfolding/unfolding and prevent protein aggregate formation that might be otherwise toxic to plant cells [25]. Here, the *Zm*HSFs were shown to interact with chaperons belonging to different families (*HSP101*, *HSP90*, *HSP40*) and genes with a well-known role in thermotolerance (*HSA32*, *HSP82*, *HSBP-2*). DREB2A is a major transcriptional activator that functions downstream of HSFA1s dependent transcriptional cascade in *Arabidopsis thaliana* [3,12,15,29]. Similarly, MBF1c is a major protein characterized by its role in regulating abiotic stress responses and growth in plants [3,82,83]. SUMO proteins are attached to their target proteins and modify their biological activities under various physiological and stress conditions [84]. In *Arabidopsis*, SUMOylation has been proposed as one of the molecular mechanisms that are responsible for the activation of HSFA1s [12]. Many HSFs in *Arabidopsis* such as *HSFA1d*, *HSFA2*, and *HSFB2B* have the potential to be SUMOylated [103]. In tomatoes, the knockout of SIZI (a SUMO ligase) reduces plant thermotolerance [104]. Here, *Zm*HSFs were shown to interact with all these proteins (DREB2A, MBF1c, and SUMO proteins), which further confirms their role in regulating the abiotic stress response.

Taken together, our present analysis provides strong support for the positive role of HSFs in the growth and development of maize by the regulation of primary metabolism. Furthermore, HSFs of maize interact with the major stress-responsive proteins and confer abiotic stress resistance.

5. Conclusions

Here, we identified a total of 25 HSFs from the maize genome through genome-wide investigation analysis. To better understand the roles of *HSF* genes in the maize genome, comprehensive in silico analysis was performed, including phylogenetic analysis, gene structure, and conserved protein motif analysis, gene duplication and evolutionary analysis, domain analysis, and physio-chemical properties, protein 3-D structure, GO and PPI network. Further, the expression profiles of *ZmHSFs* under various developmental stages, and stress conditions were studied. The results indicate that *ZmHSFs* play a major role in plant growth and stress responses. These discoveries will lay the basis for studying the roles of *ZmHSFs* genes in maize developmental processes and response to several stresses using different functional validation options, such as overexpression, knockout via CRISPR/Cas9 systems, etc.

Supplementary Materials: The following are available online at https://www.mdpi.com/article/10.3390/agronomy11112335/s1, Table S1: Protein sequences of *Zea mays* HSFs; Table S2: The protein sequences of *Arabidopsis thaliana*, *Oryza sativa*, *Brachypodium distachyon*, and *Sorghum bicolor* HSFs; Table S3: The class specific amino acid composition of *Zea mays* HSFs; Table S4: FPKM values of *Zea mays* HSFs under different growth stages across various tissues; Table S5: FPKM values of *Zea mays* HSFs under multiple abiotic stress conditions; Table S6: GO annotation file of *Zea mays* HSFs; Table S7: Protein–protein interaction network file of *Zea mays* HSFs.

Author Contributions: Conceptualization, S.H. and T.M.; methodology, S.H.; software, S.H.; formal analysis, S.H., S.R. and Y.A.; resources, A.R.; writing—original draft preparation, S.H., S.R., Y.A., A.R., J.T. and T.J.; writing—review and editing, A.R. and H.S.O.; supervision, T.M.; funding acquisition, A.R. and H.S.O. All authors have read and agreed to the published version of the manuscript.

Funding: This research received no external funding.

Institutional Review Board Statement: Not applicable.

Informed Consent Statement: Not applicable.

Data Availability Statement: Not applicable.

Conflicts of Interest: The authors declare no conflict of interest.

References

1. Shabbir, R.; Javed, T.; Afzal, I.; Sabagh, A.E.; Ali, A.; Vicente, O.; Chen, P. Modern Biotechnologies: Innovative and Sustainable Approaches for the Improvement of Sugarcane Tolerance to Environmental Stresses. *Agronomy* **2021**, *11*, 1042. [CrossRef]
2. Lippmann, R.; Babben, S.; Menger, A.; Delker, C.; Quint, M. Development of wild and cultivated plants under global warming conditions. *Curr. Biol.* **2019**, *29*, R1326–R1338. [CrossRef]
3. Haider, S.; Iqbal, J.; Naseer, S.; Yaseen, T.; Shaukat, M.; Bibi, H.; Ahmad, Y.; Daud, H.; Abbasi, N.L.; Mahmood, T. Molecular mechanisms of plant tolerance to heat stress: Current landscape and future perspectives. *Plant Cell Rep.* **2021**, *40*, 2247–2271. [CrossRef]
4. Chaturvedi, P.; Wiese, A.J.; Ghatak, A.; Záveská Drábková, L.; Weckwerth, W.; Honys, D. Heat stress response mechanisms in pollen development. *New Phytol.* **2021**, *231*, 571–585. [CrossRef]
5. Raza, A. Metabolomics: A systems biology approach for enhancing heat stress tolerance in plants. *Plant Cell Rep.* **2020**, 1–23. [CrossRef] [PubMed]
6. Sharif, R.; Raza, A.; Chen, P.; Li, Y.; El-Ballat, E.M.; Rauf, A.; Hano, C.; El-Esawi, M.A. HD-ZIP gene family: Potential roles in improving plant growth and regulating stress-responsive mechanisms in plants. *Genes* **2021**, *12*, 1256. [CrossRef]
7. Russo, S.; Marchese, A.F.; Sillmann, J.; Immé, G. When will unusual heat waves become normal in a warming Africa? *Environ. Res. Lett.* **2016**, *11*, 054016. [CrossRef]

8. Hasanuzzaman, M.; Nahar, K.; Alam, M.; Roychowdhury, R.; Fujita, M. Physiological, biochemical, and molecular mechanisms of heat stress tolerance in plants. *Int. J. Mol. Sci.* **2013**, *14*, 9643–9684. [CrossRef] [PubMed]
9. Hayes, S.; Schachtschabel, J.; Mishkind, M.; Munnik, T.; Arisz, S.A. Hot topic: Thermosensing in plants. *Plant Cell Environ.* **2021**, *44*, 2018–2033. [CrossRef]
10. Zahra, N.; Shaukat, K.; Hafeez, M.B.; Raza, A.; Hussain, S.; Chaudhary, M.T.; Akram, M.Z.; Kakavand, S.N.; Saddiq, M.S.; Wahid, A. Physiological and molecular responses to high, chilling, and freezing temperature in plant growth and production: Consequences and mitigation possibilities. In *Harsh Environment and Plant Resilience: Molecular and Functional Aspects*; Springer: New York, NY, USA, 2021; p. 235.
11. Mittler, R.; Finka, A.; Goloubinoff, P. How do plants feel the heat? *Trends Biochem. Sci.* **2012**, *37*, 118–125. [CrossRef]
12. Ohama, N.; Sato, H.; Shinozaki, K.; Yamaguchi-Shinozaki, K. Transcriptional regulatory network of plant heat stress response. *Trends Plant Sci.* **2017**, *22*, 53–65. [CrossRef]
13. Javed, T.; Shabbir, R.; Ali, A.; Afzal, I.; Zaheer, U.; Gao, S.J. Transcription factors in plant stress responses: Challenges and potential for sugarcane improvement. *Plants* **2020**, *9*, 491. [CrossRef]
14. Raza, A.; Tabassum, J.; Kudapa, H.; Varshney, R.K. Can omics deliver temperature resilient ready-to-grow crops? *Crit. Rev. Biotechnol.* **2021**, *41*, 1209–1232. [CrossRef]
15. Haider, S.; Iqbal, J.; Naseer, S.; Shaukat, M.; Abbasi, B.A.; Yaseen, T.; Zahra, S.A.; Mahmood, T. Unfolding molecular switches in plant heat stress resistance: A comprehensive review. *Plant Cell Rep.* **2021**, 1–24. [CrossRef] [PubMed]
16. Andrási, N.; Pettkó-Szandtner, A.; Szabados, L. Diversity of plant heat shock factors: Regulation, interactions, and functions. *J. Exp. Bot.* **2021**, *72*, 1558–1575. [CrossRef]
17. Sorger, P.K.; Pelham, H.R. Yeast heat shock factor is an essential DNA-binding protein that exhibits temperature-dependent phosphorylation. *Cell* **1988**, *54*, 855–864. [CrossRef]
18. Scharf, K.-D.; Heider, H.; Höhfeld, I.; Lyck, R.; Schmidt, E.; Nover, L. The tomato Hsf system: HsfA2 needs interaction with HsfA1 for efficient nuclear import and may be localized in cytoplasmic heat stress granules. *Mol. Cell. Biol.* **1998**, *18*, 2240–2251. [CrossRef]
19. Guo, M.; Liu, J.-H.; Ma, X.; Luo, D.-X.; Gong, Z.-H.; Lu, M.-H. The plant heat stress transcription factors (HSFs): Structure, regulation, and function in response to abiotic stresses. *Front. Plant Sci.* **2016**, *7*, 114. [CrossRef] [PubMed]
20. Nover, L.; Bharti, K.; Döring, P.; Mishra, S.K.; Ganguli, A.; Scharf, K.-D. Arabidopsis and the heat stress transcription factor world: How many heat stress transcription factors do we need? *Cell Stress Chaperones* **2001**, *6*, 177. [CrossRef]
21. von Koskull-Döring, P.; Scharf, K.-D.; Nover, L. The diversity of plant heat stress transcription factors. *Trends Plant Sci.* **2007**, *12*, 452–457. [CrossRef]
22. Scharf, K.-D.; Berberich, T.; Ebersberger, I.; Nover, L. The plant heat stress transcription factor (Hsf) family: Structure, function and evolution. *Biochim. Biophys. Acta (BBA) Gene Regul. Mech.* **2012**, *1819*, 104–119. [CrossRef]
23. Bharti, K.; von Koskull-Döring, P.; Bharti, S.; Kumar, P.; Tintschl-Körbitzer, A.; Treuter, E.; Nover, L. Tomato heat stress transcription factor HsfB1 represents a novel type of general transcription coactivator with a histone-like motif interacting with the plant CREB binding protein ortholog HAC1. *Plant Cell* **2004**, *16*, 1521–1535. [CrossRef] [PubMed]
24. Fragkostefanakis, S.; Simm, S.; El-Shershaby, A.; Hu, Y.; Bublak, D.; Mesihovic, A.; Darm, K.; Mishra, S.K.; Tschiersch, B.; Theres, K. The repressor and co-activator HsfB1 regulates the major heat stress transcription factors in tomato. *Plant Cell Environ.* **2019**, *42*, 874–890. [CrossRef] [PubMed]
25. Jacob, P.; Hirt, H.; Bendahmane, A. The heat-shock protein/chaperone network and multiple stress resistance. *Plant Biotechnol. J.* **2017**, *15*, 405–414. [CrossRef]
26. Czarnecka-Verner, E.; Yuan, C.-X.; Scharf, K.-D.; Englich, G.; Gurley, W.B. Plants contain a novel multi-member class of heat shock factors without transcriptional activator potential. *Plant Mol. Biol.* **2000**, *43*, 459–471. [CrossRef] [PubMed]
27. Kotak, S.; Larkindale, J.; Lee, U.; von Koskull-Döring, P.; Vierling, E.; Scharf, K.-D. Complexity of the heat stress response in plants. *Curr. Opin. Plant Biol.* **2007**, *10*, 310–316. [CrossRef]
28. Mishra, S.K.; Tripp, J.; Winkelhaus, S.; Tschiersch, B.; Theres, K.; Nover, L.; Scharf, K.-D. In the complex family of heat stress transcription factors, HsfA1 has a unique role as master regulator of thermotolerance in tomato. *Genes Develop.* **2002**, *16*, 1555–1567. [CrossRef] [PubMed]
29. Yoshida, T.; Ohama, N.; Nakajima, J.; Kidokoro, S.; Mizoi, J.; Nakashima, K.; Maruyama, K.; Kim, J.-M.; Seki, M.; Todaka, D. Arabidopsis HsfA1 transcription factors function as the main positive regulators in heat shock-responsive gene expression. *Mol. Genet. Genom.* **2011**, *286*, 321–332. [CrossRef]
30. Hahn, A.; Bublak, D.; Schleiff, E.; Scharf, K.-D. Crosstalk between Hsp90 and Hsp70 chaperones and heat stress transcription factors in tomato. *Plant Cell* **2011**, *23*, 741–755. [CrossRef] [PubMed]
31. Charng, Y.-Y.; Liu, H.-C.; Liu, N.-Y.; Chi, W.-T.; Wang, C.-N.; Chang, S.-H.; Wang, T.-T. A heat-inducible transcription factor, HsfA2, is required for extension of acquired thermotolerance in Arabidopsis. *Plant Physiol.* **2007**, *143*, 251–262. [CrossRef]
32. Lämke, J.; Brzezinka, K.; Altmann, S.; Bäurle, I. A hit-and-run heat shock factor governs sustained histone methylation and transcriptional stress memory. *EMBO J.* **2016**, *35*, 162–175. [CrossRef]
33. Nishizawa, A.; Yabuta, Y.; Yoshida, E.; Maruta, T.; Yoshimura, K.; Shigeoka, S. Arabidopsis heat shock transcription factor A2 as a key regulator in response to several types of environmental stress. *Plant J.* **2006**, *48*, 535–547. [CrossRef] [PubMed]

34. Yoshida, T.; Sakuma, Y.; Todaka, D.; Maruyama, K.; Qin, F.; Mizoi, J.; Kidokoro, S.; Fujita, Y.; Shinozaki, K.; Yamaguchi-Shinozaki, K. Functional analysis of an Arabidopsis heat-shock transcription factor HsfA3 in the transcriptional cascade downstream of the DREB2A stress-regulatory system. *Biochem. Biophys. Res. Commun.* **2008**, *368*, 515–521. [CrossRef] [PubMed]
35. Lin, K.-F.; Tsai, M.-Y.; Lu, C.-A.; Wu, S.-J.; Yeh, C.-H. The roles of Arabidopsis HSFA2, HSFA4a, and HSFA7a in the heat shock response and cytosolic protein response. *Bot. Stud.* **2018**, *59*, 1–9. [CrossRef]
36. Shim, D.; Hwang, J.-U.; Lee, J.; Lee, S.; Choi, Y.; An, G.; Martinoia, E.; Lee, Y. Orthologs of the class A4 heat shock transcription factor HsfA4a confer cadmium tolerance in wheat and rice. *Plant Cell* **2009**, *21*, 4031–4043. [CrossRef] [PubMed]
37. Guo, X.-L.; Yuan, S.-N.; Zhang, H.-N.; Zhang, Y.-Y.; Zhang, Y.-J.; Wang, G.-Y.; Li, Y.-Q.; Li, G.-L. Heat-response patterns of the heat shock transcription factor family in advanced development stages of wheat (*Triticum aestivum* L.) and thermotolerance-regulation by *TaHsfA2–10*. *BMC Plant Biol.* **2020**, *20*, 364. [CrossRef]
38. Kotak, S.; Vierling, E.; Bäumlein, H.; Koskull-0Döring, P.V. A novel transcriptional cascade regulating expression of heat stress proteins during seed development of Arabidopsis. *Plant Cell* **2007**, *19*, 182–195. [CrossRef]
39. Albihlal, W.S.; Obomighie, I.; Blein, T.; Persad, R.; Chernukhin, I.; Crespi, M.; Bechtold, U.; Mullineaux, P.M. Arabidopsis heat shock transcription factora1b regulates multiple developmental genes under benign and stress conditions. *J. Exp. Bot.* **2018**, *69*, 2847–2862. [CrossRef]
40. Guo, J.; Wu, J.; Ji, Q.; Wang, C.; Luo, L.; Yuan, Y.; Wang, Y.; Wang, J. Genome-wide analysis of heat shock transcription factor families in rice and Arabidopsis. *J. Genet. Genom.* **2008**, *35*, 105–118. [CrossRef]
41. Lin, Y.-X.; Jiang, H.-Y.; Chu, Z.-X.; Tang, X.-L.; Zhu, S.-W.; Cheng, B.-J. Genome-wide identification, classification and analysis of heat shock transcription factor family in maize. *BMC Genom.* **2011**, *12*, 76. [CrossRef]
42. Chung, E.; Kim, K.-M.; Lee, J.-H. Genome-wide analysis and molecular characterization of heat shock transcription factor family in *Glycine max*. *J. Genet. Genom.* **2013**, *40*, 127–135. [CrossRef] [PubMed]
43. Wang, F.; Dong, Q.; Jiang, H.; Zhu, S.; Chen, B.; Xiang, Y. Genome-wide analysis of the heat shock transcription factors in *Populus trichocarpa* and *Medicago truncatula*. *Mol. Biol. Rep.* **2012**, *39*, 1877–1886. [CrossRef]
44. Yang, X.; Zhu, W.; Zhang, H.; Liu, N.; Tian, S. Heat shock factors in tomatoes: Genome-wide identification, phylogenetic analysis and expression profiling under development and heat stress. *Peer J.* **2016**, *4*, e1961. [CrossRef] [PubMed]
45. Wen, F.; Wu, X.; Li, T.; Jia, M.; Liu, X.; Li, P.; Zhou, X.; Ji, X.; Yue, X. Genome-wide survey of heat shock factors and heat shock protein 70s and their regulatory network under abiotic stresses in *Brachypodium distachyon*. *PLoS ONE* **2017**, *12*, e0180352. [CrossRef]
46. Duan, S.; Liu, B.; Zhang, Y.; Li, G.; Guo, X. Genome-wide identification and abiotic stress-responsive pattern of heat shock transcription factor family in *Triticum aestivum* L. *BMC Genom.* **2019**, *20*, 257. [CrossRef] [PubMed]
47. Letunic, I.; Copley, R.R.; Schmidt, S.; Ciccarelli, F.D.; Doerks, T.; Schultz, J.; Ponting, C.P.; Bork, P. SMART 4.0: Towards genomic data integration. *Nucl. Acids Res.* **2004**, *32*, D142–D144. [CrossRef]
48. Delorenzi, M.; Speed, T. An HMM model for coiled-coil domains and a comparison with PSSM-based predictions. *Bioinformatics* **2002**, *18*, 617–625. [CrossRef]
49. Horton, P.; Park, K.-J.; Obayashi, T.; Fujita, N.; Harada, H.; Adams-Collier, C.; Nakai, K. WoLF PSORT: Protein localization predictor. *Nucl. Acids Res.* **2007**, *35*, W585–W587. [CrossRef]
50. Yu, C.S.; Chen, Y.C.; Lu, C.H.; Hwang, J.K. Prediction of protein subcellular localization. *Proteins Struct. Funct. Bioinform.* **2006**, *64*, 643–651. [CrossRef]
51. Bailey, T.L.; Boden, M.; Buske, F.A.; Frith, M.; Grant, C.E.; Clementi, L.; Ren, J.; Li, W.W.; Noble, W.S. MEME SUITE: Tools for motif discovery and searching. *Nucl. Acids Res.* **2009**, *37*, W202–W208. [CrossRef]
52. Hu, B.; Jin, J.; Guo, A.-Y.; Zhang, H.; Luo, J.; Gao, G. GSDS 2.0: An upgraded gene feature visualization server. *Bioinformatics* **2015**, *31*, 1296–1297. [CrossRef]
53. Thompson, J.D.; Higgins, D.G.; Gibson, T.J. CLUSTAL W: Improving the sensitivity of progressive multiple sequence alignment through sequence weighting, position-specific gap penalties and weight matrix choice. *Nucl. Acids Res.* **1994**, *22*, 4673–4680. [CrossRef] [PubMed]
54. Kumar, S.; Stecher, G.; Li, M.; Knyaz, C.; Tamura, K. MEGA X: Molecular evolutionary genetics analysis across computing platforms. *Mol. Biol. Evol.* **2018**, *35*, 1547–1549. [CrossRef] [PubMed]
55. Gu, Z.; Cavalcanti, A.; Chen, F.-C.; Bouman, P.; Li, W.-H. Extent of gene duplication in the genomes of Drosophila, nematode, and yeast. *Mol. Biol. Evol.* **2002**, *19*, 256–262. [CrossRef]
56. Yang, S.; Zhang, X.; Yue, J.-X.; Tian, D.; Chen, J.-Q. Recent duplications dominate NBS-encoding gene expansion in two woody species. *Mol. Genet. Genom.* **2008**, *280*, 187–198. [CrossRef]
57. Lynch, M.; Conery, J.S. The evolutionary fate and consequences of duplicate genes. *Science* **2000**, *290*, 1151–1155. [CrossRef]
58. Chen, C.; Chen, H.; Zhang, Y.; Thomas, R.H.; Frank, H.M.; He, Y.; Xia, R. TBtools: An Integrative Toolkit Developed for Interactive Analyses of Big Biological Data. *Mol. Plant* **2020**, *13*, 1194–1202. [CrossRef]
59. Stelpflug, S.C.; Sekhon, R.S.; Vaillancourt, B.; Hirsch, C.N.; Buell, C.R.; de Leon, N.; Kaeppler, S.M. An expanded maize gene expression atlas based on RNA sequencing and its use to explore root development. *Plant Genome* **2016**, *9*, 1–16. [CrossRef] [PubMed]
60. Jumper, J.; Evans, R.; Pritzel, A.; Green, T.; Figurnov, M.; Ronneberger, O.; Tunyasuvunakool, K.; Bates, R.; Žídek, A.; Potapenko, A. Highly accurate protein structure prediction with AlphaFold. *Nature* **2021**, *596*, 583–589. [CrossRef] [PubMed]

61. Szklarczyk, D.; Gable, A.L.; Nastou, K.C.; Lyon, D.; Kirsch, R.; Pyysalo, S.; Doncheva, N.T.; Legeay, M.; Fang, T.; Bork, P. The STRING database in 2021: Customizable protein–protein networks, and functional characterization of user-uploaded gene/measurement sets. *Nucleic Acids Res.* **2021**, *49*, D605–D612. [CrossRef] [PubMed]
62. Shannon, P.; Markiel, A.; Ozier, O.; Baliga, N.S.; Wang, J.T.; Ramage, D.; Amin, N.; Schwikowski, B.; Ideker, T. Cytoscape: A software environment for integrated models of biomolecular interaction networks. *Genome Res.* **2003**, *13*, 2498–2504. [CrossRef] [PubMed]
63. Consortium, G.O. The gene ontology resource: 20 years and still GOing strong. *Nucl. Acids Res.* **2019**, *47*, D330–D338.
64. Rehman, S.; Jørgensen, B.; Aziz, E.; Batool, R.; Naseer, S.; Rasmussen, S.K. Genome wide identification and comparative analysis of the serpin gene family in brachypodium and barley. *Plants* **2020**, *9*, 1439. [CrossRef] [PubMed]
65. Gaut, B.S. Evolutionary dynamics of grass genomes. *New Phytol.* **2002**, *154*, 15–28. [CrossRef]
66. Initiative, I.B. Genome sequencing and analysis of the model grass *Brachypodium distachyon*. *Nature* **2010**, *463*, 763–768.
67. Ali, A.; Javed, T.; Zaheer, U.; Zhou, J.R.; Huang, M.T.; Fu, H.Y.; Gao, S.J. Genome-Wide Identification and Expression Profiling of the bHLH Transcription Factor Gene Family in *Saccharum spontaneum* Under Bacterial Pathogen Stimuli. *Tropical Plant Biol.* **2021**, *14*, 283–294. [CrossRef]
68. Xu, X.; Yang, Y.; Liu, C.; Sun, Y.; Zhang, T.; Hou, M.; Huang, S.; Yuan, H. The evolutionary history of the sucrose synthase gene family in higher plants. *BMC Plant Biol.* **2019**, *19*, 566. [CrossRef]
69. Fragkostefanakis, S.; Röth, S.; Schleiff, E.; Scharf, K.D. Prospects of engineering thermotolerance in crops through modulation of heat stress transcription factor and heat shock protein networks. *Plant Cell Environ.* **2015**, *38*, 1881–1895. [CrossRef]
70. De Cuyper, C.; Struk, S.; Braem, L.; Gevaert, K.; De Jaeger, G.; Goormachtig, S. Strigolactones, karrikins and beyond. *Plant Cell Environ.* **2017**, *40*, 1691–1703. [CrossRef]
71. Ye, J.; Yang, X.; Hu, G.; Liu, Q.; Li, W.; Zhang, L.; Song, X. Genome-wide investigation of heat shock transcription factor family in wheat (*Triticum aestivum* L.) and possible roles in anther development. *Int. J. Mol. Sci.* **2020**, *21*, 608. [CrossRef]
72. Wu, T.-Y.; Juan, Y.-T.; Hsu, Y.-H.; Wu, S.-H.; Liao, H.-T.; Fung, R.W.; Charng, Y.-Y. Interplay between heat shock proteins HSP101 and HSA32 prolongs heat acclimation memory posttranscriptionally in Arabidopsis. *Plant Physiol.* **2013**, *161*, 2075–2084. [CrossRef]
73. Borkovich, K.; Farrelly, F.; Finkelstein, D.; Taulien, J.; Lindquist, S. hsp82 is an essential protein that is required in higher concentrations for growth of cells at higher temperatures. *Mol. Cell. Biol.* **1989**, *9*, 3919–3930.
74. Gu, L.; Jiang, T.; Zhang, C.; Li, X.; Wang, C.; Zhang, Y.; Li, T.; Dirk, L.M.; Downie, A.B.; Zhao, T. Maize HSFA2 and HSBP2 antagonistically modulate raffinose biosynthesis and heat tolerance in Arabidopsis. *Plant J.* **2019**, *100*, 128–142. [CrossRef] [PubMed]
75. Pulido, P.; Leister, D. Novel DNAJ-related proteins in Arabidopsis thaliana. *New Phytol.* **2018**, *217*, 480–490. [CrossRef]
76. Kong, F.; Deng, Y.; Wang, G.; Wang, J.; Liang, X.; Meng, Q. LeCDJ1, a chloroplast DnaJ protein, facilitates heat tolerance in transgenic tomatoes. *J. Integr. Plant Biol.* **2014**, *56*, 63–74. [CrossRef] [PubMed]
77. Kong, F.; Deng, Y.; Zhou, B.; Wang, G.; Wang, Y.; Meng, Q. A chloroplast-targeted DnaJ protein contributes to maintenance of photosystem II under chilling stress. *J. Exp. Bot.* **2014**, *65*, 143–158. [CrossRef]
78. Mizoi, J.; Shinozaki, K.; Yamaguchi-Shinozaki, K. AP2/ERF family transcription factors in plant abiotic stress responses. *Biochim. Biophys. Acta (BBA) Gene Regul. Mech.* **2012**, *1819*, 86–96. [CrossRef]
79. Suzuki, N.; Bajad, S.; Shuman, J.; Shulaev, V.; Mittler, R. The transcriptional co-activator MBF1c is a key regulator of thermotolerance in Arabidopsis thaliana. *J. Biol. Chem.* **2008**, *283*, 9269–9275. [CrossRef]
80. Jaimes-Miranda, F.; Chávez Montes, R.A. The plant MBF1 protein family: A bridge between stress and transcription. *J. Exp. Bot.* **2020**, *71*, 1782–1791. [CrossRef]
81. Hay, R.T. SUMO: A history of modification. *Mol. Cell* **2005**, *18*, 1–12. [CrossRef] [PubMed]
82. Rytz, T.C.; Miller, M.J.; McLoughlin, F.; Augustine, R.C.; Marshall, R.S.; Juan, Y.-T.; Charng, Y.-Y.; Scalf, M.; Smith, L.M.; Vierstra, R.D. SUMOylome profiling reveals a diverse array of nuclear targets modified by the SUMO ligase SIZ1 during heat stress. *Plant Cell* **2018**, *30*, 1077–1099. [CrossRef] [PubMed]
83. Strable, J.; Scanlon, M.J. Maize (Zea mays): A model organism for basic and applied research in plant biology. *Cold Spring Harb. Protoc.* **2009**, *2009*, emo132. [CrossRef]
84. Schnable, P.S.; Ware, D.; Fulton, R.S.; Stein, J.C.; Wei, F.; Pasternak, S.; Liang, C.; Zhang, J.; Fulton, L.; Graves, T.A. The B73 maize genome: Complexity, diversity, and dynamics. *Science* **2009**, *326*, 1112–1115. [CrossRef]
85. Nagaraju, M.; Sudhakar Reddy, P.; Anil Kumar, S.; Srivastava, R.K.; Kavi Kishor, P.; Rao, D.M. Genome-wide scanning and characterization of Sorghum bicolor L. heat shock transcription factors. *Curr. Genom.* **2015**, *16*, 279–291. [CrossRef]
86. Kaul, S.; Koo, H.L.; Jenkins, J.; Rizzo, M.; Rooney, T.; Tallon, L.J.; Feldblyum, T.; Nierman, W.; Benito, M.-I.; Lin, X. Analysis of the genome sequence of the flowering plant Arabidopsis thaliana. *Nature* **2000**, *408*, 796–815.
87. Young, N.D.; Debellé, F.; Oldroyd, G.E.; Geurts, R.; Cannon, S.B.; Udvardi, M.K.; Benedito, V.A.; Mayer, K.F.; Gouzy, J.; Schoof, H. The Medicago genome provides insight into the evolution of rhizobial symbioses. *Nature* **2011**, *480*, 520–524. [CrossRef] [PubMed]
88. Yu, J.; Hu, S.; Wang, J.; Wong, G.K.-S.; Li, S.; Liu, B.; Deng, Y.; Dai, L.; Zhou, Y.; Zhang, X. A draft sequence of the rice genome (*Oryza sativa* L. ssp. indica). *Science* **2002**, *296*, 79–92. [CrossRef]
89. Raza, A.; Su, W.; Gao, A.; Mehmood, S.S.; Hussain, M.A.; Nie, W.; Lv, Y.; Zou, X.; Zhang, X. Catalase (CAT) Gene Family in Rapeseed (*Brassica napus* L.): Genome-Wide Analysis, Identification, and Expression Pattern in Response to Multiple Hormones and Abiotic Stress Conditions. *Int. J. Mol. Sci.* **2021**, *22*, 4281. [CrossRef] [PubMed]

90. Van de Peer, Y.; Maere, S.; Meyer, A. The evolutionary significance of ancient genome duplications. *Nat. Rev. Genet.* **2009**, *10*, 725–732. [CrossRef]
91. Cannon, S.B.; Mitra, A.; Baumgarten, A.; Young, N.D.; May, G. The roles of segmental and tandem gene duplication in the evolution of large gene families in *Arabidopsis thaliana*. *BMC Plant Biol.* **2004**, *4*, 10. [CrossRef]
92. Liu, Y.; Jiang, H.; Chen, W.; Qian, Y.; Ma, Q.; Cheng, B.; Zhu, S. Genome-wide analysis of the auxin response factor (ARF) gene family in maize (*Zea mays*). *Plant Growth Regul.* **2011**, *63*, 225–234. [CrossRef]
93. Solé, R.V.; Fernandez, P.; Kauffman, S.A. Adaptive walks in a gene network model of morphogenesis: Insights into the Cambrian explosion. *arXiv* **2003**, arXiv:q-bio/0311013.
94. Maere, S.; De Bodt, S.; Raes, J.; Casneuf, T.; Van Montagu, M.; Kuiper, M.; Van de Peer, Y. Modeling gene and genome duplications in eukaryotes. *Proc. Natl. Acad. Sci. USA* **2005**, *102*, 5454–5459. [CrossRef]
95. Jiao, Y.; Wickett, N.J.; Ayyampalayam, S.; Chanderbali, A.S.; Landherr, L.; Ralph, P.E.; Tomsho, L.P.; Hu, Y.; Liang, H.; Soltis, P.S. Ancestral polyploidy in seed plants and angiosperms. *Nature* **2011**, *473*, 97–100. [CrossRef]
96. Rose, A. Intron-mediated regulation of gene expression. *Nucl. Pre-Mrna Process. Plants* **2008**, *326*, 277–290.
97. Liu, H.-C.; Charng, Y.-Y. Common and distinct functions of Arabidopsis class A1 and A2 heat shock factors in diverse abiotic stress responses and development. *Plant Physiol.* **2013**, *163*, 276–290. [CrossRef] [PubMed]
98. Kumar, R.R.; Goswami, S.; Singh, K.; Dubey, K.; Rai, G.K.; Singh, B.; Singh, S.; Grover, M.; Mishra, D.; Kumar, S. Characterization of novel heat-responsive transcription factor (*TaHSFA6e*) gene involved in regulation of heat shock proteins (HSPs)—A key member of heat stress-tolerance network of wheat. *J. Biotechnol.* **2018**, *279*, 1–12. [CrossRef]
99. Yokotani, N.; Ichikawa, T.; Kondou, Y.; Matsui, M.; Hirochika, H.; Iwabuchi, M.; Oda, K. Expression of rice heat stress transcription factor OsHsfA2e enhances tolerance to environmental stresses in transgenic Arabidopsis. *Planta* **2008**, *227*, 957–967. [CrossRef] [PubMed]
100. Jiang, Y.; Zheng, Q.; Chen, L.; Liang, Y.; Wu, J. Ectopic overexpression of maize heat shock transcription factor gene ZmHsf04 confers increased thermo and salt-stress tolerance in transgenic Arabidopsis. *Acta Physiol. Plant.* **2018**, *40*, 9. [CrossRef]
101. Li, G.; Zhang, Y.; Zhang, H.; Zhang, Y.; Zhao, L.; Liu, Z.; Guo, X. Characteristics and regulating role in thermotolerance of the heat shock transcription factor *ZmHsf12* from *Zea mays* L. *J. Plant Biol.* **2019**, *62*, 329–341. [CrossRef]
102. Banti, V.; Mafessoni, F.; Loreti, E.; Alpi, A.; Perata, P. The heat-inducible transcription factor *HsfA2* enhances anoxia tolerance in Arabidopsis. *Plant Physiol.* **2010**, *152*, 1471–1483. [CrossRef] [PubMed]
103. Miller, M.J.; Barrett-Wilt, G.A.; Hua, Z.; Vierstra, R.D. Proteomic analyses identify a diverse array of nuclear processes affected by small ubiquitin-like modifier conjugation in Arabidopsis. *Proc. Natl. Acad. Sci. USA* **2010**, *107*, 16512–16517. [CrossRef] [PubMed]
104. Zhang, S.; Wang, S.; Lv, J.; Liu, Z.; Wang, Y.; Ma, N.; Meng, Q. SUMO E3 ligase SlSIZ1 facilitates heat tolerance in tomato. *Plant Cell Physiol.* **2018**, *59*, 58–71. [CrossRef] [PubMed]

MDPI

Article

Genome-Wide Identification, Characterization, and Expression Analysis of TUBBY Gene Family in Wheat (*Triticum aestivum* L.) under Biotic and Abiotic Stresses

Adil Altaf [1,†], Ahmad Zada [2,†], Shahid Hussain [3], Sadia Gull [4], Yonggang Ding [1], Rongrong Tao [1], Min Zhu [1,5,*] and Xinkai Zhu [1,5,6,*]

1 Jiangsu Key Laboratory of Crop Genetics and Physiology, Jiangsu Key Laboratory of Crop Cultivation and Physiology, Wheat Research Center, College of Agriculture, Yangzhou University, Yangzhou 225009, China; er.altafadil@outlook.com (A.A.); 15949083474@163.com (Y.D.); ap_peach@163.com (R.T.)
2 College of Bioscience and Biotechnology, Yangzhou University, Yangzhou 225009, China; dh19025@yzu.edu.cn
3 College of Agriculture, Yangzhou University, Yangzhou 225009, China; shahidh91@outlook.com
4 College of Horticulture and Plant Protection, Yangzhou University, Yangzhou 225009, China; sadiagull137@yahoo.com
5 Co-Innovation Center for Modern Production Technology of Grain Crops, Yangzhou University, Yangzhou 225009, China
6 Joint International Research Laboratory of Agriculture and Agri-Product Safety, The Ministry of Education of China, Yangzhou University, Yangzhou 225009, China
* Correspondence: minzhu@yzu.edu.cn (M.Z.); xkzhu@yzu.edu.cn (X.Z.)
† These authors contributed equally to this work.

Abstract: The TUBBY gene family is a group of transcription factors found in animals and plants with many functions. *TLP* genes have a significant role in response to different abiotic stresses. However, there is limited knowledge regarding the TUBBY gene family in *T. aestivum*. Here we identified 40 *TaTLP* genes in wheat to reveal their potential function. This study found that TUBBY (*TaTLP*) genes are highly conserved in wheat. The GO analysis of *TaTLP* genes revealed their role in growth and stress responses. Promoter analysis revealed that most *TaTLPs* participate in hormone and abiotic stress responses. The heatmap analysis also showed that *TaTLP* genes showed expression under various hormonal and abiotic stress conditions. Several genes were upregulated under different hormonal and temperature stresses. The qRT-PCR analysis confirmed our hypotheses. The results clearly indicate that various *TaTLP* genes showed high expression under temperature stress conditions. Furthermore, the results showed that *TaTLP* genes are expressed in multiple tissues with different expression patterns. For the first time in wheat, we present a comprehensive *TaTLP* analysis. These findings provide valuable clues for future research about the role of *TLPs* in the abiotic stress process in plants. Overall, the research outcomes can serve as a model for improving wheat quality through genetic engineering.

Keywords: wheat; temperature stress; hormonal stress; genes; *TaTLPs*

Citation: Altaf, A.; Zada, A.; Hussain, S.; Gull, S.; Ding, Y.; Tao, R.; Zhu, M.; Zhu, X. Genome-Wide Identification, Characterization, and Expression Analysis of TUBBY Gene Family in Wheat (*Triticum aestivum* L.) under Biotic and Abiotic Stresses. *Agronomy* **2022**, *12*, 1121. https://doi.org/10.3390/agronomy12051121

Academic Editor: Dilip R. Pathee

Received: 23 March 2022
Accepted: 2 May 2022
Published: 6 May 2022

1. Introduction

Global warming has resulted in significant decreases in crop production over the last few decades [1]. Plants are exposed to numerous environmental stresses that disrupt biochemical and physiological processes [2]. Temperature, heat, drought, and salt stress directly reduce the quality and total yield [3,4]. To overcome annual yield losses in crops such as wheat, it is critical to identify and understand new sources of defense biomarkers. The TUBBY-like proteins are a family of bipartite transcription factors discovered in plants [5–7]. It was possible to trace the TUBBY-like gene family's phylogenetic history back to the earliest stages of eukaryotic evolution after discovering TUBBY-like genes in both single-celled and multicellular eukaryotes [7]. TUBBY-like proteins are distinguished from other proteins by the presence of the conserved C-terminal tubby domain, which is composed of 12 antiparallels closed β-barrel strands with a central hydrophobic α–helix [5]. A conserved N-terminal F-box domain and the C terminal tubby domain are found in the *TLP* family of plants, which is much larger than the *TLP* family found in animals [8]. The function of *TLP* genes was studied in various plants such as *Arabidopsis thaliana, Oryza sativa, Populus deltoides* [9], *Malus domestica* [8], *Zea maize* [10], *Solanum lycopersicum* [11], and cotton [12]. In *A. thaliana,* 11 TUBBY family genes were identified, whereas in *Oryza sativa* 14, *Malus domestica* 15, *Zea maize* 10, *Solanum lycopersicum* 11, and cotton 105 *TLP* genes were previously identified [8–12]. *TLP* shows different expression levels in tissues in plants in response to various environmental and hormonal stresses [7,8,13]. It was found that At*TLP3* and At*TLP9* play an essential role in abscisic acid and osmotic stress [13], whereas At*TLP9* plays a significant role in salt and drought stress [13,14]. Many TUBBY family genes showed up-regulation in *Malus domestica* in response to abiotic stresses, suggesting a substantial role of *TLP* genes in abiotic stresses [8]. Previous observation showed that Ca*TLP1* in *Cicer arietinum* plays a vital role in dehydration stress resistance, and its overexpression in tobacco offers salt and drought stress resistance [15]. Thus, *TLP*s seem to have a significant role in abiotic stress tolerance in plants. However, the function of *TLP*s and their mode of action in plants is an unexplored topic [11].

Wheat is an important crop providing sustenance to 35% of the world's population. However, unpredictable climatic conditions have stagnated wheat production in the past two to three decades. Biotic and abiotic stresses affect the growth of wheat crops and have decreased the plant's output and performance [16]. Wheat crops' evolutionary diversity allows them to adapt to different environmental conditions, although the molecular basis of this adaptation is unknown. Therefore, we were interested in the evolution of the wheat TUBBY family genes and their function in response to abiotic stress. This research aimed to understand wheat TUBBY family genes to improve wheat production, plant quality, and abiotic stress response.

2. Materials and Methods

2.1. Wheat TUBBY Family Genes Identification

We retrieved the protein sequences of TUBBY genes from the Arabidopsis [17] and wheat [18] using the Hidden Markov Model (HMM). The extracted *Triticum aestivum L.* protein sequences were analyzed using CD-search NCBI (http://www.ncbi.nlm.nih.gov/Structure/cdd/wrpsb.cgi, http://www.ebi.ac.uk/interpro/search/sequence/) and SMART (http://smart.embl-heidelberg.de/) (accessed on 7 March 2022) databases. Proteins that do not exhibit the TUBBY domain were excluded. The chemical properties of TaTUBBY proteins were examined using the Expasy online server (http://web.expasy.org/protparam/) (accessed on 7 March 2022). The CELLO2GO [19] online server was used to predict the subcellular location of TaTUBBY genes.

2.2. Phylogenetic Tree, Digital Expression, and Motif Analysis

The Mega (version 7.0) program was used to create the maximum likelihood phylogenetic tree [19]. The conserved motif in the TUBBY gene was predicted using the online MEME server (latest Version 4.12.0) (http://meme-suite.org/tools/meme) (accessed on 8 March 2022). In response to biotic and abiotic stress, the gene expression levels were determined at various stages in all available tissue. The RNA-seq data were retrieved in transcripts per million (TPM) from the expVIP wheat Expression Browser (http://www.wheat-expression.com/) (accessed on 8 March 2022) [20,21]. The abiotic stress was comprised of temperature stress ranging from 20 to 40 °C, and biotic stresses were comprised of abscisic acid (ABA), gibberellic acid (GA), and a combination of *Fusarium graminearum* (FG), ABA, and GA. The ratio of the expression value under treatment to the control was calculated to determine the regulation patterns of a given gene subjected to stress. Ratios greater than or less than 1.0 under a given treatment indicated that the stress treatment had altered gene expression levels. In contrast, a ratio equal to 1.0 showed that the treatment did not affect gene expression levels [20]. The heatmap was created using the Heml 1.0 software tool (http://hemi.biocuckoo.org/faq.php) (accessed on 8 March 2022).

2.3. Chromosomal Location and Protein-Protein Interaction of TUBBY Genes

The chromosomal location of the TUBBY genes was determined using plants from the Ensemble genomes (https://plants.ensembl.org/Triticum_aestivum/Info/Annotation/) (accessed on 9 March 2022) [20]. MAPDraw was also used to map the physical location of TUBBY genes, and nomenclature followed the order in which they appeared on the chromosomes. Analyses of Arabidopsis protein–protein interactions were conducted using the STRING online server (http://string.embl.de) (version. 10) (accessed on 9 March 2022).

2.4. Gene Structure and Conserved Motif Analysis

In the Gene Structure Display server program (http://gsds.cbi.pku.edu.cn/) (accessed on 10 March 2022), genomic and CDS sequences of *TaTLPs* genes were used to create an exon/intron map [22]. The conserved motifs in the TUBBY proteins were discovered using the online server MEME 4.11.3 (http://meme-suite.org/tools/meme) (accessed on 10 March 2022) [23].

2.5. Gene Ontology and Cis-Elements Analysis of TUBBY Family Genes

A 1.5 Kb genomic DNA sequence upstream of each identified *TaTLP* gene's start codon was obtained from the Ensemble Plants database (http://plants.ensemble.org/Triticum_aestivum) (accessed on 11 March 2022) using the Ensemble Plants search engine (ATG). The online Plant CARE (http://bioinformatics.psb.ugent.be/webtools/plantcare/html/) (accessed on 11 March 2022) database was used to identify cis-regulatory elements for all the TUBBY genes. Ontology analysis of the *TaTLP* protein sequences was performed using the Blast2GO program Ver.2.7.2 (http://www.blast2go.com) (accessed on 11 March 2022), and the groups of GO classification (molecular functions, biological process, and cellular component) were documented.

2.6. RNA Isolation and cDNA Synthesis

Total RNA was isolated from stress-exposed seedlings at selected time points, including 0 (control) and stress using the RNeasy plant mini kit (Qiagen, Redwood City, CA, USA), as per the manufacturer's instructions. The quantity and quality of isolated RNA were determined by spectrophotometry (Nanodrop 2000, Thermo Fisher Scientific, Waltham, MA, USA) and formaldehyde-based gel electrophoresis, respectively. For cDNA synthesis,1 µg of total RNA was transcribed in 20 µL using Revert Aid First Strand cDNA Synthesis Kit (Fermentas Life Sciences, Waltham, MA, USA) using oligo (dT)primers as per the manufacturer's instructions.

2.7. Expression Analysis of Different Genes

To examine the temporal expression patterns of selected genes, qRT-PCR was performed. The qRT-PCR was performed in a CFX-96 Real-time PCR Detection 4 System (Bio-Rad, Hercules, CA, USA). Reactions were conducted in a total volume of 20 pl using 50 ng of cDNA, 10 pmol of forward and reverse primers, and 10 L of 2× Sso Fast Eva GreenqPCR Supermix (Bio-Rad, Hercules, CA, USA). The cycling conditions were as per the manufacturer's protocol with a primer-specific annealing temperature. The threshold cycle (Ct) was automatically determined for each reaction using the system set with default parameters. The transcript levels were normalized to the actin transcript, and the fold differences of each amplified product in the samples were calculated using the 2-AACt method.

3. Results

3.1. Identification and Analysis of TaTLPs Genes

In the current study, 40 *TaTLP* proteins from wheat were retrieved using the Ensemble Plants (http://plants.ensemble.org/Triticum_aestivum) (accessed on 11 March 2022) database. The genes were named based on their chromosomal position from *TaTLPq1* to *TaTLP40* (Table 1). Among these, *TaTLP2, TaTLP4, TaTLP5, TALP13, TaTLP16,* and *TaTLP35* genes were located in the *Extracellular* region, *TaTLF14* and *TaTLP24* in the mitochondrial region, and *TaTLP23* was located in the chloroplast, while the remaining 31 *TaTLPs* genes were found in the nucleus (Table 1). More details regarding *TaTLPs* were also recorded, including Locus ID, Proteins, and Molecular weight

Table 1. The gene features of the wheat TUBBY gene family.

Gene Name	Locus ID	Proteins	MW	PI	SL
TaTLP1	Traes_1AL_399C1DBF5	269	-	-	N
TaTLP2	Traes_1AL_45AB9EF34	247	27,640.24	9.22	EC
TaTLP3	Traes_1BL_BDF2FCEC7	175	19,434.98	9.08	N
TaTLP4	Traes_1DL_106460E68	50	5544.55	9.42	EC
TaTLP5	Traes_1DL_9DE76F004	251	27,986.78	9.16	EC
TaTLP6	Traes_1DL_FAB396374	204	22,671.21	9.80	N
TaTLP7	Traes_2AL_436D234EE	471	52,097.90	9.28	N
TaTLP8	Traes_2AS_11876C298	203	22,666.81	9.14	N
TaTLP9	Traes_2BL_181C4AA28	472	52,073.74	9.22	N
TaTLP10	Traes_2BS_9DCC9CC7A	203	22,657.80	9.14	N
TaTLP11	Traes_2DL_79449BF6D	347	38,850.30	9.73	N
TaTLP12	Traes_2DS_AC89AEF36	203	22,661.79	9.14	N
TaTLP13	Traes_3AL_108550E28	126	14,455.70	8.34	EC
TaTLP14	Traes_3AL_731AE8008	60	6858.86	10.08	M
TaTLP15	Traes_3AL_9B4AF9950	152	17,297.97	9.37	N
TaTLP16	Traes_3AL_C3AEDD333	82	9123.34	10.72	EC
TaTLP17	Traes_3B_021E89FE5	377	-	-	N
TaTLP18	Traes_3B_02CE045341	194	22,052.95	9.33	N
TaTLP19	Traes_3B_C967E97B9	180	20,367.29	9.30	N
TaTLP20	Traes_3DL_697C6F117	211	23,933.51	9.69	N
TaTLP21	Traes_3DL_C81B58D98	279	31,811.57	9.63	EC
TaTLP22	Traes_4AL_EDE236978	440	48,793.98	9.39	N
TaTLP23	Traes_4AS_0C8542099	402	44,497.04	9.43	C
TaTLP24	Traes_4BL_7E9BC637F	559	61,782.42	9.73	M
TaTLP25	Traes_4BS_D5B5C14F6	440	48,817.02	9.39	N
TaTLP26	Traes_4DL_7D905B6BC	404	44,726.34	9.29	C
TaTLP27	Traes_4DS_620432A0D	440	48,833.02	9.39	N
TaTLP28	Traes_5AL_83533B97D	361	40,470.04	9.54	N
TaTLP29	Traes_5AL_D38708404	64	7023.98	4.75	N
TaTLP30	Traes_5BL_ECCAFFEB4	440	49,017.86	9.34	N
TaTLP31	Traes_5DL_FA0200E13	439	48,856.72	9.25	N
TaTLP32	Traes_6AL_058D829C3	374	-	-	N
TaTLP33	Traes_6AS_FB1249AB4	321	35,747.87	9.70	N
TaTLP34	Traes_6BL_9F6ACF02D	322	35,555.86	9.39	N
TaTLP35	Traes_6BS_5C281F303	177	20,314.47	9.98	EC
TaTLP36	Traes_6DL_E7A7DAE5C	368	40,756.92	9.26	N
TaTLP37	Traes_6DS_D6AD8C3ED	177	20,355.52	9.98	N
TaTLP38	Traes_7AL_52F3AE87C	373	41,569.32	9.80	N
TaTLP39	Traes_7BL_8312EAB48	441	-	-	N
TaTLP40	Traes_7DL_037F2A4F4	238	26,443.41	9.69	N

CDS: Coding Sequence, MW: Molecular Weight, SL: Sub Cellular Location, EC: Extracellular, N: Nuclear, PM: Plasma membrane, C: Chloroplast.

3.2. Phylogenetic Analysis of TaTLPs

We used the Neighbor-Joining method to construct a phylogenetic tree that included *Triticum aestivum*, *A. thaliana*, and *O. sativa TLPs* to investigate their phylogenetic relationship (Figure 1). The results showed that 40 *TaTLPs*, 15 *OsTLPs*, and 14 *AtTLPs* were clustered and further divided into three families, namely, A, B, and C. Furthermore, Family A was divided into two subfamilies, Family AI and Family AII. Family AI was the largest family containing most *TLPs* including 18 *TaTLPs*, (*TaTLP1, TaTLP6, TaTLP7, TaTLP9, TaTLP11, TaTLP16, TaTLP17, TaTLP20, TaTLP22, TaTLP25, TaTLP27, TaTLP28, TaTLP29, TaTLP30, TaTLP31, TaTLP38, TaTLP39, TaTLP40*), six *OsTLPs*, and three *AtTLPs*. The subfamily AII contained 14 *TLPs* including nine *TaTLPs* (*TaTLP2, TaTLP3, TaTLP4, TaTLP5, TaTLP13, TaTLP14, TaTLP18, TaTLP19, TaTLP21*), four *OsTLPs*, and one *AtTLP*. Family B was the second-largest family, containing 10 *TaTLPs* including (*TaTLP8, TaTLP10, TaTLP12, TaTLP15, TaTLP23, TaTLP24, TaTLP26, TaTLP33, TaTLP35, TaTLP37*), four OsTLPs, and five AtTLPs. Family C was the smallest, containing three *TaTLPs*, including (*TaTLP32, TaTLP34, TaTLP36,*) one *OsTLP*, and two *AtTLPs*. The results confirmed that the evolutionary relationships of *A. thaliana*, *O. sativa*, and *Triticum aestivum* are closer.

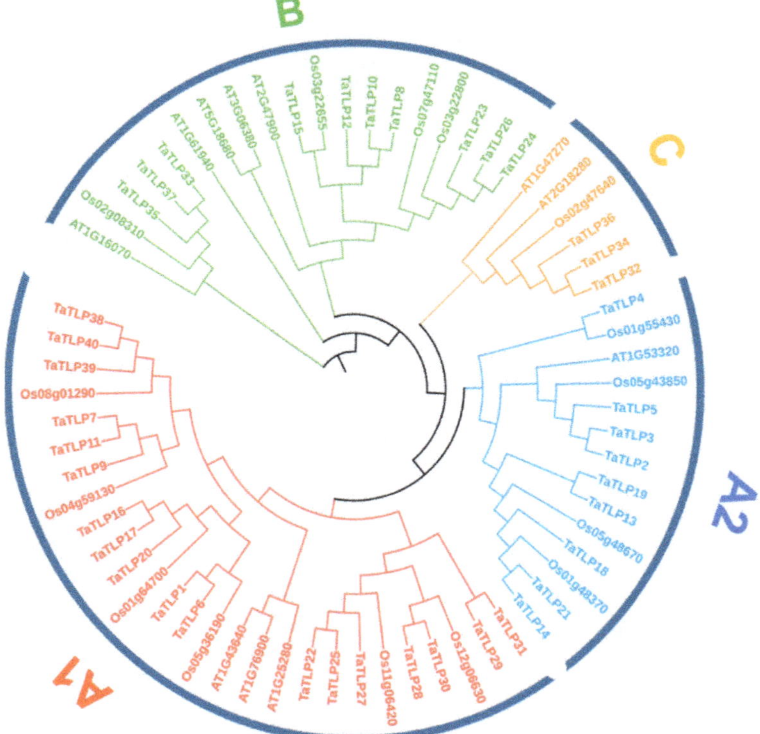

Figure 1. *TLPs* protein phylogeny in three plants: *T. aestivum*, *A. thaliana*, and *O. sativa*. MEGA 7 was used to generate the phylogenetic tree using the following parameters: Bootstrap D 1000 replicates, Neighbor-Joining method, Poisson correction. All group members are divided into four groups, each represented by an assorted color. Different labels are used to identify members of various species.

3.3. Conserved Motif Analysis of TaTLPs-Genes

A total of 10 conserved motifs were discovered using the MEME online server, and they were found to be appropriate for explaining the *TaTLPs'* structure (Figure 2). Among the 40 *TaTLPs* genes, *TaTLP7, TaTLP9, TaTLP22, TaTLP24,* and *TaTLP39* contained more than seven *TLPs* motifs.

Figure 2. The *TaTLPs* genes' conserved motifs.

3.4. Gene Ontology of TaTLP Genes

For the functional prediction of *TaTLP* genes, we used gene ontology (GO) enrichment pathway analysis to identify potential pathways. Three different processes were studied and predicted in their functional outcomes: molecular, biological, and cellular (Figure 3). The molecular prediction suggested that *TaTLP* genes participate in many activities such as signal transduction activity, hydrolase activity, lipid binding, ion binding activity, and cytoskeletal protein binding activity. The biological prediction suggested that *TaTLP* genes participate in various biological processes such as cellular protein modification, signal transduction, vesicle-mediated transport, anatomical structural development, abiotic stresses, cell differentiation, lipid metabolic process, morphogenesis, and embryo development. The cellular prediction suggested that *TaTLP* genes are located in the plasma membrane, intracellular, cytoplasm, nucleus, and extracellular region. Based on the results, it is suggested that *TaTLP* genes play an essential role in plant growth regulation by modulating biological, molecular, and cellular activities.

3.5. Protein-Protein Interaction of TaTLPs

The *TaTLP* protein prediction analysis revealed various other proteins that predictably interact with TaTLP5 (Figures 2 and 4). Thus, TaTLP5 possibly interacts closely with ATG2G20050, which plays a vital role in signal transduction, ATP binding, metal ion binding, and protein serine phosphatase activity, as highlighted by the bit score. The putative bit score of 0.837 closely interacts with our reference gene *TaTLP5*, a member of the TUBBY gene family. Similarly, it interacted with ATG2G35680, Phosphotyrosine protein phosphatase superfamily protein, and Possesses phosphate activity. Furthermore, these genes interact with AT3G12370, AT3G10330, pBRP2, AT2G45910, ENDOL9, ATGRIP, AT2G04940, and MLO1. The details are given in Table 2.

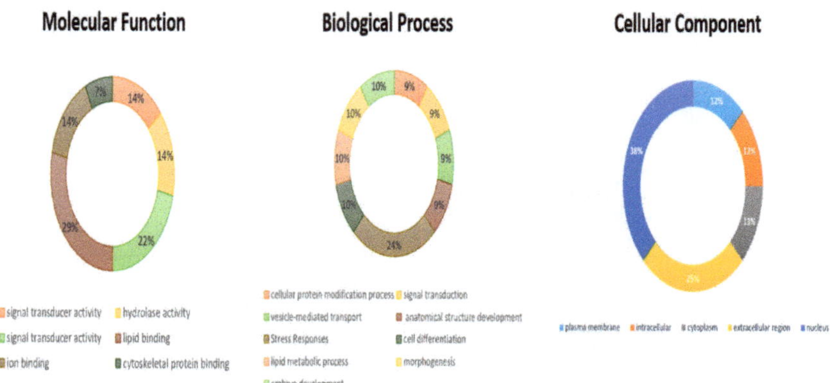

Figure 3. *TaTLP* gene GO enrichment analysis shows molecular function, biological process, and cellular component data.

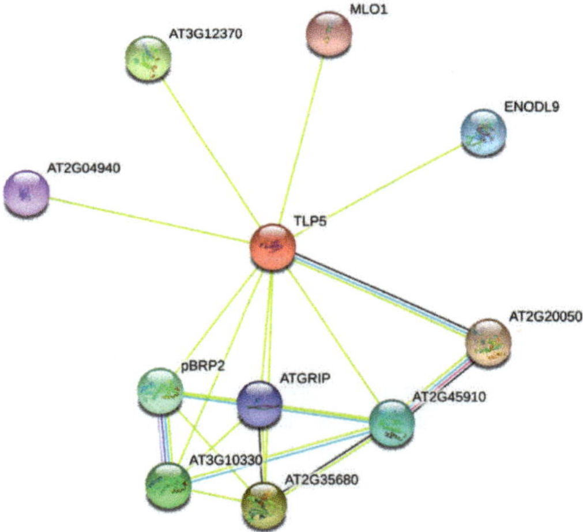

Figure 4. The predicted functional partners of *TaTLP* protein.

Table 2. Different predicted protein families interacted with *TaTLP* genes.

Gene Name	Protein Family	Putative Function	Interactive-Bit Score
ATG2G20050	Protein phosphatase 2C and cyclic nucleotide-binding	Signal transduction, ATP binding, metal ion binding, protein serine phosphatase activity	0.837
ATG2G35680	Phosphotyrosine protein phosphatase superfamily protein	Possess phosphate activity	0.698
AT3G12370	EMB3136—Ribosomal protein L10 family protein	The function is a structural protein	0.691
AT3G10330	Cyclin-like family protein	DNA-templated transcription, initiation, transcription preinitiation complex assembly	0.637
pBRP2	Plant-specific TFIIB-related protein 2	Regulation of endosperm proliferation, DNA-templated transcription, initiation	0.637
AT2G45910	U-box domain-containing protein kinase family protein	Cellular response to oxygen-containing compound, defense response to the bacterium, flower development,	0.633

Table 2. *Cont.*

Gene Name	Protein Family	Putative Function	Interactive-Bit Score
ENDOL9	Early nodulin-like protein 9	electron carrier activity	0.623
ATGRIP	Golgi-localized grip domain-containing protein	Involved in Golgi protein trafficking. AtARL1 binds directly to the GRIP domain of AtGRIP in a GTP-dependent manner.	0.616
AT2G04940	scramblase-like protein	plasma membrane phospholipid scrambling	0.603
MLO1	Transmembrane domain protein	barely mildew resistance	0.588

3.6. *TaTLPs Cis-Regulatory Elements*

According to the results of the in-silico analyses of the *TaTLP*-genes, the upstream region of the *TaTLP* genes contained 13 hormonal, stress, and growth responsive cis-regulatory elements, six of which were responsive to hormones. Seven of these elements were responsive to stress and growth-related changes (Table 3). The hormone-responsive cis-elements were ABRE, which participates in the abscisic acid responsiveness, TCA (cis-acting element involved in salicylic acid responsiveness), TATC-Box (gibberellin-responsive element), AuxRR-Core (auxin-responsive element), CGTCA (cis-acting regulatory element involved in the MeJA-responsiveness), and TGACG (Cis-acting regulatory element involved in the MeJA-responsiveness). The stress and growth responsive cis-elements were ARE (stimulate mRNA decay), ACE (cis-acting element involved in light responsiveness), G-Box (cis-acting element involved in light responsiveness), LTR (Long-terminal repeat), CAT-Box (Cis-acting element involved in meristem development), O2-Site (Cis-acting regulatory element involved in zein metabolism regulation), and MSA-Like (Cis-acting regulatory element involved in the cell cycle). The presence of these cis-elements in the promoter region of *TaTLP* genes indicates that they regulate gene expression in response to various environmental stimuli at various stages of development.

Table 3. Analysis of the diverse types and numbers of cis-acting regulatory elements involved in growth, development, stress, and hormonal response.

Category	Cis-Elements	Annotations
Hormone	ABRE	Cis-acting element involved in the abscisic acid responsiveness.
	TCA	Cis-acting element involved in salicylic acid responsiveness.
	TATC-Box	Gibberellin-responsive element.
	AuxRR-Core	Auxin-responsive element.
	CGTCA	Cis-acting regulatory element involved in the MeJA-responsiveness.
	TGACG	Cis-acting regulatory element involved in the MeJA-responsiveness.
Stress and Growth	ARE	Stimulate mRNA decay.
	ACE	Cis-acting element involved in light responsiveness.
	G-Box	Cis-acting element involved in light responsiveness.
	LTR	Long-terminal repeat.
	CAT-Box	Cis-acting element involved in meristem development.
	O2-Site	Cis-acting regulatory element involved in zein metabolism regulation.
	MSA-Like	Cis-acting regulatory element involved in the cell cycle.

3.7. *Expression Analysis of TaTLP Genes in Different Wheat Cultivars in Response to Fusarium graminum Stress*

The gene expression pattern in different wheat cultivars subjected to *Fusarium graminum* stress was drawn on the heatmap (Figure 5). The *TaTLP7, TaTLP22, TaTLP26, TaTLP27, TaTLP35,* and *TaTLP36* showed dominant expression in all cultivars compared to other *TLPs* genes, whereas *TaTLP9, TaTLP12, TaTLP20,* and *TaTLP39* showed dominant expression in annonng0771 and zhongmai66 cultivars as compared to the control and sumai3. The *TaTLP6* showed low expression levels in annonng0771 and zhongmai66 cultivars compared to the control and sumai3. The *TaTLP6* showed lower expression levels for cultivars annonng0771 and zhongmai66 than the control and sumai3 cultivar, which showed low expression levels. Similarly, *TaTLP15* and *TaTLP30* showed lower expression levels for cultivar zhongmai66, and in comparison, the control, sumai3, and annonng0771 showed higher expression levels. *TaTLP10, TaTLP37,* and *TaTLP24* displayed higher expression levels for annonng0771 and zhongmai66 cultivars, and lower expression levels were recorded for the control and sumai3. The *TaTLP1, and TaTLP2,* showed lower expression levels in sumai3 than the control, annonng0771, and zhongmai66 cultivars. The *TaTLP4* and *TaTLP31* showed high expression levels in the control and sumai3 compared to annonng0771 and zhongmai66 cultivars. Furthermore, *TaTLP8, TaTLP23, TaTLP29, TaTLP3, TaTLP25, TaTLP38, TaTLP13* and *TaTLP28* displayed high expression levels as compared to *TaTLP17, TaTLP18, TaTLP5, TaTLP16, TaTLP33, TaTLP32, TaTLP40, TaTLP19, TaTLP11, TaTLP21, TaTLP14,* and *TaTLP34* which displayed low expression levels, except for *TaTLP19, TaTLP11,* and *TaTLP21* which showed higher expression levels for annong0711 cultivar under *Fusarium graminum* stress.

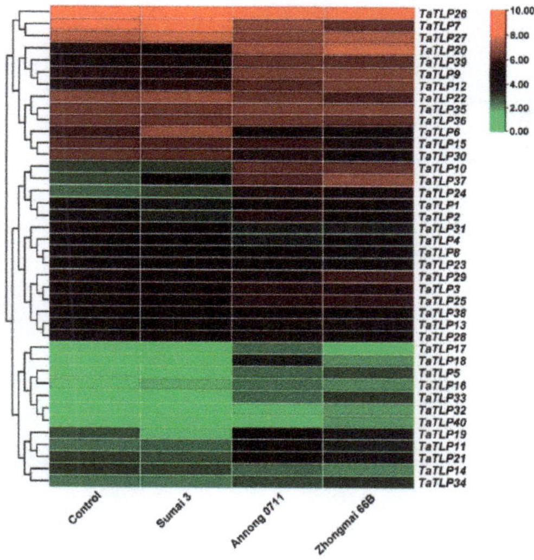

Figure 5. Heatmap representing expression analysis of *TaTLP* genes in different wheat cultivars subjected to *Fusarium graminum* stress.

3.8. Expression Analysis of TaTLP Genes in Wheat in Response to Different Temperatures

The gene expression pattern in response to temperature stress was drawn on the heatmap (Figure 6). *TaTLP*10, *TaTLP*15, *TaTLP*28, *TaTLP*2 *TaTLP*34, *TaTLP*24, *TaTLP*26, *TaTLP*16, *TaTLP*30, *TaTLP7*, *TaTLP*40, *TaTLP*20, *TaTLP*31, *TaTLP*38, *TaTLP*22, *TaTLP*14, and *TaTLP*17 displayed a high expression level under temperatures of 20 and 30 °C, whereas *TaTLP*24 and *TaTLP*22 showed the highest expression levels in comparison to other *TaTLP* genes. Similarly, *TaTLP*25 and *TaTLP*39 showed higher expression levels at 30 °C than under 20 or 40 °C.

The role of *TaTLP* proteins in response to temperature stress is of great interest. To reveal the potential role of *TaTLP* genes in response to temperature stress, we used qRT-PCR to detect *TaTLP* expression levels for 0, 3, and 6 h stress at 40 °C (Figure 7). *TaTLP* genes were found to respond positively to temperature stress. For example, *TaTLP1, TaTLP 2, TaTLP6, TaTLP8,* and *TaTLP15* expression increased at 3 and 6 h temperature stress, indicating that these genes are stress responsive in wheat. Further, *TaTLP3, TaTLP12, TaTLP16,* and *TaTLP32* expression significantly decreased under high-temperature stress. *TaTLP4, TaTLP5, TaTLP10. TaTLP17, TaTLP34,* and *TaTLP36* expression decreased under

3 h temperature stress, and, as temperature stress increased, the expression levels of these genes significantly increased.

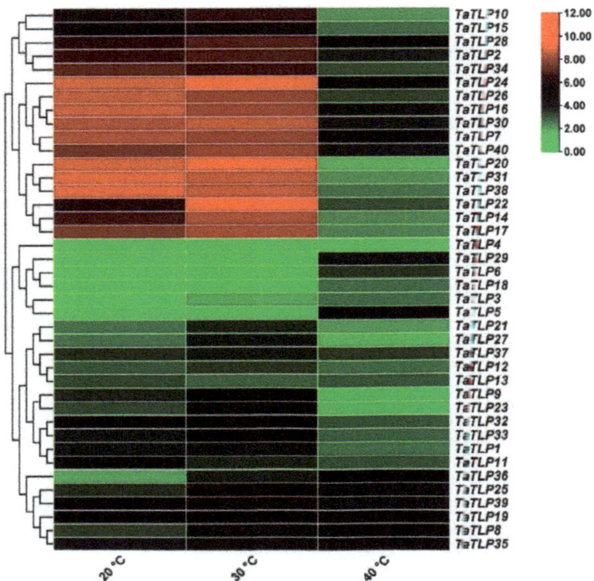

Figure 6. Heatmap representing expression analysis of *TaTLP* genes in wheat subjected to different temperatures.

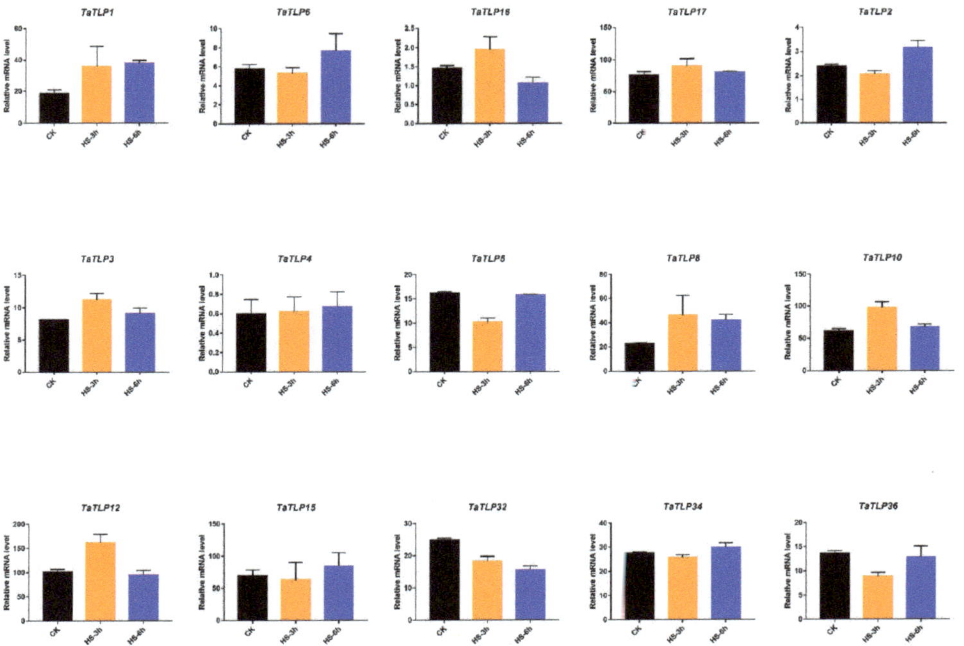

Figure 7. Expression analysis of *TaTLP* genes under high temperature at different time points.

3.9. Expression Analysis of TaTLP Genes in Wheat Subjected to Hormonal Treatment

The heatmap analysis showed different expression patterns under various hormonal treatments. *TaTLP*11, *TaTLP*2, *TaTLP*9, *TaTLP*12, *TaTLP*34, *TaTLP*13, *TaTLP*35, *TaTLP*30, *TaTLP*22, *TaTLP*23, *TaTLP*29, *TaTLP*10, *TaTLP*4, *TaTLP*37 *TaTLP*3 *TaTLP*19, *TaTLP*38, and *TaTLP*17 showed high levels of expression in comparison to other *TATLP* genes under control (CK) and a similar expression patterns were observed for ABA (abscisic acid) treatment, whereas *TaTLP*11, *TaTLP*2, *TaTLP*9, and *TaTLP*34 showed the highest expression level under CK and ABA treatments (Figure 8). Furthermore, *TaTLP*40, *TaTLP*11, *TaTLP*2, *TaTLP*9, *TaTLP*12, *TaTLP*34, *TaTLP*13, *TaTLP*35, *TaTLP*30, *TaTLP*22, *TaTLP*23, *TaTLP*3 *TaTLP*19, *TaTLP*38, *TaTLP*17, *TaTLP*1, and *TaTLP*27 showed a high expression level under GA (gibberellic acid) treatment as compared to other *TLP* genes. The expression pattern of *TaTLP* genes in response to the combination of ABA and FG (*Fusarium graminum*) treatment showed a similar expression pattern as that of GA treatment. The expression pattern of *TaTLP* genes in response to the combination of GA and FG was different; for example, *TaTLP*16, *TaTLP*25, *TaTLP*7, *TaTLP*15, *TaTLP*11, *TaTLP*12, *TaTLP*34, *TaTLP*13, *TaTLP*35, *TaTLP*22, *TaTLP*23, *TaTLP*37, *TaTLP*19, *TaTLP*38, *TaTLP*17, *TaTLP*1, and *TaTLP*27 displayed high expression levels as compared to other *TLP* genes.

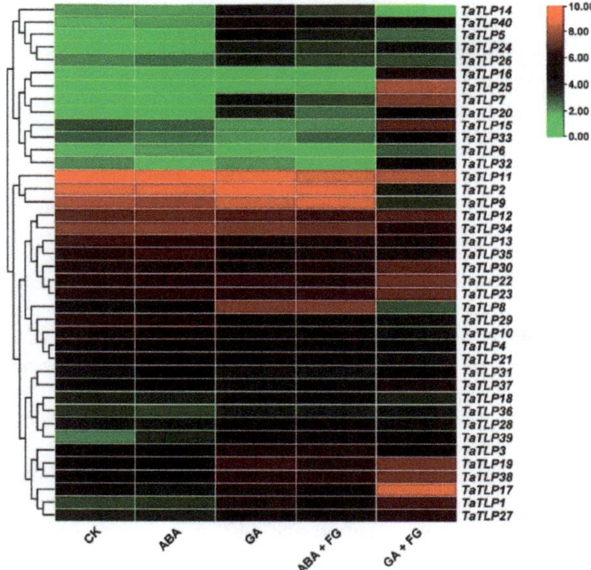

Figure 8. Heatmap represents expression analysis of *TaTLP* genes in wheat subjected to hormonal treatment and hormones + *Fusarium graminum*. ABA (abscisic acid), GA (gibberellic acid), ABA+FG (abscisic acid + *Fusarium graminum*), GA + FG (gibberellic acid + *Fusarium graminum*).

3.10. Expression Analysis of TaTLP Genes in Wheat in Response to Iron Deficiency Stress

The heatmap analysis showed varying expression patterns under iron deficiency conditions in roots and leaf tissues, as shown in Figure 9. *TaTLP7*, *TaTLP17*, *TaTLP24*, *TaTLP16*, *TaTLP22*, *TaTLP2*, *TaTLP7*, *TaTLP710*, *TaTLP7,14*, *TaTLP720*, *TaTLP32*, *TaTLP35*, and *TaTLP39* displayed the highest expressions levels as compared to other *TLP* genes in root-control and root low-Fe conditions, whereas *TaTLP*26 showed high expression as compared to root-control. The remaining *TLP* genes' expression levels were unaffected under root low-Fe conditions. Similarly, *TaTLP7*, *TaTLP17*, *TaTLP24*, *TaTLP16*, *TaTLP22*, *TaTLP10*, *TaTLP14*, and *TaTLP20* showed high expression levels as compared to other *TLPSs*, but similar expression levels between leaf-control and leaf low-Fe conditions. *TaTLP*34 showed a high expression level under the leaf low-Fe condition, whereas *TaTLP*40 and

*TaTLP*1 showed a high expression level in leaf-control compared to the leaf low-Fe condition. The remaining *TLP* gene expression levels were unaffected under leaf low-Fe conditions.

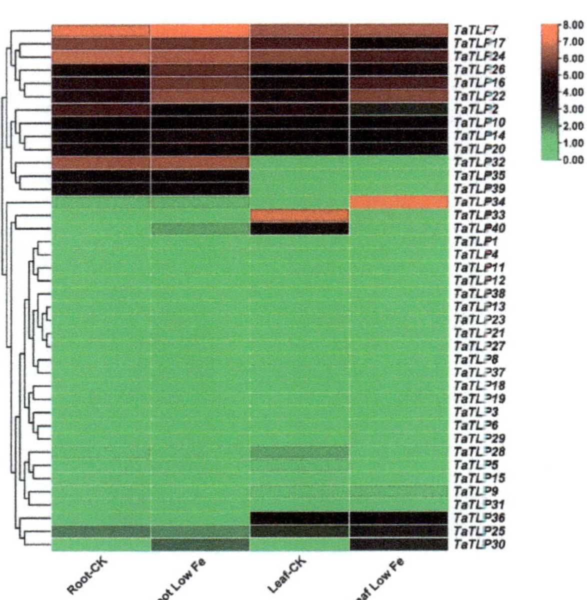

Figure 9. Heatmap representing expression analysis of *TaTLP* genes in wheat under iron deficiency stress.

3.11. Expression Analysis of TaTLP Genes in Different Wheat Tissues

To further study the responses of *TaTLP* genes against biotic and abiotic stresses, we used qRT-PCR to analyze the expression patterns in various wheat tissues (Figure 10). The results showed that *TaTLP*s were expressed in different tissues. The *TaTLP1* transcript level showed high expression in the leaf compared to other tissues. *TaTLP6* and *TaTLP16* showed significantly high expression levels in the root, leaf, and spikelet. *TaTLP17* and *TaTLP6* showed significantly high expression in the stem and leaf, respectively.

Similarly, *TaTLP3* and *TaTLP4* showed significantly high expression levels in leaf, stem, and spikelet tissues. *TaTLP5* was highly expressed in the leaf and stem compared to other tissues, whereas *TaTLP8* expression level was higher in the leaf. *TaTLP10* was expressed in almost all tissues, and the *TaTLP12* expression level was much higher in spikelets than in other tissues. *TaTLP15* and *TaTLP32* were expressed in all tissues, and the expression level was significantly high in the stem and leaf, respectively. *TaTLP34* showed significantly high expression levels in spikelets. *TaTLP36* showed a high expression level in the root, stem, and spikelet. The various *TaTLPs* were expressed in different tissues at varying levels, indicating that they may play a significant role in wheat against various biotic and abiotic stresses.

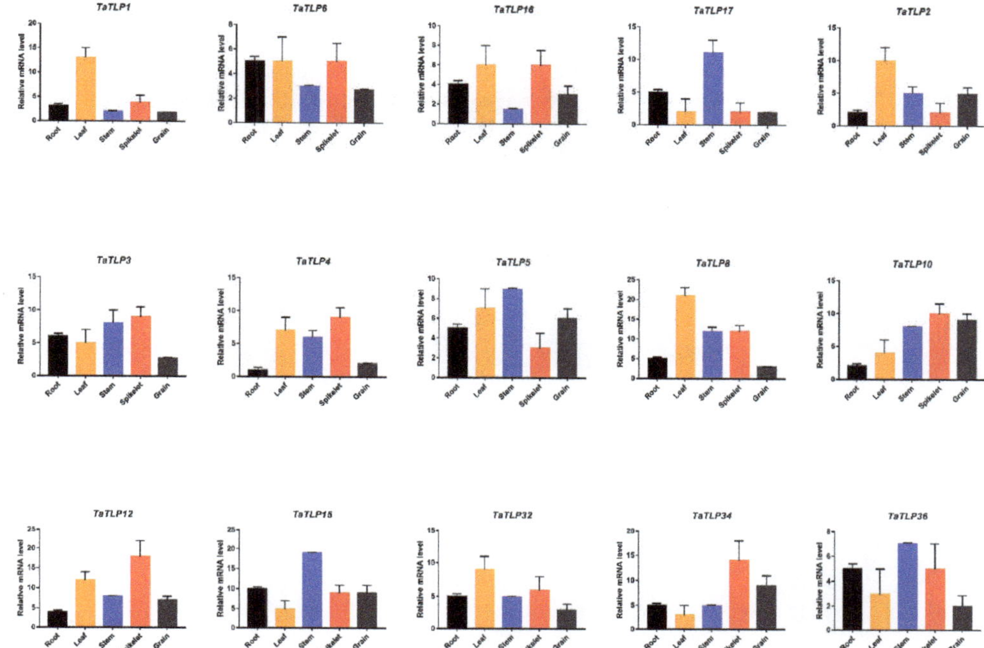

Figure 10. Expression analysis of *TaTLP* genes in different tissues of wheat.

4. Discussion

Plants contend with various environmental conditions throughout their life cycle that may interfere with their development. *TaTLP* genes are members of a gene family found in multiple animals. Plants have a smaller number of functionally studied *TLPs* than animals. *TLPs* have only been discovered in a few plant species, including Arabidopsis [13], rice [7], maize [10], *Solanum lycopersicum* [11], and cotton [12]. This study found 40 genes encoding *TLP* proteins in wheat, which is higher than the numbers found in other plants: 11 in Arabidopsis, 14 in rice, and 15 in maize. This research will advance the knowledge and understanding of their functional characteristics in the future.

4.1. TaTLP Genes Are Distributed Widely in the Wheat Genome

The hexaploid wheat, created by crossing Triticum and Aegilops, is a valuable tool for studying allopolyploidization evolution [24]. An analysis of the phylogenetic tree (Figure 1) shows that the *TaTLPs* are clustered and divided into three large subfamilies: A, B, and C. The A subfamily has two groups: AI and AII. This grouping matches previous *S. lycopersicum* reports [8,10,11]. The *TLPs* within each subfamily share a high degree of homology and have evolved close to one another [25]. Interestingly, we found that *TLPs* possess the F-box domain related to plant stress resistance [13,26,27]. This finding suggests that *TLPs* in wheat are highly conserved and may have additional functions, as demonstrated in Figure 3.

4.2. TaTLP Genes Are Thought to Be Involved in Critical Biological and Molecular Processes

The GO analysis revealed that *TaTLP* genes perform a wide range of biological, molecular, and cellular functions (Figure 3). Many genes are directly involved in cell wall biosynthesis, which is the first line of defense against abiotic and biotic factors [28]. The TUBBY gene family appears to be essential for wheat plant growth in both normal and stressful conditions. Trans-acting elements are required for any biological or molecular process in plants. Multiple signaling pathways regulate plant stress responses, and there is much overlap between the gene expression patterns induced by different stresses [29–31]. Several transcription factors influence the expression of stress-related genes in plants. Several closely related transcription factors can frequently activate or repress genes via cis-acting sequences in response to specific stresses [14,32]. In our study, many hormones (ABRE, TCA, TATC-BOX, AUXRR-Core, CGTCA, and TGACG), stress, and growth-related (ARE, ACE, G-Box, LTR, CAT-Box, O2-Site, MSA-Like) cis-elements were identified in the promoter region of *TaTLP* genes (Table 3). These elements are primarily involved in drought, low-temperature, and hormone responses [33,34].

The CGTCA and TGACG motif were found in nearly all *TLP* promoters, indicating that they were associated with the jasmonate acid response in most cases. Also found in most *TLP* promoters where the enzymes ARE (associated with anaerobic reaction) and ABRE (associated with ABA response) [7,8,13,25]. Based on these results, we suggest that *TaTLP*s may play an essential role in stress responses, but this needs further experimental verification. The PPI analysis demonstrated that *TaTLP*s interact with other essential proteins, such as ATG2G20050, which play an indispensable role in signal transduction, ATP binding, metal ion binding, and protein serine phosphatase activity. Similarly, the *TLP* gene interacted with ATG2G35680, AT3G12370, AT3G10330, pBRP2, AT2G45910, ENDOL9, ATGRIP, AT2G04940, and MLO1 (Table 2 and Figure 4).

Plant *TLP* gene families have been previously studied, and it has been discovered that multiple *TLP* genes are involved in the responses of plants to biological and abiotic stresses [7,8,10–13]. According to this, *TLP* genes can be used as candidate genes in plant resistance breeding.

4.3. TaTLP Genes Control Plant Response to Hormones and Abiotic and Biotic Stresses

To further investigate the response of *TaTLPs* to abiotic stress, the expression patterns of 40 putative *TaTLPs* in wheat were determined using a heatmap analysis and confirmed through qRT-PCR to analyze the expression patterns in various tissues and under temperature stress (Figures 7 and 10). TaTLP genes were induced to varying degrees under multiple conditions, including high temperature, GA, exogenous ABA, and low iron deficiency stress. Due to their immobility, plants face abiotic stresses. Abiotic stresses can significantly reduce crop yields by impeding their physiological and biochemical processes [35,36]. Modern research breakthroughs have relied heavily on understanding the impact of changing climate. The underlying mechanism in systematic temporal variation is complex and challenging to comprehend. Our current results showed that many *TaTLP* genes, such as *TaTLP16, TaTLP20, TaTLP22*, and *TaTLP24*, displayed upregulated expression patterns under different degrees of temperature stress. The findings of this study are consistent with those of previous studies [11,12,25]. GA is a plant hormone involved in seed germination, phase transition, flowering, fruit, and grain development [37–39]. Here, *TaTLP40, TaTLP11, TaTLP2, TaTLP9, TaTLP12, TaTLP34, TaTLP13, TaTLP35, TaTLP30, TaTLP22, TaTLP23, TaTLP3 TaTLP19, TaTLP38, TaTLP17, TaTLP1*, and *TaTLP27* showed higher expression levels under GA treatment (Figure 8). Our findings suggest that genes may be involved in GA-mediated plant growth activities, but more research is needed. Most *TaTLP* genes showed a decrease in expression in response to ABA, FG, and a combination of these factors, in addition to low iron deficiency stress.

5. Conclusions

This study identified and analyzed 40 *TLPs* in wheat (*Triticum aestivum* L.*)*. We performed a comprehensive analysis of *TaTLPs* that included gene identification, phylogenetic analysis, chromosomal location, protein–protein interactions, cis-regulatory elements, and expression analysis. Forty *TaTLPs* were identified and classified into three subfamilies based on their domain and structural characteristics. A heatmap analysis revealed the expression of *TaTLPs* in different cultivars in response to biotic and hormonal stress. The qRT-PCR analysis showed that the expression patterns under high temperature and in various wheat tissues were significantly high, suggesting that these genes may play a role in wheat resistance mediation. Stress regulation is also a complicated mechanism to comprehend. The in-silico analysis provided valuable information for future functional stress biology studies. More research is needed to fully understand the regulation and pathways of the mechanism of *TaTLPs* in wheat.

Author Contributions: Conceptualization, formal analysis, investigation, methodology, software, validation, visualization, writing—original draft, A.A.; investigation, formal analysis, software, validation, writing—review and editing, A.Z.; writing—review and editing, S.H.; writing—review and editing, S.G.; writing—review and editing, Y.D.; writing—review and editing, R.T.; conceptualization, methodology, resources, supervision, writing—review and editing, M.Z.; conceptualization, funding acquisition, methodology, project administration, supervision, writing—review and editing, X.Z. All authors have read and agreed to the published version of the manuscript.

Funding: This work was jointly supported by the earmarked fund for Jiangsu Agricultural Industry Technology System (JATS[2021]503), the Postgraduate Research & Practice Innovation Program of Jiangsu Province (XKYCX20_020), the National Key Research and Development Program of China (2018YFD0200500), the National Natural Science Foundation of China (31901433, 31771711), Jiangsu Modern Agricultural (Wheat) Industry Technology System, Pilot Projects of the Central Cooperative Extension Program for Major Agricultural Technologies, The Priority Academic Program Development of Jiangsu Higher Education Institutions, and The Science and Technology Innovation Team of Yangzhou University, Yangzhou, China.

Institutional Review Board Statement: Not applicable.

Informed Consent Statement: Not applicable.

Data Availability Statement: Not applicable.

Conflicts of Interest: The authors declare no conflict of interest.

References

1. Carmo-Silva, A.E.; Gore, M.A.; Andrade-Sanchez, P.; French, A.N.; Hunsaker, D.J.; Salvucci, M.E. Decreased CO_2 availability and inactivation of Rubisco limit photosynthesis in cotton plants under heat and drought stress in the field. *Environ. Exp. Bot.* **2012**, *83*, 1–11. [CrossRef]
2. Sharif, R.; Xie, C.; Wang, J.; Cao, Z.; Zhang, H.; Chen, P.; Li, Y. Genome-wide identification, characterization and expression analysis of HD-ZIP gene family in Cucumis sativus L. under biotic and various abiotic stresses. *Int. J. Biol. Macromol.* **2020**, *158*, 502–520. [CrossRef] [PubMed]
3. Altaf, A.; Zhu, X.; Zhu, M.; Quan, M.; Irshad, S.; Xu, D.; Aleem, M.; Zhang, X.; Gull, S.; Li, F. Effects of Environmental Stresses (Heat, Salt, Waterlogging) on Grain Yield and Associated Traits of Wheat under Application of Sulfur-Coated Urea. *Agronomy* **2021**, *11*, 2340. [CrossRef]
4. Altaf, A.; Gull, S.; Zhu, X.; Zhu, M.; Rasool, G.; Ibrahim, M.E.H.; Aleem, M.; Uddin, S.; Saeed, A.; Shah, A.Z. Study of the effect of peg-6000 imposed drought stress on wheat (*Triticum aestivum* L.) cultivars using relative water content (RWC) and proline content analysis. *Pak. J. Agric. Sci.* **2021**, *58*, 357–367.
5. Boggon, T.J.; Shan, W.S.; Santagata, S.; Myers, S.C.; Shapiro, L. Implication of tubby proteins as transcription factors by structure-based functional analysis. *Science* **1999**, *286*, 2119–2125. [CrossRef] [PubMed]
6. Carroll, K.; Gomez, C.; Shapiro, L. Tubby proteins: The plot thickens. *Nat. Rev. Mol. Cell Biol.* **2004**, *5*, 55–64. [CrossRef]
7. Liu, Q. Identification of rice TUBBY-like genes and their evolution. *FEBS J.* **2008**, *275*, 163–171. [CrossRef]
8. Xu, J.-N.; Xing, S.-S.; Zhang, Z.-R.; Chen, X.-S.; Wang, X.-Y. Genome-wide identification and expression analysis of the tubby-like protein family in the Malus domestica genome. *Front. Plant Sci.* **2016**, *7*, 1693. [CrossRef]
9. Yang, Z.; Zhou, Y.; Wang, X.; Gu, S.; Yu, J.; Liang, G.; Yan, C.; Xu, C. Genomewide comparative phylogenetic and molecular evolutionary analysis of tubby-like protein family in Arabidopsis, rice, and poplar. *Genomics* **2008**, *92*, 246–253. [CrossRef]
10. Yulong, C.; Wei, D.; Baoming, S.; Yang, Z.; Qing, M. Genome-wide identification and comparative analysis of the TUBBY-like protein gene family in maize. *Genes Genom.* **2016**, *38*, 25–36. [CrossRef]
11. Zhang, Y.; He, X.; Su, D.; Feng, Y.; Zhao, H.; Deng, H.; Liu, M. Comprehensive profiling of tubby-like protein expression uncovers ripening-related *TLP* genes in tomato (*Solanum lycopersicum*). *Int. J. Mol. Sci.* **2020**, *21*, 1000. [CrossRef] [PubMed]
12. Bano, N.; Fakhrah, S.; Mohanty, C.S.; Bag, S.K. Genome-Wide Identification and Evolutionary Analysis of Gossypium Tubby-Like Protein (*TLP*) Gene Family and Expression Analyses During Salt and Drought Stress. *Front. Plant Sci.* **2021**, *12*, 667929. [CrossRef] [PubMed]
13. Lai, C.-P.; Lee, C.-L.; Chen, P.-H.; Wu, S.-H.; Yang, C.-C.; Shaw, J.-F. Molecular analyses of the Arabidopsis TUBBY-like protein gene family. *Plant Physiol.* **2004**, *134*, 1586–1597. [CrossRef] [PubMed]

14. Bao, Y.; Song, W.-M.; Jin, Y.-L.; Jiang, C.-M.; Yang, Y.; Li, B.; Huang, W.-J.; Liu, H.; Zhang, H.-X. Characterization of Arabidopsis Tubby-like proteins and redundant function of At*TLP3* and At*TLP9* in plant response to ABA and osmotic stress. *Plant Mol. Biol.* **2014**, *86*, 471–483. [CrossRef]
15. Wardhan, V.; Jahan, K.; Gupta, S.; Chennareddy, S.; Datta, A.; Chakraborty, S.; Chakraborty, N. Overexpression of Ca*TLP1*, a putative transcription factor in chickpea (*Cicer arietinum* L.), promotes stress tolerance. *Plant Mol. Biol.* **2012**, *79*, 479–493. [CrossRef]
16. Rad, R.N.; Kadir, M.A.; Jaafar, H.Z.; Gement, D. Physiological and biochemical relationship under drought stress in wheat (*Triticum aestivum*). *Afr. J. Biotechnol.* **2012**, *11*, 1574–1578.
17. Lamesch, P.; Berardini, T.Z.; Li, D.; Swarbreck, D.; Wilks, C.; Sasidharan, R.; Muller, R.; Dreher, K.; Alexander, D.L.; Garcia-Hernandez, M. The Arabidopsis Information Resource (TAIR): Improved gene annotation and new tools. *Nucleic Acids Res.* **2012**, *40*, D1202–D1210. [CrossRef]
18. Lukaszewski, A.J.; Alberti, A.; Sharpe, A.; Kilian, A.; Stanca, A.M.; Keller, B.; Clavijo, B.J.; Friebe, B.; Gill, B.; Wulff, B. A chromosome-based draft sequence of the hexaploid bread wheat (*Triticum aestivum*) genome. *Science* **2014**, *345*, 1251788.
19. Kumar, S.; Stecher, G.; Tamura, K. MEGA7: Molecular evolutionary genetics analysis version 7.0 for bigger datasets. *Mol. Biol. Evol.* **2016**, *33*, 1870–1874. [CrossRef]
20. Ma, J.; Yang, Y.; Luo, W.; Yang, C.; Ding, P.; Liu, Y.; Qiao, L.; Chang, Z.; Geng, H.; Wang, P. Genome-wide identification and analysis of the MADS-box gene family in bread wheat (*Triticum aestivum* L.). *PLoS ONE* **2017**, *12*, e0181443. [CrossRef]
21. Borrill, P.; Ramirez-Gonzalez, R.; Uauy, C. expVIP: A customizable RNA-seq data analysis and visualization platform. *Plant Physiol.* **2016**, *170*, 2172–2186. [CrossRef] [PubMed]
22. Guo, A.-Y.; Zhu, Q.-H.; Chen, X.; Luo, J.-C. GSDS: A gene structure display server. *Yi Chuan = Hered.* **2007**, *29*, 1023–1026. [CrossRef]
23. Bailey, T.L.; Boden, M.; Buske, F.A.; Frith, M.; Grant, C.E.; Clementi, L.; Ren, J.; Li, W.W.; Noble, W.S. MEME SUITE: Tools for motif discovery and searching. *Nucleic Acids Res.* **2009**, *37*, W202–W208. [CrossRef] [PubMed]
24. Feldman, M.; Levy, A.A. Genome evolution due to allopolyploidization in wheat. *Genetics* **2012**, *192*, 763–774. [CrossRef]
25. Wang, M.; Xu, Z.; Kong, Y. The tubby-like proteins kingdom in animals and plants. *Gene* **2018**, *642*, 16–25. [CrossRef]
26. Xu, J.; Xing, S.; Sun, Q.; Zhan, C.; Liu, X.; Zhang, S.; Wang, X. The expression of a tubby-like protein from Malus domestica (Md*TLP7*) enhances abiotic stress tolerance in Arabidopsis. *BMC Plant Biol.* **2019**, *19*, 60. [CrossRef]
27. Gagne, J.M.; Downes, B.P.; Shiu, S.-H.; Durski, A.M.; Vierstra, R.D. The F-box subunit of the SCF E3 complex is encoded by a diverse superfamily of genes in Arabidopsis. *Proc. Natl. Acad. Sci. USA* **2002**, *99*, 11519–11524. [CrossRef]
28. Endler, A.; Kesten, C.; Schneider, R.; Zhang, Y.; Ivakov, A.; Froehlich, A.; Funke, N.; Persson, S. A mechanism for sustained cellulose synthesis during salt stress. *Cell* **2015**, *162*, 1353–1364. [CrossRef]
29. Durrant, W.E.; Rowland, O.; Piedras, P.; Hammond-Kosack, K.E.; Jones, J.D. cDNA-AFLP reveals a striking overlap in race-specific resistance and wound response gene expression profiles. *Plant Cell* **2000**, *12*, 963–977. [CrossRef]
30. Schenk, P.M.; Kazan, K.; Wilson, I.; Anderson, J.P.; Richmond, T.; Somerville, S.C.; Manners, J.M. Coordinated plant defense responses in Arabidopsis revealed by microarray analysis. *Proc. Natl. Acad. Sci. USA* **2000**, *97*, 11655–11660. [CrossRef]
31. Chen, W.; Provart, N.J.; Glazebrook, J.; Katagiri, F.; Chang, H.-S.; Eulgem, T.; Mauch, F.; Luan, S.; Zou, G.; Whitham, S.A. Expression profile matrix of Arabidopsis transcription factor genes suggests their putative functions in response to environmental stresses. *Plant Cell* **2002**, *14*, 559–574. [CrossRef] [PubMed]
32. Karam, B.; Rhonda, C.; Luis, O. Transcription factors in plant defense and stress response. *Curr Opin Plant. Biol* **2002**, *5*, 430–436.
33. He, Y.; Li, W.; Lv, J.; Jia, Y.; Wang, M.; Xia, G. Ectopic expression of a wheat MYB transcription factor gene, TaMYB73, improves salinity stress tolerance in Arabidopsis thaliana. *J. Exp. Bot.* **2012**, *63*, 1511–1522. [CrossRef] [PubMed]
34. Li, W.; Cui, X.; Meng, Z.; Huang, X.; Xie, Q.; Wu, H.; Jin, H.; Zhang, D.; Liang, W. Transcriptional regulation of Arabidopsis MIR168a and argonaute1 homeostasis in abscisic acid and abiotic stress responses. *Plant. Physiol.* **2012**, *158*, 1279–1292. [CrossRef] [PubMed]
35. Pan, J.; Sharif, R.; Xu, X.; Chen, X. Waterlogging response mechanisms in plants: Research progress and prospects. *Front. Plant Sci.* **2020**, *11*, 2319.
36. Raza, A.; Tabassum, J.; Kudapa, H.; Varshney, R.K. Can omics deliver temperature resilient ready-to-grow crops? *Crit. Rev. Biotechnol.* **2021**, *41*, 1209–1232. [CrossRef]
37. Gupta, R.; Chakrabarty, S. Gibberellic acid in plant: Still a mystery unresolved. *Plant. Signal. Behav.* **2013**, *8*, e25504. [CrossRef]
38. Hedden, P. The current status of research on gibberellin biosynthesis. *Plant. Cell Physiol.* **2020**, *61*, 1832–1849. [CrossRef]
39. Yu, C.-S.; Cheng, C.-W.; Su, W.C.; Chang, K.-C.; Huang, S.-W.; Hwang, J.-K.; Lu, C.-H. CELLO2GO: A web server for protein subCELlular LOcalization prediction with functional gene ontology annotation. *PLoS ONE* **2014**, *9*, e99368. [CrossRef]

Article

Phenylalanine Ammonia-Lyase (PAL) Genes Family in Wheat (*Triticum aestivum* L.): Genome-Wide Characterization and Expression Profiling

Fatima Rasool [1,2], Muhammad Uzair [2], Muhammad Kashif Naeem [2], Nazia Rehman [2], Amber Afroz [3], Hussain Shah [4] and Muhammad Ramzan Khan [1,2,*]

1 Genome Editing & Sequencing Lab, National Centre for Bioinformatics, Quaid-i-Azam University, Islamabad 45320, Pakistan; fatimarasool014@yahoo.com
2 National Institute for Genomics and Advanced Biotechnology, National Agricultural Research Centre, Park Road, Islamabad 45500, Pakistan; uzairbreeder@gmail.com (M.U.); kashifuaar@gmail.com (M.K.N.); naziarehman96@yahoo.com (N.R.)
3 Department of Biochemistry and Biotechnology, University of Gujrat, Gujrat 50700, Pakistan; dramber.afroz@uog.edu.pk
4 Pakistan Agriculture Research Council, Park Road, Islamabad 44000, Pakistan; hussainshah05@yahoo.com
* Correspondence: mrkhan@parc.gov.pk; Tel.: +92-51-8443705

Citation: Rasool, F.; Uzair, M.; Naeem, M.K.; Rehman, N.; Afroz, A.; Shah, H.; Khan, M.R. Phenylalanine Ammonia-Lyase (PAL) Genes Family in Wheat (*Triticum aestivum* L.): Genome-Wide Characterization and Expression Profiling. *Agronomy* **2021**, *11*, 2511. https://doi.org/10.3390/agronomy11122511

Academic Editors: Channapatna S. Prakash, Ali Raza, Xiling Zou and Daojie Wang

Received: 18 October 2021
Accepted: 20 November 2021
Published: 10 December 2021

Abstract: Phenylalanine ammonia-lyase (PAL) is the first enzyme in the phenylpropanoid pathway and plays a vital role in adoption, growth, and development in plants but in wheat its characterization is still not very clear. Here, we report a genome-wide identification of *TaPAL* genes and analysis of their transcriptional expression, duplication, and phylogeny in wheat. A total of 37 *TaPAL* genes that cluster into three subfamilies have been identified based on phylogenetic analysis. These *TaPAL* genes are distributed on 1A, 1B, 1D, 2A, 2B, 2D, 4A, 5B, 6A, 6B, and 6D chromosomes. Gene structure, conserved domain analysis, and investigation of *cis*-regulatory elements were systematically carried out. Chromosomal rearrangements and gene loss were observed by evolutionary analysis of the orthologs among *Triticum urartu*, *Aegilops tauschii*, and *Triticum aestivum* during the origin of bread wheat. Gene ontology analysis revealed that *PAL* genes play a role in plant growth. We also identified 27 putative miRNAs targeting 37 *TaPAL* genes. The high expression level of *PAL* genes was detected in roots of drought-tolerant genotypes compared to drought-sensitive genotypes. However, very low expressions of *TaPAL10, TaPAL30, TaPAL32, TaPAL3,* and *TaPAL28* were recorded in all wheat genotypes. Arogenate dehydratase interacts with TaPAL29 and has higher expression in roots. The analysis of all identified genes in RNA-seq data showed that they are expressed in roots and shoots under normal and abiotic stress. Our study offers valuable data on the functioning of *PAL* genes in wheat.

Keywords: phenylalanine ammonia-lyase (PAL); phylogenetic analysis; expression profiling; gene structure; drought stress

1. Introduction

Phenylalanine ammonia-lyase (PAL) produces precursors of various secondary metabolites, including lignin, phytoalexin, and phenolic compounds. This gene family is also associated with the production of the first enzyme of the phenylpropanoid pathway [1–3]. *PAL* genes have a molecular mass in the range of 270–330 kilodalton (kDa) and are present in higher plants, yeast, some bacteria, and fungi. However, these genes are not found in animals because they have another histidine ammonia lyase (HAL) [4]. The PAL family encodes a variety of protective compounds such as components of the cell wall, flavonoids, phytoalexins, and furanocoumarin [5,6]. The conversion of L-phenylalanine to cinnamic acid, linking primary metabolism with secondary metabolism catalyzed by the PAL enzymes, also plays an essential role in phenylpropanol biosynthesis, a speed-limiting step in phenylpropanol metabolism [1].This metabolic pathway is involved in the production of various natural products (phytoalexin, hydroxycinnamic acids, flavonoids, etc.), and is also reported as a role player in phenolic glycoside and benzene compound synthesis, which are part of several enzyme-regulated reactions [1,2,7–10]. Thus, phenylpropanoids play a critical role for the growth, development, and survival of vascular plants [1]. *PAL* activity is induced dramatically in reply to various stimuli such as tissue wounding, pathogenic attack, light, low temperature, and hormonal triggers [5,11].

The first plant PAL was found in *Petroselinum crispum* in crystal forms [12]. The *PAL* encoding genes are typically discovered as small gene families comprising one to five members [13,14]. During the evolution of higher plants, PAL diversified into different functions. Both HAL and PAL have different primary protein sequences, but they perform similar functions *in vivo*. It was thought that PAL is formed from HAL when the fungi and plants separated from other kingdoms [15,16]. There are two (the first is horizontal gene transfer (HGT) and the second is gene duplication) methods of evolution are reported. Studies showed that gene duplication is the major method of evolution and gymnosperms are thought to be the ancestors of angiosperms [16,17]. For instance, four *PAL* gene family members in *Arabidopsis thaliana* [18,19], five in *Populus trichocarpa* [20], three in *Scutellaria baicalensis* [21], and three in *Coffeac anephora* [22] have been recognized and functionally described. Nevertheless, some studies have indicated more than five *PAL* genes in certain plants. Moreover, five separate *PAL* genes were recognized in *Pinus taeda* [23]. Furthermore, as many as thirteen *PAL* genes were discovered in *Cucumis sativus* [24], twelve in *Citrullus lanatus* [24], thirteen in *Cucumis melo*, and sixteens in *Vitis vinifera* [25].

Wheat (*Triticum aestivum*) is an important source of starch, protein, and minerals in the diet for more than 35% of the world's inhabitants. It is grown on a variety of soil and in a range of environmental conditions [26]. To prevent environmental stresses, the wheat plant has evolved multiple plant protection systems [27]. Previous studies showed the involvement of the *PAL* gene family in coping with the environmental stresses by activating the transcriptional processes. The *PAL* gene family is responsible for the adaptation and resistance of plants to unfavorable biotic and abiotic environmental conditions. It also controls the expression and inhibition of genes to amend different biochemical pathways. Our research explored, a theoretical way. the functional characterization, and differential expression analysis of the *PAL* gene family engaged in the root development of six different wheat genotypes. This study carries immense importance in understanding the stress tolerance mechanisms in wheat and the role of the *PAL* gene family in the same.

2. Materials and Methods

2.1. Retrieval of Protein Sequences Containing the PAL Gene Family in Triticum aestivum

Two methods were applied to retrieve the phenylalanine ammonia-lyase (PAL) domain-containing sequences in wheat. The first method searched the *PAL* gene family members in the *Triticum aestivum* by inputting the keywords "Phenyl ammonium lyase (PAL)" in the Ensembl plants database (http://plants.ensembl.org/Triticum_aestivum/ (accessed on 16 March 2021) [28], while in the second method, the search for wheat *PAL* genes was conducted using *Arabidopsis thaliana PAL* genes (*At3g53260*, *At2g37040*, *At3g10340*, and *At5g04230*) as reference/query to BLASTP [29] against wheat protein database International Wheat Genome Sequencing Consortium (IWGSC) (V2.1), and *Triticum aestivum* chromosome 3B RELEASE 1.0 (http://wheat-urgi.versailles.inra.fr/, accessed on 16 April 2021). Based on more than 75% sequence identity and E-value $\leq$ 1e-10, the wheat *PAL* gene family was identified. Unique non-redundant wheat PAL gene family members were identified by performing multiple sequence alignments using the ClustalW tool [30], and redundant gene sequences were removed. Further, the Pfam [31] and SMART (http://smart.embl-heidelberg.de/, accessed on 18 May 2021) databases [32] were used for the identification and confirmation of PAL-conserved domains.

2.2. Gene Structure and Conserved Domain Analysis of TaPAL Genes

The online Gene Structure Display Server GSDS 2.0 (http://gsds.gao-lab.org/, accessed on 20 May 2021) [33] was used to examine the gene structure by comparing the open reading frame (ORF) sequence with the corresponding genomic sequences. The conserved motifs of TaPAL protein analysis were determined by MEME Suit (Multiple EM for Motif Elicitation) Version 4.12.0 (http://meme-suite.org/tools/meme, accessed on 25 May 2021) [34] using the following parameters: the number of motifs to be found was ten, and the motif width was kept between 10 and 200; site distribution was set at zero or one occurrence per sequence (thus each sequence was allowed to contain at most one occurrence of each motif). The chromosomal location was drawn on respective chromosomes. The molecular weight (g/mol), isoelectric point, protein charge, and the subcellular location were retrieved from UniProt (https://www.uniprot.org, accessed on 28 May 2021) [35]. The conserved domains of TaPAL proteins were examined using the Unipro UGENE software package [36], which joined the sequences into alignment by the ClustalW algorithm and displayed conservation in the form of color patterns differentiating each amino acid based on physiochemical properties. Protein domain analysis was also performed by using TaPAL1 protein sequence and SMART database containing Pfam domain search option (https://pfam.xfam.org/, accessed on 18 November 2021), and confirmed through the InterPro (https://www.ebi.ac.uk/interpro/, accessed on 18 November 2021) database [37].

2.3. Phylogenetic Identification

To retrieve the protein sequence containing the PAL domain, 37 protein sequences of wheat were used as queries to BLASTP against the *Triticum urartu*, *Solanum tuberosum*, and *Hordeum vulgare*. The protein sequences with more than 70% sequence identity were downloaded from NCBI (https://www.ncbi.nlm.nih.gov/, accessed on 16 June 2021). The *PAL* genes of *Arabidopsis thaliana*, *Zea mays*, and *Oryza sativa* were retrieved from Ensembl (http://plants.ensembl.org, accessed on 18 June 2021). Molecular Evolutionary Genetics Analysis (MEGA version X) [38] was used to infer the evolutionary history of TaPAL by the maximum-likelihood (ML) method and 1000 bootstrap replicates were used. While the gene duplication was calculated by using MCScanX in Tbtools [39].

2.4. Synteny Analysis

The visualization sequence identity and synteny analysis of the *PAL* family genes were performed using Tbtools [40]. These analyses were used to study the sequence similarity patterns [41].

2.5. miRNA Prediction in Wheat PAL Family Genes

miRNA prediction was carried out as previously described [42]. In detail, all the genome sequences of *TaPAL* genes were submitted against the available reference of miRNA sequences using the psRNATarget Server (https://www.zhaolab.org/psRNATarget/, accessed on 14 September 2021) with default setting [43]. While the visualization of interaction was carried out with the help of Cytoscape software (https://cytoscape.org/, accessed on 14 September 2021) by following the default setting [44].

2.6. Promoter and Gene Ontology (GO) Enrichment Analysis

The upstream 1 kb nucleotide sequence from the start codon was retrieved for promoter analysis of all the 37 *TaPAL* genes using the Ensembl Plants database (http://plants.ensembl.org/Triticum_aestivum/, accessed on 19 June 2021). Subsequently, these were also subjected to identification of the already-defined motif by using the PLACE *cis*-regulatory element database [2,45]. These databases also helped to obtain five *cis*-regulatory elements (CACTFTPPCA1, CATTBOX1, ARR1AT, CGCGBOXAT, and WBOXNTERF) and their location. GO analysis of TaPAL protein sequences was conducted by using online tool gProfiler (https://biit.cs.ut.ee/gprofiler/gost, accessed on 20 June 2021) with default parameters [46].

2.7. Protein–Protein Interaction

Protein–protein interactions of wheat were analyzed by using the STRING online server (http://string.embl.de, accessed on 14 September 2021) with the default setting [47].

2.8. Analysis of RNA-Seq Base expression profiling

Six different wheat varieties (Table 1) were used to analyze *PAL* RNA-seq base expression profiling of the *TaPAL* genes. All the wheat varieties were grown under normal conditions at the National Agricultural Research Centre (NARC), Islamabad, Pakistan. The root samples (each data point pooled from eight plants) were collected as previously described [48,49] from 35-day-old seedlings of all six wheat varieties and were frozen in liquid nitrogen prior to storage at −80 °C until use. Total RNA of the above-prepared samples was isolated using the Gene JET™ Plant RNA Purification Mini Kit (Catalog # K0801). Illumina HiSeq2500 platform was used for paired-end (PE) sequencing of wheat RNA samples. The quality of raw data was checked with the help of FastQC (http://www.bioinformatics.bbsrc.ac.uk/projects/fastqc/, accessed on 25 June 2021). Trimming of reads (quality scores < 20) was done with the help of the Trimmomatic tool [50]. The HISAT2 (version 2.0.5) (http://ccb.jhu.edu/software/hisat2/faq.shtml, accessed on 28 June 2021) tool with default settings [51] was used for constructing a transcriptome map based on the genome reference of wheat (ftp://ftp.ensemblgenomes.org/pub/release-25/plants/fasta/triticum_aestivum/dna/, accessed on 4 July 2021). The transcripts were assembled with String Tie software [52], while the NOIseq package was used to find the expression level of genes and transcripts and to draw the graph of the genes. The NOIseq package [53] was used to calculate the FPKM (fragments per kilobase per million) mapped. The genes with FPKM values greater than one were retained for subsequent analyses.

Table 1. Wheat genotypes used for RNA-seq analysis.

Genotypes	Characteristics
Batis	Winter wheat (drought-susceptible) from Germany (INRES)
Blue Silver	Spring wheat (drought-susceptible) was released in Pakistan in 1971 by the NARC
Chakwal-50	Released in 2008 as drought-tolerant, high-yielding, and disease-resistant wheat grown in rain-fed areas of Pakistan by BARI
Local White	Drought-tolerant land race, grown in dryer areas of Pakistan (NARC)
Syn-22	Synthetic wheat (drought-tolerant) from The Netherlands (INRES)
UZ-11-CWA-8	Drought-tolerant line collected from Uzbekistan (INRES)

The expression levels were also analyzed at different stages in root and shoot tissues in response to abiotic stress (drought, heat, combination of both, Supplementary Sheet S1). The RNA-seq data was retrieved in transcripts per million (TPM) from the expVIP [54] wheat expression browser (http://www.wheat-expression.com/, accessed on 6 August 2021). To check the expression patterns of a given *PAL* gene subjected to abiotic stress, the ratio of the expression under treatment to the control was calculated (ratio ≥ 1 = altered under stress; ratio ≤ 1 = un-altered under stress). Finally, a heatmap was constructed by R package pheatmap (version 1.7) [55].

3. Results

3.1. Common Wheat PAL Gene Characterization and Identification

Systematic approaches were used to identify and characterize *TaPAL* genes from the *T. aestivum* genome using various genomic resources and tools. Finally, 37 full-length coding *PAL* genes were identified. The results evidenced that the 37 sequences containing *PAL*-HAL domains belonged to the *PAL* gene family (Supplementary File S1). The detailed information on these identified genes, including gene ID, chromosomal location, start and ending genomic position, protein length, and other related information, are summarized in Table 2. The stability of protein can be checked through the number of amino acids. The peptide length of deduced TaPAL proteins ranged from 498 (*TaPAL37*) to 714 (*TaPAL7*) amino acids with corresponding molecular weights ranging from 52.78 to 77.34 kDa with an average weight 74.97 kDa. Their predicted isoelectric (IP) points varied from 5.76 (*TaPAL35*) to 7.57 (*TaPAL31*), indicating that different TaPAL proteins function in different microenvironments. This IP value was used to measure the net charge on the proteins. The proteins with IP < 7 were considered as acidic and IP > 7 as basic. Thirty-five of the 37 *TaPAL* genes were acidic in nature. In addition, analysis of the subcellular localization of the *T. aestivum* indicated that all 37 PAL transcripts are localized in the cytoplasm (Table 2).

3.2. Localization of TaPAL Genes on the Chromosomes

The predicted *TaPAL* genes were localized on *Triticum aestivum* chromosomes. For this purpose, 37 *TaPAL* genes were mapped to chromosomes of common wheat based on physical positions, as shown in Figure 1. Our results showed that the distribution pattern of the *TaPAL* genes was different on each chromosome. The maximum number (six) of *PAL* genes were present on chromosome 1B and 2B followed by chromosome 2A (five genes), chromosome 2D (four genes), and chromosomes 1A, 1D, 6B, and 6D (three genes each), while chromosome 5B had two genes. The remaining chromosomes (4A and 6A) each contained a single gene. The shortest chromosome was 6D with three genes, while the largest chromosome was 2B containing six genes. Some genes were far away from each other, and some genes were in cluster form, which indicates that these may contains a single QTL. Thus, the chromosomal localization studies revealed an uneven distribution of the 37 candidate genes on all the chromosomes of *T. aestivum* (Figure 1). In nature there are two types of duplication involved in evolution. The first is first tandem duplication (among two or more genes on the same chromosome) and the second is segmental duplication (among different chromosomes and the same clades). *TaPAL25/TaPAL7*, *TaPAL33/TaPAL29*, *TaPAL37/TaPAL18*, *TaPAL25/TaPAL14*, *TaPAL33/TaPAL11*, *TaPAL37/TaPAL27*, *TaPAL18/TaPAL27*, *TaPAL1/TaPAL14*, *TaPAL29/TaPAL11*, *TaPAL34/TaPAL22*, *TaPAL12/TaPAL23*, *TaPAL36/TaPAL35*, *TaPAL34/TaPAL16*, *TaPAL22/TaPAL16*, *TaPAL24/TaPAL3*, *TaPAL24/TaPAL19*, *TaPAL28/TaPAL26*, *TaPAL32/TaPAL30*, and *TaPAL3/TaPAL19* are segmentally duplicated in wheat.

3.3. Identification of Conserved Protein Domains and Motifs in TaPAL Proteins

The MEME server was used to analyze the conserved domains (motifs) within the TaPAL gene family. The MEME program resulted in the identification of 10 conserved motifs of the 37 TaPAL members (Figure 2). The length of the predicted motifs ranged from 40 to 49 amino acids. Motif 9 and motif 10 were primarily present in all genes except the *TaPAL13*, *TaPAL37*, and *TaPAL31* genes. Figure 2B shows that motifs 8, 9, and 10 were not present on the *TaPAL37* gene. Furthermore, motifs 1–7 were conserved in all groups of the phylogenetic tree. The SMART database's result indicated that all *TaPAL* genes contain a well-conserved aromatic lyase domain (PF00221) [56]. This domain (e-value 2.4e-153) starts from 56 aa and ends at 536 aa (Figure 2D and Supplementary File S2).

Table 2. Detailed information of *TaPAL* genes in wheat.

Sr.#	Gene ID	Name	Subcellular Location	AA	MW (g/mol)	IP	Charge	Chrom	Gene Position Start–End (bp)	St
1	TraesCS1B02G048400	*TaPAL1*	Cytoplasm	707	76,625.34	6.2111	−3.5	1B	28,442,914–28,445,758	R
2	TraesCS1B02G048300	*TaPAL2*	Cytoplasm	713	77,090.98	6.0532	−5.0	1B	28,373,087–28,375,944	R
3	TraesCS6B02G258600	*TaPAL3*	Cytoplasm	712	76,561.87	6.2119	−3.0	6B	465,682,950–465,685,480	R
4	TraesCS2A02G381000	*TaPAL4*	Cytoplasm	713	77,163.92	5.9516	−6.0	2A	624,446,955–624,449,699	F
5	TraesCS2D02G204700	*TaPAL5*	Cytoplasm	707	76,132.96	6.2107	−3.0	2D	156,707,284–156,709,989	F
6	TraesCS2B02G398400	*TaPAL6*	Cytoplasm	713	77,233.31	6.5058	0	2B	565,209,774–565,212,553	F
7	TraesCS1B02G048100	*TaPAL7*	Cytoplasm	714	77,341.05	5.9303	−6.5	1B	28,305,997–28,308,847	R
8	TraesCS2B02G224300	*TaPAL8*	Cytoplasm	706	76,220.04	6.3114	−2.0	2B	214,312,779–214,315,434	F
9	TraesCS2D02G377500	*TaPAL9*	Cytoplasm	713	77,188.05	5.9739	−5.5	2D	481,960,998–481,964,109	F
10	TraesCS4A02G401300	*TaPAL10*	Cytoplasm	713	77,087.87	6.1487	−4.0	4A	675,283,030–675,285,827	R
11	TraesCS1D02G039500	*TaPAL11*	Cytoplasm	714	77,180.97	6.3467	−2.0	1D	19,022,361–19,025,014	F
12	TraesCS2A02G380900	*TaPAL12*	Cytoplasm	713	77,261.17	6.1497	−4.0	2A	624,394,141–624,397,057	F
13	TraesCS5B02G468400	*TaPAL13*	Cytoplasm	552	58,944.16	6.5575	0.5	5B	641,833,347–641,835,469	F
14	TraesCS1D02G039400	*TaPAL14*	Cytoplasm	707	76,681.33	5.9509	−6.0	1D	19,012,091–19,020,939	R
15	TraesCS2D02G377600	*TaPAL15*	Cytoplasm	709	76,866.84	6.0796	−4.5	2D	481,988,232–481,990,914	F
16	TraesCS2D02G377200	*TaPAL16*	Cytoplasm	713	77,097.96	6.2412	−3.0	2D	481,599,475–481,602,192	F
17	TraesCS5B02G468300	*TaPAL17*	Cytoplasm	707	76,488.16	6.4644	−0.5	5B	641,822,078–641,824,730	F
18	TraesCS1B02G122800	*TaPAL18*	Cytoplasm	708	76,047.86	6.4086	−1.0	1B	148,413,936–148,416,442	R
19	TraesCS6D02G212500	*TaPAL19*	Cytoplasm	712	76,757.91	5.9498	−6.0	6D	300,896,671–300,900,381	R
20	TraesCS1B02G048200	*TaPAL20*	Cytoplasm	714	77,250.05	6.2977	2.5	1B	28,322,038–28,324,849	R
21	TraesCS2A02G381100	*TaPAL21*	Cytoplasm	713	77,133.01	6.0528	−5.0	2A	624,458,106–624,460,956	F
22	TraesCS2B02G398000	*TaPAL22*	Cytoplasm	713	77,170.99	6.1485	−4.0	2B	564,920,332–564,923,399	F
23	TraesCS2B02G398200	*TaPAL23*	Cytoplasm	713	77,216.16	6.0265	−5.5	2B	565,076,277–565,079,194	F
24	TraesCS6A02G222700	*TaPAL24*	Cytoplasm	712	76,580.83	6.3123	−2.0	6A	416,648,337–416,650,876	F
25	TraesCS1A02G037700	*TaPAL25*	Cytoplasm	707	76,566.32	6.0782	−4.5	1A	20,924,887–20,932,614	R

Table 2. *Cont.*

Sr.#	Gene ID	Name	Subcellular Location	AA	MW (g/mol)	IP	Charge	Chrom	Gene Position Start–End (bp)	St
26	TraesCS6D02G212200	*TaPAL26*	Cytoplasm	704	75,619.36	6.211	−3.0	6D	300,326,934–300,333,023	R
27	TraesCS1D02G103500	*TaPAL27*	Cytoplasm	708	76,233.04	6.211	−3.0	1D	93,102,756–93,105,441	R
28	TraesCS6B02G258400	*TaPAL28*	Cytoplasm	704	75,605.29	6.1061	−4.0	6B	465,148,031–465,154,487	R
29	TraesCS1B02G048500	*TaPAL29*	Cytoplasm	714	77,314.11	6.096	−5.0	1B	28,483,566–28,486,349	F
30	TraesCS6D02G212400	*TaPAL30*	Cytoplasm	714	77,167.39	6.1081	−4.0	6D	300,888,767–300,893,342	R
31	TraesCS2B02G398100	*TaPAL31*	Cytoplasm	520	56,049.19	7.5761	8.5	2B	565,039,719–565,041,383	F
32	TraesCS6B02G258500	*TaPAL32*	Cytoplasm	714	77,108.33	6.2111	−3.0	6B	465,675,009–465,679,798	R
33	TraesCS1A02G037800	*TaPAL33*	Cytoplasm	714	77,228.05	6.3463	−2.0	1A	20,935,423–20,938,081	R
34	TraesCS2A02G380800	*TaPAL34*	Cytoplasm	713	77,015.89	6.2406	−3.0	2A	624,359,166–624,362,152	F
35	TraesCS2B02G224000	*TaPAL35*	Cytoplasm	707	76,207.83	5.7689	−7.0	2B	214,280,956–214,283,508	F
36	TraesCS2A02G196400	*TaPAL36*	Cytoplasm	668	71,935.46	7.4015	7.0	2A	166,254,570–166,256,576	F
37	TraesCS1A02G094900	*TaPAL37*	Cytoplasm	498	52,784.16	6.7764	2.0	1A	90,162,001–90,163,497	R

AA, amino acid; MW, molecular weight (g/mol); IP, isoelectric point; Chrom, chromosome; St, strand; R. reverse strand; F, forward strand.

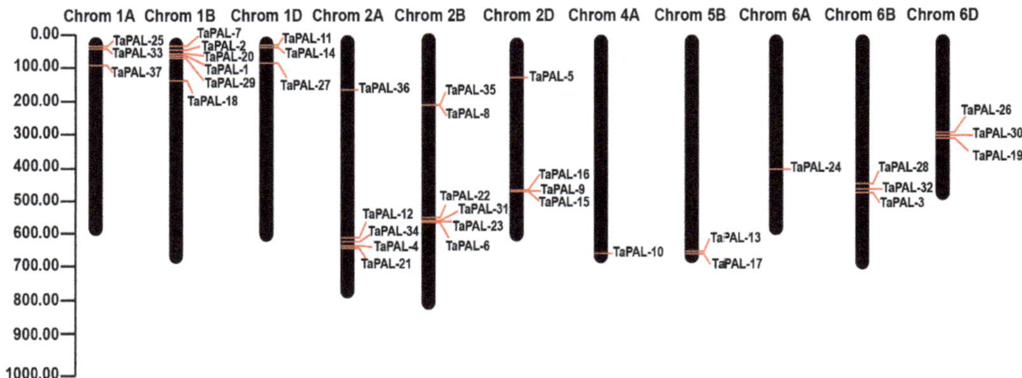

Figure 1. Localization of 37 *TaPAL* genes on wheat chromosomes. Each bar starts from the top, contains the chromosome number, and ends at the bottom. Genes numbers are according to Table 2. The size of the chromosome and the position of *TaPAL* are represented by the vertical scale in megabasepairs (Mb).

3.4. Gene Structure Analysis

For the determination of intron and exon numbers and their positions, all the coding and genomic sequences of *TaPAL* members were aligned. The analysis of *TaPAL* genes illustrated variations in exon–intron structure. Nine *TaPAL* genes (*TaPAL3, TaPAL5, TaPAL8, TaPAL18, TaPAL24, TaPAL27, TaPAL35, TaPAL36*, and *TaPAL37*) contained no introns in their ORFs (Figure 3). The present findings agreed with the previous studies, which reported that seven *CsPAL* genes contained no intron [26]. Parallel to our results, Vogt [5] also mentioned that none of the nine *ClPAL* genes contained introns. The ORFs of the 26 *TaPAL* genes (*TaPAL1-2, TaPAL4, TaPAL6-7, TaPAL9-12, TaPAL14-17, TaPAL20-23, TaPAL26*, and *TaPAL27-34*) were interrupted by a single intron (length ranged from 99 bp to 138 bp), whereas the intronic length of four genes (*TaPAL28, TaPAL32, TaPAL26* and *TaPAL30*) varied from 1055 bp to 1618 bp (Figure 3). Dong et al. [25] reported a similar exon–intron pattern in *AtPAL1* and *AtPAL2* [6], and *NtPALl*. Our results also explained that one additional intron was detected in *TaPAL25* and *TaPAL13*. Finally, the length of exon 2 was highly conserved in all the *TaPAL* genes with one intron.

3.5. Gene Ontology of PAL Genes

For the functional prediction of *PAL*-genes, we conducted GO annotation analysis. In silico functional prediction was carried out and results showed that there were three types of processes involved—biological processes (BPs), molecular processes (MPs), and cellular processes (CPs) (Figure 4). The BPs suggested that *PAL* genes are actively involved in the different metabolic activities and biosynthesis of different organic substances. Furthermore, CPs prediction clarified that almost all the *PAL* genes reside in the cytoplasm and could be involved in regulation of metabolic processes. Meanwhile, MPs showed that *PAL* genes have the enzymatic ability. Such findings clearly indicate that *PAL* genes play role in plant growth by modulating the BPs, MPs, and CPs.

3.6. MicroRNA-Targeting TaPAL Genes

We discovered 27 putative miRNAs targeting 37 *TaPAL* genes to create an interaction network using Cytoscape software to better understand the underlying regulatory mechanism of miRNAs involved in the regulation of *PALs* (Figure 5 and Supplementary Sheet S4). In the connection distribution and regulation network, we found that *TaPAL26* is one of the most-targeted *PAL* genes of wheat. The tae-miR1119 targets the wheat genes *TaPAL7, TaPAL32, TaPAL24, TaPAL3, TaPAL19, TaPAL30, TaPAL26, TaPAL27, TaPAL18, TaPAL9,*

TaPAL22, *TaPAL23*, *TaPAL21*, *TaPAL6*, *TaPAL16*, *TaPAL15*, *TaPAL33*, *TaPAL31*, and *TaPAL4*. Our results also indicated that miRNA tae-miR9781 target *TaPAL22* and *TaPAL34*. Both these genes have low expression in shoots. Furthermore, the miRNAs tae-miR1119, tae-miR398, tae-miR444a, tae-miR444b, and tae-miR9664-3p targeting *TaPAL29* have high expression in root tissues.

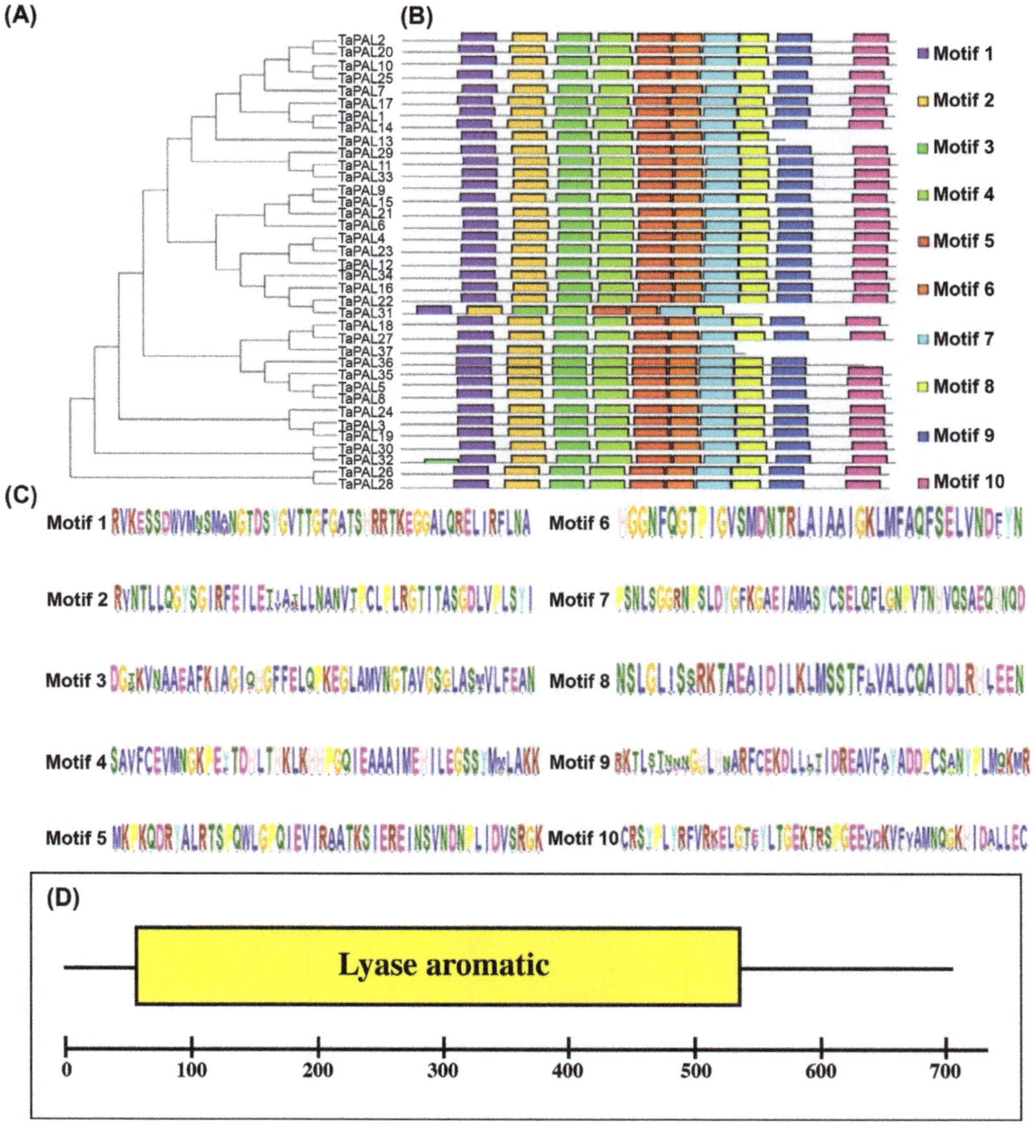

Figure 2. Phylogenetic tree and conserved domain analysis of *TaPAL* genes. (**A**) Thirty-seven TaPAL proteins sequences were used for the construction of phylogenetic tree, and 1000 replicates were also used for bootstrap test. (**B**) Ten conserve motifs with different lengths are shown. Motifs codes are presented at the right of the figure, with different colors. (**C**) Sequence logos of these motifs are presented. (**D**) Protein domain analysis showing the Pfam lyase aromatic domain.

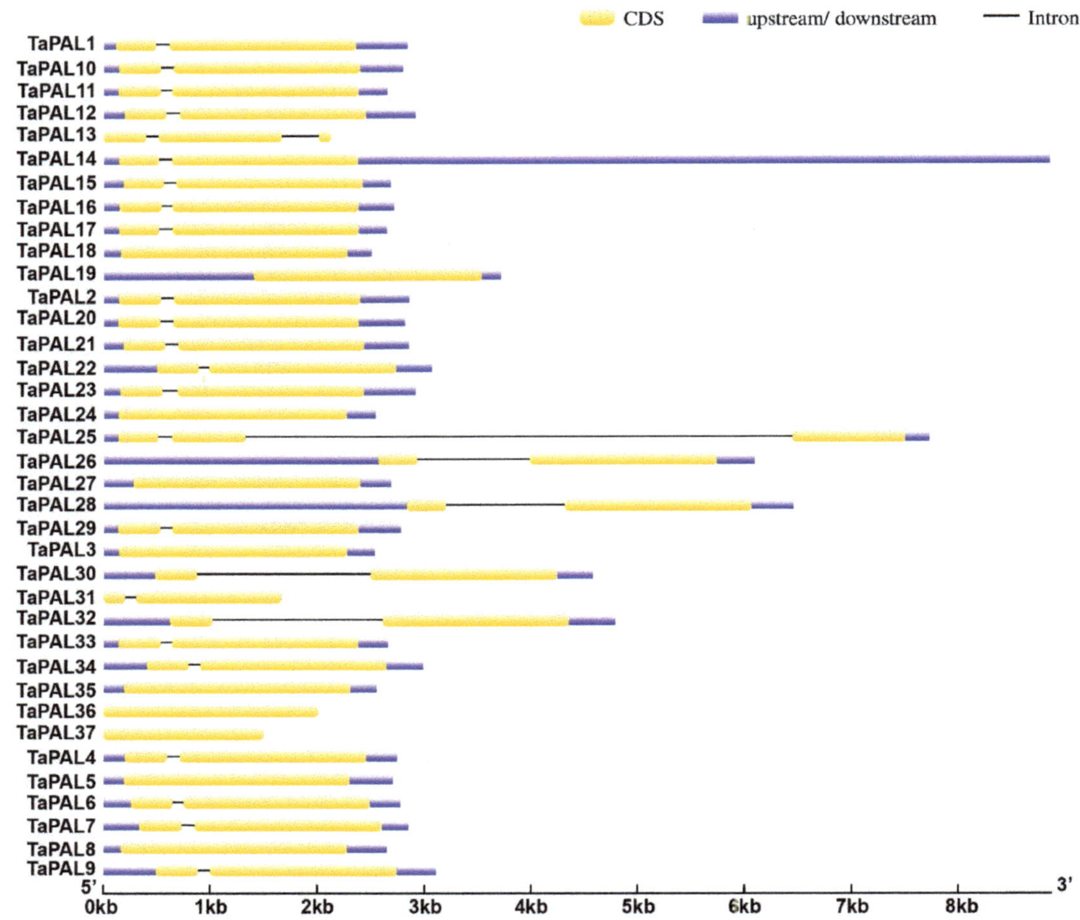

Figure 3. Gene structure analysis. Exon and intron arrangement-based gene structure of *TaPAL* genes.

3.7. Promoter Analysis

The promoter sequence is known as a regulatory element that controls gene expression and regulation [7–9]. The promoters are also called *cis*-acting regulatory DNA elements. Their location can be retrieved from the PLACE database (Table 3). Three regulatory elements, TCA-element, CGTAC-motif, and ABRE (abscisic acid or ABA responses), were identified for *TaPAL* genes. The TCA-element, CGTAC-motif, and ABRE-motifs are associated with SA responses, MeJA, and ABA, respectively.

Additionally, the expression of the *TaPAL* gene family is closely related to light, which was confirmed by the presence of MRE light-responsive element, G-Box, GT1-motif, AE-box, ATC-motif, C-box, CAG-motif, I-box, Sp1, Box 4, and ACE on some member *TaPAL* gene families. The putative TATA box was present on the upstream sequences from the start codon ATG on all *TaPAL* genes. Moreover, *TaPAL* gene promoters also contained several phytohormone-responsive elements, including ABRE, AuxRE (auxin-response elements), and GARE (gibberellin (GA) responses). The promoter of *TaPAL* genes also contained MBS (drought induction) and LTR repetitive sequences (cold stress) related to stress-response regulatory elements. *TaPAL21*, *TaPAL35*, *TaPAL8*, *TaPAL5*, *TaPAL16*, *TaPAL15*, *TaPAL28*, *TaPAL32*, *TaPAL26*, *TaPAL25*, *TaPAL18*, *TaPAL27*, *TaPAL36*, *TaPAL34*, and *TaPAL12* contained a single copy of MBS *cis*-regulatory element, while *TaPAL22* contained two copies of MBS *cis*-regulatory element. It was also observed that only *TaPAL37*, *TaPAL29*, *TaPAL34*, *TaPAL35*, *TaPAL31*, *TaPAL24*, *TaPAL28*, *TaPAL3*, and *TaPAL19* contained the LTR *cis*-regulatory element. Additionally, the upstream regulatory sequences of the *TaPAL34* and *TaPAL16* genes contained the TC-rich repeat, which is related to the defense mechanism. These results suggested that the *TaPAL* gene family members may play an important role in the survival of plants under various environmental stresses. These *cis*-regulatory elements (promoters) receive stimuli from the environment via complex mechanisms and induce gene expression and regulation in response to various abiotic and biological stresses.

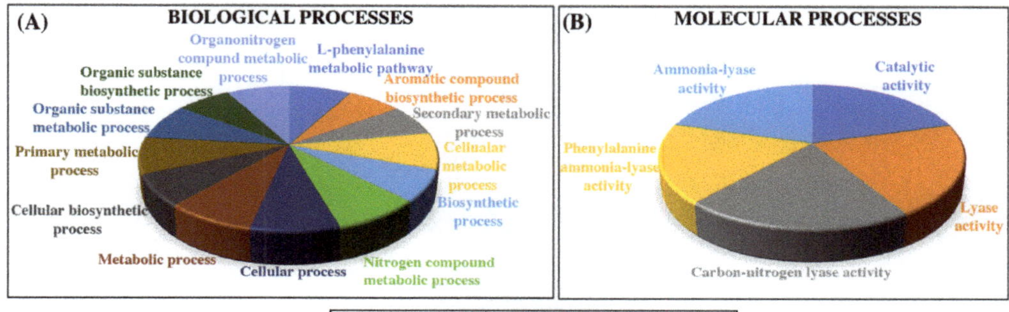

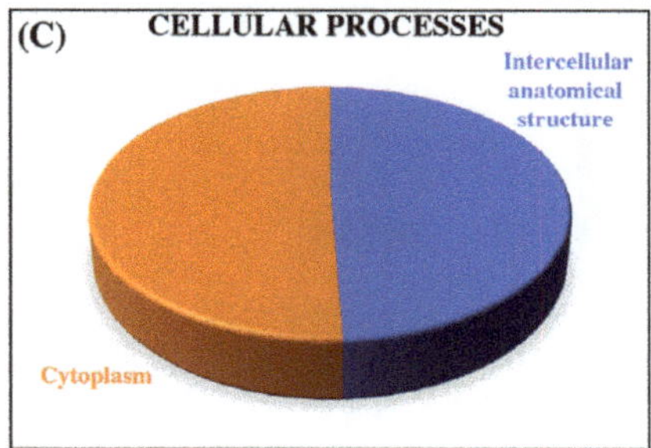

Figure 4. GO analysis of *PAL* genes. The data show (**A**) biological processes, (**B**) molecular processes, and (**C**) cellular processes.

3.8. Protein–Protein Interaction of TaPAL

The TaPAL protein predicted analysis showed an array of other proteins which co-regulate with TaPAL29 (Traes_1BS_BD86C90A7.1) (Figure 6). Arogenate dehydratase (Traes_5BL_7B0ED7548.1), which is a key enzyme involved in synthesis of L-phenylalanine from L-arogenate, showed interaction with our reference gene. The bit-score of 0.895 showed that optimum interaction with our reference gene TaPAL29, which is a member of the *TaPAL* genes, while rest of the gene was uncharacterized.

3.9. Phylogenetic Analysis of the PAL Gene Family

Of the 37 *TaPAL* genes identified in this study, four *PAL* genes from *Arabidopsis thaliana*, nine *PAL* genes from *Oryza sativa*, and eight *PAL* genes from *Zea mays* were used to construct a maximum-likelihood-approach tree using MEGA X to determine the evolutionary relationships (Figure 7). The resultant phylogenetic tree based on protein sequence similarities divided PAL proteins into four major clades or groups represented in different colors. The first three groups represent the monocots, while the fourth group shows the dicots. Overall group I exhibited 30 *TaPAL* genes (*TaPAL35-37*, *TaPAL1-2*, *TaPAL4-18*, *TaPAL20-23*, *TaPAL25*, *TaPAL27*, *TaPAL29*, TaPAL31, and *TaPAL33-37*) and one *PAL* gene from each rice and maize. Group II possessed five *TaPAL* genes (*TaPAL3*, *TaPAL19*, *TaPAL24*, *TaPAL30*, and *TaPAL32*) that were found to be more closely associated with the genes of *Z. mays* genes as compared to *O. sativa* genes. Moreover, group III illustrated two *TaPAL* genes (*TaPAL28* and *TaPAL26*) that were closely associated with *PAL* genes of *O. sativa* (*OsPAL1*, *OsPAL5*, and *OsPAL6*) versus that of *Z. mays* (*ZmPAL1*). The *AtPALs* genes which are dicots and made a separate group IV.

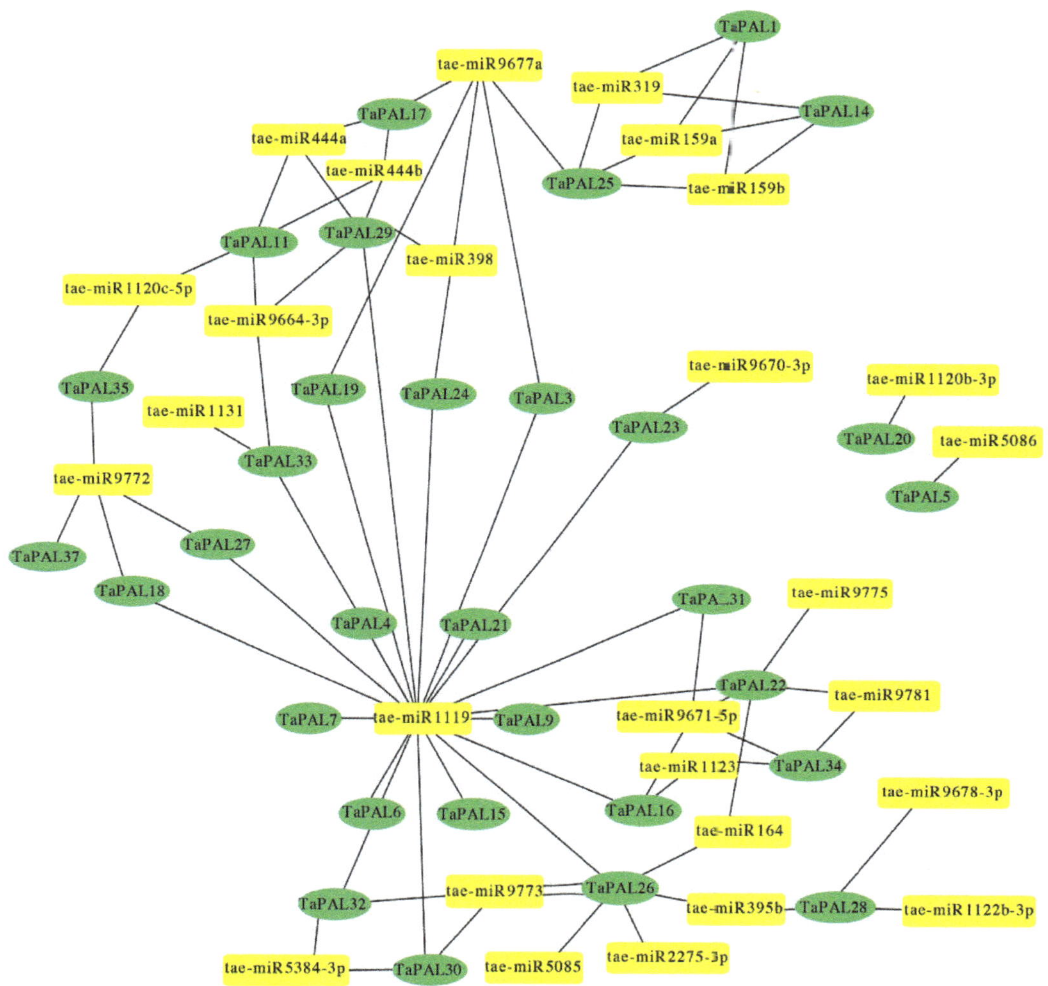

Figure 5. Regulatory network relationship between the miRNA and their targeted *TaPAL* genes.

Table 3. *Cis*-regulatory elements involved in plant growth regulation, stress, and hormonal responses.

	Site Name	Functions
Hormone	ABRE	*cis*-acting element involved in abscisic acid responsiveness
	ACE	*cis*-acting element involved in light responsiveness
	CCAAT-box	MYBHv1 binding site
	CGTCA-motif	*cis*-acting regulatory element involved in MeJA-responsiveness
	GARE-motif	Gibberellin-responsive element
	GC-motif	Enhancer-like element involved in anoxic-specific inducibility
	P-box	Gibberellin-responsive element and part of a light-responsive element
	TCA-element	*cis*-acting element involved in salicylic acid responsiveness
	TGA-element	Auxin-responsive element
	TGACG-motif	*cis*-acting regulatory element involved in MeJA-responsiveness

Table 3. *Cont.*

	Site Name	Functions
Stress and Growth	A-box	*cis*-acting regulatory element
	AE-box	Part of a module for light response
	Box 4	Part of a conserved DNA module involved in light responsiveness
	ARE	*cis*-acting regulatory element essential for anaerobic induction
	ATC-motif	Part of a conserved DNA module involved in light responsiveness
	ATCT-motif	Part of a conserved DNA module involved in light responsiveness
	C-box	*cis*-acting regulatory element involved in light responsiveness
	CAAT-box	Common *cis*-acting element in promoter and enhancer regions
	CAG-motif	Part of a light response element
	CAT-box	*cis*-acting regulatory element related to meristem expression
	chs-CMA2a	Part of a light responsive element
	chs-Unit 1 ml	Part of a light responsive element
	Circadian	*cis*-acting regulatory element involved in circadian control
	G-box	cis-acting regulatory element involved in light responsiveness
	GATA-motif	Part of a light-responsive element
	GCN4-motif	cis-regulatory element involved in endosperm expression
	GT1-motif	Light-responsive element
	GTGGC-motif	Part of a light-responsive element
	I-box	Part of a light-responsive element
	LTR	*cis*-acting element involved in low-temperature responsiveness
	MBS	MYB binding site involved in drought-inducibility
	O2-site	*cis*-acting regulatory element involved in zein metabolism regulation
	Sp1	Light-responsive element
	TATA-box	Core promoter element around -30 of transcription start
	TC-rich repeats	*cis*-acting element involved in defense and stress responsiveness
	TCT-motif	Part of a light responsive element

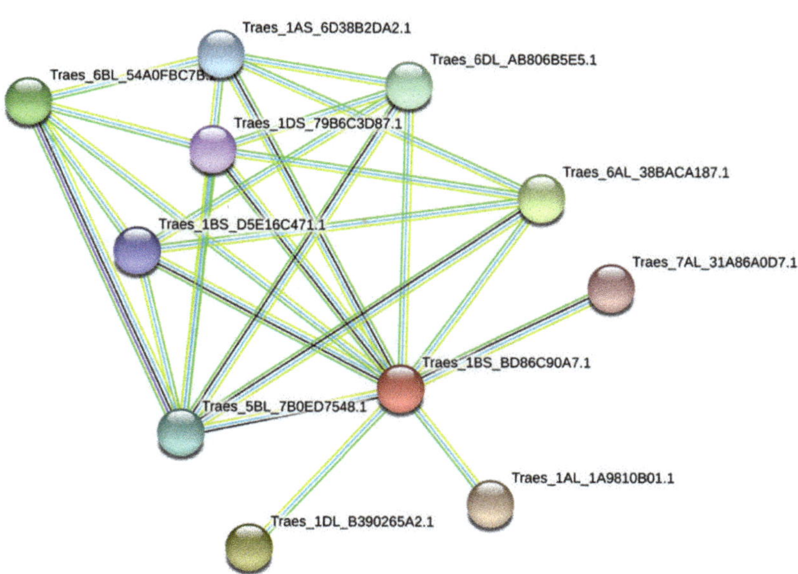

Figure 6. The predicted functional partners of TaPAL29.

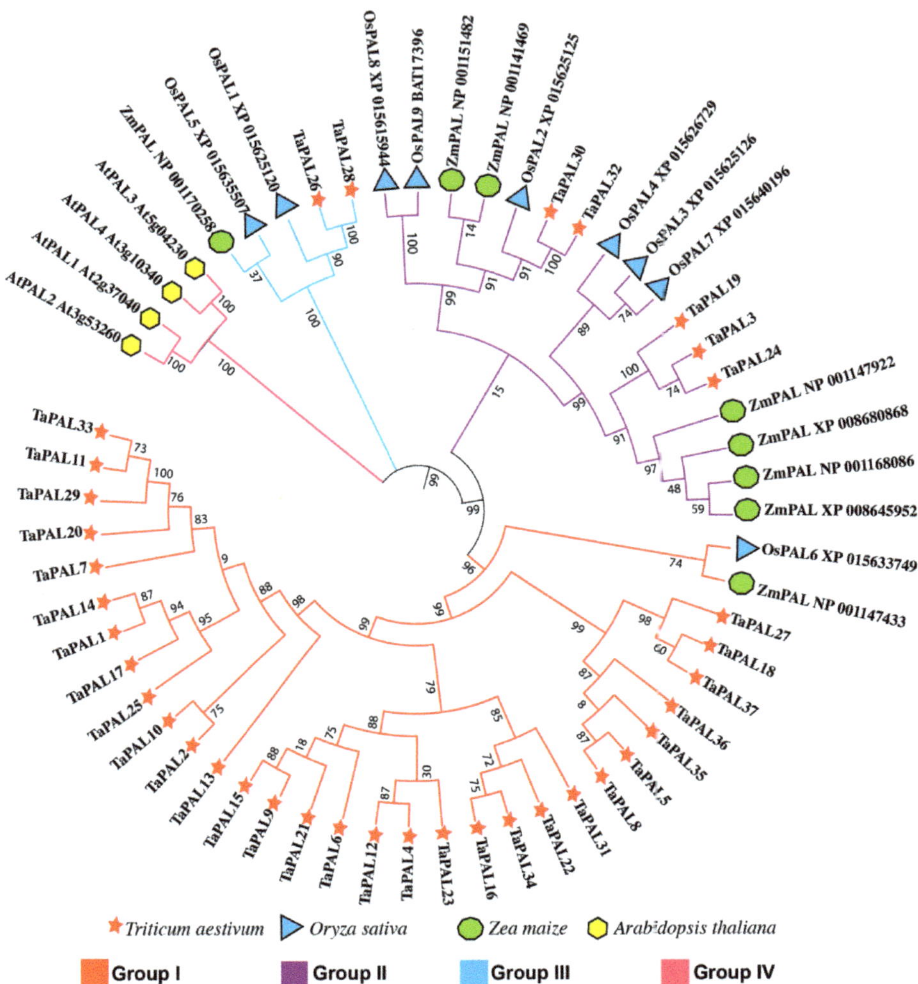

Figure 7. Comparative phylogenetic tree of PAL genes between *Triticum aestivum* (*Ta*), *Oryza sativa* (*Os*), *Zea maize* (*Zm*), and *Arabidopsis thaliana* (*At*). One thousand replicates were used for bootstrap test and the percentage of replication is presented next to the branches.

To investigate the ancestral relationship of the *PAL* gene family in *T. aestivum* with its ancestral species, the phylogenetic analysis also showed all the *PAL* genes from *Hordeum vulgare*, *Solanum tuberosum*, and *Triticum urartu*. *Hordeum vulgare* was domesticated from its wild relative, *Hordeum spontaneum*, while *Triticum urartu* is the progenitor of tetraploid *Triticum turgidum* and hexaploid *Triticum aestivum*. The ancestral plants had 8, 10, and 11 *PAL* genes, respectively. Common wheat *PAL* genes showed maximum association with *HvPAL* (*H. vulgare*), followed by *TuPAL* (*T. urartu*), as shown in Supplementary Figure S1.

To study the origin and evolutionary relationship of *Triticum aestivum* (*tr*), *Aegilops tauschii* (*ae*), *Triticum turgidum* (*tg*), and *Triticum dicocoides* (*td*) PAL protein sequences, a comparative synteny analysis was conducted (Figure 8 and Supplementary sheet S2). The proteins from four species were closely associated and showed higher similarity in evolutionary correlation analysis. It was noted that *TaPAL* genes on chromosome trchr6D have some evolutionary origins in common wheat with genes on chromosomes td6B, td6A, and ae6D. We identified that 10 genes of *Aegilops tauschii* are duplicated with *TaPAL33*,

TaPAL37, *TaPAL27*, *TaPAL34*, *TaPAL36*, *TaPAL22*, *TaPAL23*, *TaPAL35*, *TaPAL26*, and *TaPAL30*. Sixteen genes of *Triticum dicocoides* are orthologs with *TaPAL* genes of wheat, and *TaPAL26* is twice duplicated in *PAL* genes of *Triticum dicocoides*. Nineteen orthologous gene pairs of *Triticum turgidum* and wheat were identified. More than two of the orthologous gene pairs of *TaPAL 37*, *TaPAL36*, *TaPAL35*, *TaPAL34*, and *TaPAL26* were identified in *Triticum turgidum*. Seven paralogous pairs of *TaPAL* genes were identified.

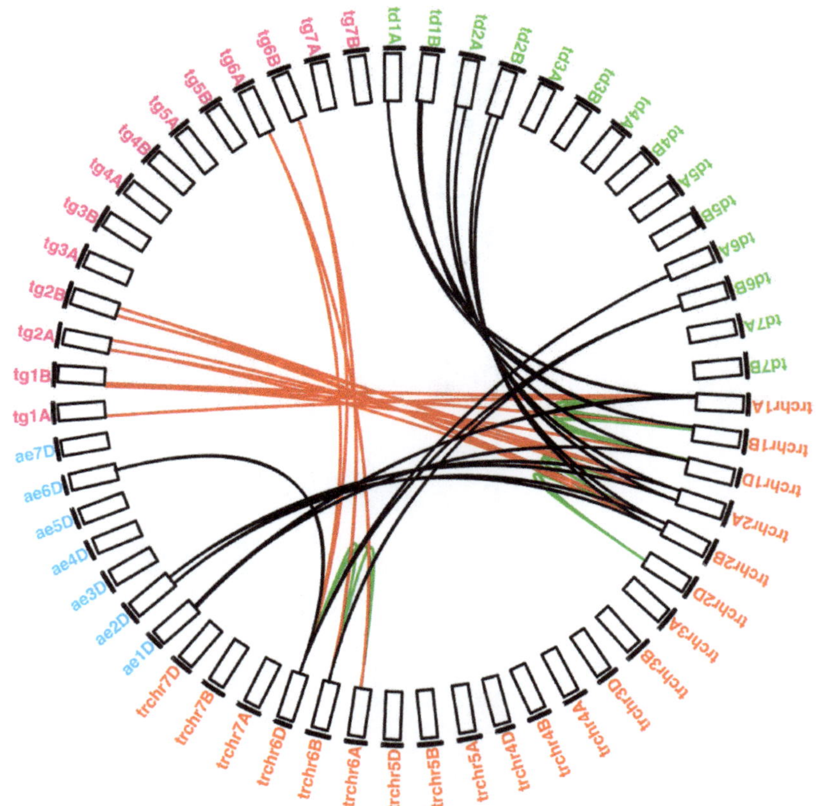

Figure 8. Evolutionary relationship of *PAL* genes between *Triticum aestivum* (*tr*, red), *Aegilops tauschii* (*ae*, blue), *Triticum turgidum* (*tg*, pink), and *Triticum dicocoides* (*td*, green). White bars represent chromosomes.

3.10. In Silico Expression Profile Analysis of PAL Gene Family in Six Genotypes of Wheat

Gene expression analysis helps to probe the potential role and functions of a gene family [57]. Comparative gene expression analysis was used to elucidate the physiological function of different *PAL* gene family members. The in-silico expression profiling was done on the roots of different wheat genotypes (Figure 9). For this purpose, RNA-seq-normalized data were analyzed and based on FPKM values, a heatmap was constructed for diverse *TaPAL* genes. The expression of *TaPAL* genes was variable in different wheat genotypes. *TaPAL35*, *TaPAL31*, *TaPAL23*, *TaPAL22*, *TaPAL8*, *TaPAL5*, and *TaPAL6* were only expressed in Local White. Very low expression of *TaPAL10*, *TaPAL30*, *TaPAL32*, *TaPAL3*, and *TaPAL28* were recorded in all wheat genotypes. Nonetheless, they may have tissue-specific expression, such as in seeds, or their expression may be induced only by certain environmental stresses. *TaPAL11*, *TaPAL14*, *TaPAL12*, *TaPAL34*, *TaPAL4*, *TaPAL21*, *TaPAL19*, *TaPAL24*, and *TaPAL36* were highly expressed in UZ-11-CWA-8. Similarly, *TaPAL27*, *TaPAL16*, *TaPAL9*, and *TaPAL15* were highly expressed in Chakwal-50. Overall, *TaPAL* genes showed a higher expression pattern in roots of drought-tolerant genotypes as compared to drought-sensitive genotypes of wheat.

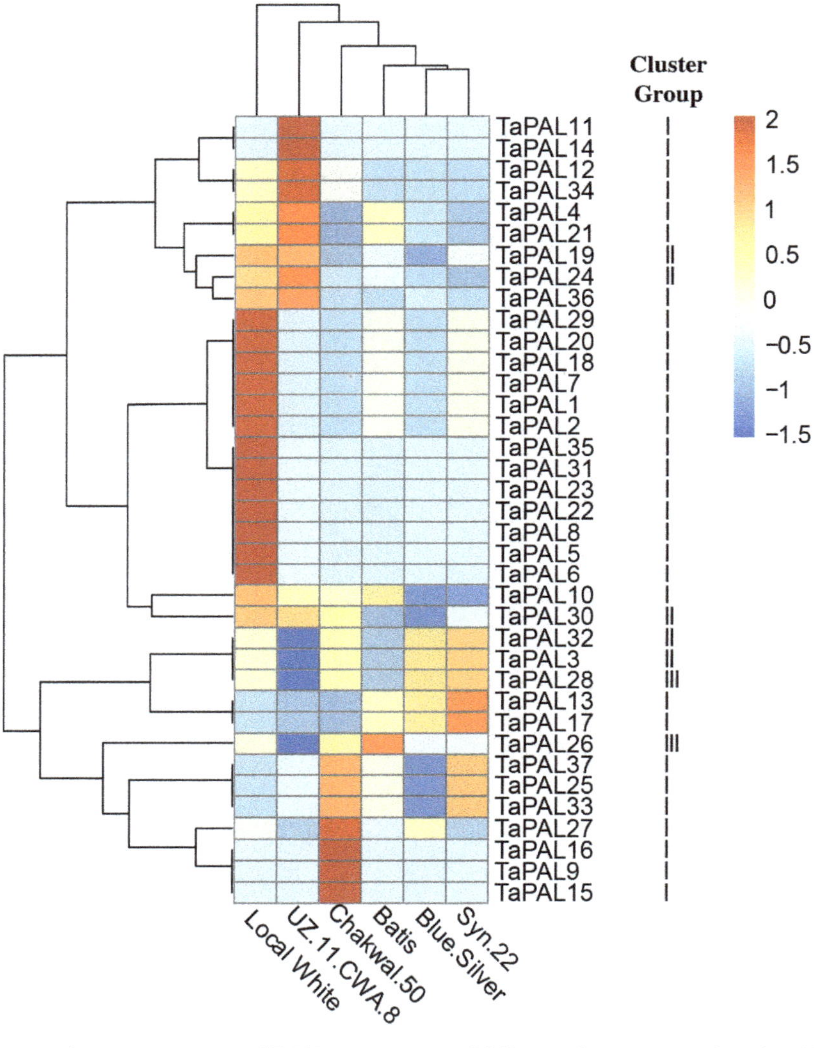

Figure 9. Heatmap of expression pattern of *TaPAL* genes in roots of different wheat genotypes based on FPKM values. Cluster grouping shown as in Figure 7. Color scheme showing the intensity of expression (blue, low expression; red, high expression, and Z score was used).

3.11. Expression of PAL Gene Family under Abiotic Stress

The expression of *TaPAL* genes under abiotic stresses such as drought (DH), heat stress (HS), and phosphorous deficiency (PS) at various stages and in various tissues were explored (Figure 10 and Supplementary Sheet S3). The expression levels of *TaPAL37*, *TaPAL36*, *TaPAL35*, *TaPAL33*, *TaPAL29*, *TaPAL25*, *TaPAL24*, *TaPAL17*, *TaPAL14*, *TaPAL11*, *TaPAL7*, *TaPAL3*, and *TaPAL4* were upregulated in roots. Similarly, the same trend was shown by *TaPAL10*, *TaPAL27*, *TaPAL15*, *TaPAL6*, and *TaPAL21* under drought stress, while *TaPAL23*, *TaPAL20*, *TaPAL19*, *TaPAL2*, and *TaPAL13* were regulated non-significantly under all conditions.

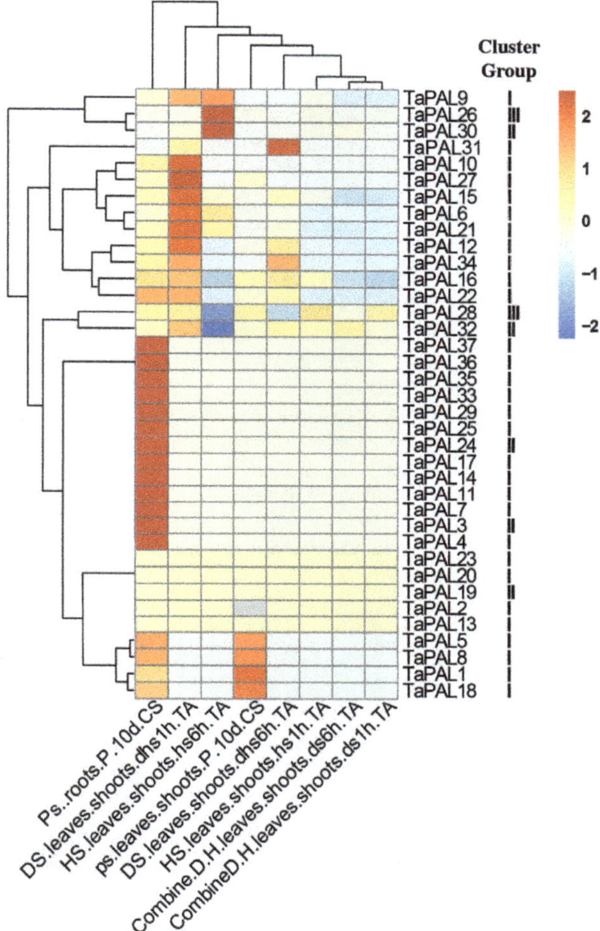

Figure 10. Expression profile of *TaPAL* genes under abiotic stresses. Cluster grouping shown as in Figure 7. Expression values are represented by color scale. Red and blue colors represent the gene expression of up-regulated and down-regulated genes, respectively; yellow shows expression was unregulated. Details are presented in Supplementary Sheet S1.

4. Discussion

Wheat is the main crop for half of the world's population. Wheat faces various types of biotic and abiotic stresses. It has been suggested that *phenylalanine ammonia-lyase (PAL)* genes are essential for plant growth, development, adaptation, and mitigation responses to various environmental and pathogens stresses by producing secondary metabolites regulating plant growth response [11,58,59]. Phenylpropanoids are plant-based organic compounds, which are produced from the amino acids phenylalanine and tyrosine. PAL serves as the first enzyme in the phenylpropanoid pathway and in flavonoid biosynthesis that catalyzes the deamination of phenylalanine [1,24,45,60,61]. Recently, these enzymes have been reported by many researchers in different crops, including *Juglans regia* [62], *Citrus reticulata* [63], *Citrullus lanatus* [64], and *Medicago truncatula* [65]. This study was an investigation of *PAL* in wheat.

The *PAL* family is a very large, multigene family. The family includes ten putative members in maize [66], four members in *Arabidopsis* [19] and tobacco [67], and more than

20 copies in tomato and potato [68]. In the present study, we demonstrated that common wheat (*Triticum aestivum*) has 37 genes of the *PAL* family, a significantly higher number than the above-mentioned species. However, the increase and decrease of *PAL* genes present among species (*Z. mays*, *A. thaliana*, and *O. sativa*) is random [6]. Our results showed that the number of *PAL* genes in *T. aestivum* far exceeds the four *AtPALs*, seven *OsPALs*, twelve *JrPALs*, and six *ZmPALs*, suggesting that whole-genome duplication, small-scale segmental duplications, local tandem duplications, or a combination of these duplication events may have caused this expansion in *T. aestivum* [7,69,70]. The duplicated *PAL* genes in this study were mapped to 11 chromosomes (Figure 1). This diversity of chromosomal distribution indicates that these genes have diverse function. The duplication events might have caused the expansion and dispersion of *PAL* genes giving rise to potential sources of functional variability in common wheat. Gene duplication events may have caused the significant increase in *PAL* genes in *T. aestivum*, as stated in recent studies on different species [16,42,65]. The isolation and identification of *PAL* genes in *T. aestivum* is critical because of their importance in adaption and stress resistance [1,71,72]. The activity of *PAL* genes in response to cold stress of *Juglans regia* (walnut) suggested that the *PAL* gene family in *T. aestivum* is also involved in providing resistance against cold, drought, salt, and disease [70,72]. Similarly, this study also indicates that the expression of *TaPAL* genes is higher in drought-tolerant wheat genotypes as compared to sensitive genotypes. Furthermore, we also checked the subcellular location of TaPAL. Our results showed that the 37 PAL genes are localized to the cytoplasm [62,73,74].

Conserved motifs referred to a part of proteins that is functionally important. The motifs were selected from the PLACE database and conservation patterns were retrieved from MEME suite (Figure 2 and Table 3) and UGENE depicted that the protein structure of the *PAL*-gene family has been highly conserved. The *PAL*-gene family, including *Z. mays*, *A. thaliana*, *O. sativa*, *H. vulgare*, and *T. urartu* plant species, contained all the conserved domains indicating that the *PAL*-gene family remained highly conserved during evolution and took long-term speciation and duplication events to evolve; thus, the results demonstrated its importance in antiretroviral effects. It was evident that the key domain is phenyl ammonium lyase/aromatic lyase, which exists in all families and ancestral species, suggesting a structural similarity between proteins of the *PAL* gene family.

The intron–exon gene structure gives clues for gene evolution [10]. In parallel to the gene number, the structure of the *TaPAL* genes in *Triticum aestivum* has experienced developmental/evolutionary modifications. Out of 37 *TaPAL* genes, ten *TaPAL* genes (*TaPAL3*, *TaPAL5*, *TaPAL8*, *TaPAL18*, *TaPAL19*, *TaPAL24*, *TaPAL27*, *TaPAL35*, *TaPAL36*, and *TaPAL37*) have no intron in their coding regions, two of the *TaPAL* genes (*TaPAL25*, and *TaPAL13*) are interrupted by two introns in their ORFs, while 25 *TaPAL* genes have one intron in their ORFs (Figure 3). Recent studies stated that the duplicated genes showed structural divergence, which is very prevalent in the generation of functionally distinct paralogs. This structural divergence has played a key role in the evolution of duplicated genes compared to non-duplicated genes [69]. The *PAL*-gene structural-data analysis showed a significant variation in the evolution of the *PAL* family of common wheat, walnut, and poplar.

For the functional prediction of *TaPAL* genes we did the GO enrichment analysis (Figure 4). In silico prediction indicated that *TaPAL* genes were involved in numerous developmental processes by regulating biological processes (BPs), molecular processes (MPs), and cellular process (CPs), and showed response against environmental stresses. Many previous studies also reported that microRNAs respond to stress stimuli through regulation of gene expression [42]. *TaPAL* is highly expressed in roots as compared to shoot tissues against abiotic stress. The miRNAs tae-miR1119, tae-miR398, tae-miR444a, tae-miR444b, and tae-miR9664-3p targeting *TaPAL29* have high expression in root tissues (Figures 5 and 10). Previously it has been reported that plant miRNAs play a role in response to environmental stress. In bread wheat under the drought stress, different miRNAs such as miR159, and miR395 were found to be differentiated [75]. Similarly, *VM-*

milR37 plays role in pathogenicity through regulation of the *VmGPX* gene [76]. In another study, miR164 regulated the salinity tolerance in maize [77]. We also checked the protein–protein interaction of TaPAL29 with other co-regulated proteins. Results showed that arogenate dehydratase belongs to the class lyases and is a key enzyme that catalyzes the reaction of L-arogenate into L-phenylalanine [78] and shows interaction with the TaPAL29 (Figure 6).

Phylogenetic analysis, both with ancestral and family species, proposed that the evolution trajectories are like family species (*Z. mays*, *A. thaliana*, and *O. sativa)* and suggested that the *PAL* gene family converge to a single ancestor. This ancestor might be involved in the evolution of plants with respect to adaptation and resistance. Previously it has been reported that during the evolution of PAL, lineage-specific duplication (to promote the diversity of multi-gene families) occurs in *Arabidopsis* and other species [79]. The close paralogs of each PAL gene clustered together phylogenetically into clades in *T. aestivum*, *A. thaliana*, *O. sativa*, and *Z. mays* (Figure 7). In contrast, the PALs from *T. aestivum* and *Z. mays* clustered together along with some of the *O. sativa* genes (*OsPAL1*, *OsPAL5*, *OsPAL6*, and *OsPAL8*), indicating that the expansion of the common wheat PAL gene family might have occurred after the divergence of eurosids I and eurosids II (approximately 100 million years ago) which was reported by [62,80]. Based on phylogenetic analysis, our 37 *TaPAL* genes were separated into three different groups as in tea plant (*Camellia sinensis)* [79] and in other woody plants (*Juglans regia* L., *Salix babylonica*, *Ornithogalum saundersiae*, and *Populus trichocarpa*) they cluster into two groups [18,21,42,81]. TaPALs showed no expansion events as in *Cucumis sativus* [26]. The *PAL* gene family has significant similarities and dissimilarities among various plant species, i.e., *ZmPAL3-5* and *OsPAL2-4*. Among *TaPAL* genes, *TaPAL13*, *TaPAL31*, *TaPAL36*, and *TaPAL37* showed a slight difference in sequence as compared to other 33 *PAL* genes of *T. aestivum* (common wheat), which indicated an 80% similarity score in syntenic analysis. This relationship demonstrated that PALs with comparable evolutionary status might play a similar role in plant development, which enabled us to examine the elements of PALs from different families such as Poaceae via utilizing a comparative genomic approach.

PAL gene is strictly involved in controlling the pre- and post-transcriptional stages, which is considered a doorway to the initiation of the phenylpropanoid pathway. Differential expression patterns for *PAL* genes in higher plants was observed. Moreover, the *PAL* genes in common wheat (*T. aestivum*) show distinct patterns of expression in roots. The genes *TaPAL11*, *TaPAL14*, *TaPAL12*, *TaPAL29*, *TaPAL20*, *TaPAL7*, *TaPAL1*, *TaPAL2*, *TaPAL9*, *TaPAL15*, and *TaPAL16* exhibited high expression levels in roots of drought-tolerant genotypes as compared to drought-susceptible genotypes (Figure 9). These variations in expression level were attributed to the differences in proteins and gene structures, as shown in Figures 2 and 3. The *PAL* family genes showed diverse expression patterns, which indicated that a complex regulation of the PAL-mediated phenylpropanoid pathways existed during the development of drought-tolerant and drought-sensitive wheat genotypes (Figure 9). A similar expression pattern of the *PAL* gene family has also been reported in walnut and barrel clover [62,82]. *Cis*-regulatory elements are also present upstream of the TaPALs (Table 3). Some of the TaPALs from the same evolutionary cluster co-express under stress conditions. This might be due to the presence of *Cis* elements [16]. Similarly, *GdPAL5* is also reported to be an auxin producer which activates plant defense mechanisms during the abiotic stress [83]. Different gene family members usually display abundance disparities in different tissues or under distinct stresses [84].

To overcome the problem of changing climatic conditions of abiotic stress including heat and drought stress on wheat, there is a need to explore the transcriptome profile of this gene family. This study used transcriptomic information of various tissues, at various stages, as shown in Figure 10. The transcript levels of *TaPAL37*, *TaPAL36*, *TaPAL35*, *TaPAL33*, *TaPAL29*, *TaPAL25*, *TaPAL24*, *TaPAL17*, *TaPAL14*, *TaPAL11*, *TaPAL7*, *TaPAL3*, and *TaPAL4* were upregulated in roots. The expression levels of *TaPAL* genes were consistent with previous studies, showing that expression of *TaPAL* genes is higher in roots as compared to other tissues of plants such as *Hordeum vulagare* [85], *Solanum tuberosum* [86], *Arabidopsis thaliana* [19], and *Juglans regia* [87]. The higher expression of the *TaPAL* gene family in drought-tolerant genotypes as compared to drought-sensitive genotypes may be due to high level of lignification, which is part of normal root development [88]. Furthermore, publicly available transcriptomic data which we used was validated by qRT-PCR [89,90].

5. Conclusions

In this study, we have identified 37 *TaPAL* gene family members, which were distributed onto 11 chromosomes. These *TaPAL* genes were found to be involved in drought-stress response mechanisms as they showed high expression in root tissues. Since a few *PAL* genes are reported in wheat, this is the first detailed study of the *PAL* gene family in wheat. We also find 27 putative miRNAs targeting *TaPAL* genes. Some questions are still to be answered, such as what is the exact role of each *TaPAL* gene, and how is the expression of each *TaPAL* controlled in different phases of development and in reaction to distinct stress or hormone signals? Therefore, to further our knowledge of the *TaPAL* family, more molecular, biochemical, and physiological studies are expected. Due to the potential roles of *TaPAL* in the growth of common wheat (*T. aestivum*), it may provide prospective targets for molecular high-quality grain breeding.

Supplementary Materials: The following are available online at https://www.mdpi.com/article/10.3390/agronomy11122511/s1, Figure S1: Ancestral relationship (phylogenetic tree) of *PAL* genes between *Triticum aestivum (Ta)*, *Hordeum vulgare (Hv)*, *Solanum tuberosum (St)*, and *Triticum urartu (Tu)*; File S1: Protein sequences of identified TaPAL in this study; File S2: Multiple sequence alignment in wheat; Supplementary Sheet S1:The details of the materials and treatments for the retrieved expression values; Sheet S2:List of genes in wheat to explore the gene duplication within the TaPAL gene family and ancestral species of wheat; Sheet S3: Expression level of genes in different conditions; Sheet S4:Putative miRNAs targeting the *TaPAL* genes.

Author Contributions: Conceptualization, F.R. and M.R.K.; data curation, F.R., M.U., M.K.N., N.R., A.A. and H.S.; formal analysis, F.R. and N.R.; funding acquisition, M.R.K.; methodology, M.R.K.; resources, M.R.K.; software, F.R.; supervision, M.R.K.; validation, M.U. and M.R.K.; visualization, M.U.; writing—original draft, F.R.; writing—review and editing, F.R., M.U. and M.R.K. All authors have read and agreed to the published version of the manuscript.

Funding: This research was funded by ALP, PARC, Pakistan.

Institutional Review Board Statement: Not applicable.

Informed Consent Statement: Not applicable.

Data Availability Statement: The data and materials presented in this study are mentioned in the main text as well as in the supplementary files, further data will be provided on request from the corresponding author.

Acknowledgments: We acknowledge the assistance of NIGAB, NARC, Islamabad, Pakistan for infrastructural support.

Conflicts of Interest: The authors declare no conflict of interest. The funders had no role in the design of the study; in the collection, analyses, or interpretation of data; in the writing of the manuscript, or in the decision to publish the results.

References

1. Huang, J.; Gu, M.; Lai, Z.; Fan, B.; Shi, K.; Zhou, Y.-H.; Yu, J.-Q.; Chen, Z. Functional analysis of the Arabidopsis *PAL* gene family in plant growth, development, and response to environmental stress. *Plant Physiol.* **2010**, *153*, 1526–1538. [CrossRef] [PubMed]
2. Wu, P.; Guo, Q.-q.; Qin, Z.-w. The fungicide propamocarb increases lignin by activating the phenylpropanoid pathway in *Cucumis sativus* L. *Hortic. Environ. Biotechnol.* **2016**, *57*, 511–518. [CrossRef]
3. Raza, A.; Su, W.; Hussain, M.A.; Mehmood, S.S.; Zhang, X.; Cheng, Y.; Zou, X.; Lv, Y. Integrated Analysis of Metabolome and Transcriptome Reveals Insights for Cold Tolerance in Rapeseed (*Brassica napus* L.). *Front. Plant Sci.* **2021**, *12*, 1796. [CrossRef]
4. Schwede, T.F.; Rétey, J.; Schulz, G.E. Crystal structure of histidine ammonia-lyase revealing a novel polypeptide modification as the catalytic electrophile. *Biochemistry* **1999**, *38*, 5355–5361. [CrossRef]

5. Vogt, T. Phenylpropanoid biosynthesis. *Mol. Plant* **2010**, *3*, 2–20. [CrossRef] [PubMed]
6. Cochrane, F.C.; Davin, L.B.; Lewis, N.G. The Arabidopsis phenylalanine ammonia lyase gene family: Kinetic characterization of the four PAL isoforms. *Phytochemistry* **2004**, *65*, 1557–1564. [CrossRef]
7. Pascual, M.B.; El-Azaz, J.; de la Torre, F.N.; Cañas, R.A.; Avila, C.; Cánovas, F.M. Biosynthesis and metabolic fate of phenylalanine in conifers. *Front. Plant Sci.* **2016**, *7*, 1030. [CrossRef]
8. Lei, L.; Zhou, S.-L.; Ma, H.; Zhang, L.-S. Expansion and diversification of the SET domain gene family following whole-genome duplications in *Populus trichocarpa*. *BMC Evol. Biol.* **2012**, *12*, 1–17. [CrossRef]
9. Dixon, R.A.; Paiva, N.L. Stress-induced phenylpropanoid metabolism. *Plant Cell* **1995**, *7*, 1085. [CrossRef]
10. La Camera, S.; Gouzerh, G.; Dhondt, S.; Hoffmann, L.; Fritig, B.; Legrand, M.; Heitz, T. Metabolic reprogramming in plant innate immunity: The contributions of phenylpropanoid and oxylipin pathways. *Immunol. Rev.* **2004**, *198*, 267–284. [CrossRef]
11. Kim, D.S.; Hwang, B.K. An important role of the pepper phenylalanine ammonia-lyase gene (PAL1) in salicylic acid-dependent signalling of the defence response to microbial pathogens. *J. Exp. Bot.* **2014**, *65*, 2295–2306. [CrossRef] [PubMed]
12. Ritter, H.; Schulz, G.E. Structural basis for the entrance into the phenylpropanoid metabolism catalyzed by phenylalanine ammonia-lyase. *Plant Cell* **2004**, *16*, 3426–3436. [CrossRef]
13. Purwar, S.; Sundaram, S.; Sinha, S.; Gupta, A.; Dobriyall, N.; Kumar, A. Expression and in silico characterization of Phenylalanine ammonium lyase against karnal bunt (*Tilletia indica*) in wheat (*Triticum aestivum*). *Bioinformation* **2013**, *9*, 1013. [CrossRef] [PubMed]
14. Rawal, H.; Singh, N.; Sharma, T. Conservation, divergence, and genome-wide distribution of *PAL* and *POX* A gene families in plants. *Int. J. Genom.* **2013**, *2013*, 678969.
15. MacDonald, M.J.; D'Cunha, G.B. A modern view of phenylalanine ammonia lyase. *Biochem. Cell Biol.* **2007**, *85*, 273–282. [CrossRef] [PubMed]
16. Wu, Z.; Gui, S.; Wang, S.; Ding, Y. Molecular evolution and functional characterisation of an ancient phenylalanine ammonia-lyase gene (*NnPAL1*) from *Nelumbo nucifera*: Novel insight into the evolution of the PAL family in angiosperms. *BMC Evol. Biol.* **2014**, *14*, 1–14. [CrossRef]
17. Chaw, S.-M.; Zharkikh, A.; Sung, H.-M.; Lau, T.-C.; Li, W.-H. Molecular phylogeny of extant gymnosperms and seed plant evolution: Analysis of nuclear 18S rRNA sequences. *Mol. Biol. Evol.* **1997**, *14*, 56–68. [CrossRef]
18. de Jong, F.; Hanley, S.J.; Beale, M.H.; Karp, A. Characterisation of the willow phenylalanine ammonia-lyase (PAL) gene family reveals expression differences compared with poplar. *Phytochemistry* **2015**, *117*, 90–97. [CrossRef]
19. Wanner, L.A.; Li, G.; Ware, D.; Somssich, I.E.; Davis, K.R. The phenylalanine ammonia-lyase gene family in *Arabidopsis thaliana*. *Plant Mol. Biol.* **1995**, *27*, 327–338. [CrossRef]
20. Raes, J.; Rohde, A.; Christensen, J.H.; Van de Peer, Y.; Boerjan, W. Genome-wide characterization of the lignification toolbox in Arabidopsis. *Plant Physiol.* **2003**, *133*, 1051–1071. [CrossRef]
21. Shi, R.; Sun, Y.-H.; Li, Q.; Heber, S.; Sederoff, R.; Chiang, V.L. Towards a systems approach for lignin biosynthesis in *Populus trichocarpa*: Transcript abundance and specificity of the monolignol biosynthetic genes. *Plant Cell Physiol.* **2010**, *51*, 144–163. [CrossRef] [PubMed]
22. Xu, H.; Park, N.I.; Li, X.; Kim, Y.K.; Lee, S.Y.; Park, S.U. Molecular cloning and characterization of phenylalanine ammonia-lyase, cinnamate 4-hydroxylase and genes involved in flavone biosynthesis in *Scutellaria baicalensis*. *Bioresour. Technol.* **2010**, *101*, 9715–9722. [CrossRef] [PubMed]
23. Lepelley, M.; Mahesh, V.; McCarthy, J.; Rigoreau, M.; Crouzillat, D.; Chabrillange, N.; de Kochko, A.; Campa, C. Characterization, high-resolution mapping and differential expression of three homologous PAL genes in *Coffea canephora* Pierre (Rubiaceae). *Planta* **2012**, *236*, 313–326. [CrossRef] [PubMed]
24. Bagal, U.R.; Leebens-Mack, J.H.; Lorenz, W.W.; Dean, J.F. The phenylalanine ammonia lyase (PAL) gene family shows a gymnosperm-specific lineage. In *Proceedings of the BMC Genomics*; Springer: Berlin/Heidelberg, Germany, 2012; pp. 1–9.
25. Dong, C.-j.; Ning, C.; ZHANG, Z.-g.; SHANG, Q.-m. Phenylalanine ammonia-lyase gene families in cucurbit species: Structure, evolution, and expression. *J. Integr. Agric.* **2016**, *15*, 1239–1255. [CrossRef]
26. Shang, Q.-M.; Li, L.; Dong, C.-J. Multiple tandem duplication of the phenylalanine ammonia-lyase genes in *Cucumis sativus* L. *Planta* **2012**, *236*, 1093–1105. [CrossRef]
27. Souri, Z.; Karimi, N.; Sandalio, L.M. Arsenic hyperaccumulation strategies: An overview. *Front. Cell Dev. Biol.* **2017**, *5*, 67. [CrossRef]
28. Bolser, D.; Staines, D.M.; Pritchard, E.; Kersey, P. Ensembl plants: Integrating tools for visualizing, mining, and analyzing plant genomics data. In *Plant Bioinformatics*; Springer: Berlin/Heidelberg, Germany, 2016; pp. 115–140.
29. Madden, T. The BLAST sequence analysis tool. In *NCBI Handbook*, 2nd ed.; National Center for Biotechnology Information (US): Bethesda, MD, USA, 2013; Volume 2, pp. 425–436.
30. Larkin, M.A.; Blackshields, G.; Brown, N.P.; Chenna, R.; McGettigan, P.A.; McWilliam, H.; Valentin, F.; Wallace, I.M.; Wilm, A.; Lopez, R. Clustal W and Clustal X version 2.0. *Bioinformatics* **2007**, *23*, 2947–2948. [CrossRef]
31. Bateman, A.; Coin, L.; Durbin, R.; Finn, R.D.; Hollich, V.; Griffiths-Jones, S.; Khanna, A.; Marshall, M.; Moxon, S.; Sonnhammer, E.L. The Pfam protein families database. *Nucleic Acids Res.* **2004**, *32*, D138–D141. [CrossRef]
32. Schultz, J.; Copley, R.R.; Doerks, T.; Ponting, C.P.; Bork, P. SMART: A web-based tool for the study of genetically mobile domains. *Nucleic Acids Res.* **2000**, *28*, 231–234. [CrossRef]

33. Hu, B.; Jin, J.; Guo, A.-Y.; Zhang, H.; Luo, J.; Gao, G. GSDS 2.0: An upgraded gene feature visualization server. *Bioinformatics* **2015**, *31*, 1296–1297. [CrossRef]
34. Bailey, T.L.; Boden, M.; Buske, F.A.; Frith, M.; Grant, C.E.; Clementi, L.; Ren, J.; Li, W.W.; Noble, W.S. MEME SUITE: Tools for motif discovery and searching. *Nucleic Acids Res.* **2009**, *37*, W202–W208. [CrossRef]
35. Consortium, U. UniProt: A hub for protein information. *Nucleic Acids Res.* **2015**, *43*, D204–D212. [CrossRef] [PubMed]
36. Okonechnikov, K.; Golosova, O.; Fursov, M.; Team, U. Unipro UGENE: A unified bioinformatics toolkit. *Bioinformatics* **2012**, *28*, 1166–1167. [CrossRef] [PubMed]
37. Blum, M.; Chang, H.-Y.; Chuguransky, S.; Grego, T.; Kandasaamy, S.; Mitchell, A.; Nuka, G.; Paysan-Lafosse, T.; Qureshi, M.; Raj, S. The InterPro protein families and domains database: 20 years on. *Nucleic Acids Res.* **2021**, *49*, D344–D354. [CrossRef] [PubMed]
38. Kumar, S.; Tamura, K.; Nei, M. MEGA: Molecular evolutionary genetics analysis software for microcomputers. *Bioinformatics* **1994**, *10*, 189–191. [CrossRef] [PubMed]
39. Wang, Y.; Tang, H.; DeBarry, J.D.; Tan, X.; Li, J.; Wang, X.; Lee, T.-h.; Jin, H.; Marler, B.; Guo, H. MCScanX: A toolkit for detection and evolutionary analysis of gene synteny and collinearity. *Nucleic Acids Res.* **2012**, *40*, e49. [CrossRef] [PubMed]
40. Chen, C.; Chen, H.; He, Y.; Xia, R. TBtools, a toolkit for biologists integrating various biological data handling tools with a user-friendly interface. *BioRxiv* **2018**, 289660. [CrossRef]
41. Bektas, Y.; Eulgem, T. Synthetic plant defense elicitors. *Front. Plant Sci.* **2015**, *5*, 804. [CrossRef] [PubMed]
42. Yan, F.; Li, H.; Zhao, P. Genome-Wide Identification and transcriptional expression of the PAL Gene family in common Walnut (*Juglans regia* L.). *Genes* **2019**, *10*, 46. [CrossRef] [PubMed]
43. Dai, X.; Zhao, P.X. psRNATarget: A plant small RNA target analysis server. *Nucleic Acids Res.* **2011**, *39*, W155–W159. [CrossRef]
44. Shannon, P.; Markiel, A.; Ozier, O.; Baliga, N.S.; Wang, J.T.; Ramage, D.; Amin, N.; Schwikowski, B.; Ideker, T. Cytoscape: A software environment for integrated models of biomolecular interaction networks. *Genome Res.* **2003**, *13*, 2498–2504. [CrossRef]
45. Higo, K.; Ugawa, Y.; Iwamoto, M.; Korenaga, T. Plant cis-acting regulatory DNA elements (PLACE) database: 1999. *Nucleic Acids Res.* **1999**, *27*, 297–300. [CrossRef] [PubMed]
46. Raudvere, U.; Kolberg, L.; Kuzmin, I.; Arak, T.; Adler, P.; Peterson, H.; Vilo, J. g: Profiler: A web server for functional enrichment analysis and conversions of gene lists (2019 update). *Nucleic Acids Res.* **2019**, *47*, W191–W198. [CrossRef] [PubMed]
47. Mering, C.v.; Huynen, M.; Jaeggi, D.; Schmidt, S.; Bork, P.; Snel, B. STRING: A database of predicted functional associations between proteins. *Nucleic Acids Res.* **2003**, *31*, 258–261. [CrossRef]
48. Iquebal, M.A.; Sharma, P.; Jasrotia, R.S.; Jaiswal, S.; Kaur, A.; Saroha, M.; Angadi, U.; Sheoran, S.; Singh, R.; Singh, G. RNAseq analysis reveals drought-responsive molecular pathways with candidate genes and putative molecular markers in root tissue of wheat. *Sci. Rep.* **2019**, *9*, 1–18.
49. Zou, C.; Wang, P.; Xu, Y. Bulked sample analysis in genetics, genomics and crop improvement. *Plant Biotechnol. J.* **2016**, *14*, 1941–1955. [CrossRef]
50. Bolger, A.M.; Lohse, M.; Usadel, B. Trimmomatic: A flexible trimmer for Illumina sequence data. *Bioinformatics* **2014**, *30*, 2114–2120. [CrossRef]
51. Kim, D.; Paggi, J.M.; Park, C.; Bennett, C.; Salzberg, S.L. Graph-based genome alignment and genotyping with HISAT2 and HISAT-genotype. *Nat. Biotechnol.* **2019**, *37*, 907–915. [CrossRef]
52. Pertea, M.; Pertea, G.M.; Antonescu, C.M.; Chang, T.-C.; Mendell, J.T.; Salzberg, S.L. StringTie enables improved reconstruction of a transcriptome from RNA-seq reads. *Nat. Biotechnol.* **2015**, *33*, 290–295. [CrossRef]
53. Tarazona, S.; Furió-Tarı, P.; Ferrer, A.; Conesa, A. *NOISeq: Differential Expression in RNA-Seq—Bioconductor*; Department of Statistics: TU Dortmund, Germany, 2013.
54. Borrill, P.; Ramirez-Gonzalez, R.; Uauy, C. expVIP: A customizable RNA-seq data analysis and visualization platform. *Plant Physiol.* **2016**, *170*, 2172–2186. [CrossRef]
55. Kolde, R.; Kolde, M.R. Package 'pheatmap'. *R Package* **2015**, *1*, 790.
56. Hossain, M.S.; Rasel Ahmed, M.; Ullah, W.; Honi, U.; Tareq, M.Z.; Sarker, M.S.A.; Ahmed, B.; Islam, M.S. Phenylalanine ammonia-lyase gene family (PAL): Genome wide characterization and transcriptional expression in jute (*Corchorus olitorius*). *J. Biosci. Agric. Res.* **2020**, *26*, 2185–2191. [CrossRef]
57. Darzentas, N. Circoletto: Visualizing sequence similarity with Circos. *Bioinformatics* **2010**, *26*, 2620–2621. [CrossRef] [PubMed]
58. Jin, Q.; Yao, Y.; Cai, Y.; Lin, Y. Molecular cloning and sequence analysis of a phenylalanine ammonia-lyase gene from Dendrobium. *PLoS ONE* **2013**, *8*, e62352. [CrossRef] [PubMed]
59. Li, G.; Wang, H.; Cheng, X.; Su, X.; Zhao, Y.; Jiang, T.; Jin, Q.; Lin, Y.; Cai, Y. Comparative genomic analysis of the PAL genes in five Rosaceae species and functional identification of Chinese white pear. *PeerJ* **2019**, *7*, e8064. [CrossRef]
60. Dong, C.-J.; Shang, Q.-M. Genome-wide characterization of phenylalanine ammonia-lyase gene family in watermelon (*Citrullus lanatus*). *Planta* **2013**, *238*, 35–49. [CrossRef]
61. Rushton, P.J.; Reinstadler, A.; Lipka, V.; Lippok, B.; Somssich, I.E. Synthetic plant promoters containing defined regulatory elements provide novel insights into pathogen-and wound-induced signaling. *Plant Cell* **2002**, *14*, 749–762. [CrossRef]
62. Kaur, A.; Pati, P.K.; Pati, A.M.; Nagpal, A.K. In-silico analysis of cis-acting regulatory elements of pathogenesis-related proteins of *Arabidopsis thaliana* and *Oryza sativa*. *PLoS ONE* **2017**, *12*, e0184523. [CrossRef]

63. Yang, H.; Dong, T.; Li, J.; Wang, M. Molecular cloning, expression, and subcellular localization of a PAL gene from *Citrus reticulata* under iron deficiency. *Biol. Plant.* **2016**, *60*, 482–488. [CrossRef]
64. Kong, W.; Ding, L.; Cheng, J.; Wang, B. Identification and expression analysis of genes with pathogen-inducible cis-regulatory elements in the promoter regions in *Oryza sativa*. *Rice* **2018**, *11*, 1–12. [CrossRef]
65. Ren, W.; Wang, Y.; Xu, A.; Zhao, Y. Genome-wide identification and characterization of the Phenylalanine Ammonia-lyase (PAL) gene family in *Medicago truncatula*. *Legume Res. Int. J.* **2019**, *42*, 461–466. [CrossRef]
66. Yuan, W.; Jiang, T.; Du, K.; Chen, H.; Cao, Y.; Xie, J.; Li, M.; Carr, J.P.; Wu, B.; Fan, Z. Maize phenylalanine ammonia-lyases contribute to resistance to Sugarcane mosaic virus infection, most likely through positive regulation of salicylic acid accumulation. *Mol. Plant Pathol.* **2019**, *20*, 1365–1378. [CrossRef] [PubMed]
67. Fukasawa-Akada, T.; Kung, S.-d.; Watson, J.C. Phenylalanine ammonia-lyase gene structure, expression, and evolution in Nicotiana. *Plant Mol. Biol.* **1996**, *30*, 711–722. [CrossRef]
68. Han, H.; Woeste, K.E.; Hu, Y.; Dang, M.; Zhang, T.; Gao, X.-X.; Zhou, H.; Feng, X.; Zhao, G.; Zhao, P. Genetic diversity and population structure of common walnut (*Juglans regia*) in China based on EST-SSRs and the nuclear gene phenylalanine ammonia-lyase (PAL). *Tree Genet. Genomes* **2016**, *12*, 1–12. [CrossRef]
69. Cheng, X.; Wang, S.; Xu, D.; Liu, X.; Li, X.; Xiao, W.; Cao, J.; Jiang, H.; Min, X.; Wang, J. Identification and analysis of the GASR gene family in common wheat (*Triticum aestivum* L.) and characterization of *TaGASR34*, a gene associated with seed dormancy and germination. *Front. Genet.* **2019**, *10*, 980. [CrossRef] [PubMed]
70. Reichert, A.I.; He, X.-Z.; Dixon, R.A. Phenylalanine ammonia-lyase (PAL) from tobacco (*Nicotiana tabacum*): Characterization of the four tobacco PAL genes and active heterotetrameric enzymes. *Biochem. J.* **2009**, *424*, 233–242. [CrossRef]
71. Hamberger, B.; Ellis, M.; Friedmann, M.; de Azevedo Souza, C.; Barbazuk, B.; Douglas, C.J. Genome-wide analyses of phenylpropanoid-related genes in *Populus trichocarpa*, *Arabidopsis thaliana*, and *Oryza sativa*: The Populus lignin toolbox and conservation and diversification of angiosperm gene families. *Botany* **2007**, *85*, 1182–1201.
72. Kaur, H.; Salh, P.; Singh, B. Role of defense enzymes and phenolics in resistance of wheat crop (*Triticum aestivum* L.) towards aphid complex. *J. Plant Interact.* **2017**, *12*, 304–311. [CrossRef]
73. Kumar, A.; Batra, R.; Gahlaut, V.; Gautam, T.; Kumar, S.; Sharma, M.; Tyagi, S.; Singh, K.P.; Balyan, H.S.; Pandey, R. Genome-wide identification and characterization of gene family for RWP-RK transcription factors in wheat (*Triticum aestivum* L.). *PLoS ONE* **2018**, *13*, e0208409. [CrossRef]
74. Chang, A.; Lim, M.-H.; Lee, S.-W.; Robb, E.J.; Nazar, R.N. Tomato phenylalanine ammonia-lyase gene family, highly redundant but strongly underutilized. *J. Biol. Chem.* **2008**, *283*, 33591–33601. [CrossRef]
75. Akdogan, G.; Tufekci, E.D.; Uranbey, S.; Unver, T. miRNA-based drought regulation in wheat. *Funct. Integr. Genom.* **2016**, *16*, 221–233. [CrossRef]
76. Feng, H.; Xu, M.; Gao, Y.; Liang, J.; Guo, F.; Guo, Y.; Huang, L. Vm-milR37 contributes to pathogenicity by regulating glutathione peroxidase gene *VmGP* in Valsa mali. *Mol. Plant Pathol.* **2021**, *22*, 243–254. [CrossRef]
77. Shan, T.; Fu, R.; Xie, Y.; Chen, Q.; Wang, Y.; Li, Z.; Song, X.; Li, P.; Wang, B. Regulatory mechanism of maize (*Zea mays* L.) miR164 in salt stress response. *Russ. J. Genet.* **2020**, *56*, 835–842. [CrossRef]
78. Yamada, T.; Matsuda, F.; Kasai, K.; Fukuoka, S.; Kitamura, K.; Tozawa, Y.; Miyagawa, H.; Wakasa, K. Mutation of a rice gene encoding a phenylalanine biosynthetic enzyme results in accumulation of phenylalanine and tryptophan. *Plant Cell* **2008**, *20*, 1316–1329. [CrossRef]
79. Wu, Y.; Wang, W.; Li, Y.; Dai, X.; Ma, G.; Xing, D.; Zhu, M.; Gao, L.; Xia, T. Six phenylalanine ammonia-lyases from *Camellia sinensis*: Evolution, expression, and kinetics. *Plant Physiol. Biochem.* **2017**, *118*, 413–421. [CrossRef] [PubMed]
80. Wada, K.C.; Mizuuchi, K.; Koshio, A.; Kaneko, K.; Mitsui, T.; Takeno, K. Stress enhances the gene expression and enzyme activity of phenylalanine ammonia-lyase and the endogenous content of salicylic acid to induce flowering in pharbitis. *J. Plant Physiol.* **2014**, *171*, 895–902. [CrossRef] [PubMed]
81. Wang, Z.-B.; Chen, X.; Wang, W.; Cheng, K.-D.; Kong, J.-Q. Transcriptome-wide identification and characterization of *Ornithogalum saundersiae* phenylalanine ammonia lyase gene family. *RSC Adv.* **2014**, *4*, 27159–27175. [CrossRef]
82. Cao, Y.; Meng, D.; Abdullah, M.; Jin, Q.; Lin, Y.; Cai, Y. Genome wide identification, evolutionary, and expression analysis of VQ genes from two Pyrus species. *Genes* **2018**, *9*, 224. [CrossRef] [PubMed]
83. Tufail, M.A.; Touceda-González, M.; Pertot, I.; Ehlers, R.-U. Gluconacetobacter diazotrophicus PAL5 Enhances Plant Robustness Status under the Combination of Moderate Drought and Low Nitrogen Stress in *Zea mays* L. *Microorganisms* **2021**, *9*, 870. [CrossRef]
84. Christopoulos, M.V.; Tsantili, E. Participation of phenylalanine ammonia-lyase (PAL) in increased phenolic compounds in fresh cold stressed walnut (*Juglans regia* L.) kernels. *Postharvest Biol. Technol.* **2015**, *104*, 17–25. [CrossRef]
85. Kervinen, T.; Peltonen, S.; Utriainen, M.; Kangasjärvi, J.; Teeri, T.H.; Karjalainen, R. Cloning and characterization of cDNA clones encoding phenylalanine ammonia-lyase in barley. *Plant Sci.* **1997**, *123*, 143–150. [CrossRef]
86. Joos, H.J.; Hahlbrock, K. Phenylalanine ammonia-lyase in potato (*Solanum tuberosum* L.) Genomic complexity, structural comparison of two selected genes and modes of expression. *Eur. J. Biochem.* **1992**, *204*, 621–629. [CrossRef] [PubMed]
87. Xu, F.; Deng, G.; Cheng, S.; Zhang, W.; Huang, X.; Li, L.; Cheng, H.; Rong, X.; Li, J. Molecular cloning, characterization and expression of the phenylalanine ammonia-lyase gene from *Juglans regia*. *Molecules* **2012**, *17*, 7810–7823. [CrossRef] [PubMed]

88. Dixon, R.A.; Maxwell, C.A.; Ni, W.; Oommen, A.; Paiva, N.L. Genetic manipulation of lignin and phenylpropanoid compounds involved in interactions with microorganisms. In *Genetic Engineering of Plant Secondary Metabolism*; Springer: Berlin/Heidelberg, Germany, 1994; pp. 153–178.
89. Pearce, S.; Vazquez-Gross, H.; Herin, S.Y.; Hane, D.; Wang, Y.; Gu, Y.Q.; Dubcovsky, J. WheatExp: An RNA-seq expression database for polyploid wheat. *BMC Plant Biol.* **2015**, *15*, 299. [CrossRef] [PubMed]
90. Ma, J.; Ding, P.; Qin, P.; Liu, Y.-X.; Xie, Q.; Chen, G.; Li, W.; Jiang, Q.; Chen, G.; Lan, X.-J. Structure and expression of the *TaGW7* in bread wheat (*Triticum aestivum* L.). *Plant Growth Regul.* **2017**, *82*, 281. [CrossRef]

 *agronomy*

Review

MicroRNA and cDNA-Microarray as Potential Targets against Abiotic Stress Response in Plants: Advances and Prospects

Tariq Pervaiz [1,2], Muhammad Waqas Amjid [3], Ashraf El-kereamy [2], Shi-Hui Niu [1,*] and Harry X. Wu [1,4,5,*]

1 Beijing Advanced Innovation Center for Tree Breeding by Molecular Design, National Engineering Laboratory for Tree Breeding, College of Biological Sciences and Technology, Beijing Forestry University, Beijing 100083, China; tariqzoqi2009@gmail.com
2 Department of Botany and Plant Sciences, University of California Riverside, Riverside, CA 22963, USA; ashrafe@ucr.edu
3 State Key Laboratory of Crop Genetics and Germplasm Enhancement, Cotton Germplasm Enhancement and Application Engineering Research Center (Ministry of Education), Nanjing Agricultural University, Nanjing 210095, China; waqasamjid@hotmail.com
4 Umeå Plant Science Centre, Department of Forest Genetics and Plant Physiology, Swedish University of Agricultural Sciences, Linnaeus väg 6, SE-901 83 Umeå, Sweden
5 CSIRO National Research Collection Australia, Black Mountain Laboratory, Canberra, ACT 2601, Australia
* Correspondence: arrennew@bjfu.edu.cn (S.-H.N.); harry.wu@slu.se (H.X.W.)

Abstract: Abiotic stresses, such as temperature (heat and cold), salinity, and drought negatively affect plant productivity; hence, the molecular responses of abiotic stresses need to be investigated. Numerous molecular and genetic engineering studies have made substantial contributions and revealed that abiotic stresses are the key factors associated with production losses in plants. In response to abiotic stresses, altered expression patterns of miRNAs have been reported, and, as a result, cDNA-microarray and microRNA (miRNA) have been used to identify genes and their expression patterns against environmental adversities in plants. MicroRNA plays a significant role in environmental stresses, plant growth and development, and regulation of various biological and metabolic activities. MicroRNAs have been studied for over a decade to identify those susceptible to environmental stimuli, characterize expression patterns, and recognize their involvement in stress responses and tolerance. Recent findings have been reported that plants assign miRNAs as critical post-transcriptional regulators of gene expression in a sequence-specific manner to adapt to multiple abiotic stresses during their growth and developmental cycle. In this study, we reviewed the current status and described the application of cDNA-microarray and miRNA to understand the abiotic stress responses and different approaches used in plants to survive against different stresses. Despite the accessibility to suitable miRNAs, there is a lack of simple ways to identify miRNA and the application of cDNA-microarray. The elucidation of miRNA responses to abiotic stresses may lead to developing technologies for the early detection of plant environmental stressors. The miRNAs and cDNA-microarrays are powerful tools to enhance abiotic stress tolerance in plants through multiple advanced sequencing and bioinformatics techniques, including miRNA-regulated network, miRNA target prediction, miRNA identification, expression profile, features (disease or stress, biomarkers) association, tools based on machine learning algorithms, NGS, and tools specific for plants. Such technologies were established to identify miRNA and their target gene network prediction, emphasizing current achievements, impediments, and future perspectives. Furthermore, there is also a need to identify and classify new functional genes that may play a role in stress resistance, since many plant genes constitute an unexplained fraction.

Keywords: abiotic stress tolerance; drought stress; salinity stress; cold stress; miRNA target gene expression; adaptation

Citation: Pervaiz, T.; Amjid, M.W.; El-kereamy, A.; Niu, S.-H.; Wu, H.X. MicroRNA and cDNA-Microarray as Potential Targets against Abiotic Stress Response in Plants: Advances and Prospects. *Agronomy* **2022**, *12*, 11. https://doi.org/10.3390/agronomy12010011

Academic Editor: Alfonso Albacete

Received: 17 November 2021
Accepted: 15 December 2021
Published: 22 December 2021

1. Introduction

Plants are subjected to a wide range of abiotic stresses that are primarily hostile to plant growth, leading to plant death worldwide. Abiotic stresses have an extensive impact on various physiological, molecular, and metabolic responses. Much progress has been made in unravelling the complex stress response mechanisms, particularly in identifying stress-responsive genes with the help of biotechnological tools [1,2]. MicroRNAs (miRNAs), play a critical role in post-transcriptional regulation through base-pairing with other miRNA targets, including transcription factors (TFs) [1,3]. Understanding the role of miRNAs in abiotic stresses may be helpful in the development of innovative ways for improving plant responses against abiotic stresses. MicroRNAs are involved in multiple cellular and metabolic pathways under abiotic stresses, such as flowering, morphogenesis, signal transduction [4–6], and gene feedback regulation [7]. MicroRNAs are a group of single-stranded non-protein-coding short length RNA of approximately 18–25 nucleotides in length with a highly conserved class [8–10]. MicroRNAs are formed by antecedence with distinctive stem-loop assemblies [11]. In the plants, miRNAs are important regulators of gene expression at various stages of plant development; for instance, 959 founding members representing 178 miRNA families were identified in rapeseed (*Brassica napus*), earth mosses (*Physcomitrella patens*), arabidopsis (*Arabidopsis thaliana*), maize (*Zea mays*), black cottonwood (*Populus trichocarpa*), barrel clover (*Medicago truncatula*), rice (*Oryza sativa*), soybean (*Glycine max*), sorghum (*Sorghum bicolor*), and sugar cane (*Saccharum officinarum*) [12,13] (Tables 1 and 2). Usually, intronic miRNAs are coordinately expressed in host plant miRNAs, suggesting that they are also initiated from mutual transcripts. Host gene expression by situ analysis was used to probe the temporal and spatial localization of intronic miRNAs. These non-coding small RNAs are proposed to perform crucial roles in plant adaptation and immunity to adverse environmental conditions [14,15].

Table 1. Examples of miRNAs identified in model plants under drought, cold and salinity stresses.

Stress Condition	Plant Species	Inducible Genes	Known Responsive miRNAs	Functions	References
Drought stress	*Arabidopsis thaliana*	Rd29A (At5g52310) *CCAAT-binding transcription factors*	miR164, miR169, miR389, miR393, miR396, miR397, miR402	Pathogen immune response Drought tolerance Oxidative stress tolerance Pathogen immunity response Syncytium formation response to parasitic nematodes	[16–19]
	Medicago truncatula	*CCAAT Binding Factor (CBF)* *Growth Regulating Factor (GRF)* *Cu/Zn superoxide dismutases (CSD1, CSD2)* *TIR-NBS-LRR domain protein*	miR169, miR396 miR398, miR2118	Drought tolerance Syncytium formation response to parasitic nematodes Oxidative stress tolerance Photoperiod-sensitive male sterility	[16,20]
	Oryza sativa	SalT (LOC_Os01g24710) *TIR1* OsLEA3 (LOC_Os05g46480)	miR393 miR402	Salt/cold tolerance	[6,17,18,21]
Cold stress	*Arabidopsis thaliana*	*Rd29A* (At5g52310) *CBF3* (At4g25480)	*miR165, miR172, miR169, miR396, miR397, miR402*	Drought/cold tolerance Drought tolerance Heat stress tolerance	[16,17]
	Oryza Sativa	*OsWRKY71* (LOC_Os02g08440) *OsMAPK2*(LOC_Os03g17700) Os05g47550, Os03g42280 Os01g73250, Os12g16350 Os03g19380	*miR319, miR389, miR393,* miR1320, miR1435 miR1884b, CHY1 CP12-2	Drought/salt tolerance Cold tolerance Pathogen immunity response	[17,21–23]

Table 1. *Cont.*

Stress Condition	Plant Species	Inducible Genes	Known Responsive miRNAs	Functions	References
Salinity stress	*Arabidopsis thaliana*	*Rd29A* (At5g52310) *COR15A* (At2g42540)	*miR389, miR393,*	Oxidative stress tolerance Heat stress tolerance	[24]
	Populus trichocarpa	Dihydropyrimidinase	miR162, miR164, miR166, miR167, miR168, miR172, miR395, miR396	Pathogen immune response Drought tolerance Drought/cold tolerance Sulfate-deficiency response	[25–27]
	Glycine max		miR1507a, miR395	Sulfate-deficiency response	[28]
	Oryza sativa	*SalT* (LOC_Os01g24710) *OsLEA3* (LOC_Os05g46480)	miR156, miR158, miR159, miR397, miR398, miR482.2, miR530a, miR1445	Drought tolerance Pathogen immune response Heat stress tolerance	[22,29–31]
	Zea mays		*miR402*	Seed germination and seedling growth of Arabidopsis under stress	[18]

Table 2. Microarray analysis of genes involved in the drought, salinity and cold stress responses in Arabidopsis.

Phenotype of Mutants	Genes	Function	AGI Code	Coded Proteins	Microarrays	
Increased tolerance to drought	*AtPARP2*	DNA repair	At2g31320	Poly (ADPribose) polymerase	24K Affymetrix	[32–34]
Hypersensitive to drought stress	*AHK1/ ATHK1*	positive regulator of drought and salt stress responses	At2g17820	Histidine kinase	22K Agilent	[32,35,36]
Increased tolerance to drought stress	*AREB1/ ABF2*	regulate the ABRE-dependent expression	At1g45249	bZIP TF	22K Agilent	[33,37,38]
Increased tolerance to salt stress	*AtbZIP60*	encodes a predicted protein of 295 aa	At1g42990	bZIP TF	44K Agilent	[37,39]
Increased tolerance to drought stress	*AtMYB60*	regulates stomatal movements and plant drought tolerance	At1g08810	MYB TF	7K cDNA	[40]
Increased sensitivity to drought stress	*AtMYB41*	control of primary metabolism and negative regulation	At4g28110	MYB TF	24K Affymetrix	[41,42]
Increased tolerance to drought and salt stress	*AHK2*	positive regulators for cytokinin signaling	At5g35750	Histidine kinase	Agilent	[35,36]
Increased tolerance to drought and salt stress	*AHK3*	perception of cytokinin, downstream signal transduction	At1g27320	Histidine kinase	22K Agilent	[35,36]
Increased tolerance to drought and freezing stress	*DREB1A/ CBF3*	stress-inducible transcription factor	ERF/AP2 TF	ERF/AP2 TF	1.3K cDNA	[43]
Increased tolerance to drought stress	*DREB2A*	heat shock-stress responses.	At5g05410	ERF/AP2 TF	22K Agilent 7K cDNA	[44]
Hypersensitive to salt	*HOS10*	coordinating factor for responses to abiotic stress and for growth and development.	At1g35515	MYB TF	24K Affymetrix	[32,45]
Increased tolerance to drought stress	*ZFHD1*	mediates all the protein-protein interactions	At1g69600	Zinc finger HD TF	22K Agilent	[36,39]

Numerous miRNAs/target gene expression modules are responsive to abiotic stresses in arabidopsis; therefore, altering the molecular profile of certain expression modules might help plants adapt to abiotic stresses [46,47]. To date, miRNAs have become an important field of intense study in recent years. Functional analysis of conserved miRNAs revealed their association with numerous developmental and biological processes. They regulate diverse metabolic events, including meristem boundary formation, organ separation and auxin signaling, the transition from the vegetative to the reproductive stage (juvenile-to-adult), and stress tolerance (Figure 1). The first reported miRNA in *Arabidopsis thaliana* to regulate the auxin signaling pathway was miR398, and miR398 was the first-ever reported miRNA related to stress tolerance. At the same time, the expression of miR398 was down-regulated under various oxidative activities and environmental stresses (Figure 1) [48,49], which further validate the substantial involvement of miRNAs in adverse environmental conditions [15]. MicroRNAs are significantly hardboard during plant development by negative gene expressions at the post-transcriptional level [50,51], and hence are considered as a popular molecular tool in modern biotechnology to study signal transduction, environmental extremes, response to stresses, protein degradation, biogenesis, and pathogen incursion [50,52,53]. Recently, several miRNAs have been mutually recognized by experimental and computational tactics in many crops [54]. In contrast, hundreds of identified miRNAs are documented as conserved across several species, suggesting that miRNAs might be used to develop abiotic stress tolerance in plants through genetic modifications [52,55].

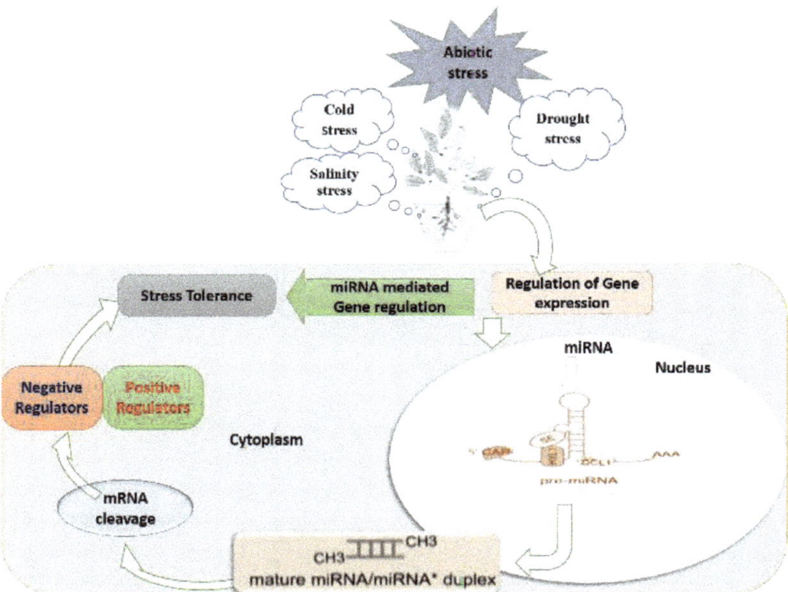

Figure 1. Schematic summary of miRNA-mediated regulatory mechanisms under abiotic stress in plant cells, with the particular formation process of miRNAs and miRNA mediated gene regulation: (1) miRNA gene is transcribed to a long sequence of primary miRNA (pri-miRNA). Primary miRNAs (pri-miRNAs) are transcribed from nuclear-encoded MIR genes by RNA polymerase II (Pol II), leading to precursor transcripts with a characteristic hairpin structure. (2) The pri-miRNA is cleaved to a stem-loop intermediate called miRNA precursor or pre-miRNA.

The second important function of miRNAs is in post-transcriptional regulation by targeting mRNAs for repressing or cleavage translation [16]. Many detrimental environmental factors adversely affect the plant's metabolic activities which, as a result, inhibit plant growth and development. However, it is quite challenging to differentiate and quantify the impact of various stresses on the plants through visual identification of hazardous factors, such as ozone, wound, and drought. Therefore, the development of sensitive and reliable techniques for diagnostics based on determining altering genes expression in DNA microarray is required [56]. Thus, the use of high-throughput sequencing (HTS) and genome tilling miRNA are focused on discovering the function of epigenetic mechanisms in ecological adaptation and genome idiomatic expressions. However, epigenomics, expression-pattern, and functional characterization urge us to elucidate the communal regulatory pathways by miRNAs that control abiotic stress resistance in plants [57]. Small RNA cloning and high-throughput deep sequencing technologies can obtain the expression profiles of both known and unknown miRNAs. The study of post-transcriptional regulation is also crucial in improving stress tolerance and suggesting next-generation targeting for classical breeding and genetic improvement.

DNA microarrays are a commonly developed tool in functional genomics. Analysis of the microarray expression profiles is a positive approach to improve in-depth understanding of genes involved in regulatory networks and signal transduction associated with resistance against multiple abiotic stresses [58,59]. With the continued progress of genome sequencing, DNA-microarray technology has become the pioneer in biotechnology and has bridged the gap between functional genomics and sequencing data. Microarrays are classified into two main classes according to the nature of immobilized probes: (1) DNA microarrays created with DNA-fragments which are normally produced by employing PCR techniques [60–62] and spotted cDNA-microarrays (most commonly used) and (2) oligonucleotide microarrays produced with longer (up to 120-mer) or shorter (10 to 40-mer) oligonucleotides premeditatedly corresponding to explicit coding targets. These cDNA-microarrays have certain advantages, particularly for regulating gene expression patterns. However, oligonucleotide-microarrays are restricted to low sequence complication array elements. The hybridization specificity for a compound probe is amended with arrays containing DNA fragments that are significantly longer than oligonucleotides [61,63]. The spotted cDNA-microarray was the earliest and widely used technology, which comprised several PCR-amplified probes of cDNA-fragments dropped, cross-linked, and dried in a matrix pattern of spots on a treated glass surface. The targets for these samples are preferentially identified cDNA solutions derived from reverse-transcribed mRNAs obtained from two cell samples populations [64,65]. There are two modifications to the DNA array series that may contain cDNAs that are immobilized to a firm base, such as oligonucleotides or glass/nylon membranes, that are perceived on glass slides (20 to 80-mer) [63]. The most hotly debated topics are the data normalization techniques, the purpose of which is to reduce the sample variations resulting from the technical features of microarray processing that may obscure biological differences in a specific experiment [66]. The review presents a perspective analysis and bridges the gap between previous and recent advancements in MicroRNAs and cDNA-Microarray as potent targets to cope with abiotic adversities in plants.

2. MicroRNAs and Microarray Target Prediction against Abiotic Stress

Perusing plant stress responses is an inclusive concern, which has been threatened by global warming and other abiotic factors. Currently, numerous miRNAs related to stress-responses have been identified as being triggered under high salinity, low temperature, and drought [58,67,68] (Tables 1–3, Figure 2). The stress-induced miRNAs depend upon the type of stress, tissues or organs, and plant genotype. Stress-sensitive miRNAs can either be negative regulators by downregulation or positive regulators by upregulation of the accumulation of positive regulators [57]. MicroRNA regulates gene modulation in a sequence-specific mode and plays a significant role against stress. Understanding and recognizing abiotic stress-associated microRNAs can help to establish schemes and improve tolerance against extreme stress [69,70]. Various advancements in miRNA identification—for example, deep sequencing, cloning, and prediction by bioinformatics methods, including miRNA-regulated network, miRNA target prediction, miRNA identification, expression profile, features (disease or stress, biomarker) association, tools based on machine learning algorithms, NGS, and tools specific for plants—have been developed to study the expression patterns of miRNA against stress [70–72]. High-throughput sequencing (HTS) evaluated the miRNA landscape of Arabidopsis entire seedlings subjected to heat, drought, and salinity stress, and 121, 123, and 118 miRNAs with a larger than 2-fold changed abundance, respectively, were discovered [46]. cDNA-microarray includes 3628 distinctive sequences retrieved from the Yukon ecotype of *Thellungiella salsuginea*, earlier stress-induced cDNA libraries, and reported transcript profiles in response to simulated drought, cold, and salinity [73]. Many stress-inducible genes are responsible for low temperature and dehydration; their sequences have been used to prepare cDNA-microarray with descriptive exposure of the *T. salsuginea* genome developed with stress-associated gene expression [41,73,74]. In addition, microarray revealed a larger number of stress-related genes (1886) as differentially regulated in *RGA1* mutants [75]. Using full-length cDNA or Gene Chips array transcription profiling experiments on *A. thaliana* reveals an extensive alteration occurrence in transcription against salinity, cold, and drought stress [74,76] (Table 2).

Table 3. miRNAs regulated by drought stress, salinity stress, and cold stress in plants.

Stress Condition	Plant Species	miRNA	Key Functions	Response	References
Drought stress	*Medicago truncatula*	miR398a,b miR408 miR399k miR2089 miR2111a-f,h-s miR2111g miR4414a	Oxidative stress tolerance Salt/drought/cold/oxidative osmotic-stress responses Phosphate-deficiency response	Up-regulated	[20,77–80]
		miR398b,c miR2111u,v miR5274b miR1510a-3p, 5p miR1510a	Heat stress tolerance Drought responsive Oxidative-stress tolerance triggering phasiRNA production from numerous NB-LRRs	Down-regulated	[77,79,80]
	Glycine max	miR5554a-c	Drought responsive		[79]
Salinity stress	*Glycine max*	miR169d miR395a miR395b,c miR1510a-5p miR1520d,e,l,n,q	Drought tolerance Sulfate-deficiency response triggering phasiRNA production from numerous NB-LRRs	Up-regulated	[20,81,82]
		gma-miR159b,c gma-miR169b,c gma-miR1520c	Pathogen immune response Drought/Salt tolerance	Down-regulated	[82]
	Phaseolus vulgaris	pvu-miR159.2	Plant–nematode interaction		[31]
Cold stress	*Phaseolus vulgaris*	pvu-miR2118	regulate the expression of genes encoding the TIR-NBS-LRR resistance protein	Up-regulated	

Cold- or drought-inducing genes were clustered based on the RNA gel blot and microarray analyses. The clusters were (1) cold-specific, (2) cold-inducible, and (3) drought-specific inducible genes. Recently, microRNAs have appeared as gene expression regulators that have also been associated with stress responses. However, the association between stress responses and miRNA expression is just beginning to be unfolded and documented. Fourteen stress-inducible miRNAs were established using microarray, in which the results of three main environmental stresses in Arabidopsis were plotted. Of them, 10 were cold regulated and had high salinity, while four were detected for drought miRNAs [83,84] (Tables 1 and 2). Seki M., et al. [43] reported 20 genes related with cold and drought-inducible genes, five which were drought-specific, and four novel genes, including *FL5-2D23*, *FL5-3J4*, *FL2-56*, and *FL6-55*, and two genes that were cold-specific inducible, including a novel (*FL5-90*) gene. Additionally, in rice, two siRNAs were previously reported as miR441 and miR446 [70,85,86]. They were testified to be down-regulated due to water deficiency; miR169g is the individual gene tempted by the scarcity of water which belongs to the miR169 family (Table 1). Moreover, the miRNAs responsive to abiotic stress inducements were comprised of 21 miRNAs belonging to 11 miRNA families which were up-regulated by UV-B stress in Arabidopsis [51,87,88].

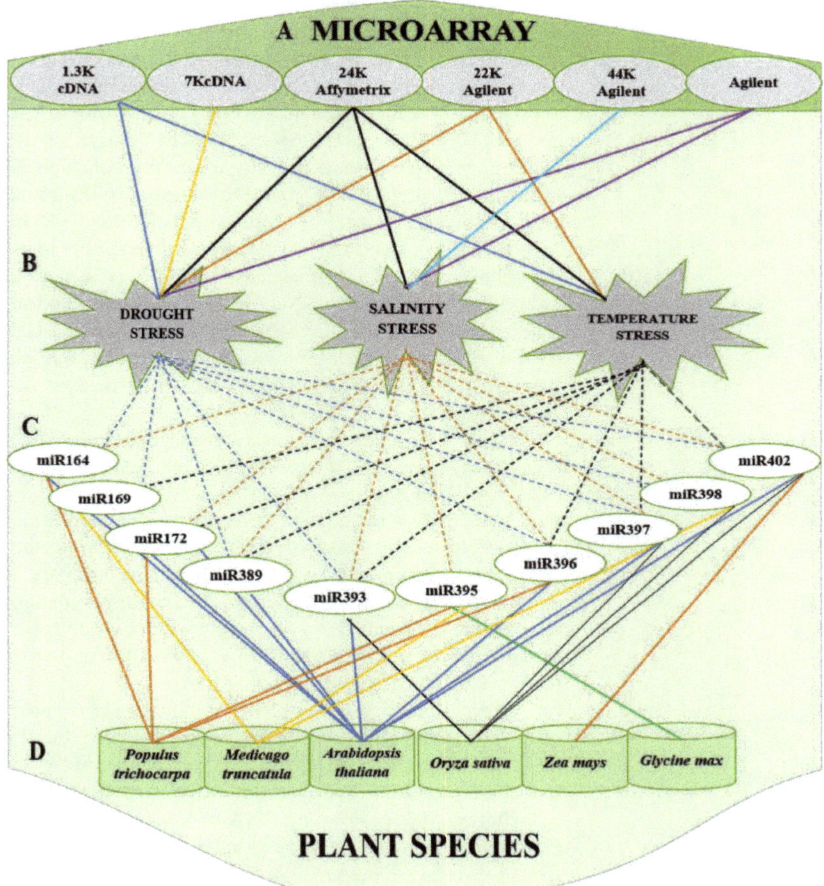

Figure 2. Summary of commonly used (**A**) microarrays (cDNA, Affymetrix, and Agilent) to stress and (**B**) miRNAs, categorized based on the stress, that respond to drought stress, salinity and temperature stress and (**C**) miRNAs reported in (**D**) plant species: *Populus trichocarpa*, *Medicago truncatula*, *Arabidopsis thaliana*, *Oryza sativa*, *Zea mays* and *Glycine max*.

High-throughput sequencing (HTS) microarray techniques have been employed for gene expression profiling under environmental stresses [42,89–91]. Several members of stress-regulated gene families were reported, such as *bZIP* to drought, *AP2* family to drought and cold, *MYB* to dehydration, *NAC* and *bHLH* to drought, ABA, and salinity, and *zinc finger* to drought and cold [92–94]. In addition, up-to-date, numerous drought-sensitive genes have been acknowledged in populous and pine [95,96].

3. Drought Responsive miRNAs and cDNA-Microarray

Drought stress is the foremost ecological factor that profoundly influences plant growth and development. Drought or soil water scarcity and perturbations is a main abiotic stress condition that causes yield reduction or complete crop loss [69]. It may be enduring in climatic zones with low or random water accessibility, due to meteorological changes during plant growth [97,98]. Therefore, preliminary physiological modifications in drought stress lead to radical gene expression variations [99]. A transcriptomic study in *Pinus taeda* was conducted in order to understand how plants were treated for mild drought and recuperation cycles [100]. To understand the role of microRNA, an oligonucleotide microarray was employed to control a rice microRNA expression profile against drought stress. Furthermore, it was confirmed that mir169g was stimulated by drought along with the mir169 family, and the introduction of mir169g was more prominent in roots than in shoots [16,93]. Among the miR169 family, only miR169g in *Oryza sativa* was regulated by drought [16]. Many genes associated with drought stress responses have been identified (Table 3) through cloning and characterization of cDNAs [101]. The examination of gene regulation through the drought stress response illuminated the roles of genes involved against abiotic stresses [67]. Moreover, microRNAs induced in drought was identified, and *mir169g* was reported as the only family member of mir169 which was induced under drought condition. The presence of *mir169g* was more pronounced in plant roots than in shoots. Several microRNAs in rice were modified against stress conditions on the microarray. RNA-seq analysis revealed two adjacent Dehydration-responsive elements (*DREs*) upstream of the *MIR169g*. *Mir169g* was substantially up-regulated and mir169 was the only family member caused via drought. The expression of *mir169g* might be directly synchronized by the *CBF/DREBs* [16,41].

Water uptake mechanisms are improved under stress, and the crop cells can confer drought avoidance to retain water and regulate the water deficit. The molecular response of higher plant mechanisms to water stress was analyzed by identifying various genes that are sensitive to drought stress at the transcriptional levels [102]. Comprehensive study on transcriptome analysis has presented important evidence on gene expression and pathways expressed differently in cotton cultivars, which are useful in developing drought tolerance [15,42]. MicroRNAs are known to significantly regulate the function against stress, but miRNAs associated with drought have not been recognized (Figure 2). Moreover, it is unclear that miRNAs could contribute to drought lenience capabilities in some plants (such as cowpeas) [103,104].

Expression microarrays provided novel insights into the physiological and metabolic pathways of dehydration tolerance, which led to the detection of candidate genes that might be helpful to speed up the breeding of tolerant varieties [99,100], and exhibited a photosynthetic acclimatization trend in response to moderate drought. Because of the novelty of the technology, performing DNA-microarray experiments remains a challenge [105,106]. *PHENOPSIS* was developed as an automated controlled drought screen to measure various Arabidopsis accessions efficiency and identify resistant ecotypes [107]. cDNA-microarrays have been designed for aquaporins (AQPs) to determine the expression patterns of 35 Arabidopsis *AQPs* in roots, flowers, and leaves. however, no leaf specific *AQPs* were identified. Plasma membrane intrinsic protein (PIP) transcripts were reported, usually down-regulated under moderate drought in the leaves, apart from *AtPIP2;5-6* and *AtPIP1*; 4, which were expressed constitutively and were unaffected by drought stress [108]. Liu, et al. [84] reported seven drought regulated miRNAs by microarray analysis in *Arabidopsis thaliana* (miR167, miR165, miR31, miR156, miR168, miR171, and miR396) and confirmed this by spotting their expression patterns in their promoter sequences and analyzing the cis-elements. Moreover, an additional subset of c.150 gene expression was discovered during recovery from the stress. Identifying co-regulated gene groups has made it possible to identify common sequence patterns between promoters of certain genes and to detect transcription factors that control their expression [30,67,76] (Tables 1 and 2).

The plant stress-responsive pathways are not linear, but are dynamically integrated circuits consisting of several passages involving various tissues, cellular compartments, cofactors, and signaling molecules to organize a precise response to particular signals [109,110]. Microarray research showed that transgenic drought resistance was associated with several stress tolerant pathway genes, such as *DREB1A/CBF3*, *RD29A*, and *COR15A*, and was up-regulated. Protein phosphorylation/dephosphorylation is the main signaling event, which is being stimulated by osmotic stress. Arabidopsis 2-Oligo Microarray (Agilent) was used to analyze transcription profiles of the *SRK2C* gene, and protein kinase activated by osmotic stress (Table 1, Figure 2) [111].

4. miRNAs and cDNA-Microarray Associated with Cold Stress

Cold stress (frost and chilling) decreases crop yields worldwide through tissue degradation and delayed growth. Most temperate plants have evolved cold resistance through cold-acclimatization [112]. Signaling pathways were being used in response to winter stress. The functional genes transform reactions, and reposts suggest that the signaling pathways for leaf senescence and plant defense responses may overlap [113]. The most characteristic region of cold-stress responsive genes includes transcription factors, such as *CBF/DREB* and stress-inducible candidate genes, identified as *KIN* (cold-induced), *COR* (cold-regulated), and *LTI* genes (induced by low temperature) or *RD* (dehydration) [114]. Several *HSPs* (heat shock proteins) are also reported for their functions against cold stress. *HSPs*, which perform as molecular chaperons, play an important regulatory function in protecting from stress by restoring normal protein conformation and thus maintaining cellular homeostasis in plants [115]. The number of the miRNA target genes in expression is intricate during stress and plant growth. These miRNAs are co-regulated by both developmental signals and ecological factors (Table 3). The cold-responsive miRNAs were detected by microarray analysis in *Arabidopsis thaliana* (miR165, miR31, miR156, miR168, miR171, miR396) and recommended by identifying their expression patterns in their promoter sequences and evaluating the cis-components (Table 3, Figure 1) [116,117]. Furthermore, high-intensity light (HL) responsive genes were assessed with the drought-inducible genes reported with a similar microarray system, which exposed an impenetrable intersection between drought and HL-induced genes. Moreover, 10 genes were identified as being involved in the regulation by HL, drought, salinity, and cold stress (Tables 1 and 2). These genes are comprised of *ERD10*, *RD29A*, *KIN1*, *LEA14*, *COR15a*, and *ERD7*, and most of them are considered to be concerned in the defense of cellular components [78,118,119]. Along with the HL-inducible genes, some are also identified and encouraged by other stresses (heat, drought, and cold), including *AtGolS*, *LEA*, *RAB*, *RD*, *COR*, *ERD*, *HSP*, *KIN*, lipid-transfer proteins, and *fibrillins* [76,120,121].

DNA microarrays almost in all genes of the unicellular *Synechocystis sp PCC6803* were used to investigate the gene expression sequential software [122]. A cDNA-microarray was used to test the profile expression in cold stress, and 328 temperature-regulated transcripts were reported. *OsMYB3R-2* was studied further and was shown to be a dominant regulator against stress [123]. In this study, there was an attempt to use a 3.1K cDNA-microarray to express the cold-regulated transcripts in the *Capsicum annuum*. Several TFs, including the *EREBP* (*CaEREBP-C1* to *C4*) family of four genes, a protein of the ring domain, a *bZIP* protein (CaBZ1), RVA1, a WRKY (CaWRKY1), and HSF1 protein have been observed among the cold stress-regulated genes. These genes included *CaBZ1*, *CaEREBP-C3*, *NtPRp27*, the *SAR8.2* protein precursor, putative trans-activator factor, malate hydrogenase, putative protein of auxin-repressed, xyloglu-canendo-1, 4-D-gucanase precursor, LEA protein 5 (*LEA5*), homologous *DNAJ* protein, *PR10* and Stns *LTP* [124,125]. cDNA microarray z1300 full-length cDNAs were used in Arabidopsis to identify cold stress-inducing genes and target genes of *DREB1A/CBF3*. Six genes were documented based on microarray and, in RNA gel blot analyses, it was observed that a novel *DREB1A* controls cold- and drought-inducible genes [43,126]. Furthermore, microarray with full-length cDNA was performed by 1300 full-length cDNAs and cDNA microarray to discover cold-induced genes. Previous reposts exhibited the target genes of *DREB1A/CBF3* and stress-inducible gene expressions were controlled by transcription factors [76]; in contrast, stress-sensitive genes' expressions were reported as specific to the growth stage [42]. Full-length cDNA microarray is convenient for analyzing the Arabidopsis gene expression patterns under cold stress, and can also be used to identify the functional genes of stress-related TFs that are likely to act as DNA elements by merging the genomic sequence data with the expression data [76,127]. Additionally, cold stress is also induced by the increase in the proline content in plants (osmoprotectant). Microarray and RNA gel blot research found that the proline can induce the expression of several genes with the proline-responsive elements in their promoters (PRE, ACTCAT) [120,127,128]. Microarray analysis was carried out to detect the cold-inducible

AP2 gene family transcription factor *RAV1* [129], which could control plant growth under stress. *RAV1* is down-regulated by epibrassinolide, and transgenic Arabidopsis overexpressing *RAV1* exhibits a rosette leaf and adjacent root growth retardation, although the early-flowering phenotype showed antisense to RAV1 plants [130,131].

5. miRNAs and cDNA-Microarray Response to Salinity Stress

Salt intrusion from saline soils and irrigation water is one of the most severe and harmful risks to reduce agricultural production and adverse effects on cultivated land and the geographical distribution of plant species [70,132,133], coupled with oxidative stress [134]. The most imperative cations in saline soils are calcium, potassium, magnesium, and sodium, and the main anions in saline soils are chloride, bicarbonate, sulfate, nitrate, and carbonates. Other electrolytes causative to salinity are borane, molybdenum, strontium, silicon dioxide, aluminum cation, and barium ion [135,136]. Higher concentrations of sodium chloride (NaCl) typically affect plant development, metabolism, and physiology at various metabolic phases (ion toxicity, nutrient imbalance, and oxidative stress) [70,137]. Despite such advances in scientific research, it remains unclear about the underlying molecular mechanism of salinity responses in plants. However, based on the combination of microarray and inhibition subtractive hybridization (SSH), changes in the transcriptome profile caused by salt induction were studied and evaluated [138]. Investigation of complete transcriptomics suggests that these processes, such as the synthesis of osmolytes and ion carriers and the regulation of transcription and translation mechanisms, have distinctive reactions under salinity stress. In particular, the introduction of transcripts of specific TFs, ribosomal genes, RNA-binding proteins, and translation initiation and elongation factors has been testified [139,140].

Using cDNA microarray in *Synechocystis*, 19 genes were reported to be instantaneously regulated under salinity stress. The salt- and osmo-regulated genes, and some putative sensor molecules, have been implicated during salinity stress signaling [35]. Several differentially regulated miRNAs have been reported against salinity stress. In *A. thaliana*, several microRNAs are regulated against salinity stress, such as miR156, miR158, miR159, miR165, miR167, miR168, miR169, miR171, miR319, miR393, miR394, miR396, and miR397 (Table 3, Figure 2) [84]. In *Populus trichocarpa*, miR1445, miR1447, miR1446a-e, miR530a, and miR171l-n were down-regulated (Table 3) [141]. Arenas-Huertero et al. [31] reported, in *Proteus vulgaris*, the production of miRS1 and miR159.2 expression in response to salinity. Furthermore, miR169g and family members of miR169n were induced in saline-rich conditions [142]. However, there is a need to discover and annotate novel functional genes which have a probable function against salinity stress. Subsequently, a large number of genes in plants still have unknown functions [143]. Recent studies revealed that specific down-regulation of the bacterial-type *phosphoenolpyruvate carboxylase* (*PEPC*) gene *Atppc4* by artificial microRNA enhanced the salinity tolerance in *A. thaliana*. The increased salinity tolerance might be linked to enhanced PEPC activity [10,144]. Transcript control for salinity-tolerant rice with microarrays, like 1728 cDNAs from salinity-stressed roots libraries, was studied in response to high salinity (Table 3) [144–146].

A tiling path microarray was used to examine the high-throughput expression profiling patterns under various environmental stresses for all of the known miRNAs [16,70] (Tables 1 and 4). The analysis revealed that the effects of miRNAs under low-temperature, drought, and high salinity with miRNA chips represent, approximately, all of the reported miRNAs cloned or recognized in *A. thaliana* (L.). High salinity stress agitates homeostasis in water potential. Extreme changes in water homeostasis and ions lead to molecular breakdown, stunted growth, and even the death of cells or whole plants [16,147].

Table 4. Software and tools used for the detection of plant miRNA and cDNA microarray data analysis.

Software and Tools	Function	Website	Reference	Accessed
Software and tools used for detection of plant miRNA and data analysis				
MiPred	Random forest (RF)-based miRNA predictor, which can distinguish between real and pseudo-miRNA precursors	http://server.malab.cn/MiPred/	[72]	5 November 2021
miBridge	Algorithm and database	http://sitemaker.umich.edu/mibridge/home	[148]	5 November 2021
miRTar	A novel rule-based model learning method for cell line specific microRNA target prediction	http://miRTar.mbc.nctu.edu.tw	[72]	5 November 2021
PolymiRTS	Linking polymorphisms in microRNAs and their target sites	http://compbio.uthsc.edu/miRSNP	[149]	25 November 2021
miRGator	microRNA portal for deep sequencing, expression profiling and mRNA targeting	http://mirgator.kobic.re.kr	[150]	10 November 2021
Bowtie	Aligns efficiently, and short-read aligners	http://bowtie-bio.sourceforge.net	[72]	5 November 2021
miRBase	Provides handy and useful ID conversion tools	http://www.mirbase.org/	[72]	25 November 2021
miRDB	miRNA target databases	http://www.mirdb.org	[151]	25 November 2021
mirDIP	Integrative database of microRNA target predictions	http://ophid.utoronto.ca/mirDIP	[152]	25 November 2021
miRanda	Predict or collect miRNA targets	http://34.236.212.39/microrna/home.do	[72]	25 November 2021
RNAhybrid	microRNA target prediction	https://bibiserv.cebitec.uni-bielefeld.de/rnahybrid	[72]	8 November 2021
miTALOS	Analyzes tissue specific microRNA function.	http://mips.helmholtz-muenchen.de/mitalos	[153]	5 November 2021
RNA22	microRNA target predictions	https://cm.jefferson.edu/rna22	[154]	5 November 2021
psRNATarget	Small RNA target analysis server	http://plantgrn.noble.org/psRNATarget/	[155]	5 November 2021
miRandola	Curated knowledge base of non-invasive biomarkers	http://mirandola.iit.cnr.it/	[155]	5 November 2021
ChIPBase	Decoding transcriptional regulatory networks of non-coding RNAs and protein-coding genes from ChIP-seq data	http://rna.sysu.edu.cn/chipbase/	[155,156]	1 October 2021
MirGeneDB	Curated miRNA gene database	http://mirgenedb.org/	[157]	28 November 2021
TarHunter	Predicting conserved microRNA targets and target mimics in plants	http://tarhunter.genetics.ac.cn	[158]	28 November 2021
TissueAtlas	Tissue specificity miRNA database	https://ccb-web.cs.uni-saarland.de/tissueatlas/	[72]	28 November 2021
miRNAme Converter	miRNA ID converter	http://163.172.134.150/miRNAmeConverter-shiny	[159]	28 November 2021

Table 4. *Cont.*

Software and Tools	Function	Website	Reference	Accessed
Software and tools used for detection of plant microarray and data analysis				
Array Designer	Design primers and probes for oligo and cDNA expression microarrays.	http://www.premierbiosoft.com/dnamicroarray/index.html	[160]	1 November 2021
Stanford Microarray Database SMD	Stores raw and normalized data from microarray experiments	http://smd-www.stanford.edu//download/	[161]	1 November 2021
eArray	Designing Agilent arrays	http://earray.chem.agilent.com/earray/login.do	[160]	1 November 2021
Significance Analysis of Microarrays	Adjustments for multiple testing, statistical analysis for discrete, quantitative, and time series data, gene set enrichment analysis	http://www-stat.stanford.edu/~tibs/SAM/	[162]	5 November 2021
Visual OMP	Design software for RNA, DNA, single or multiple probe design, microarrays, Taq Manassays, genotyping, single and multiplex PCR, secondary structure simulation, sequencing, genotyping.	http://www.dnasoftware.com/Products/VisualOMP	[160]	5 November 2021
caArray	Open-source, web and programmatically accessible microarray data management system that supports the annotation of microarray	http://caarray.nci.nih.gov/		5 November 2021
Gene Expression Model Selector	Diagnostic models and biomarker discovery	http://www.gems-system.org/	[163]	18 November 2021
Gene index	Gene Index Project is to use the available EST and gene sequences, along with the reference genomes, to provide an inventory of likely genes and variants.	http://compbio.dfci.harvard.edu/tgi/plant.html	[160]	5 November 2021
Genesis	Java package of tools to simultaneously visualize and analyze a whole set of gene expression experiments	http://genome.tugraz.at/genesisclient/genesisclient_description.shtml		18 November 2021
RMA Express	Standalone GUI program for Windows, OS X and Linux to compute gene expression summary values for Affymetrix	http://rmaexpress.bmbolstad.com http://www.r-project.org http://www.bioconductor.org		18 November 2021
dCHIP	Model-based expression analysis for Affymetrix gene expression arrays	http://www.dchip.org	[164]	18 November 2021
TM4	Microarray Data Manager (MADAM), TIGR Spotfinder, Microarray Data Analysis System (MIDAS), and Multi experiment Viewer (MeV)	http://www.tm4.org/	[164]	18 November 2021

Table 4. *Cont.*

Software and Tools	Function	Website	Reference	Accessed
Able Image Analyser	Software for image analysis. It enables dimensional measurements: distance, area, angle in digital images	http://able.mulabs.com	[160]	18 November 2021
ImaGene	Unique, robust, room-temperature preservation solutions for nucleic acids, biospecimens and bioreagents for in the living ectors	http://www.biodiscovery.com/index/imagene	[160]	13 November 2021
Spotfinder	Custom-designed cDNA array, the chips are scanned using a microarray scanner	http://www.tm4.org/spotfinder.html	[164]	18 November 2021
SNOMAD	Web-based tool and has various normalization options for two-channel and single-channel experiments	http://pevsnerlab.kennedykrieger.org/snomadinput.html	[164]	18 November 2021
Multiexperimet Viewer	Cloud-based application supporting analysis, visualization, and stratification of large genomic data	http://www.tm4.org/mev.html		18 November 2021
Onto-Express and Pathway-Express	Automatically translates DE gene transcripts from microarray experiments into functional profiles characterizing the impact of the condition studied	http://vortex.cs.wayne.edu/projects.htm	[164]	13 November 2021
DAVID/EASE	Database for annotation, visualization and integrated discovery (DAVID) is an online tool for annotation and functional analysis. Expression analysis systematic sxplorer (EASE)	http://david.abcc.ncifcrf.gov	[164]	13 November 2021

Oligo-DNA microarrays were developed in common wheat, and these microarrays were designed to include approximately 32,000 distinctive genes characterized by several expressed sequence tags (ESTs). To classify the salinity-stress responsive genes, the expression profiles of transcripts that responded to stress were examined using microarrays. It was concluded that 5996 genes were verified by more than a 2-fold change in expression. These genes were categorized into twelve groups based on gene expression patterns [165]. Transcription-regulator activity, DNA binding, and the genes' assigned transcription factor functions were preferentially classified as immediate response genes. In wheat, candidate genes were identified as involved in salinity-stress tolerance [165,166]. These genes are active in the regulation of transcription [112,143] and the signal transduction that is engaged in metabolic pathways [167] or acting as ion transporters [168]. cDNA library in yeast (*Saccharomyces cerevisiae*) was examined using a synthetic medium augmented with excessive salt concentrations (900 mM). A few clones showed comparatively improved growth. The notorious clones bore the *Guanyl transferase* (OsMPG1) mannose-1-phosphate gene [133]. Extreme salinity stress was significantly linked with the transcription factors of four tomato genes from the family of *zinc finger*. There has been prior evidence of the relationship between *zinc finger* transcription factors and plant salinity tolerance [169,170]. Overexpression of *OSISAP1* in transgenic tobacco resulted in tolerance to salinity, dehydration, and cold stress in the new sprouts [171].

A microarray containing 384 genes associated with stress responses was used in *Medicago truncatula* genotypes (*Jemalong A17* and *108-R*) to compare rooting gene expression during salt stress. The homolog of flora *TFIIIA*-related TF, *MtZpt2*-1, and *COLD-REGULATEDA1* genes were known to regulate the previous genes and were acknowledged in *Jemalong A17* stress-tolerant genotypes. Two *MtZpt2* Transcription factors (*MtZpt2-1* and *MtZpt2-2*) have shown increased expression in the roots compared to *108-R* [172]. Salinity stress is attributed to diverse stresses that persuade overlapping patterns in gene expression. For example, in an investigation of 8100 *A. thaliana* genes, approximately 2400 genes were reported to have a widespread expression in exposure to salt, oxidative and cold stress [92]. In addition, 23 genes were reported against NaCl stress. This also accounted for a small percentage of DEGs, including encoding transcription factors *WOX2* and *BZIP3*, calcium-binding protein *CML42*, ubiquitin-protein ligase *UBC17*, and *IDA-like 5* protein [92]. Most prominently, synthesized *isiA* encoded a novel chlorophyll (*Chl*)-binding protein [173] (Table 3).

6. Potential Role of Bioinformatics in the Prediction of miRNA and cDNA Microarray

Next-generation sequencing methods are crucial in gene expression profiling, epigenomics, genomics, and transcriptomics. These tools can sequence multiple DNA molecules within a short period. The recent introduction of innovative "-omics" technologies, such as metabolomics, proteomics, and genomics allows for analyzing and identifying the genetic elements that contribute to system complexity [72,90,174,175]. Bioinformatics tools developed for miRNA prediction include miRNA target prediction, analysis, and structure prediction. For example, miRanda, RNAhybrid, RNA22, and TarHunter detect miRNA expression and perform analysis based on miRNA-Seq data (Table 4). Existing plant miRNA prediction tools lack a cross-species conservation filter and eTM prediction function. TarHunter features a strict cross-species conservation filter and the capability of predicting eTMs [158]. Despite ongoing progress, bioinformatics prediction of microRNA targets remains difficult, since current tools have a lack of accuracy and sensitivity. [72,176]. Microarrays are an effective method for determining the quantity of RNA in a sample. Since microarray data have computational complexity and contain hundreds of genes, statistical and bioinformatics methods are required for data interpretation [160]. These specialized tools provide statistical analysis, sample comparisons, and functional interpretation of data generated in a series following visualization and normalization in a microarray study, such as Array Designer, eArray, Visual OMP, caArray, and dCHIP (Table 4). The software, including Able Image Analyser, Gene pix pro-6.0, and GeneChip operating software, are used for analyzing images in order to obtain the intensity at each spot and quantify the expression for each transcript. Additionally, this also provides different types of discoveries by comparing gene expression data with already reported biological information, such as protein–protein interactions, pathway analysis, transcription factor binding sites, and network analysis tools, including Array Designer, eArray, Significance Analysis of Microarrays, Gene Expression, and Model Selector (Table 4) [164].

7. Conclusions and Future Perspectives

MicroRNAs (miRNAs) have been considered a potential target in genetic engineering against abiotic stresses in plants. Thus, miRNAs can also be utilized in the initial monitoring and transmission of abiotic stresses, and to elucidate the genetic and physiological responses against stress in plants. This review summarized current developments and the history of miRNAs and microarray with diverse functions in several stress responses, predominantly abiotic stresses. Many traditional approaches have identified significant numbers of miRNAs in plants from various organisms. Microarray-based genomic technologies for ecological studies have received great attention, particularly in plants, to disclose the role of stress-responsive loci in plants. DNA microarrays provide a novel insight into the cell and provide a solution for several problems from the viewpoint of analytical calculation, despite the inconceivable amount of work done in the last two decades to reduce the different sources of uncertainty on the subsequent measurements. The review will provide valuable insight to plant researchers, especially plant breeders and stress physiologists, to design a comprehensive strategy to cope with environmental stresses.

The elucidation of miRNA responses to abiotic stresses may lead to the development of technologies for the early detection of plants' environmental stressors. MicroRNAs and cDNA-microarrays are powerful targets for engineering abiotic stress tolerance in transgenic plants. The field of bioinformatics is developing rapidly, and it is inevitable to progress in plant genomics and breeding without integrating the latest bioinformatics tools. Multiple advanced sequencing and bioinformatics tools were established to identify miRNAs and their target gene network and prediction. As the understanding of the function of miRNAs under stress deepens, the potential use of miRNA mediated genes to enhance plant tolerance will also increase. In the future, the large-scale microarrays might be substituted with small biosensors which contain a unique or a small number of novel microbes deposited on an electronic platform. We would like to conclude by illustrating the existing gap between the detection of stress-regulated miRNAs and microarray to validate their role. In conclusion, we recommend the utilization of miRNAs for the identification and classification of new functional genes conferring a significant functional role in stress tolerance and to exploit the unexplained fraction of genes further.

Author Contributions: T.P. and M.W.A.: have made substantial contributions to formulate, Drafting and retrieve data, A.E.-k.; revised and improved the technical language S.-H.N.; Revised and improved the MS, H.X.W.; Final approval and revised for publication. All authors have read and agreed to the published version of the manuscript.

Funding: This work is funded by the National Natural Science Foundation of China (31870651).

Institutional Review Board Statement: Not applicable.

Informed Consent Statement: Not applicable.

Data Availability Statement: Not applicable.

Conflicts of Interest: The authors declare no conflict of interest.

References

1. Pervaiz, T.; Liu, S.-W.; Uddin, S.; Amjid, M.W.; Niu, S.-H.; Wu, H.X. The Transcriptional Landscape and Hub Genes Associated with Physiological Responses to Drought Stress in Pinus tabuliformis. *Int. J. Mol. Sci.* **2021**, *22*, 9604. [CrossRef]
2. Raza, A. Eco-physiological and biochemical responses of rapeseed (Brassica napus L.) to abiotic stresses: Consequences and mitigation strategies. *J. Plant Growth Regul.* **2021**, *40*, 1368–1388. [CrossRef]
3. Li, C.; Zhang, B. MicroRNAs in control of plant development. *J. Cell. Physiol.* **2016**, *231*, 303–313. [CrossRef]
4. Latef, A.A.H.A.; Hashem, A.; Rasool, S.; Abd_Allah, E.F.; Alqarawi, A.; Egamberdieva, D.; Jan, S.; Anjum, N.A.; Ahmad, P. Arbuscular mycorrhizal symbiosis and abiotic stress in plants: A review. *J. Plant Biol.* **2016**, *59*, 407–426. [CrossRef]
5. Vishwakarma, K.; Upadhyay, N.; Kumar, N.; Yadav, G.; Singh, J.; Mishra, R.K.; Kumar, V.; Verma, R.; Upadhyay, R.; Pandey, M. Abscisic acid signaling and abiotic stress tolerance in plants: A review on current knowledge and future prospects. *Front. Plant Sci.* **2017**, *8*, 161. [CrossRef] [PubMed]
6. Meng, Y.; Ma, X.; Chen, D.; Wu, P.; Chen, M. MicroRNA-mediated signaling involved in plant root development. *Biochem. Biophys. Res. Commun.* **2010**, *393*, 345–349. [CrossRef] [PubMed]
7. Begara-Morales, J.C.; Chaki, M.; Valderrama, R.; Mata-Pérez, C.; Padilla, M.N.; Barroso, J.B. The function of S-nitrosothiols during abiotic stress in plants. *J. Exp. Bot.* **2019**, *70*, 4429–4439. [CrossRef]
8. Dunn, L.E.; Ivens, A.; Netherton, C.L.; Chapman, D.A.; Beard, P.M. Identification of a functional small non-coding RNA encoded by African swine fever virus. *J. Virol.* **2020**, *94*, 865147. [CrossRef] [PubMed]
9. Borges, F.; Martienssen, R.A. The expanding world of small RNAs in plants. *Nat. Rev. Mol. Cell Biol.* **2015**, *16*, 727–741. [CrossRef]
10. Wang, C.; Han, J.; Liu, C.; Kibet, K.N.; Kayesh, E.; Shangguan, L.; Li, X.; Fang, J. Identification of microRNAs from Amur grape (Vitis amurensis Rupr.) by deep sequencing and analysis of microRNA variations with bioinformatics. *BMC Genom.* **2012**, *13*, 122. [CrossRef]
11. Bartel, D.P. MicroRNAs: Genomics, biogenesis, mechanism, and function. *Cell* **2004**, *116*, 281–297. [CrossRef]
12. Islam, W.; Adnan, M.; Huang, Z.; Lu, G.-D.; Chen, H.Y. Small RNAS from seed to mature plant. *Crit. Rev. Plant Sci.* **2019**, *38*, 117–139. [CrossRef]

13. Wang, Y.; Wang, Q.; Gao, L.; Zhu, B.; Ju, Z.; Luo, Y.; Zuo, J. Parsing the regulatory network between small RNAs and target genes in ethylene pathway in tomato. *Front. Plant Sci.* **2017**, *8*, 527. [CrossRef]
14. Slaby, O.; Laga, R.; Sedlacek, O. Therapeutic targeting of non-coding RNAs in cancer. *Biochem. J.* **2017**, *474*, 4219–4251. [CrossRef]
15. Nejat, N.; Mantri, N. Emerging roles of long non-coding RNAs in plant response to biotic and abiotic stresses. *Crit. Rev. Biotechnol.* **2018**, *38*, 93–105. [CrossRef] [PubMed]
16. Zhao, B.; Liang, R.; Ge, L.; Li, W.; Xiao, H.; Lin, H.; Ruan, K.; Jin, Y. Identification of drought-induced microRNAs in rice. *Biochem. Biophys. Res. Commun.* **2007**, *354*, 585–590. [CrossRef]
17. Sunkar, R.; Zhu, J.-K. Novel and stress-regulated microRNAs and other small RNAs from Arabidopsis. *Plant Cell* **2004**, *16*, 2001–2019. [CrossRef] [PubMed]
18. Wang, Y.; Long, L.-H. Identification and isolation of the coldresistance related miRNAs in Pisum sativum Linn. *J. Liaoning Norm. Univ.* **2010**, *2*, 027.
19. Ebert, M.S.; Neilson, J.R.; Sharp, P.A. MicroRNA sponges: Competitive inhibitors of small RNAs in mammalian cells. *Nat. Methods* **2007**, *4*, 721–726. [CrossRef]
20. Sunkar, R.; Chinnusamy, V.; Zhu, J.; Zhu, J.-K. Small RNAs as big players in plant abiotic stress responses and nutrient deprivation. *Trends Plant Sci.* **2007**, *12*, 301–309. [CrossRef]
21. Raffaele, S.; Mongrand, S.; Gamas, P.; Niebel, A.; Ott, T. Genome-wide annotation of remorins, a plant-specific protein family: Evolutionary and functional perspectives. *Plant Physiol.* **2007**, *145*, 593–600. [CrossRef]
22. Zhou, X.; Jeker, L.T.; Fife, B.T.; Zhu, S.; Anderson, M.S.; McManus, M.T.; Bluestone, J.A. Selective miRNA disruption in T reg cells leads to uncontrolled autoimmunity. *J. Exp. Med.* **2008**, *205*, 1983–1991. [CrossRef]
23. Rabbani, M.A.; Maruyama, K.; Abe, H.; Khan, M.A.; Katsura, K.; Ito, Y.; Yoshiwara, K.; Seki, M.; Shinozaki, K.; Yamaguchi-Shinozaki, K. Monitoring expression profiles of rice genes under cold, drought, and high-salinity stresses and abscisic acid application using cDNA microarray and RNA gel-blot analyses. *Plant Physiol.* **2003**, *133*, 1755–1767. [CrossRef] [PubMed]
24. Moons, A.; Prinsen, E.; Bauw, G.; Van Montagu, M. Antagonistic effects of abscisic acid and jasmonates on salt stress-inducible transcripts in rice roots. *Plant Cell* **1997**, *9*, 2243–2259. [PubMed]
25. Boualem, A.; Laporte, P.; Jovanovic, M.; Laffont, C.; Plet, J.; Combier, J.P.; Niebel, A.; Crespi, M.; Frugier, F. MicroRNA166 controls root and nodule development in Medicago truncatula. *Plant J.* **2008**, *54*, 876–887. [CrossRef] [PubMed]
26. Yang, F.; Yu, D. Overexpression of Arabidopsis MiR396 enhances drought tolerance in transgenic tobacco plants. *Acta Bot. Yunnanica* **2009**, *31*, 421–426. [CrossRef]
27. Subramanian, S.; Fu, Y.; Sunkar, R.; Barbazuk, W.B.; Zhu, J.-K.; Yu, O. Novel and nodulation-regulated microRNAs in soybean roots. *BMC Genom.* **2008**, *9*, 1–14. [CrossRef]
28. Zeng, H.Q.; Zhu, Y.Y.; Huang, S.Q.; Yang, Z.M. Analysis of phosphorus-deficient responsive miRNAs and cis-elements from soybean (Glycine max L.). *J. Plant Physiol.* **2010**, *167*, 1289–1297. [CrossRef] [PubMed]
29. Cui, H.; Zhai, J.; Ma, C. miRLocator: Machine learning-based prediction of mature microRNAs within plant pre-miRNA sequences. *PLoS ONE* **2015**, *10*, e0142753. [CrossRef]
30. Yu, Y.; Wu, G.; Yuan, H.; Cheng, L.; Zhao, D.; Huang, W.; Zhang, S.; Zhang, L.; Chen, H.; Zhang, J.; et al. Identification and characterization of miRNAs and targets in flax (Linum usitatissimum) under saline, alkaline, and saline-alkaline stresses. *BMC Plant Biol.* **2016**, *16*, 124. [CrossRef] [PubMed]
31. Arenas-Huertero, C.; Pérez, B.; Rabanal, F.; Blanco-Melo, D.; De la Rosa, C.; Estrada-Navarrete, G.; Sanchez, F.; Covarrubias, A.A.; Reyes, J.L. Conserved and novel miRNAs in the legume Phaseolus vulgaris in response to stress. *Plant Mol. Biol.* **2009**, *70*, 385–401. [CrossRef] [PubMed]
32. Gu, Z.; Pan, W.; Chen, W.; Lian, Q.; Wu, Q.; Lv, Z.; Cheng, X.; Ge, X. New perspectives on the plant PARP family: Arabidopsis PARP3 is inactive, and PARP1 exhibits predominant poly (ADP-ribose) polymerase activity in response to DNA damage. *BMC Plant Biol.* **2019**, *19*, 364. [CrossRef]
33. Rasheed, S.; Bashir, K.; Matsui, A.; Tanaka, M.; Seki, M. Transcriptomic Analysis of Soil-Grown Arabidopsis thaliana Roots and Shoots in Response to a Drought Stress. *Front. Plant Sci.* **2016**, *7*, 180. [CrossRef] [PubMed]
34. Vanderauwera, S.; De Block, M.; Van de Steene, N.; Van De Cotte, B.; Metzlaff, M.; Van Breusegem, F. Silencing of poly (ADP-ribose) polymerase in plants alters abiotic stress signal transduction. *Proc. Natl. Acad. Sci. USA* **2007**, *104*, 15150–15155. [CrossRef] [PubMed]
35. Marin, K.; Suzuki, I.; Yamaguchi, K.; Ribbeck, K.; Yamamoto, H.; Kanesaki, Y.; Hagemann, M.; Murata, N. Identification of histidine kinases that act as sensors in the perception of salt stress in Synechocystis sp. PCC 6803. *Proc. Natl. Acad. Sci. USA* **2003**, *100*, 9061–9066. [CrossRef]
36. Tran, L.-S.P.; Urao, T.; Qin, F.; Maruyama, K.; Kakimoto, T.; Shinozaki, K.; Yamaguchi-Shinozaki, K. Functional analysis of AHK1/ATHK1 and cytokinin receptor histidine kinases in response to abscisic acid, drought, and salt stress in Arabidopsis. *Proc. Natl. Acad. Sci. USA* **2007**, *104*, 20623–20628. [CrossRef] [PubMed]
37. Seki, M.; Narusaka, M.; Ishida, J.; Nanjo, T.; Fujita, M.; Oono, Y.; Kamiya, A.; Nakajima, M.; Enju, A.; Sakurai, T. Monitoring the expression profiles of 7000 Arabidopsis genes under drought, cold and high-salinity stresses using a full-length cDNA microarray. *Plant J.* **2002**, *31*, 279–292. [CrossRef]

38. Fujita, Y.; Fujita, M.; Satoh, R.; Maruyama, K.; Parvez, M.M.; Seki, M.; Hiratsu, K.; Ohme-Takagi, M.; Shinozaki, K.; Yamaguchi-Shinozaki, K. AREB1 is a transcription activator of novel ABRE-dependent ABA signaling that enhances drought stress tolerance in Arabidopsis. *Plant Cell* **2005**, *17*, 3470–3488. [CrossRef]
39. Fujita, M.; Mizukado, S.; Fujita, Y.; Ichikawa, T.; Nakazawa, M.; Seki, M.; Matsui, M.; Yamaguchi-Shinozaki, K.; Shinozaki, K. Identification of stress-tolerance-related transcription-factor genes via mini-scale Full-length cDNA Over-eXpressor (FOX) gene hunting system. *Biochem. Biophys. Res. Commun.* **2007**, *364*, 250–257. [CrossRef]
40. Cominelli, E.; Galbiati, M.; Vavasseur, A.; Conti, L.; Sala, T.; Vuylsteke, M.; Leonhardt, N.; Dellaporta, S.L.; Tonelli, C. A guard-cell-specific MYB transcription factor regulates stomatal movements and plant drought tolerance. *Curr. Biol.* **2005**, *15*, 1196–1200. [CrossRef]
41. Wang, R.; Okamoto, M.; Xing, X.; Crawford, N.M. Microarray analysis of the nitrate response in Arabidopsis roots and shoots reveals over 1,000 rapidly responding genes and new linkages to glucose, trehalose-6-phosphate, iron, and sulfate metabolism. *Plant Physiol.* **2003**, *132*, 556–567. [CrossRef] [PubMed]
42. Padmalatha, K.V.; Dhandapani, G.; Kanakachari, M.; Kumar, S.; Dass, A.; Patil, D.P.; Rajamani, V.; Kumar, K.; Pathak, R.; Rawat, B. Genome-wide transcriptomic analysis of cotton under drought stress reveal significant down-regulation of genes and pathways involved in fibre elongation and up-regulation of defense responsive genes. *Plant Mol. Biol.* **2012**, *78*, 223–246. [CrossRef]
43. Seki, M.; Narusaka, M.; Abe, H.; Kasuga, M.; Yamaguchi-Shinozaki, K.; Carninci, P.; Hayashizaki, Y.; Shinozaki, K. Monitoring the expression pattern of 1300 Arabidopsis genes under drought and cold stresses by using a full-length cDNA microarray. *Plant Cell* **2001**, *13*, 61–72. [CrossRef] [PubMed]
44. Sakuma, Y.; Maruyama, K.; Osakabe, Y.; Qin, F.; Seki, M.; Shinozaki, K.; Yamaguchi-Shinozaki, K. Functional analysis of an Arabidopsis transcription factor, DREB2A, involved in drought-responsive gene expression. *Plant Cell* **2006**, *18*, 1292–1309. [CrossRef]
45. Zhu, J.; Verslues, P.E.; Zheng, X.; Lee, B.-H.; Zhan, X.; Manabe, Y.; Sokolchik, I.; Zhu, Y.; Dong, C.-H.; Zhu, J.-K. HOS10 encodes an R2R3-type MYB transcription factor essential for cold acclimation in plants. *Proc. Natl. Acad. Sci. USA* **2005**, *102*, 9966–9971. [CrossRef] [PubMed]
46. Pegler, J.L.; Oultram, J.M.J.; Grof, C.P.L.; Eamens, A.L. Profiling the Abiotic Stress Responsive microRNA Landscape of Arabidopsis thaliana. *Plants* **2019**, *8*, 58. [CrossRef]
47. Sharif, R.; Raza, A.; Chen, P.; Li, Y.; El-Ballat, E.M.; Rauf, A.; Hano, C.; El-Esawi, M.A. HD-ZIP gene family: Potential roles in improving plant growth and regulating stress-responsive mechanisms in plants. *Genes* **2021**, *12*, 1256. [CrossRef]
48. De la Rosa, C.; Covarrubias, A.A.; Reyes, J.L. A dicistronic precursor encoding miR398 and the legume-specific miR2119 coregulates CSD1 and ADH1 mRNAs in response to water deficit. *Plant Cell Environ.* **2019**, *42*, 133–144. [CrossRef]
49. Lu, Q.; Guo, F.; Xu, Q.; Cang, J. LncRNA improves cold resistance of winter wheat by interacting with miR398. *Funct. Plant Biol.* **2020**, *47*, 544–557. [CrossRef]
50. Dykes, I.M.; Emanueli, C. Transcriptional and post-transcriptional gene regulation by long non-coding RNA. *Genom. Proteom. Bioinform.* **2017**, *15*, 177–186. [CrossRef]
51. Jeong, D.-H.; Green, P.J. The role of rice microRNAs in abiotic stress responses. *J. Plant Biol.* **2013**, *56*, 187–197. [CrossRef]
52. Behbahani, G.D.; Ghahhari, N.M.; Javidi, M.A.; Molan, A.F.; Feizi, N.; Babashah, S. MicroRNA-mediated post-transcriptional regulation of epithelial to mesenchymal transition in cancer. *Pathol. Oncol. Res.* **2017**, *23*, 1–12. [CrossRef]
53. Haider, M.S.; Jogaiah, S.; Pervaiz, T.; Yanxue, Z.; Khan, N.; Fang, J. Physiological and transcriptional variations inducing complex adaptive mechanisms in grapevine by salt stress. *Environ. Exp. Bot.* **2019**, *162*, 455–467. [CrossRef]
54. Yu, H.; Song, C.; Jia, Q.; Wang, C.; Li, F.; Nicholas, K.K.; Zhang, X.; Fang, J. Computational identification of microRNAs in apple expressed sequence tags and validation of their precise sequences by miR-RACE. *Physiol. Plant.* **2011**, *141*, 56–70. [CrossRef] [PubMed]
55. Michlewski, G.; Cáceres, J.F. Post-transcriptional control of miRNA biogenesis. *Rna* **2019**, *25*, 1–16. [CrossRef]
56. González, A.; Ramos, J.; De Paz, J.F.; Corchado, J.M. Obtaining relevant genes by analysis of expression arrays with a multi-agent system. In *Proceedings of the 9th International Conference on Practical Applications of Computational Biology and Bioinformatics*; Springer: Singapore, 2015; pp. 137–146.
57. Wani, S.H.; Kumar, V.; Khare, T.; Tripathi, P.; Shah, T.; Ramakrishna, C.; Aglawe, S.; Mangrauthia, S.K. miRNA applications for engineering abiotic stress tolerance in plants. *Plant Cell* **2020**, *75*, 1081. [CrossRef]
58. Ghorbani, R.; Alemzadeh, A.; Razi, H. Microarray analysis of transcriptional responses to salt and drought stress in Arabidopsis thaliana. *Heliyon* **2019**, *5*, e02614. [CrossRef] [PubMed]
59. Raza, A.; Su, W.; Hussain, M.A.; Mehmood, S.S.; Zhang, X.; Cheng, Y.; Zou, X.; Lv, Y. Integrated Analysis of Metabolome and Transcriptome Reveals Insights for Cold Tolerance in Rapeseed (Brassica napus L.). *Front. Plant Sci.* **2021**, *12*, 12. [CrossRef]
60. Chun, J.Y. Detection of Target Nucleic Acid Sequences Using Dual-Labeled Immobilized Probes on Solid Phase. European Patent EP2630262A2, 10 May 2017.
61. Jiménez Meneses, P. Study of Substrate Modulation and Bioreceptor Anchoring for the Development of High Performance Microarrays. Ph.D. Thesis, Universitat Politècnica de València, Valencia, Spain, 2020.
62. Raza, A. Metabolomics: A systems biology approach for enhancing heat stress tolerance in plants. *Plant Cell Rep.* **2020**. [CrossRef]
63. Viemann, D.; Schulze-Osthoff, K.; Roth, J. Potentials and pitfalls of DNA array analysis of the endothelial stress response. *Biochim. Biophys. Acta (BBA)-Mol. Cell Res.* **2005**, *1746*, 73–84. [CrossRef] [PubMed]

64. Sasik, R.; Woelk, C.; Corbeil, J. Microarray truths and consequences. *J. Mol. Endocrinol.* **2004**, *33*, 1–9. [CrossRef] [PubMed]
65. Fredonnet, J.; Foncy, J.; Cau, J.-C.; Séverac, C.; François, J.M.; Trévisiol, E. Automated and multiplexed soft lithography for the production of low-density DNA microarrays. *Microarrays* **2016**, *5*, 25. [CrossRef] [PubMed]
66. Jaksik, R.; Iwanaszko, M.; Rzeszowska-Wolny, J.; Kimmel, M. Microarray experiments and factors which affect their reliability. *Biol. Direct.* **2015**, *10*, 46. [CrossRef] [PubMed]
67. Shinozaki, K.; Yamaguchi-Shinozaki, K. Gene networks involved in drought stress response and tolerance. *J. Exp. Bot.* **2007**, *58*, 221–227. [CrossRef]
68. Jain, M. Emerging role of metabolic pathways in abiotic stress tolerance. *J. Plant Biochem. Physiol.* **2013**, *1*, 10–4172. [CrossRef]
69. Barrera-Figueroa, B.E.; Wu, Z.; Liu, R. Abiotic stress-associated microRNAs in plants: Discovery, expression analysis, and evolution. *Front. Biol.* **2013**, *8*, 189–197. [CrossRef]
70. Lotfi, A.; Pervaiz, T.; Jiu, S.; Faghihi, F.; Jahanbakhshian, Z.; Khorzoghi, E.G.; Fang, J. Role of microRNAs and their target genes in salinity response in plants. *Plant Growth Regul.* **2017**, *82*, 377–390. [CrossRef]
71. Li, X.; Wang, X.; Zhang, S.; Liu, D.; Duan, Y.; Dong, W. Identification of soybean microRNAs involved in soybean cyst nematode infection by deep sequencing. *PLoS ONE* **2012**, *7*, e39650. [CrossRef]
72. Chen, L.; Heikkinen, L.; Wang, C.; Yang, Y.; Sun, H.; Wong, G. Trends in the development of miRNA bioinformatics tools. *Brief Bioinform.* **2019**, *20*, 1836–1852. [CrossRef]
73. Wong, C.E.; Li, Y.; Labbe, A.; Guevara, D.; Nuin, P.; Whitty, B.; Diaz, C.; Golding, G.B.; Gray, G.R.; Weretilnyk, E.A. Transcriptional profiling implicates novel interactions between abiotic stress and hormonal responses in Thellungiella, a close relative of Arabidopsis. *Plant Physiol.* **2006**, *140*, 1437–1450. [CrossRef]
74. Liu, Y.; Ji, X.; Zheng, L.; Nie, X.; Wang, Y. Microarray analysis of transcriptional responses to abscisic acid and salt stress in Arabidopsis thaliana. *Int. J. Mol. Sci.* **2013**, *14*, 9979–9998. [CrossRef] [PubMed]
75. Jangam, A.P.; Pathak, R.R.; Raghuram, N. Microarray Analysis of Rice d1 (RGA1) Mutant Reveals the Potential Role of G-Protein Alpha Subunit in Regulating Multiple Abiotic Stresses Such as Drought, Salinity, Heat, and Cold. *Front. Plant Sci.* **2016**, *7*, 11. [CrossRef] [PubMed]
76. Seki, M.; Narusaka, M.; Kamiya, A.; Ishida, J.; Satou, M.; Sakurai, T.; Nakajima, M.; Enju, A.; Akiyama, K.; Oono, Y. Functional annotation of a full-length Arabidopsis cDNA collection. *Science* **2002**, *296*, 141–145. [CrossRef] [PubMed]
77. Trindade, I.; Capitão, C.; Dalmay, T.; Fevereiro, M.P.; Dos Santos, D.M. miR398 and miR408 are up-regulated in response to water deficit in Medicago truncatula. *Planta* **2010**, *231*, 705–716. [CrossRef]
78. Fang, Z.; Zhang, X.; Gao, J.; Wang, P.; Xu, X.; Liu, Z.; Shen, S.; Feng, B. A buckwheat (Fagopyrum esculentum) DRE-binding transcription factor gene, FeDREB1, enhances freezing and drought tolerance of transgenic Arabidopsis. *Plant Mol. Biol. Report.* **2015**, *33*, 1510–1525. [CrossRef]
79. Wang, T.; Chen, L.; Zhao, M.; Tian, Q.; Zhang, W.-H. Identification of drought-responsive microRNAs in Medicago truncatula by genome-wide high-throughput sequencing. *BMC Genom.* **2011**, *12*, 1–11. [CrossRef]
80. Mantri, N.; Basker, N.; Ford, R.; Pang, E.; Pardeshi, V. The role of micro-ribonucleic acids in legumes with a focus on abiotic stress response. *Plant Genome* **2013**, *6*. [CrossRef]
81. Li, H.; Dong, Y.; Yin, H.; Wang, N.; Yang, J.; Liu, X.; Wang, Y.; Wu, J.; Li, X. Characterization of the stress associated microRNAs in Glycine max by deep sequencing. *BMC Plant Biol.* **2011**, *11*, 1–12. [CrossRef]
82. Dong, Z.; Shi, L.; Wang, Y.; Chen, L.; Cai, Z.; Wang, Y.; Jin, J.; Li, X. Identification and dynamic regulation of microRNAs involved in salt stress responses in functional soybean nodules by high-throughput sequencing. *Int. J. Mol. Sci.* **2013**, *14*, 2717–2738. [CrossRef]
83. Zhang, B. MicroRNA: A new target for improving plant tolerance to abiotic stress. *J. Exp. Bot.* **2015**, *66*, 1749–1761. [CrossRef]
84. Liu, H.-H.; Tian, X.; Li, Y.-J.; Wu, C.-A.; Zheng, C.-C. Microarray-based analysis of stress-regulated microRNAs in Arabidopsis thaliana. *Rna* **2008**, *14*, 836–843. [CrossRef] [PubMed]
85. Meyers, B.C.; Axtell, M.J.; Bartel, B.; Bartel, D.P.; Baulcombe, D.; Bowman, J.L.; Cao, X.; Carrington, J.C.; Chen, X.; Green, P.J. Criteria for annotation of plant MicroRNAs. *Plant Cell* **2008**, *20*, 3186–3190. [CrossRef] [PubMed]
86. Ta, K.N.; Sabot, F.; Adam, H.; Vigouroux, Y.; De Mita, S.; Ghesquière, A.; Do, N.V.; Gantet, P.; Jouannic, S. miR2118-triggered phased siRNAs are differentially expressed during the panicle development of wild and domesticated African rice species. *Rice* **2016**, *9*, 10. [CrossRef] [PubMed]
87. Yan, H.; Kikuchi, S.; Neumann, P.; Zhang, W.; Wu, Y.; Chen, F.; Jiang, J. Genome-wide mapping of cytosine methylation revealed dynamic DNA methylation patterns associated with genes and centromeres in rice. *Plant J.* **2010**, *63*, 353–365. [CrossRef]
88. Zhou, S.; Wei, S.; Boone, B.; Levy, S. Microarray analysis of genes affected by salt stress in tomato. *Afr. J. Environ. Sci. Technol.* **2007**, *1*, 14–26.
89. Cohen, D.; Bogeat-Triboulot, M.-B.; Tisserant, E.; Balzergue, S.; Martin-Magniette, M.-L.; Lelandais, G.; Ningre, N.; Renou, J.-P.; Tamby, J.-P.; Le Thiec, D. Comparative transcriptomics of drought responses in Populus: A meta-analysis of genome-wide expression profiling in mature leaves and root apices across two genotypes. *BMC Genom.* **2010**, *11*, 630. [CrossRef]
90. Pervaiz, T.; Lotfi, A.; Haider, M.S.; Haifang, J.; Fang, J. High Throughput Sequencing Advances and Future Challenges. *J. Plant Biochem. Physiol.* **2017**, *5*, 188. [CrossRef]
91. Raza, A.; Tabassum, J.; Kudapa, H.; Varshney, R.K. Can omics deliver temperature resilient ready-to-grow crps? *Crit. Rev. Biotechnol.* **2021**, *41*, 1–24. [CrossRef]

92. Liu, M.; Shi, J.; Lu, C. Identification of stress-responsive genes in Ammopiptanthus mongolicususing ESTs generated from cold-and drought-stressed seedlings. *BMC Plant Biol.* **2013**, *13*, 88. [CrossRef]
93. Mittal, S.; Banduni, P.; Mallikarjuna, M.G.; Rao, A.R.; Jain, P.A.; Dash, P.K.; Thirunavukkarasu, N. Structural, Functional, and Evolutionary Characterization of Major Drought Transcription Factors Families in Maize. *Front. Chem.* **2018**, *6*, 177. [CrossRef]
94. Pervaiz, T.; Songtao, J.; Faghihi, F.; Haider, M.S.; Fang, J. Naturally occurring anthocyanin, structure, functions and biosynthetic pathway in fruit plants. *J. Plant Biochem. Physiol.* **2017**, *5*, 1–9. [CrossRef]
95. Street, N.R.; Skogström, O.; Sjödin, A.; Tucker, J.; Rodríguez-Acosta, M.; Nilsson, P.; Jansson, S.; Taylor, G. The genetics and genomics of the drought response in Populus. *Plant J.* **2006**, *48*, 321–341. [CrossRef]
96. Lorenz, W.W.; Sun, F.; Liang, C.; Kolychev, D.; Wang, H.; Zhao, X.; Cordonnier-Pratt, M.-M.; Pratt, L.H.; Dean, J.F. Water stress-responsive genes in loblolly pine (Pinus taeda) roots identified by analyses of expressed sequence tag libraries. *Tree Physiol.* **2006**, *26*, 1–16. [CrossRef] [PubMed]
97. Rosegrant, M.W.; Cline, S.A. Global food security: Challenges and policies. *Science* **2003**, *302*, 1917–1919. [CrossRef] [PubMed]
98. Du, M.; Ding, G.; Cai, Q. The transcriptomic responses of Pinus massoniana to drought stress. *Forests* **2018**, *9*, 326. [CrossRef]
99. Kathiresan, A.; Lafitte, H.; Chen, J.; Mansueto, L.; Bruskiewich, R.; Bennett, J. Gene expression microarrays and their application in drought stress research. *Field Crops Res.* **2006**, *97*, 101–110. [CrossRef]
100. Vásquez-Robinet, C.; Watkinson, J.I.; Sioson, A.A.; Ramakrishnan, N.; Heath, L.S.; Grene, R. Differential expression of heat shock protein genes in preconditioning for photosynthetic acclimation in water-stressed loblolly pine. *Plant Physiol. Biochem.* **2010**, *48*, 256–264. [CrossRef]
101. Contour-Ansel, D.; Torres-Franklin, M.L.; Cruz De Carvalho, M.H.; D'Arcy-Lameta, A.; Zuily-Fodil, Y. Glutathione reductase in leaves of cowpea: Cloning of two cDNAs, expression and enzymatic activity under progressive drought stress, desiccation and abscisic acid treatment. *Ann. Bot.* **2006**, *98*, 1279–1287. [CrossRef] [PubMed]
102. Hasegawa, P.M.; Bressan, R.A.; Zhu, J.-K.; Bohnert, H.J. Plant cellular and molecular responses to high salinity. *Annu. Rev. Plant Biol.* **2000**, *51*, 463–499. [CrossRef] [PubMed]
103. Barrera-Figueroa, B.E.; Gao, L.; Diop, N.N.; Wu, Z.; Ehlers, J.D.; Roberts, P.A.; Close, T.J.; Zhu, J.-K.; Liu, R. Identification and comparative analysis of drought-associated microRNAs in two cowpea genotypes. *BMC Plant Biol.* **2011**, *11*, 127. [CrossRef] [PubMed]
104. Han, M.; Lu, X.; Yu, J.; Chen, X.; Wang, X.; Malik, W.A.; Wang, J.; Wang, D.; Wang, S.; Guo, L.; et al. Transcriptome Analysis Reveals Cotton (Gossypium hirsutum) Genes That Are Differentially Expressed in Cadmium Stress Tolerance. *Int. J. Mol. Sci.* **2019**, *20*, 1479. [CrossRef] [PubMed]
105. Harb, A.; Krishnan, A.; Ambavaram, M.M.; Pereira, A. Molecular and physiological analysis of drought stress in Arabidopsis reveals early responses leading to acclimation in plant growth. *Plant Physiol.* **2010**, *154*, 1254–1271. [CrossRef]
106. Muyal, J.P.; Singh, S.K.; Fehrenbach, H. DNA-microarray technology: Comparison of methodological factors of recent technique towards gene expression profiling. *Crit. Rev. Biotechnol.* **2008**, *28*, 239–251. [CrossRef] [PubMed]
107. Granier, C.; Aguirrezabal, L.; Chenu, K.; Cookson, S.J.; Dauzat, M.; Hamard, P.; Thioux, J.J.; Rolland, G.; Bouchier-Combaud, S.; Lebaudy, A. PHENOPSIS, an automated platform for reproducible phenotyping of plant responses to soil water deficit in Arabidopsis thaliana permitted the identification of an accession with low sensitivity to soil water deficit. *New Phytol.* **2006**, *169*, 623–635. [CrossRef]
108. Alexandersson, E.; Fraysse, L.; Sjövall-Larsen, S.; Gustavsson, S.; Fellert, M.; Karlsson, M.; Johanson, U.; Kjellbom, P. Whole gene family expression and drought stress regulation of aquaporins. *Plant Mol. Biol.* **2005**, *59*, 469–484. [CrossRef]
109. Dombrowski, J.E. Salt stress activation of wound-related genes in tomato plants. *Plant Physiol.* **2003**, *132*, 2098–2107. [CrossRef]
110. Vishwakarma, K.; Mishra, M.; Patil, G.; Mulkey, S.; Ramawat, N.; Pratap Singh, V.; Deshmukh, R.; Kumar Tripathi, D.; Nguyen, H.T.; Sharma, S. Avenues of the membrane transport system in adaptation of plants to abiotic stresses. *Crit. Rev. Biotechnol.* **2019**, *39*, 861–883. [CrossRef]
111. Umezawa, T.; Yoshida, R.; Maruyama, K.; Yamaguchi-Shinozaki, K.; Shinozaki, K. SRK2C, a SNF1-related protein kinase 2, improves drought tolerance by controlling stress-responsive gene expression in Arabidopsis thaliana. *Proc. Natl. Acad. Sci. USA* **2004**, *101*, 17306–17311. [CrossRef]
112. Chinnusamy, V.; Zhu, J.; Zhu, J.-K. Cold stress regulation of gene expression in plants. *Trends Plant Sci.* **2007**, *12*, 444–451. [CrossRef]
113. Xue-Xuan, X.; Hong-Bo, S.; Yuan-Yuan, M.; Gang, X.; Jun-Na, S.; Dong-Gang, G.; Cheng-Jiang, R. Biotechnological implications from abscisic acid (ABA) roles in cold stress and leaf senescence as an important signal for improving plant sustainable survival under abiotic-stressed conditions. *Crit. Rev. Biotechnol.* **2010**, *30*, 222–230. [CrossRef] [PubMed]
114. Benedict, C.; Geisler, M.; Trygg, J.; Huner, N.; Hurry, V. Consensus by democracy. Using meta-analyses of microarray and genomic data to model the cold acclimation signaling pathway in Arabidopsis. *Plant Physiol.* **2006**, *141*, 1219–1232. [CrossRef] [PubMed]
115. Wang, Z.-L.; Li, P.-H.; Fredricksen, M.; Gong, Z.-Z.; Kim, C.; Zhang, C.; Bohnert, H.J.; Zhu, J.-K.; Bressan, R.A.; Hasegawa, P.M. Expressed sequence tags from Thellungiella halophila, a new model to study plant salt-tolerance. *Plant Sci.* **2004**, *166*, 609–616. [CrossRef]
116. Ciarmiello, L.F.; Woodrow, P.; Fuggi, A.; Pontecorvo, G.; Carillo, P. *Plant Genes for Abiotic Stress. Abiotic Stress in Plants—Mechanisms and Adaptations*. 2011, pp. 283–308. Available online: https://www.intechopen.com/chapters/18407 (accessed on 22 December 2020).

117. Kurikesu, I.; Anuja, T.; Gangaprasad, A.; Nair, A.J. Regulation of micrornas during biotic and abiotic stress. *Bull. Pure Appl. Sci. -Bot.* **2018**, *37*, 49–55. [CrossRef]
118. Kimura, M.; Yamamoto, Y.Y.; Seki, M.; Sakurai, T.; Sato, M.; Abe, T.; Yoshida, S.; Manabe, K.; Shinozaki, K.; Matsui, M. Identification of Arabidopsis genes regulated by high light-stress using cDNA microarray. *Photochem. Photobiol.* **2003**, *77*, 226–233.
119. Albert, E. Genetic and Genomic Determinants of Response to Water Deficit in Tomato (Solanum lycopersicum) and Impact on Fruit Quality. Ph.D. Thesis, Université d'Avignon, Avignon, France, 2017.
120. Satoh, R.; Nakashima, K.; Seki, M.; Shinozaki, K.; Yamaguchi-Shinozaki, K. ACTCAT a novel cis-acting element for proline-and hypoosmolarity-responsive expression of the ProDH gene encoding proline dehydrogenase in Arabidopsis. *Plant Physiol.* **2002**, *130*, 709–719. [CrossRef]
121. Guo, M.; Liu, J.H.; Ma, X.; Luo, D.X.; Gong, Z.H.; Lu, M.H. The Plant Heat Stress Transcription Factors (HSFs): Structure, Regulation, and Function in Response to Abiotic Stresses. *Front. Plant Sci.* **2016**, *7*, 114. [CrossRef] [PubMed]
122. Hihara, Y.; Kamei, A.; Kanehisa, M.; Kaplan, A.; Ikeuchi, M. DNA microarray analysis of cyanobacterial gene expression during acclimation to high light. *Plant Cell* **2001**, *13*, 793–806. [CrossRef]
123. Dai, X.; Xu, Y.; Ma, Q.; Xu, W.; Wang, T.; Xue, Y.; Chong, K. Overexpression of an R1R2R3 MYB gene, OsMYB3R-2, increases tolerance to freezing, drought, and salt stress in transgenic Arabidopsis. *Plant Physiol.* **2007**, *143*, 1739–1751. [CrossRef] [PubMed]
124. Hwang, E.-W.; Kim, K.-A.; Park, S.-C.; Jeong, M.-J.; Byun, M.-O.; Kwon, H.-B. Expression profiles of hot pepper (Capsicum annuum) genes under cold stress conditions. *J. Biosci.* **2005**, *30*, 657–667. [CrossRef]
125. da Silva, J.A.T.; Yücel, M. Revealing response of plants to biotic and abiotic stresses with microarray technology. *Genes Genomes Genom.* **2008**, *2*, 14–48.
126. Ravikumar, G.; Manimaran, P.; Voleti, S.R.; Subrahmanyam, D.; Sundaram, R.M.; Bansal, K.C.; Viraktamath, B.C.; Balachandran, S.M. Stress-inducible expression of AtDREB1A transcription factor greatly improves drought stress tolerance in transgenic indica rice. *Transgenic Res.* **2014**, *23*, 421–439. [CrossRef]
127. Leviatan, N.; Alkan, N.; Leshkowitz, D.; Fluhr, R. Genome-wide survey of cold stress regulated alternative splicing in Arabidopsis thaliana with tiling microarray. *PLoS ONE* **2013**, *8*, e66511. [CrossRef]
128. Oono, Y.; Seki, M.; Nanjo, T.; Narusaka, M.; Fujita, M.; Satoh, R.; Satou, M.; Sakurai, T.; Ishida, J.; Akiyama, K. Monitoring expression profiles of Arabidopsis gene expression during rehydration process after dehydration using ca. 7000 full-length cDNA microarray. *Plant J.* **2003**, *34*, 868–887. [CrossRef]
129. Maruyama, K.; Sakuma, Y.; Kasuga, M.; Ito, Y.; Seki, M.; Goda, H.; Shimada, Y.; Yoshida, S.; Shinozaki, K.; Yamaguchi-Shinozaki, K. Identification of cold-inducible downstream genes of the Arabidopsis DREB1A/CBF3 transcriptional factor using two microarray systems. *Plant J.* **2004**, *38*, 982–993. [CrossRef]
130. Hu, H.; Dai, M.; Yao, J.; Xiao, B.; Li, X.; Zhang, Q.; Xiong, L. Overexpressing a NAM, ATAF, and CUC (NAC) transcription factor enhances drought resistance and salt tolerance in rice. *Proc. Natl. Acad. Sci. USA* **2006**, *103*, 12987–12992. [CrossRef] [PubMed]
131. Song, X.; Li, Y.; Hou, X. Genome-wide analysis of the AP2/ERF transcription factor superfamily in Chinese cabbage (Brassica rapa ssp. pekinensis). *BMC Genom.* **2013**, *14*, 573. [CrossRef] [PubMed]
132. Lata, R.; Chowdhury, S.; Gond, S.K.; White, J.F., Jr. Induction of abiotic stress tolerance in plants by endophytic microbes. *Lett. Appl. Microbiol.* **2018**, *66*, 268–276. [CrossRef]
133. Kumar, R.; Mustafiz, A.; Sahoo, K.K.; Sharma, V.; Samanta, S.; Sopory, S.K.; Pareek, A.; Singla-Pareek, S.L. Functional screening of cDNA library from a salt tolerant rice genotype Pokkali identifies mannose-1-phosphate guanyl transferase gene (OsMPG1) as a key member of salinity stress response. *Plant Mol. Biol.* **2012**, *79*, 555–568. [CrossRef]
134. Miller, G.; Suzuki, N.; Ciftci-Yilmaz, S.; Mittler, R. Reactive oxygen species homeostasis and signalling during drought and salinity stresses. *Plant Cell Environ.* **2010**, *33*, 453–467. [CrossRef]
135. Hu, Y.; Burucs, Z.; von Tucher, S.; Schmidhalter, U. Short-term effects of drought and salinity on mineral nutrient distribution along growing leaves of maize seedlings. *Environ. Exp. Bot.* **2007**, *60*, 268–275. [CrossRef]
136. Mangrauthia, S.K.; Agarwal, S.; Sailaja, B.; Madhav, M.S.; Voleti, S. MicroRNAs and their role in salt stress response in plants. In *Salt Stress in Plants*; Springer: Berlin/Heidelberg, Germany, 2013; pp. 15–46.
137. Vinocur, B.; Altman, A. Recent advances in engineering plant tolerance to abiotic stress: Achievements and limitations. *Curr. Opin. Biotechnol.* **2005**, *16*, 123–132. [CrossRef]
138. Ouyang, B.; Yang, T.; Li, H.; Zhang, L.; Zhang, Y.; Zhang, J.; Fei, Z.; Ye, Z. Identification of early salt stress response genes in tomato root by suppression subtractive hybridization and microarray analysis. *J. Exp. Bot.* **2007**, *58*, 507–520. [CrossRef] [PubMed]
139. Sahi, C.; Singh, A.; Kumar, K.; Blumwald, E.; Grover, A. Salt stress response in rice: Genetics, molecular biology, and comparative genomics. *Funct. Integr. Genom.* **2006**, *6*, 263–284. [CrossRef] [PubMed]
140. Zhu, Y.N.; Shi, D.Q.; Ruan, M.B.; Zhang, L.L.; Meng, Z.H.; Liu, J.; Yang, W.C. Transcriptome analysis reveals crosstalk of responsive genes to multiple abiotic stresses in cotton (Gossypium hirsutum L.). *PLoS ONE* **2013**, *8*, e80218.
141. Chan, Z.; Bigelow, P.J.; Loescher, W.; Grumet, R. Comparison of salt stress resistance genes in transgenic Arabidopsis thaliana indicates that extent of transcriptomic change may not predict secondary phenotypic or fitness effects. *Plant Biotechnol. J.* **2012**, *10*, 284–300. [CrossRef] [PubMed]
142. Khraiwesh, B.; Zhu, J.-K.; Zhu, J. Role of miRNAs and siRNAs in biotic and abiotic stress responses of plants. *Biochim. Biophys. Acta (BBA)-Gene Regul. Mech.* **2012**, *1819*, 137–148. [CrossRef]

143. Kumari, S.; Singh, P.; Singla-Pareek, S.L.; Pareek, A. Heterologous expression of a salinity and developmentally regulated rice cyclophilin gene (OsCyp2) in E. coli and S. cerevisiae confers tolerance towards multiple abiotic stresses. *Mol. Biotechnol.* **2009**, *42*, 195–204. [CrossRef]
144. Wu, T.; Yang, C.; Ding, B.; Feng, Z.; Wang, Q.; He, J.; Tong, J.; Xiao, L.; Jiang, L.; Wan, J. Microarray-based gene expression analysis of strong seed dormancy in rice cv. N22 and less dormant mutant derivatives. *Plant Physiol. Biochem.* **2016**, *99*, 27–38. [CrossRef] [PubMed]
145. Kawasaki, S.; Borchert, C.; Deyholos, M.; Wang, H.; Brazille, S.; Kawai, K.; Galbraith, D.; Bohnert, H.J. Gene expression profiles during the initial phase of salt stress in rice. *Plant Cell* **2001**, *13*, 889–905. [CrossRef] [PubMed]
146. Kong, W.; Zhong, H.; Gong, Z.; Fang, X.; Sun, T.; Deng, X.; Li, Y. Meta-analysis of salt stress transcriptome responses in different rice genotypes at the seedling stage. *Plants* **2019**, *8*, 64. [CrossRef] [PubMed]
147. Zhu, J.K. Salt and drought stress signal transduction in plants. *Annu. Rev. Plant Biol.* **2002**, *53*, 247–273. [CrossRef]
148. Lee, I.; Ajay, S.S.; Yook, J.I.; Kim, H.S.; Hong, S.H.; Kim, N.H.; Dhanasekaran, S.M.; Chinnaiyan, A.M.; Athey, B.D. New class of microRNA targets containing simultaneous 5′-UTR and 3′-UTR interaction sites. *Genome Res.* **2009**, *19*, 1175–1183. [CrossRef]
149. Bhattacharya, A.; Ziebarth, J.D.; Cui, Y. PolymiRTS Database 3.0: Linking polymorphisms in microRNAs and their target sites with human diseases and biological pathways. *Nucleic Acids Res.* **2014**, *42*, D86–D91. [CrossRef]
150. Cho, S.; Jang, I.; Jun, Y.; Yoon, S.; Ko, M.; Kwon, Y.; Choi, I.; Chang, H.; Ryu, D.; Lee, B.; et al. MiRGator v3.0: A microRNA portal for deep sequencing, expression profiling and mRNA targeting. *Nucleic Acids Res.* **2013**, *41*, D257. [CrossRef]
151. Wang, X. Improving microRNA target prediction by modeling with unambiguously identified microRNA-target pairs from CLIP-ligation studies. *Bioinformatics* **2016**, *32*, 1316–1322. [CrossRef] [PubMed]
152. Tokar, T.; Pastrello, C.; Rossos, A.E.M.; Abovsky, M.; Hauschild, A.C.; Tsay, M.; Lu, R.; Jurisica, I. mirDIP 4.1-integrative database of human microRNA target predictions. *Nucleic Acids Res.* **2018**, *46*, D360–D370. [CrossRef]
153. Preusse, M.; Theis, F.J.; Mueller, N.S. miTALOS v2: Analyzing Tissue Specific microRNA Function. *PLoS ONE* **2016**, *11*, e0151771. [CrossRef] [PubMed]
154. Loher, P.; Rigoutsos, I. Interactive exploration of RNA22 microRNA target predictions. *Bioinformatics* **2012**, *28*, 3322–3323. [CrossRef] [PubMed]
155. Dai, X.; Zhuang, Z.; Zhao, P.X. psRNATarget: A plant small RNA target analysis server (2017 release). *Nucleic Acids Res.* **2018**, *46*, W49–W54. [CrossRef]
156. Zhou, K.R.; Liu, S.; Sun, W.J.; Zheng, L.L.; Zhou, H.; Yang, J.H.; Qu, L.H. ChIPBase v2.0: Decoding transcriptional regulatory networks of non-coding RNAs and protein-coding genes from ChIP-seq data. *Nucleic Acids Res.* **2017**, *45*, D43–D50. [CrossRef] [PubMed]
157. Fromm, B.; Billipp, T.; Peck, L.E.; Johansen, M.; Tarver, J.E.; King, B.L.; Newcomb, J.M.; Sempere, L.F.; Flatmark, K.; Hovig, E.; et al. A Uniform System for the Annotation of Vertebrate microRNA Genes and the Evolution of the Human microRNAome. *Annu. Rev. Genet.* **2015**, *49*, 213–242. [CrossRef] [PubMed]
158. Ma, X.; Liu, C.; Gu, L.; Mo, B.; Cao, X.; Chen, X. TarHunter, a tool for predicting conserved microRNA targets and target mimics in plants. *Bioinformatics* **2018**, *34*, 1574–1576. [CrossRef]
159. Haunsberger, S.J.; Connolly, N.M.; Prehn, J.H. miRNAmeConverter: An R/bioconductor package for translating mature miRNA names to different miRBase versions. *Bioinformatics* **2017**, *33*, 592–593. [CrossRef] [PubMed]
160. Page, G.P.; Coulibaly, I. Bioinformatic tools for inferring functional information from plant microarray data: Tools for the first steps. *Int. J. Plant Genom.* **2008**, *2008*, 147563. [CrossRef] [PubMed]
161. Demeter, J.; Beauheim, C.; Gollub, J.; Hernandez-Boussard, T.; Jin, H.; Maier, D.; Matese, J.C.; Nitzberg, M.; Wymore, F.; Zachariah, Z.K.; et al. The Stanford Microarray Database: Implementation of new analysis tools and open source release of software. *Nucleic Acids Res.* **2007**, *35*, D766–D770. [CrossRef]
162. Tusher, V.G.; Tibshirani, R.; Chu, G. Significance analysis of microarrays applied to the ionizing radiation response. *Proc. Natl. Acad. Sci. USA* **2001**, *98*, 5116–5121. [CrossRef]
163. Alonso-Betanzos, A.; Bolón-Canedo, V.; Morán-Fernández, L.; Sánchez-Maroño, N. A Review of Microarray Datasets: Where to Find Them and Specific Characteristics. *Methods Mol. Biol.* **2019**, *1986*, 65–85.
164. Mehta, J.P.; Rani, S. Software and tools for microarray data analysis. *Methods Mol. Biol.* **2011**, *784*, 41–53.
165. Kawaura, K.; Mochida, K.; Ogihara, Y. Genome-wide analysis for identification of salt-responsive genes in common wheat. *Funct. Integr. Genom.* **2008**, *8*, 277–286. [CrossRef]
166. Fernandez, P.; Di Rienzo, J.; Fernandez, L.; Hopp, H.E.; Paniego, N.; Heinz, R.A. Transcriptomic identification of candidate genes involved in sunflower responses to chilling and salt stresses based on cDNA microarray analysis. *BMC Plant Biol.* **2008**, *8*, 11. [CrossRef]
167. Singla-Pareek, S.L.; Yadav, S.K.; Pareek, A.; Reddy, M.K.; Sopory, S.K. Enhancing salt tolerance in a crop plant by overexpression of glyoxalase II. *Transgenic Res.* **2008**, *17*, 171–180. [CrossRef] [PubMed]
168. Uddin, M.I.; Qi, Y.; Yamada, S.; Shibuya, I.; Deng, X.P.; Kwak, S.S.; Kaminaka, H.; Tanaka, K. Overexpression of a new rice vacuolar antiporter regulating protein OsARP improves salt tolerance in tobacco. *Plant Cell Physiol.* **2008**, *49*, 880–890. [CrossRef] [PubMed]
169. Sakamoto, K.; Göransson, O.; Hardie, D.G.; Alessi, D.R. Activity of LKB1 and AMPK-related kinases in skeletal muscle: Effects of contraction, phenformin, and AICAR. *Am. J. Physiol. Endocrinol. Metab.* **2004**, *287*, E310–E317. [CrossRef]

170. Hichri, I.; Muhovski, Y.; Žižková, E.; Dobrev, P.I.; Gharbi, E.; Franco-Zorrilla, J.M.; Lopez-Vidriero, I.; Solano, R.; Clippe, A.; Errachid, A. The Solanum lycopersicum WRKY3 transcription factor SlWRKY3 is involved in salt stress tolerance in tomato. *Front. Plant Sci.* **2017**, *8*, 1343. [CrossRef] [PubMed]

171. Mukhopadhyay, S.; Mandal, S.K.; Bhaduri, S.; Armstrong, W.H. Manganese clusters with relevance to photosystem II. *Chem. Rev.* **2004**, *104*, 3981–4026. [CrossRef]

172. De Lorenzo, L.; Merchan, F.; Blanchet, S.; Megías, M.; Frugier, F.; Crespi, M.; Sousa, C. Differential expression of the TFIIIA regulatory pathway in response to salt stress between Medicago truncatula genotypes. *Plant Physiol.* **2007**, *145*, 1521–1532. [CrossRef] [PubMed]

173. Havaux, M.; Guedeney, G.; Hagemann, M.; Yeremenko, N.; Matthijs, H.C.; Jeanjean, R. The chlorophyll-binding protein IsiA is inducible by high light and protects the cyanobacterium Synechocystis PCC6803 from photooxidative stress. *FEBS Lett.* **2005**, *579*, 2289–2293. [CrossRef]

174. Upadhyay, J.; Joshi, R.; Singh, B.; Bohra, A.; Vijayan, R.; Bhatt, M.; Bisht, S.S.; Wani, S.H. Application of Bioinformatics in Understanding of Plant Stress Tolerance. *Plant Bioinform.* **2017**, 347–374.

175. Raza, A.; Razzaq, A.; Mehmood, S.S.; Hussain, M.A.; Wei, S.; He, H.; Zaman, Q.U.; Xuekun, Z.; Yong, C.; Hasanuzzaman, M. Omics: The way forward to enhance abiotic stress tolerance in Brassica napus L. *GM Crops Food* **2021**, *12*, 251–281. [CrossRef] [PubMed]

176. Quillet, A.; Saad, C.; Ferry, G.; Anouar, Y.; Vergne, N.; Lecroq, T.; Dubessy, C. Improving Bioinformatics Prediction of microRNA Targets by Ranks Aggregation. *Front. Genet.* **2019**, *10*, 1330. [CrossRef]

 agronomy

Article

The Role of Salicylic Acid in Mitigating the Adverse Effects of Chilling Stress on "Seddik" Mango Transplants

Ibrahim Hmmam [1,*], Amr E. M. Ali [1], Samir M. Saleh [2], Nagwa Khedr [1] and Abdou Abdellatif [1]

[1] Pomology Department, Faculty of Agriculture, Cairo University, 4, El-Gamaa St., Giza 12613, Egypt; amr.ali@agr.cu.edu.eg (A.E.M.A.); nagwahamdy5@yahoo.com (N.K.); abdo.abdullatif@agr.cu.edu.eg (A.A.)
[2] Central Laboratory for Agricultural Climate (CLAC), Agriculture Research Center (ARC), 6, Michael Bakhom St., Dokki, Giza 12411, Egypt; samir.saleh@arc.sci.eg
* Correspondence: ibrahim.s.hmmam@agr.cu.edu.eg; Tel.: +20-1006985497

Abstract: Salicylic acid (SA) was sprayed on "Seddik" mango transplants at concentrations of 0, 0.5, 1, and 1.5 mM. Then, the mango transplants were subjected to 72 h of chilling stress at 4 ± 1 °C, followed by a six-day recovery under greenhouse conditions. Untreated transplants exposed to chilling stress represented the positive control, while those not exposed were the negative control. SA-pretreated mango transplants were compared to the positive and negative controls, evaluating physiological and biochemical changes. The SA concentration of 1.5 mM L^{-1} was the most efficient in mitigating chilling injury (CI) in mango transplants by maintaining the integrity of the leaves' cell membrane and minimizing electrolyte leakage (EL), specifically after six days of recovery. SA increased photosynthetic pigment content, total sugar content, and 2, 2-diphenyl-1-picrylhydrazyl (DPPH) radical scavenging activity and decreased proline and total phenolic content in the "Seddik" mango transplants' leaves. After exposure to chilling stress, the antioxidant enzymes' internal activities in SA-pretreated chilled mango transplants improved, especially on the sixth day of recovery, compared to the negative control; the transplants nearly attained normal growth levels. Thus, SA can protect plants against the adverse effects of chilling stress.

Keywords: antioxidant enzymes; chilling stress; climate; *Mangifera indica*; salicylic acid; Seddik; transplant

Citation: Hmmam, I.; Ali, A.E.M.; Saleh, S.M.; Khedr, N.; Abdellatif, A. The Role of Salicylic Acid in Mitigating the Adverse Effects of Chilling Stress on "Seddik" Mango Transplants. *Agronomy* **2022**, *12*, 1369. https://doi.org/10.3390/agronomy12061369

Academic Editor: Juan Jose Rios

Received: 25 April 2022
Accepted: 30 May 2022
Published: 6 June 2022

1. Introduction

Mango (*Mangifera indica* L.), a tropical tree, is highly sensitive to low temperatures and needs protection against chilling damage. Mango grows very well in a warm, frost-free environment with a specified winter dry season. Mango's optimum temperature ranges from 24 to 26.7 °C, with a minimum temperature of 10–12 °C, and chilling injury (CI) symptoms occur below these temperatures [1]. Currently, mango is experiencing CI during winter in Egypt, which affects its normal growth and development. According to the Central Laboratory for Agricultural Climate (CLAC) data, the minimum temperatures during the winter season have fluctuated over the last decade. This has caused cold symptoms in plants, specifically in tropical and subtropical fruit trees. Several cytological, biochemical, physiological, and molecular activities are altered due to chilling stress. They include photosynthesis, plasma membrane permeability, water status, osmotic balance, antioxidant activities, and other processes [2]. Many plant species have accrued SA when exposed to chilling stress. Salicylic acid (SA), an endogenous plant hormone, plays a vital role in growth, photosynthetic activity, and pigment content and has crucial physiological and biological roles in normal and stressed plants' metabolism [3–6]. SA application improves CI tolerance in many fruit tree species, including apple [7], apricot [8], banana [9], citrus [10], guava [11], kiwifruit [12], loquat [13], mango [14], peach [15], cactus pear [16], plum [17], and pomegranate [18,19]. The vast majority of SA applications has the goal to alleviate CI to fruits during cold storage, and SA use to ease CI on fruit trees is very limited.

Therefore, this study aimed to investigate SA role in mitigating the effects of chilling stress on Egyptian "Seddik" mango cultivar transplants.

2. Materials and Methods

This study was conducted during the 2021/2022 growing season at the Department of Pomology, Faculty of Agriculture, Cairo University, Egypt. All chemicals used in this study were reagent-grade and purchased from many worldwide suppliers (Bio Basic Inc., Markham, ON, Canada; Caisson Laboratories Inc., North Logan, UT, USA; Chem-Lab, Zedelgem, Belgium; Loba Chemie Pvt. Ltd., Maharashtra, India; Merck Ltd., Darmstadt, Germany; SDFCL, Chennai, India, and Sigma-Aldrich, Taufkirchen, Germany). Unless otherwise specified, all solutions were prepared in distilled water.

2.1. Greenhouse, Chilling Chamber Preparation, and Plant Materials

One-year-old mango (*Mangifera indica* L., cv. "Seddik") transplants grafted on "Sukkary" rootstock were used. The selected plants had uniform sizes, received the recommended water amounts to prevent the development of water deficit, were supplemented with macro- and microelements, and were treated with suitable pesticides to prevent other biotic stresses. A greenhouse was prepared for mango transplants' growth and recovery under normal growth conditions (25 °C/20 °C, day/night) before and after exposure to chilling. A chilling room was prepared to expose the mango transplants to chilling stress at 4 ± 1 °C. This temperature was determined based on the last decade's data (Figure 1) provided by the Central Laboratory for Agricultural Climate (CLAC).

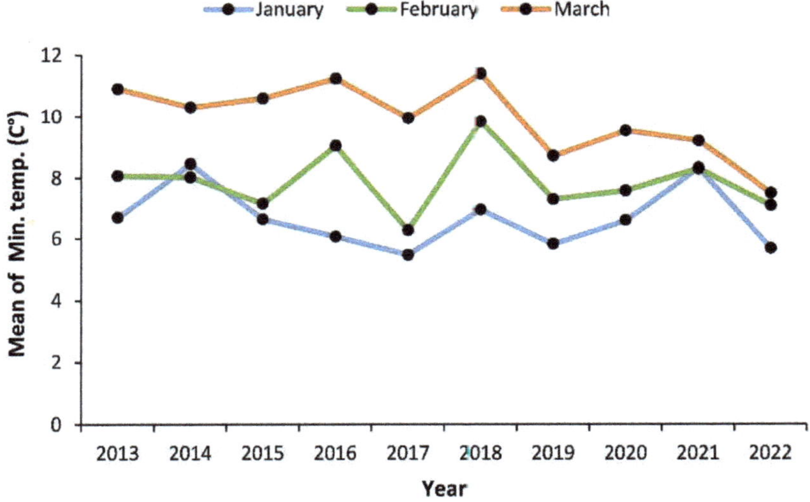

Figure 1. Minimum mean January, February, and March temperatures (open-air temperature) during the last decade in Giza district, Egypt.

2.2. Exogenous Salicylic Acid Treatments and Chilling Stress Induction

Salicylic acid (SA) was sprayed 48 and 24 h before the chilling treatment for 3 d at different concentrations (0, 0.5, 1, and 1.5 mM) on uniform juvenile mango transplants of "Seddik" cultivar. SA was dissolved in a NaOH solution (0.002 N); the pH was adjusted to 6.8 before the treatment. Then, Tween 80 was added at 0.1% as a surfactant. Untreated (treated with tap water) transplants exposed to chilling represented the positive control, while untreated transplants not exposed to chilling and kept under greenhouse conditions (25 °C/20 °C, day/night) throughout the entire period were the negative control.

For chilling exposure, the transplants were transferred to the Agriculture Development Systems (ADS) project's growth chamber, Faculty of Agriculture, Cairo University, at

4 ± 1 °C for 72 consecutive hours, followed by a period of six days of recovery under greenhouse conditions (25 °C/20 °C, day/night).

2.3. *Measurements*

Leaf samples were collected from the third to fifth leaf from the top of the transplant at 0 and 6 days after the culmination of the period of chilling exposure at 4 ± 1 °C for 72 consecutive hours. All treatments were assessed by the following measurements:

2.3.1. Defoliation

Defoliation percentage was estimated visually as the mean number of fallen leaves from each transplant with and without treatment after three weeks of chilling and compared to the initial mean number of leaves in each transplant.

2.3.2. Chilling Injury Index

Chilling injury (CI) was visually assessed after three weeks of exposure to chilling by employing the following scale: normal, no visible symptoms; trace, small necrotic areas on leaves but without growth restrictions (less than 5% of necrotic leaf area); slight, small necrotic areas on leaves (less than 15% of necrotic leaf area); moderate, well-defined necrotic areas on leaves (less than 30% of necrotic leaf area); and severe, extensive necrotic areas with severe growth restrictions (more than 50% of necrotic leaf area, plant still alive). By assigning a score of 1, 2, 3, 4, or 5, respectively, to each group, the average injury for each treatment was calculated [20].

2.3.3. Chlorophyll Content

Fresh leaf samples (0.25 g) were randomly chosen from three plants per replicate and homogenized in 20 mL of acetone (80% *v*/*v*). The absorbance was measured using a spectrophotometer (JENWAY 6300, Staffordshire, UK) at 663 and 646 nm, and the chlorophyll content was determined using the following equations [21]:

$$\text{Chl a (mg g}^{-1}\text{ fw)} = 12.21 \times \text{A663} - 2.81 \times \text{A646}$$

$$\text{Chl b (mg g}^{-1}\text{ fw)} = 20.13 \times \text{A646} - 5.03 \times \text{A663}$$

$$\text{Total chlorophyll content (mg g}^{-1}\text{ fw)} = \text{Chl a} + \text{Chl b}$$

2.3.4. Chlorophyll Stability Index (CSI)

The CSI was determined by heating fresh leaf samples (0.25 g) in 20 mL of distilled water at 56 °C in a water bath for 30 min [22]. Normal and heated leaf samples were homogenized in 80% acetone, then the total chlorophylls content was computed as above; the CSI was determined by using the following formula:

$$\text{CSI (\%)} = (\text{Total chlorophyll without heating} - \text{Total chlorophyll after heating}) \times 100$$

2.3.5. Electrolyte Leakage (EL) and Membrane Stability Index (MSI)

Electrolyte leakage was determined to evaluate membrane permeability, following the procedure of Guo et al. [23] with some modifications. Ten leaf discs of randomly selected plants per replicate were taken from the youngest fully expanded leaf. Then, the discs were placed in a 50 mL falcon tube and washed three times with distilled water to remove surface contamination. Next, the discs were placed in a 50 mL falcon tube containing 20 mL of deionized water (Aquinity^2 P^{10}, MembraPure GmbH, Hennigsdorf, Germany) and incubated, at room temperature, for 24 h. The bathing solution's (EC1) electrical conductivity (EC) was read after incubation using an electrical conductivity meter (BALRAMA, Digital EC Meter, New Delhi, India). Afterward, the same samples were placed into a boiling water bath for 20 min, and a second reading (EC2) was carried out after cooling the solution to room temperature. The EL was expressed as a percent value using the following formula:

$$EL (\%) = (EC1/EC2) \times 100$$

The membrane stability index (MSI) was computed based on electrolyte leakage data and expressed as a percent value using the following formula:

$$MSI (\%) = [1 - (EC1/EC2)] \times 100$$

2.3.6. Total Sugar Content

The total sugar content was determined utilizing the phenol–sulfuric acid method, according to Dubois et al. [24]. To this aim, 0.5 g of fresh leaf was homogenized in 20 mL of 80% ethanol (v/v). One mL ethanolic solution was mixed with 1 mL of 5% phenol dissolved in water (v/v), followed by the addition of 5 mL of concentrated sulfuric acid. The absorbance was read at 490 nm by a spectrophotometer (JENWAY 6300, Staffordshire, UK). A standard curve was generated, employing a pure glucose solution, and the total sugar content was expressed in mg glucose equivalent g^{-1} of fresh weight.

2.3.7. Total Phenolic Content

A fresh leaf sample (0.5 g) was extracted with 20 mL of 80% (v/v) methanol and then used for determining the total phenolic content. Total phenols were determined with a spectrophotometer, employing the modified Folin–Ciocalteu colorimetric method [25]. The methanolic extract (1 mL) was diluted 1:10, then mixed with 1 mL of Folin–Ciocalteu reagent in a test tube and allowed to stand for 6 min, followed by the addition of 5 mL of 1M Na_2CO_3 (w/v). Then, 3 mL of distilled water was added. The samples were incubated for 90 min at room temperature in the dark, and the absorbance at 760 nm was measured using the JENWAY 6300 spectrophotometer, Staffordshire, UK. The results were expressed as mg gallic acid (GAE) g^{-1} fw.

2.3.8. Proline Content

Proline content was determined using the Bates's method [26]. Leaf tissue (0.5 g) was homogenized in 10 mL of 3% aqueous sulfosalicylic acid for 10 min, followed by filtration. Two milliliters of the filtrate were mixed with 2 mL of acid ninhydrin and 2 mL of glacial acetic acid, then placed in a boiling water bath for 1 h at 90 °C, and the reaction was completed in an ice bath. The developed color was extracted in 4 mL of toluene, and the intensity of the reaction mixture was determined spectrophotometrically (JENWAY 6300, Staffordshire, UK) at the wavelength of 520 nm. Proline concentration was determined from a standard curve and calculated on a fresh weight basis as follows: μmoles proline/g of fresh weight = [(μg proline/mL × mL toluene)/115.5 μg/μmole]/[(g sample)/5].

2.3.9. DPPH Free Radical Scavenging Assay

The antioxidant activity was assessed by the 1,1-diphenyl-2-picrylhydrazyl (DPPH) free radical scavenging method [27,28]. We added 2.4 mL of 0.1 mM DPPH to 1.6 mL of methanolic leaf extract, vortexed the mixture, and incubated it at room temperature in the dark. The absorbance of the samples was measured after 30 min at 517 nm using a spectrophotometer (JENWAY 6300, Staffordshire, UK). The percentage of DPPH scavenging activity was calculated as % inhibition of DPPH = (A_{517} control −A_{517} sample/A_{517} control) × 100.

2.3.10. Extraction and Determination of Antioxidant Enzymes Activity

We used about 0.5 g of fresh young mango leaves to determine the concentration of superoxide dismutase (SOD), catalase (CAT), peroxidase (POX), and polyphenol oxidase (PPO) enzymes. Crude enzyme extracts were prepared using ground fresh leaves samples in liquid nitrogen that were homogenized in 10 mL of phosphate buffer pH 6.8 (0.1 M) and then centrifuged at 2 °C for 20 min at 20,000 rpm in a refrigerated centrifuge. The clear supernatant was the crude enzyme extract [29].

- *Superoxide Dismutase (SOD) Activity*

SOD (EC 1.15.1.1) activity was determined by measuring the inhibition of the auto-oxidation of pyrogallol by employing a method described by Marklund and Marklund [30]. We used a 10 mL reaction mixture consisting of 3.6 mL of distilled water, 0.1 mL of enzyme extract, 5.5 mL of 50 mM phosphate buffer (pH 7.8), and 0.8 mL of 3 mM pyrogallol (dissolved in 10 mM HCl). The pyrogallol reduction rate was measured at 325 nm with a UV–VIS spectrophotometer (PD 303, Apel Co., Ltd., Saitama, Japan). One unit of enzyme activity is the amount of the enzyme leading to 50% inhibition of the auto-oxidation rate of pyrogallol at 25 °C [31].

- *Catalase (CAT) Activity*

Catalase (E.C.1.11.1.6) activity was assayed based on Chen et al. [32]. The reaction mixture with a final volume of 10 mL, containing 40 μL of enzyme extract, was added to 9.96 mL of H_2O_2 phosphate buffer pH 7.0 (0.16 mL of 30% H_2O_2 to 100 mL of 50 mM phosphate buffer). Catalase activity was measured as the change in H_2O_2 absorbance in 60 s against a buffer blank at 250 nm using a UV–VIS spectrophotometer (PD 303, Apel Co., Ltd., Saitama, Japan). The blank sample was made by using buffer instead of the enzyme extract. One unit of enzyme activity is the amount of the enzyme reducing 50% of H_2O_2 in 60 s at 25 °C [31].

- *Peroxidase (POX) Activity*

Peroxidase (EC 1.11.1.7) activity was assayed in a reaction mixture containing 5.8 mL of 50 mM phosphate buffer pH 7.0, 0.2 mL of the enzyme extract, and 2 mL of 20 mM H_2O_2. The change in optical density was determined spectrophotometrically (PD 303, Apel Co., Ltd., Saitama, Japan) within 60 s at 470 nm at 25 °C after adding 2 mL of 20 mM pyrogallol [33]. One unit of enzyme activity is the amount of the enzyme catalyzing one micromole of H_2O_2 per minute at 25 °C [31].

- *Polyphenol Oxidase (PPO) Activity*

Polyphenol oxidase (EC 1.10.3.1) activity was determined using 125 μmol of phosphate buffer (pH 6.8), 100 μmol of pyrogallols, and 2 mL of enzyme extract. After incubating the mixture for 5 min at 25 °C, the reaction was stopped by adding 1 mL of 5% H_2SO_4. The developed color was read spectrophotometrically at 430 nm (PD 303, Apel Co., Ltd., Saitama, Japan) [34].

2.4. Statistical Analysis

The experiment used a completely randomized design (CRD) with three replicates; each replicate contained four mango transplants. The statistical analysis was performed using the R software, version 4.0.5, R Core Team, Vienna, Austria [35]. The main treatment effects at each sampling time were analyzed, and the means were compared by Duncan's multiple range tests [36] at a significance level of 0.05. Pearson's correlation coefficient was also calculated to specify associations between any measured parameters at each sampling time in response to chilling and SA application.

3. Results and Discussion

The effects of the exogenous SA application on chilled "Seddik" mango transplants were observed. Defoliation percentages for all treatments, after 21 days of exposure to chilling stress, were significantly high ($p \leq 0.05$) compared to those of the negative control (8.33%), but the chilled transplants pretreated with 1.5 mM L^{-1} SA yielded an acceptable defoliation percentage (23.33%). The positive control revealed the highest defoliation percentage (45.14%) after 21 days of exposure to chilling stress (Figure 2A). Chilling temperatures (lower than 10 °C) cause many physiological changes in chilling-sensitive plants, inducing CI and even mortality in tropical and subtropical species [37].

The plants exhibited a steady increase in leaf fall in the early stages of cold stress exposure and, as days progressed, they became even more defoliated [38]. Similarly, the chilling injury index (CII) of all chilled transplants not treated (positive control) or pretreated with SA was significantly higher ($p \leq 0.05$) than that of the negative control (Figure 2B). Thus, SA pretreatment at 1 and 1.5 mM alleviated mango transplants' CI symptoms (Figure 2E,F, respectively) compared with the positive control (Figure 2D). In fact, the application led to a vital reduction in CI incidence [2,39]. Exposure of the mango transplants to chilling stress for 72 h critically affected plant photosynthetic pigments, chlorophyll a, b, and total pigments, even in transplants pretreated with SA. However, after six days of recovery, the SA-treated mango transplants, specifically, those treated with SA at 1.5 mM, showed pigment values similar to those of the negative control (normal growth conditions). Hence, SA application mitigated the chilled mango transplants' chilling stress during the recovery period (Figure 3A–C). The lowest total chlorophyll values were 8.99 and 8.87 mg g^{-1} for positive control treatment after zero and six days of chilling stress, respectively. Similarly, the chlorophyll stability index (CSI) was the lowest for the untreated transplants (positive control) exposed to chilling stress; SA increased mango leaves' CSI compared with the positive control (Figure 3D).

Figure 2. Defoliation percentage for each treatment after 21 days of chilling stress (**A**), chilling injury index (**B**), chilling injury (CI) symptoms in mango transplants, negative control (**C**), CI symptoms in chilled mango transplants, positive control (**D**), CI symptoms in chilled mango transplants pretreated with 1 mM SA (**E**) and 1.5 mM SA (**F**). Bars with different letters represent significantly different data at 95% confidence, as determined by Duncan's Multiple Range Test. Error bars represent the standard deviation.

Under low-temperature conditions, chlorophyll-degrading enzymes' (chlorophyllases) activity increases, and their biosynthesis is inhibited, leading to a decrease in chlorophyll content in chilled plants [40,41].

The leaves' low chlorophyll content at low temperatures can be interpreted as a lack of photosynthetic efficiency [42,43]. Moreover, the reduction in photosynthetic capacity at low temperatures is associated with a decrease in PSII quantum efficiency, the primary target of damage at low temperatures [1,44]. Chilling damage occurs when membranes acquire more saturated fatty acids due to the exposure to low temperatures [44–46]. In Figure 4A, it is discernible that the transplants exposed to chilling stress showed an increase in electrolyte leakage percentage after zero (52.99%) or six days (64.51%) compared to those pretreated with SA. The negative control recorded the lowest electrolyte leakage percentage values after zero (28.49%) and six days (28.05%) of exposure to chilling stress.

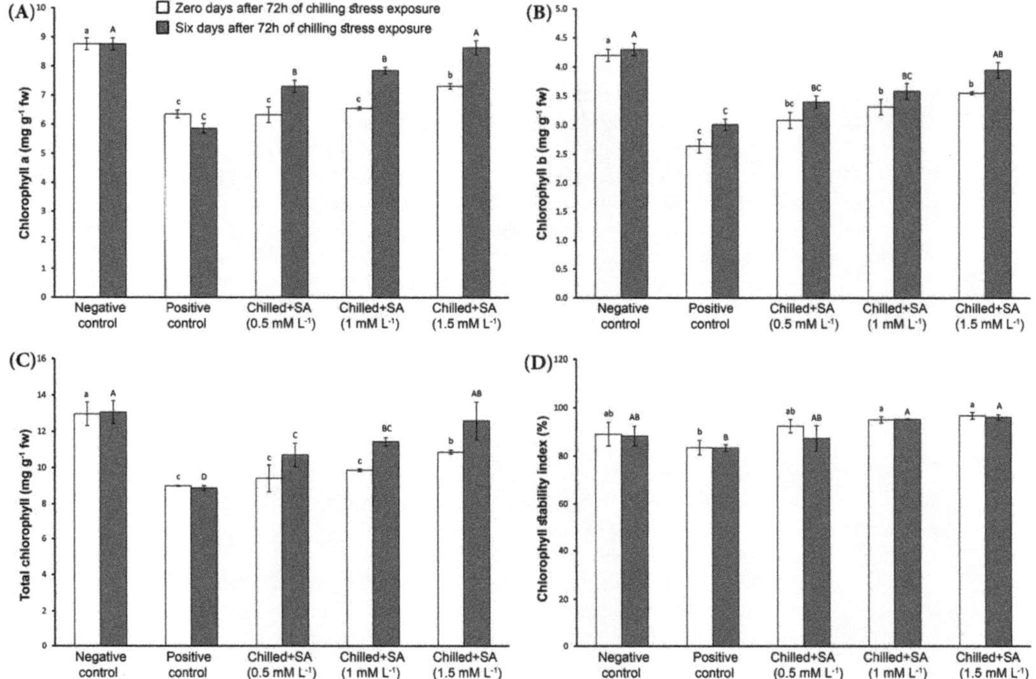

Figure 3. Changes in leaf chlorophyll content; chlorophyll a content (**A**); chlorophyll b content (**B**); total chlorophyll (**C**), and chlorophyll stability index (**D**) in mango transplants under the studied treatments. Different lower-case letters indicate statistical differences between treatments at zero days after 72 h of chilling stress exposure, while the upper-case letters indicate significant differences between treatments at six days at 95% confidence, as determined by Duncan's Multiple Range Test. Error bars represent the standard deviation.

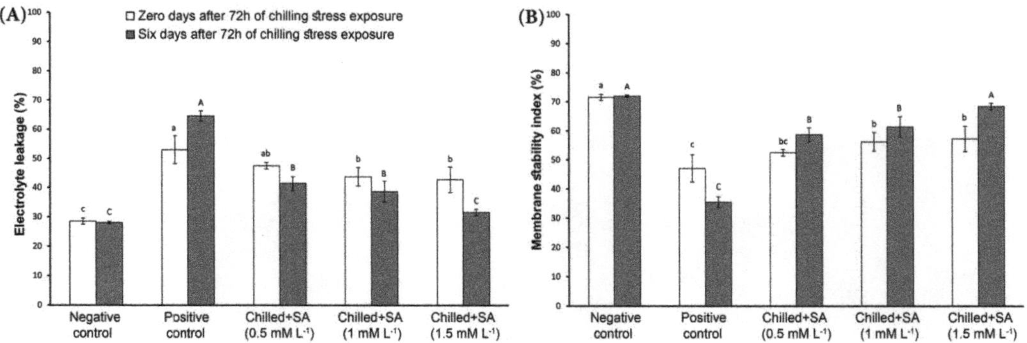

Figure 4. Changes in electrolyte leakage (**A**) and membrane stability index (**B**) of mango transplants' leaves under the studied treatments. Different lower-case letters indicate statistical differences between treatments at zero days after 72 h of chilling stress exposure, while the upper-case letters indicate significant differences between treatments at six days at 95% confidence, as determined by Duncan's Multiple Range Test. Error bars represent the standard deviation.

Contrarily, the membrane stability index was the highest for the negative control compared to the positive control, with the lowest values (47.01 and 35.49%) after zero and six days of chilling stress exposure, respectively (Figure 4B). Usually, electrolytes leakage is used to assess the chilling damage [47]. Guinn [48] suggested that the increase in electrolyte leakage is likely due to chilling-induced water stress. Furthermore, increased electrolyte leakage from chilled plants was attributed to membrane deterioration and corresponded to the presence of leaked inorganic and organic ions [46,49–52]. SA role in maintaining the fatty acids' content and ratio in the cell membranes could explain its protection of the cell membrane structure [53]. Differences in total sugar content were not statistically significant but accrued gradually with increased SA concentrations, specifically, after six days of recovery, in the chilled mango transplants compared with the positive control (Figure 5A). Generally, plants amass many relevant solutes, such as soluble sugars and amino acids, in response to cold and other osmotic stresses [54–56]. During the recovery period, the highest significant ($p \leq 0.05$) total phenolic content was recorded in the positive control, with 3.76 mg g^{-1}, and the lowest ($p \leq 0.05$) in the negative control, with 1.81 mg g^{-1}. The same trend was also evident after exposure to chilling stress for all treatments (Figure 5B). The transplants pretreated with increased SA concentrations before being subjected to chilling had a lower proline content after zero or six days of stress exposure, as shown in Figure 5C. Rivero et al. [57] observed that mango tissues accrued phenolic compounds under cold stress. A decreased amount of phenolics was observed in SA-pretreated transplants, depicting the effect of SA in alleviating CI in mango under chilling. This finding agrees with Han et al. [58]. However, Wongsheree et al. [45] found that total phenolic compounds in lemon basil leaves, whether young or mature, were not affected by chilling stress exposure at 4 °C. Under cold stress, exogenous SA application caused an increase of soluble carbohydrates in *Phaselous vulgaris* [59]. SA treatment substantially increased solutes and total soluble sugars, and these osmolytes promoted cryostability in the cell membranes, protecting the plants from cold stress [60,61]. Moreover, the stress conditions increased proline metabolism, attributable to an increase in proline biosynthesis enzymes (pyrolline-5-carboxylate reductase and -glutamyl kinase) [62,63]. Sayyari et al. [64] reported that SA ameliorated CI by inhibiting proline accumulation; the variation in proline content in response to chilling stress primarily depended on the plant genotype [64,65].

Concerning DPPH radical-scavenging activity, SA application positively impacted the chilled mango transplants during recovery by enhancing the function of DPPH. It increased with increases in SA concentration, and the lowest value was measured for the positive control (Figure 5D). The induction of DPPH scavenging activity in chilled mango transplants by SA depended on the concentration applied. This was also observed for banana [66], mango [67,68], and lemons [69] when cold-exposed fruits were treated with SA. Under chilling conditions, SA pre-treatment reduced SOD activity in mango leaves (Figure 6A). After 72 h of exposure to chilling stress, SOD activity significantly ($p \leq 0.05$) decreased compared to that in the negative control (0.81 U g^{-1}) but significantly ($p \leq 0.05$) increased with the gradual increase of SA concentration (0.65, 0.93, and 1.00 U g^{-1} with 0.5, 1, and 1.5 mM SA, respectively). The positive control exhibited the lowest significant ($p \leq 0.05$) value (0.56 U g^{-1}). This trend was also evident for the recovery period with fewer responses, whereas the negative control showed the highest (0.80 U g^{-1}) and the lowest values (0.38 U g^{-1}). The SOD values in chilled transplants pretreated with SA were 0.42, 0.61, and 0.64 U g^{-1} with 0.5, 1, and 1.5 mM SA, respectively. Similarly, chilling stress (positive control) significantly ($p \leq 0.05$) reduced CAT activity during recovery compared to the negative control, but SA pre-treatment gradually increased CAT activity ($p \leq 0.05$) in chilled mango transplants (Figure 6B). Under chilling conditions, POX activity in mango leaves was significantly ($p \leq 0.05$) high for the positive control, reaching the highest value (0.73 U g^{-1}) compared to the negative control (0.104 U g^{-1}). As expected, SA pre-treatment gradually decreased POX activity after six days of chilling stress to its normal level (0.104 U g^{-1}) with 1.5 mM L^{-1} SA treatment (Figure 6C).

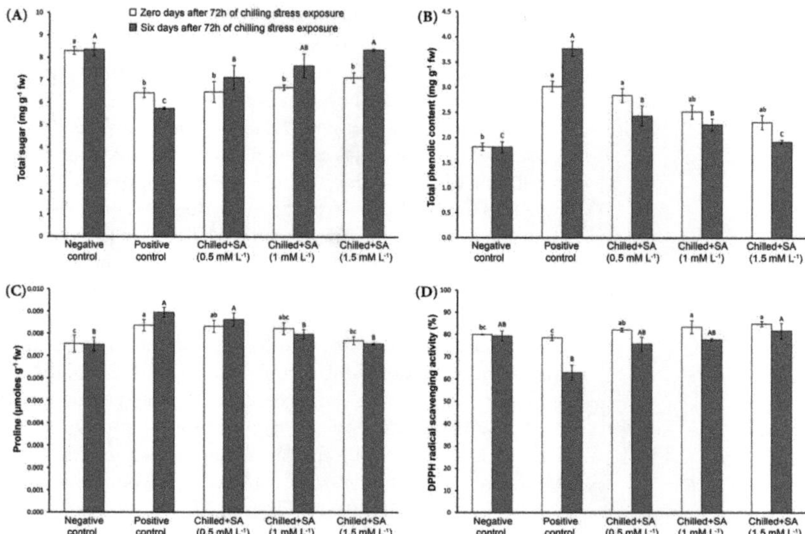

Figure 5. Changes in total sugar (**A**), total phenolic content (**B**), proline content (**C**), and DPPH (**D**) in mango transplant leaves. Different lower-case letters indicate statistical differences between treatments at zero days after 72 h of chilling stress exposure, while the upper-case letters indicate significant differences between treatments at six days at 95% confidence, as determined by Duncan's Multiple Range Test. Error bars represent the standard deviation.

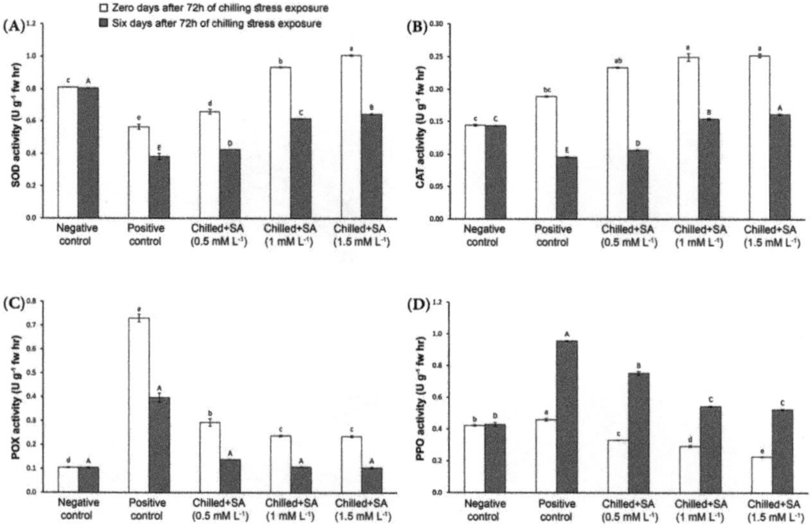

Figure 6. Changes in the activities of superoxide dismutase (**A**), catalase (**B**), peroxidase (**C**), and polyphenol oxidase (**D**) in mango transplant leave. Different lower-case letters indicate statistical differences between treatments at zero days after 72 h of chilling stress exposure, while the various upper-case letters indicate significant differences between treatments at six days at 95% confidence, as determined by Duncan's Multiple Range Test. Error bars represent the standard deviation.

Moreover, PPO activity in chilled mango leaves was also significantly ($p \leq 0.05$) high after exposure to chilling. However, SA gradually decreased the enzyme activity by about half compared to the negative control. After six days of exposure to chilling stress, PPO activity in chilled mango leaves was significantly ($p \leq 0.05$) higher, and the SA treatments significantly ($p \leq 0.05$) reduced PPO activity until it approximately returned to the normal level compared to the negative control (Figure 6D).

The untreated mango leaves exhibited lower SOD and CAT enzyme activities after six days of chilling stress exposure than the treated ones. It implies less protection against membrane oxidation in untreated leaves [45]. Accordingly, a positive correlation between SOD and CAT enzyme activities was detected after zero ($r = 0.34$) and six days ($r = 0.81$) of chilling stress exposure (Figure 7A,B). Generally, SA treatment effectively alleviated the chilled mango transplants' CI compared to the negative control, reaching the normal levels after six days of stress (recovery). These findings are aligned with those of Chen et al. [9] and Khademi et al. [66], who found that SA treatment effectively reduced banana CI by maintaining membrane integrity and improving antioxidants' activity. SOD and CAT enzymatic activities exhibited the same trend. SOD, the primary line of defense against ROS-induced oxidative damages, catalyzes the dismutation of two superoxide radicals into H_2O_2 and O_2. CAT converts H_2O_2 into H_2O and O_2. Still, POX enables H_2O_2 oxidation and yields water and another oxidizing molecule. This was evident in the highly positive significant correlation ($p < 0.001$) between POX enzyme activity and EL ($r = 0.77$) after 72 h of exposure to chilling stress (Figure 7A). PPO causes tissue browning in most horticultural crops by oxidizing phenolic compounds to quinones. CI alleviation by SA application decreased PPO activity and increased antioxidant systems and specific bioactive chemicals' concentration, as reported for banana [70], cherry [71], litchi [72], pomegranate [18], and wax apple [73]. After six days of exposure to chilling, PPO had significant ($p < 0.001$) correlations with EL, total phenol, and proline (Figure 7B). Meanwhile, it showed strong negative correlations with MSI, plant photosynthetic pigments (chlorophyll a, b, and total), SOD, and CAT. SA treatment of cold-stressed plants altered the activities of different enzymatic antioxidants such as CAT, SOD, and POX [74,75]. In the correlation analysis, a significant positive correlation ($p < 0.001$) between CAT, chlorophyll a, b, MSI, SOD, and total sugar (Figure 7B) was found. SA was shown to enable some crops to recover from cold damage by regulating antioxidative mechanisms [39,59,76,77].

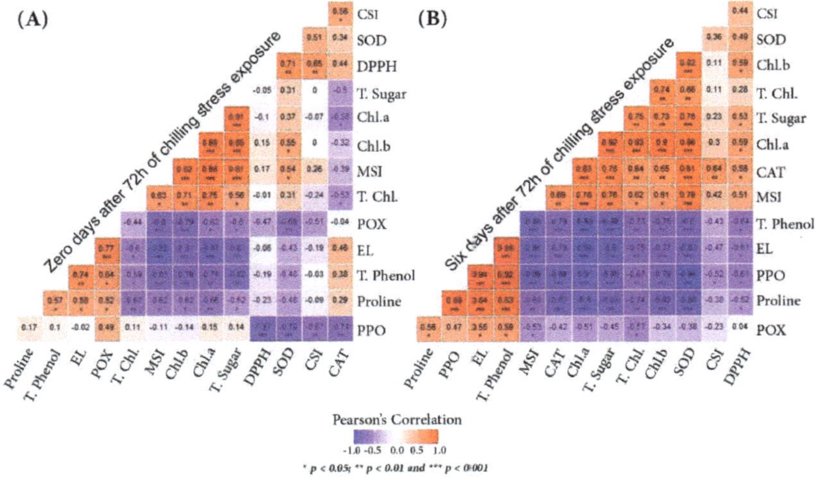

Figure 7. Pearson's correlation coefficient analysis for each sampling time, (**A**) for day zero and (**B**) for day six after 72 h of chilling stress exposure. R is presented in different colors; the legend shows the color range of different R values with * $p < 0.05$, ** $p < 0.01$, and *** $p < 0.001$ as indicators of statistical significance.

4. Conclusions

The results revealed that exogenous SA application minimized the adverse effects of chilling in local mango transplants of the "Seddik" cultivar, as evidenced by reduced membrane damage and enhanced endogenous production of photosynthetic pigments, total sugar, and DPPH. SA also reduced the content of total phenolic compounds and proline and regulated the activity of antioxidant enzymes, including SOD, CAT, PPO, and POX, especially during the recovery period. These findings suggest that exogenous SA treatment can help mango transplants recover from chilling stress damage.

Author Contributions: Conceptualization, I.H.; methodology, I.H., A.E.M.A., N.K. and A.A; investigation, I.H. and A.A; resources, S.M.S. and A.A.; formal analysis, I.H.; writing—original draft preparation, I.H.; writing—review and editing, I.H. and A.A.; visualization, I.H. and S.M.S.; validation, A.E.M.A. and N.K.; supervision, I.H.; project administration, I.H.; funding acquisition, I.H. All authors have read and agreed to the published version of the manuscript.

Funding: This research was funded and supported by a research grant from the Science, Technology and Innovation Funding Authority (STIFA), under STDF-RG, No. 35967.

Institutional Review Board Statement: Not applicable.

Informed Consent Statement: Not applicable.

Data Availability Statement: Not applicable.

Acknowledgments: The authors would like to express special gratitude to the Science and Technology Development Fund (STDF) for funding and to the Faculty of Agriculture, Cairo University, for the use of its facilities during this work. Current name: Science, Technology and Innovation Funding Authority (STIFA).

Conflicts of Interest: The authors declare no conflict of interest.

References

1. Allen, D.J.; Ort, D.R. Impacts of chilling temperatures on photosynthesis in warm-climate plants. *Trends Plant Sci.* **2001**, *6*, 36–42. [CrossRef]
2. Saleem, M.; Fariduddin, Q.; Janda, T. Multifaceted role of salicylic acid in combating cold stress in plants: A review. *J. Plant Growth Regul.* **2021**, *40*, 464–485. [CrossRef]
3. Cheng, F.; Lu, J.; Gao, M.; Shi, K.; Kong, Q.; Huang, Y.; Bie, Z. Redox signaling and CBF-responsive pathway are involved in salicylic acid-improved photosynthesis and growth under chilling stress in watermelon. *Front. Plant. Sci.* **2016**, *7*, 1519. [CrossRef] [PubMed]
4. Ding, P.; Ding, Y. Stories of salicylic acid: A plant defense hormone. *Trends Plant. Sci.* **2020**, *25*, 549–565. [CrossRef] [PubMed]
5. Poór, P. Effects of salicylic acid on the metabolism of mitochondrial reactive oxygen species in plants. *Biomolecules* **2020**, *10*, 341. [CrossRef] [PubMed]
6. Mishra, A.K.; Baek, K.H. Salicylic acid biosynthesis and metabolism: A divergent pathway for plants and bacteria. *Biomolecules* **2021**, *11*, 705. [CrossRef] [PubMed]
7. Hadian-Deljou, M.; Esna-Ashari, M.; Sarikhani, H. Effect of pre-and post-harvest salicylic acid treatments on quality and antioxidant properties of 'Red Delicious' apples during cold storage. *Adv. Hortic. Sci.* **2017**, *31*, 31–38. [CrossRef]
8. Ezzat, A.; Ammar, A.; Szabó, Z.; Holb, I. Salicylic acid treatment saves quality and enhances antioxidant properties of apricot fruit. *Hortic. Sci.* **2017**, *44*, 73–81. [CrossRef]
9. Chen, L.; Zhao, X.; Wu, J.E.; He, Y.; Yang, H. Metabolic analysis of salicylic acid-induced chilling tolerance of banana using NMR. *Food Res. Int.* **2020**, *128*, 108796. [CrossRef]
10. Supapvanich, S.; Polpakdee, R.; Wongsuwan, P. Chilling injury alleviation and quality maintenance of lemon basil by preharvest salicylic acid treatment. *Emir. J. Food Agric.* **2015**, *27*, 801–807. [CrossRef]
11. Killadi, B.; Lenka, J.; Chaurasia, R.; Shukla, D.K. Effect of salicylic acid and acylated salicylic acid on the shelf-life of guava (*Psidium guajava* L.) fruit during storage. *Int. J. Chem. Stud.* **2019**, *7*, 1848–1852.
12. Zhang, Y.; Chen, K.; Zhang, S.; Ferguson, I. The role of salicylic acid in postharvest ripening of kiwifruit. *Postharvest Biol. Technol.* **2003**, *28*, 67–74. [CrossRef]
13. Cai, C.; Li, X.; Chen, K. Acetylsalicylic acid alleviates chilling injury of postharvest loquat (*Eriobotrya japonica* Lindl.) fruit. *Eur. Food Res. Technol.* **2006**, *223*, 533–539. [CrossRef]
14. Reddy, S.V.R.; Sharma, R.R.; Srivastava, M.; Kaur, C. Effect of pre-harvest application of Salicylic acid on the postharvest behavior of 'Amrapali' mango fruits during storage. *Indian J. Hortic.* **2016**, *73*, 405–409. [CrossRef]

15. Yang, Z.; Cao, S.; Zheng, Y.; Jiang, Y. Combined salicyclic acid and ultrasound treatments for reducing the chilling injury on peach fruit. *J. Agric. Food Chem.* **2012**, *60*, 1209–1212. [CrossRef]
16. Al-Qurashi, A.D.; Awad, M.A. Postharvest salicylic acid treatment reduces chilling injury of 'Taify' cactus pear fruit during cold storage. *J. Food Agric. Environ.* **2012**, *10*, 120–124.
17. Davarynejad, G.H.; Zarei, M.; Nasrabadi, M.E.; Ardakani, E. Effects of salicylic acid and putrescine on storability, quality attributes and antioxidant activity of plum cv. 'Santa Rosa'. *J. Food Sci. Technol.* **2015**, *52*, 2053–2062. [CrossRef]
18. Sayyari, M.; Babalar, M.; Kalantari, S.; Serrano, M.; Valero, D. Effect of salicylic acid treatment on reducing chilling injury in stored pomegranates. *Postharvest Biol. Technol.* **2009**, *53*, 152–154. [CrossRef]
19. Güneş, N.T.; Yildiz, M.; Variş, B.; Horzum, Ö. Postharvest salicylic acid treatment influences some quality attributes in air-stored pomegranate fruit. *J. Agric. Sci.* **2020**, *26*, 499–506. [CrossRef]
20. Wang, C.Y. Modification of chilling susceptibility in seed¬lings of cucumber and zucchini squash by the bio regulator paclobutrazol (PP333). *Sci. Hortic.* **1985**, *26*, 293–298. [CrossRef]
21. Lichtenthaler, H.K.; Wellburn, A.R. Determinations of total carotenoids and chlorophylls a and b of leaf extracts in different solvents. *Biochem. Soc. Trans.* **1983**, *11*, 591–592. [CrossRef]
22. Thakur, A. Use of easy and less expensive methodology to rapidly screen fruit crops for drought tolerance. *Acta Hortic.* **2004**, *662*, 231–235. [CrossRef]
23. Guo, Z.; Huang, M.; Lu, S.; Yaqing, Z.; Zhong, Q. Differential response to paraquat induced oxidative stress in two rice cultivars on antioxidants and chlorophyll a fluorescence. *Acta Physiol. Plant* **2007**, *29*, 39–46. [CrossRef]
24. Dubois, M.; Gilles, K.A.; Hamilton, J.K.; Rebers, P.T.; Smith, F. Colorometric method for determination of sugars and related substances. *Analyt. Chem.* **1956**, *28*, 350–356. [CrossRef]
25. Singleton, V.L.; Rossi, J.A. Colorimetry of total phenolics with phosphomolybdic-phosphotungstic acid reagents. *Am. J. Enol. Vitic.* **1965**, *16*, 144–158.
26. Bates, L.S.; Waldren, R.P.; Teare, I.D. Rapid determination of free proline for water-stress studies. *Plant Soil* **1973**, *39*, 205–207. [CrossRef]
27. Blois, M.S. Antioxidant determinations by the use of a stable free radical. *Nature* **1958**, *181*, 1199–1200. [CrossRef]
28. Rahman, M.M.; Islam, M.B.; Biswas, M.; Alam, A.K. In vitro antioxidant and free radical scavenging activity of different parts of Tabebuia pallida growing in Bangladesh. *BMC Res. Notes* **2015**, *8*, 621. [CrossRef]
29. Mukherjee, S.P.; Choudhuri, M.A. Implication of water stress-induced changes in the level of endogenous ascorbic acid and hydrogen peroxide in Vigna seedlings. *Physiol. Plant* **1983**, *58*, 166–170. [CrossRef]
30. Marklund, S.; Marklund, G. Involvement of the superoxide anion radical in the autoxidation of pyrogallol and a convenient assay for superoxide dismutase. *Eur. J. Biochem.* **1974**, *47*, 469–474. [CrossRef]
31. Kong, F.X.; Hu, W.; Chao, S.Y.; Sang, W.L.; Wang, L.S. Physiological responses of the lichen Xanthoparmelia mexicana to oxidative stress of SO_2. *Environ. Exp. Bot.* **1999**, *42*, 201–209. [CrossRef]
32. Chen, Y.; Cao, X.X.D.; Lu, Y.; Wang, X.R. Effect of rare earth metal ions and their EDTA complex on antioxidant enzymes of fish liver. *Bull. Environ. Contam. Toxicol.* **2000**, *65*, 357–365. [CrossRef] [PubMed]
33. Pütter, J. Peroxidases. In *Methods of Enzymatic Analysis*, 2nd ed.; Bergmeyer, H.U., Ed.; Verlag Chemie Weinheim, Academic Press: New York, NY, USA, 1974; Volume 2, pp. 685–690. [CrossRef]
34. Kar, M.; Mishra, D. Catalase, peroxidase and polyphenol oxidase activities during rice leaf senescence. *Plant Physiol.* **1976**, *57*, 315–319. [CrossRef] [PubMed]
35. R Core Team. *R: A Language and Environment for Statistical Computing*; R Foundation for Statistical Computing: Vienna, Austria, 2021. Available online: https://www.R-project.org/ (accessed on 29 January 2022).
36. Duncan, D.B. Multiple range and multiple F tests. *Biometrics* **1955**, *11*, 1–42. [CrossRef]
37. Lukatkin, A.S.; Brazaityte, A.; Bobinas, C.; Duchovskis, P. Chilling injury in chilling-sensitive plants: A review. *Agriculture* **2012**, *99*, 111–124.
38. Chang, Y.C.; Lin, T.C. Temperature effects on fruit development and quality performance of nagami kumquat (*Fortunella margarita* [Lour.] Swingle). *Hort. J.* **2020**, *89*, 351–358. [CrossRef]
39. Kang, G.; Wang, C.; Sun, G.; Wang, Z. Salicylic acid changes activities of H2O2-metabolizing enzymes and increases the chilling tolerance of banana seedlings. *Environ. Exp. Bot.* **2003**, *50*, 9–15. [CrossRef]
40. Reddy, M.P. Changes in pigment composition. Hill reaction activity and saccharides metabolism in bajra (*Penisetum typhoides*) leaves under NaCl salinity. *Photosynthetica* **1986**, *20*, 50–55.
41. Wise, R.R.; Naylor, A.W. Chilling-enhanced photooxidation: Evidence for the role of singlet oxygen and superoxide in the breakdown of pigments and endogenous antioxidants. *Plant Physiol.* **1987**, *83*, 278–282. [CrossRef]
42. Schapendonk, A.H.C.M.; Dolstra, O.; Van Kooten, O. The use of chlorophyll fluorescence as a screening method for cold tolerance in maize. *Photosynth. Res.* **1989**, *20*, 235–247. [CrossRef]
43. Kingston-Smith, A.H.; Foyer, C.H. Bundle sheath proteins are more sensitive to oxidative damage than those of the mesophyll in maize leaves exposed to paraquat or low temperatures. *J. Exp. Bot.* **2000**, *51*, 123–130. [CrossRef] [PubMed]
44. Durner, E.F. *Principles of Horticultural Physiology*; CABI Publishing: Oxford, UK, 2013; p. 391.
45. Wongsheree, T.; Ketsa, S.; van Doorn, W.G. The relationship between chilling injury and membrane damage in lemon basil (Ocimum × citriodourum) leaves. *Postharvest Biol. Technol.* **2009**, *51*, 91–96. [CrossRef]

46. Gadallah, F.M.; El-Yazal, S.; Abdel-Samad, G.A. Physiological changes in leaves of some mango cultivars as response to exposure to low temperature degrees. *Hortic. Int. J.* **2019**, *3*, 266–273. [CrossRef]
47. McKay, H.M. Electrolyte leakage from fine roots of conifer seedlings: A rapid index of plant vitality following cold storage. *Can. J. For. Res.* **1992**, *22*, 1371–1377. [CrossRef]
48. Guinn, G. Chilling injury in cotton seedlings: Changes in permeability of cotyledons. *Crop. Sci.* **1971**, *11*, 101–102. [CrossRef]
49. Chinnusamy, V.; Zhu, J.; Zhu, J.K. Cold stress regulation of gene expression in plants. *Trends Plant Sci.* **2007**, *12*, 444–451. [CrossRef]
50. El-Moniem, E.A.A.; Ismail, O.M.; Shaban, A.E.A. Changes in leaves component of some mango cultivars in relation to exposure to low temperature degrees. *World J. Agric. Res.* **2010**, *6*, 212–217.
51. Gadallah, F.M.; El-Yazal, M.A.S.; Abdel-Samad, G.A.; Sayed, A.A. Leakage compositional changes accompanying to exposure of some mango cultivars to low temperature under field conditions. *Int. J. Curr. Microbiol. App. Sci.* **2020**, *9*, 1448–1463. [CrossRef]
52. Sayed, A.A.; Gadallah, F.M.; El-Yazal, M.A.S.; Abdel-Samad, G.A. Impact of exposure to low temperature degrees in field conditions on leaf pigments and chlorophyll fluorescence in leaves of mango trees. *JHPR* **2020**, *10*, 30–45. [CrossRef]
53. Popova, L.P.; Maslenkova, L.T.; Ivanova, A.; Stoinova, Z. Role of salicylic acid in alleviating heavy metal stress. In *Environmental Adaptations and Stress Tolerance of Plants in the Era of Climate Change*; Ahmad, P., Prasad, M., Eds.; Springer: New York, NY, USA, 2012; pp. 447–466. [CrossRef]
54. Rhodes, D.; Nadolska-Orczyk, A.; Rich, P.J. Salinity, osmolytes and compatible solutes. In *Salinity: Environment-Plants-Molecules*; Läuchli, A., Lüttge, U., Eds.; Springer: Dordrecht, The Netherlands, 2002; pp. 181–204. [CrossRef]
55. Ruelland, E.; Zachowski, A. How plants sense temperature. *Environ. Exp. Bot.* **2010**, *69*, 225–232. [CrossRef]
56. Kamal, A.; Kumar, V.; Muthukumar, M.; Bajpai, A. Morphological indicators of salinity stress and their relation with osmolyte associated redox regulation in mango cultivars. *J. Plant Biochem. Biotechnol.* **2021**, *30*, 918–929. [CrossRef]
57. Rivero, R.M.; Ruiz, J.M.; Garcıa, P.C.; Lopez-Lefebre, L.R.; Sánchez, E.; Romero, L. Resistance to cold and heat stress: Accumulation of phenolic compounds in tomato and watermelon plants. *Plant Sci.* **2001**, *160*, 315–321. [CrossRef]
58. Han, J.; Tian, S.P.; Meng, X.H.; Ding, Z.S. Response of physiologic metabolism and cell structures in mango fruit to exogenous methyl salicylate under low temperature stress. *Physiol. Plant* **2006**, *128*, 125–133. [CrossRef]
59. Soliman, M.H.; Alayafi, A.A.; El Kelish, A.A.; Abu-Elsaoud, A.M. Acetylsalicylic acid enhance tolerance of *Phaseolus vulgaris* L. to chilling stress, improving photosynthesis, antioxidants and expression of cold stress responsive genes. *Bot. Stud.* **2018**, *59*, 1–17. [CrossRef]
60. Gusta, L.V.; Wisniewski, M. Understanding plant cold hardiness: An opinion. *Physiol. Plant* **2013**, *147*, 4–14. [CrossRef]
61. Luo, Y.L.; Su, Z.L.; Bi, T.J.; Cui, X.L.; Lan, Q.Y. Salicylic acid improves chilling tolerance by affecting antioxidant enzymes and osmoregulators in sacha inchi (*Plukenetia volubilis*). *Rev. Bras. Bot.* **2014**, *37*, 357–363. [CrossRef]
62. Zhu, B.; Su, J.; Chang, M.; Verma, D.P.S.; Fan, Y.L.; Wu, R. Overexpression of a Δ1-pyrroline-5-carboxylate synthetase gene and analysis of tolerance to water-and salt stress in transgenic rice. *Plant Sci.* **1998**, *139*, 41–48. [CrossRef]
63. Forlani, G.; Trovato, M.; Funck, D.; Signorelli, S. Regulation of Proline Accumulation and Its Molecular and Physiological Functions in Stress Defence. In *Osmoprotectant-Mediated Abiotic Stress Tolerance in Plants*; Hossain, M., Kumar, V., Burritt, D., Fujita, M., Mäkelä, P., Eds.; Springer: Cham, Switzerland, 2019; pp. 73–97. [CrossRef]
64. Sayyari, M.; Ghanbari, F.; Fatahi, S.; Bavandpour, F. Chilling tolerance improving of watermelon seedling by salicylic acid seed and foliar application. *Not. Sci. Biol.* **2013**, *5*, 67–73. [CrossRef]
65. Bastam, N.; Baninasab, B.; Ghobadi, C. Improving salt tolerance by exogenous application of salicylic acid in seedlings of pistachio. *Plant. Growth Regul.* **2013**, *69*, 275–284. [CrossRef]
66. Khademi, O.; Ashtari, M.; Razavi, F. Effects of salicylic acid and ultrasound treatments on chilling injury control and quality preservation in banana fruit during cold storage. *Sci. Hortic.* **2019**, *249*, 334–339. [CrossRef]
67. Junmatong, C.; Faiyue, B.; Rotarayanont, S.; Uthaibutra, J.; Boonyakiat, D.; Saengnil, K. Cold storage in salicylic acid increases enzymatic and non-enzymatic antioxidants of Nam Dok Mai No. 4 mango fruit. *Sci. Asia* **2015**, *41*, 12–21. [CrossRef]
68. Awad, M.A.; Al-Qurashi, A.D. Postharvest salicylic acid and melatonin dipping delay ripening and improve quality of 'Sensation' Mangoes. *Philipp. Agric. Sci.* **2021**, *104*, 34–44.
69. Siboza, X.I.; Bertling, I. The effects of methyl jasmonate and salicylic acid on suppressing the production of reactive oxygen species and increasing chilling tolerance in 'Eureka'lemon [*Citrus limon* (L.) Burm. F.]. *J. Hortic. Sci. Biotechnol.* **2013**, *88*, 269–276. [CrossRef]
70. Chen, J.Y.; He, L.H.; Jiang, Y.M.; Kuang, J.F.; Lu, C.B.; Joyce, D.C.; Macnish, A.; He, Y.X.; Lu, W.J. Expression of PAL and HSPs in fresh-cut banana fruit. *Environ. Exp. Bot.* **2009**, *66*, 31–37. [CrossRef]
71. Dokhanieh, A.Y.; Aghdam, M.S.; Fard, J.R.; Hassanpour, H. Postharvest salicylic acid treatment enhances antioxidant potential of cornelian cherry fruit. *Sci. Hortic.* **2013**, *154*, 31–36. [CrossRef]
72. Kumari, P.; Barman, K.; Patel, V.B.; Siddiqui, M.W.; Kole, B. Reducing postharvest pericarp browning and preserving health promoting compounds of litchi fruit by combination treatment of salicylic acid and chitosan. *Sci. Hortic.* **2015**, *197*, 555–563. [CrossRef]
73. Supapvanich, S.; Mitsang, P.; Youryon, P.; Techavuthiporn, C.; Boonyaritthongchai, P.; Tepsorn, R. Postharvest quality maintenance and bioactive compounds enhancement in 'Taaptimjaan' wax apple during short-term storage by salicylic acid immersion. *Hortic. Environ. Biotechnol.* **2018**, *59*, 373–381. [CrossRef]

74. Mutlu, S.; Karadağoğlu, Ö.; Atici, Ö.; Nalbantoğlu, B. Protective role of salicylic acid applied before cold stress on antioxidative system and protein patterns in barley apoplast. *Biol. Plant* **2013**, *57*, 507–513. [CrossRef]
75. Mo, Y.; Gong, D.; Liang, G.; Han, R.; Xie, J.; Li, W. Enhanced preservation effects of sugar apple fruits by salicylic acid treatment during postharvest storage. *J. Sci. Food Agric.* **2008**, *88*, 2693–2699. [CrossRef]
76. Mutlu, S.; Atıcı, Ö.; Nalbantoğlu, B.; Mete, E. Exogenous salicylic acid alleviates cold damage by regulating antioxidative system in two barley (*Hordeum vulgare* L.) cultivars. *Front. Life Sci.* **2016**, *9*, 99–109. [CrossRef]
77. Ignatenko, A.; Talanova, V.; Repkina, N.; Titov, A. Exogenous salicylic acid treatment induces cold tolerance in wheat through promotion of antioxidant enzyme activity and proline accumulation. *Acta Physiol. Plant* **2019**, *41*, 1–10. [CrossRef]

Article

Seed Priming with Sulfhydral Thiourea Enhances the Performance of *Camelina sativa* L. under Heat Stress Conditions

Ejaz Ahmad Waraich [1,*], Muhammad Ahmad [1], Walid Soufan [2], Muhammad Taimoor Manzoor [1], Zahoor Ahmad [3], Muhammad Habib-Ur-Rahman [4] and Ayman EL Sabagh [5,*]

1 Department of Agronomy, University of Agriculture Faisalabad, Faisalabad 38040, Pakistan; ahmadbajwa516@gmail.com (M.A.); taimoorpansota95@gmail.com (M.T.M.)
2 Plant Production Department, College of Food and Agriculture Sciences, King Saud University, P.O. Box 2460, Riyadh 11451, Saudi Arabia; wsoufan@ksu.edu.sa
3 Department of Botany, Bahawalpur Campus, University of Central Punjab, Bahawalpur 63100, Pakistan; zahoorahmadbwp@gmail.com
4 Institute of Crop Science and Resource Conservation (INRES), Crop Science, University of Bonn, 53115 Bonn, Germany; mhabibur@uni-bonn.de
5 Department of Agronomy, Faculty of Agriculture, Kafrelsheikh University, Kafrelsheikh 33516, Egypt
* Correspondence: uaf_ewarraich@yahoo.com (E.A.W.); ayman.elsabagh@agr.kfs.edu.eg (A.E.S.)

Abstract: Temperature is a key factor influencing plant growth and productivity; however, temperature fluctuations can cause detrimental effects on crop growth. This study aimed to assess the effect of seed priming on *Camelina sativa* L. under heat stress. Experimental treatments were comprised of; seed priming including, no-priming, hydropriming (distilled water priming), and osmopriming (thiourea applications at 500 ppm), heat stress (control = 20 °C and heat stress = 32 °C), and camelina varieties (7126 and 8046). Heat stress hammered crop growth as relative water content and photosynthetic rate were reduced by 35.9% and 49.05% in 7126, respectively, and 25.6% and 41.2% in 8046 as compared with control-no thiourea applied. However, osmopriming with thiourea improved the root and shoot length, and biomass production compared to control–no application under heat stress, with more improvement in variety 8046 as compared with 7126. Moreover, the maximum values of gas exchange and water relations were recorded at thiourea priming and no stress as compared with no-priming under heat stress that helped to improve seed yield by 12% in 7126 and 15% in 8046, respectively. Among the varieties, camelina variety 8046 showed better performance than 7126 by producing higher seed yield especially when subjected to thiourea priming. In conclusion, thiourea seed priming helped the plants to mitigate the adverse effects of heat stress by upregulating plant physiological attributes that lead to maintain camelina seed yield.

Keywords: gas exchange; hydropriming; osmoprimimg; water relations; yield

Citation: Waraich, E.A.; Ahmad, M.; Soufan, W.; Manzoor, M.T.; Ahmad, Z.; Habib-Ur-Rahman, M.; Sabagh, A.E. Seed Priming with Sulfhydral Thiourea Enhances the Performance of *Camelina sativa* L. under Heat Stress Conditions. *Agronomy* **2021**, *11*, 1875. https://doi.org/10.3390/agronomy11091875

Academic Editors: Channapatna S. Prakash, Ali Raza, Xiling Zou and Daojie Wang

Received: 15 August 2021
Accepted: 16 September 2021
Published: 17 September 2021

1. Introduction

The world's population has been projected to cross 10 billion by 2050, which will significantly increase the demand for food supply [1], while climate change has already threatened food safety. Abiotic stresses have reduced the productivity of the staple crops, which has multiplied the existing challenge of food and nutritional security [2,3]. According to NASA [4,5], the first decade of the 21st century was the hottest in human history, which had huge impacts on agriculture productivity. Various climatic models predict that there will be a significant rise in the Earth's average annual temperature due to the increasing CO_2 concentration [6] that will lead to a significant reduction in crop yield and reduced the farmer's income drastically [7]. Heat stress could impart numerous phenological, morphological, and physicochemical changes in crop plants. Considering all the climatic challenges, heat stress at seed formation had the primary role in affecting final yield and quality of oilseed crops.

In the wake of rising temperature, the emphasis should be on crop health [8] because crops require optimal temperature for proper development, and camelina is no exception. The effects of high temperature may hamper the performance of the photosynthetic apparatus that could lead to reduce carbon assimilation to reduce crop growth. Chemical signaling mechanisms in the thylakoids and carbon metabolism are more prone to heat stress damages [9], as high-temperature damages photosystem II that affects the electron transport chain and glycolate pathway due to the overproduction of reactive oxygen species (ROS; such as hydrogen peroxide, H_2O_2) [10]. High temperature severely influences the mineral and water transport system of plant tissues, which results in mineral deficiency and decreased turgidity [11,12]. Crop yield may decrease up to 10–15% due to each degree (°C) rise in temperature above the optimum [13]. Plants have developed several mechanisms to reduce ROS levels in plant cells [14] by activating enzymatic and non-enzymatic scavenging systems [15]. Heat stress at any crop stage can cause substantial yield losses in Brassica crops [16]. Camelina has shown a substantial reduction in photosynthetic efficiency and crop yield when grown under high-temperature stress [17]. Ahmad et al. [16] have reported a reduction in the performance of camelina under heat-stressed conditions due to the impairment of plant physiological attributes. Temperature above 32 °C is critical at the reproductive growth stage and tends to decreased crop yield by the pod abortion, decreasing the number of seeds per plant, and seed weight [17]. Innovative and sustainable methods need to be introduced to improve the performance of crops under heat stress. Recently, thiourea (TU) emerged as one of the effective approaches to enhance high-temperature stress tolerance in plants by regulating metabolic balance, plant growth, and development [18].

Seed priming is a controlled hydration technique that accelerates the key metabolites for osmotic up-regulation [19]. Priming is one of the most feasible and economic technologies enabling the efficient uptake of nutrients, boosting water use efficiency, breaking seed dormancy, promoting early maturity, and improving crop physiology that ensures successful crop production [20,21]. It has been reported that osmopriming enhanced the antioxidant defense system under stressful conditions [22] to improve the defense system against heat stress. In addition, TU modulates the activity of numerous biological compounds such as plant growth regulators, polyamines, enzymes, mineral nutrients, and produces many derivatives, which have the potential to mitigate heat stress damages [23]. It has a major role in the production of proteins, vitamins, enzymes, and chlorophyll in plant cells and tissues from vegetative growth to maturity [24]. Interaction of TU-containing compounds with various biological compounds produces specific derivatives essential for enhancing thermo-tolerance by modulating the ROS scavenging system [25]. The applications of TU may upregulate enzymatic activities in different plant parts that help to remove the ROS, by activating the ascorbate-glutathione cycle to alleviate heat stress [26,27]. Nonetheless, optimal quantity is vital for TU-induced increase in antioxidant activities that contributed to reverse the high temperature stress [28,29]. The heat stress-induced damages can be ameliorated by the pre-sowing seed treatments with different chemical agents, which may be useful for the generation of heat-stress tolerance in plants.

Camelina [*Camelina sativa* L. Crantz] is an emerging oilseed crop with unique characteristics in relation to its high adaptability against abiotic stresses [30]. Its oil constitutes a vital product for the bio-based industries, as its distinctive composition permits multiple applications [31]. It is a rediscovered oil crop that belongs to the family *Brassicaceae*, having a seed oil content of 26–43% [32,33] with high percentage of unsaturated fatty acids. The seeds of camelina are unique compared to other members of *Brassicaceae* due to their high amount of polyunsaturated fatty acids and low level of erucic acid. Regular ingestion of camelina oil reduces the level of cholesterol in blood and the presence of tocopherols prevents the oil from rancidity [34]. According to Zubr [35], Camelina can be adapted to various environmental conditions including limited water conditions, high and low temperatures, etc. Camelina is popularly known as false flax, which is typically a cool-season crop with a temperature that seldomly exceeds 30 °C [36]; however, it can be grown in the winter and spring seasons [37].

However, the role of TU has been documented under abiotic stress tolerance, while the role of TU to alleviate heat stress damages in camelina needs further investigation. The study was hypothesized that TU priming regulates the heat stress tolerance in camelina. The objective of the study was to evaluate the impact of seed priming techniques on growth and yield parameters on physiological basis in camelina under heat stress conditions.

2. Materials and Methods

2.1. Crop Husbandry

The pot experiment was laid out under completely randomized design (CRD) with factorial arrangements and three replications. Camelina seeds (10 seeds) were sown in plastic pots (36/24 cm) containing 5 kg of sand, while each pot was considered as a biological replicate. Camelina seeds were obtained from the Stress Physiology Laboratory, Department of Agronomy, University of Agriculture, Faisalabad. Sand was sieved to opt-out all the contaminants, and then field capacity was calculated through proper procedure. The experiment was comprised of three factors: (a) seed priming; TU_0 = control-no priming, TU_1 = hydropriming (water priming), and TU_2 = osmopriming (TU primming at 500 ppm), (b) heat stress; control—20/18 °C day/night and heat stress—32/22 °C day and night at 65 days after sowing (DAS), and (c) camelina varieties (7126 and 8046). The screening experiment was done at stress physiology laboratory, which led us to select the one resistant (8046) and one susceptible (7126) variety for this study. The Hoagland solution ($NH_4H_2PO_4$ 1 mM; $Ca(NO_3)_2{\cdot}4H_2O$ 4 mM; KNO_3 6 mM; $MnCl_2{\cdot}4H_2O$ 9.1 µM; H_3BO_3 46.2 µM; $CuSO_4{\cdot}5H_2O$ 0.3 µM; $ZnSO_4{\cdot}7H_2O$ 0.8 µM; $MgSO_4{\cdot}7H_2O$ 2 mM; Fe–Na_2–EDTA 0.1 mM.) was applied for the nutritional requirements of camelina. The application of Hoagland solution was done at the time of sowing and topped up after every fortnight.

Seeds of both varieties were separately soaked for the hydro-priming and osmopriming (TU solution) for 6 h. For hydro-priming, seeds were soaked in the distilled water for 6 h while TU (500 ppm) solution was used for the osmopriming, and continuous aeration was provided by using an aquarium pump to avoid anxious conditions. The experiment was comprised of 36 pots and grouped into two sets containing 18 pots in each, which were grown under the same conditions until heat stress was applied on one set. Stress was induced just before the onset of the flowering stage by increasing the temperature of the growth room from 20 °C (control) to 32 °C (heat stress). The experiment was performed in a growth room having a mechanized unit of cooling, heating, light (∼12,000 lux), and humidifier/dehumidifier adjustment systems. Relative humidity (70%) was maintained in the growth rooms and water was provided regularly for achieving the field capacity to prevent drought stress. Then, the temperature was gradually increased by 2 °C each day to avoid any heat shock to seedlings till it reached 32/22 °C day/night. The stress lasts for 10 days as it reached to maximum temperature and came back with the same way. Gas exchange attributes were measured at 76 DAS after imposition of stress, while growth parameters were measured at 80 DAS, while seed yield and related parameters were measured at 108 DAS.

2.2. Growth Parameters

For the measurement of growth parameters, two uniform plants were randomly selected from each biological replicate. Plant height and root length was measured using a meter rod from the surface of the soil to the tip of the plant. Pots were filled with water to gently uproot the randomly selected plants and averaged. The length of five roots was taken by using a meter rod from the uprooted plants and averaged. These samples were cleaned and washed with distilled water, and root and shoot were separated by a pair of scissors. After cleaning, the fresh weight of root and shoot was taken by using a digital balance (Uni Block AUX220, Shimadzu Corporation, Kyoto, Japan). These samples were oven-dried (Memmert-110, Schawabach, Germany) at 70 °C for 72 h to take dry weight by using a digital balance.

2.3. Gas Exchange Parameters

Different physiological traits including photosynthetic rate (*A*), transpiration rate (*E*), stomatal conductance (*gs*), and internal CO_2 concentration (*Ci*) were measured by using an open system, portable infrared gas analyzer (IRGA) (LCA-4 ADC (USA)). The fully expanded young leaves of three plants selected randomly from each pot were used to measure these attributes. Measurements were made between 9:00 a.m. and 10:00 a.m. to opt-out of the effect of high temperature. The following adjustments were made for these measurements; leaf surface area 6.25 cm^2, ambient CO_2 concentration (326 $\mu molmol^{-1}$), the temperature of the leaf chamber ranged from 31.5 to 37.8 °C, ambient pressure (P) 98.2 k Pa, chamber gas flow rate (V) 408 mL min^{-1}, the molar flow of air per unit leaf area (Us) 409.5 $molm^{-2}s^{-1}$, the water vapor pressure in the chamber (ref.) ranged 21.2–24 mbar, and PAR at leaf surface was maximum up to 1181 μ mol $m^{-2}s^{-1}$.

2.4. Water Relations

To observe leaf water potential, the top third youngest and fully expanded leaf of camelina plants was harvested from each treatment. Scholander-type pressure chamber (ARIMAD-2, ELE-International, Tokyo, Japan) according to the method defined by Ahmad et al. [17].

2.5. Yield and Related Attributes

Yield component, i.e., a number of silique per plant and number of seed per silique were measured from two tagged plants per pot and averaged. A 1000 seed weight was taken from two representative plants from each biological replicate and seeds were taken to measure 1000 seed weight by using a digital balance (Uni Block AUX220, Shimadzu Corporation, Kyoto, Japan). Each pot was manually harvested, seeds were separated, and the seed yield per pot was obtained using a digital balance.

2.6. Statistical Analysis

Data collected were statistically analyzed through analysis of variance technique using Statistix 10.1 (Analytical Software, Statistix, Tallahassee, FL, USA). Fisher's analysis of variance was used to compare the treatment means at a 5% probability level [38]. Graphical representation was done by using SigmaPlot 10.0.

3. Results

3.1. Growth Parameters

Analysis of variance showed that TU supplementation significantly affected the growth parameters in camelina varieties under different environments compared to control–no TU (Table 1). Heat stress reduced the growth attributes in camelina varieties, while more reduction was noted in camelina with no TU supplementation under heat stress. Plant height was reduced by 23.7% and 30.5% in 8046 and 7126, respectively, under heat stress as compared to control-no stress. Different growth parameters such as plant height, root length, root-shoot length, and their fresh and dry weight were significantly improved with TU priming under normal as well as heat stress conditions (Table 1). Nevertheless, plant height was improved by 27.5% in 8046 and 19.4% in 7126, respectively, which showed higher improvement in variety 8046 compared with 7126 with TU priming compared to control-no TU. Results showed that seed priming with TU performed better in all the mentioned characters and improved these growth characters under normal temperature (control) as well as under high-temperature stress conditions. In addition, camelina variety 8046 performed better for all growth parameters with osmopriming under normal and heat stress conditions compared to 7126 (Table 1). Maximum values of growth attributes including plant height (64.5 cm) and root length (17.2 cm) were observed at TU_2 (TU priming), T_1 (22 °C), and V_2 (8046), while minimum values of growth attributes including plant height (33.6 cm) and root length (7.38 cm) were observed at TU_0 (control-no priming), T_2 (32 °C), and V_1 (7126). Among the seed priming, hydro-priming showed an increase of 8.99% in plant height and TU-seed priming showed an increase of 23.6% in plant height as compared to control-no priming.

Table 1. Impact of thiourea priming and heat stress on growth parameters of two camelina varieties.

Varieties (V)	Heat Stress (T)	Thiourea (TU) Applications	Plant Height (cm)	Root Length (cm)	Shoot Fresh Weight (g)	Root Fresh Weight (g)	Shoot Dry Weight (g)	Root Dry Weight (g)
7126	Control	TU_0	52.7 ± 1.25 ^{d}	9.19 ± 0.15 ^{f}	4.04 ± 0.06 ^{d}	1.41 ± 0.004 ^{e}	0.56 ± 0.02 ^{d}	0.08 ± 0.003 ^{f}
		TU_1	55.5 ± 0.56 ^{c}	10.3 ± 0.51 ^{e}	4.61 ± 0.21 ^{c}	1.47 ± 0.02 ^{c,d}	0.60 ± 0.01 ^{c}	0.09 ± 0.002 ^{d,e}
		TU_2	58.5 ± 2.17 ^{b}	11.9 ± 0.61 ^{c}	5.35 ± 0.20 ^{b}	1.50 ± 0.005 ^{c}	0.72 ± 0.01 ^{b}	0.10 ± 0.002 ^{b}
	Heat stress	TU_0	33.6 ± 1.00 ^{g}	7.37 ± 0.17 ^{g}	2.13 ± 0.10 ^{h}	1.11 ± 0.01 ^{i}	0.21 ± 0.002 ^{g}	0.04 ± 0.001 ^{i}
		TU_1	37.6 ± 1.21 ^{f}	8.91 ± 1.12 ^{f}	2.47 ± 0.05 ^{g}	1.15 ± 0.01 ^{h}	0.22 ± 0.01 ^{g}	0.05 ± 0.001 ^{h}
		TU_2	45.2 ± 1.42 ^{e}	11.6 ± 0.22 ^{c,d}	2.92 ± 0.40 ^{f}	1.21 ± 0.01 ^{g}	0.29 ± 0.01 ^{f}	0.09 ± 0.00 ^{f}
8046	Control	TU_0	54.5 ± 0.15 ^{c,d}	12.09 ± 1.17 ^{c}	4.27 ± 0.02 ^{d}	1.45 ± 0.002 ^{d,e}	0.63 ± 0.01 ^{c}	0.09 ± 0.002 ^{e}
		TU_1	60.6 ± 0.62	13.2 ± 1.57 ^{b}	5.13 ± 0.14 ^{b}	1.55 ± 0.02 ^{b}	0.72 ± 0.02 ^{b}	0.10 ± 0.002 ^{c}
		TU_2	64.5 ± 0.86 ^{a}	17.2 ± 0.50 ^{a}	6.01 ± 0.17 ^{a}	1.60 ± 0.04 ^{a}	0.82 ± 0.004 ^{a}	0.13 ± 0.005 ^{a}
	Heat stress	TU_0	39.2 ± 0.66 ^{f}	10.7 ± 0.05 ^{d,e}	2.74 ± 0.03 ^{f,g}	1.15 ± 0.01 ^{h}	0.29 ± 0.01 ^{f}	0.06 ± 0.00 ^{h}
		TU_1	46.6 ± 1.69 ^{e}	13.1 ± 0.53 ^{b}	2.96 ± 0.04 ^{f}	1.22 ± 0.01 ^{g}	0.32 ± 0.01 ^{f}	0.07 ± 0.00 ^{g}
		TU_2	55.0 ± 2.00 ^{c}	16.3 ± 0.74 ^{a}	3.56 ± 0.03 ^{e}	1.29 ± 0.02 ^{f}	0.47 ± 0.03 ^{e}	0.09 ± 0.00 ^{d}

Values (mean ± standard deviation, n = 3), TU_0 = No thiourea priming, TU_1 = Water priming, TU_2 = Thiourea priming; LSD = least significant difference; values sharing same case letter or without lettering for a parameter do not differ significantly ($p \leq 0.05$) by the LSD test.

Among the interactions, TU × T was significant for plant height, root and shoot fresh weight, and shoot dry weight. The interaction, TU × V was significant for plant height, root length, root fresh weight, and shoot dry weight in camelina. The interaction, V × T was significant for plant height, shoot fresh and dry weight.

3.2. Physiological Parameters

3.2.1. Gas Exchange Attributes

Analysis of variance showed that TU priming significantly influenced the physiological parameters under different environmental conditions (Table 2). Among the gas exchange attributes, photosynthetic rate and stomatal conductance were decreased, while transpiration and intercellular CO_2 rates were increased under heat stress conditions as compared to control–no stress. The photosynthetic rate was decreased by 41.2% in 8046 and 49.2% in 7126, respectively, while stomatal conductance was decreased by 19.4% in 8046 and 34.6% in camelina under heat stress over control–no stress. Maximum photosynthetic rate (6.94 µmol CO_2 m^{-2} s^{-1}) and stomatal conductance (0.079 mol H_2O m^{-2} s^{-1}) were noted with osmopriming, while a lower value of these attributes (2.04 µmol CO_2 m^{-2} s^{-1}, and 0.04 mol H_2O m^{-2} s^{-1}, respectively) was noted with control–no priming. Nevertheless, photosynthetic rate was improved by 44.3% in 8046 and 42% in 7126 with TU priming compared to control–no TU. Transpiration rate (0.68 mol H_2O m^{-2} s^{-1}) was increased, and internal CO_2 rate (344.4 µmol CO_2 mol^{-1}) was decreased with TU seed priming compared to control–no TU. In relation to camelina varieties, the 8046 variety was more tolerant to heat stress conditions compared to 7126 variety (Table 2). Among the varieties, higher values of photosynthetic rate (5.11 µmol CO_2 m^{-2} s^{-1}) were noted in 8046, while lower values (3.67 µmol CO_2 m^{-2} s^{-1}) were noted in 7126. Maximum values of gas exchange attributes including photosynthetic rate (7.45 µmol CO_2 m^{-2} s^{-1}) were observed at TU_2 (TU priming), T_1 (22 °C), and V_2, while minimum values of gas exchange attributes including photosynthetic rate (2.05 µmol CO_2 m^{-2} s^{-1}) were observed at TU_0 (control–no priming), T_2 (32 °C), and V_1 (7126). Among the seed priming, hydro-priming showed an increase of 18.07% in photosynthetic rate and TU–seed priming showed an increase of 43.3% in photosynthetic rate as compared to control–no priming

Table 2. Impact of thiourea priming and heat stress on gas exchange attributes of two camelina varieties.

Varieties (V)	Heat Stress (T)	Thiourea (TU) Applications	Photosynthetic Rate (μmol H_2O m^{-2} s^{-1})	Transpiration Rate (mmol m^{-2} s^{-1})	Stomatal Conductance (mmol m^{-2} s^{-1})	Intercellular CO_2 Concentration (μmol m^{-2} s^{-1})
7126	Control	TU_0	4.05 ± 0.02^{f}	0.41 ± 0.00^{h}	0.05 ± 0.00^{e}	278.1 ± 1.43^{e}
		TU_1	4.75 ± 0.06^{e}	0.47 ± 0.00^{g}	0.05 ± 0.00^{d}	267.8 ± 1.72^{f}
		TU_2	5.78 ± 0.06^{c}	0.50 ± 0.02^{f}	0.06 ± 0.00^{c}	254.3 ± 2.15^{g}
	Heat stress	TU_0	2.05 ± 0.03^{j}	0.43 ± 0.00^{h}	0.04 ± 0.00^{h}	344.3 ± 3.72^{a}
		TU_1	2.51 ± 0.02^{j}	0.48 ± 0.01^{g}	0.04 ± 0.00^{g}	331.0 ± 2.75^{b}
		TU_2	2.87 ± 0.00^{h}	0.49 ± 0.00^{e}	0.04 ± 0.00^{f}	311.0 ± 2.75^{c}
8046	Control	TU_0	5.42 ± 0.37^{d}	0.56 ± 0.00^{e}	0.06 ± 0.00^{c}	248.4 ± 5.37^{g}
		TU_1	6.42 ± 0.37^{b}	0.63 ± 0.01^{c}	0.07 ± 0.00^{b}	237.1 ± 3.19^{h}
		TU_2	7.45 ± 0.01^{a}	0.67 ± 0.01^{ab}	0.07 ± 0.00^{a}	224.2 ± 6.10^{i}
	Heat stress	TU_0	3.07 ± 0.06^{h}	0.60 ± 0.00^{d}	0.04 ± 0.00^{g}	329.2 ± 3.51^{b}
		TU_1	3.53 ± 0.11^{g}	$0.64 \pm 0.03^{b,c}$	0.04 ± 0.00^{f}	311.4 ± 2.76^{c}
		TU_2	4.76 ± 0.08^{e}	0.68 ± 0.00^{a}	0.05 ± 0.00^{d}	296.4 ± 7.28^{d}

Values (mean $\pm$ standard error, $n = 3$), TU_0 = No thiourea priming, TU_1 = Water priming, TU_2 = Thiourea priming; LSD = least significant difference; values sharing same case letter or without lettering for a parameter do not differ significantly ($p \leq 0.05$) by the LSD test.

Among the interactions, all interactions were significant for the photosynthetic rate. TU $\times$ T was significant for photosynthetic rate and intercellular CO_2 rates. The T $\times$ V interaction remained significant for intercellular CO_2 rates. The higher-order interaction TU $\times$ T $\times$ V was significant for stomatal conductance.

3.2.2. Water Relations

Seed priming significantly affected the plant water relations under heat stress (Figures 1 and 2). Heat stress reduced the plant water relations including water potential, osmotic potential, pressure potential, and relative water content as compared to control–no stress. Water potential decreased by 30% in 8046 and 33.7% in 7126, respectively, and leaf relative water content was decreased by 25.6% in 8046 and 35.9% in 7126, respectively, under heat stress over control–no stress. High values of water potential (−0.80 MPa), osmotic potential, (−1.33 MPa), pressure potential (0.48 MPa), and relative water content (86.5%), respectively, were noted with osmo-priming (TU priming) compared to control–no TU applied (Figures 1 and 2). Nevertheless, the relative water content was improved by 13.6% and pressure potential was increased by 29.5% with TU priming compared to control–no TU. In relation to camelina varieties, the 8046 variety was more tolerant to heat stress conditions compared to 7126 variety (Figures 1 and 2). In addition, higher values of water potential (−1.00 MPa) were noted in 8046, while lower values (−1.09 MPa) were noted in 7126. Maximum values of gas exchange attributes including water potential (−0.79 MPa) were observed at TU_2 (TU priming), T_1 (22 °C) and V_2, while minimum values of water relations including water potential (−1.32 MPa) were observed at TU_0 (control–no priming), T_2 (32 °C), and V_1 (7126). Among the seed priming, hydro-priming showed an increase of 15.05% in relative water content and TU–seed priming showed an increase of 60.9% in relative water content as compared to control–no priming.

Among the interactions, TU $\times$ T was significant for osmotic potential. The T $\times$ V was significant for water potential and relative water content. The higher-order interaction TU $\times$ T $\times$ V was significant for water potential.

3.3. Yield and Related Attributes

Analysis of variance showed that TU seed priming significantly affected the yield attributes in camelina varieties under high temperature stress (Table 1). Heat stress decreased the growth attributes in camelina varieties, while more reduction was noted with no-TU applications (Table 3).

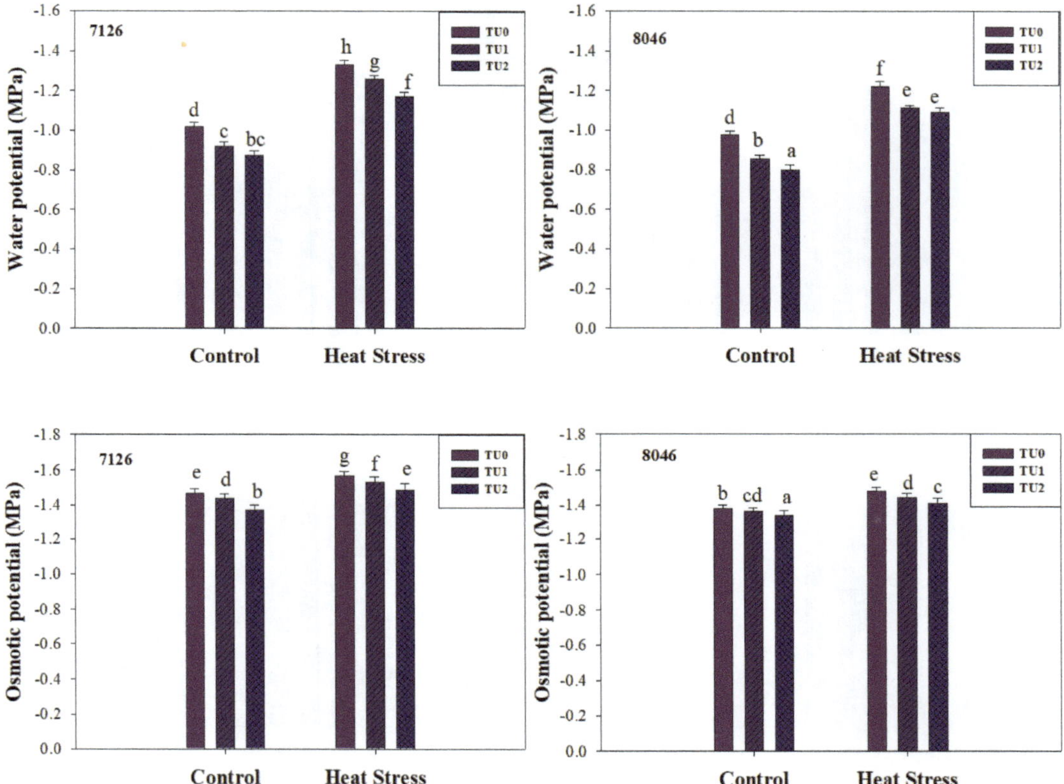

Figure 1. Impact of thiourea priming (TU_0 = control–no application, TU_1 = Water priming, TU_2 = Thiourea priming (500 ppm) and heat stress (32 °C) on water potential (−MPa) and osmotic potential (−MPa) on camelina genotypes 7126 and 8046. Error bars above means indicate the ±S.E. Means sharing the same letter in both varieties do not differ significantly at $p \leq 0.05$.

Thousand seed weight was reduced by 28% in 8046 and 33.5% in 7126, respectively, under heat stress conditions as compared to control–no stress. Yield traits such as the number of silicle per plant, number of seeds per silicle, thousand seed weight, and yield per pot were significantly higher under TU treatment as compared to control–no TU under heat stress conditions (Table 3). However, seed weight was improved by 41.2%, and seed yield was improved by 65.6% with TU priming compared to control–no TU. Among the varieties, seed yield was improved by 63% in 8046 and 58.7% in 7126 which showed that seed yield was improved 11.7% more in 8046 as compared 7126 (Table 3). Maximum values of yield attributes including seed yield per pot (3.91 g) was observed at TU_2 (TU priming), T_1 (22 °C), and V_2, while minimum values of yield attributes including photosynthetic rate (0.59 g) was observed at TU_0 (control–no priming), T_2 (32 °C), and V_1 (7126). Among the seed priming, hydro-priming showed an increase of 15.05% in seed yield and TU–seed priming showed an increase of 60.9% in seed yield as compared to control-no priming.

Among the interactions, TU × T was significant for the number of silicle/plant, number of seeds/silicle, and seed yield. The interaction, TU × V was significant for number of seeds/silicle and T × V was significant for seed yield. The interaction, TU × T × V was significant for seed yield.

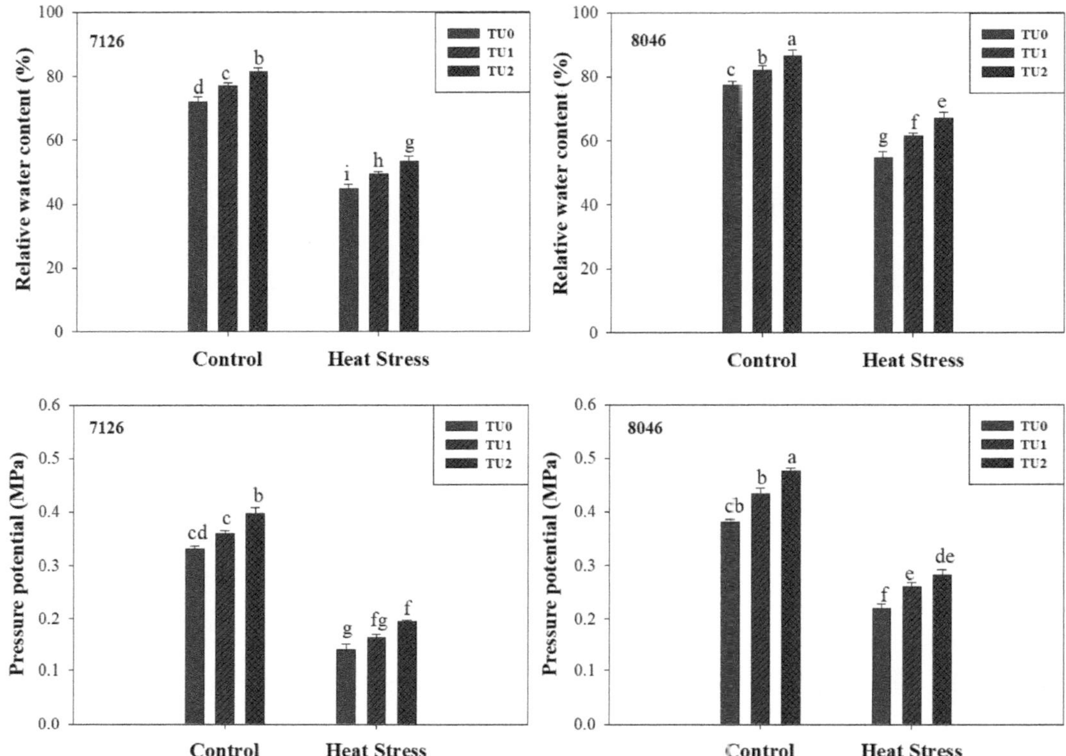

Figure 2. Impact of thiourea priming (TU_0 = control–no application, TU_1 = Water priming, TU_2 = Thiourea priming (500 ppm) and heat stress (32 °C) on pressure potential (MPa) and relative water content (MPa) on camelina genotypes 7126 and 8046. Error bars above means indicate the ±S.E. Means sharing the same letter in both varieties do not differ significantly at $p \leq 0.05$.

Table 3. Impact of thiourea priming and heat stress on growth and yield parameters of two camelina varieties.

Varieties (V)	Heat Stress (T)	Thiourea (TU) Applications	No. of Silicle Plant^{-1}	No. of Seeds Silicle^{-1}	1000-Seed Weight (g)	Seed Yield Pot^{-1} (g)
7126	Control	TU_0	33.3 ± 1.52^{e}	9.73 ± 0.64^{e}	0.76 ± 0.04^{ef}	2.45 ± 0.06^{d}
		TU_1	40.6 ± 0.57^{c}	11.7 ± 0.68^{c}	0.87 ± 0.05^{cd}	$2.92 \pm 0.17^{b,c}$
		TU_2	45.6 ± 1.15^{b}	13.6 ± 0.57^{b}	1.00 ± 0.03^{b}	3.81 ± 0.24^{a}
	Heat stress	TU_0	23.0 ± 1.00^{g}	7.42 ± 0.46^{h}	0.48 ± 0.01^{h}	0.57 ± 0.02^{h}
		TU_1	27.0 ± 1.00^{f}	$8.30 \pm 0.26^{f,g}$	0.53 ± 0.01^{g}	0.62 ± 0.01^{h}
		TU_2	32.6 ± 0.58^{e}	9.66 ± 0.58^{e}	0.73 ± 0.01^{f}	1.31 ± 0.24^{f}
8046	Control	TU_0	38.0 ± 1.03^{d}	10.5 ± 0.25^{d}	$0.81 \pm 0.01^{d,e}$	$2.66 \pm 0.04^{c,d}$
		TU_1	44.0 ± 1.00^{b}	13.3 ± 0.57^{b}	0.89 ± 0.02^{c}	2.96 ± 0.15^{b}
		TU_2	50.6 ± 1.52^{a}	15.3 ± 0.35^{a}	1.08 ± 0.05^{a}	3.89 ± 0.28^{a}
	Heat stress	TU_0	28.1 ± 1.25^{f}	$7.97 \pm 0.16^{g,h}$	0.53 ± 0.01^{h}	$0.72 \pm 0.03^{g,h}$
		TU_1	33.0 ± 2.64^{e}	$9.00 \pm 0.05^{e,f}$	0.60 ± 0.02^{g}	0.96 ± 0.09^{g}
		TU_2	36.3 ± 0.57^{d}	12.3 ± 0.57^{c}	0.84 ± 0.02^{cd}	1.61 ± 0.18^{e}

Values (mean ± standard error, n = 3), LSD = least significant difference; values sharing same case letter or without lettering for a parameter do not differ significantly ($p \leq 0.05$) by the LSD test.

4. Discussion

Crop productivity is mainly dependent upon environmental growth factors, which suggested that the ever rising CO_2 concentration in the atmosphere is the major reason for climate change that causes temperature fluctuations at an alarming rate, leading to the imposition of heat stress. The suboptimal temperature at any crop growth stage imparts deleterious impacts on crops productivity including camelina. In this study, heat stress showed negative effects on crop growth and yield attributes as it reduced the photosynthetic efficiency and stomatal conductance due to reduction in leaf water status in camelina, while TU supplementation remained effective in ameliorating the negative impacts of heat stress in camelina varieties (Tables 1–3; Figures 1 and 2).

The hypothesis of the study has been accepted as results indicated that the TU supplementation has played important role to ameliorate the negative impacts of heat stress by seed priming (TU_0 = No thiourea priming, TU_1 = Water priming, TU_2 = Thiourea priming), which lead to improving the seed yield. High temperature hampered plant growth by causing a reduction in morpho-physiological attributes in camelina varieties. Among the seed priming techniques tested in this study, TU-osmopriming (TU priming at 500 ppm) remained an effective technique to improve the performance of camelina varieties under heat stress, as TU application boosted stomatal conductance that might be attributed to an increase in photosynthetic rate, which conferred the resistance against heat stress through physiological regulations (Tables 1–3 and Figures 1 and 2). Indeed, TU priming ameliorated the negative impacts of heat stress, regulated plant growth, improved the water status to facilitate the stomatal conductance, and also acted as a compatible osmolyte, which assisted plants to cope with heat stress. The present study confirmed a significant reduction in plant height, roots, and shoot lengths along with their fresh and dry weights under heat stress, while more reduction was observed in the control treatment (no TU-priming).

Temperature above threshold level increased the physiological activities, which consequently fastened the crop growth rate that further reduced the growth period of crops [39,40]. This reduction in the growth period led to the premature completion of phenological plant parts that could damage the final yield. Results showed that heat stress reduced the physiological attributes including photosynthetic rate, stomatal conductance, transpiration rate, and intercellular CO_2 concentration under control–no priming as compared to TU osmopriming. Results have shown that the transpiration rate was increased due to high temperature stress, which leads to disturb plant water status; however, it modulated higher water loss, which was primarily caused by heat stress. Heat stress reduced the photosynthetic rate, which is more prone to heat damages that lead to reduce crop growth and grain yield [41,42]. Heat stress negatively affected the reproductive growth that disturbed the seed formation process which may reduce the number of seed pod^{-1} and seed weight [43,44], because high-temperature stress affects the source-sink relationship and explain the differences in seed yield [45]. Among the gas exchange attributes, results exhibited that heat stress restricted the rate of CO_2 assimilation as indicated by lower photosynthetic efficiency, transpiration rate, and stomatal conductance, while intercellular CO_2 concentration was increased, which showed the negative effect of heat stress on the stomatal component of photosynthesis; nevertheless, it also hampered the assimilation of absorbed CO_2. The reduction in the photosynthetic efficiency resulted in reducing the assimilation formation and translocation towards the sink, which led to reduce the seed yield and related attributes [18]. Heat stress at the reproductive stage could damage the seed formation processes including flowering and seed set that hampered the rate of grain filling and grain yield [46]. High temperature directly affected the crop water relations as water potential, osmotic potential, pressure potential, and relative water content were decreased by 31.9, 6.72, 46.9, and 30.6% as compared to control–no stress with more reduction in no-TU treatment (Figures 1 and 2). Reduction in plant water status effects the turgidity of the cell that directly effects the elongation of the cell, which leads to reduce the crop growth.

Pre-sowing seed treatment was the foundation for the early activation of seed metabolism that in combination with other elements might be helpful to the proper vegetative growth and higher seed yield. The survival of plants under stress conditions could be possible by the supplementation of stress alleviating chemical compounds [47,48], like TU, that can potentially upregulate the plant defense to improve plant tolerance under stressed conditions [49]. The applications of TU manifolds growth regulatory roles in plant species including camelina varieties. Our results depicted that exogenous application of TU as a seed priming treatment improved the seed yield of both camelina varieties under heat stress conditions compared to control (Table 3). The TU priming significantly improved the yield attributes including plant height, root and shoot lengths along with their fresh and dry weights indicating higher biomass accumulation triggered by TU which is in line with the findings of Asthir et al. [40]. The sulfhydryl TU not only increased the root length but also increased branching in roots under heat stress that tends to increase the root fresh and dry weight. The ascribed TU-induced increase in growth could be due to the mediation of a number of important metabolic functions.

Thus, seed priming with TU offered a promising and economical solution for improving crop resistance against heat stress [17]. Plants treated with TU exhibited maximum biomass accumulation as compared to control–no TU applied, indicating the positive role of TU in boosting plant growth by alleviating the adverse effects of heat stress (Table 1). In addition, osmoprimed crops could timely complete all the phenological events, and this phenological plasticity can be helpful when integrated with high-temperature stress to avoid their negative effects on crop growth and development during early and later reproductive stages without yield penalty. TU supplementation reduced heat stress-induced oxidative stress by upregulating the important phenomenon of photosynthesis, and assimilate translocation which was also reported by Patade et al. [27] to enhance the defense system in camelina plants to impart heat stress. The TU-supplementation improved the seed weight and consequently gave higher seed yield per pot in both varieties under heat stress conditions as compared to control–no TU priming at the same conditions. This can be attributed to the improvement in plant metabolism, which enabled the plant defense against heat stress [50]. In the current study, TU priming (500 ppm) helped to increase photosynthetic rate and stomatal conductance [48]; however, the net CO_2 assimilation rate was more with TU priming than control–no TU applied.

The possible reason for this variation in photosynthetic rate may be that TU application increased the leaf growth which in turn up-folded the photosynthetic rate by increasing the harvesting of photosynthetically active light (Table 2). Thiourea supplementation at any growth stage and through any methods of applications may improve the photosynthetic apparatus in plants that helped the plants to maintain photosynthetic rate [51,52]. In the present study, exogenous use of TU as pre-sowing seed treatment was found effective in decreasing the damages caused by heat stress. Additionally, the TU-applied reduction in intercellular CO_2 concentration rate might be due to the effective role of TU metabolites in the regulation of activities of antioxidative enzymes (Table 2). Available reports support the present findings where seed priming with different compounds was found effective in improving the plant physiological attributes [53–55]. Our results are in line with Orman and Kaplan [56], who reported that TU application increased the biomass of tomato plants by 6–8% grown in sandy loam soil. TU at either stage performed well to alleviate heat induced damages, while, TU supplementation at vegetative stage improved plant height, root length, and dry weight compared to TU applied at the vegetative growth stage [48,57]. In addition, TU supplementation upregulated the plant water relations which played significant role to improve the stomatal conductance as compared to no-TU application in line with the findings of Ahmad et al. [18].

The results have shown variability among two varieties of camelina under heat-stressed conditions, as 8046 has shown more resistance to the deleterious effects of heat stress as compared to variety 7126. Variety 8046 has shown better performance in relation to plant growth attributes as compared to 7126, as plant height and root length were maximum in 8046 and minimum in 7126 (Table 1). The impact of heat stress-induced damages is cultivar-specific depending on the extent of tolerance based on various tolerance mechanisms including the cellular oxidative defenses in terms of the enzymatic and non-enzymatic antioxidant compounds [58–60]. The endogenous content of photosynthetic pigments with CO_2 assimilation rate was also affected in heat susceptible variety (7126) as compared to heat resistant variety (8046), which is in line with the findings of [61]. Almeselmani et al. [62] and Balla et al. [63] have also noted that the activities of plant defense system upregulated the plant physiological attributes including photosynthetic rate and stomatal conductance while decreasing intercellular CO_2 concentration intolerant variety (8046), but downregulated insensitive variety (7126) under high-temperature stress. However, variety 8046 was not so affected and sustained its biomass, photosynthetic rate, plant water status, and seed yield as compared to 7126 (Figures 1 and 2).

5. Conclusions

Heat stress imparted deleterious effects on photosynthesis and plant water status that led to reduce the plant growth and yield in camelina varieties. Nevertheless, osmoriming with thiourea (500 ppm) improved the growth and yield of both varieties of camelina under normal and heat-stressed conditions. Thiourea priming upregulated the plant water relations to regulate stomatal conductance and photosynthetic efficiency, which added to improve crop yield. It was also inferred that camelina variety 8046 performed better against heat stress as compared to 7126 grown under high-temperature stress. Overall, this study provides a good understanding for scientists to find out the actual physiological mechanisms behind the thiourea induced heat stress tolerance mechanism in camelina that will be a roadmap for the further investigation at cellular level.

Author Contributions: Conceptualization, E.A.W., M.A. and M.T.M.; methodology, M.T.M., M.A.; software, A.E.S.; validation, E.A.W., Z.A., M.H.-U.-R., W.S. and A.E.S.; formal analysis, Z.A.; investigation, M.A., M.T.M., Z.A.; resources, EA.W and A.E.S.; data curation, W.S., A.E.S., M.A.; writing—original draft preparation, E.A.W., M.A., M.T.M. and A.E.S.; writing—review and editing, M.A., M.H.-U.-R., W.S. and A.E.S. and supervision, E.A.W.; funding acquisition, W.S. All authors have read and agreed to the published version of the manuscript.

Funding: This Research was funded by the Researchers Supporting Project number (RSP-2021/390), King Saud University, Riyadh, Saudi Arabia.

Institutional Review Board Statement: Not applicable.

Informed Consent Statement: Not applicable.

Data Availability Statement: Data is contained within the article.

Acknowledgments: This research was funded by the Researchers Supporting Project number (RSP-2021/390), King Saud University, Riyadh, Saudi Arabia.

Conflicts of Interest: The authors declare no conflict of interest.

References

1. Faisal, M.; Iqbal, M.A.; Aydemir, S.K.; Hamid, A.; Rahim, N.; El Sabagh, A.; Khaliq, A.; Siddiqui, M.H. Exogenously foliage applied micronutrients efficacious impact on achene yield of sunflower under temperate conditions. *Pak. J. Bot.* **2020**, *52*, 1215–1221. [CrossRef]
2. Mahajan, S.; Tuteja, N. Cold, salinity and drought stresses: An overview. *Arch. Biochem. Biophys.* **2005**, *444*, 139–158. [CrossRef]
3. IPCC. *IPCC Expert Meeting Report: Towards New Scenarios for Analysis of Emissions, Climate Change, Impacts, and Response Strategies*; IPCC Secretariat: Geneva, Switzerland, 2007.
4. NASA. What Is Climate Change? Available online: https://climatekids.nasa.gov/climate-change-meaning/ (accessed on 14 May 2019).
5. Olesen, J.E.; Bindi, M. Consequences of climate change for European agricultural productivity, land use and policy. *Eur. J. Agron.* **2002**, *16*, 239–262. [CrossRef]
6. Ahmad, Z.; Waraich, E.A.; Ahmad, R.; Shahbaz, M. Modulation in water relations, chlorophyll contents and antioxidants activity of maize by foliar phosphorus application under drought stress. *Pak. J. Bot.* **2017**, *49*, 11–19.
7. Stern, N. *Stern Review: The Economics of Climate Change*; Cambridge University Press: Cambridge, UK, 2006.
8. Wise, R.R.; Olson, A.J.; Schrader, S.M.; Sharkey, T.D. Electron transport is the functional limitation of photosynthesis in field-grown Pima cotton plants at high temperature. *Plant Cell Environ.* **2004**, *27*, 717–724. [CrossRef]
9. Ribeiro, R.V.; Machado, E.C.; Oliveira, R.F. Temperature response of photosynthesis and its interaction with light intensity insweet orange leaf discs under non-photorespiratory condition. *Ciênc. Agrotecnol.* **2006**, *30*, 670–678. [CrossRef]

10. Rout, G.R.; Das, A.B. *Molecular Stress Physiology of Plants*; Springer: New Delhi, India, 2013; ISBN 978-81-322-0806-8.
11. Ahanger, M.A.; Akram, N.A.; Ashraf, M.; Alyemeni, M.N.; Wijaya, L.; Ahmad, P. Plant responses to environmental stresses—Fromgene to biotechnology. *AoB Plants.* **2017**, *1*, 9.
12. Kumar, S.; Kaur, R.; Kaur, N.; Bhandhari, K.; Kaushal, N.; Gupta, K.; Bains, T.S.; Nayyar, H. Heat-stress induced inhibition ingrowth and chlorosis in mungbean (Phaseolus aureus Roxb.) is partly mitigated by ascorbic acid application and is related toreduction in oxidative stress. *Acta Physiol. Plant.* **2011**, *33*, 2091–2101. [CrossRef]
13. Rai, A.C.; Singh, M.; Shah, K. Effect of water withdrawal on formation of free radical, proline accumulation and activities of antioxidant enzymes in ZAT12-transformed transgenic tomato plants. *Plant Physiol. Biochem.* **2012**, *61*, 108–114. [CrossRef]
14. Hasanuzzaman, M.; Hossain, M.A.; Silva, J.A.T.D.; Fujita, M. Plant response and tolerance to abiotic oxidative stress: Antiox-idant defense is a key factor. In *Crop Stress and Its Management: Perspectives and Strategies*; Venkateswarlu, B., Shanker, A.K., Shanker, C., Maheswari, M., Eds.; Springer: Dordrecht, The Netherlands, 2012; pp. 261–315.
15. Morrison, M.J.; Stewart, D.W. Heat stress during flowering in summer Brassica. *Crop Sci.* **2002**, *42*, 797–803. [CrossRef]
16. Carmo-Silva, A.E.; Gore, M.A.; Andrade-Sanchez, P.; French, A.N.; Hunsaker, D.J.; Salvucci, M.E. Decreased CO_2 availability and inactivation of Rubisco limit photosynthesis in cotton plants under heat and drought stress in the field. *Environ. Exp. Bot.* **2012**, *83*, 1–11. [CrossRef]
17. Ahmad, M.; Waraich, E.A.; Tanveer, A.; Anwar-Ul-Haq, M. Foliar Applied Thiourea Improved Physiological Traits and Yield of Camelina and Canola Under Normal and Heat Stress Conditions. *J. soil Sci. Plant Nutr.* **2021**, *21*, 1666–1678. [CrossRef]
18. Chen, S.; Stefanova, K.; Siddique, K.H.M.; Cowling, W.A. Transient daily heat stress during the early reproductive phase disrupts pod and seed development in *Brassica napus* L. *Food Energy Secur.* **2021**, *10*, 262. [CrossRef]
19. Ihsan, M.Z.; Daur, I.; Alghabari, F.; Alzamanan, S.; Rizwan, S.; Ahmad, M.; Waqas, M.; Shafqat, W. Heat stress and plant development: Role of sulphur metabolites and management strategies. *Acta Agric. Scand. Sect. B—Plant Soil Sci.* **2019**, *69*, 332–342. [CrossRef]
20. Dawood, M.G. Stimulating Plant Tolerance against Abiotic Stress Through Seed Priming. In *Advances in Seed Priming*; Springer International Publishing: Cham, Switzerland, 2018; pp. 147–183.
21. Ahmad, Z.; Waraich, E.A.; Barutçular, C.; Alharby, H.; Bamagoos, A.; Kizilgeci, F.; Öztürk, F.; Hossain, A.; Bayoumi, Y.; El Sabagh, A. Enhancing drought tolerance in Camelina sativa L. and Canola napus L. through application of selenium. *Pak. J. Bot.* **2020**, *52*, 1927–1939. [CrossRef]
22. Dutta, P. Seed Priming: New Vistas and Contemporary Perspectives. In *Advances in Seed Priming*; Springer: Singapore, 2018; pp. 3–22.
23. Nathawat, N.S.; Nair, J.S.; Kumawat, S.M.; Yadava, N.S.; Singh, G.; Ramaswamy, N.K.; Sahu, M.P.; D'Souza, S.F. Effect of seed soaking with thiols on the antioxidant enzymes and photosystem activities in wheat subjected to water stress. *Biol. Plant.* **2007**, *51*, 93–97. [CrossRef]
24. Hasanuzzaman, M.; Bhuyan, M.H.M.B.; Mahmud, J.A.; Nahar, K.; Mohsin, S.M.; Parvin, K.; Fujita, M. Interaction of sulfur with phytohormones and signaling molecules in conferring abiotic stress tolerance to plants. *Plant Signal. Behav.* **2018**, *13*, e1477905. [CrossRef] [PubMed]
25. Skudra, I.; Ruza, A. Effect of Nitrogen and Sulphur Fertilization on Chlorophyll Content in Winter Wheat. *Rural. Sustain. Res.* **2017**, *37*, 29–37. [CrossRef]
26. Pandey, M.; Srivastava, A.K.; D'Souza, S.F.; Penna, S. Thiourea, a ROS scavenger, regulates source-to-sink relationship to enhance crop yield and oil content in *Brassica juncea* (L.). *PLoS ONE* **2013**, *8*, e73921.
27. Patade, V.Y.; Nikalje, G.C.; Srivastava, S. Role of Thiourea in Mitigating Different Environmental Stresses in Plants. In *Protective Chemical Agents in the Amelioration of Plant Abiotic Stress: Biochemical and Molecular Perspectives*; Roychoudhury, A., Tripathi, D.K., Eds.; John Wiley & Sons: Hoboken, NJ, USA, 2020; pp. 467–482. [CrossRef]
28. Akladious, S.A. Influence of thiourea application on some physiological and molecular criteria of sunflower (*Helianthus annuus* L.) plants under conditions of heat stress. *Protoplasma* **2013**, *251*, 625–638. [CrossRef]
29. Ahmad, M.; Waraich, E.A.; Zulfiqar, U.; Ullah, A.; Farooq, M. Thiourea application improves heat tolerance in camelina (*Camelina sativa* L. Crantz) by modulating gas exchange, antioxidant defense and osmoprotection. *Ind. Crop. Prod.* **2021**, *170*, 113826. [CrossRef]
30. Hao, Z.; Singh, V.P. Drought characterization from a multivariate perspective: A review. *J. Hydrol.* **2015**, *527*, 668–678. [CrossRef]
31. Righini, D.; Zanetti, F.; Martínez-Force, E.; Mandrioli, M.; Toschi, T.G.; Monti, A. Shifting sowing of camelina from spring to autumn enhances the oil quality for bio-based applications in response to temperature and seed carbon stock. *Ind. Crop. Prod.* **2019**, *137*, 66–73. [CrossRef]
32. Righini, D.; Zanetti, F.; Monti, A. The bio-based economy can serve as the springboard for camelina and crambe to quit the limbo. *OCL* **2016**, *23*, D504. [CrossRef]
33. Gesch, R.; Dose, H.; Forcella, F. Camelina growth and yield response to sowing depth and rate in the northern Corn Belt USA. *Ind. Crop. Prod.* **2017**, *95*, 416–421. [CrossRef]
34. Murphy, E.J. Camelina (Camelina sativa). In *Industrial Oil Crops*; Elsevier BV: Amsterdam, The Netherlands, 2016; pp. 207–230.
35. Zubr, J. Qualitative variation of Camelina sativa seed from different locations. *Ind. Crop. Prod.* **2003**, *17*, 161–169. [CrossRef]

36. Dixon, G.R. *Vegetable Brassicas and Related Crucifers (No. 14)*; CABI: Wallingford, UK, 2007.
37. Zanetti, F.; Eynck, C.; Christou, M.; Krzyżaniak, M.; Righini, D.; Alexopoulou, E.; Stolarski, M.J.; van Loo, E.N.; Puttick, D.; Monti, A. Agronomic performance and seed quality attributes of Camelina (*Camelina sativa* L. crantz) in multi-environment trials across Europe and Canada. *Ind. Crops Prod.* **2017**, *107*, 602–608. [CrossRef]
38. Steel, R.G.D.; Torrie, J.H.; Deekey., D.A. Principles and Procedures of Statistics. In *A Biometrical Approach*; McGraw Hill Book. Int. Co.: New York, NY, USA, 1997; pp. 400–428.
39. Fahad, S.; Hussain, S.; Saud, S.; Hassan, S.; Ihsan, Z.; Shah, A.N.; Wu, C.; Yousaf, M.; Nasim, W.; Alharby, H.; et al. Exogenously Applied Plant Growth Regulators Enhance the Morpho-Physiological Growth and Yield of Rice under High Temperature. *Front. Plant Sci.* **2016**, *7*, 1250. [CrossRef] [PubMed]
40. Asthir, B.; Thapar, R.; Farooq, M.; Bains, N.S. Exogenous application of thiourea improves the performance of late sown wheat by inducing terminal heat resistance. *Int. J. Agric. Biol.* **2013**, *15*, 1337–1342.
41. Allakhverdiev, S.I.; Kreslavski, V.D.; Klimov, V.V.; Los, D.; Carpentier, R.; Mohanty, P. Heat stress: An overview of molecular responses in photosynthesis. *Photosynth. Res.* **2008**, *98*, 541–550. [CrossRef] [PubMed]
42. Fahad, S.; Hussain, S.; Saud, S.; Hassan, S.; Chauhan, B.; Khan, F.; Ihsan, M.Z.; Ullah, A.; Wu, C.; Bajwa, A.; et al. Responses of Rapid Viscoanalyzer Profile and Other Rice Grain Qualities to Exogenously Applied Plant Growth Regulators under High Day and High Night Temperatures. *PLoS ONE* **2016**, *11*, e0159590. [CrossRef]
43. Jiang, Y. Effect of heat stress on pollen development and seed set of field pea (*Pisum sativum* L.). Ph.D. Thesis, University of Saskatchewan, Saskatoon, SK, Canada, 2016.
44. Farooq, M.; Bramley, H.; Palta, J.; Siddique, K. Heat Stress in Wheat during Reproductive and Grain-Filling Phases. *Crit. Rev. Plant Sci.* **2011**, *30*, 491–507. [CrossRef]
45. Weymann, W.; Böttcher, U.; Sieling, K.; Kage, H. Effects of weather conditions during different growth phases on yield formation of winter oilseed rape. *Field Crop. Res.* **2015**, *173*, 41–48. [CrossRef]
46. Afzal, I.; Basra, S.; Shahid, M.; Farooq, M.; Saleem, M. Priming enhances germination of spring maize (*Zea mays* L.) under cool conditions. *Seed Sci. Technol.* **2008**, *36*, 497–503. [CrossRef]
47. Farooq, M.; Wahid, A.; Kobayashi, N.; Fujita, D.; Basra, S.M.A. Plant drought stress: Effects, mechanisms and management. *Agron. Sustain. Dev.* **2009**, *29*, 185–212. [CrossRef]
48. Ali, M.M.; Shafique, M.W.; Gull, S.; Naveed, W.A.; Javed, T.; Yousef, A.F.; Mauro, R.P. Alleviation of Heat Stress in Tomato by Exogenous Application of Sulfur. *Horticulturae* **2021**, *7*, 21. [CrossRef]
49. Singh, S.; Rathore, P. Influence of phosphorus and thiourea on yield and economics of greengram [Vigna radiata var. aureus (L.) Wilczek]. *Res. Crop.* **2003**, *4*, 210–212.
50. Ahmad, M.; Waraich, E.A.; Hussain, S.; Ayyub, C.M.; Ahmad, Z.; Zulfiqar, U. Improving Heat Stress Tolerance in Camelina sativa and *Brassica napus* Through Thiourea Seed Priming. *J. Plant Growth Regul.* **2021**, 1–17. [CrossRef]
51. Rezaei, M.; Bagherian, F. Influence of planting date and sulfur coating in seed coating solution on cotton (*Gossypium hirsutum* L.) seeds: Physiological traits. *Iran. J. Plant Physiol.* **2013**, *4*, 917–923.
52. Brooks, A. Effects of Phosphorus Nutrition on Ribulose-1,5-Bisphosphate Carboxylase Activation, Photosynthetic Quantum Yield and Amounts of Some Calvin-Cycle Metabolites in Spinach Leaves. *Funct. Plant Biol.* **1986**, *13*, 221–237. [CrossRef]
53. Ibrahim, E.A. Seed priming to alleviate salinity stress in germinating seeds. *J. Plant Physiol.* **2016**, *192*, 38–46. [CrossRef] [PubMed]
54. Kaya, C.; Ashraf, M.; Sonmez, O.; Tuna, A.L.; Polat, T.; Aydemir, S. Exogenous application of thiamin promotes growth and antioxidative defense system at initial phases of development in salt-stressed plants of two maize cultivars differing in salinity tolerance. *Acta Physiol. Plant.* **2014**, *37*, 1741 . [CrossRef]
55. Mobin, M.; Khan, M.N.; Abbas, Z.K.; Ansari, H.R.; Al-Mutairi, K. Significance of sulfur in heat stressed cluster bean (*Cymopsis tetragonoloba* L. Taub) genotypes: Responses of growth, sugar and antioxidative metabolism. *Arch. Agron. Soil Sci.* **2016**, *63*, 288–295. [CrossRef]
56. Orman, S.; Kaplan, M. Effects of elemental sulphur and farmyard manure on pH and salinity of calcareous sandy loam soil and some nutrient elements in tomato plant. *J. Agric. Sci. Technol.* **2011**, *5*, 20–26.
57. Sohag, A.A.M.; Arif, T.-U.; Brestic, M.; Afrin, S.; Sakil, A.; Hossain, T.; Hossain, M.A.; Hossain, A. Exogenous salicylic acid and hydrogen peroxide attenuate drought stress in rice. *Plant Soil Environ.* **2020**, *66*, 7–13. [CrossRef]
58. Ali, Q.; Daud, M.; Haider, M.Z.; Ali, S.; Rizwan, M.; Aslam, N.; Noman, A.; Iqbal, N.; Shahzad, F.; Deeba, F.; et al. Seed priming by sodium nitroprusside improves salt tolerance in wheat (Triticum aestivum L.) by enhancing physiological and biochemical parameters. *Plant. Physiol. Biochem.* **2017**, *119*, 50–58. [CrossRef]
59. Javeed, H.; Ali, M.; Skalicky, M.; Nawaz, F.; Qamar, R.; Rehman, A.; Faheem, M.; Mubeen, M.; Iqbal, M.; Rahman, M.; et al. Lipoic Acid Combined with Melatonin Mitigates Oxidative Stress and Promotes Root Formation and Growth in Salt-Stressed Canola Seedlings (*Brassica napus* L.). *Molecules* **2021**, *26*, 3147. [CrossRef] [PubMed]
60. Ahmad, Z.; Anjum, S.; Skalicky, M.; Waraich, E.; Tariq, R.M.S.; Ayub, M.; Hossain, A.; Hassan, M.; Brestic, M.; Islam, M.S.; et al. Selenium Alleviates the Adverse Effect of Drought in Oilseed Crops Camelina (*Camelina sativa* L.) and Canola (*Brassica napus* L.). *Molecules* **2021**, *26*, 1699. [CrossRef]

61. Marchand, F.L.; Mertens, S.; Kockelbergh, F.; Beyens, L.; Nijs, I. Performance of High Arctic tundra plants improved during but deteriorated after exposure to a simulated extreme temperature event. *Glob. Chang. Biol.* **2005**, *11*, 2078–2089. [CrossRef]
62. Almeselmani, M.; Deshmukh, P.S.; Sairam, R.K. High temperature stress tolerance in wheat genotypes: Role of antioxidant defence enzymes. *Acta Agron. Hung.* **2009**, *57*, 1–14. [CrossRef]
63. Balla, K.; Bencze, S.; Janda, T.; Veisz, O. Analysis of heat stress tolerance in winter wheat. *Acta Agron. Hung.* **2009**, *57*, 437–444. [CrossRef]

Article

Effects of Environmental Stresses (Heat, Salt, Waterlogging) on Grain Yield and Associated Traits of Wheat under Application of Sulfur-Coated Urea

Adil Altaf [1], Xinkai Zhu [1,2,3,*], Min Zhu [1,2,*], Ma Quan [1], Sana Irshad [4], Dongyi Xu [1], Muhammad Aleem [5], Xinbo Zhang [1], Sadia Gull [6], Fujian Li [1], Amir Zaman Shah [6] and Ahmad Zada [7]

1. Jiangsu Key Laboratory of Crop Genetics and Physiology, Jiangsu Key Laboratory of Crop Cultivation and Physiology, Wheat Research Center, College of Agriculture, Yangzhou University, Yangzhou 225009, China; er.altafadil@outlook.com (A.A.); mq_agriculture@163.com (M.Q.); xudongyi0923@163.com (D.X.); zhxb202@126.com (X.Z.); fjli_agriculture@163.com (F.L.)
2. Co-Innovation Center for Modern Production Technology of Grain Crops, Yangzhou University, Yangzhou 225009, China
3. Joint International Research Laboratory of Agriculture and Agri-Product Safety, The Ministry of Education of China, Yangzhou University, Yangzhou 225009, China
4. School of Environmental Studies, China University of Geosciences, Wuhan 430074, China; sanairshad55@gmail.com
5. College of Environment, Hohai University, Nanjing 210098, China; mmaleem@hotmail.com
6. College of Horticulture and Plant Protection, Yangzhou University, Yangzhou 225009, China; sadiagull137@yahoo.com (S.G.); dh18052@yzu.edu.cn (A.Z.S.)
7. College of Bioscience and Biotechnology, Yangzhou University, Yangzhou 225009, China; dh19025@yzu.edu.cn

* Correspondence: xkzhu@yzu.edu.cn (X.Z.); minzhu@yzu.edu.cn (M.Z.)

Citation: Altaf, A.; Zhu, X.; Zhu, M.; Quan, M.; Irshad, S.; Xu, D.; Aleem, M.; Zhang, X.; Gull, S.; Li, F.; et al. Effects of Environmental Stresses (Heat, Salt, Waterlogging) on Grain Yield and Associated Traits of Wheat under Application of Sulfur-Coated Urea. *Agronomy* **2021**, *11*, 2340. https://doi.org/10.3390/agronomy11112340

Academic Editor: Juan M. Ruiz

Received: 26 October 2021
Accepted: 15 November 2021
Published: 19 November 2021

Abstract: Abiotic stresses, such as heat, salt, waterlogging, and multiple-stress environments have significantly reduced wheat production in recent decades. There is a need to use effective strategies for overcoming crop losses due to these abiotic stresses. Fertilizer-based approaches are readily available and can be managed in all farming communities. This research revealed the effects of sulfur-coated urea (SCU, 130 kg ha^{-1}, release time of 120 days) on wheat crops under heat, salt, waterlogging, and combined-stress climatic conditions. The research was done using a completely randomized design with three replicates. The results revealed that SCU at a rate of 130 kg of N ha^{-1} showed a significantly ($p \leq 0.05$) high SPAD value (55) in the case of waterlogging stress, while it was the lowest (31) in the case of heat stress; the control had a SPAD value of 58. Stress application significantly ($p \leq 0.05$) reduced the leaf area and was the highest in control (1898 cm^2), followed by salt stress (1509 cm^2), waterlogging (1478 cm^2), and heat stress (1298 cm^2). A significantly ($p \leq 0.05$) lowest crop yield was observed in the case of heat stress (3623.47 kg ha^{-1}) among all stresses, while it was 10,270 kg ha^{-1} in control and was reduced up to 35% after the application of heat stress. Among all stresses, the salt stress showed the highest crop yield of 5473.16 kg ha^{-1}. A significant correlation was observed among growth rate, spike length, yield, and physiological constraints with N content in the soil. The SCU fertilizer was the least effective against heat stress but could tolerate salt stress in wheat plants. The findings suggested the feasibility of adding SCU as an alternative to normal urea to alleviate salt stresses and improve wheat crop growth and yield traits. For heat stress tolerance, the applicability of SCU with a longer release period of ~180 days is recommended as a future prospect for study.

Keywords: abiotic stresses; winter wheat; photosynthetic activity; morphological parameters; controlled-release nitrogen fertilizer

1. Introduction

Feeding the world's growing population necessitates having a greater focus on the efficient and particular use of scarce resources, such as fertilizers. The yield of crops decreases due to various abiotic parameters, such as drought stress, salt stress, late sowing, poor seed quality, climate change, and lack of fertilizer [1–5]. High-temperature stress is the primary environmental issue that confines the yield of wheat. For every 1 °C increase in average temperature from 23 °C, the wheat yield will decrease by about 10% [6]. Heat stress significantly reduces the photosynthetic rate, chlorophyll content, leaf areas, and grain weight, and ultimately crop yield per hector is reduced to half [7–9]. Besides this, >40% of the world's total wheat area is facing abiotic stress. The productivity of wheat is often unfavorably affected by salt stress, which has been linked to slow growth, changes in reproductive behavior, variations in enzymatic activity, damage to photosynthesis, injury to the ultrastructure of cell components, serotonin deficiency, and oxidative stress [10]. Many studies in the past have revealed the adverse effects of salt stress on the physiological traits of numerous plants, including cardoon genotypes [11], pepper [12], *Vicia faba* [13], and *Olea europea* L. [14]. Besides this, drought stress reduces morphological traits, such as leaf size and vegetative growth; physiological traits, such as reduction in photosynthesis and stomatal conductance; and the transpiration rate [15]. Drought stress has also been observed to alter the biochemical and physiological responses of wheat [7]. On the other hand, waterlogging is also a significant factor influencing the yield and quality of wheat. Waterlogging stress reduces yield, number of ears per square meter, grain weight, protein content, and levels of chlorophyll a and b while increasing proline levels [16]. Similarly, waterlogging induced a stress-activated antioxidant response system in *Phalaris arundinacea* [17]. In another study, long-term waterlogging stress affected photosynthetic traits such as leaf area, stomatal density, and stomatal conductance in apple cultivars [18]. So, the wheat yield has been reduced in recent years, resulting in price volatility and food insecurity. It has been proposed that wheat production must be increased by 60% to fulfill the needs of 9 billion people by 2050 [4,5]. This will necessitate an increase in annual wheat production of at least 1.6%, which will require resistance to abiotic and biotic stresses and enhanced input use efficiency.

Many approaches have been used to reduce the deleterious effects of abiotic stress in plants; one of these approaches is the use of nitrogen (N) fertilizer. The rational use of chemical fertilizers is essential for wheat production, food security, and the environment Nitrogen has been reported to enhance wheat crop yields [19]. Nitrate is a communal form of N that exists in cell vacuoles and is reduced by nitrate and nitrite reductase activities in the cytoplasm. Leaves contain chlorophyll, which is responsible for photosynthesis. When N is readily available in the soil solution, the nitrogen use efficiency of the plant is critical [20]. The excessive use of nitrogen fertilizer causes environmental pollution, as well as economic losses. The unwise use of nitrogen fertilizers can cause crop lodging and reduce economic yields. Thus, studying nitrogen for a good yield and time is inevitable for the wheat crop, especially once grown under stress conditions [21].

At the same time, loss of N is the main threat of environmental pollution, which causes health problems. The volatilization of ammonia in urea fertilizer is up to 65%, depending on the environment and soil characteristics. Nitrate pollution produces serious health problems for humans and animals. The use of nitrogen higher than the crop requirement may be the reason for a low nitrogen utilization rate and nitrogen loss in the soil [19]. Hence, in order to reduce the loss of nitrogen under abiotic stress and increase the yield, it is recommended to use the 4R principle (right time, right amount, right source, and right place) for fertilization [22]. In this regard, slow-release nitrogen fertilizers can improve the tolerance of abiotic stresses [19,21]. Slow-release nitrogen fertilizer technology could be used to decrease water and environmental pollution [22]. Slow-release fertilizers contain a semipermeable layer of various essential oils, as well as secondary and significant nutrients, and control particle water solubility by slowing the hydrolysis procedure of water-soluble fertilizers.

One of the good slow-release nitrogen fertilizers is sulfur-coated urea (SCU), which promotes wheat growth and development. The wheat crop has a positive correlation between "S" and "N" elements [23]. The S element is a secondary and fungicide with acidic possessions that neutralize the alkalinity of soil [24]. As a result, the excessive application of N without the S coating material results in the extreme leaching of N [21]. As a result, the prudent application of nitrogen fertilizers and nitrogen sources reduces strength while increasing crop yield under stress.

Few studies have narrated the effects of SCU on wheat under comparative examination of heat, salt stress, waterlogging and combined stress conditions. A study addressing the comparative effect of slow-release nitrogen fertilizer on three stresses (salt, waterlogging, and heat) has not been conducted before, hence the novelty of our study. There is a need to conduct a detailed study based on various abiotic conditions under controlled nitrogen fertilization. To address all the issues mentioned above, the current study aimed to determine the effects of salt stress, waterlogging, and heat stresses on the physiological attributes and wheat crop yield. Furthermore, the current study also focused on improving wheat growth and development, as well as viable soil management using SCU (with a nitrogen release period of 120 days) under heat, salt stress, waterlogging, and combined stresses. Another objective of the current study was to assess the effect of the N source and release rate on wheat production and abiotic stress tolerance. The effectiveness of slow-release SCU against various abiotic stresses was determined along with control wheat with the same SCU fertilizer. This study also aimed to determine the key physical parameters affecting crop yield once the stresses were applied.

2. Materials and Methods

2.1. Experimental Site and Environmental Conditions:

Pot experiments were carried out during the wheat growing season in 2020–2021 (2021) at the Agricultural Experiment Station (32°39′ N, 119°42′ E) of Agricultural College, Yangzhou University, in China. Winter wheat (Yangmai 25) was grown in the pot field. Each experiment consisted of five developmental phases (overwintering, jointing, booting, flowering, and maturity), with 6 seeds of Yangmai 25 genotype grown in individual pots (Figure S2). Four stresses were chosen: waterlogging, salt stress, heat, and the combined effect of these three stresses, which were monitored at different growth stages (Figure S3). Due to the abundant rainfall in this area, we did not need to irrigate during the wheat growing season. However, if necessary, the pots were irrigated with tap water accordingly to retain the field capacity required for healthy wheat growth.

The experiment was a completely randomized design with SCU treatment (control—SCU only) as the main plot and stress as the subplot. There were four stress treatments (waterlogging, salt stress, heat, and combined). There were five plots in the experiment (Figure S2). Each treatment was conducted in triplicate and was subjected to the same field management. The N rate adopted in the experiment was 130 kg ha^{-1}, consistent with the optimum N rate in all the treatments, including control (CK). Phosphate (114 kg ha^{-1} P_2O_5) and potassium (62 kg ha^{-1} K_2O) fertilizers were applied once before sowing conferring to the pre-soil analysis report. Basal fertilizer was applied at a depth of 10–15 cm [22]. A total of six seeds of Yangmai 25 were sown in a 10 kg pot filled with a standard potting mixture [25] (Table S2).

The slow-release sulfur-coated urea (SCU; release time 120 days) was used as a stress-alleviation substance at a recommended dose of 130 kg ha^{-1} in all treatments, including the control. Seeds were sown on 8 November 2020, while sampling was performed at the overwintering stage (28 December 2020), jointing stage (12 March 2021), booting stage (30 Mar 2021), flowering stage (17 April 2021), and maturity stage (28 May 2021). The setups for the different abiotic stress applications were established as follows.

For the heat stress, plants were upraised with control plants until the flowering stage and then moved to the heat stress chamber (cryogenic room) for the remainder of the growth period (up to maturity). The plants were kept under a 16 h photoperiod duration and provided a light intensity of 350 µmol m^{-2} s^{-1} generated by metal halide lamps. The temperature treatment was as follows: the night temperature was kept at 20 °C, whereas the day temperature was gradually increased to 33 °C and held for 8 h, then subsequently decreased to 20 °C. The relative humidity (%RH) was kept at 64–68% and 76% during the day and night, respectively, in the stress chamber (Figures S1 and S3).

For waterlogging stress, plants were grown with control plants, and stress was applied from germination to the overwintering stage (35 days). Waterlogging was applied using water from a nearby water service by flooding the pots allocated for the waterlogging treatments. The soil was moisturized using water above field capacity using incessant flooding, generally every day, to produce an oxygen-deficient atmosphere. For this purpose, pots were placed in a water basin (i.e., used for nursery rice growing). There was a storage tank on one side and a drainage valve on the other side. The standing water level was maintained at ~12 cm. Water was replaced after 4–7 days according to weather and water conditions. For replacing water, a drainage valve was opened according to the storage tank valve so that the water level could be maintained. The soil moisture content was measured by an oven-drying method using 1 g of soil sample at 105 °C overnight [26]. In control, the soil moisture content was <85%, while it was from 85 to 100% for the stressed plants (Figures S1 and S3).

For salt stress, a 15 mM sodium chloride (NaCl) solution was used as a source, while plants were grown with control plants under the same conditions. The salt stress was applied after the jointing stage and continued until maturation (Figures S1 and S3).

For combined stress, plants were grown in waterlogged stress, and after the jointing stage, salt stress was applied. After the flowering stage, these plants were transferred to the heating chamber, and they continued to grow there until the maturation stage (Figure S3). In total, the experimental setup consisted of 1 genotype × 3 pots × 4 abiotic stresses × 5 developmental stages × 1 fertilizer (SCU: for all treatments including control) × 6 seeds per pot.

A control treatment was also established in triplicate in a separate plot containing the same SCU fertilizer but at a recommended dose of 130 kg ha^{-1} (Figure S2). A separate pot experiment without any stress and fertilizer was also conducted under the same conditions to compare the nitrogen accumulation. Seeds were sown on 8 November 2020, and the wheat crop was harvested on 28 May 2021. During this study period, samples were collected at 5 stages: (1) overwintering stage: 28 December 2020; (2) jointing stage: 12 March 2021; (3) booting stage: 30 March 2021; (4) flowering stage: 17 April 2021; and (5) maturity stage: 28 May 2021.

2.2. Procedure/Protocols for Growth and Yield Parameters

Each experimental unit was measured for yield and yield-related characteristics, such as plant height, seeds per spike, total dry matter (kg ha^{-1}), number of tillers per plant, grain yield (kg ha^{-1}), average seeds weight per spike, and harvest index. Portable chlorophyll meters (SPAD-502, Konica Minolta, Osaka, Japan) and mobile photosynthesis systems (LI-6400, LI-COR Biosciences, Lincoln, NE 68504, USA) were used to collect data on physiological parameters, including chlorophyll content percent and net photosynthetic rate [26,27]. Wheat plants were harvested at various stages of development to measure fresh and dry biomass. An electrical weight balance was used to measure the fresh weight of leaves and stems. Then, oven drying of leaves and stems was done for 48 h (up to constant weight) at 70 °C, and the dry weight was calculated [24]. A leaf area meter (Model, CI-202, CID Bio-Science, Inc., 1554 NE 3rd Avenue, Camas, WA 98607, USA) was used to calculate the leaf area. At maturity, plants from three pots were collected from each experimental component, and various yield components were quantified. The final yield and biomass of the entire experimental unit were assessed separately, and on the rationale of dry biomass production, transformed into kg ha^{-1}. The activity of photosynthetic properties, fluorescence, photosynthetic rate, stomatal conductance, the intercellular CO_2 concentration of the plants, transpiration rate, and water use efficiency were all measured with a portable system (LI-6400, Li-Cor Inc., USA). Water use efficiency was also calculated (photosynthetic rate divided by transpiration rate) following the methodology of a previous study [24]. All analyses were performed in triplicate, and mean values were calculated.

2.3. Estimation of Plant NPK

The plant N content was measured at 20, 60, and 120 days using the standard procedure described by Watson et al. [24]. A digestion tube was filled with the plant-dried sample (1.0 g). Then, in a digestion block, 15 mL of concentrated H_2SO_4 and 1 g digestion mixture (K_2SO_4 + $CuSO_4$ @ 9:1) were combined, and the tubes were heated for 2 h at 450 °C. After heating, the color of the solution was changed from transparent to yellowish-green, visible in the digestion tubes. A distillate unit produced the required volume for the distillation process. Then, the material was placed in a receiver containing 4% boric acid (25 mL). After that, a few drops of the indicator were added; the purple color then changed to golden yellow through the distillation process. The subsequent distillates were then titrated with 0.1 N H_2SO_4, resulting in purple as an endpoint from a golden yellow shade [24].

The following spectrophotometer and the spectrophotometric vanadium phosphomolybdate processes were used to estimate the P percent in the plant following the standard protocol of a previous study [28]. The spectrophotometer was used to run the standard of P samples. Following this, the P concentration of plant samples was assessed using the yellow color procedure. The digested plant samples were used to distill water, and a coloring reagent was placed in a flask. The flask was then placed at room temperature for about 30–35 min, with the development of color over the subsequent time. The P percent in plant samples was then calculated using just a spectrophotometer at 420 nm using the standard method [29]. The flame photometer procedure was used to measure the K concentration using a method developed by a previous study [30].

2.4. Statistical Analysis

Statistical analyses were performed with SPSS 16.0 (SPSS Inc., Chicago, IL, USA) [31]. All results are presented as the means of three replicates. Data from each sampling stage were analyzed separately and were subjected to one-way analysis of variance (ANOVA) followed by means comparison using Duncan's multiple range test (DMRT) using a statistically significant level of $p < 0.05$. In the correlation analysis, the nonparametric Spearman test was used for the correlation of different traits.

3. Results

Wheat (*Triticum aestivum* L.), a staple cereal crop in various world regions, is a major cereal crop subjected to many biotic and abiotic stresses such as heat, salt, waterlogging, and sometimes, a combination of stresses. These stresses have a global impact on crop yields. Farmers have used various mechanisms to combat a vast array of biotic and abiotic stresses. One of these aspects is fertilizers that affect the nutrient availability to the plant. The experiment was conducted on winter wheat crop Yangmai 25 genotype for a five-stage (germination to maturity) growth period amended with SCU having a release time of 120 days. The sample and experimental analysis were conducted at every stage. The effect of SCU or stress alleviation effect of SCU was analyzed in terms of growth and physiological parameters. The study showed exciting aspects of each type of stress that are presented below.

3.1. Plant Growth Study under Abiotic Stress Amended with SCU

3.1.1. Waterlogging Stress and Wheat Growth during the Study

The waterlogging stress was applied up to the overwintering stage; initially, stress did not affect plant growth significantly ($p \leq 0.05$). It was evident in the change of florescence value and photosynthetic rate (Figure 1a,b), which were 0.794 and 14.58 μmol CO_2 m^{-2} s^{-1}, respectively, at the start of the experiment. At the flowering stage, a significant ($p \leq 0.05$)

change of ~5% was observed in the florescence value, and the photosynthetic rate was decreased by 7%, compared to the control. This was also related to the number of tillers and leaf area of the plants (Figures 2 and 3), which were also significantly ($p < 0.05$) lower than the control. At the flowering stage, the plant had four tillers in the case of stress, while the control had five; the leaf area of the control plant was 1898 cm^2, while it was 1330 cm^2 in stress conditions. The photosynthetic rate corresponded to the SPAD value, which showed a significant ($p < 0.05$) decrease at the flowering stage (Figure 4). At the flowering stage, it was ~57 in the case of the control and ~50 once waterlogging stress was applied, i.e., a decrease of 10% was observed. The SPAD value is a measure of the chlorophyll value. Both leaf area and photosynthetic rate were negatively affected by stress, consequently reducing total dry matter accumulation rates and final yields. The same trend of fresh and dry biomass was observed in our study, where stress application considerably ($p \leq 0.05$) reduced the biomass content of the wheat plant (Figures 5 and 6). This significant ($p \leq 0.05$) decrease in the photosynthetic rate was also correlated with the stomatal conductance (Figure 7b), transpiration rate (Figure 7a), and water use efficiency (Figure 7c). In April 2021, stressed wheat plants showed 9, 10, 8.8, and 9.6% lower stomatal conductance, transpiration rate, intercellular CO_2, and water use efficiency, respectively, compared to the control plant.

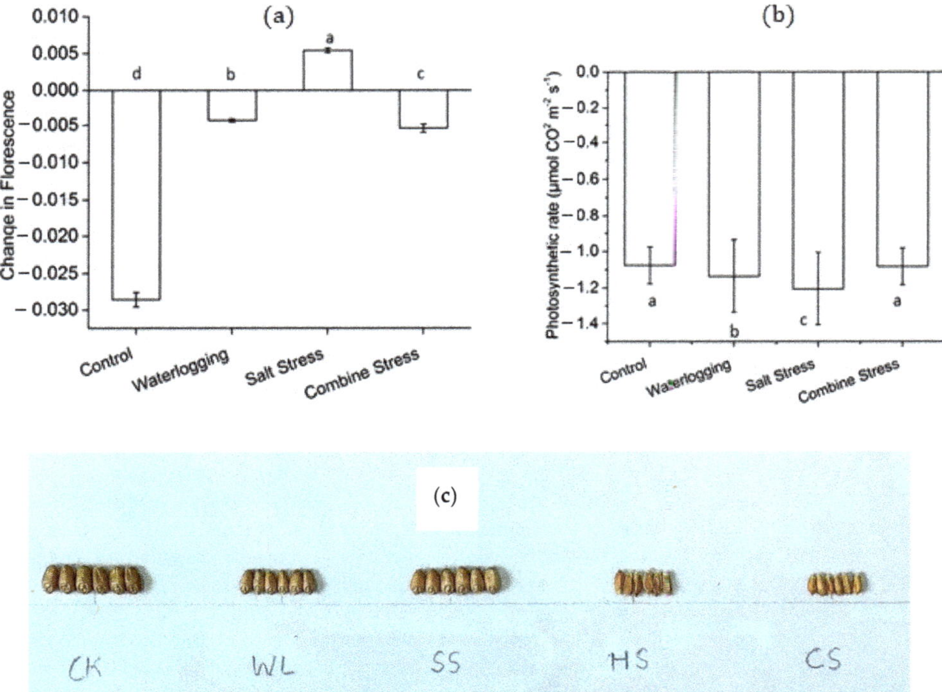

Figure 1. Change in the (**a**) fluorescence and (**b**) photosynthetic activity of the wheat plants from 15 April 2020 to 22 April 2021 (flowering stage) grown under different abiotic stresses. (**c**) The difference in the grain sizes after harvest. Note: reported values are the means and standard deviations of triplicates for each treatment (exact values are given in Table S1). Lowercase letters show the significant differences among treatments according to one-way ANOVA and DMRT, while the significance level was $p \leq 0.05$. CK: control; WL: waterlogging; SS: salt stress; HS: heat stress; CS: combined stress.

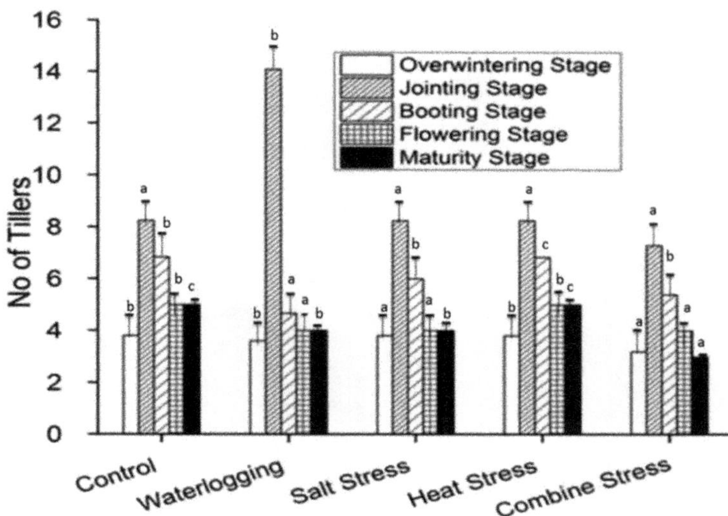

Figure 2. Number of tillers during five-stage growth of wheat plants grown under different abiotic stresses. Reported values are the means and standard deviations of triplicates for each treatment. Note: sampling date at overwintering stage was 28 December 2020; at jointing stage was 12 March 2021; at booting stage was 30 March 2021; at the flowering stage was 17 April 2021; and at maturity stage was 28 May 2021. Lowercase letters show the significant differences within treatments according to one-way ANOVA and DMRT, while the significance level was $p \leq 0.05$.

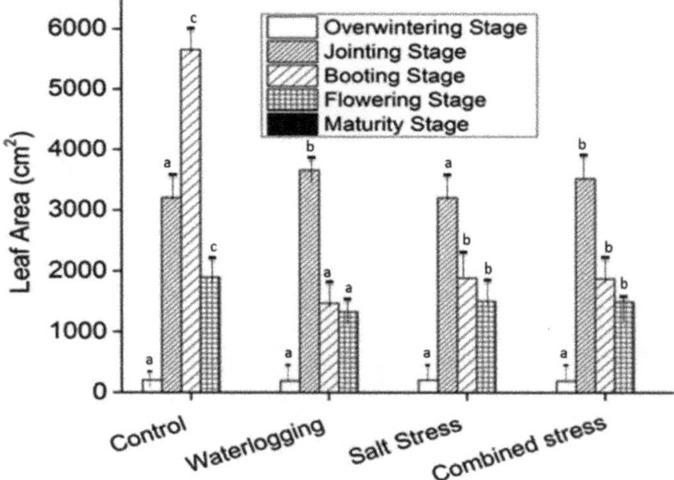

Figure 3. Leaf area of wheat plants during the five-stage plant growth (until flowering stage) grown under different abiotic stresses. Reported values are the means and standard deviations of triplicates for each treatment. Note: sampling date at overwintering stage was 28 December 2020; at jointing stage was 12 March 2021; at booting stage was 30 March 2021; at the flowering stage was 17 April 2021; and at maturity stage was 28 May 2021. Lowercase letters show the significant differences among treatments according to one-way ANOVA and DMRT, while the significance level was $p \leq 0.05$.

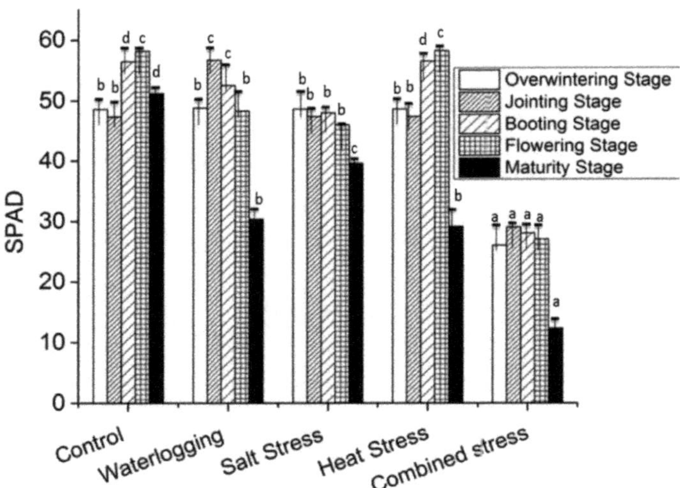

Figure 4. SPAD value of wheat plants during the five-stage growth of plants grown under different abiotic stresses. Reported values are the means and standard deviations of triplicates for each treatment. Note: sampling date at overwintering stage was 28 December 2020; at jointing stage was 12 March 2021; at booting stage was 30 March 2021; at the flowering stage was 17 April 2021; and at maturity stage was 28 May 2021. Lowercase letters show the significant differences among treatments according to one-way ANOVA and DMRT, while the significance level was $p \leq 0.05$.

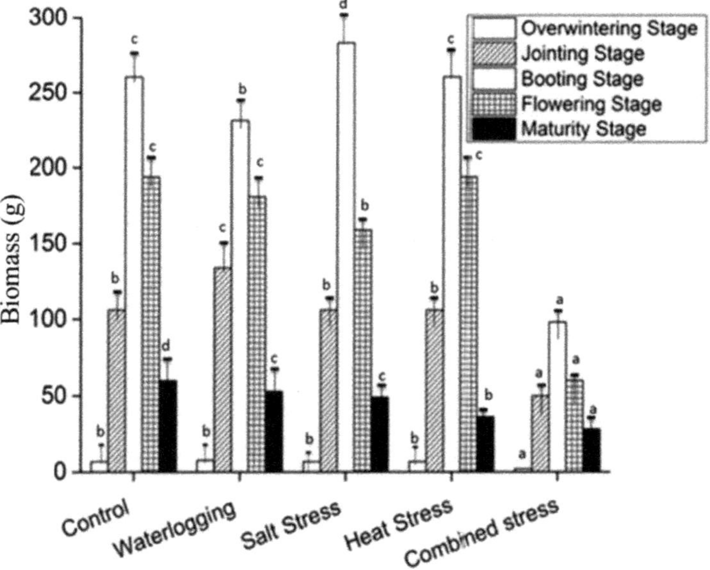

Figure 5. Fresh biomass of wheat plants during five-stage growth of plants grown under different abiotic stresses. Reported values are the means and standard deviations of triplicates for each treatment. Note: sampling date at overwintering stage was 28 December 2020; at jointing stage was 12 March 2021; at booting stage was 30 Mar 2021; at the flowering stage was 17 April 2021; and at maturity stage was 28 May 2021. Lowercase letters show the significant differences within treatments according to one-way ANOVA and DMRT, while the significance level was $p \leq 0.05$.

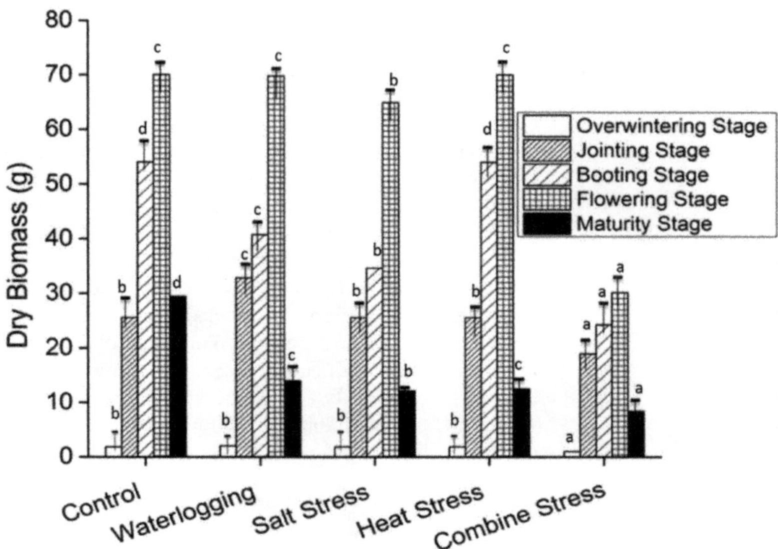

Figure 6. Dry biomass of wheat plants during the five-stage growth of plants grown under different abiotic stresses. Reported values are the means and standard deviations of triplicates for each treatment. Note: sampling date at overwintering stage was 28 December 2020; at jointing stage was 12 March 2021; at booting stage was 30 Mar 2021; at the flowering stage was 17 April 2021; and at maturity stage was 28 May 2021. Lowercase letters show the significant differences among treatments according to one-way ANOVA and DMRT, while the significance level was $p \leq 0.05$.

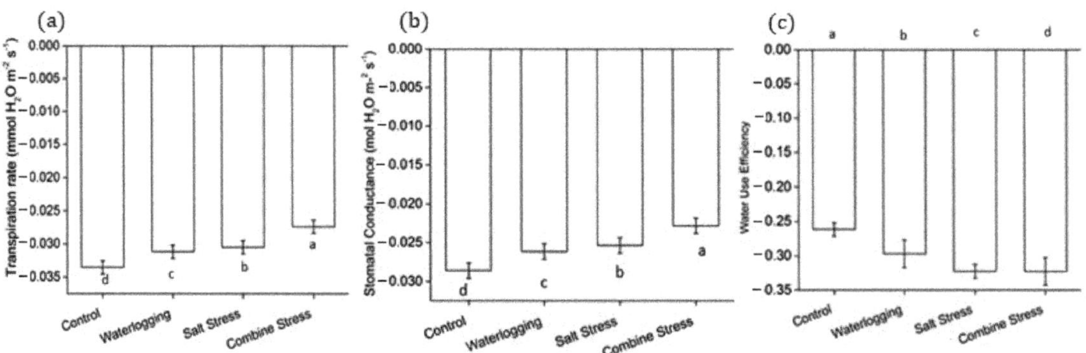

Figure 7. (**a**) Transpiration rate; (**b**) stomata conductance; and (**c**) water use efficiency of the wheat plants from 15 April 2020 to 22 April 2021 (flowering stage) grown under different abiotic stresses. Water use efficiency was measured in µmol CO_2 m^{-2} s^{-1}/mmol H_2O m^{-2} s^{-1}. Note: Reported values are the means and standard deviations of triplicates for each treatment (exact values are given in Table S1). Lowercase letters show the significant differences among treatments according to one-way ANOVA and DMRT, while the significance level was $p \leq 0.05$.

3.1.2. Salt Stress and Wheat Growth during the Study

The salt stress was applied after the jointing stage; initially, it was less disturbing for the plant, but the wheat growth was significantly ($p \leq 0.05$) reduced after the booting stage. It was evident in the change in fluorescence value and photosynthetic rate (Figure 1a,b), which were 0.794 and 14.58 µmol CO_2 m^{-2} s^{-1}, respectively, at the start of the experiment in control. A significant decrease of 2 and 4% was observed in the case of florescence value and photosynthetic rate, respectively. It was also correlated with the number of tillers. Tillers and leaf area of the plant (Figures 2 and 3) were 6.8 and 3655 cm^2 at the booting stage, respectively, and both were significantly ($p \leq 0.05$) reduced at the flowering stage. The plant had four tillers in case of stress while the control had 5; the leaf area at this stage was 1898 cm^2, while it was 1590 cm^2 in salt stress conditions. The photosynthetic rate corresponded to the SPAD value, which showed a significant ($p \leq 0.05$) decrease at the flowering stage (Figure 4). It was ~57 in the case of the control at the flowering stage, while it was reduced to ~46 once salt stress was applied. SPAD value is the measure of chlorophyll value. Both leaf area and photosynthetic rate were decreased significantly ($p \leq 0.05$) by stress, resulting in significant reductions in total dry matter accumulation rates and final yields. The same trend of fresh and dry biomass was observed in our study, in which the application of stress considerably ($p \leq 0.05$) reduced the biomass content of the wheat plant (Figures 5 and 6). This trend of photosynthetic rate was also correlated with the stomatal conductance (Figure 7b), transpiration rate (Figure 7a), and water use efficiency (Figure 7c). In April 2021, stressed wheat plants showed 8.9, 12, 9.8, and 8.9% lower stomatal conductance, transpiration rate, intercellular CO_2, and water use efficiency, respectively than control plants.

3.1.3. Heat Stress and Wheat Growth during the Study

The heat stress was applied after the flowering stage; it was significantly ($p \leq 0.05$) most disturbing to the plant, perhaps due to the unavailability of nitrogen, as release time for most of the nitrogen in the case of SCU, is 120 days. The trend of fresh and dry biomass explains this in our study, in which stress application considerably ($p \leq 0.05$) reduced the biomass content of the wheat plant (Figure 5). At the maturation phase, the biomass for control was ~55 g, while it was reduced to 25 g after heat stress. For all growth parameters, control plants grew better than stressed ones. However, the stress affected the wheat plant comparatively less until after the first ~110 days of the experiment (until the booting stage).

In contrast, afterward, growth was significantly affected ($p \leq 0.05$). It can be explained in terms of the nitrogen release period of the SCU, which was ~120 days. Combined stress had quite the same action as heat stress.

3.2. Plant Stress Response and Yield Aspects

3.2.1. Effect of Waterlogging Stress on Yield of the Wheat Plant

Table 1 shows the effect of waterlogging stress on the yield of the wheat plant. The crop yield kg ha^{-1} was reduced significantly ($p \leq 0.05$) to 6034.5, which was ~40% less than the control group (10,270 kg ha^{-1}). It was 60% of the growth without stress. It is explainable in terms of the 1000-grain weight, which was reduced significantly ($p \leq 0.05$) from 56 g in control to 46 g once the stress was applied.

Table 1. Influence of abiotic stresses on the wheat yields after harvest.

Treatments/ Parameters	1000 Grain Weight (g)	Harvest Index (%)	Yield (Mann/Acre)	Yield (kg ha^{-1})
Control	56.75 ^{a} ± 2.30	44.09 ^{d} ± 1.31	103.90 ^{a} ± 2.65	10270.10 ^{a} ± 14
Waterlogging	46.95 ^{c} ± 2.12	54.41 ^{b} ± 1.45	61.05 ^{b} ± 1.31	6034.50 ^{b} ± 08
Salt stress	48.50 ^{b} ± 2.10	56.96 ^{a} ± 1.23	55.37 ^{c} ± 1.11	5473.16 ^{c} ± 08
Heat stress	21.95 ^{d} ± 1.10	36.54 ^{e} ± 0.93	36.66 ^{d} ± 1.09	3623.47 ^{d} ± 09
Combined stress	22.80 ^{d} ± 0.98	34.82 ^{f} ± 1.03	23.36 ^{f} ± 0.78	2309.10 ^{f} ± 11

Note: lowercase letters show the significant differences among treatments according to one-way ANOVA and DMRT at a significance level of $p \leq 0.05$.

3.2.2. Effect of Salt Stress on Yield of the Wheat Plant

The effect of salt stress on the yield of the wheat plant is given in Table 1. The crop yield kg ha^{-1} was reduced significantly ($p \leq 0.05$) to 5473.16 kg ha^{-1}. It was not significantly ($p \leq 0.05$) less, but just half of the growth without stress, which was further correlated with the 1000-grain weight, which was reduced significantly ($p \leq 0.05$) from 56 g in control to 48.50 g once the salt stress was applied.

3.2.3. Effect of Heat Stress on Yield of the Wheat Plant

In the case of heat stress, the crop yield was 3623.47 kg ha^{-1}, while in control, it was 10270 kg ha^{-1} (Table 1). In other words, the crop yield was reduced by 75% after heat stress application. This was further related to the 1000-grain weight, which was reduced significantly ($p \leq 0.05$) from 56 g in control to 21.50 g once the stress was applied. The heat stress was significantly ($p \leq 0.05$) more deadly for the plant than salt and waterlogging stress, as shown in Table 1.

3.2.4. Effect of Combined Stress on Yield of the Wheat Plant

Table 1 shows the effect of combined stress on the yield of the wheat plant. The crop yield kg ha^{-1} was reduced significantly ($p \leq 0.05$) from 10,270 in the control group to 2309.10 in the stressed one. The 1000-grain weight was significantly ($p \leq 0.05$) reduced from 56.75 g in control to 22.80 g in stress conditions.

The correlation matrix (Table 2) shows the negative correlation between spike number and grain weight. The leaf area, photosynthesis, SPAD, tiller, and spike weight were positively correlated. The PCA matrix and plot (Table 3 and Figure 8) show the spike length, spike weight, and leaf area as the most prominent parameters affecting plant yield once the leaf area was reduced, which might have been due to the waterlogging stress. As a result, photosynthesis was reduced, and ultimately, spike weight was compromised. On the other hand, once spike weight was compromised, ultimately, grain weight was reduced, and finally, the plant yield was reduced.

Table 2. Correlation between morphological attributes of the wheat plants grown under abiotic stresses.

Correlation	PH	T	LA	SPAD	S.P	SL	SW	S.S	GW.S	G.S	IGW	TGW	GY.P	B.P	HI	Ymann.ac	Ykg.ha	Pearson	Color
PH		0.357	0.148	0.89	0.293	−0.075	0.531	−0.051	0.481	0.064	0.782	0.789	0.649	0.634	0.24	0.649	0.649	−1	
T	0.185		0.816	0.513	0.891	0.521	0.396	0.519	0.307	0.233	0.292	0.276	0.591	0.652	−0.002	0.591	0.591	−0.5	
LA	0.086	0.802		0.344	0.605	0.858	0.685	0.845	0.609	0.72	0.248	0.206	0.719	0.746	0.054	0.719	0.719	0	
SPAD	0.943	0.463	0.257		0.321	−0.049	0.663	−0.05	0.67	0.108	0.939	0.934	0.736	0.623	0.585	0.737	0.737	0.5	
S.P	0.12	0.904	0.598	0.359		0.439	0.175	0.435	0.049	0.083	0.098	0.113	0.469	0.601	−0.274	0.469	0.469	1	
SL	0.029	0.679	0.943	0.143	0.478		0.554	0.997	0.435	0.866	−0.108	−0.145	0.556	0.67	−0.326	0.555	0.555		
SW	0.371	0.494	0.829	0.429	0.239	0.714		0.534	0.981	0.799	0.737	0.708	0.94	0.844	0.45	0.94	0.94		
S.S	−0.029	0.626	0.899	0.058	0.485	0.986	0.638		0.408	0.854	−0.127	−0.165	0.538	0.666	−0.368	0.538	0.538		
GW.S	0.429	0.216	0.657	0.371	0	0.6	0.943	0.551		0.735	0.788	0.758	0.88	0.732	0.607	0.88	0.88		
G.S	0.429	0.216	0.657	0.371	0	0.6	0.943	0.551	1		0.188	0.148	0.688	0.697	0	0.687	0.687		
IGW	0.883	0.334	0.412	0.883	0.123	0.294	0.736	0.194	0.736	0.736		0.997	0.738	0.557	0.763	0.739	0.739		
TGW	0.943	0.185	0.257	0.886	0	0.2	0.6	0.116	0.657	0.657	0.971		0.726	0.547	0.751	0.726	0.726		
GY.P	0.486	0.617	0.714	0.6	0.478	0.486	0.886	0.406	0.771	0.771	0.765	0.6		0.957	0.315	1	1		
B.P	0.143	0.648	0.829	0.257	0.598	0.657	0.829	0.638	0.714	0.714	0.441	0.257	0.886		0.03	0.957	0.957		
HI	0.429	0	−0.029	0.486	−0.239	−0.257	0.371	−0.406	0.314	0.314	0.618	0.543	0.486	0.086		0.315	0.315		
Ymann.ac	0.486	0.617	0.714	0.6	0.478	0.486	0.886	0.406	0.771	0.771	0.765	0.6	1	0.886	0.486		1		
Ykg.ha	0.486	0.617	0.714	0.6	0.478	0.486	0.886	0.406	0.771	0.771	0.765	0.6	1	0.886	0.486	1			
Spearman Values	−1	−0.5	0	0.5	1														
Color																			

PH: plant height; T: tillers (FS); LA: leaf area (FS), SPAD: SPAD (15DAFS); S.P: spikes/plant; SL: spike length; SW: spike weight; S.S: spikelet/spike; GW.S: grain weight/spike; G.S: grains/spike; IGW: individual grain weight; TGW: thousand-grain weight; GY.P: grain yield/plant; B.P: biomass/plant; HI: harvest index; Ymann.ac: yield (Mann/acre); Ykg.ha: yield (kg ha^{-1}); FS: flowering stage; DAFS: days after flowering stage.

Table 3. Principal component analysis between morphology and abiotic stress responses.

Pattern Matrix	Component 1	Component 2
SL	1.091	−0.424
S.S	1.091	−0.441
LA	0.942	
G.S	0.865	
B.P	0.749	0.361
T	0.662	
S.P	0.628	
GY.P	0.587	0.584
Ymann.ac	0.587	0.584
Ykg.ha	0.587	0.584
TGW		1.07
IGW		1.062
SPAD		0.973
HI	−0.463	0.882
PH		0.834
GW.S	0.371	0.685
SW	0.523	0.595

Note: PH: plant height; T: tillers (FS); LA: leaf area (FS); SPAD: SPAD (15DAFS); S.P: spikes/plant; SL: spike length; SW: spike weight; S.S: spikelet/spike; GW.S: grain weight/spike; G.S: grains/spike; IGW: individual grain weight; TGW: thousand-grain weight; GY.P: grain yield/plant; B.P: biomass/plant; HI: harvest index; Ymann.ac: yield (Mann/acre); Ykg.ha: yield (kg ha^{-1}); FS: flowering stage; DAFS: days after flowering stage.

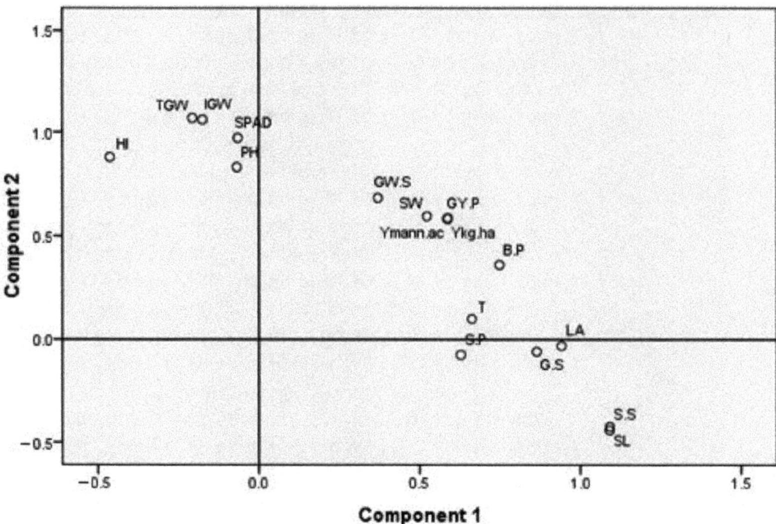

Figure 8. PCA plot showing chief components of plant stress responses. Note: PH: plant height; T: tillers (FS); LA: leaf area (FS); SPAD: SPAD (15DAFS); S.P: spikes/plant; SL: spike length; SW: spike weight; S.S: spikelet/spike; GW.S: grain weight/spike; G.S: grains/spike; IGW: individual grain weight; TGW: thousand-grain weight; GY.P: grain yield/plant; B.P: biomass/plant; HI: harvest index; Ymann.ac: yield (Mann/acre); Ykg.ha: yield (kg ha^{-1}); FS: flowering stage; DAFS: days after flowering stage.

3.2.5. NPK in the Plant and Stress Response

Fertilizer is an essential resource for plant growth. The results of NPK in wheat plants grown under different abiotic stresses and with SCU application are shown in Table 4. It was revealed that NPK was highest in control and lowest in the case of heat and combined stresses. At the heat stress application stage, the plant was already at the flowering stage, and %N was significantly ($p \leq 0.05$) reduced from 2.8 to 1.8%, P was reduced from 0.31 mg/kg to 0.24 mg/kg, and %K was reduced 1.2 to 0.98%. A 20% decrease in %K was observed in waterlogging stress, and it was the same in all treatments. Combined stressed accumulated the least nitrogen (1.33%) and P (0.98%) in plants among all treatments (Table 4).

Table 4. Phosphorous, potassium, and nitrogen accumulation in wheat plants grown under different abiotic stresses.

Treatments	P in Plant	K in Plant	N in Soil	N in Plant
	(mg/kg)	(%)	(%)	(%)
Control	0.31 ^{d} ± 0.01	1.2 ^{d} ± 0.12	0.08 ^{a} ± 0.001	2.8 ^{d} ± 0.2
Waterlogging	0.26 ^{c} ± 0.02	1.13 ^{c} ± 0.11	0.09 ^{b} ± 0.002	2.01 ^{c} ± 0.2
Salt stress	0.25 ^{b} ± 0.02	1.11 ^{b} ± 0.13	0.08 ^{a} ± 0.002	1.99 ^{c} ± 0.1
Heat stress	0.24 ^{b} ± 0.01	1.09 ^{b} ± 0.09	0.07 ^{a} ± 0.003	1.89 ^{b} ± 0.12
Combined Stress	0.21 ^{a} ± 0.01	0.98 ^{a} ± 0.01	0.09 ^{a} ± 0.001	1.33 ^{a} ± 0.3

Note: lowercase letters show the significant differences among treatments according to one-way ANOVA and DMRT at a significance level of $p \leq 0.05$.

4. Discussion

In the current experiment, the effect of different abiotic stresses was analyzed individually and in a combined form under the influence of SCU. It was found that the spike weight and root network were affected by all the stresses, but heat stress significantly ($p \leq 0.05$) reduced crop yield (Table 1). Our results of crop yield reduction due to heat stress were following the previous studies conducted on wheat and soya bean [7–9]. It was also found that the nitrogen release period was key to the stress alleviation for the plant at a growth point of ~120 days (that is, nitrogen release period of the fertilizer), as the plant was able to grow in all stressed environments. However, after this phase (at maturity and late flowering), plant growth was significantly ($p \leq 0.05$) reduced. Our studies observed correlations between growth parameters, such as spike number and grain weight leaf area, photosynthesis, SPAD, tiller, and spike weight (Table 2). Studies in the past also revealed that different growth parameters were related to each other, and changes in one growth parameter could cause alterations in the other parameters of the wheat plant [32]. The PCA matrix and plot (Table 3 and Figure 8) show the spike length, spike weight, and leaf area as the most prominent parameters affecting crop yield. Many studies have suggested the dry weight of wheat seedlings as the best principle for measuring the stress resistance attribute of the genotype [33,34]. A further detailed discussion of these results is given below.

The stress (waterlogging, salt stress, and heat) tolerance in the wheat plants is a matter of growth tolerance in plants, which initiate vigorous root systems and proliferate abundantly [1,2,33]. Our experiments also showed fewer tillers and reduced photosynthetic rate, fluorescence, dry biomass, leaf area, SPAD value, and crop yield, especially for the heat and salt stresses. However, these effects were moderate in waterlogging stress, especially for the first 120 days (Table 1, Figures 2–4 and 6). Salt stress was already reported to adversely affect plants' growth by Yu et al. [10]. In another study, Domico et al. [11] also reported reduced metabolic activity in the cardoon plant when subjected to long-term and short-term salt stress. The work of Pezo et al. further supported our results [12], reporting the reduced crop yield and pepper seed quality under salt stress.

Similarly, our results of reduced photosynthetic rate, leaf area, dry biomass, and yield were consistent with previous research conducted on wheat [7] and soya bean [8]. Waterlogging is undoubtedly reported to negatively impact a plant's physiological and biochemical response [16–18]. However, few studies also reported the innate adaptability of a plant against waterlogging stress [35]. It might be attributed to the capability of a plant to grow in hypoxic conditions that are mainly based on the root system tolerance of a plant [18,35]. It was hypothesized that the Yangmai 25 was able to build a good root network. That is why it was less affected by waterlogging stress than other stresses. In the first phase of waterlogging stress, the root growth might have been rapid or close to that of the control plants in Yangmai 25, but after ~120 days, the growth of the roots might have been compromised, which resulted in the decreased crop yield (Table 1). Once the root growth is clasped, the shoot growth can never recover to the control value for any genotypes [35]. Malik et al. [36] also suggested root growth recovery as a key to waterlogging stress tolerance. Similar results were also revealed by Ahmed et al. [37] while working on mung bean under waterlogging stress.

Now let us find out why the plant tolerated the stress for the first 120 days. The answer lies in the nitrogen release time of the SCU used in the current study, which was 120 days. In the first phase, wheat could withstand all the stresses due to the soil's high accessibility of nutrients and water availability [33,34]. Our findings disclosed that crops reached maturity phases a bit early; this might be due to controlled release/coated fertilizer [20]. The experiment showed that the N released by SCU attained maximum nitrogen content in the control plants, followed by waterlogging, salt stress, combined, and then heat stress. Our results agreed with Praharaj et al. [38] and Joshi et al. [39], who worked on pulses and rice, respectively. These studies found that adequate irrigation and controlled nitrogen supplies could induce extra productivity in plants in abiotic stress conditions. The coated urea also helped the wheat plant attain better grain weight (1000-grain weight) than the control (Table 1), but fertilizer was the least effective in heat stress. Similar results of enhanced grain weight in the wheat plant were observed by Ghafoor et al. [22] while applying SCU to alleviate stress on the plant in an arid climate.

In our experiment, heat stress significantly ($p \leq 0.05$) reduced plant growth among all stresses (Figures 2–6). This can be explained in terms of the heat sensitivity of the plant [8]. High-temperature stress is the leading environmental factor that limits wheat yield. Wheat yield decreases by 10% for every 1 °C increase above the mean temperature of 23 °C [6]. High-temperature stress affects more than 40% of the world's wheat area every year. It reduces wheat yield through chronic stress resulting from prolonged, relatively high temperatures up to 32 °C, or through heat-shock caused by abruptly but comparatively brief exposure to 33 °C and above. High temperatures cause changes in the physiological, biochemical, and molecular components of wheat crops. The high temperature might have initially accelerated the thylakoid membrane breakdown, resulting in electrolyte leakage and disruption of all electrochemical processes, particularly photosystem II (PS II)- and cytochrome f/b6-mediated reactions have resulted in a drastic decrease in the photosynthesis rate [40,41]. Wheat's PS II is more exposed to extreme temperature stress, as it is a winter season crop instead of a warm-season crop, such as rice and pearl millet (*Pennisetum glaucum*) [42].

One argument can be further made based on photophosphorylation; high-temperature stress also tends to cause a halt in photophosphorylation due to thylakoid membrane damage [43]. The rate of photosynthetic CO_2 assimilation was lower after stress application than in control (Figure 7b). This has previously been reported in various plant studies. The decrease in net photosynthesis could be attributed to changes in leaf water potential (Figure 7c), stomatal conductance (Figure 7b), the amount or activity of photosynthetic enzymes, and chlorophyll (Figure 4). According to some experiments, one possible factor in reducing photosynthesis in plants grown under heat stress is the accumulation of carbohydrates in leaves, indicating a feedback inhibition of photosynthesis [43,44]. Other studies have suggested that abiotic stress, particularly salt stress, reduces net respiratory activity in the roots, asserting a feedback mechanism that uses photosynthesis and inhibits plant growth [45,46]. Based on our findings, it appeared that stomatal closure may have contributed to the decreased photosynthetic rate in this experiment, particularly once the nitrogen source, i.e., SCU, was exhausted. Our results are supported by past literature, in which it was found that the nitrogen release timing of the fertilizer was key to the stress survival and nitrogen release from the slow-release fertilizer help in the recovery of the wheat plant once exposed to heat [8], salt stress [11], and waterlogging stress [17].

The wheat plant was able to grow under stress for the first 120 days. The controlled release of nitrogen fertilizer was key to the wheat plant growth, especially in waterlogged and salt stress. Many studies found that a higher grain yield can be attained by applying controlled-release SCU fertilization [23]. SCU increased grain yield efficiently by lowering rhizosphere pH, and the results were consistent with findings of a previous study [47]. Our findings also showed that using SCU N at a rate of 130 kg ha^{-1} resulted in better growth in the first 120 days in terms of leaf area, biomass gain, photosynthesis rate, fluorescence, SPAD value, etc. However, grain yield, number of grains per spike, grain weight, and harvest index were compromised. These results are supported by past studies that reported the effects of abiotic stress on cardoon [11], wheat genotype [32], and other cash crops [35].

The earlier phase of the experiment (120 days) revealed that controlled-release fertilizers increased total N percent with equal N level application vs. later stages once the nitrogen source, i.e., SCU, was exploited [22]. After stress application, grain yield was decreased by 9.58 to 11.21%, and N uptake was reduced by 19.06 to 23.94 % (Table 4). This was because protein contains nitrogen as an essential constituent, and N is involved in all vital processes of plants. For this reason, nitrogen application is both necessary and unavoidable for crop production [20]. The optimal soil N content increases photosynthetic processes, leaf area production, leaf area duration, and net assimilation rate [33,37]. Since crop yields have increased globally due to increased N use and good management practices [19], all plants, including cereals, oilseeds, fiber, and sugar-producing plants, require a balanced amount of nitrogen for vigorous growth and development in a larger harvest with higher quality. Nitrogen fertilization has also improved Pakistani crops' growth and yield parameters for crops such as wheat, rice, sugarcane, and cotton. Wheat growth and yield parameters such as plant height (cm), number of tillers (m^{-2}), number of spikelets ($spike^{-1}$), grains ($spike^{-1}$), and 1000-grain weight have been improved by nitrogen fertilization. Ali et al. [48] also demonstrated that coated urea fertilizer with higher nitrate contents and neem nitrification increased grain yield. Hence, SCU is effective under abiotic stresses once it can control nitrogen release but is no longer effective once the limit is reached. It is suggested to use a more advanced fertilizer with a better nitrogen release period of about 160 days.

Furthermore, screening should be done in soil rather than potting mix. The adverse effects in Vertosol soil were much more apparent and more representative of the actual situation on farms. The chlorophyll fluorescence model proved to be the most suitable for large-scale programs when choosing wheat genotypes for abiotic stress tolerance, requiring only a few seconds per sample. More specific studies at the cellular and tissue levels are needed to understand the fundamental physiological mechanisms fully.

5. Conclusions

The SCU fertilizer was applied at a recommended rate (130 kg ha^{-1}) to increase wheat stress tolerance. The experiment presented the positive possessions of SCU on wheat growth and development, physiological conditions, and nitrogen accumulation under different abiotic stress conditions. After 120 days, all stress types significantly ($p < 0.05$) reduced plant growth (leaf area, dry biomass, SPAD value, and the number of tillers). However, the crop yield was most compromised in the cases of heat and combined stress. The heat stress showed the lowest grain yield of 3623.47 kg ha^{-1}, while waterlogging stress showed a better yield of 6034.5 kg ha^{-1} in all stresses.

SCU, used in the current study, has a controlled nitrogen release time that meets nitrogen requirements for up to 120 days only. Hence, the wheat plant was able to tolerate salt and waterlogging stress to some extent. However, once the nitrogen source was exhausted at the time of heat stress, the plant could not tolerate heat stress. Therefore, it is suggested to use SCU with a longer release time to provide nitrogen until the wheat reaches the harvesting stage. In this way, wheat growers, especially farmers in developing countries, can use sustainable ecosystem practices in the salt-affected soils. There is a need to conduct studies on SCU with a release period of ~180 days, particularly in heat stress conditions. Another future recommendation for the study is to analyze the nitrogen losses and ecosystem benefits while using slow-release SCU instead of common nitrogen fertilizers. Future research studies, such as modeling ecosystem services and N loss under various crop and climate change scenarios, may also indicate the agricultural system's sustainability.

Supplementary Materials: The following are available online at https://www.mdpi.com/article/10.3390/agronomy11112340/s1, Figure S1. Setups for the waterlogging stress (a), control/salt stress (b), and heat stress (c) experiments. Figure S2. Schematic representation of the experimental design. Figure S3. Time bar graph of the whole study. Table S1. Photosynthetic attributes of the wheat plant. Table S2. Properties of the soil used in the study.

Author Contributions: Conceptualization, formal analysis, investigation, methodology, software, validation, visualization, writing—original draft, A.A.; conceptualization, funding acquisition, methodology, project administration, supervision, writing—review and editing, X.Z. (Xinkai Zhu); conceptualization, methodology, resources, supervision, writing—review and editing, M.Z.; investigation, M.Q.; formal analysis, writing—review and editing, S.I.; investigation, D.X.; formal analysis, writing—review and editing, M.A.; writing—review and editing, X.Z. (Xinbo Zhang); writing—review and editing, S.G.; writing—review and editing, F.L.; writing—review and editing, A.Z.S.; writing—review and editing, A.Z. All authors have read and agreed to the published version of the manuscript.

Funding: This work was jointly supported by the earmarked fund for Jiangsu Agricultural Industry Technology System (JATS[2021]503), the National Key Research and Development Program of China (2018YFD0200500), the National Natural Science Foundation of China (31901433, 31771711), Jiangsu Modern Agricultural (Wheat) Industry Technology System, Pilot Projects of the Central Cooperative Extension Program for Major Agricultural Technologies, The Priority Academic Program Development of Jiangsu Higher Education Institutions, and The Science and Technology Innovation Team of Yangzhou University, Yangzhou, China.

Institutional Review Board Statement: Not applicable.

Informed Consent Statement: Not applicable.

Conflicts of Interest: The authors declare no conflict of interest.

References

1. Ding, J.; Huang, Z.; Zhu, M.; Li, C.; Zhu, X.; Guo, W. Does cyclic water stress damage wheat yield more than a single stress? *PLoS ONE* **2018**, *13*, e0195535. [CrossRef] [PubMed]
2. Liu, L.; Xia, Y.; Liu, B.; Chang, C.; Xiao, L.; Shen, J.; Tang, L.; Cao, W.; Zhu, Y. Individual and combined effects of jointing and booting low-temperature stress on wheat yield. *Eur. J. Agron.* **2020**, *113*, 125989. [CrossRef]
3. Anosheh, H.P.; Emam, Y.; Ashraf, M.; Foolad, M. Exogenous application of salicylic acid and chlormequat chloride alleviates negative effects of drought stress in wheat. *Adv. Stud. Biol.* **2012**, *4*, 501–520.
4. Searchinger, T.; Hanson, C.; Ranganathan, J.; Lipinski, B.; Waite, R.; Winterbottom, R.; Dinshaw, A.; Heimlich, R.; Boval, M.; Chemineau, P. *Creating a Sustainable Food Future: A Menu of Solutions to Feed Nearly 10 Billion People by 2050*; Final Report; WRI: Washington, DC, USA, 2019.
5. Altaf, A.; Gull, S.; Zhu, X.; Zhu, M.; Rasool, G.; Ibrahim, M.E.H.; Aleem, M.; Uddin, S.; Saeed, A.; Shah, A.Z. Study of the effect of peg-6000 imposed drought stress on wheat (*Triticum aestivum* L.) cultivars using relative water content (RWC) and proline content analysis. *Pak. J. Agric. Sci.* **2021**, *58*, 357–367.
6. Prasad, P.V.; Pisipati, S.; Ristic, Z.; Bukovnik, U.; Fritz, A. Impact of nighttime temperature on physiology and growth of spring wheat. *Crop. Sci.* **2008**, *48*, 2372–2380. [CrossRef]
7. Sattar, A.; Sher, A.; Ijaz, M.; Ul-Allah, S.; Rizwan, M.S.; Hussain, M.; Jabran, K.; Cheema, M.A. Terminal drought and heat stress alter physiological and biochemical attributes in flag leaf of bread wheat. *PLoS ONE* **2020**, *15*, e0232974.
8. Cohen, I.; Zandalinas, S.I.; Fritschi, F.B.; Sengupta, S.; Fichman, Y.; Azad, R.K.; Mittler, R. The impact of water deficit and heat stress combination on the molecular response, physiology, and seed production of soybean. *Physiol. Plant.* **2021**, *172*, 41–52. [CrossRef]
9. Zahra, N.; Shaukat, K.; Hafeez, M.B.; Raza, A.; Hussain, S.; Chaudhary, M.T.; Akram, M.Z.; Kakavand, S.N.; Saddiq, M.S.; Wahid, A. Physiological and Molecular Responses to High, Chilling, and Freezing Temperature in Plant Growth and Production: Consequences and Mitigation Possibilities. In *Harsh Environment and Plant Resilience*; Springer: Berlin, Germany, 2021; p. 235.
10. Yu, G.; Zhang, X.; Ma, H. Changes in the physiological parameters of SbPIP1-transformed wheat plants under salt stress. *Int. J. Genom.* **2015**, *2015*, 384356.

11. Docimo, T.; De Stefano, R.; Cappetta, E.; Piccinelli, A.L.; Celano, R.; De Palma, M.; Tucci, M. Physiological, biochemical, and metabolic responses to short and prolonged saline stress in two cultivated cardoon genotypes. *Plants* **2020**, *9*, 554. [CrossRef]
12. Pezo, C.; Valdebenito, S.; Flores, M.F.; Oyanedel, E.; Vidal, K.; Neaman, A.; Peñaloza, P. Impact of Mother Plant Saline Stress on the Agronomical Quality of Pepper Seeds. *J. Soil Sci. Plant Nutr.* **2020**, *20*, 2600–2605. [CrossRef]
13. Benidire, L.; El Khalloufi, F.; Oufdou, K.; Barakat, M.; Tulumello, J.; Ortet, P.; Heulin, T.; Achouak, W. Phytobeneficial bacteria improve saline stress tolerance in Vicia faba and modulate microbial interaction network. *Sci. Total Environ.* **2020**, *729*, 139020. [CrossRef]
14. Galicia-Campos, E.; Ramos-Solano, B.; Montero-Palmero, M.; Gutierrez-Mañero, F.J.; García-Villaraco, A. Management of Plant Physiology with Beneficial Bacteria to Improve Leaf Bioactive Profiles and Plant Adaptation under Saline Stress in *Olea europea* L. *Foods* **2020**, *9*, 57. [CrossRef]
15. Bhusal, N.; Lee, M.; Han, A.R.; Han, A.; Kim, H.S. Responses to drought stress in Prunus sargentii and Larix kaempferi seedlings using morphological and physiological parameters. *For. Ecol. Manag.* **2020**, *465*, 118099. [CrossRef]
16. Wu, L.; Han, X.; Islam, S.; Zhai, S.; Zhao, H.; Zhang, G.; Cui, G.; Zhang, F.; Han, W.; You, X. Effects of sowing mode on lodging resistance and grain yield in winter wheat. *Agronomy* **2021**, *11*, 1378. [CrossRef]
17. Wang, X.; He, Y.; Zhang, C.; Tian, Y.-a.; Lei, X.; Li, D.; Bai, S.; Deng, X.; Lin, H. Physiological and transcriptional responses of Phalaris arundinacea under waterlogging conditions. *J. Plant Physiol.* **2021**, *261*, 153428. [CrossRef]
18. Bhusal, N.; Kim, H.S.; Han, S.-G.; Yoon, T.-M. Photosynthetic traits and plant–water relations of two apple cultivars grown as bi-leader trees under long-term waterlogging conditions. *Environ. Exp. Bot.* **2020**, *176*, 104111. [CrossRef]
19. Zörb, C.; Ludewig, U.; Hawkesford, M.J. Perspective on wheat yield and quality with reduced nitrogen supply. *Trends Plant Sci.* **2018**, *23*, 1029–1037. [CrossRef]
20. Agami, R.A.; Alamri, S.A.; Abd El-Mageed, T.; Abousekken, M.; Hashem, M. Role of exogenous nitrogen supply in alleviating the deficit irrigation stress in wheat plants. *Agric. Water Manag.* **2018**, *210*, 261–270. [CrossRef]
21. Gooding, M.; Pinyosinwat, A.; Ellis, R. Responses of wheat grain yield and quality to seed rate. *J. Agric. Sci.* **2002**, *138*, 317–331. [CrossRef]
22. Ghafoor, I.; Habib-ur-Rahman, M.; Ali, M.; Afzal, M.; Ahmed, W.; Gaiser, T.; Ghaffar, A. Slow-release nitrogen fertilizers enhance growth, yield, NUE in wheat crop and reduce nitrogen losses under an arid environment. *Environ. Sci. Pollut. Res.* **2021**, *28*, 43528–43543. [CrossRef]
23. Eghbali Babadi, F.; Yunus, R.; Abbasi, A.; Masoudi Soltani, S. Response surface method in the optimization of a rotary pan-equipped process for increased efficiency of slow-release coated urea. *Processes* **2019**, *7*, 125. [CrossRef]
24. Watson, M. *Plant Analysis Handbook: A Practical Sampling, Preparation, Analysis, and Interpretation Guide*; Micro Macro International Inc.: Athens, GA, USA, 1992; Volume 1, p. 82.
25. Chen, Z.; Zhou, M.; Newman, I.A.; Mendham, N.J.; Zhang, G.; Shabala, S. Potassium and sodium relations in salinised barley tissues as a basis of differential salt tolerance. *Funct. Plant Biol.* **2007**, *34*, 150–162. [CrossRef] [PubMed]
26. Wang, Z.-M.; Wei, A.-L.; Zheng, D.-M. Photosynthetic characteristics of non-leaf organs of winter wheat cultivars differing in ear type and their relationship with grain mass per ear. *Photosynthetica* **2001**, *39*, 239–244. [CrossRef]
27. Lu, Y.; Yan, Z.; Li, L.; Gao, C.; Shao, L. Selecting traits to improve the yield and water use efficiency of winter wheat under limited water supply. *Agric. Water Manag.* **2020**, *242*, 106410. [CrossRef]
28. Jixin, W.C.L. Spectraphotometry with Vanadium Phosphomolybdate-Nile Blue Multicomponent Complex and Determination of Vanadium in Monomineral Silicate. *Chin. J. Anal. Chem.* **1984**, *10*, 37–40.
29. Olsen, S.R. *Estimation of Available Phosphorus in Soils by Extraction with Sodium Bicarbonate*; US Department of Agriculture: Washington, DC, USA, 1954.
30. WHEELER, O.H.; MATEOS, J.L. The Ultraviolet Absorption of Isolated Double Bonds1. *J. Org. Chem.* **1956**, *21*, 1110–1112. [CrossRef]
31. d Steel, R.G.; Torrie, J.H. *Principles and Procedures of Statistics: A Biometrical Approach*; McGraw-Hill: New York, NY, USA, 1986.
32. Schirrmann, M.; Giebel, A.; Gleiniger, F.; Pflanz, M.; Lentschke, J.; Dammer, K.-H. Monitoring agronomic parameters of winter wheat crops with low-cost UAV imagery. *Remote Sens.* **2016**, *8*, 706. [CrossRef]
33. Ahmed, H.G.M.-D.; Zeng, Y.; Yang, X.; Anwaar, H.A.; Mansha, M.Z.; Hanif, C.M.S.; Ikram, K.; Ullah, A.; Alghanem, S.M.S. Conferring drought-tolerant wheat genotypes through morpho-physiological and chlorophyll indices at seedling stage. *Saudi J. Biol. Sci.* **2020**, *27*, 2116–2123. [CrossRef]
34. Ahmed, H.G.M.-D.; Sajjad, M.; Li, M.; Azmat, M.A.; Rizwan, M.; Maqsood, R.H.; Khan, S.H. Selection criteria for drought-tolerant bread wheat genotypes at seedling stage. *Sustainability* **2019**, *11*, 2584. [CrossRef]
35. Ahmed, F.; Rafii, M.; Ismail, M.R.; Juraimi, A.S.; Rahim, H.; Asfaliza, R.; Latif, M.A. Waterlogging tolerance of crops: Breeding, mechanism of tolerance, molecular approaches, and future prospects. *BioMed Res. Int.* **2013**, *2013*, 963525. [CrossRef]
36. Malik, A.I.; Colmer, T.D.; Lambers, H.; Schortemeyer, M. Changes in physiological and morphological traits of roots and shoots of wheat in response to different depths of waterlogging. *Funct. Plant Biol.* **2001**, *28*, 1121–1131. [CrossRef]
37. Ahmed, S.; Nawata, E.; Sakuratani, T. Effects of waterlogging at vegetative and reproductive growth stages on photosynthesis, leaf water potential and yield in mungbean. *Plant Prod. Sci.* **2002**, *5*, 117–123. [CrossRef]
38. Praharaj, C.; Singh, U.; Singh, S.; Singh, N.; Shivay, Y. Supplementary and life-saving irrigation for enhancing pulses production, productivity and water-use efficiency in India. *Indian J. Agron.* **2016**, *61*, 249–261.

39. Joshi, R.; Sahoo, K.K.; Tripathi, A.K.; Kumar, R.; Gupta, B.K.; Pareek, A.; Singla-Pareek, S.L. Knockdown of an inflorescence meristem-specific cytokinin oxidase–OsCKX2 in rice reduces yield penalty under salinity stress condition. *Plant Cell Environ.* **2018**, *41*, 936–946. [CrossRef]
40. Sarkar, J.; Chakraborty, B.; Chakraborty, U. Plant growth promoting rhizobacteria protect wheat plants against temperature stress through antioxidant signalling and reducing chloroplast and membrane injury. *J. Plant Growth Regul.* **2018**, *37*, 1396–1412. [CrossRef]
41. Pospíšil, P. Production of reactive oxygen species by photosystem II as a response to light and temperature stress. *Front. Plant Sci.* **2016**, *7*, 1950. [CrossRef] [PubMed]
42. Brestic, M.; Zivcak, M. PSII fluorescence techniques for measurement of drought and high temperature stress signal in crop plants: Protocols and applications. In *Molecular Stress Physiology of Plants*; Springer: Berlin, Germany, 2013; pp. 87–131.
43. Narayanan, S. Membrane Fluidity and Compositional Changes in Response to High Temperature Stress in Wheat. In *Physiological, Molecular, and Genetic Perspectives of Wheat Improvement*; Springer: Cham, Switzerland, 2021; pp. 115–123.
44. Morsy, M.R.; Jouve, L.; Hausman, J.-F.; Hoffmann, L.; Stewart, J.M. Alteration of oxidative and carbohydrate metabolism under abiotic stress in two rice (Oryza sativa L.) genotypes contrasting in chilling tolerance. *J. Plant Physiol.* **2007**, *164*, 157–167. [CrossRef] [PubMed]
45. Hasanuzzaman, M.; Bhuyan, M.; Zulfiqar, F.; Raza, A.; Mohsin, S.M.; Mahmud, J.A.; Fujita, M.; Fotopoulos, V. Reactive oxygen species and antioxidant defense in plants under abiotic stress: Revisiting the crucial role of a universal defense regulator. *Antioxidants* **2020**, *9*, 681. [CrossRef] [PubMed]
46. Demirel, U.; Morris, W.L.; Ducreux, L.J.; Yavuz, C.; Asim, A.; Tindas, I.; Campbell, R.; Morris, J.A.; Verrall, S.R.; Hedley, P.E. Physiological, biochemical, and transcriptional responses to single and combined abiotic stress in stress-tolerant and stress-sensitive potato genotypes. *Front. Plant Sci.* **2020**, *11*, 169. [CrossRef]
47. Shivay, Y.; Pooniya, V.; Prasad, R.; Pal, M.; Bansal, R. Sulphur-coated urea as a source of sulphur and an enhanced efficiency of nitrogen fertilizer for spring wheat. *Cereal Res. Commun.* **2016**, *44*, 513–523. [CrossRef]
48. Ali, K.; Munsif, F.; Zubair, M.; Hussain, Z.; Shahid, M.; Din, I.U.; Khan, N. Management of organic and inorganic nitrogen for different maize varieties. *Sarhad J. Agric.* **2011**, *27*, 525–529.

Review

Role of Glycine Betaine in the Thermotolerance of Plants

Faisal Zulfiqar [1,*], Muhammad Ashraf [2] and Kadambot H. M. Siddique [3,*]

[1] Department of Horticultural Sciences, Faculty of Agriculture and Environment, The Islamia University of Bahawalpur, Bahawalpur 63100, Pakistan
[2] Institute of Molecular Biology and Biotechnology, The University of Lahore, Lahore 54000, Pakistan; ashrafbot@yahoo.com
[3] The UWA Institute of Agriculture, The University of Western Australia, Perth, WA 6001, Australia
* Correspondence: ch.faisal.zulfiqar@gmail.com (F.Z.); kadambot.siddique@uwa.edu.au (K.H.M.S.)

Abstract: As global warming progresses, agriculture will likely be impacted enormously by the increasing heat stress (HS). Hence, future crops, especially in the southern Mediterranean regions, need thermotolerance to maintain global food security. In this regard, plant scientists are searching for solutions to tackle the yield-declining impacts of HS on crop plants. Glycine betaine (GB) has received considerable attention due to its multiple roles in imparting plant abiotic stress resistance, including to high temperature. Several studies have reported GB as a key osmoprotectant in mediating several plant responses to HS, including growth, protein modifications, photosynthesis, gene expression, and oxidative defense. GB accumulation in plants under HS differs; therefore, engineering genes for GB accumulation in non-accumulating plants is a key strategy for improving HS tolerance. Exogenous application of GB has shown promise for managing HS in plants, suggesting its involvement in protecting plant cells. Even though overexpressing GB in transgenics or exogenously applying it to plants induces tolerance to HS, this phenomenon needs to be unraveled under natural field conditions to design breeding programs and generate highly thermotolerant crops. This review summarizes the current knowledge on GB involvement in plant thermotolerance and discusses knowledge gaps and future research directions for enhancing thermotolerance in economically important crop plants.

Keywords: abiotic stresses; heat stress; high temperature; osmolytes; heat tolerance; transgenic plants

Citation: Zulfiqar, F.; Ashraf, M.; Siddique, K.H.M. Role of Glycine Betaine in the Thermotolerance of Plants. *Agronomy* **2022**, *12*, 276. https://doi.org/10.3390/agronomy12020276

Academic Editors: Channapatna S. Prakash, Ali Raza, Xiling Zou and Daojie Wang

Received: 18 November 2021
Accepted: 20 January 2022
Published: 21 January 2022

1. Introduction

Temperature can adversely affect the normal functioning of plant metabolism [1,2]. In the last few decades, climate change-induced rising temperatures have beome a major challenge for modern crop production, especially in southern Mediterranean regions [3]. Thus, efforts to achieve maximum crop yields to ensure food security for an ever-increasing human population remain challenging. Global food security could be jeopardized if mitigation efforts are not implemented aggressively [4]. Any fluctuation in an environmental cue such as temperature can directly affect the production of temperature-sensitive crops, and hence food security, causing considerable losses to growers and other stakeholders. According to the Inter-Governmental Panel on Climate Change, global temperatures increased by 0.74 °C within a century (1906–2005), which was ascribed to ongoing anthropogenic activities and their resultant greenhouse gases [5]. The terrestrial surface temperature is expected to increase by a further 1–6 °C by 2050, with arable areas projected to see the greatest increases [6]. There is little doubt that the ongoing climate change will affect all societies inhabiting the globe [7,8], decreasing crop production, and threatening food security. Therefore, strategies are needed to ameliorate its effects on modern agriculture.

Heat stress is a condition of unfavorable temperature causing irreversible damage to plants [9]. High-temperature stress adversely affects plant physio-biochemical and molecular characteristics, resulting in poor plant growth and development [10]. At the physiological level, heat stress negatively influences photosynthesis by adversely affecting the oxygen evolving complex, photosystem II, RuBisCo, and energy-(ATP) producing processes [11,12]. Furthermore, heat stress-induced disturbance of the electron transport chain leads to excessive reactive oxygen species (ROS) production in different cell organelles, such as mitochondria and chloroplasts, causing severe damage to DNA and cell membranes by inducing lipid peroxidation and ultimately causing cell death 13]. Increasing the capacity of heat stress-induced excessive ROS scavenging is considered an efficient defense strategy for ameliorating heat stress in plants [13]. A plant's thermotolerance ability is attributed to enhanced plasma membrane thermostability and reduced toxic ROS levels [14]. Plants have naturally adapted various defense mechanisms to counteract harsh environmental conditions such as heat stress. These defense mechanisms include an antioxidant machinery, osmolyte accumulation, maintenance of membrane integrity, and increased biosynthesis of heat-shock proteins (HSPs) by upregulating their associated genes' expression [15,16]. These defense mechanisms are involved in cellular defense against heat stress. Osmolyte accumulation has a significant role in mediating stress tolerance in plants (reviewed in Zulfiqar et al. [17]). Various studies have reported enhanced GB accumulation in plants under heat stress, revealing the positive role of this osmolyte in heat stress tolerance [18–22]. Sorwong and Sakhonwasee [23] stated that exogenous GB application alleviated the heat stress-induced reduction in leaf gas exchange traits.

There are no critical reviews on the role of GB in heat stress tolerance. Here, we summarize the fundamental impact of GB in inducing heat stress tolerance in economically important crops.

2. GB Structure and Biosynthesis in Plants

Glycinebetaine is an *N,N,N*-trimethylglycine quaternary ammonium compound that is naturally produced in numerous living organisms, including plants [24]. At physiological pH, GB is electrically a neutral molecule, but dipolar in nature. There are two pathways related to GB biosynthesis in plants [25]. The initiating metabolites for these pathways are choline and glycine [26]. GB biosynthesis in plants occurs via N methylation of glycine or the oxidation of choline [27]. GB biosynthesis, particularly in higher plants, is a two-step pathway beginning with choline, which is catalyzed by a ferredoxin-dependent Rieske-type protein, namely, choline monooxygenase (CMO), and by a soluble NAD+-dependent enzyme [28,29]. Betaine aldehyde is oxidized by NAD^+-dependent betaine aldehyde dehydrogenase (*BADH*) to produce GB. Both *BADH* and CMO generally exist in chloroplast stroma [24,30] (Figure 1).

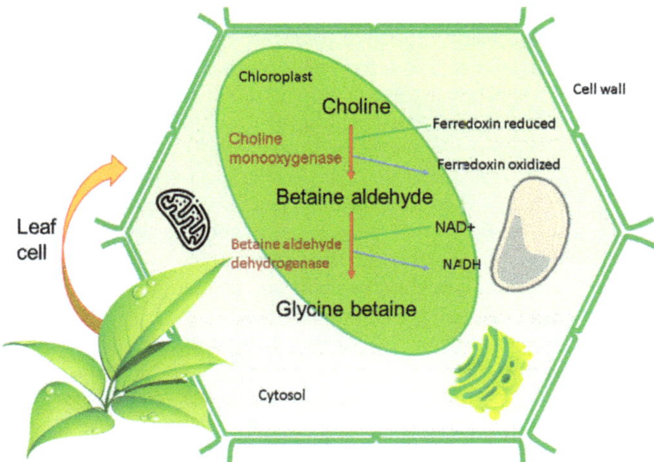

Figure 1. Biosynthesis of GB in plant cells (Figure adapted from Sakamoto and Murata [24]; Ashraf and Foolad [31]).

3. Glycine Betaine-Accumulating and -Non-Accumulating Plants

Glycine betaine is a vital compatible osmolyte that plays multifarious roles in plant growth and metabolism. However, plant species differ in naturally accumulating GB. It is now evident that GB synthesis occurs in both chloroplasts and cytosol [32,33], but chloroplastic GB has been positively related to stress tolerance, whereas cytosolic GB has not shown such a relationship [32]. Thus, the presence of high amounts of GB in a plant may not necessarily account for its enhanced stress tolerance. Metabolic restriction of GB synthesis in plants has been ascribed to the availability of the substrate (choline) [33]. Since choline occurs in the cytosol [34], its transport to chloroplasts through appropriate transporters is essential for GB synthesis [34,35]. GB-non-accumulating plants have either limited amounts of intrinsic choline or poor activity of choline transporters in the chloroplast envelope [35]. Thus, genetic engineers need this information to generate transgenic lines with high GB-accumulating ability.

Plants accumulate various amounts of GB; naturally accumulating plants under normal and stress conditions are categorized as GB accumulators, while non-accumulating plants do not increase GB level under stress [36]. Table 1 lists GB accumulators and non-accumulators. Natural GB accumulators accrue a certain amount of GB solely under stressful cues [18,21]. For example, Alhaithloul et al. [22] studied the responses of *Catharanthus roseus* and *Mentha piperita* under HS and drought stress, individually and in combination. They reported that the level of GB increased by 46% and 58% for *C. roseous* and *M. piperita*, respectively, in response to HS, and more so under combined heat and drought stress. This evolutionary adaptation enables plants to survive under a varied range of climatic conditions. Screening plants for their ability to accumulate osmolytes, particularly GB, offers a way to target such plants for acquiring GB biosynthesis-related genes to develop GB-producing plants.

Table 1. Glycine betaine accumulating crops.

GB Accumulators	Accumulating Condition	References
Plant families with known naturally high accumulation of GB: Asteraceae, Chenopodiaceae, Poaceae, and Solanaceae	Different types of stresses	[37–39]
Spinach (*Spinacia oleracea*)	Naturally accumulates under non-stress conditions; GB levels increase under stress conditions	[31,32,36]
Sugar beet (*Beta vulgaris*)	Naturally accumulates under non-stress conditions; GB levels increase under stress conditions	[31,32,36]
Barley (*Hordeum vulgare*)	Naturally accumulates under non-stress conditions; GB levels increase under stress conditions	[31,36,40]
Wheat (*Triticum aestivum*)	Naturally accumulates under non-stress conditions; GB levels increase under stress conditions	[31,36,40]
Sorghum (*Sorghum bicolor*)	Naturally accumulates under non-stress conditions; GB levels increase under stress conditions	[41,42]
Maize (*Zea mays*)	Naturally accumulates under non-stress conditions; GB levels increase under stress conditons	[36,43–45]
	GB-non-accumulators	
Rice (*Oryza sativa*)	Non-stress and stress conditions	[31,36,46,47]
Mustard (*Brassica* spp.)	Non-stress and stress conditions	[46]
Arabidopsis (*Arabidopsis thaliana*)	Non-stress and stress conditions	[33,36,46]
Tobacco (*Nicotiana tabacum*)	Non-stress and stress conditions	[33,46]
Tomato (*Solanum lycopersicum*)	Non-stress and stress conditions	[33,36,46,48]
Potato (*Solanum tuberosum*)	Non-stress and stress conditions	[36,46]

4. Mechanisms of GB-Mediated Thermotolerance

Glycine betaine is a compatible osmolyte that likely plays an important role in osmoregulation in plants subjected to extreme environmental cues, including high-temperature stress) [21,49]. Additionally, it is likely that GB activates signaling molecules such as calcium-dependent protein kinases (CDPKs) and mitogen-activated protein kinases (MAPKs) [50], which could activate stress-responsive and heat-shock transcription factor (HSF) genes [51,52]. The activated stress-responsive genes may boost the natural defense system by enhancing the activities of enzymatic antioxidants, such as superoxide dismutase (SOD), catalase (CAT), and peroxidase (POD), which may alleviate the negative impact of uncontrolled ROS causing oxidative damage triggered by heat stress)[53] (Figure 1). The elimination/reduction of ROS may keep biological membranes intact [54]. Furthermore, activated HSF genes may lead to the synthesis and activation of HSPs [55]. Most HSPs can also act as chaperones, which can prevent heat-induced aggregation of proteins [56]. The role of HSPs in plant thermotolerance has been elucidated in several comprehensive reviews [56,57]. GB can also significantly prevent photoinhibition by stabilizing the structure of the O_2-evolving center (PSII) [19,58]. Thus, overall, GB can stabilize photosynthesis in heat-stressed plants, promoting growth under heat stress.

It is now evident that high temperatures cause many metabolic changes in plants that involve intricate reprogramming of cellular activities to safeguard organellar ultrastructures and functions under heat stress [59]. Although some promising roles of GB are depicted in Figure 1 for counteracting heat-induced physiological disorders, intensive research is needed to elucidate how and to what extent GB can regulate some key processes involved in plant thermotolerance, other than those highlighted in Figure 2. For example, very little information is available on the crosstalk between GB and other biomolecules, including various plant growth regulators.

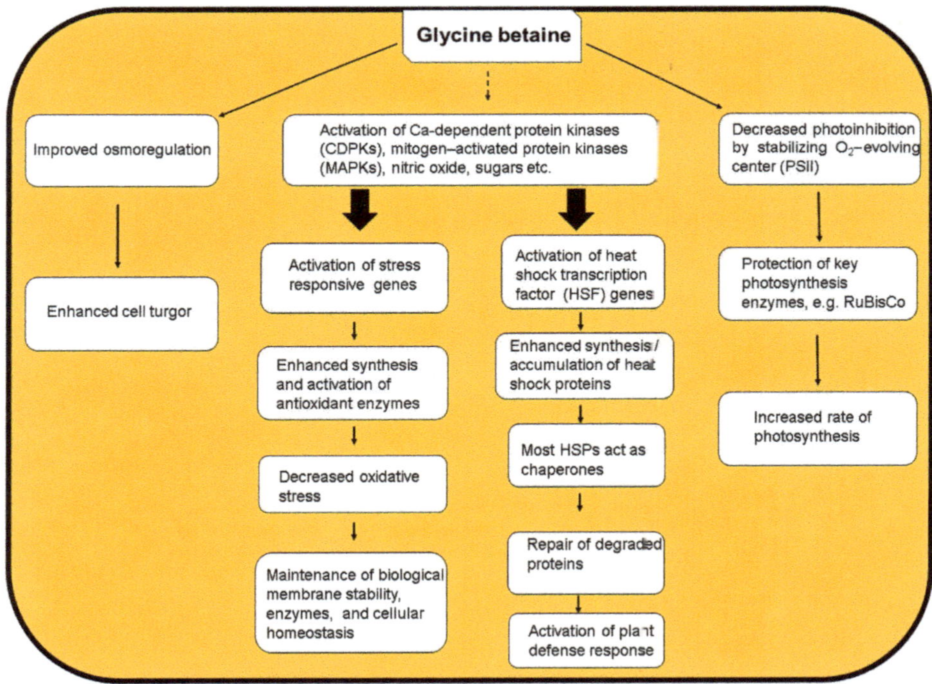

Figure 2. Proposed mechanism of glycine betaine-mediated thermotolerance in plants.

5. Improvement in Heat Tolerance through Exogenous Application of GB

Exogenous application of GB improves thermotolerance in many plants by enhancing their growth and yield via counteracting metabolic maladjustments caused by HS (Table 2). For example, while appraising the role of exogenous GB application on heat-stressed tomato plants, Li et al. [60] reported enhanced seed germination, expression of heat-shock genes, and accumulation of heat-shock protein 70 (HSP70). Likewise, exogenous GB supplementation likely controls many other key metabolic processes in heat-stressed plants. For example, exogenously applied GB protected photosystem II (PSII) in heat-stressed plants of *Hordeum vulgare* [61] and *Triticum aestivum* [20] and decreased the relative membrane permeability and leakage of ions such as Ca^{2+}, K^+, and NO_3^- in heat-stressed barley seedlings [62]. Furthermore, GB supplied to sprouting sugarcane nodal buds under HS markedly reduced H_2O_2 generation, increased K^+ and Ca^{2+} contents, and increased the levels of endogenous GB, free proline, and soluble sugars, enhancing the overall growth [63]. Sorwong and Sakhonwasee [23] stated that exogenous GB supplementation alleviated the heat stress-induced reduction in CO_2 assimilation rate, stomatal conductance, relative water content, and transpiration rate in marigold. The high-temperature-induced oxidative stress in marigold was mitigated due to reduced levels of H_2O_2, peroxide, superoxide, and lipid peroxidation [23]. Exogenous application of GB during mid-flowering in heat-stressed tomato in the field increased fruit yield [64]. In apple, the application of GB enhanced photosynthesis under individual HS or drought stress and combined stresses [65]. In a three-year field study, Chowdhury et al. [66] evaluated the role of GB and potassium nitrate in heat-stressed late-sown wheat; these osmolytes improved grain yield under heat stress compared to the control. Hence, it is clear that exogenously applied GB mediates HS. However, future studies should focus on field-based heat stress evaluations of different crops.

Table 2. Effect of exogenously applied GB on the regulation of different physio-biochemical attributes in heat-stressed plants.

Crop	Heat Stress Range	GB Concentration Applied	Exogenously Applied GB-Induced Regulation of Different Attributes in Heat-Stressed Plants	Reference
Tomato (*Lycopersicon esculentum*)	34 °C	0, 0.1, 1, and 5 mM GB	• Improved seed germination • Enhanced expression of heat-shock genes and accumulation of HSPs	[60]
Barley (*Hordeum vulgare*)	45 °C	10 mM	• Protective effect on oxygen-evolving complex by increasing connectivity among PSII antennae, inducing greater PSII stability	[61]
Wheat (*Triticum aestivum*)	25/20 °C day/night	100 mM	• Maintenance of higher chlorophyll content, PSII photochemical activity, and net photosynthetic rate • Reversed the decline in *psbA* gene transcription • Accelerated endogenous accumulation of GB in leaves • Decreased photoinhibition by D1 protein synthesis	[20]
Wheat (*T. aestivum*)	30–38 °C	100 and 50 mM	• Improved yield • Marginally improved the relative membrane permeability	[66]
Barley (*H. vulgare*)	40/32 °C day/night	10, 20, 30, 40, and 50 mM	• Improved growth, photosynthesis, and water relations • Decrease ion leakage	[62]

Table 2. *Cont.*

Crop	Heat Stress Range	GB Concentration Applied	Exogenously Applied GB-Induced Regulation of Different Attributes in Heat-Stressed Plants	Reference
Sugarcane (*Saccharum* spp.)	42 °C	20 mM	• Improved bud sprouting • Decreased H_2O_2 production • Improved soluble sugar accumulation • Improved K^+ and Ca^{2+} content • Enhanced the endogenous levels of osmolytes	[63]
Marigold (*Tagetes erecta*)	39/29 °C day/night	0.5 and 1 mM	• Improved leaf gas exchange traits • Decreased ROS accumulation	[23]

Abbreviations: HSPs: Heat-shock proteins, PSII: Photosystem II, ROS: Reactive oxygen species, *psbA*: PSII protein D1 precursor gene.

6. Genetic Engineering for Enhanced Thermotolerance

Developing transgenic plants for thermotolerance is a cost-effective and efficient biotechnological approach for achieving optimum agricultural production under the changing climate scenario [17]. Genes involved in encoding GB biosynthetic enzymes in different organisms and plants have been cloned to produce transgenic plants overexpressing one or more of these genes to enhance endogenous GB production, improving HS tolerance [67,68] (Table 3). For example, Zhang et al. [68] compared the thermotolerance ability of two transgenic tomato lines containing the betaine aldehyde dehydrogenase (*BADH*) and choline oxidase (*COD*) genes responsible for GB synthesis. They observed that *codA* transgenic plants had higher GB levels, CO_2 assimilation rate, and photosystem II (PSII) photochemical activity and lower accumulation of H_2O_2, $O_2^{\bullet-}$, and malondialdehyde (MDA) than wild-type (WT) plants. In addition, the codA transgenic line had higher heat-response gene expression, heat-shock protein 70 (HSP70) accumulation, and expression of a mitochondrial small heat-shock protein (MT-sHSP), heat-shock cognate 70 (HSC70), and heat-shock protein 70 (HSP70) than WT plants under HS. In another study, Yang et al. [69] reported GB accumulation in tobacco by introducing the *BADH* gene in tobacco, which increased tolerance to high-temperature stress and improved photosynthesis. While transferring the *BADH* gene from spinach to tomato, Li et al. [67] reported enhanced accumulation of GB, antioxidant activity, and photosynthetic capacity by improving D1 protein content, which could repair heat stress-induced damage to PSII. Furthermore, transgenic tomato accumulated less MDA and ROS (H_2O_2 and $O_2^{\bullet-}$), reducing oxidative stress relative to the WT. Reduced oxidative stress and restored PSII from HS-induced enhanced photoinhibition occurred in transgenic tobacco plants transformed with the *BADH* gene relative to the WT [70]. The role of *BADH* in xerophyllic *Ammopiptanthus nanus* under severe stress was confirmed by transferring the *BADH* gene of this plant into *E. coli* treated with 700 mM NaCl at 55 °C; the transgenic bacteria showed tolerance to these combined stresses [71]. The above studies reveal the positive role of GB-related genes in providing stress tolerance in plants. The introgression of GB synthesis-related genes could enhance endogenous GB accumulation to protect plants from heat-induced oxidative stress.

Numerous studies have been conducted on engineering GB biosynthesis [33]. Performance of single-gene-based transgenics under field conditions may not be the same as that reported under controlled or semi-controlled conditions. Thus, the development of transgenic lines by transforming with multiple genes (pyramiding of genes) for enhanced GB biosynthesis under heat stress is plausible.

Table 3. Genetic engineering for enhanced GB accumulation and improved thermotolerance in different crops.

Gene Trans-formed	Donor/Source	Gene Action	Transgenic Plant Species	Stress Condition	Transgenic Plant Response	References
BADH and *codA*	Spinach (*Spinacia oleracea* L.) as a donor of the *BADH* gene; binary vector pCG/codA for chloroplast-targeted expression of the *codA* gene	Genes related to key enzymes involved in GB synthesis	Tomato (*Solanum lycopersicum*)	Two months after transplanting, plants were exposed to 42 °C for 0–8 h in a growth chamber	• Enhanced GB levels in leaves • Improved CO_2 assimilation and photosystem II (PSII) photochemical activity • Transgenic plants, especially those containing *codA*, accumulated low ROS levels and thus exhibited reduced oxidative stress • Enhanced expression of heat-response genes and accumulation of heat-shock protein 70 (HSP70) • Enhanced thermotolerance	[68]
BADH	Spinach (*S. oleracea* L.)	Gene for betaine aldehyde dehydrogenase	Tobacco (*Nicotiana tabacum*)	Two-month-old seedlings subjected to various temperatures (25–50 °C) for 4 h in a growth chamber	• Increased GB accumulation in vivo increased PSII tolerance under heat stress • Enhanced GB accumulation in vivo improved the thermostability of PSII reaction centers • Reversed heat-induced PSII photoinhibition • Accumulated low levels of ROS, improving oxidative defense system	[70]
BADH	Spinach (*S. oleracea* L.)	Gene for betaine aldehyde dehydrogenase	Tomato (*S. lycopersicum*)	Six-week-old seedlings placed in a growth chamber at 42 °C for 0, 2, 4, or 6 h	• Higher GB accumulation • Enhanced chlorophyll fluorescence • Increased tolerance to heat-induced photoinhibition • Improved D1 protein content • Enhanced antioxidant enzyme activities • Reduced oxidative stress	[67]

Table 3. *Cont.*

Gene Transformed	Donor/Source	Gene Action	Transgenic Plant Species	Stress Condition	Transgenic Plant Response	References
BADH	Spinach (*S. oleracea* L.)	—do—	Tomato (*Solanum lycopersicum*)	Two-month-old plants exposed to varying temperature regimes (25–45 °C) for 2 h	• Enhanced CO_2 assimilation • Enhanced photosynthesis related to RuBisCo activase-mediated activation of RuBisCo • Enhanced thermotolerance	[69]
BADH	Garden orache (*Atriplex hortensis* L.)	–do–	Wheat (*Triticum aestivum*)	Individual and combined heat stress (40 °C) and drought stress as PEG-6000 (osmotic potential about −1.88 MPa) for 3 h in an artificial chamber	• Improved photosynthesis • Improved antioxidant activities/levels • Improved water status	[72]
codA	Spinach (*S. oleracea* L.)	Gene encodescholine oxidase (COD), a key enzyme for GB synthesis	Rice (*Oryza sativa*)	Plants grown at 28/13 °C for 5 weeks	• Enhanced endogenous GB accumulation • Improved heat and salt stress tolerance	[47]

7. Conclusions and Prospects

Under the changing climate threat, strategies are needed to alleviate the adverse impacts of harsh environmental stresses such as HS on plants. The most expedient strategy is to use the plant's natural defense system for withstanding these stresses. Under HS, many plants naturally accumulate GB, a compound known to mediate HS tolerance via osmoregulation, photosynthetic mechanisms, and signaling molecules, such as CDPKs, MAPKs, nitric oxide (NO), and sugars, which activate stress-responsive genes and HSF genes.

As discussed above, GB biosynthesis has different roles in different organelles; for example, chloroplastic GB is actively involved in stress tolerance, while cytosolic GB lacks such functionality. As a result, high levels of GB in plants are not the only factor enhancing stress tolerance. The principal substrate for GB synthesis is choline, which occurs in the cytosol. Choline transport to the chloroplast takes place via choline transporters. Several problems occur in GB-non-accumulating plants: (1) a limited amount of endogenous choline exists and (2) choline transporters present on the chloroplast membrane do not transport the required level of choline to chloroplasts. Thus, molecular biologists require this information for different crops to develop transgenic lines/cultivars with enhanced GB-accumulating ability.

Plants that do not naturally accumulate GB under HS have less HS tolerance than those that do. Various strategies have been used to increase GB accumulation in these plants to improve their tolerance against HS. Exogenous supplementation of GB in different forms, such as seed priming or plant priming, has enhanced endogenous GB levels and thus improved plant growth and development under HS. Genetic engineering could be used to introduce biosynthetic pathway-related genes into GB-deficient species from plants or microorganisms. While various studies have demonstrated the importance of GB in thermotolerance, very few have revealed the molecular roles of GB in thermotolerance. Moreover, transgenic lines generated for different crops have been based on a single gene transformation, with marked progress in terms of enhanced GB accumulation coupled with improved thermotolerance. However, all these studies have been undertaken under semi-controlled or controlled conditions, and the developed transgenic lines have not been tested under natural field conditions. Thus, further research is needed to generate thermotolerant lines/cultivars for different crops threatened by the rising temperatures of climate change. The effectiveness of exogenous GB application should be tested at the field level.

Author Contributions: Conceptualization, F.Z. and M.A.; writing—original draft preparation; F.Z. writing—review and editing, F.Z., M.A. and K.H.M.S. All authors have read and agreed to the published version of the manuscript.

Funding: This research received no external funding.

Institutional Review Board Statement: Not applicable.

Informed Consent Statement: Not applicable.

Data Availability Statement: Not applicable.

Conflicts of Interest: The authors declare no conflict of interest.

References

1. Geange, S.R.; Arnold, P.A.; Catling, A.A.; Coast, O.; Cook, A.M.; Gowland, K.M.; Leigh, A.; Notarnicola, R.F.; Posch, B.C.; Venn, S.E.; et al. The thermal tolerance of photosynthetic tissues: A global systematic review and agenda for future research. *New Phytol.* **2020**, *229*, 2497–2513. [CrossRef]
2. Gil, K.E.; Park, C.M. Thermal adaptation and plasticity of the plant circadian clock. *New Phytol.* **2018**, *221*, 1215–1229. [CrossRef] [PubMed]
3. Wang, X.; Zhao, C.; Müller, C.; Wang, C.; Ciais, P.; Janssens, I.; Peñuelas, J.; Asseng, S.; Li, T.; Elliott, J.; et al. Emergent constraint on crop yield response to warmer temperature from field experiments. *Nat. Sustain.* **2020**, *3*, 908–916. [CrossRef]
4. Sadok, W.; Lopez, J.R.; Smith, K.P. Transpiration increases under high-temperature stress: Potential mechanisms, trade-offs and prospects for crop resilience in a warming world. *Plant Cell Environ.* **2021**, *44*, 2102–2116. [CrossRef]

5. IPCC. Climate change 2014: Impacts, adaptation and vulnerability. In *Working Group II Contribution to the Fifth Assessment Report of the Intergovernmental Panel on Climate Change*; Cambridge University Press: Cambridge, UK, 2014.
6. Ferguson, J.N.; Tidy, A.C.; Murchie, E.H.; Wilson, Z.A. The potential of resilient carbon dynamics for stabilizing crop reproductive development and productivity during heat stress. *Plant Cell Environ.* **2021**, *44*, 2066–2089. [CrossRef]
7. O'Neill, B.C.; Carter, T.R.; Ebi, K.; Harrison, P.A. Achievements and needs for the climate change scenario framework. *Nat. Clim. Chang.* **2020**, *10*, 1074–1084. [CrossRef] [PubMed]
8. Hertel, T.W.; de Lima, C.Z. Climate impacts on agriculture: Searching for keys under the streetlight. *Food Policy* **2020**, *95*, 101954. [CrossRef]
9. Wahid, A.; Gelani, S.; Ashraf, M.; Foolad, M.R. Heat tolerance in plants: An overview. *Environ. Exp. Bot.* **2007**, *61*, 199–223. [CrossRef]
10. Fahad, S.; Bajwa, A.A.; Nazir, U.; Anjum, S.A.; Farooq, A.; Zohaib, A.; Sadia, S.; Nasim, W.; Adkins, S.; Saud, S.; et al. Crop production under drought and heat stress: Plant responses and management options. *Front Plant Sci.* **2017**, *8*, 1147. [CrossRef]
11. Tan, S.L.; Yang, Y.J.; Liu, T.; Zhang, S.B.; Huang, W. Responses of photosystem I compared with photosystem II to combination of heat stress and fluctuating light in tobacco leaves. *Plant Sci.* **2020**, *292*, 110371. [CrossRef]
12. Parrotta, L.; Aloisi, I.; Faleri, C.; Romi, M.; Del Duca, S.; Cai, G. Chronic heat stress affects the photosynthetic apparatus of *Solanum lycopersicum* L. cv Micro-Tom. *Plant Physiol. Biochem.* **2020**, *154*, 463–475. [CrossRef] [PubMed]
13. Suzuki, N.; Katano, K. Coordination between ROS regulatory systems and other pathways under heat stress and pathogen attack. *Front Plant Sci.* **2018**, *9*, 490. [CrossRef] [PubMed]
14. Chen, W.L.; Yang, W.J.; Lo, H.F.; Yeh, D.M. Physiology, anatomy, and cell membrane thermostability selection of leafy radish (*Raphanus sativus* var. oleiformis Pers.) with different tolerance under heat stress. *Sci. Hortic.* **2014**, *179*, 367–375. [CrossRef]
15. Akter, N.; Islam, M.R. Heat stress effects and management in wheat. A review. *Agron. Sustain. Dev.* **2017**, *37*, 37. [CrossRef]
16. Waters, E.R.; Vierling, E. Plant small heat shock proteins–evolutionary and functional diversity. *New Phytol.* **2020**, *227*, 24–37. [CrossRef]
17. Zulfiqar, F.; Akram, N.A.; Ashraf, M. Osmoprotection in plants under abiotic stresses: New insights into a classical phenomenon. *Planta* **2020**, *251*, 3. [CrossRef] [PubMed]
18. Storey, R.; Ahmad, N.; Wyn Jones, R.G. Taxonomic and ecological aspects of the distribution of glycinebetaine and related compounds in plants. *Oecologia* **1977**, *27*, 319–332. [CrossRef]
19. Allakhverdiev, S.I.; Los, D.A.; Mohanty, P.; Nishiyama, Y.; Murata, N. Glycinebetaine alleviates the inhibitory effect of moderate heat stress on the repair of photosystem II during photoinhibition. *Biochim. Biophys. Acta (BBA)-Bioenerg.* **2007**, *1767*, 1363–1371. [CrossRef]
20. Wang, Y.; Liu, S.; Zhang, H.; Zhao, Y.; Zhao, H.; Liu, H. Glycine betaine application in grain filling wheat plants alleviates heat and high light-induced photoinhibition by enhancing the psbA transcription and stomatal conductance. *Acta Physiol. Plant* **2014**, *36*, 2195–2202. [CrossRef]
21. Annunziata, M.G.; Ciarmiello, L.F.; Woodrow, P.; Dell'Aversana, E.; Carillo, P. Spatial and temporal profile of glycine betaine accumulation in plants under abiotic stresses. *Front. Plant Sci.* **2019**, *10*, 230. [CrossRef]
22. Alhaithloul, H.A.; Soliman, M.H.; Ameta, K.L.; El-Esawi, M.A.; Elkelish, A. Changes in ecophysiology, osmolytes, and secondary metabolites of the medicinal plants of *Mentha piperita* and *Catharanthus roseus* subjected to drought and heat stress. *Biomolecules* **2020**, *10*, 43. [CrossRef]
23. Sorwong, A.; Sakhonwasee, S. Foliar application of glycine betaine mitigates the effect of heat stress in three marigold (*Tagetes erecta*) cultivars. *Hort. J.* **2015**, *48*, 161–171. [CrossRef]
24. Sakamoto, A.; Murata, N. The role of glycine betaine in the protection of plants from stress: Clues from transgenic plants. *Plant Cell Environ.* **2002**, *25*, 163–171. [CrossRef]
25. Hanson, A.D.; Rhodes, D. ^{14}C tracer evidence for synthesis of choline and betaine via phosphoryl base intermediates in salinized sugarbeet leaves. *Plant Physiol.* **1983**, *71*, 692–700. [CrossRef]
26. Weretilnyk, E.A.; Bednarek, S.; McCue, K.F.; Rhodes, D.; Hanson, A.D. Comparative biochemical and immunological studies of the glycine betaine synthesis pathway in diverse families of dicotyledons. *Planta* **1989**, *178*, 342–352. [CrossRef]
27. Chen, T.H.; Murata, N. Glycinebetaine: An effective protectant against abiotic stress in plants. *Trends Plant Sci.* **2008**, *13*, 499–505. [CrossRef] [PubMed]
28. Brouquisse, R.; Weigel, P.; Rhodes, D.; Yocum, C.F.; Hanson, A.D. Evidence for a ferredoxin-dependent choline monooxygenase from spinach chloroplast stroma. *Plant Physiol.* **1989**, *90*, 322–329. [CrossRef] [PubMed]
29. Hibino, T.; Waditee, R.; Araki, E.; Ishikawa, H.; Aoki, K.; Tanaka, Y.; Takabe, T. Functional characterization of choline monooxygenase, an enzyme for betaine synthesis in plants. *J. Biol. Chem.* **2002**, *277*, 41352–41360. [CrossRef] [PubMed]
30. Rathinasabapathi, B.; Burnet, M.; Russell, B.L. Choline monooxygenase, an unusual iron-sulfur enzyme catalyzing the first step of glycine betaine synthesis in plants: Prosthetic group characterization and cDNA Cloning. *Proc. Natl. Acad. Sci. USA* **1997**, *94*, 3454–3458. [CrossRef] [PubMed]
31. Ashraf, M.; Foolad, M.R. Roles of glycine betaine and proline in improving plant abiotic stress resistance. *Environ. Exp. Bot.* **2007**, *59*, 206–216. [CrossRef]
32. Chen, T.H.; Murata, N. Enhancement of tolerance of abiotic stress by metabolic engineering of betaines and other compatible solutes. *Curr. Opin. Plant Biol.* **2002**, *5*, 250–257. [CrossRef]

33. Mansour, M.M.F.; Ali, E.F. Glycinebetaine in saline conditions: An assessment of the current state of knowledge. *Acta Physiol. Plant.* **2017**, *39*, 56. [CrossRef]
34. Chen, T.H.; Murata, N. Glycinebetaine protects plants against abiotic stress: Mechanisms and biotechnological applications. *Plant Cell Environ.* **2011**, *34*, 1–20. [CrossRef] [PubMed]
35. Huang, J.; Hirji, R.; Adam, L.; Rozwadowski, K.L.; Hammerlindl, J.K.; Keller, W.A.; Selvaraj, G. Genetic engineering of glycinebetaine production toward enhancing stress tolerance in plants: Metabolic limitations. *Plant Physiol.* **2000**, *122*, 747–756. [CrossRef] [PubMed]
36. Rhodes, D.; Hanson, A. Quaternary ammonium and tertiary sulfonium compounds in higher plants. *Annu. Rev. Plant Physiol. Plant Mol. Biol.* **1993**, *44*, 357–384. [CrossRef]
37. Wyn Jones, R.G.; Storey, R. Betaines. In *The Physiology and Biochemistry of Drought Resistance in Plants*; Paleg, L.G., Aspinall, D., Eds.; Academic Press: Cambridge, MA, USA, 1981; pp. 171–204.
38. Ladyman, J.A.R. The accumulation of glycinebetaine in barley in relation to water stress. *Diss. Abs. Bull.* **1982**, *43*, 1320.
39. Wani, S.H.; Singh, N.B.; Haribhushan, A.; Mir, J.I. Compatible solute engineering in plants for abiotic stress tolerance role of glycine betaine. *Curr. Genom.* **2013**, *14*, 157–165. [CrossRef]
40. Tian, F.; Wang, W.; Liang, C.; Wang, X.; Wang, G.; Wang, W. Overaccumulation of glycine betaine makes the function of the thylakoid membrane better in wheat under salt stress. *Crop J.* **2017**, *5*, 73–82. [CrossRef]
41. Weimberg, R.; Lerner, H.R.; Poljakoff-Mayber, A. Changes in growth and water soluble solute concentrations in Sorghum bicolor stressed with sodium and potassium. *Physiol. Plant.* **1984**, *62*, 472–480. [CrossRef]
42. Yang, W.-J.; Rich, P.J.; Axtell, J.D.; Wood, K.V.; Bonham, C.C.; Ejeta, G.; Mickelbart, M.V.; Rhodes, D. Genotypic variation for glycine betaine in sorghum. *Crop Sci.* **2003**, *43*, 162–169. [CrossRef]
43. McCue, R.F.; Hanson, A.D. Drought and salt tolerance: Towards understanding and application. *Tibtech* **1990**, *8*, 358–362. [CrossRef]
44. Kishitani, S.; Watanabe, K.; Yasuda, S.; Arakawa, K.; Takabe, T.J.P.C. Accumulation of glycinebetaine during cold acclimation and freezing tolerance in leaves of winter and spring barley plants. *Plant Cell Environ.* **1994**, *17*, 89–95. [CrossRef]
45. Quan, R.; Shang, M.; Zhang, H.; Zhao, Y.; Zhang, J. Engineering of enhanced glycine betaine synthesis improves drought tolerance in maize. *Plant Biotechnol. J.* **2004**, *2*, 477–486. [CrossRef] [PubMed]
46. McNeil, S.D.; Nuccio, M.L.; Hanson, A.D. Betaines and related osmoprotectants. Targets for metabolic engineering of stress resistance. *Plant Physiol.* **1999**, *120*, 945–949. [CrossRef]
47. Shirasawa, K.; Takabe, T.; Takabe, T.; Kishitani, S. Accumulation of glycinebetaine in rice plants that overexpress choline monooxygenase from spinach and evaluation of their tolerance to abiotic stress. *Ann. Bot.* **2006**, *98*, 565–571. [CrossRef] [PubMed]
48. Mäkelä, P.; Munns, R.; Colmer, T.D.; Condon, A.G.; Peltonen-Sainio, P. Effect of foliar applications of glycinebetaine on stomatal conductance, abscisic acid and solute concentrations in leaves of salt-or drought-stressed tomato. *Funct. Plant Biol.* **1998**, *25*, 655–663. [CrossRef]
49. Al-Huqail, A.; El-Dakak, R.M.; Sanad, M.N.; Badr, R.H.; Ibrahim, M.M.; Soliman, D.; Khan, F. Effects of climate temperature and water Stress on plant growth and accumulation of antioxidant compounds in sweet basil (*Ocimum basilicum* L.) leafy vegetable. *Scientifica* **2020**, *2020*, 3808909. [CrossRef]
50. Hemantaranjan, A.; Bhanu, A.N.; Singh, M.N.; Yadav, D.K.; Patel, P.K.; Singh, R.; Katiyar, D. Heat stress responses and thermotolerance. *Adv. Plants Agric. Res.* **2014**, *1*, 62–70. [CrossRef]
51. Divya, K.; Bhatnagar-Mathur, P.; Sharma, K.K.; Reddy, P.S. Heat shock proteins (Hsps) mediated signalling pathways during abiotic stress conditions. In *Plant Signaling Molecules*; Woodhead Publishing: Shaston, UK, 2019; pp. 499–516.
52. Hashemi-Petroudi, S.H.; Nematzadeh, G.; Mohammadi, S.; Kuhlmann, M. Expression pattern analysis of heat shock transcription factors (HSFs) gene family in *Aeluropus littoralis* under salinity stress. *Environ. Stresses Crop Sci.* **2020**, *13*, 571–581.
53. Zulfiqar, F.; Ashraf, M. Bioregulators: Unlocking their potential role in regulation of the plant oxidative defense system. *Plant Mol. Biol.* **2021**, *105*, 11–41. [CrossRef]
54. Giri, J. Glycinebetaine and abiotic stress tolerance in plants. *Plant Signal. Behav.* **2011**, *6*, 1746–1751. [CrossRef]
55. Andrási, N.; Pettkó-Szandtner, A.; Szabados, L. Diversity of plant heat shock factors: Regulation, interactions, and functions. *J. Exp. Bot.* **2020**, *72*, 1558–1575. [CrossRef] [PubMed]
56. Jacob, P.; Hirt, H.; Bendahmane, A. The heat-shock protein/chaperone network and multiple stress resistance. *Plant Biotechnol. J.* **2017**, *15*, 405–414. [CrossRef]
57. Reddy, P.S.; Chakradhar, T.; Reddy, R.A.; Nitnavare, R.B.; Mahanty, S.; Reddy, M.K. Role of heat shock proteins in improving heat stress tolerance in crop plants. In *Heat Shock Proteins and Plants*; Springer: Cham, Switzerland, 2016; pp. 283–307.
58. Yang, G.; Rhodes, D.; Joly, R.J. Effects of high temperature on membrane stability and chlorophyll fluorescence in glycinebetaine-deficient and glycinebetaine-containing maize lines. *Funct. Plant Biol.* **1996**, *23*, 437–443. [CrossRef]
59. Jagadish, S.K.; Way, D.A.; Sharkey, T.D. Plant heat stress: Concepts directing future research. *Plant Cell Environ.* **2021**, *44*, 1992–2005. [CrossRef] [PubMed]
60. Li, S.; Li, F.; Wang, J.; Zhang, W.; Meng, Q.; Chen, T.H.H.; Murata, N.; Yang, X. Glycinebetaine enhances the tolerance of tomato plants to high temperature during germination of seeds and growth of seedlings. *Plant Cell Environ.* **2011**, *34*, 1931–1943. [CrossRef] [PubMed]

61. Oukarroum, A.; El Madidi, S.; Strasser, R.J. Exogenous glycine betaine and proline play a protective role in heat-stressed barley leaves (*Hordeum vulgare* L.): A chlorophyll a fluorescence study. *Plant Biosyst.—Int. J. Deal. All Asp. Plant Biol.* **2012**, *146*, 1037–1043.
62. Wahid, A.; Shabbir, A. Induction of heat stress tolerance in barley seedlings by pre-sowing seed treatment with glycinebetaine. *Plant Growth Regul.* **2005**, *46*, 133–141. [CrossRef]
63. Rasheed, R.; Wahid, A.; Farooq, M.; Hussain, I.; Basra, S.M. Role of proline and glycinebetaine pretreatments in improving heat tolerance of sprouting sugarcane (*Saccharum* sp.) buds. *Plant Growth Regul.* **2011**, *65*, 35–45. [CrossRef]
64. Makela, P.; Jokinen, K.; Kontturi, M.; Peltonen-Sainio, P.; Pehu, E.; Somersalo, S. Foliar application of glycine betaine—A novel product from sugar beet as an approach to increase tomato yield. *Ind. Crops Prod.* **1998**, *7*, 139–148. [CrossRef]
65. Wang, G.; Wang, J.; Xue, X.; Lu, C.; Chen, R.; Wang, L.; Han, X. Foliar spraying of glycine betaine lowers photosynthesis inhibition of Malus hupehensis leaves under drought and heat stress. *Int. J. Agric. Biol.* **2020**, *23*, 1121–1128.
66. Chowdhury, A.R.; Ghosh, M.; Lal, M.; Pal, A.; Hazra, K.K.; Acharya, S.; Chaurasiya, A.; Pathak, S.K. Foliar spray of synthetic osmolytes alleviates terminal heat stress in late-sown wheat. *Int. J. Plant Prod.* **2020**, *14*, 321–333. [CrossRef]
67. Li, M.; Li, Z.; Li, S.; Guo, S.; Meng, Q.; Li, G.; Yang, X. Genetic Engineering of glycine betaine biosynthesis reduces heat-enhanced photoinhibition by enhancing antioxidative defense and alleviating lipid peroxidation in tomato. *Plant Mol. Biol. Rep.* **2014**, *32*, 42–51. [CrossRef]
68. Zhang, T.; Li, Z.; Li, D.; Li, C.; Wei, D.; Li, S.; Liu, Y.; Chen, T.H.H.; Yang, X. Comparative effects of glycinebetaine on the thermotolerance in codA-and *BADH*-transgenic tomato plants under high temperature stress. *Plant Cell Rep.* **2020**, *39*, 1525–1538. [CrossRef]
69. Yang, X.H.; Liang, Z.; Lu, C.M. Genetic engineering of the biosynthesis of glycinebetaine enhances photosynthesis against high temperature stress in transgenic tobacco plants. *Plant Physiol.* **2005**, *138*, 299–309. [CrossRef] [PubMed]
70. Yang, X.; Wen, X.; Gong, H.; Lu, Q.; Yang, Z.; Tang, Y.; Liang, Z.; Lu, C. Genetic engineering of the biosynthesis of glycinebetaine enhances thermotolerance of photosystem II in tobacco plants. *Planta* **2007**, *225*, 719–733. [CrossRef]
71. Yu, H.Q.; Wang, Y.G.; Yong, T.M.; She, Y.H.; Fu, F.L.; Li, W.C. Heterologous expression of betaine aldehyde dehydrogenase gene from *Ammopiptanthus nanus* confers high salt and heat tolerance to *Escherichia coli*. *Gene* **2014**, *549*, 77–84. [CrossRef] [PubMed]
72. Wang, G.P.; Li, F.; Zhang, J.; Zhao, M.R.; Hui, Z.; Wang, W. Overaccumulation of glycine betaine enhances tolerance of the photosynthetic apparatus to drought and heat stress in wheat. *Photosynthetica* **2010**, *48*, 30–41. [CrossRef]

MDPI

Review

Heat Stress in Cotton: A Review on Predicted and Unpredicted Growth-Yield Anomalies and Mitigating Breeding Strategies

Sajid Majeed [1], Iqrar Ahmad Rana [2], Muhammad Salman Mubarik [2], Rana Muhammad Atif [1], Seung-Hwan Yang [3], Gyuhwa Chung [3,*], Yinhua Jia [4], Xiongming Du [4], Lori Hinze [5] and Muhammad Tehseen Azhar [6,7,*]

1 Department of Plant Breeding and Genetics, University of Agriculture, Faisalabad 38400, Pakistan; sajidmajeedpbg@gmail.com (S.M.); dratif@uaf.edu.pk (R.M.A.)
2 Centre of Agricultural Biochemistry and Biotechnology, University of Agriculture, Faisalabad 38400, Pakistan; iqrar_rana@uaf.edu.pk (I.A.R.); msmubarik@gmail.com (M.S.M.)
3 Department of Biotechnology, Chonnam National University, Chonnam 59626, Korea; ymichigan@jnu.ac.kr
4 State Key Laboratory of Cotton Biology, Institute of Cotton Research Chinese Academy of Agricultural Science, Anyang 455000, China; jiayinhua_0@sina.com (Y.J.); duxiongming@caas.cn (X.D.)
5 US Department of Agriculture, Agricultural Research Service, College Station, TX 77845, USA; lori.hinze@usda.gov
6 School of Agriculture Sciences, Zhengzhou University, Zhengzhou 450000, China
7 Institute of Molecular Biology and Biotechnology, Bahauddin Zakariya University, Multan 60800, Pakistan
* Correspondence: chung@chonnam.ac.kr (G.C.); tehseenazhar@gmail.com (M.T.A.)

Citation: Majeed, S.; Rana, I.A.; Mubarik, M.S.; Atif, R.M.; Yang, S.-H.; Chung, G.; Jia, Y.; Du, X.; Hinze, L.; Azhar, M.T. Heat Stress in Cotton: A Review on Predicted and Unpredicted Growth-Yield Anomalies and Mitigating Breeding Strategies. *Agronomy* **2021**, *11*, 1825. https://doi.org/10.3390/agronomy11091825

Academic Editors: Channapatna S. Prakash, Ali Raza, Xiling Zou and Daojie Wang

Received: 4 August 2021
Accepted: 9 September 2021
Published: 12 September 2021

Publisher's Note: MDPI stays neutral with regard to jurisdictional claims in published maps and institutional affiliations.

Abstract: The demand for cotton fibres is increasing due to growing global population while its production is facing challenges from an unpredictable rise in temperature owing to rapidly changing climatic conditions. High temperature stress is a major stumbling block relative to agricultural production around the world. Therefore, the development of thermo-stable cotton cultivars is gaining popularity. Understanding the effects of heat stress on various stages of plant growth and development and its tolerance mechanism is a prerequisite for initiating cotton breeding programs to sustain lint yield without compromising its quality under high temperature stress conditions. Thus, cotton breeders should consider all possible options, such as developing superior cultivars through traditional breeding, utilizing molecular markers and transgenic technologies, or using genome editing techniques to obtain desired features. Therefore, this review article discusses the likely effects of heat stress on cotton plants, tolerance mechanisms, and possible breeding strategies.

Keywords: breeding; climate change; genetics; heat stress; upland cotton

1. Introduction

Upland cotton (*Gossypium hirsutum*) is a multipurpose cash crop. The lint of cotton is the major product of this crop and is used by the textile industry for cloth manufacturing. Cotton is cultivated on an area of about 34.1 million hectares with a production of 126.5 million bales, and it is grown in more than 35 countries [1]. India is the largest grower and producer of cotton, with production of ~6.1 million tons of cotton every year. Other leading cotton producing countries include China, United States, Brazil, Pakistan, Turkey, and Uzbekistan. China is the largest consumer of cotton, with consumption of ~7.60 million tons of cotton annually [2]

From sowing to harvesting, the cotton crop faces numerous problems including infestation of insect pests, diseases, heat, drought, cold and salinity stresses, trash during picking, and post-harvest management problems [3–6]. Each of these causes significant reduction in yield and quality of cotton fibres. Therefore, comprehensive research on each aspect is required in order to understand these problems. The present discussion focuses on high temperature stress and on minimizing the losses due to this abiotic stress. Thus, a detailed review about the effects of heat stress on cotton plants and possible strategies for its mitigation is described in the following paragraphs.

Heat stress is often called high temperature stress. It is one of the limiting factors in crop productivity. Heat stress is defined as a condition when the temperature is high enough for a sufficient period of time to cause irreversible damage to plant development and functions [7]. A sudden increase of 5–7 °C in maximum temperature for a few days with a corresponding increase in ambient minimum temperature causes "high temperature stress" in plants. The temperature requirement varies from species to species, and it also depends upon time of exposure, intensity of exposure, air or soil temperature, night/day temperature, and age of the plant. Therefore, a particular temperature cannot be defined as a cardinal point for heat stress. Generally, cool season plant species are more vulnerable to heat stress than compared to warm season crops [7,8]. Moreover, plants of similar species adapted in different climatic regions have different degrees of temperature tolerance. For example, cotton grown in the United States and China is considered to be under heat stress when temperatures increase above 38 °C, while in Pakistan and India this temperature is considered optimum and temperatures greater than 46 °C are considered as heat [9–11]. Cotton responses to heat stress are presented in detail below.

Traditional breeding strategies have been utilized to incorporate heat stress tolerance into upland cotton. Most of the improved cultivars and breeding lines are the outcome of purely classical breeding as very little molecular and genomic tools have been used to date. Such breeding efforts were based on intensive selection, which reduced genetic variability in cotton. Mutagenic agents have been used to create variability and have resulted in the release of high-yielding cultivars, including NIAB-78 as a successful example from Pakistan [12,13]. Due to the increased resources needed to screen large populations, reduced frequency of desirable alleles, and pleiotropic effects, mutation breeding has gradually been replaced with marker assisted breeding and site-specific mutagenesis technologies. The use of advanced genomics and biotechnological tools has also become important as the challenges to cotton production escalating. Later in this review, advanced breeding tools are discussed that can be utilized to mitigate the effect of changing climatic conditions, especially high temperature stress.

2. Effects of High Temperature Stress on Cotton

High temperatures in arid and semi-arid regions of the world are inducing negative impacts on growth, development, and productivity of several field crops [14]. Heat stress can cause damage to a cotton plant in almost every stage of its life, but it is reported that the reproductive stages of cotton are more sensitive to high temperature than compared to vegetative growth stages [15]. Both day and night temperatures play an important role in determining yield potential in crop plants, but high night temperature reduces yield and causes significant damage, while the role of high day temperature is secondary [16]. The adverse effects of high night/day temperature on different plant stages are shown in Figure 1.

2.1. Effects on Germination of Cotton Seed

Successful germination of seed and development of seedlings requires a good soil environment, especially soil moisture content and soil temperature. These requirements vary from plant species to species [17,18]. Generally, high temperature results in poor seed germination in field crops. The germination of seed is a complex physiological process that depends upon the activity of several cellular organelles and enzymes. These enzymes need to be produced in a continuous manner to perform various activities related to metabolic processes. For instance, in maize at high temperatures, these enzymes denature or slow down these activities, which results in reduced metabolism processes [19]. Moreover, high soil temperature also increases the rate of transpiration, which reduces the water availability to seed. Less availability of water also slows down the seed germination [20]. The optimal temperature of 28 to 30 °C is needed for germination of cotton seed. As the temperature decreases, the rate of seed germination also decreases, and very poor germination can be observed at temperatures <20 °C. Similarly, increases in temperature from the optimal range (≥38 °C) have also been reported to result in decreased germination of cotton seed [21].

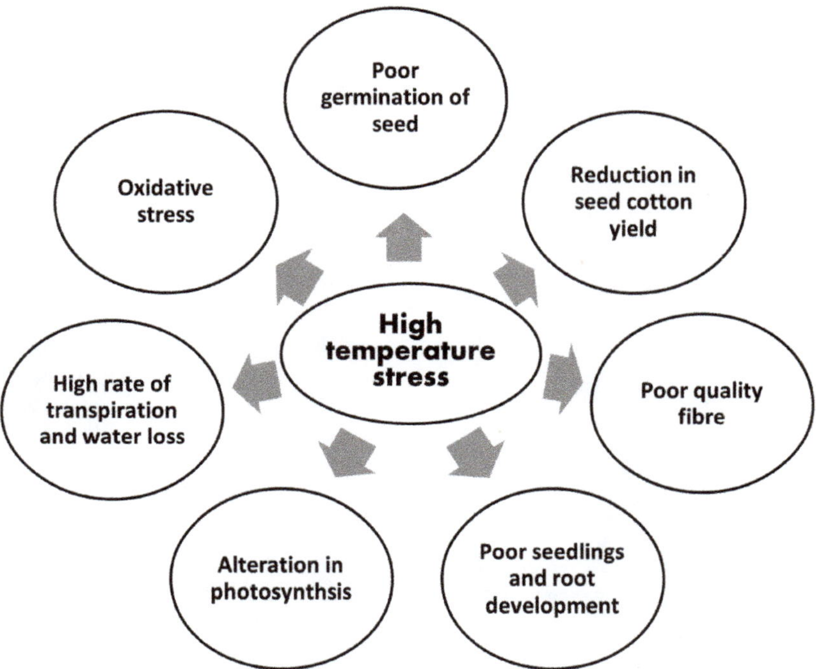

Figure 1. Salient adverse effects of high temperature stress on cotton plants.

2.2. Effects on Early Vegetative Growth Stages

In addition to germination, uniform and rapid emergence of seedlings from soil is necessary to achieve vigorous growth of plants and high yield [22]. High temperature stress during the early stages of plant growth affects the emergence of seedlings and results in the development of poor seedlings. In a previous experiment, researchers grew seedlings of cotton under various temperature regimes, i.e., 38 °C and 32 °C for 8 days. They recorded 152 mm shoot and 173 mm root growth at 32 °C temperature, while at 38 °C only 50 mm shoot and 86 mm root growths were observed [17]. The emergences and growths of cotton seedlings were also assessed for four different temperatures, i.e., 20, 30, 40, and 50 °C. The maximum emergence of seedling with vigorous growth was observed at 30 °C. The genotypes developed for hot tropical environments with higher seed weight showed vigorous seedlings even at 40 °C than compared to genotypes with smaller seed size and weight. Cotton seedlings did not emerge at 50 °C [23].

Along with the above ground parts of plants, roots are also severely affected by high temperature stress. The damaging of roots minimizes their uptake of nutrients and water from soil that can disturb the entire physiological mechanism of the plant and limit its productivity. The cotton plant has a taproot system. Studies showed that cotton roots develop poorly under high soil temperature and result in a poor crop stand [24]. A study was conducted on 10 upland cotton genotypes in order to determine the heat stress ability of roots by measuring related parameters. After 120 min of seed germinations, the seedlings were subjected to seven different temperature regimes between 25 to 45 °C. The results revealed that all the genotypes showed normal root growth up to 35 °C. The roots' growth declined when temperature exceeded from 35 °C, and irreversible damage to roots of all the cultivars occurred at 45 °C [25].

2.3. Effects on Lateral Vegetative Growth Stages

Upland cotton is generally sown at the start of the summer season in tropical and subtropical parts of the world. Therefore, seedlings of cotton experience lower temperatures than compared to later lateral growth stages. The cotton crop normally experiences its highest temperature at bud initiation. This is the first flowering stage that causes instability in production [26,27]. A significant reduction in the number of fruiting branches has been observed under heat stress [28]. Dropping of floral buds, flowers, and bolls are common when the temperature increases above average. The retention of fruits on fruiting branches declined rapidly at temperatures increased to ≥33 °C in Mississippi, USA [29]. The shedding of fruiting bodies is a natural mechanism of the cotton plant for decreasing the fruit load in order to adjust the supply and demand balance of nutrients and water in the plant. Generally, the cotton plant drops ~50% premature bolls, but this percentage increases with increases in temperature. Researchers claimed that retention of bolls until harvest is the bigger challenge than compared to increasing the total number of fruiting bodies or bolls per plant. They indicated that the best possible strategy to increase cotton production is to minimize flower and boll shedding [30].

Boll weight and boll size are also reduced under heat stress conditions. Temperatures greater than 40 °C result in shortening the boll maturation period and formation of smaller sized bolls having less weight than bolls developed under normal temperatures [31]. An experiment was carried out with 23 accessions of cotton, which were grown under normal and heat stress conditions for two years. The boll weight of all the genotypes was reduced under the heat stress condition [32]. The reduction in boll weight of cotton cultivars grown under high temperatures was also observed by other researchers [33,34]. It was found that foliar application of ascorbic acid increased the heat tolerant ability of plants and minimized losses due to heat stress. Zeiher and Brown conducted several experiments under various environments, including greenhouse, growth chamber, and field environments. They concluded that boll size, fruit retention percentage, and number of seeds per boll were also reduced when temperatures were above 30 °C [29,35–37].

2.4. Effects on Yield and Quality of Fibre

Fibre production upon maturity is the ultimate goal of growers. Both the quantity and quality of fibre determine the end value of the cotton crop. The yield of seed cotton is a complex trait, which is the result of various morphological and physiological features of the plant. It also has a strong correlation with antioxidant activities. Both of these parameters are polygenic in nature and highly influenced by environmental conditions [38]. The negative association of fibre quantity with its quality is another challenge for breeders in terms of improving both features simultaneously [39]. Numerous studies have reported the effect of high temperature on fibre yield and quality [40–43]. An experiment was conducted on three cotton cultivars grown in nine diverse environments in order to investigate the effects of heat stress. The results revealed that lint yield and quality attributes were severely affected by high temperature. It was further concluded that cotton plants can perform well within specific temperature ranges as low temperature also adversely affected the cotton yield. The minimum threshold temperature of 22 °C was recorded to maintain seed cotton yield [44]. Another experiment revealed that both short and long term increases in day temperature decrease the biomass of cotton fruits, which ultimately results in low yield and poor quality fibres [45]. An increase of 2–3 °C temperature from the optimal temperature (32 °C) in Nanjing, China, resulted in a decrease of 10% of biomass, while yield declined by 40%. The results also revealed that the micronaire value of fibre increased under heat stress, which results in coarse fibres with less strength [46]. The developmental process of cotton fibres is sensitive to increases in temperature. Such conditions also reduce the time required for boll maturation that results in short fibres. High temperature conditions lasting up to 5 days did not change fibre quality, but the prevalence of these conditions for more than a week may cause irreversible damage to cellulose and significant reduction in fibre quality [47].

2.5. Effect on Floral Parts

Information about pollen viability and stigma receptivity is important for increasing productivity because effective pollination is essential for fruiting and seed setting in crop plants [48]. The germination and length of pollen tubes in cotton were monitored in 12 cultivars grown under different temperature regimes. The results indicated that maximum pollen germination and longest pollen tube length were observed at 32 °C. The pollen failed to germinate at 47 °C, while no pollen tubes formed above 44 °C [49]. Another experiment revealed that length of the pollen tube decreases as temperature increases above 32 °C, while germination of pollen decreases above 37 °C. Therefore, it can be suggested that the formation of the pollen tube is more sensitive to high temperature stress than compared to pollen germination. *G. barbadense* was found to be more sensitive to heat stress. A lower in vitro pollen germination percentage was recorded for heat treated pollen of pima cotton than compared to upland cotton [50]. The length of the filament was significantly reduced when cotton flowers were exposed to high temperature stress, and this resulted in the appearance of an elongated stigma. The actual length of stigma remained the same, but the length of filaments were reduced to the extent that the stigma appeared to be very long, and the process of self-pollination was badly affected [29]. The loss of receptivity of the stigma under high temperature is also reported in sweet cherry and peach [51,52]. The penetration of pollen grains to the ovule via the pollen tube ensures the successful fertilization process, and the penetration of pollen tube through the stigma, style, and ovule at high temperatures was reduced, which resulted in poor seed setting [53]. The effects of heat stress on various reproductive phases of plants are illustrated in Figure 2.

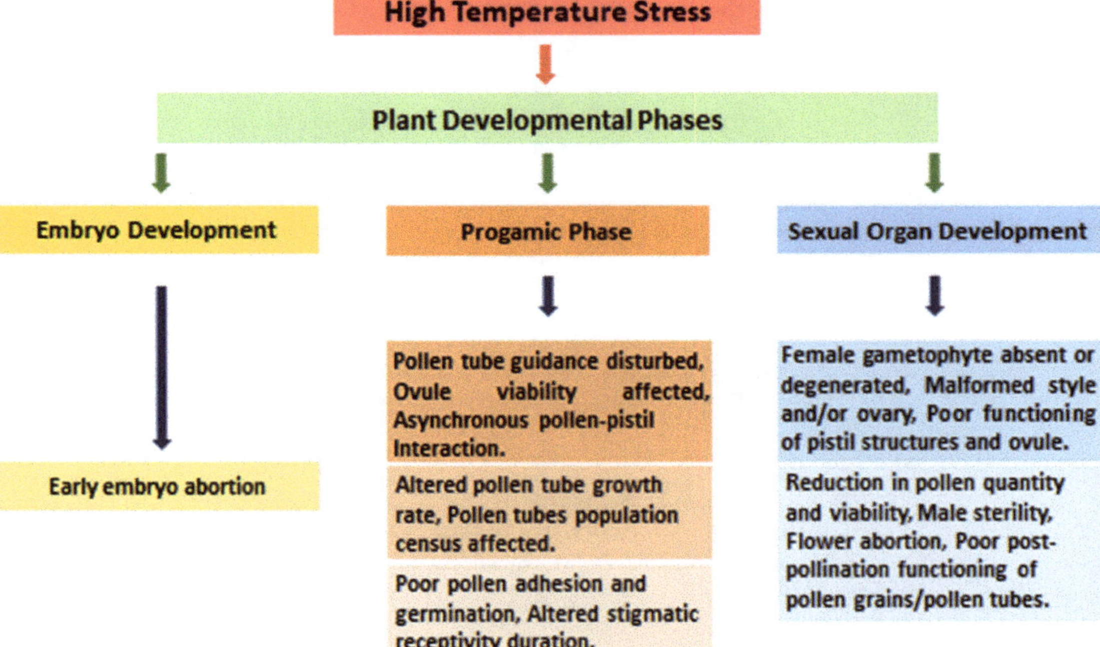

Figure 2. Effects of high temperature stress on reproductive phases of plant.

2.6. Effects on Physiological Attributes

Plant growth and development is based on physiological processes such as photosynthesis, membrane permeability, and stomatal conductance [54]. Heat stress adversely affects physiological attributes of plants and limits productivity [55]. The rate of photosynthesis is significantly reduced or even inhibited at high temperatures. The deactivation of ribulose 1,5 bisphosphate carboxylase/oxygenase (Rubisco) and increase in ionic conductance of thylakoid membranes are the primary causes of photosynthesis reduction or inhibition in cotton during high temperature conditions [56,57]. Chlorophyll content also decreases when cotton plants are exposed to high temperature that results in a decrease in the rate of photosynthesis [58]. Heat stress changes the permeability of membranes and alters cell differentiation and elongation by causing injuries to cellular membranes and deforming the organization of microtubules and cytoskeleton [59]. Higher cell membrane thermostability is positively associated with heat tolerance in cotton. Therefore, a number of experiments have been conducted to screen for heat tolerant accessions on the basis of cell membrane thermostability [60–62]. Stomatal conductance is directly related to water relations and photosynthesis in plants. High temperature causes opening of stomata and an increase in stomatal conductance that results in a decrease in the water potential of leaf. High stomatal conductance also increases the rate of transpiration and intercellular CO_2. Stomatal conductance increases up to 40% in most of plant species when temperature rises from 30 to 40 °C [63]. The advantage of higher stomatal conductance is associated with cooling of leaves, which provides tolerance to heat stress [64]. Experimental results have shown that upland cotton has more stomatal conductance and higher rate of photosynthesis under high temperature conditions than compared to pima cotton [65]. Studies revealed that differences in cotton accessions and species for stomatal conductance are under genetic control. Thus, this trait can be improved through breeding and selection [66–68].

3. Mechanisms of Heat Tolerance

3.1. Antioxidant Activity in Response to Oxidative Stress

Crop plants face environmental stresses which alter the various metabolic activities in order to ensure balance between production and consumption of energy through oxidation and reduction reactions [69]. This change in metabolism alters the concentration of various molecules. Likewise, metabolic imbalances during high temperature stress promote the extra-accumulation of reactive oxygen species (ROS) into the cellular compartment of plants [70]. ROSs are highly reactive chemicals formed from O_2. Excessive production and over-accumulation of ROS in plant cells cause irreversible damage to its organelles through oxidative stress. However, a balanced amount of ROS is required for normal activities such as detoxification of poisonous substances, antimicrobial phagocytosis, and apoptosis. ROS also benefits plants by acting as signaling molecules for activation of numerous genes related to stress tolerance, cell proliferation, seed germination, growth of root hairs, and cell senescence [71,72]. The over-accumulation of ROS during stress conditions results in the oxidation damage of vital molecules such as DNA, proteins, and lipids. This condition is termed as oxidative stress in plants [73]. ROS includes free radicals such as the hydroxyl radical (OH) and superoxide anion (O_2^-), as well as non-radicals such as singlet oxygen ($^1O_2$) and hydrogen peroxide (H_2O_2). These species are produced by excitation and reduction in intra-cellular oxygen (O_2). The schematic diagram of ROS production is illustrated in Figure 3.

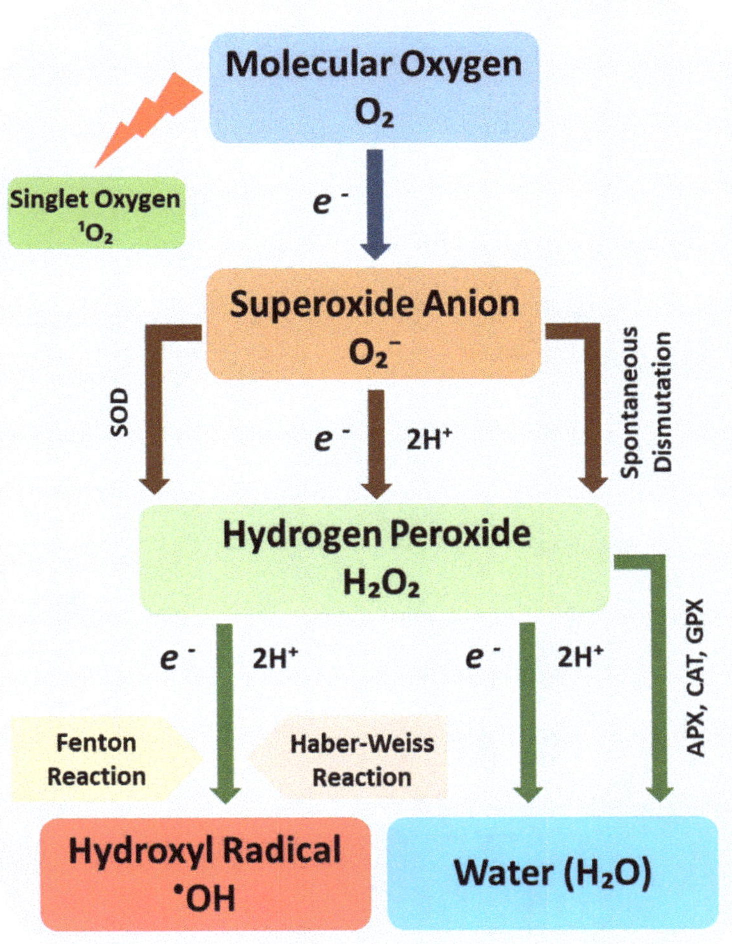

Figure 3. Schematic diagram of ROS production in plants.

The concentration of ROS and scavenging capacity of antioxidants in cotton is considered as a selection criterion for heat tolerant accessions [74]. An experiment was conducted using two cotton cultivars, and heat stress was applied gradually from 30 to 45 °C on 30 days old seedlings. The results revealed a 206 to 248% increase in hydrogen peroxide content and a 40 to 170% increase in lipid peroxidation under heat stress conditions. The concentration of non-enzymatic antioxidants also increases with increases in temperature, while the activity of enzymatic antioxidants such as superoxide dismutase (SOD), catalase (CAT), peroxidase (POX), and ascorbate peroxidase (APX) increase 56–70%, 37–69%, 43–91%, and 22–27%, respectively. It was concluded that genotypic differences exist in cultivars for ROS production and antioxidants response. Higher levels of antioxidants and lower levels of ROS during high temperature are an indication of heat stress tolerance [75]. In another study, the cotton plants were grown at two temperature regimes, i.e., 38 and 45 °C. The results indicated non-significant differences in the concentration of hydrogen peroxide at both temperatures, while the concentration of proline decreased rapidly and significantly as the temperature increased from 30 to 45 °C. The activity of

SOD declined at 45 °C while the activity of CAT, POX, and APX increased with the increase in temperature [76]. It is found that the exogenous application of hydrogen peroxide on cotton plants triggers the activity of SOD and CAT. It was further concluded that foliar applications of H_2O_2 on field grown cotton can enhance the heat tolerant ability without compromising yield [77]. The effect of high night temperature on biochemistry of leaf and pistil was studied in upland cotton cultivar. The results indicated that glutathione reductase activity in leaves is increased with an increase in night temperature, while no change in the concentration of glutathione reductase was observed in pistils. This shows that the antioxidant mechanism of pistil or floral parts is less sensitive to changes in mean night temperature than compared to leaves or vegetative parts of the cotton plant [78].

3.2. Role of Heat Shock Proteins

Heat shock proteins (HSPs) are important contributors to cellular homeostasis under heat stress. These molecular chaperons are upregulated during high temperature conditions and perform various activities in order to maintain the integrity of the cell [79]. On the basis of molecular weight, HSPs are divided into five major groups, i.e., small HSPs, HSP60, HSP70, HSP90, and HSP100 [80]. The details of each group are summarized in Table 1. Each HSP group is unique in nature and specific in function. Small HSPs (sHSPs) have low molecular weight (12–40 kDa) and are the most diverse in nature with respect to cellular location, function, and sequence similarities [81]. These sHSPs bind to non-native proteins, prevent non-native aggregation through hydrophobic interactions, and facilitate their refolding by ATP-dependent chaperons such as ClpB/DnaK [82]. Almost all sHSPs have an α-crystallin domain, which forms a dodecamer double ring and helps in the folding of proteins [83]. Previous work has revealed that the expression of the sHSP coding gene, i.e., *Hsp 17.7*, is directly related to thermal stress tolerance in plants [84]. A quantitative expression analysis of *GHSP26* (a small HSP coding gene in cotton) indicated that the leaves of cotton have 100-fold increased concentration of proteins encoded by this gene during water deficit conditions [85].

Table 1. Characterization of various groups of HSP in plants.

Group (Sub-Families)	Representative Members	Intracellular Localization	Major Role	Reference
sHSP				
I	Hsp17.6	Cytosol	Stabilization of non-native proteins and prevent aggregation	[86,87]
II	Hsp17.9	Cytosol		
III	Hsp21, Hsp26.2 [5]	Chloroplast		
IV	Hsp22	Endoplasmic reticulum		
V	Hsp23 [5]	Mitochondria		
VI	Hsp22.3	Membrane		
HSP60				
Group I	Cpn60 [2]	Chloroplast, mitochondria	Folding of proteins	[88,89]
Group II	CCT [3]	Cytosol		
HSP70				
DnaK	Hsp/Hsc70	Cytosol	Protein import, signal transduction, transcriptional activation, assist refolding and prevent aggregation of proteins	[90–93]
	Hsp70	Chloroplast, mitochondria		
	Bip [1]	Endoplasmic reticulum		
Hsp110/SSE	Hsp91	Cytosol		
HSP90	AtHsp90-1	Cytosol	Facilitate in genetic buffering and maturation of signaling molecules	[94–96]
	AtHsp90-5	Chloroplast		
	AtHsp90-6	Mitochondria		
	AtHsp90-7	Endoplasmic reticulum		
HSP100				
Class I	ClpB, ClpA/C, ClpD	Cytosol, mitochondria	Unfolding and disaggregation of proteins	[97,98]
Class II	ClpM, ClpN, ClpX, ClpY	Chloroplast		

HSP60 is generally known as a mitochondrial chaperon or chaperonin 60. It plays two essential roles in mitochondria during high temperature conditions, i.e., maintenance of the unfolded state of proteins for their transportation across the inner mitochondrial membrane and the folding of important proteins into a matrix [88]. HSP60 is also involved in assisting proteins that help in photosynthesis such as Rubisco [99]. Studies revealed that a mutation in *Chaperonin-60α* gene that codes for HSP60 protein causes a defection in chloroplasts, which ultimately results in poor seedlings and embryo development in *Arabidopsis* plant [100]. However, deletion of this gene results in cell death [101]. It has been experimentally verified that transgenic tobacco plants with reduced *Cpn60β* (chaperonin 60β) exhibited phenotypic defects such as delayed flowering, stunted growth, and leaf chlorosis [102]. HSP70s are considered important cellular machinery involved in the folding of proteins and in preventing their aggregation [103]. The overexpression of HSP70s is an indication of heat tolerance in plants. It is reported that *HSP70* genes of cotton play essential roles during fibre development. The inhibition of these genes results in the retardation of fibre elongation. The inhibition of *HSP70* genes results in oxidative stress by elevating the level of H_2O_2, which causes damage to epidermal layers of the ovule [104]. HSP70 proteins also act as signaling molecules for transcriptional activation and de-activation [91].

HSP90 proteins are quite distinct from other chaperons because most of them are substrates involved in signal transduction, such as signaling kinases and hormone receptors [105]. They also manage the folding of proteins [106]. HSP90s are among the most abundant proteins of the cell (1–2% of the total), are constitutive in nature, and act together with HSP70s as multi-chaperone machinery. The expression of HSP90 proteins increases significantly during hot conditions [107]. HSP100 belongs to the AAA ATPase family and performs various functions such as unfolding and disaggregation of proteins [108]. In addition to heat stress tolerance, HSP100 also performs housekeeping functions in the cell, including the development of chloroplasts [109,110].

3.3. Small RNAs Activity in Regulating Heat Stress

With the advancement in high throughput sequencing technologies, plant researchers have revealed various roles of microRNA (miRNA) and small interfering RNA (siRNA) during biotic and abiotic stress conditions. These small RNAs are actively involved in the degradation of mRNA and prevent translation of various proteins in plant cells [111–113]. The regulatory roles of various miRNAs have been characterized during high temperature conditions in many plant species, including chestnuts and *Arabidopsis*, e.g., the miR156 and miR157 families comprise 17 and four miRNAs, respectively. These miRNAs upregulate during hot temperature conditions and target *SPL* genes, which are essential for floral development in plants. Thus, flowering is controlled by these miRNAs when plants are exposed to high temperature conditions [114,115]. Over-production of miR157 targets *SPL* genes in cotton during heat stress results in a reduction in flowers and smaller sized bolls with fewer seeds [116]. The similar role of these miRNA families is reported in *Brassica rapa* [117], citrus [118], and *Arabidopsis* [119].

Experiments revealed that the expression of miR159 is down-regulated in heat tolerant genotypes of wheat upon exposure to hot temperature regimes. This miRNA acts as a negative regulator of *MYB* transcriptional factors [120]. The major role of the auxin response factors (*ARFs*) gene family is the regulation of auxin levels in plants. The over-expression of miR160 in cotton increases its susceptibility to high temperature stress by suppressing the expression of *ARF* genes [121]. It is found that the expression of miR162 increases up to 15-fold during drought conditions. Increases in concentration of miR162 under heat and salinity stresses is reported in cotton and rice [122,123]. This microRNA controls the transcription of numerous genes by targeting zinc finger proteins (ZFPs) and acts as a regulator of dicer such as proteins [124,125]. The expression level of miR164 was observed to decrease 0.3-fold under heat stress conditions. It targets *HSP17* genes and also regulates the expression of various genes essential for mitogen activated protein kinase (*MAPK*) mediated signaling pathways and the activation of *NAC* transcriptional factors in wheat, rice, and alfalfa [126–128]. It is reported that expression of miR171 increased several folds upon exposure of the plant to high temperature. It targets *GRAS* genes, which are involved in various developmental processes, i.e., flowering time, floral meristem determinacy, plant height, and leaf architecture in cotton [129,130].

The regulation of flowering and floral organs is controlled by *AP2* genes [131]. This gene family is regulated by miR172, as reported in roses [132]. Thus, up-regulation and downregulation of these miRNAs is directly related to flowering timing and the transition of floral and vegetative phases in rice and *Arabidopsis* during high temperature conditions [133,134]. The miR390 of cotton controls the formation of lateral roots by targeting *ARF* genes [135]. The miR393 is considered a regulator of auxin receptors [136]. Overexpression causes a delay in flowering and results in poor development of roots. Thus, decreased levels of miR393 are an indicator of stress tolerance in cotton and rice [137,138]. F-box proteins perform various activities during stress conditions such as degradation of proteins, rolling, and senescence of leaves [139,140]. It was found that miR394 prevents the translation of F-box genes family transcripts during abiotic stresses to maintain the optimal levels of proteins for normal functioning of plants, particularly in rice and *Arabidopsis* [141,142]. The miR395 regulates *APS* genes in response to various abiotic stresses. The major function of this gene family is the assimilation of sulphate [143].

4. Breeding Strategies for High Temperature Stress Tolerance

4.1. Conventional Breeding Approaches

Assessment of germplasm is a prerequisite for breeding stress tolerance. Numerous experiments have been conducted to identify heat stress tolerant genotypes from the available genepool. Moreover, the utilization of crop wild relatives is also gaining popularity in plant breeding due to their novel features that are lacking in domesticated cultivars. Most of these novel features are related to biotic and abiotic environmental stress. It is recommended to screen related wild species and relatives in order to have a diverse gene pool [144]. Although gene transfer from wild to cultivated species encounters numerous problems and is not always possible without recombinant DNA technology, the rapidly evolving technologies in plant sciences have made it quite possible to transfer genes among many species, as is discussed below [145]. After the identification of a suitable gene and trait, the next step is to transfer it to a desirable genotype or to purify the identified plant through selection. For this purpose, single plant selection, bulk selection, and pedigree selection are among the most widely used classical breeding methods in cotton [146,147]. These methods are used in cotton improvement along with molecular breeding tools for quick and efficient screening and genetic gain.

4.2. Molecular and Biotechnological Approaches

In addition to conventional screening and breeding approaches, molecular markers and biotechnological tools are also useful for improving stress tolerance of cotton genotypes [148]. Numerous markers such as amplified fragment length polymorphism (AFLP) and randomly amplified polymorphic DNA (RAPD) markers have been successfully utilized screening cotton genotypes for heat tolerance in the past [149,150]. Currently, simple sequence repeats (SSRs) and single nucleotide polymorphism (SNPs) are widely used markers for identifying quantitative traits loci (QTLs) related to stress tolerance in cotton [151,152]. The experiments were conducted by using heat tolerant and susceptible cultivars to determine heat responsive genes in upland cotton. Twenty-five expressed sequence tags (ESTs) were sequenced to study the homology of genes. The expression level of a few ESTs was also quantified using real time PCR. The results indicated that expression of folylpolyglutamate synthase (*FPGS3*) and IAA-ala hydrolase (*IAR3*) coding genes was significantly up-regulated during long-term and short-term high temperature stress. The expression of two non-annotated ESTs, i.e., *GhHS128* and *GhHS126*, was also found to be up-regulated under hot conditions. Thus, it was suggested that these two non-annotated ESTs are heat tolerant candidate genes [153].

In order to investigate the molecular mechanism of high temperature stress tolerance, the expression of some heat responsive genes was quantified through real time PCR in tolerant and susceptible upland cotton cultivars. The genes belong to various groups, i.e., *HSFA1b* and *HSFA2* are heat stress transcriptional factors; *HSP101*, *HSP70-1*, and *GHSP26* code for heat shock proteins; *ANNAT8* is involved in calcium signaling; and *APX1* controls antioxidant activity. The level of *GHSP26* increased in all genotypes, while the expression of *HSP101* and *HSP70-1* increased several-fold only in the seedlings of heat tolerant cultivars. The expression of *APX1* increased significantly in a heat-tolerant cultivar (VH-260), indicating the involvement of antioxidant activity in conferring heat tolerance. No significant change in the expression of *ANNAT8* was observed in heat susceptible cultivars. The expression of *HSFA2* and *HSFA1b* was several folds higher in leaves and ovaries of heat tolerant accessions than compared to heat susceptible accessions [154]. In order determine the SNP markers linked to the mitochondrial small heat shock protein (*MTsHSP*), a study was conducted by using accessions belonging to various cotton species, i.e., *G. aridum*, *G. sturtianum*, *G. gossypioides*, *G. stocksii*, *G. arboreum*, *G. laxum*, and *G. herbaceum*. Approximately 21 SNPs were identified for this gene by using PCR cloning and sequencing techniques, which could be useful for cotton improvement [155]. Transcriptomic analysis of 82 genes belonging to the *GhHSP20* family revealed their involvement in developmental and physiological processes of cotton. Most of them were regulatory in nature and expressed only under hot conditions, while eight genes were found to be involved in conferring tolerance for multiple stresses, namely heat, drought, and salinity [156].

Rapid advancements in applied genomics have resulted in useful tools for plant improvement. For example, markers linked to known genes or QTLs can be used for marker-assisted selection (MAS), as well as for genomic selection. Genomic selection assists the breeders in utilizing the molecular markers in the absence of phenotypic data. It can reduce the time for cultivar development through more efficient selection of progeny in early generations. An experiment was conducted with 550 recombinant inbred lines of cotton, and six fiber quality traits were evaluated using genomic selection. A total of 6292 markers were obtained through genotyping by sequencing. It was revealed that genomic selection could potentially predict genomic estimated breeding value in upland cotton fiber quality attributes [157]. Association mapping is also an effective technique for cotton improvement when information on population structure and linkage disequilibrium (LD) is available. This method is quite useful for reducing the laborious work involved in screening large populations [158]. Genome-wide association studies (GWAS) represent a powerful approach for identifying the locations of genetic factors that underlie complex traits. GWAS has been successfully implemented in cotton for the identification of single nucleotide polymorphism (SNP) loci and candidate genes for various attributes [159].

4.3. Transgenic Approaches

Transgenic techniques have also been extensively used to improve cotton cultivars for better tolerance with respect to high temperature stress. Recently, heat shock protein 70 (*AsHSP70*) from *Agave sisalana* was transformed in cotton through the *Agrobacterium* mediated transformation method. Expression studies showed the higher expression of transformed gene in different plant tissues under high temperature. Additionally, transgenic cotton plants exhibited improved performance for the measurable physiological and biochemical indicators [160]. In another study, over-expression of both *AVP1* and *OsSIZ1* genes in cotton improves lint yield compared to wild-type cotton under combined drought and heat stress conditions, and it does not have any negative affect on overall cotton yield when there is no stress. Furthermore, transgenic cotton plants also had 72% more photosynthetic rates two hours before the outset of heat stress and 108% higher photosynthetic rates during heat stress [161]. The function of *Arabidopsis* heat shock protein 101 (*AtHSP101*) is well known in heat tolerance at a vegetative stage, and the overexpression of this protein in cotton (Coker-312) clearly exhibited the increased germination percentage and enhanced pollen tube elongation under high temperatures than compared to non-transgenic cotton [162]. Therefore, improved heat tolerance of reproductive systems in transgenic cotton is crucial for enhanced yield on a sustainable basis in the face of climate change. Conversely, the *Arabidopsis* SUMO E3 ligase (*AtSIZ1*) is a key gene for plant heat stress response, as *AtSIZ1* mutant plants exhibited increased susceptibility to high temperatures. In cotton, the overexpression of *OsSIZ1* gene, a rice homolog of *AtSIZ1*, conferred tolerance to both drought and heat stresses than compared with non-transgenic plants by demonstrating enhanced net photosynthesis rate and improved growth and development [163]. In another study, research revealed that the ectopic expression of Arabidopsis stress associated gene (*AtSAP5*) in transgenic cotton (Coker-312) protects several components of carbon gain and growth under extreme drought and associated heat stress conditions [164].

However, in order to render transgenic cotton more acceptable and efficient, it is necessary to focus on improving transformation efficiency. In a nutshell, these studies indicate that integrating heat stress-related genes in cotton is a viable strategy for engineering high-temperature tolerant cotton cultivars that could significantly improve lint yields in marginal environments, resulting in a sustainable cotton production. The effectiveness of heat stress responsive genes, for which its expression could be successful without the accompanying yield penalties, would determine the end degree of success.

4.4. CRISPR-Cas Mediated Genome Editing

Tolerance relative to high temperatures is mostly regulated by many genes based on the degree of stress and the stress tolerance mechanism. It will be more difficult to target a tolerance mechanism that is controlled by multiple genes. Plant heat stress response is precisely regulated by a complex web of TFs from distinct families. These TFs improve plant heat stress tolerance by modulating the expression of several stress responsive genes, either individually or in conjunction with other regulatory factors. There are numerous successful genetic engineering applications inducing heat stress tolerance in plants using heat stress TFs and HSPs genes [165,166]. However, genetically modified crop plants are subject to stringent regulatory requirements, which may cause lab research to be delayed in reaching the market. As an alternative to traditional transgenic approaches, recently emerged CRISPR-Cas-mediated genome editing allows researchers to alter, modify or swap alleles, and insert or silence gene(s) in a predefined manner [167].

High temperatures alter the expression pattern of several plant genes either by upregulating or down-regulating them. Although our understanding of differentially expressed genes in response to drought and salt stress has expanded, relatively less focus has been made on studying heat stress associated genes in cotton. Studying the expression pattern of heat stress responsive genes in cotton under long-term heat stress clearly showed that expressions of *HS126*, *HS128*, *FPGS*, *TH1*, and *IAR3* genes increased under high temperature. In contrast, the expressions of *ABCC3*, *CIPK*, *CTL2*, *LSm8*, and *RPS14* genes were down-regulated [153]. Therefore, targeted modulation of these up-regulated and down-regulated genes in cotton using the CRISPR-Cas system would be an exciting opportunity for combating the negative impact of heat stress. Moreover, multiple HSPs and TFs associated with heat stress sensitive genes have been proposed as potential candidates for improving plant heat tolerance [168]. Therefore, understanding the exact role of these genetic regulators paves the way for the development of enhanced heat tolerance, while maintaining overall plant resilience. In maize plants, peak photosynthesis has been observed between 20 and 32 °C, and the subsequent increase in temperature caused a decrease in net photosynthesis rate depending on plant growth stage [169]. CRISPR/Cas9 mediated disruption of the heat-stress sensitive albino-1 (*HSA1*) gene in rice exhibits higher heat sensitivity compared to wild plants [170]. The *slagamous-like 6* (*Slagl6*) gene was reported as a potential gene in the study of facultative parthenocarpy. Thus, researchers have successfully developed heat-tolerant parthenocarpic tomato fruit by mutating the *SlAGL6* gene using the state-of-the-art CRISPR-Cas9 system [171]. In addition, thermosensitive male sterile maize lines have also been developed by mutating the *thermosensitive genic male-sterile 5* (*TMS5*) gene using CRISPR-Cas9 editing [172].

CRISPR-Cas9 has been modified and exploited for a range of new functions, including controlling gene regulation by activating and suppressing target gene expression using CRISPR activation and interference systems [173]. Positive gene regulators associated with HSPs and stress related TFs could be activated through the CRISPR activation system with high specificity. Moreover, negative regulators could be knocked out by using the CRISPR interference system. In one of study, the *BZR1* gene was up-regulated and repressed using CRISPR activator and interference systems. The results show that the overexpression of the *BZR1* gene enhances H_2O_2 production and recovery of thermo tolerance in rice, while plants with suppression of the gene show impaired production of H_2O_2 in apoplast and reduced heat tolerance [174]. Previously, the roles of *MAP3Ks* remained poorly understood in cotton. Recently, it has been reported that *MAP3K65* gene expression is induced by multiple signaling molecules, pathogen infection, and heat stress. This gene enhances susceptibility to pathogen infection and heat stress by negatively modulating growth and development related processes. Moreover, silencing *GhMAP3K65* enhanced resistance to pathogen infection and heat stress in cotton. Therefore, *GhMAP3K65* is a potential candidate gene to target with the CRISPR-Cas9 genome editing system in order to engineer heat tolerance in cotton [175]. In conclusion, maneuvering positive and negative regulators of heat stress signaling molecules in cotton could, thus, be exploited to develop new cotton cultivars tolerant to extreme temperatures.

5. Conclusions and Future Directions

High temperature exerts a negative impact on cotton growth and yield in all stages of plant growth. It reduces lint quantity and quality by impeding normal plant biological processes and pathways. Understanding the heat tolerance mechanism and molecular characterization of related genes is essential for developing stress tolerant cultivars for sustainable cotton production under changing climatic conditions. Undoubtedly, the heat stress TFs, HSPs, and other genes are important for maintaining the secondary structure of proteins, and rapid sensing of heat is also critical to induce the protective mechanisms against high temperature stress. Traditional breeding approaches for developing stress tolerance are being complemented by new technologies, including state-of-the-art genome editing tools, speed breeding approaches, and various omics tools, which may decrease the time needed to develop cotton cultivars with increased heat tolerance. For engineering high temperature stress tolerance, the CRISPR-Cas system is considered a non-genetically modified (nGM) approach, thus allowing for the expansion of scientific community efforts to introduce heat stress tolerance in future cotton cultivars for all cotton growing regions. Furthermore, the increasing demand of high-quality lint yield in a rapidly growing world population will be confronted through concept of speed breeding. It will aid quick generation advancement in order to shorten the overall growth cycle and accelerate cotton breeding programs. In addition, recent advances in sequencing technologies have made significant strides in successfully sequencing complex cotton genome. Subsequently, applications of various omics approaches have pointedly enhanced our understanding of cotton physiology and the functions of genes in response to heat stress. Therefore, differently expressed genes, proteins, and metabolites identified through different omics tools can be used as a potential biomarker to develop high temperature-resilient cotton cultivars.

Author Contributions: Conceptualization, M.T.A., S.M., L.H., Y.J. and M.S.M.; writing and draft preparation, S.M., M.T.A. and M.S.M. writing, review and editing, M.T.A., R.M.A., S.-H.Y. and G.C.; supervision, X.D., I.A.R. and M.T.A. All authors have read and agreed to the published version of the manuscript.

Funding: This research received no external funding.

Institutional Review Board Statement: Not applicable.

Informed Consent Statement: Not applicable.

Acknowledgments: The presented study is part of research proposal "Genotyping and development of heat stress tolerant cot-ton germplasm having enhanced quality traits" No. 964, and the authors are grateful to the Center for Advanced Studies and Punjab Agricultural Research Board (CAS-PARB), Pakistan, for providing funds for this study.

Conflicts of Interest: The authors declare no conflict of interest.

Disclaimer: USDA is an equal opportunity provider and employer.

References

1. Meyer, L.A. The World and US Cotton Outlook for 2019/20. 2019. Available online: https://www.usda.gov/sites/default/files/documents/Leslie_Meyer.pdf (accessed on 12 May 2021).
2. Khan, M.A.; Wahid, A.; Ahmad, M.; Tahir, M.T.; Ahmed, M.; Ahmad, S.; Hasanuzzaman, M. World cotton production and consumption: An overview. *Cotton Prod. Uses* **2020**, 1–7. [CrossRef]
3. Zahra, N.; Shaukat, K.; Hafeez, M.B.; Raza, A.; Hussain, S.; Chaudhary, M.T.; Wahid, A. Physiological and molecular responses to high chilling and freezing temperature in plant growth and production: Consequences and mitigation possibilities. In *Harsh Environment and Plant Resilience: Molecular and Functional Aspects*; Springer: New York, NY, USA, 2021; p. 235.
4. Zhu, Y.N.; Shi, D.Q.; Ruan, M.B.; Zhang, L.L.; Meng, Z.H.; Liu, J.; Yang, W.C. Transcriptome analysis reveals crosstalk of responsive genes to multiple abiotic stresses in cotton (*Gossypium hirsutum* L.). *PLoS ONE* **2013**, *8*, e80218.
5. Downes, S.; Walsh, T.; Tay, W.T. Bt resistance in Australian insect pest species. *Curr. Opin. Insect Sci.* **2016**, *15*, 78–83. [CrossRef]
6. Van der Sluijs, M.J.; Hunter, L. A review on the formation, causes, measurement, implications and reduction of neps during cotton processing. *Text. Prog.* **2016**, *48*, 221–323. [CrossRef]
7. Hall, A.E.; Botany and Plant Sciences Department University of California, Riverside. Heat Stress and Its Impact. 2001. Available online: https://plantstress.com/heat/ (accessed on 13 June 2021).
8. Dhyani, K.; Ansari, M.W.; Rao, Y.R.; Verma, R.S.; Shukla, A.; Tuteja, N. Comparative physiological response of wheat genotypes under terminal heat stress. *Plant Signal. Behav.* **2013**, *8*, e24564. [CrossRef] [PubMed]
9. Phillips, J.B. Cotton Response to High Temperature Stress During Reproductive Development. 2012. Available online: https://scholarworks.uark.edu/etd/394 (accessed on 22 July 2021).
10. Raza, A.; Ahmad, M. Analysing the Impact of Climate Change on Cotton Productivity in Punjab and Sindh, Pakistan. 2015. Available online: https://mpra.ub.uni-muenchen.de/72867/ (accessed on 7 July 2021).
11. Zahid, K.R.; Ali, F.; Shah, F.; Younas, M.; Shah, T.; Shahwar, D.; Hassan, W.; Ahmad, Z.; Qi, C.; Lu, Y. Response and tolerance mechanism of cotton *Gossypium hirsutum* L. to elevated temperature stress: A review. *Front. Plant Sci.* **2016**, *7*, 937. [CrossRef]
12. Shamsuzzaman, K.; Hamid, M.; Azad, M.; Hussain, M.; Majid, M. Varietal improvement of cotton (*Gossypium hirsutum*) through mutation breeding. In *Improvement of New and Traditional Industrial Crops by Induced Mutations and Related Biotechnology*; International Atomic Energy Agency: Vienna, Austria, 2003; pp. 81–94.
13. Ahloowalia, B.; Maluszynski, M.; Nichterlein, K. Global impact of mutation-derived varieties. *Euphytica* **2004**, *135*, 187–204. [CrossRef]
14. Challinor, A.; Wheeler, T.; Craufurd, P.; Slingo, J. Simulation of the impact of high temperature stress on annual crop yields. *Agric. For. Meteorol.* **2005**, *135*, 180–189. [CrossRef]
15. Snider, J.L.; Oosterhuis, D.M.; Skulman, B.W.; Kawakami, E.M. Heat stress-induced limitations to reproductive success in *Gossypium hirsutum*. *Physiol. Plant.* **2009**, *137*, 125–138. [CrossRef]
16. Khan, A.H.; Min, L.; Ma, Y.; Wu, Y.; Ding, Y.; Li, Y.; Xie, S.; Ullah, A.; Shaban, M.; Manghwar, H. High day and night temperatures distinctively disrupt fatty acid and jasmonic acid metabolism, inducing male sterility in cotton. *J. Exp. Bot.* **2020**, *71*, 6128–6141. [CrossRef]
17. Nabi, G.; Mullins, C. Soil temperature dependent growth of cotton seedlings before emergence. *Pedosphere* **2008**, *18*, 54–59. [CrossRef]
18. Zia, S.; Khan, M.A. Effect of light, salinity, and temperature on seed germination of Limonium stocksii. *Can. J. Bot.* **2004**, *82*, 151–157. [CrossRef]
19. Riley, G.J. Effects of high temperature on the germination of maize (*Zea mays* L.). *Planta* **1981**, *151*, 68–74. [CrossRef] [PubMed]
20. Burke, J.J.; Upchurch, D.R. Leaf temperature and transpirational control in cotton. *Environ. Exp. Bot.* **1989**, *29*, 487–492. [CrossRef]
21. Krzyzanowski, F.C.; Delouche, J.C. Germination of cotton seed in relation to temperature. *Rev. Bras. Sementes* **2011**, *33*, 543–548. [CrossRef]
22. Parera, C.A.; Cantliffe, D.J. Presowing seed priming. *Hortic. Rev.* **1994**, *16*, 109–141.
23. Raphael, J.; Gazola, B.; Nunes, J.G.; Macedo, G.C.; Rosolem, C.A. Cotton germination and emergence under high diurnal temperatures. *Crop Sci.* **2017**, *57*, 2761–2769. [CrossRef]
24. Carmo-Silva, A.E.; Gore, M.A.; Andrade-Sanchez, P.; French, A.N.; Hunsaker, D.J.; Salvucci, M.E. Decreased CO2 availability and inactivation of Rubisco limit photosynthesis in cotton plants under heat and drought stress in the field. *Environ. Exp. Bot.* **2012**, *83*, 1–11. [CrossRef]
25. Sethar, M.A.; Laidman, D.; Pahoja, V.M.; Chachar, Q.; Mirjat, M.A. Thermotolerance in Cotton Roots at Early Stage of Emergence. *Pak. J. Biol. Sci.* **2001**, *4*, 361–364. [CrossRef]

26. Reddy, K.R.; Hodges, H.F.; McKinion, J.M. A comparison of scenarios for the effect of global climate change on cotton growth and yield. *Funct. Plant Biol.* **1997**, *24*, 707–713. [CrossRef]
27. Kamal, M.; Saleem, M.; Shahid, M.; Awais, M.; Khan, H.; Ahmed, K. Ascorbic acid triggered physiochemical transformations at different phenological stages of heat-stressed Bt cotton. *J. Agron. Crop Sci.* **2017**, *203*, 323–331. [CrossRef]
28. Saleem, M.F.; Kamal, M.A.; Anjum, S.A.; Shahid, M.; Raza, M.A.S.; Awais, M. Improving the performance of Bt-cotton under heat stress by foliar application of selenium. *J. Plant Nutr.* **2018**, *41*, 1711–1723. [CrossRef]
29. Brown, P. Cotton Heat Stress. 2008. Available online: https://cals.arizona.edu/azmet/az1448.pdf (accessed on 14 July 2021).
30. Tariq, M.; Yasmeen, A.; Ahmad, S.; Hussain, N.; Afzal, M.N.; Hasanuzzaman, M. Shedding of fruiting structures in cotton: Factors, compensation and prevention. *Trop. Subtrop. Agroecosystems* **2017**, *20*, 251–262.
31. Karademir, E.; Karademir, Ç.; Ekinci, R.; Başbağ, S.; Başal, H. Screening cotton varieties (*Gossypium hirsutum* L.) for heat tolerance under field conditions. *Afr. J. Agric. Res.* **2012**, *7*, 6335–6342.
32. Rauf, S.; Khan, T.M.; Naveed, A.; Munir, H. Modified path to high lint yield in upland cotton (*Gossypium hirsutum* L.) under two temperature regimes. *Turk. J. Biol.* **2007**, *31*, 119–126.
33. Kamal, M.; Saleem, M.; Wahid, M.; Shakeel, A. Effects of ascorbic acid on membrane stability and yield of heatstressed BT cotton. *J. Anim. Plant Sci.* **2017**, *27*, 192–199.
34. Singh, K.; Wijewardana, C.; Gajanayake, B.; Lokhande, S.; Wallace, T.; Jones, D.; Reddy, K.R. Genotypic variability among cotton cultivars for heat and drought tolerance using reproductive and physiological traits. *Euphytica* **2018**, *214*, 57. [CrossRef]
35. Zeiher, C.A.; Brown, P.W.; Silvertooth, J.C.; Matumba, N.; Mitton, N. The Effect of Night Temperature on Cotton Reproductive Development. 1994. Available online: https://repository.arizona.edu/bitstream/handle/10150/209598/370096-089-096.pdf?sequence=1 (accessed on 27 June 2021).
36. Zeiher, C.; Matumba, N.; Brown, P.; Silvertooth, J. Response of upland cotton to elevated night temperatures. II. Results of controlled environmental studies. In Proceedings of the Beltwide Cotton Conferences, San Antonio, TX, USA, 4–7 January 1995.
37. Brown, P.; Zeiher, C. A model to estimate cotton canopy temperature in the desert southwest. In Proceedings of the Beltwide Cotton Conferences, San Diego, CA, USA, 5–9 January 1998; National Cotton Council of America: Memphis, TN, USA, 1998.
38. Sabagh, A.E.; Hossain, A.; Islam, M.S.; Barutcular, C.; Ratnasekera, D.; Gormus, O.; Amanet, K.; Mubeen, M.; Nasim, W.; Fahad, S. Drought and heat stress in cotton (*Gossypium hirsutum* L.): Consequences and their possible mitigation strategies. In *Agroomic Crops*; Springer: New York, NY, USA, 2020. [CrossRef]
39. Yaqoob, M.; Fiaz, S.; Ijaz, B. Correlation analysis for yield and fiber quality traits in upland cotton. *Commun. Plant Sc.* **2016**, *6*, 55–60.
40. Peng, S.; Krieg, D.; Hicks, S. Cotton lint yield response to accumulated heat units and soil water supply. *Field Crops Res.* **1989**, *19*, 253–262. [CrossRef]
41. Reddy, K.R.; Davidonis, G.H.; Johnson, A.S.; Vinyard, B.T. Temperature regime and carbon dioxide enrichment alter cotton boll development and fiber properties. *Agron. J.* **1999**, *91*, 851–858. [CrossRef]
42. Pettigrew, W. The effect of higher temperatures on cotton lint yield production and fiber quality. *Crop Sci.* **2008**, *48*, 278–285. [CrossRef]
43. Abro, S.; Rajput, M.T.; Khan, M.A.; Sial, M.A.; Tahir, S.S. Screening of cotton (*Gossypium hirsutum* L.) genotypes for heat tolerance. *Pak. J. Bot* **2015**, *47*, 2085–2091.
44. Liakatas, A.; Roussopoulos, D.; Whittington, W. Controlled-temperature effects on cotton yield and fibre properties. *J. Agric. Sci.* **1998**, *130*, 463–471. [CrossRef]
45. Li, X.; Shi, W.; Broughton, K.; Smith, R.; Sharwood, R.; Payton, P.; Bange, M.; Tissue, D.T. Impacts of growth temperature, water deficit and heatwaves on carbon assimilation and growth of cotton plants (*Gossypium hirsutum* L.). *Environ. Exp. Bot.* **2020**, *179*, 104204. [CrossRef]
46. He, X.-Y.; Zhou, Z.-G.; Dai, Y.-J.; Qiang, Z.-Y.; Chen, B.-L.; Wang, Y.-H. Effect of increased temperature in boll period on fiber yield and quality of cotton and its physiological mechanism. *Yingyong Shengtai Xuebao* **2013**, *24*, 12.
47. Bo, X.; Zhou, Z.-G.; Guo, L.-T.; Xu, W.-Z.; Zhao, W.-Q.; Chen, B.-L.; Meng, Y.-L.; Wang, Y.-H. Susceptible time window and endurable duration of cotton fiber development to high temperature stress. *J. Integr. Agric.* **2017**, *16*, 1936–1945.
48. Abdelgadir, H.; Johnson, S.; Van Staden, J. Pollen viability, pollen germination and pollen tube growth in the biofuel seed crop *Jatropha curcas* (Euphorbiaceae). *S. Afr. J. Bot.* **2012**, *79*, 132–139. [CrossRef]
49. Kakani, V.; Reddy, K.; Koti, S.; Wallace, T.; Prasad, P.; Reddy, V.; Zhao, D. Differences in in vitro pollen germination and pollen tube growth of cotton cultivars in response to high temperature. *Ann. Bot.* **2005**, *96*, 59–67. [CrossRef]
50. Burke, J.J.; Velten, J.; Oliver, M.J. In vitro analysis of cotton pollen germination. *Agron. J.* **2004**, *96*, 359–368. [CrossRef]
51. Hedhly, A.; Hormaza, J.; Herrero, M. The effect of temperature on stigmatic receptivity in sweet cherry (*Prunus avium* L.). *Plant Cell Environ.* **2003**, *26*, 1673–1680. [CrossRef]
52. Hedhly, A.; Hormaza, J.; Herrero, M. The effect of temperature on pollen germination, pollen tube growth, and stigmatic receptivity in peach. *Plant Biol.* **2005**, *7*, 476–483. [CrossRef]
53. Snider, J.L.; Oosterhuis, D.M. Heat stress and pollen-pistil interactions. In *Flowering and Fruiting in Cotton*; Publ. Cotton Foundation: Memphis, TN, USA, 2012; pp. 59–78.
54. Gago, J.; de Menezes Daloso, D.; Figueroa, C.M.; Flexas, J.; Fernie, A.R.; Nikoloski, Z. Relationships of leaf net photosynthesis, stomatal conductance, and mesophyll conductance to primary metabolism: A multispecies meta-analysis approach. *Plant Physiol.* **2016**, *171*, 265–279. [CrossRef] [PubMed]

55. Hasanuzzaman, M.; Nahar, K.; Alam, M.; Roychowdhury, R.; Fujita, M. Physiological, biochemical, and molecular mechanisms of heat stress tolerance in plants. *Int. J. Mol. Sci.* **2013**, *14*, 9643–9684. [CrossRef] [PubMed]
56. Crafts-Brandner, S.; Law, R. Effect of heat stress on the inhibition and recovery of the ribulose-1, 5-bisphosphate carboxylase/oxygenase activation state. *Planta* **2000**, *212*, 67–74. [CrossRef] [PubMed]
57. Schrader, S.; Wise, R.; Wacholtz, W.; Ort, D.R.; Sharkey, T. Thylakoid membrane responses to moderately high leaf temperature in Pima cotton. *Plant Cell Environ.* **2004**, *27*, 725–735. [CrossRef]
58. Karademir, E.; Karademir, C.; Sevilmis, U.; Basal, H. Correlations between canopy temperature, chlorophyll content and yield in heat tolerant cotton (*Gossypium hirsutum* L.) genotypes. *Fresenius Environ. Bull.* **2018**, *27*, 5230–5237.
59. Bita, C.; Gerats, T. Plant tolerance to high temperature in a changing environment: Scientific fundamentals and production of heat stress-tolerant crops. *Front. Plant Sci.* **2013**, *4*, 273. [CrossRef]
60. Bibi, A.; Oosterhuis, D.; Brown, R.; Gonias, E.; Bourland, F. The physiological response of cotton to high temperatures for germplasm screening. *Summ. Ark. Cotton Res.* **2003**, *521*, 87–93.
61. Rahman, H.; Malik, S.A.; Saleem, M. Heat tolerance of upland cotton during the fruiting stage evaluated using cellular membrane thermostability. *Field Crops Res.* **2004**, *85*, 149–158. [CrossRef]
62. Khan, N.; Ahmad, I.; Azhar, M.T. Genetic variation in relative cell injury for breeding upland cotton under high temperature stress. *Turk. J. Field Crops* **2017**, *22*, 152–159. [CrossRef]
63. Urban, J.; Ingwers, M.; McGuire, M.A.; Teskey, R.O. Stomatal conductance increases with rising temperature. *Plant Signal. Behav.* **2017**, *12*, e1356534. [CrossRef] [PubMed]
64. Lu, Z.; Percy, R.G.; Qualset, C.O.; Zeiger, E. Stomatal conductance predicts yields in irrigated Pima cotton and bread wheat grown at high temperatures. *J. Exp. Bot.* **1998**, 453–460. [CrossRef]
65. Lu, Z.; Chen, J.; Percy, R.G.; Zeiger, E. Photosynthetic rate, stomatal conductance and leaf area in two cotton species (*Gossypium barbadense* and *Gossypium hirsutum*) and their relation with heat resistance and yield. *Funct. Plant Biol.* **1997**, *24*, 693–700. [CrossRef]
66. Lu, Z.; Quiñones, M.A.; Zeiger, E. Temperature dependence of guard cell respiration and stomatal conductance co-segregate in an F2 population of Pima cotton. *Funct. Plant Biol.* **2000**, *27*, 457–462. [CrossRef]
67. Khan, A.I.; Khan, I.A.; Sadaqat, H.A. Heat tolerance is variable in cotton (*Gossypium hirsutum* L.) and can be exploited for breeding of better yielding cultivars under high temperature regimes. *Pak. J. Bot.* **2008**, *40*, 2053–2058.
68. Rahman, H.; Murtaza, N.; Shah, K.; Qayyum, A.; Ullah, I.; Malik, W. Genetic variation for stomatal conductance in upland cotton as influenced by heat-stressed and non-stressed growing regimes. *Acta Agron. Hung.* **2008**, *56*, 11–19. [CrossRef]
69. Suzuki, N.; Mittler, R. Reactive oxygen species and temperature stresses: A delicate balance between signaling and destruction. *Physiol. Plant.* **2006**, *126*, 45–51. [CrossRef]
70. Suzuki, N.; Koussevitzky, S.; Mittler, R.; Miller, G. ROS and redox signalling in the response of plants to abiotic stress. *Plant Cell Environ.* **2012**, *35*, 259–270. [CrossRef]
71. Considine, M.J.; María Sandalio, L.; Helen Foyer, C. Unravelling how plants benefit from ROS and NO reactions, while resisting oxidative stress. *Ann. Bot.* **2015**, *116*, 469–473. [CrossRef]
72. Singh, R.; Singh, S.; Parihar, P.; Mishra, R.K.; Tripathi, D.K.; Singh, V.P.; Chauhan, D.K.; Prasad, S.M. Reactive oxygen species (ROS): Beneficial companions of plants' developmental processes. *Front. Plant Sci.* **2016**, *7*, 1299. [CrossRef]
73. Mittler, R. Oxidative stress, antioxidants and stress tolerance. *Trends Plant Sci.* **2002**, *7*, 405–410. [CrossRef]
74. Majeed, S.; Malik, T.A.; Rana, I.A.; Azhar, M.T. Antioxidant and physiological responses of upland cotton accessions grown under high-temperature regimes. *Iran. J. Sci. Technol. Trans. A Sci.* **2019**, *43*, 2759–2768. [CrossRef]
75. Sekmen, A.H.; Ozgur, R.; Uzilday, B.; Turkan, I. Reactive oxygen species scavenging capacities of cotton (*Gossypium hirsutum*) cultivars under combined drought and heat induced oxidative stress. *Environ. Exp. Bot.* **2014**, *99*, 141–149. [CrossRef]
76. Gür, A.; Demirel, U.; Özden, M.; Kahraman, A.; Çopur, O. Diurnal gradual heat stress affects antioxidant enzymes, proline accumulation and some physiological components in cotton (*Gossypium hirsutum* L.). *Afr. J. Biotechnol.* **2010**, *9*, 1008–1015.
77. Sarwar, M.; Saleem, M.F.; Ullah, N.; Rizwan, M.; Ali, S.; Shahid, M.R.; Alamri, S.A.; Alyemeni, M.N.; Ahmad, P. Exogenously applied growth regulators protect the cotton crop from heat-induced injury by modulating plant defense mechanism. *Sci. Rep.* **2018**, *8*, 17086. [CrossRef]
78. Loka, D.A.; Oosterhuis, D.M. Effect of high night temperatures during anthesis on cotton (*Gossypium hirsutum* L.) pistil and leaf physiology and biochemistry. *Aust. J. Crop Sci.* **2016**, *10*, 741. [CrossRef]
79. Wang, W.; Vinocur, B.; Shoseyov, O.; Altman, A. Role of plant heat-shock proteins and molecular chaperones in the abiotic stress response. *Trends Plant Sci.* **2004**, *9*, 244–252. [CrossRef]
80. Kotak, S.; Larkindale, J.; Lee, U.; von Koskull-Döring, P.; Vierling, E.; Scharf, K.-D. Complexity of the heat stress response in plants. *Curr. Opin. Plant Biol.* **2007**, *10*, 310–316. [CrossRef]
81. Waters, E.R.; Lee, G.J.; Vierling, E. Evolution, structure and function of the small heat shock proteins in plants. *J. Exp. Bot.* **1996**, *47*, 325–338. [CrossRef]
82. Mogk, A.; Schlieker, C.; Friedrich, K.L.; Schönfeld, H.-J.; Vierling, E.; Bukau, B. Refolding of substrates bound to small Hsps relies on a disaggregation reaction mediated most efficiently by ClpB/DnaK. *J. Biol. Chem.* **2003**, *278*, 31033–31042. [CrossRef]
83. Van Montfort, R.L.; Basha, E.; Friedrich, K.L.; Slingsby, C.; Vierling, E. Crystal structure and assembly of a eukaryotic small heat shock protein. *Nat. Struct. Mol. Biol.* **2001**, *8*, 1025. [CrossRef]

84. Malik, M.K.; Slovin, J.P.; Hwang, C.H.; Zimmerman, J.L. Modified expression of a carrot small heat shock protein gene, Hsp17. 7, results in increased or decreased thermotolerance. *Plant J.* **1999**, *20*, 89–99. [CrossRef]
85. Maqbool, A.; Zahur, M.; Irfan, M.; Qaiser, U.; Rashid, B.; Husnain, T. Identification, characterization and expression of drought related alpha-crystalline heat shock protein gene (*GHSP26*) from Desi cotton. *Crop Sci.* **2007**, *47*, 2437–2444. [CrossRef]
86. Lee, G.J.; Vierling, E. A small heat shock protein cooperates with heat shock protein 70 systems to reactivate a heat-denatured protein. *Plant Physiol.* **2000**, *122*, 189–198. [CrossRef]
87. Haslbeck, M.; Vierling, E. A first line of stress defense: Small heat shock proteins and their function in protein homeostasis. *J. Mol. Biol.* **2015**, *427*, 1537–1548. [CrossRef]
88. Boston, R.S.; Viitanen, P.V.; Vierling, E. Molecular chaperones and protein folding in plants. In *Post-Transcriptional Control of Gene Expression in Plants*; Springer: Berlin/Heidelberg, Germany, 1996; pp. 191–222.
89. Hartl, F.U.; Bracher, A.; Hayer-Hartl, M. Molecular chaperones in protein folding and proteostasis. *Nature* **2011**, *475*, 324. [CrossRef]
90. Sung, D.Y.; Kaplan, F.; Guy, C.L. Plant Hsp70 molecular chaperones: Protein structure, gene family, expression and function. *Physiol. Plant.* **2001**, *113*, 443–451. [CrossRef]
91. Scarpeci, T.E.; Zanor, M.I.; Valle, E.M. Investigating the role of plant heat shock proteins during oxidative stress. *Plant Signal. Behav.* **2008**, *3*, 856–857. [CrossRef] [PubMed]
92. Li, H.; Liu, S.S.; Yi, C.Y.; Wang, F.; Zhou, J.; Xia, X.J.; Shi, K.; Zhou, Y.H.; Yu, J.Q. Hydrogen peroxide mediates abscisic acid-induced HSP 70 accumulation and heat tolerance in grafted cucumber plants. *Plant Cell Environ.* **2014**, *37*, 2768–2780. [CrossRef] [PubMed]
93. Rehman, A.; Atif, R.M.; Qyyum, A.; Du, X.; Hinze, L.; Azhar, M.T. Genome-wide identification and characterization of HSP70 gene family in four species of cotton. *Genomics* **2020**, *112*, 4442–4453. [CrossRef] [PubMed]
94. Krishna, P.; Gloor, G. The Hsp90 family of proteins in Arabidopsis thaliana. *Cell Stress Chaperones* **2001**, *6*, 238. [CrossRef]
95. Kadota, Y.; Shirasu, K. The HSP90 complex of plants. *Biochim. Et Biophys. Acta (BBA)-Mol. Cell Res.* **2012**, *1823*, 689–697. [CrossRef]
96. Sable, A.; Rai, K.M.; Choudhary, A.; Yadav, V.K.; Agarwal, S.K.; Sawant, S.V. Inhibition of heat shock proteins HSP90 and HSP70 induce oxidative stress, suppressing cotton fiber development. *Sci. Rep.* **2018**, *8*, 3620. [CrossRef] [PubMed]
97. Zietkiewicz, S.; Krzewska, J.; Liberek, K. Successive and synergistic action of the Hsp70 and Hsp100 chaperones in protein disaggregation. *J. Biol. Chem.* **2004**, *279*, 44376–44383. [CrossRef] [PubMed]
98. Winkler, J.; Tyedmers, J.; Bukau, B.; Mogk, A. Hsp70 targets Hsp100 chaperones to substrates for protein disaggregation and prion fragmentation. *J. Cell Biol* **2012**, *198*, 387–404. [CrossRef]
99. Hemmingsen, S.M.; Woolford, C.; van der Vies, S.M.; Tilly, K.; Dennis, D.T.; Georgopoulos, C.P.; Hendrix, R.W.; Ellis, R.J. Homologous plant and bacterial proteins chaperone oligomeric protein assembly. *Nature* **1988**, *333*, 330. [CrossRef]
100. Apuya, N.R.; Yadegari, R.; Fischer, R.L.; Harada, J.J.; Zimmerman, J.L.; Goldberg, R.B. The Arabidopsis embryo mutant schlepperless has a defect in the chaperonin-60α gene. *Plant Physiol.* **2001**, *126*, 717–730. [CrossRef]
101. Ishikawa, A.; Tanaka, H.; Nakai, M.; Asahi, T. Deletion of a chaperonin *60β* gene leads to cell death in the Arabidopsis lesion initiation 1 mutant. *Plant Cell Physiol.* **2003**, *44*, 255–261. [CrossRef]
102. Zabaleta, E.; Oropeza, A.; Assad, N.; Mandel, A.; Salerno, G.; Herrera-Estrella, L. Antisense expression of chaperonin *60β* in transgenic tobacco plants leads to abnormal phenotypes and altered distribution of photoassimilates. *Plant J.* **1994**, *6*, 425–432. [CrossRef]
103. Miemyk, J. The 70 kDa stress-related proteins as molecular chaperones. *Trends Plant Sci.* **1997**, *2*, 180–187. [CrossRef]
104. Zhang, Y.; Wang, M.; Chen, J.; Rong, J.; Ding, M. Genome-wide analysis of HSP70 superfamily in *Gossypium raimondii* and the expression of orthologs in *Gossypium hirsutum*. *Yi Chuan Hered.* **2014**, *36*, 921–933.
105. Young, J.C.; Moarefi, I.; Hartl, F.U. Hsp90: A specialized but essential protein-folding tool. *J. Cell Biol.* **2001**, *154*, 267. [CrossRef] [PubMed]
106. Buchner, J. Hsp90 & Co.—A holding for folding. *Trends Biochem. Sci.* **1999**, *24*, 136–141.
107. Zhang, Z.; Quick, M.K.; Kanelakis, K.C.; Gijzen, M.; Krishna, P. Characterization of a plant homolog of hop, a cochaperone of hsp90. *Plant Physiol.* **2003**, *131*, 525–535. [CrossRef] [PubMed]
108. Latterich, M.; Patel, S. The AAA team: Related ATPases with diverse functions. *Trends Cell Biol.* **1998**, *8*, 65–71. [CrossRef]
109. Lee, U.; Rioflorido, I.; Hong, S.W.; Larkindale, J.; Waters, E.R.; Vierling, E. The Arabidopsis ClpB/Hsp100 family of proteins: Chaperones for stress and chloroplast development. *Plant J.* **2007**, *49*, 115–127. [CrossRef]
110. Mishra, R.C.; Grover, A. ClpB/Hsp100 proteins and heat stress tolerance in plants. *Crit. Rev. Biotechnol.* **2016**, *36*, 862–874. [CrossRef]
111. Barrera-Figueroa, B.E.; Gao, L.; Wu, Z.; Zhou, X.; Zhu, J.; Jin, H.; Liu, R.; Zhu, J.-K. High throughput sequencing reveals novel and abiotic stress-regulated microRNAs in the inflorescences of rice. *BMC Plant Biol.* **2012**, *12*, 132. [CrossRef]
112. Barakat, A.; Sriram, A.; Park, J.; Zhebentyayeva, T.; Main, D.; Abbott, A. Genome wide identification of chilling responsive microRNAs in Prunus persica. *BMC Genom.* **2012**, *13*, 481. [CrossRef]
113. Huang, J.; Yang, M.; Zhang, X. The function of small RNAs in plant biotic stress response. *J. Integr. Plant Biol.* **2016**, *58*, 312–327. [CrossRef] [PubMed]
114. Wang, J.-W.; Czech, B.; Weigel, D. miR156-regulated SPL transcription factors define an endogenous flowering pathway in Arabidopsis thaliana. *Cell* **2009**, *138*, 738–749. [CrossRef] [PubMed]

115. Chen, G.; Li, J.; Liu, Y.; Zhang, Q.; Gao, Y.; Fang, K.; Cao, Q.; Qin, L.; Xing, Y. Roles of the GA-mediated SPL Gene Family and miR156 in the Floral Development of Chinese Chestnut (*Castanea mollissima*). *Int. J. Mol. Sci.* **2019**, *20*, 1577. [CrossRef] [PubMed]
116. Liu, N.; Tu, L.; Wang, L.; Hu, H.; Xu, J.; Zhang, X. MicroRNA 157-targeted SPL genes regulate floral organ size and ovule production in cotton. *BMC Plant Biol.* **2017**, *17*, 7. [CrossRef]
117. Yu, X.; Wang, H.; Lu, Y.; de Ruiter, M.; Cariaso, M.; Prins, M.; van Tunen, A.; He, Y. Identification of conserved and novel microRNAs that are responsive to heat stress in Brassica rapa. *J. Exp. Bot.* **2011**, *63*, 1025–1038. [CrossRef]
118. Song, C.; Fang, J.; Li, X.; Liu, H.; Chao, C.T. Identification and characterization of 27 conserved microRNAs in citrus. *Planta* **2009**, *230*, 671–685. [CrossRef] [PubMed]
119. Stief, A.; Altmann, S.; Hoffmann, K.; Pant, B.D.; Scheible, W.-R.; Bäurle, I. Arabidopsis miR156 regulates tolerance to recurring environmental stress through SPL transcription factors. *Plant Cell* **2014**, *26*, 1792–1807. [CrossRef] [PubMed]
120. Wang, Y.; Sun, F.; Cao, H.; Peng, H.; Ni, Z.; Sun, Q.; Yao, Y. TamiR159 directed wheat TaGAMYB cleavage and its involvement in anther development and heat response. *PLoS ONE* **2012**, *7*, e48445. [CrossRef]
121. Ding, Y.; Ma, Y.; Liu, N.; Xu, J.; Hu, Q.; Li, Y.; Wu, Y.; Xie, S.; Zhu, L.; Min, L. micro RNA s involved in auxin signalling modulate male sterility under high-temperature stress in cotton (*Gossypium hirsutum*). *Plant J.* **2017**, *91*, 977–994. [CrossRef]
122. Wang, M.; Wang, Q.; Zhang, B. Response of miRNAs and their targets to salt and drought stresses in cotton (*Gossypium hirsutum* L.). *Gene* **2013**, *530*, 26–32. [CrossRef]
123. Sailaja, B.; Anjum, N.; Prasanth, V.V.; Sarla, N.; Subrahmanyam, D.; Voleti, S.; Viraktamath, B.; Mangrauthia, S.K. Comparative study of susceptible and tolerant genotype reveals efficient recovery and root system contributes to heat stress tolerance in rice. *Plant Mol. Biol. Rep.* **2014**, *32*, 1228–1240. [CrossRef]
124. Ruan, M.-B.; Zhao, Y.-T.; Meng, Z.-H.; Wang, X.-J.; Yang, W.-C. Conserved miRNA analysis in *Gossypium hirsutum* through small RNA sequencing. *Genomics* **2009**, *94*, 263–268. [CrossRef]
125. Boopathi, M.; Sathish, S.; Kavitha, P.; Dachinamoorthy, P.; Ravikesavan, R. Comparative miRNAome analysis revealed numerous conserved and novel drought responsive miRNAs in cotton (*Gossypium* spp.). *Cotton Genom. Genet.* **2016**. [CrossRef]
126. Fang, Y.; Xie, K.; Xiong, L. Conserved miR164-targeted NAC genes negatively regulate drought resistance in rice. *J. Exp. Bot.* **2014**, *65*, 2119–2135. [CrossRef]
127. Kumar, R.R.; Pathak, H.; Sharma, S.K.; Kala, Y.K.; Nirjal, M.K.; Singh, G.P.; Goswami, S.; Rai, R.D. Novel and conserved heat-responsive microRNAs in wheat (*Triticum aestivum* L.). *Funct. Integr. Genom.* **2015**, *15*, 323–348. [CrossRef] [PubMed]
128. Matthews, C.; Arshad, M.; Hannoufa, A. Alfalfa response to heat stress is modulated by microRNA156. *Physiol. Plant.* **2019**, *165*, 830–842. [CrossRef] [PubMed]
129. Zhao, J.; He, Q.; Chen, G.; Wang, L.; Jin, B. Regulation of non-coding RNAs in heat stress responses of plants. *Front. Plant Sci.* **2016**, *7*, 1213. [CrossRef]
130. Zhang, B.; Liu, J.; Yang, Z.E.; Chen, E.Y.; Zhang, C.J.; Zhang, X.Y.; Li, F.G. Genome-wide analysis of GRAS transcription factor gene family in *Gossypium hirsutum* L. *BMC Genom.* **2018**, *19*, 348. [CrossRef]
131. Aukerman, M.J.; Sakai, H. Regulation of flowering time and floral organ identity by a microRNA and its *APETALA2*-like target genes. *Plant Cell* **2003**, *15*, 2730–2741. [CrossRef]
132. François, L.; Verdenaud, M.; Fu, X.; Ruleman, D.; Dubois, A.; Vandenbussche, M.; Bendahmane, A.; Raymond, O.; Just, J.; Bendahmane, M. A miR172 target-deficient AP2-like gene correlates with the double flower phenotype in roses. *Sci. Rep.* **2018**, *8*, 12912. [CrossRef]
133. Wu, G.; Park, M.Y.; Conway, S.R.; Wang, J.-W.; Weigel, D.; Poethig, R.S. The sequential action of miR156 and miR172 regulates developmental timing in Arabidopsis. *Cell* **2009**, *138*, 750–759. [CrossRef] [PubMed]
134. Zhu, Q.-H.; Upadhyaya, N.M.; Gubler, F.; Helliwell, C.A. Over-expression of miR172 causes loss of spikelet determinacy and floral organ abnormalities in rice (*Oryza sativa*). *BMC Plant Biol.* **2009**, *9*, 149. [CrossRef] [PubMed]
135. Yin, Z.; Han, X.; Li, Y.; Wang, J.; Wang, D.; Wang, S.; Fu, X.; Ye, W. Comparative analysis of cotton small RNAs and their target genes in response to salt stress. *Genes* **2017**, *8*, 369. [CrossRef] [PubMed]
136. Wójcik, A.M.; Gaj, M.D. miR393 contributes to the embryogenic transition induced in vitro in Arabidopsis via the modification of the tissue sensitivity to auxin treatment. *Planta* **2016**, *244*, 231–243. [CrossRef]
137. Jeong, D.-H.; Green, P.J. The role of rice microRNAs in abiotic stress responses. *J. Plant Biol.* **2013**, *56*, 187–197. [CrossRef]
138. Xie, F.; Wang, Q.; Sun, R.; Zhang, B. Deep sequencing reveals important roles of microRNAs in response to drought and salinity stress in cotton. *J. Exp. Bot.* **2014**, *66*, 789–804. [CrossRef]
139. Wang, H.; Huang, J.; Lai, Z.; Xue, Y. F-box proteins in flowering plants. *Chin. Sci. Bull.* **2002**, *47*, 1497–1501. [CrossRef]
140. Xu, G.; Ma, H.; Nei, M.; Kong, H. Evolution of F-box genes in plants: Different modes of sequence divergence and their relationships with functional diversification. *Proc. Natl. Acad. Sci. USA* **2009**, *106*, 835–840. [CrossRef]
141. Song, J.B.; Gao, S.; Sun, D.; Li, H.; Shu, X.X.; Yang, Z.M. miR394 and LCR are involved in Arabidopsis salt and drought stress responses in an abscisic acid-dependent manner. *BMC Plant Biol.* **2013**, *13*, 210. [CrossRef]
142. Liu, Q.; Yang, T.; Yu, T.; Zhang, S.; Mao, X.; Zhao, J.; Wang, X.; Dong, J.; Liu, B. Integrating small RNA sequencing with QTL mapping for identification of miRNAs and their target genes associated with heat tolerance at the flowering stage in rice. *Front. Plant Sci.* **2017**, *8*, 43. [CrossRef]
143. Zhang, B. MicroRNA: A new target for improving plant tolerance to abiotic stress. *J. Exp. Bot.* **2015**, *66*, 1749–1761. [CrossRef]
144. Majeed, S.; Chaudhary, M.T.; Hulse-Kemp, A.M.; Azhar, M.T. Introduction: Crop Wild Relatives in Plant Breeding. In *Wild Germplasm for Genetic Improvement in Crop Plants*; Elsevier: Amsterdam, The Netherlands, 2021; pp. 1–18.

145. Raza, A.; Tabassum, J.; Kudapa, H.; Varshney, R.K. Can omics deliver temperature resilient ready-to-grow crops? *Crit. Rev. Biotechnol.* **2021**, 1–24. [CrossRef]
146. Percy, R.G. Comparison of bulk F2 performance testing and pedigree selection in thirty Pima cotton populations. *J. Cotton Sci.* **2003**, *7*, 123.
147. Tokatlidis, I.; Tsikrikoni, C.; Lithourgidis, A.; Tsialtas, J.; Tzantarmas, C. Intra-cultivar variation in cotton: Response to single-plant yield selection at low density. *J. Agric. Sci.* **2011**, *149*, 197–204. [CrossRef]
148. Mubarik, M.S.; Ma, C.; Majeed, S.; Du, X.; Azhar, M.T. Revamping of Cotton Breeding Programs for Efficient Use of Genetic Resources under Changing Climate. *Agronomy* **2020**, *10*, 1190. [CrossRef]
149. Badigannavar, A.; Myers, G.O.; Jones, D.C. Molecular diversity revealed by AFLP markers in upland cotton genotypes. *J. Crop Improv.* **2012**, *26*, 627–640. [CrossRef]
150. Mohamed, H.; Abdel-Hamid, A. Molecular and biochemical studies for heat tolerance on four cotton genotypes. *Rom. Biotechnol. Lett.* **2013**, *18*, 8823–8831.
151. Du, L.; Cai, C.; Wu, S.; Zhang, F.; Hou, S.; Guo, W. Evaluation and exploration of favorable QTL alleles for salt stress related traits in cotton cultivars (*G. hirsutum* L.). *PLoS ONE* **2016**, *11*, e0151076. [CrossRef]
152. Majeed, S.; Rana, I.A.; Atif, R.M.; Zulfiqar, A.; Hinze, L.; Azhar, M.T. Role of SNPs in determining QTLs for major traits in cotton. *J. Cotton Res.* **2019**, *2*, 5. [CrossRef]
153. Demirel, U.; Gür, A.; Can, N.; Memon, A. Identification of heat responsive genes in cotton. *Biol. Plant.* **2014**, *58*, 515–523. [CrossRef]
154. Zhang, J.; Srivastava, V.; Stewart, J.M.; Underwood, J. Heat-tolerance in Cotton Is Correlated with Induced Overexpression of Heat-Shock Factors, Heat-Shock Proteins, and General Stress Response Genes. *J. Cotton Sci.* **2016**, *20*, 253–262.
155. Shaheen, T.; Asif, M.; Zafar, Y. Single nucleotide polymorphism analysis of *MT-SHSP* gene of *Gossypium arboreum* and its relationship with other diploid cotton genomes, *G. hirsutum* and *Arabidopsis thaliana*. *Pak. J. Bot.* **2009**, *41*, 117–183.
156. Ma, W.; Zhao, T.; Li, J.; Liu, B.; Fang, L.; Hu, Y.; Zhang, T. Identification and characterization of the *GhHsp20* gene family in *Gossypium hirsutum*. *Sci. Rep.* **2016**, *6*, 32517. [CrossRef]
157. Islam, M.S.; Fang, D.D.; Jenkins, J.N.; Guo, J.; McCarty, J.C.; Jones, D.C. Evaluation of genomic selection methods for predicting fiber quality traits in Upland cotton. *Mol. Genet. Genom.* **2020**, *295*, 67–79. [CrossRef] [PubMed]
158. Baytar, A.A.; Peynircioğlu, C.; Sezener, V.; Basal, H.; Frary, A.; Frary, A.; Doğanlar, S. Genome-wide association mapping of yield components and drought tolerance-related traits in cotton. *Mol. Breed.* **2018**, *38*, 74. [CrossRef]
159. Su, J.; Pang, C.; Wei, H.; Li, L.; Liang, B.; Wang, C.; Song, M.; Wang, H.; Zhao, S.; Jia, X. Identification of favorable SNP alleles and candidate genes for traits related to early maturity via GWAS in upland cotton. *BMC Genom.* **2016**, *17*, 687. [CrossRef]
160. Batcho, A.A.; Sarwar, M.B.; Rashid, B.; Hassan, S.; Husnain, T. Heat shock protein gene identified from Agave sisalana (As HSP70) confers heat stress tolerance in transgenic cotton (*Gossypium hirsutum*). *Theor. Exp. Plant Physiol.* **2021**, *33*, 141–156. [CrossRef]
161. Esmaeili, N.; Cai, Y.; Tang, F.; Zhu, X.; Smith, J.; Mishra, N.; Hequet, E.; Ritchie, G.; Jones, D.; Shen, G. Towards doubling fibre yield for cotton in the semiarid agricultural area by increasing tolerance to drought, heat and salinity simultaneously. *Plant Biotechnol. J.* **2021**, *19*, 462. [CrossRef]
162. Burke, J.J.; Chen, J. Enhancement of reproductive heat tolerance in plants. *PLoS ONE* **2015**, *10*, e0122933. [CrossRef]
163. Mishra, N.; Sun, L.; Zhu, X.; Smith, J.; Prakash Srivastava, A.; Yang, X.; Pehlivan, N.; Esmaeili, N.; Luo, H.; Shen, G. Overexpression of the rice SUMO E3 ligase gene OsSIZ1 in cotton enhances drought and heat tolerance, and substantially improves fiber yields in the field under reduced irrigation and rainfed conditions. *Plant Cell Physiol.* **2017**, *58*, 735–746. [CrossRef]
164. Hozain, M.d.; Abdelmageed, H.; Lee, J.; Kang, M.; Fokar, M.; Allen, R.D.; Holaday, A.S. Expression of *AtSAP5* in cotton up-regulates putative stress-responsive genes and improves the tolerance to rapidly developing water deficit and moderate heat stress. *J. Plant Physiol.* **2012**, *169*, 1261–1270. [CrossRef]
165. Ali, S.; Rizwan, M.; Arif, M.S.; Ahmad, R.; Hasanuzzaman, M.; Ali, B.; Hussain, A. Approaches in enhancing thermotolerance in plants: An updated review. *J. Plant Growth Regul.* **2020**, *39*, 456–480. [CrossRef]
166. Meriç, S.; Ayan, A.; Atak, Ç. Molecular Abiotic Stress Tolerans Strategies: From Genetic Engineering to Genome Editing Era. In *Abiotic Stress in Plants*; IntechOpen: London, UK, 2020.
167. Mubarik, M.S.; Khan, S.H.; Ahmad, A.; Khan, Z.; Sajjad, M.; Khan, I.A. Disruption of Phytoene Desaturase Gene using Transient Expression of Cas9: gRNA Complex. *Int. J. Agric. Biol.* **2016**, *18*. [CrossRef]
168. Bhatnagar-Mathur, P.; Vadez, V.; Sharma, K.K. Transgenic approaches for abiotic stress tolerance in plants: Retrospect and prospects. *Plant Cell Rep.* **2008**, *27*, 411–424. [CrossRef]
169. Ruiz-Vera, U.M.; Siebers, M.H.; Drag, D.W.; Ort, D.R.; Bernacchi, C.J. Canopy warming caused photosynthetic acclimation and reduced seed yield in maize grown at ambient and elevated [CO_2]. *Glob. Chang. Biol.* **2015**, *21*, 4237–4249. [CrossRef] [PubMed]
170. Qiu, Z.; Kang, S.; He, L.; Zhao, J.; Zhang, S.; Hu, J.; Zeng, D.; Zhang, G.; Dong, G.; Gao, Z. The newly identified heat-stress sensitive albino 1 gene affects chloroplast development in rice. *Plant Sci.* **2018**, *267*, 168–179. [CrossRef] [PubMed]
171. Klap, C.; Yeshayahou, E.; Bolger, A.M.; Arazi, T.; Gupta, S.K.; Shabtai, S.; Usadel, B.; Salts, Y.; Barg, R. Tomato facultative parthenocarpy results from Sl AGAMOUS-LIKE 6 loss of function. *Plant Biotechnol. J.* **2017**, *15*, 634–647. [CrossRef]
172. Li, J.; Zhang, H.; Si, X.; Tian, Y.; Chen, K.; Liu, J.; Chen, H.; Gao, C. Generation of thermosensitive male-sterile maize by targeted knockout of the ZmTMS5 gene. *J. Genet. Genom. Yi Chuan Xue Bao* **2017**, *44*, 465–468. [CrossRef]

173. Mubarik, M.S.; Khan, S.H.; Sajjad, M.; Raza, A.; Hafeez, M.B.; Yasmeen, T.; Rizwan, M.; Ali, S.; Arif, M.S. A manipulative interplay between positive and negative regulators of phytohormones: A way forward for improving drought tolerance in plants. *Physiol. Plant.* **2021**. [CrossRef] [PubMed]
174. Yin, W.; Xiao, Y.; Niu, M.; Meng, W.; Li, L.; Zhang, X.; Liu, D.; Zhang, G.; Qian, Y.; Sun, Z. ARGONAUTE2 enhances grain length and salt tolerance by activating BIG GRAIN3 to modulate cytokinin distribution in rice. *Plant Cell* **2020**, *32*, 2292–2306. [CrossRef]
175. Zhai, N.; Jia, H.; Liu, D.; Liu, S.; Ma, M.; Guo, X.; Li, H. GhMAP3K65, a cotton Raf-like MAP3K gene, enhances susceptibility to pathogen infection and heat stress by negatively modulating growth and development in transgenic *Nicotiana benthamiana*. *Int. J. Mol. Sci.* **2017**, *18*, 2462. [CrossRef]

agronomy

MDPI

Article

Intra-Plant Variability for Heat Tolerance Related Attributes in Upland Cotton

Aneeq ur Rehman [1], Iqrar Ahmad Rana [2,†], Sajid Majeed [1], Muhammad Tanees Chaudhary [1], Mujahid Zulfiqar [1], Seung-Hwan Yang [3,*], Gyuhwa Chung [3], Yinhua Jia [4], Xiongming Du [4], Lori Hinze [5] and Muhammad Tehseen Azhar [6,7,*]

1 Department of Plant Breeding and Genetics, University of Agriculture, Faisalabad 38040, Pakistan; aneequrrehman57@gmail.com (A.u.R.); sajidmajeedpbg@gmail.com (S.M.); tanees_ch227@yahoo.com (M.T.C.); mujahidzulifiqar786@gmail.com (M.Z.)

2 Centre of Agricultural Biochemistry and Biotechnology, University of Agriculture, Faisalabad 38040, Pakistan; iqrar_rana@uaf.edu.pk

3 Department of Biotechnology, Chonnam National University, Gwangju 59626, Korea; chung@chonnam.ac.kr

4 State Key Laboratory of Cotton Biology, Institute of Cotton Research of the Chinese Academy of Agricultural Sciences, Anyang 455000, China; jiayinhua_0@sina.com (Y.J.); duxiongming@caas.cn (X.D.)

5 Southern Plains Agricultural Research Center, USDA, Agricultural Research Service, College Station, TX 77845, USA; lori.hinze@usda.gov

6 Institute of Molecular Biology and Biotechnology, Bahauddin Zakariya University, Multan 60800, Pakistan

7 School of Agriculture Sciences, Zhengzhou University, Zhengzhou 450052, China

* Correspondence: ymichigan@jnu.ac.kr (S.-H.Y.); tehseenazhar@gmail.com (M.T.A.)

† Equal Co-First Authorship.

Citation: Rehman, A.u.; Rana, I.A.; Majeed, S.; Chaudhary, M.T.; Zulfiqar, M.; Yang, S.-H.; Chung, G.; Jia, Y.; Du, X.; Hinze, L.; et al. Intra-Plant Variability for Heat Tolerance Related Attributes in Upland Cotton. *Agronomy* **2021**, *11*, 2375. https://doi.org/10.3390/agronomy11122375

Academic Editors:Channapatna S. Prakash, Ali Raza, Xiling Zou, Daojie Wang and Sheng Chen

Received: 25 September 2021
Accepted: 19 November 2021
Published: 23 November 2021

Abstract: Abiotic stress, particularly heat stress, affects various parts of the cotton plant and ultimately impacts the seed cotton yield. Different portions of a single cotton plant of a cultivar exhibit variable responses to stress during reproductive and vegetative phases. To test this hypothesis, physiological and morphological traits related to heat stress were observed for two flowering positions in 13 genotypes of upland cotton. These genotypes were sown in field conditions in triplicate following a randomized complete block design. Data were collected for pollen germination, pollen viability, cell membrane thermostability, chlorophyll content, boll weight, and boll retention for both the top and bottom branches of each genotype. The collected data were analyzed for the identification of variability within and between genotypes for these two flowering positions. Tukey's test was applied to estimate the significance of differences between genotypes and positions within each genotype. Results showed that the two positions within the same plant statistically varied from each other. The bottom branches of the genotypes performed significantly better for all traits measured except boll weight. The genotype AA-933 performed best for pollen germination and boll retention, while CYTO-608 exhibited maximum pollen viability in both the bottom and top flower positions compared with other genotypes. Overall, MNH-1016 and CIM-602 showed better cell membrane thermostability and chlorophyll content, respectively. This intra-plant variability can be further exploited in breeding programs to enhance the stress tolerance capabilities of the resulting varieties.

Keywords: genetic variability; *Gossypium hirsutum*; intra-plant variation; heat tolerance

1. Introduction

Cotton is a Kharif season crop grown mainly for feed, fiber, and oil in the Punjab and Sindh regions of Pakistan. These are considered hot regions since the temperature reaches 47 °C during the growing season. Environmental stresses such as heat, and drought affect cotton plants by impeding normal physiological processes which lead to morphological abnormalities and yield reduction [1]. Plants mostly invest their defense in the most valuable sections, such as reproductive parts under various stress conditions. Cotton production is vulnerable to abiotic stresses, particularly during the growth stages of blooming and boll formation, which have become more frequent as our climate changes [2]. Any stress during this stage abruptly reduces the yield. Numerous efforts have been made to understand the physiological, molecular, and genetic pathways of the cotton plant related to sustaining yield under stress conditions [3,4].

The reproductive efficiency of the cotton crop is negatively impacted by temperatures above 32 °C in a variety of ways, including reduced metabolism as well as suppression of photosynthesis, pollination, fertilization, and crop growth rate [5]. Heat and drought stress causes male gametes to undergo metabolic and structural changes that result in meiotic abnormalities or premature spore abortion [6]. It also results in poor pollen germination and short pollen tube growth in cotton [7,8]. It was reported that pollen germination is better in flowers that have been pollinated under the canopy of the plant as compared to flowers that are directly exposed to sunlight and pollinated during high-temperature stress [9].

Yield reduction is also associated with certain changes in metabolic and biochemical pathways in plant cells, i.e., excessive accumulation of reactive oxygen species (ROS) such as H_2O_2, singlet oxygen, hydroxyl ions, etc., during stress conditions [10]. As a result of a dramatic accumulation of ROS during stress, programmed cell death has been observed in developing pollen grains [11]. Hence, ROS scavenging through the action of antioxidants in anthers has a role in maintaining pollen viability under abiotic stress [12]. Under high temperature or water deficit conditions, the role of the cell membrane in maintaining cell osmotic balance may be impeded due to leakage of electrolytes [13]. In cotton, temperatures over 35 °C increased membrane leakage and reduced leaf size [14]. High canopy temperature adversely affects the chlorophyll content in leaf tissues and lowers the rate of photosynthesis and carbohydrate production [15]. Reduction in carbohydrate content is also associated with decreased lint yield [16]. Cotton plants will shed bolls when they are stressed, thus boll retention drops significantly under harsh environmental conditions [17,18].

It was noticed that different portions of a single cotton plant of a cultivar exhibit variable responses to stress during reproductive and vegetative phases. Although every cell in a plant has the same genetic material, the different behavior might be due to epigenetic [19,20] or other effects. Every cell expresses itself according to the stimulus received from the environment. Young leaves are more resistant to insect damage compared to old ones [21]. So, every part of the plant faces a different environment. As a result, these positions phenotypically behave differently. Cultivars also differ in canopy shape and intra-plant morphological features. Moreover, cultivars are grown in the same region exhibit variation among them. Environmental and genotypic effects both contribute to the phenotype. Therefore, the objective of this study was to identify inter- and intra-cultivar variability for physiological as well as morphological attributes associated with the yield of seed cotton under heat stress conditions.

2. Materials and Methods

2.1. Genotypes and Experimental Design

This experiment was performed in the field area of the Department of Plant Breeding and Genetics, the University of Agriculture, Faisalabad located at 31.4504° N, 73.1350° E, Pakistan. Thirteen genotypes of cotton were collected from the germplasm units of the Central Cotton Research Institute (CCRI), Multan; Cotton Research Institute (CRI), Multan; Cotton Research Station (CRS), Faisalabad; and other institutes in Pakistan listed in Table 1. These genotypes have different genetic backgrounds, have genetic variability, and grow well in the ecological niche present in the field area for this experiment. Cotton genotypes were sown on 16 May 2019 in three replications under a randomized complete block design (RCBD). Plots were single rows, 10 feet (3.1 m) long with a plant-to-plant distance of 12 inches (30 cm). Distance between rows was 30 inches (76 cm). All agronomic practices, including thinning, irrigation, weeding, and plant protection measures were performed at the appropriate crop growth stage according to cotton production technology approved for the Punjab province by the Directorate of Agriculture to maintain a healthy plant population.

Table 1. List of 13 cotton genotypes of *G. hirsutum* L. evaluated for heat tolerance.

Sr #	Genotype Name	Origin	Prominent Characteristics
1	CRS-2	Advance strain	Spreading growth habit, creamy yellow pollens, heat tolerant.
2	VH-377	CRS Vehari	Medium leaf pubescence, creamy pollen color, Good fiber quality.
3	FH-215	CRS Faisalabad	Resistant to CLCuV and pink bollworm, moderate pubescence on leaves, semi-erect branches.
4	CIM-343	CCRI Multan	Heat and drought tolerant, high yielding Bt-variety [22]
5	CIM-602	CCRI Multan	Early maturity, high lint percentage, and heat tolerant Bt-variety [23]
6	MNH-1016	CRI Multan	Semi erect branches, stem pigmentation, creamy pollen color, round shape boll, tolerant to CLCuV, high yielding Bt variety
7	MNH-1026	CRI Multan	Medium compact growth habit, semi-erect branches, creamy pollen, oblong boll shape, CLCuV tolerant, white fiber color, high yielding Bt variety
8	NIBGE-2	NIBGE Faisalabad	Resistant to Multan and Burewala strain of CLCuV, Drought resistant, Spreading growth habit. [24]
9	N-777	NIAB Faisalabad	High-density planting cotton, tolerant to heat and CLCuV-B strain. [25]
10	N-1048	NIAB Faisalabad	Tolerant to CLCuV, Spreading growth habit.
11	CYTO-124	CCRI Multan	Highly CLCuV tolerant, Non-Bt interspecific variety [23]
12	CYTO-608	CCRI Multan	Non-Bt interspecific variety [23]
13	AA-933	Ali Akbar group, Multan	Heat tolerant, good fiber quality, resistant to CLCuV, yellow pollen color. Spreading growth habit.

CRS = Cotton Research Station, CCRI = Central Cotton Research Institute, CRI = Cotton Research Institute, NIBGE = National Institute for Biotechnology and Genetic Engineering, NIAB = Nuclear Institute for Agriculture and Biology.

2.2. Data Collection

Heat tolerance measurements were taken during the growing season when 50% of the crop was flowering. Each plant was divided into two equal parts by measuring plant height in such a way that the bottom portion was under the shade of the plant canopy while the upper portion was exposed to direct sunlight. Flowers in the top part of the plant were exposed to direct sunlight while flowers on the bottom part of the plant were under the leaf canopy and received indirect sunlight. Heat tolerance-related parameters including pollen germination (PG), pollen viability (PV), and cell membrane thermostability (CMT) were assessed under in vitro conditions while boll retention and boll weight were measured in vivo.

Flowers that showed dehiscence of anthers were collected from the field and immediately transported to the laboratory where pollen grains were deposited on pollen germination media. The media was prepared following the method explained by Burke et al. [9] with little modifications. The solid germination medium consisted of 2% (w/v) agarose (Product no. A4718, Sigma Aldrich, Merck, Darmstadt Germany), 25% (w/v) sucrose (Product no. S0389, Sigma Aldrich, Merck, Germany), 0.52 mM KNO_3 (Product no. P8291, Sigma Aldrich, Merck, Darmstadt Germany), 3.06 mM $MnSO_4$ (Product no. M7899, Sigma Aldrich, Merck, Darmstadt Germany), 1.66 mM H_3BO_3 (Product no. B6768, Sigma Aldrich, Merck, Darmstadt Germany), 0.42 mM $MgSO_4{\cdot}7H_2O$ (Product no. M2643, Sigma Aldrich, Merck, Darmstadt Germany) and 1.0 µM A_3 gibberellic acid (Product no. G7645, Sigma Aldrich, Merck, Darmstadt Germany). The pH of the germination medium was brought to 7.6 before adding sucrose and agarose. The medium was autoclaved and poured into Petri plates (100 × 15 mm, Product no. P5856, Sigma Aldrich, Merck, Darmstadt Germany) under a laminar flow hood to avoid contamination. Plates were wrapped with cling film tape then placed in a refrigerator until used. Pollen grains with pollen tube lengths greater than the diameter of the pollen grains themselves were considered to be germinated (Figure 1). Percent pollen germination was estimated using the following equation:

$$\text{Pollen germination } (\%) = \frac{\text{Number of germinated pollen grains}}{\text{Total number of pollen grains}} \times 100$$

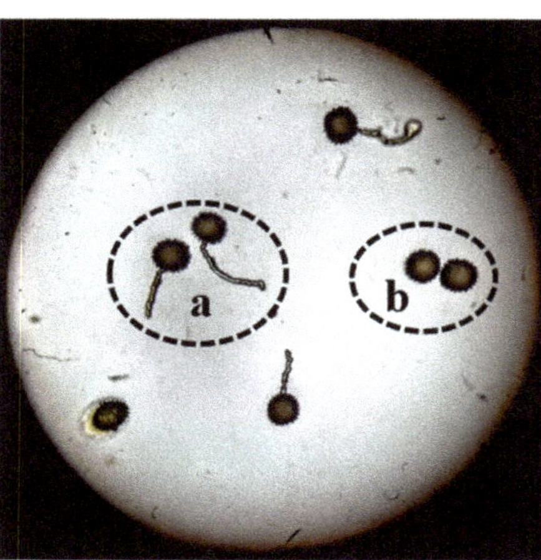

Figure 1. An example of pollen tube growth. A pollen grain that has germinated its pollen tube is labeled as 'a' while non-germinating pollen grains are labeled as 'b'.

The triphenyl-tetrazolium chloride (TTC) test was used to test the viability of pollen grains [26]. Flowers that showed dehiscence of anthers were taken into the laboratory to test pollen viability. Fresh pollen grains were sprinkled on a glass slide (76 × 26 mm) by gently tapping the flower. Two to three drops of 0.5% 2,3,5-triphenyl tetrazolium chloride (Product no. 17779, Millipore, Merck, Germany) were added in a 15% sucrose solution (Product no. S0389, Sigma Aldrich, Merck, Germany). The slide was covered with a coverslip (20 × 20 mm) to prevent desiccation and then placed under sunlight for 60 min at 30–37 °C. After this exposure, slides were observed under a light microscope (Model XSZ 107BN, Manufacturer: Zenith Lab Inc., Zhejiang China). The pollen grains that changed to red color after exposure to the TTC solution were considered viable while non-viable pollen remained yellowish in color (Figure 2). Pollen viability percentage was estimated using the following equation:

$$\text{Pollen viability } (\%) = \frac{\text{Number of viable pollen grains}}{\text{Total number of pollen grains}} \times 100$$

Two leaves from the top of the plant and two leaves from the bottom of the plant were selected for measuring CMT following the protocol of Sullivan [27] and using the following equation:

$$\text{Cell membrane thermostability } (\%) = \left[\frac{1 - T_1/T_2}{1 - C_1/C_2}\right] \times 100$$

where, the subscripts 1 and 2 refer to the 1st and 2nd electrical conductivity (EC) readings, respectively, and T and C refer to the EC of heat-treated (T) and control (C) sets of test tubes. The EC value was measured by a portable EC meter (FieldScout EC 110 Meter).

Boll weight, boll retention percentage, chlorophyll content, and canopy temperature were measured at harvest. For boll weight, all bolls from plants within the plot were harvested and weighed using an analytical balance (least count = 0.01 g). Total boll weight was divided by the total number of selected bolls to get the average weight of an individual boll. Boll retention percentage was estimated as the number of fruiting positions on the plant that had bolls divided by the total number of fruiting positions. To measure boll retention, all fruiting squares were labeled 60 days after sowing (DAS). One hundred days after sowing, the number of labeled bolls was counted. Boll retention was calculated as follows:

$$\text{Boll retention} = \frac{\text{Number of labeled bolls 100 DAS}}{\text{Number of labeled fruiting squares 60 DAS}} \times 100$$

The leaf chlorophyll content was measured using a "SPAD 502 Plus" (Konica Minolta, Japan) chlorophyll meter which works on the principle of red and blue light absorption (therefore, the SPAD measurement has no units). Top and bottom canopy temperatures of upland cotton genotypes was measured using an infrared crop temperature meter (Model: 2956, Spectrum technologies, Inc., Plainfield, NJ, USA) at crop maturity (Table 2).

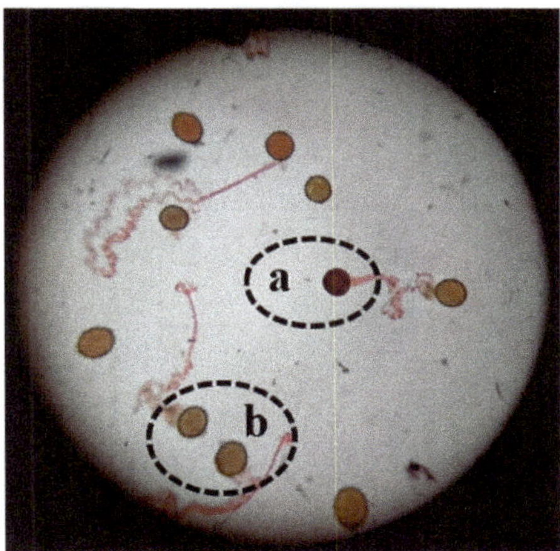

Figure 2. An example of results from the triphenyl-tetrazolium chloride (TTC) test. Viable pollen has changed to a red color (labeled as 'a') while non-viable pollen does not change color (labeled as 'b').

Table 2. Top and bottom canopy temperature in 13 cotton genotypes grown under field conditions in 2019 in Faisalabad, Pakistan.

Genotypes	Top Temp. (°C)	Bottom Temp. (°C)
CRS-2	35	33
VH-377	37	36
FH-215	35	33
CIM-343	37	35
CIM-602	35	34
MNH-1016	34	32
MNH-1026	35	33
NIBGE-2	36	34
N-777	37	35
N-1048	35	33
CYTO-124	35	33
CYTO-608	36	34
AA-933	36	35

2.3. Data Analysis

Analysis of variance was conducted with replication, genotype, and position as main effects. The interaction effect of genotype and position was also analyzed to identify sources of variation [28]. Statistix 8.1 (An Software, 2003) was used to calculate ANOVA and Tukey's test [29]. Tukey's test was applied to test the significant difference of variation between the cotton genotypes and variation between the two positions for selected traits [30]. Cluster analysis was carried out using the statistical software package of Minitab ver.17.

3. Results

Genotypes were significantly different for all recorded parameters, and plant position was also significantly different for all parameters except boll weight (Table 3). Of the sources of variation, plant position had the largest effect on PV, PG, chlorophyll content, and boll retention. The effect of genotype was largest for the cell membrane thermostability and boll weight parameters. The mean values for each measure of heat tolerance in the top and bottom plant positions of the genotypes are provided in Table 4.

Table 3. Mean squares for six measures of heat tolerance in cotton grown under field conditions.

Source of Variation	DF	PV	PG	Chl. Cont.	CMT	Boll Wt.	Boll Ret.
Replication	2	166.88 **	51.50 NS	46.67 NS	10.14 NS	0.13 NS	2.79 NS
Genotype	12	199.58 **	173.82 **	1218.46 **	3142.60 **	0.80 **	18.10 **
Position	1	1456.01 **	873.35 **	1813.86 **	2807.12 **	0.28 NS	304.88 **
Genotype × Position	12	19.37 NS	6.18 NS	65.06 NS	91.64 **	0.10 NS	6.41 NS
Error	50	21.02	26.83	41.93	30.79	0.09	6.27
Total	77						

** $p < 0.01$ and NS = Nonsignificant; DF = Degree of freedom; PV = Pollen viability; PG = Pollen germination; Chl. Cont. = Chlorophyll content; CMT = Cell membrane thermostability; Boll Wt. = Boll weight; Boll Ret. = Boll retention.

Table 4. Mean values with standard errors for six measures of heat tolerance in the top and bottom positions of 13 cotton genotypes grown in 2019 in Faisalabad, Pakistan.

Genotypes	Positions	PV (%)	PG (%)	Chl. Cont.	CMT (%)	Boll Wt. (g)	Boll Ret. (%)
CRS-2	Top	30.33 ± 2.03	24.00 ± 2.31	47.70 ± 3.65	73.80 ± 4.91	3.14 ± 0.09	29.11 ± 2.73
	Bottom	34.33 ± 2.33	28.67 ± 2.96	50.87 ± 3.88	88.75 ± 5.95	3.30 ± 0.17	36.45 ± 0.75
VH-377	Top	25.33 ± 2.91	14.67 ± 3.18	62.20 ± 2.11	78.68 ± 2.52	2.80 ± 0.20	30.51 ± 0.26
	Bottom	33.00 ± 4.36	21.00 ± 1.73	72.37 ± 3.46	86.61 ± 1.61	3.30 ± 0.07	32.48 ± 0.85
FH-215	Top	26.67 ± 1.45	17.67 ± 3.18	48.80 ± 2.93	72.81 ± 0.94	2.61 ± 0.15	28.97 ± 2.20
	Bottom	31.33 ± 2.03	25.67 ± 2.85	65.40 ± 2.95	75.95 ± 2.21	2.84 ± 0.06	32.88 ± 2.73
CIM-343	Top	23.33 ± 0.88	15.00 ± 2.65	51.20 ± 1.95	72.53 ± 4.36	3.26 ± 0.08	24.85 ± 1.19
	Bottom	32.00 ± 1.15	23.00 ± 4.16	60.23 ± 2.45	79.39 ± 1.79	3.40 ± 0.11	30.79 ± 1.27
CIM-602	Top	21.00 ± 1.73	17.00 ± 1.15	96.90 ± 4.39	81.68 ± 3.48	2.19 ± 0.06	26.79 ± 1.92
	Bottom	31.00 ± 4.16	22.33 ± 1.20	103.8 ± 2.39	91.31 ± 1.19	2.36 ± 0.16	31.34 ± 0.28
MNH-1016	Top	21.00 ± 1.73	10.33 ± 1.86	50.10 ± 4.72	88.67 ± 2.56	2.45 ± 0.19	30.53 ± 2.24
	Bottom	28.67 ± 2.03	15.67 ± 2.03	56.37 ± 2.86	92.50 ± 0.74	2.71 ± 0.08	32.01 ± 2.11
MNH-1026	Top	29.67 ± 1.67	19.00 ± 3.06	55.20 ± 3.02	83.19 ± 0.50	2.50 ± 0.13	30.21 ± 1.27
	Bottom	43.00 ± 3.61	25.33 ± 5.78	60.25 ± 3.01	88.82 ± 1.98	2.36 ± 0.20	31.95 ± 0.29
NIBGE-2	Top	19.00 ± 2.08	10.00 ± 1.53	51.27 ± 2.59	18.24 ± 2.98	2.58 ± 0.24	27.07 ± 1.32
	Bottom	31.67 ± 3.48	16.00 ± 2.08	57.63 ± 1.87	32.54 ± 3.51	2.42 ± 0.24	32.17 ± 0.66
N-777	Top	30.33 ± 3.18	17.67 ± 3.93	55.80 ± 5.42	78.88 ± 4.63	2.77 ± 0.33	29.16 ± 1.43
	Bottom	38.33 ± 3.28	23.33 ± 3.18	61.50 ± 4.85	88.16 ± 0.65	2.63 ± 0.09	34.42 ± 0.20
N-1048	Top	36.00 ± 2.89	9.00 ± 2.08	51.87 ± 1.30	41.47 ± 3.46	2.22 ± 0.26	24.08 ± 1.47
	Bottom	39.67 ± 2.91	12.67 ± 2.40	74.67 ± 4.64	70.03 ± 2.81	1.83 ± 0.05	30.35 ± 0.91
CYTO-124	Top	26.67 ± 4.26	17.67 ± 1.76	45.70 ± 5.23	22.95 ± 2.78	2.49 ± 0.16	31.71 ± 0.87
	Bottom	42.00 ± 5.51	29.00 ± 5.03	51.99 ± 4.56	36.99 ± 2.66	2.76 ± 0.09	33.72 ± 0.91
CYTO-608	Top	40.33 ± 4.26	22.00 ± 2.65	70.93 ± 2.86	83.47 ± 2.90	2.34 ± 0.13	28.35 ± 1.11
	Bottom	47.00 ± 4.04	31.00 ± 3.79	87.13 ±7.27	92.35 ± 1.47	2.58 ± 0.22	30.69 ± 0.90
AA-933	Top	32.67 ± 1.76	24.00 ± 3.06	68.00 ± 2.09	83.88 ± 2.50	2.38 ± 0.11	30.17 ± 1.72
	Bottom	42.67 ± 1.45	31.33 ± 4.06	78.83 ± 4.42	92.87 ± 2.70	2.77 ± 0.40	33.64 ± 0.56

PV = Pollen viability; PG = Pollen germination; Chl. Cont. = Chlorophyll content; CMT = Cell membrane thermostability; Boll Wt. = Boll weight; Boll Ret. = Boll retention.

3.1. Physiological Traits

The viability and germination of pollen grains were higher in the bottom part of the plant as compared to pollen that developed in flowers on the top part of the plant (Figure 3). The maximum viability of pollen grains in the bottom position was seen in CYTO-608 (47%) followed by MNH-1026 (43%), and the lowest value was observed in MNH-1016 (28.67%). The top flowers of CYTO-608 shown in Figure 4 and N-1048 showed the highest pollen viability measures of 40.33% and 36%, respectively, while MNH-1016 and CIM-602 had the lowest pollen viability of 21% for both genotypes (Table 4). Tukey's mean comparison test for pollen viability revealed non-significant differences for the genotypes CYTO-608, N-1048, and AA-933. On average, these genotypes performed well in both top and bottom positions (Table 5). Mean values of pollen viability across all genotypes to compare top and bottom positions of plants revealed a significant difference between top and bottom positions. Pollen viability at the bottom position exhibited more value compared to the top position (Figure 3).

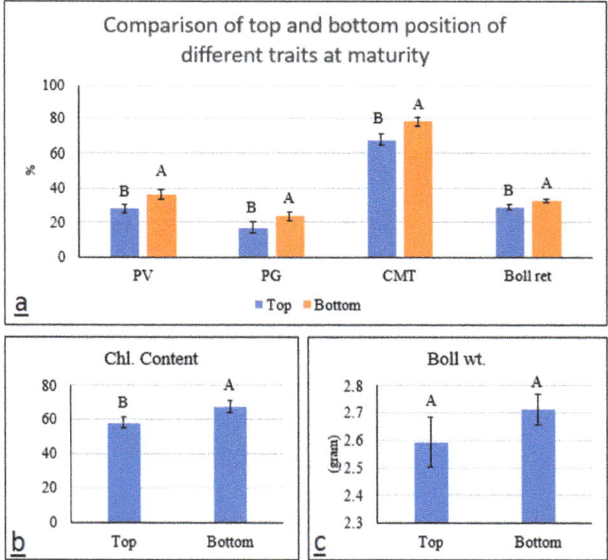

Figure 3. Means averaged across all genotypes to compare top and bottom positions of plants for different traits at maturity. The error bars are standard errors. The letters show Tukey's mean comparison where different letters show significant differences between top and bottom positions for each trait. (**a**) PV (Pollen viability), PG (Pollen germination), CMT (Cell membrane thermostability) and boll ret (boll retention), (**b**) Chl. Content (Chlorophyll content), (**c**) Boll wt (Boll weight).

The highest pollen germination from flowers at the bottom position was observed in AA-933 (31.33%) while in the top position, both AA-933 and CRS-2 showed 24% pollen germination. The lowest value for this parameter was observed in N-1048 with 12.67% and 9% germination in the bottom and top positions, respectively (Table 4). Overall, pollen from flowers that bloomed on top parts of the plant showed less germination when compared to pollen from bottom flowers (Figure 3). Pollen germination estimates were also lower than pollen viability estimates. Tukey's mean test revealed a non-significant difference between CRS-2, Cyto-608, and AA-933. It was observed that these genotypes performed well in both viability and germination tests, except CRS-2 which showed good pollen germination (Table 5). This indicates that most of the viable pollen of CRS-2 did germinate. Variation in pollen tube length was also observed as shown in Figure 5.

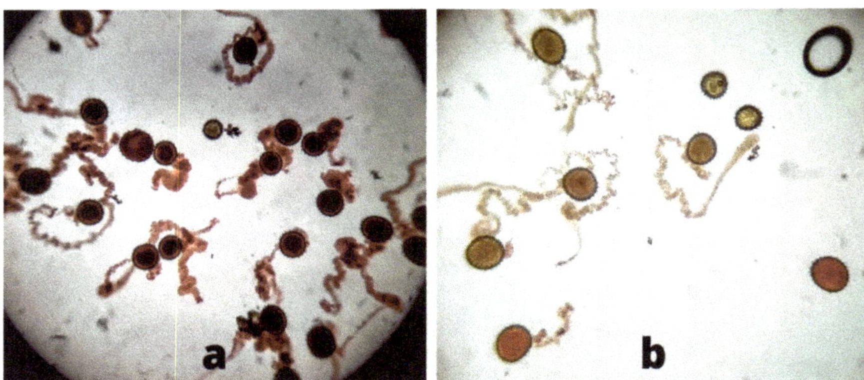

Figure 4. Results from staining pollen with triphenyl-tetrazolium chloride (TTC). (**a**) Pollen from flowers from the bottom positions of CYTO-608 shows the highest pollen viability. (**b**) Pollen from flowers from the top positions of NIBGE-2 shows the lowest pollen viability.

Table 5. Mean values for six measures of heat tolerance in 13 cotton genotypes grown in 2019 in Faisalabad, Pakistan.

Genotypes	PV (%)	PG (%)	Chl. Cont.	CMT (%)	Boll Wt. (g)	Boll Ret. (%)
CRS-2	32.33 ^{BCDE}	26.33 ^{A}	49.28 ^{F}	81.28 ^{ABC}	3.21 ^{AB}	32.78 ^{A}
VH-377	29.17 ^{BCDE}	17.83 ^{ABC}	67.28 ^{BCD}	82.64 ^{ABC}	3.05 ^{ABC}	31.50 ^{AB}
FH-215	29.00 ^{BCDE}	21.67 ^{AB}	57.10 ^{DEF}	74.38 ^{C}	2.73 ^{ABCD}	30.92 ^{AB}
CIM-343	27.67 ^{CDE}	19.00 ^{ABC}	55.72 ^{DEF}	75.96 ^{BC}	3.33 ^{A}	27.82 ^{AB}
CIM-602	26.00 ^{DE}	19.67 ^{ABC}	100.35 ^{A}	86.64 ^{A}	2.27 ^{DE}	28.57 ^{AB}
MNH-1016	24.83 ^{E}	13.00 ^{BC}	53.23 ^{EF}	90.59 ^{A}	2.58 ^{CDE}	31.27 ^{AB}
MNH-1026	36.33 ^{ABC}	22.17 ^{AB}	57.73 ^{DEF}	86.01 ^{AB}	2.43 ^{DE}	31.08 ^{AB}
NIBGE-2	25.33 ^{DE}	13.00 ^{BC}	54.45 ^{DEF}	25.39 ^{E}	2.50 ^{CDE}	29.62 ^{AB}
N-777	34.33 ^{BCD}	20.50 ^{ABC}	58.65 ^{CDEF}	85.61 ^{AB}	2.69 ^{BCD}	31.79 ^{AB}
N-1048	37.83 ^{AB}	10.83 ^{C}	65.35 ^{CDE}	55.75 ^{D}	2.03 ^{E}	31.96 ^{AB}
CYTO-124	34.33 ^{BCD}	23.33 ^{AB}	48.85 ^{F}	29.97 ^{E}	2.62 ^{BCDE}	32.67 ^{AB}
CYTO-608	43.67 ^{A}	26.50 ^{A}	79.03 ^{B}	85.81 ^{AB}	2.46 ^{CDE}	29.52 ^{AB}
AA-933	37.67 ^{AB}	27.67 ^{A}	71.33 ^{BC}	88.38 ^{A}	2.58 ^{CDE}	27.71 ^{B}

PV = Pollen viability; PG = Pollen germination; Chl. Cont. = Chlorophyll content; CMT = Cell membrane thermostability; Boll Wt. = Boll weight; Boll Ret. = Boll retention. Means with the same letters in each column are not significantly different according to Tukey's test.

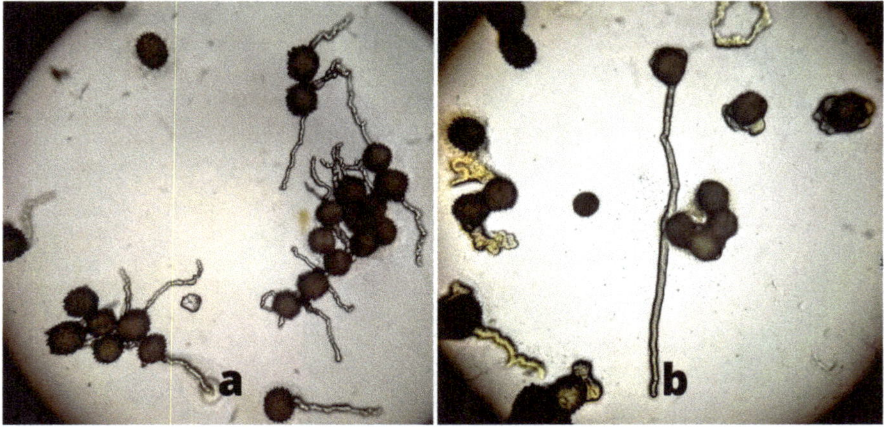

Figure 5. The observed variation in pollen tube germination where (**a**) Bottom flowers from AA-933 showed maximum pollen germination (**b**) Top flowers from N-1048 showed lowest pollen germination.

Cell membrane thermostability (CMT) and chlorophyll content were also significantly different for genotypes and plant positions (Table 3). The CMT values for top leaves (67.96%) were lower than values for bottom leaves (78.17%) (Figure 3). At the bottom of the plant, the maximum value for CMT was recorded for the genotype AA-933 (92.87%) followed by MNH-1016, CYTO-608, and CIM-602 which presented 92.50%, 92.35%, and 92.31% CMT, respectively (Table 4). Leaves of the bottom branches had more chlorophyll content as compared to leaves from the top branches (Figure 3). The genotypes CIM-602, CYTO-608, and AA-933 showed the highest chlorophyll contents in bottom branches (103.8, 87.13, and 78.83, respectively) while CIM-602 also had the highest chlorophyll content in leaves of top branches (96.9) (Table 4).

3.2. Morphological Traits

Genotypes varied significantly for boll weight, but non-significant differences were observed between the top and bottom plant portions for this trait (Table 3). On average, the largest boll weight at the bottom position was observed for CIM-343 (3.40 g) followed by CRS-2 and VH-377 (3.30 g for each) while N-1048 exhibited the lowest boll weight (Table 4). Tukey's test revealed non-significant differences among CIM-343, CRS-2, and VH-377 genotypes for boll weight (Table 5). Boll retention percentage was significantly different for genotypes and positions (Table 3). Lower boll retention was observed in the top position branches as compared to bottom branches (Figure 3). In the bottom branches, the genotype CRS-2 had maximum boll retention (36.45%) followed by N-777 (34.42%) and CYTO-124 (33.72%). The minimum boll retention was observed in N-1048 for both portions of the plant. Boll retention was also low in the top branches of CIM-343 (Table 4). It was noted that genotypes with high pollen germination retained more bolls.

3.3. Cluster Analysis

All the genotypes were clustered using pollen germination, pollen viability, cell membrane thermostability, chlorophyll content, boll weight, and boll retention at high-temperature stress as variables. The dendrogram showed three clusters with a minimum of 33.33% similarity level. The highest Euclidean distance was found between clusters 2 and 3 (57.76) and lowest between clusters 1 and 2 (27.95) as presented in Table 6.

Table 6. The distance among the various cluster centroid of cotton genotypes under high temperature.

	Cluster 1	Cluster 2	Cluster 3
Cluster 1	0	27.9541	45.5783
Cluster 2		0	57.7641
Cluster 3			0

The clusters were divided into two groups Group Y and Group Z. Group Y included two clusters named cluster 1 and cluster 2 while Group Z included only one cluster named cluster 3. Cluster 1 included seven genotypes named CRS-2, MNH-1026, N-777, MNH-1016, VH-377, FH-215, and CIM-343 which represent 53.85% of total genotypes. Cluster 2 includes three genotypes named CIM-602, Cyto-608 and AA-933 represent 23.07% of total genotypes. In cluster 3, three genotypes are included named NIBGE-2, Cyto-124, and N-1048 representing 23.07% of the total genotype (Figure 6). The genotypes within each cluster exhibited similar behavior based on six traits used in this study. The genotypes in cluster 1 showed good performance based on boll weight. The genotypes grouped in cluster 2 are characterized by high pollen germination, pollen viability, chlorophyll content, and CMT. This indicated that the genotypes grouped in cluster 2 could be selected for the breeding program. The genotypes of cluster 3 were grouped by the lowest value of pollen germination, pollen viability, CMT, and boll weight (Table 7).

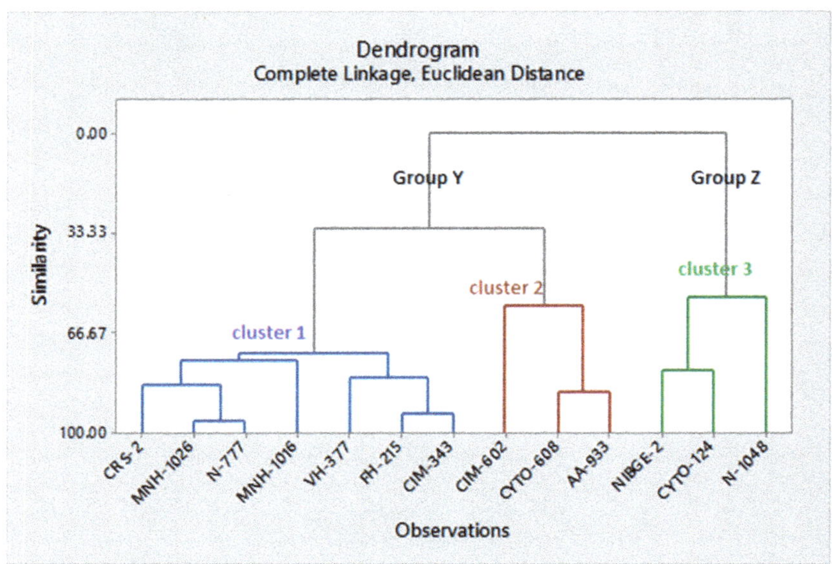

Figure 6. Cluster analysis of thirteen accessions of upland cotton evaluated for high-temperature regimes.

Table 7. Means of clusters of 13 cotton genotypes of all observed tr under high-temperature stress.

Variable	Cluster 1	Cluster 2	Cluster 3	Grand Centroid
No. of genotypes	7	3	3	13
PV (%)	30.5229	35.78	32.4967	32.1915
PG (%)	20.0714	24.6133	15.72	20.1154
Chl. Cont.	56.9986	83.57	56.2167	62.95
CMT (%)	82.3529	86.9433	37.0367	72.9546
Boll Wt. (g)	2.86	2.4367	2.3833	2.6523
Boll Ret. (%)	31.0229	28.6	31.4167	30.5546

4. Discussion

Higher pollen germination and viability percentages from flowers under the canopy of the plant as compared to flowers in direct sunlight were observed in this study. As the temperature in the experimental region in Pakistan rises to 47 °C during the time of cotton flowering, this damages the lipid as well as protein parts of the pollen membrane, thus resulting in decreases in pollen viability [12]. Pollen viability was determined by analyzing dehydrogenase enzyme activity in the pollen grains—if the enzyme is active, viable pollen grains change to a red color after TTC staining. However, there may be damage to the pollen grains that reduces germination despite this enzyme activity. It has been reported that the distribution of cell organelles such as mitochondria, vacuoles, and endoplasmic reticulum of pollen cells become disturbed under high temperatures. Lipid and starch granules are also reduced in pollen cells during heat stress [8].

In our study, lower pollen germination was observed as compared to pollen viability. Most pollen could not develop the pollen tube required for germination, likely due to metabolic or structural abnormalities of pollen grains [31]. Drought or heat stress significantly lowers carbohydrate metabolism in the pistil, resulting in a lower energy supply to the pollen tube in the style, thus leading to a failure of fertilization [32]. Under excessively high temperatures, heat shock proteins (HSPs) work to stabilize proteins that were damaged when exposed to stressful conditions. As the expression of HSPs varies between genotypes, some genotypes showed good pollen germination even in the top portion of the plant [33].

Genotypic variability for CMT has been previously reported [3,34]. Here, we have observed CMT differences between plant positions. The significant differences among cultivars are due to several factors including cuticle thickness, secondary metabolites, and heat shock proteins [35–37]. Lower CMT estimates in top leaves were due to sunlight exposure. The membranes of leaves facing direct sunlight in high-temperature conditions were more prone to damage. Sun rays cause oxidative damages to both lipid and protein parts of the cellular membranes and cause the leakage of electrolytes [38]. UV radiation from the sun causes irreversible damage to plant pigments [39]. It causes conformational changes in the structure of nucleic acids, proteins, and macromolecules in the cell and degrades the chlorophyll pigment [40,41]. Heat stress directly affects the flow of fluid through the cell membrane as relative electrical conductivity increases with temperature [42,43].

Since the chlorophyll contents under the canopy were higher as compared to the top position, it has been assumed that chlorophyll loses its integrity under direct sunlight. In addition to direct sunlight, the higher temperature in the top portion of the plant also causes chlorophyll damage [44]. Heat stress that denatures thylakoid membranes results in a loss of chlorophyll [45]. Moreover, the enzymes required for the synthesis of chlorophyll and its normal activity were also denatured under high-temperature conditions [46]. As a result, photosynthetic activity was reduced in the top portion of plants. On average, the genotypes AA-933 and CYTO-608 had good heat tolerance features in bottom positions; therefore, these genotypes would be useful as parents in a breeding program

Ascorbic acid has the potential to mitigate the negative effects of stress. It acts in ROS scavenging and maintains the integrity of membranes, including the thylakoid membrane [47]. So, ascorbic acid could be used to overcome the heat stress problem. The cell membrane thermostability of cotton crops can be improved significantly by applying the foliar application of 40 mg L^{-1} ascorbic acid [48].

Boll weight is positively associated with seed cotton yield. It is a complex polygenic trait that depends upon numerous factors namely, the weight of seed, seed size, protein and oil content within the seed, and cellulose deposition during fiber development and maturity [49]. It is one of the most important characters linked to improved yield, and significant variation for this trait has been reported in germplasm [50,51]. Although the genotypes used in this experiment were significantly different for boll weight, no significant differences for this trait were recorded between the top and bottom portions of the same genotypes. Retention of bolls during the developmental period varied significantly between the top and bottom branches. The bottom branches tend to hold more bolls as compared to the top branches. It was noted that the genotypes with higher pollen viability and germination also retained more bolls. This study revealed that the heat tolerance ability of the genotypes was associated with boll retention while heat stress has been considered one of the major factors in bolls dropping before maturity [52,53]. Thus, high temperature in the top portion of a plant due to direct exposure to sunlight can explain retaining a lower number of bolls in this portion of the plant.

The variability was found between the genotypes as shown in Table 3. Cluster analysis has revealed that CIM-602, Cyto-608, and AA-933 grouped in cluster 2 performed well and these genotypes could be used further in any breeding program. Since all genotypes are grown in the Punjab region of Pakistan, these are therefore acclimatized to this environment. These genotypes share some common, as well as different phenotypes, which showed variability based on six traits used in this study. The variability was also observed in the cotton genotypes cultivated in the Punjab region of Pakistan by khan [54].

This study provides an understanding of the role of flowering in the top and bottom portions of the cotton plant in response to high-temperature stress because high temperature is a major factor in reducing yield. It is assumed that by increasing the vegetative growth and leaf surface area, the shading effect can be increased. The spreading-type behavior of the cotton plant could be able to produce more shading. The shading effect will reduce canopy temperature and hence yield could be increased. Likewise, screening for early maturing cultivars and for having more branches on the bottom part of the plant could be beneficial because the bottom branches have shown more productivity than top branches. Keeping in view the importance of the study, another study may be conducted to assess the temperature of the microenvironment i.e., the temperature of leaf, bud, and/or boll at the top and bottom regions of each genotype, followed by correlation analysis with each trait to understand the relationship of various traits during heat stress.

5. Conclusions

Both top and bottom branches of the cotton plant exhibited variable responses for physiological and morphological traits. Significant genotypic variability for these attributes was also observed. The bottom branches of the genotypes performed better for all the recorded parameters except boll weight which was non-significant for both positions. The high temperature was found to disrupt plant physiology and morphology more on top position flowers. Further study of the shading effect is an objective for future breeding programs. A focus on increasing vegetative growth, leaf surface area, and a more spreading growth pattern would increase the canopy size and allow for shading to improve pollen germination and pollen tube growth. Further, research focusing on increased resilience to high temperature itself would allow top portions of a cotton plant to deliver a higher yield thus increasing overall yields.

Author Contributions: Conceptualization, M.T.A., I.A.R. and S.-H.Y.; methodology, M.T.A., A.u.R., M.Z., M.T.C. and S.M.; software, A.u.R., X.D., Y.J.; validation, M.T.A., S.M. and I.A.R. formal analysis, A.u.R. and M.T.C.; investigation, M.T.A., L.H. and A.u.R.; data curation, M.Z. and A.u.R., writing—original draft preparation, A.u.R. and M.T.C. and S.M.; writing—review and editing, M.T.A., I.A.R., G.C., A.u.R., S.M. and L.H.; supervision, M.T.A.; funding acquisition, S.-H.Y. All authors have read and agreed to the published version of the manuscript.

Funding: This work was carried out with the support of "Basic Science Research Program through the National Research Foundation of Korea (NRF) funded by the Ministry of Education (NRF-2021R1F1A10554820).

Institutional Review Board Statement: Not applicable.

Informed Consent Statement: Not applicable.

Acknowledgments: The authors are grateful to the Lab. Incharge for provision of necessary equipment and chemicals at the Center for Advanced Studies, University of Faisalabad (Pakistan). USDA is an equal opportunity provider and employer.

Conflicts of Interest: The authors declare no conflict of interest.

References

1. Dabbert, T.; Gore, M.A. Challenges and perspectives on improving heat and drought stress resilience in cotton. *J. Cotton Sci.* **2014**, *18*, 393–409.
2. Oosterhuis, D.M. Day or night high temperatures: A major cause of yield variability. *Cotton Grow.* **2002**, *46*, 8–9.
3. Abro, S.; Rajput, M.T.; Khan, M.A.; Sial, M.A.; Tahir, S.S. Screening of cotton (*Gossypium hirsutum* L.) genotypes for heat tolerance. *Pak. J. Bot.* **2015**, *47*, 2085–2091.
4. Salman, M.; Majeed, S.; Rana, I.A.; Atif, R.M.; Azhar, M.T. Novel breeding and biotechnological approaches to mitigate the effects of heat stress on cotton. In *Recent Approaches in Omics for Plant Resilience to Climate Change*; Wani, S., Ed.; Springer: Berlin/Heidelberg, Germany, 2019; pp. 251–277, ISBN 978-3-030-21686-3.
5. Snider, J.L.; Oosterhuis, D.M.; Skulman, B.W.; Kawakami, E.M. Heat-stress induced limitations to reproductive success in *Gossypium hirsutum* L. *Physiol. Plant* **2009**, *137*, 125–138. [CrossRef] [PubMed]
6. De Storme, N.; Geelen, D. The impact of environmental stress on male reproductive development in plants: Biological processes and molecular mechanisms. *Plant Cell Environ.* **2014**, *37*, 1–18. [CrossRef]
7. Snider, J.L.; Oosterhuis, D.M.; Kawakami, E.M. Diurnal pollen tube growth rate is slowed by high temperature in field-grown *Gossypium hirsutum* pistils. *J. Plant Physiol.* **2011**, *168*, 441–448. [CrossRef] [PubMed]
8. Song, G.; Chen, Q.; Tang, C. The effects of high-temperature stress on the germination of pollen grains of upland cotton during square development. *Euphytica* **2014**, *200*, 175–186. [CrossRef]
9. Burke, J.J.; Velten, J.; Oliver, M.J. In vitro analysis of cotton pollen germination. *Agron. J.* **2004**, *96*, 359–368. [CrossRef]
10. Foyer, C.H.; Bloom, A.; Queval, G.; Noctor, G. Photo respiratory metabolism: Genes, mutants, energetics and redox signaling. *Annu. Rev. Plant. Biol.* **2009**, *60*, 455–484. [CrossRef]

11. Jiang, P.; Zhang, X.; Zhu, Y.; Zhu, W.; Xie, H.; Wang, X. Metabolism of reactive oxygen species in cotton cytoplasmic male sterility and its restoration. *Plant Cell Rep.* **2007**, *26*, 1627–1634. [CrossRef]
12. Majeed, S.; Malik, T.A.; Rana, I.A.; Azhar, M.T. Antioxidant and physiological responses of upland cotton accessions grown under high-temperature regimes. *Iran. J. Sci. Technol. Trans. A Sci.* **2019**, *43*, 2759–2768. [CrossRef]
13. Salman, M.; Khan, A.A.; Rana, I.A.; Maqsood, R.H.; Azhar, M.T. Genetic architecture of relative cell injury and some yield related parameters in *Gossypium hirsutum* L. *Turk. J. Field Crop.* **2016**, *21*, 246–253. [CrossRef]
14. Bibi, A.C.; Oosterhuis, D.M.; Gonias, E.D. Photosynthesis, quantum yield of photosystem II and membrane leakage as affected by high temperatures in cotton genotypes. *J. Cotton Sci.* **2008**, *12*, 150–159.
15. Karademir, E.; Karademir, C.; Sevilmiş, U.; Başal, H. Correlations between canopy temperature, chlorophyll content and yield in heat tolerant cotton (*Gossypium hirsutum* L.) genotypes. *Fresenius Environ. Bull.* **2018**, *27*, 5230–5237.
16. Pettigrew, W. Environmental effects on cotton fiber carbohydrate concentration and quality. *Crop. Sci.* **2001**, *41*, 1108–1113. [CrossRef]
17. Guinn, G. *Causes of Square and Boll Shedding in Cotton*; US Department of Agriculture, Agricultural Research Service: Washington DC, USA, 1982.
18. Raper, T.; Oosterhuis, D.; Pilon, C.; Burke, J.; Coomer, T. Evaluation of 1-methylcyclopropene to reduce ethylene driven yield reductions in field-grown cotton. In *Summaries of Arkansas Cotton Research*; Oosterhuis, D.M., Ed.; University of Arkansas System Division of Agriculture: Little Rock, AR, USA, 2013; Volume 91, pp. 91–95.
19. O'Connor, C.M.; Adams, J.U.; Fairman, J. How do cells decode genetic information into functional proteins? In *Essentials of Cell Biology*; NPG Education: Cambridge, MA, USA, 2010; Volume 1, p. 54.
20. Angers, B.; Castonguay, E.; Massicotte, R. Environmentally induced phenotypes and DNA methylation: How to deal with unpredictable conditions until the next generation and after. *Mol. Ecol.* **2010**, *19*, 1283–1295. [CrossRef]
21. Anderson, P.; Agrell, J. Within-plant variation in induced defence in developing leaves of cotton plants. *Oecologia* **2005**, *144*, 427–434. [CrossRef]
22. PSC Approves 2 New Bt Cotton Varieties Developed by CCRI Multan with Production Potential of 50 Munds per Acre. Available online: https://www.urdupoint.com/en/pakistan/psc-approves-2-new-bt-cotton-varieties-develo-1154689.html (accessed on 13 October 2021).
23. Central Cotton Research Institute, Multan. Cotton Varieties. Available online: http://www.ccri.gov.pk/varieties.html# (accessed on 14 October 2021).
24. Rahman, M.U.; Zafa, Y. Registration of 'NIBGE-2' cotton. *J. Plant Regist.* **2007**, *1*, 113. [CrossRef]
25. Research Divisions, Achievements, Nuclear Institute for Agriculture and Biology, Faisalabad. Available online: http://www.niab.org.pk/mutation.htm (accessed on 14 October 2021).
26. Nortin, J. Testing of plum pollen viability with tetrazolium salts. *Proc. Amer. Soc. Hort. Sci.* **1966**, *89*, 132–134.
27. Sullivan, C.Y. Mechanisms of Heat and Drought Resistance in Grain Sorghum and Methods of Measurement. In *Sorghum in Seventies*; Oxford & IBH Pub. Co.: New Delhi, India, 1972.
28. Steel, R.G.; Torrie, J.H.; Dickey, D.A. *Principles and Procedures of Statistics, a Biometrical Approach*; McGraw-Hill: New York, NY, USA, 1997.
29. Statistix 8.1. User's Manual. Analytical Software, Tallahassee. 2003. Available online: https://statistix.informer.com/8.1/ (accessed on 28 February 2020).
30. Haynes, W. Tukey's test. In *Encyclopedia of Systems Biology*; Dubitzky, W., Wolkenhauer, O., Cho, K.H., Yokota, H., Eds.; Springer: New York, NY, USA, 2013; pp. 2303–2304.
31. Song, G.; Jiang, C.; Ge, X.; Chen, Q.; Tang, C. Pollen thermotolerance of upland cotton related to anther structure and HSP expression. *Agron. J.* **2015**, *107*, 1269–1279. [CrossRef]
32. Hu, W.; Liu, Y.; Loka, D.A.; Zahoor, R.; Wang, S.; Zhou, Z. Drought limits pollen tube growth rate by altering carbohydrate metabolism in cotton (*Gossypium hirsutum*) pistils. *Plant. Sci.* **2019**, *286*, 108–117. [CrossRef]
33. Wang, J.; Sun, N.; Deng, T.; Zhang, L.; Zuo, K. Genome-wide cloning, identification, classification and functional analysis of cotton heat shock transcription factors in cotton (*Gossypium hirsutum*). *BMC Genom.* **2014**, *15*, 1–19. [CrossRef] [PubMed]
34. Rahman, H.; Malik, S.A.; Saleem, M. Heat tolerance of upland cotton during the fruiting stage evaluated using cellular membrane thermostability. *Field Crops Res.* **2004**, *85*, 149–158. [CrossRef]
35. Hadacek, F. Secondary metabolites as plant traits: Current assessment and future perspectives. *Crit. Rev. Plant Sci.* **2002**, *21*, 273–322. [CrossRef]
36. Goodwin, S.M.; Jenks, M.A. Plant cuticle function as a barrier to water loss. *Plan. Abiotic Stress* **2005**, *21*, 14–36.
37. Grigorova, B.; Vaseva, I.; Demirevska, K.; Feller, U. Combined drought and heat stress in wheat: Changes in some heat shock proteins. *Biol. Plant* **2011**, *55*, 105–111. [CrossRef]
38. Melgar, J.C.; Guidi, L.; Remorini, D.; Agati, G.; Degl'innocenti, E.; Castelli, S.; Baratto, M.C.; Faraloni, C.; Tattini, M. Antioxidant defences and oxidative damage in salt-treated olive plants under contrasting sunlight irradiance. *Tree Physiol.* **2009**, *29*, 1187–1198. [CrossRef]
39. Petrović, S.; Zvezdanović, J.; Marković, D. Chlorophyll degradation in aqueous mediums induced by light and UV-B irradiation: An UHPLC-ESI-MS study. *Radiat. Phys. Chem.* **2017**, *141*, 8–16. [CrossRef]

40. Bassman, J.H. Ecosystem consequences of enhanced solar ultraviolet radiation: Secondary plant metabolites as mediators of multiple trophic interactions in terrestrial plant communities. *Photochem. Photobiol.* **2004**, *79*, 382–398. [CrossRef]
41. Salama, H.M.; Al Watban, A.A.; Al-Fughom, A.T. Effect of ultraviolet radiation on chlorophyll, carotenoid, protein and proline contents of some annual desert plants. *Saudi J. Biol. Sci.* **2011**, *18*, 79–86. [CrossRef]
42. Cottee, N.S.; Tan, D.K.Y.; Bange, M.P.; Cothren, J.T.; Campbell, L.C. Multi-level determination of heat tolerance in cotton (*Gossypium hirsutum* L.) under field conditions. *Crop. Sci.* **2010**, *50*, 2553–2564. [CrossRef]
43. Jamil, A.; Khan, S.J.; Ullah, K. Genetic diversity for cell membrane thermostability, yield and quality attributes in cotton (*Gossypium hirsutum* L.). *Genet. Resour. Crop. Evol.* **2020**, *67*, 1405–1414. [CrossRef]
44. Yang, G.; Rhodes, D.; Joly, R.J. Effects of high temperature on membrane stability and chlorophyll fluorescence in glycinebetaine-deficient and glycinebetaine-containing maize lines. *Funct. Plant Biol.* **1996**, *23*, 437–443. [CrossRef]
45. Ristic, Z.; Bukovnik, U.; Prasad, P.V. Correlation between heat stability of thylakoid membranes and loss of chlorophyll in winter wheat under heat stress. *Crop. Sci.* **2007**, *47*, 2067–2073. [CrossRef]
46. Gosavi, G.; Jadhav, A.; Kale, A.; Gadakh, S.; Pawar, B.; Chimote, V. Effect of heat stress on proline, chlorophyll content, heat shock proteins and antioxidant enzyme activity in sorghum (*Sorghum bicolor*) at seedlings stage. *Indian J. Biotechnol.* **2014**, *13*, 356–363.
47. Dolatabadian, A.; Sanavi, A.M.; Harifi, M.S. Effect of ascorbic acid (Vitamin C) leaf feeding on antioxidant enzymes activity, proline accumulation and lipid peroxidation of canola (*Brassica napus* L.) under salt stress condition. *J. Sci. Tech. Agri. Nat. Res.* **2009**, *13*, 611–620.
48. Kamal, M.A.; Saleem, M.F.; Wahid, M.A.; Shakeel, A. Effects of ascorbic acid on membrane stability and yield of heat stressed Bt cotton. *J. Anim. Plant. Sci.* **2017**, *27*, 192–199.
49. Zhang, Z.; Shang, H.; Shi, Y.; Huang, L.; Li, J.; Ge, Q.; Gong, J.; Liu, A.; Chen, T.; Wang, D. Construction of a high-density genetic map by specific locus amplified fragment sequencing (SLAF-seq) and its application to Quantitative Trait Loci (QTL) analysis for boll weight in upland cotton (*Gossypium hirsutum.*). *BMC Plant Biol.* **2016**, *16*, 1–18. [CrossRef]
50. Campbell, B.; Chee, P.; Lubbers, E.; Bowman, D.; Meredith, W., Jr.; Johnson, J.; Fraser, D.; Bridges, W.; Jones, D. Dissecting genotype × environment interactions and trait correlations present in the Pee Dee cotton germplasm collection following seventy years of plant breeding. *Crop. Sci.* **2012**, *52*, 690–699. [CrossRef]
51. Mubarik, M.S.; Ma, C.; Majeed, S.; Du, X.; Azhar, M.T. Revamping of cotton breeding programs for efficient use of genetic resources under changing climate. *Agronomy* **2020**, *10*, 1190. [CrossRef]
52. Reddy, K.; Hodges, H.; Reddy, V. Temperature effects on cotton fruit retention. *Agron. J.* **1992**, *84*, 26–30. [CrossRef]
53. Liu, Z.; Yuan, Y.L.; Liu, S.Q.; Yu, X.N.; Rao, L.Q. Screening for high-temperature tolerant cotton cultivars by testing in vitro pollen germination, pollen tube growth and boll retention. *J. Integr. Plant Biol.* **2006**, *48*, 706–714. [CrossRef]
54. Khan, A.I.; Khan, I.A.; Awan, F.S.; Sadaqat, H.A.; Bahadur, S. Estimation of genetic distance based on RAPDs between 11 cotton accessions varying in heat tolerance. *Genet. Mol. Res.* **2011**, *10*, 96–101. [CrossRef] [PubMed]

Review

Comprehensive Understanding of Selecting Traits for Heat Tolerance during Vegetative and Reproductive Growth Stages in Tomato

Kwanuk Lee [1], Sherzod Nigmatullayevich Rajametov [1], Hyo-Bong Jeong [1], Myeong-Cheoul Cho [1], Oak-Jin Lee [1], Sang-Gyu Kim [1], Eun-Young Yang [1,*] and Won-Byoung Chae [2,*]

1 National Institute of Horticultural & Herbal Science, Rural Development Administration, Wanju 55365, Korea; kwanuk01@korea.kr (K.L.); sherzod_2004@list.ru (S.N.R.); bong9846@korea.kr (H.-B.J.); chomc@korea.kr (M.-C.C.); ojlee6524@korea.kr (O.-J.L.); kimsg9@korea.kr (S.-G.K.)

2 Department of Environmental Horticulture, Dankook University, Cheonan 31116, Korea

* Correspondence: yangyang2@korea.kr (E.-Y.Y.); wbchae75@dankook.ac.kr (W.-B.C.); Tel.: +82-63-238-6613 (E.-Y.Y.); +82-41-550-3642 (W.-B.C.)

Abstract: Climate change is an important emerging issue worldwide; the surface temperature of the earth is anticipated to increase by 0.3 °C in every decade. This elevated temperature causes an adverse impact of heat stress (HS) on vegetable crops; this has been considered as a crucial limiting factor for global food security as well as crop production. In tomato plants, HS also causes changes in physiological, morphological, biochemical, and molecular responses during all vegetative and reproductive growth stages, resulting in poor fruit quality and low yield. Thus, to select genotypes and develop tomato cultivars with heat tolerance, feasible and reliable screening strategies are required that can be adopted in breeding programs in both open-field and greenhouse conditions. In this review, we discuss previous and recent studies describing attempts to screen heat-tolerant tomato genotypes under HS that have adopted different HS regimes and threshold temperatures, and the association of heat tolerance with physiological and biochemical traits during vegetative and reproductive growth stages. In addition, we examined the wide variety of parameters to evaluate the tomato's tolerance to HS, including vegetative growth, such as leaf growth parameters, plant height and stem, as well as reproductive growth in terms of flower number, fruit set and yield, and pollen and ovule development, thereby proposing strategies for the development of heat-tolerant tomato cultivars in response to high temperature.

Keywords: climate change; heat tolerance; heat stress regimes; tomato breeding program; vegetative and reproductive growth stages; heat tolerance traits

Citation: Lee, K.; Rajametov, S.N.; Jeong, H.-B.; Cho, M.-C.; Lee, O.-J.; Kim, S.-G.; Yang, E.-Y.; Chae, W.-B. Comprehensive Understanding of Selecting Traits for Heat Tolerance during Vegetative and Reproductive Growth Stages in Tomato. *Agronomy* **2022**, *12*, 834. https://doi.org/10.3390/agronomy12040834

Academic Editors: Channapatna S. Prakash, Ali Raza, Xiling Zou and Daojie Wang

Received: 17 February 2022
Accepted: 26 March 2022
Published: 29 March 2022

1. Introduction

Industrialization and urbanization have continuously accelerated climate change, as seen in the increase in air temperature to date [1,2]. The global average temperature is predicted to increase by approximately 0.3 °C in every decade, thereby rising by 1–4 °C from the year 2081 to 2100 compared to the temperature recorded from 1986 to 2005, according to the Intergovernmental Panel on Climatic Change (IPCC) [3,4]. An elevated temperature of approximately 3–4 °C may result in a drastic reduction in crop yields by a maximum of 35% in Asia, Africa, and the Middle East [5].

High temperature (HT) is closely associated with heat stress (HS), which is one of the major abiotic stresses, including temperature, drought, salinity, and flooding [5]. HS is a detrimental abiotic factor that influences global crop productivity by compromising crop growth during the vegetative and reproductive growth stages [5–7]. In general, HS is considered to be an elevation of temperature over a threshold level for a substantial amount of time, leading to irreversible impairments in crop growth and development [8].

A transiently elevated temperature of 10–15 °C above the optimum temperature for plant growth and development is termed as "heat" or "thermal" shock [8,9]. HS is a complex process with many factors, including the intensity and total duration of HT and the speed of the temperature increase [9,10]. For example, the frequency and period of HT can affect the occurrence and intensity of HS in a certain climatic zone during the day or night. In definition, HS tolerance is termed as the survival capability of a plant to grow, develop, and/or produce economic yields in response to HT [8,10].

Tomato (*Solanum lycopersicum*) plants, which belong to the *Solanaceae* family, are grown in a wide variety of climate conditions and areas from tropical to temperate regions. The tomato was introduced to Europe in the 16th century and then spread to the Mediterranean [11]. It is the second most important vegetable crop worldwide after the potato [12]. The cultivation area has reached almost 5,000,000 hectares and worldwide tomato yields amount to 181,000,000 metric tons (FAO, http://www.fao.org/faostat/ (accessed on 7 February 2022)). Tomatoes are consumed fresh and are also a major ingredient in a wide array of cuisines, as well as for sauces and juices [13]. In particular, the tomato fruit is a rich source of vitamins A and C and antioxidants, including lycopene, β-carotene, phenolics, and minerals but is low in calories [14–16], contributing to the maintenance of human health.

Climate change is likely to significantly reduce crop yield in the future [17–19], and a daily temperature that is a few degrees above the average can significantly affect vegetative and reproductive parameters, such as seedling growth, plant height (PH), leaf length and width, pollen grain viability, and the fertility of the female parts [20–22]. Its optimal average day and night temperatures range from 21 to 30 °C and from 18 to 21 °C, respectively [23]. Tomato production is frequently threatened by HS in diverse cultivated regions [24,25]. HT and drought stress are likely to have a negative impact on the growth, development, and yield of tomato plants in fields, leading to reduced production in its main producing regions from the year 2050 to 2100 [22].

The effect of HS on tomato plants has chiefly been evaluated during the reproductive stage owing to the importance of its fruits, as well as its sensitivity at the reproductive stage [26–28]. When the day temperature is above optimum, this induces developmental disorders in the flower organs (stamen and ovule) and in fruit development (fruit set, weight and quality, and seed number) [26–31]. In addition, a study has shown that the treatment of transient HS (>45 °C, 20 min) results in programmed cell death (PCD) in tomato fruits via DNA fragmentation and cytochrome c release, and induces caspase-like enzyme activity [32]. Vegetative growth parameters are likewise influenced under HS in many crops [8]. Non-photochemical quenching (NPQ), chlorophyll content, photosynthetic rate, transpiration rate, stomatal conductance, and CO_2 assimilation are negatively affected by HS, resulting in a reduced growth rate [19,26,33,34].

The evaluated traits and factors in HS conditions are significantly varied among genotypes and growth stages in tomato plants, suggesting that the correlations among these traits and factors can vary during the vegetative and reproductive growth stages [35,36]. It is important to find the association in terms of HT tolerance between germination or seedling stages and flowering stages in tomato plants [37–39] for enhancing the speed of tomato breeding by the early selection of heat-tolerant genotypes as an indirect selection. However, substantial knowledge gaps exist in our understanding of the correlation between vegetative and reproductive growth-related traits or factors under HS.

The development of heat-tolerant tomato cultivars is crucial for adapting to elevated temperatures at present and in the future [5,20], but some bottlenecks and difficulties exist in identifying and applying the traits associated with heat tolerance in breeding programs. There has been a different understanding of the definition of plant responses to HT and a lack of in-depth understanding of the genetic basis and architecture regarding the heat tolerance mechanism during the vegetative and reproductive growth stages [1,20,40]. These problems have led to a deficiency of common screening methods and protocols on evaluating heat-tolerant traits, as well as in screening and selecting heat-tolerant genotypes to date. Consequently, the identification of heat-tolerant genotypes seems not to be reproducible and reliable to some degree [41]. In this review, we provide an overview of the knowledge of the response of tomato plants to HT, which includes diverse HS regimes, as well as trait associations and key factors related to heat tolerance during the vegetative and reproductive growth stages. Lastly, we examine some promising traits associated with heat tolerance in tomato plants, with a discussion of recent correlation studies using a large number of genotypes.

2. Types of Heat Stress Regimes in Tomato Plants

The concept of HS is projected to assess temperature intensity and duration and the speed of increments in temperatures [8]. The application of proper HS treatment is a key point when screening heat-tolerant plants. Yeh et al. [42] and Mesihovic et al. [41] reported four major HS regimes for screening the heat-tolerant germplasm in *Arabidopsis* and crop species, suggesting that directly applied HS (DAHS) should be applied for basal thermo-tolerance, pre-induced HS for acquired thermo-tolerance (ATT), gradient HS for ATT, and mild chronic HS (MCHS) for mild heat thermo-tolerance (MHTT) for ATT in greenhouse and open-field conditions [41]. Two HS regimes, DAHS and MCHS, are mainly utilized in screening for and studying the heat tolerance of tomato plants in response to HT at physiological and molecular levels. In general, DAHS is applied for a short period of screening (from an hour to a day) with a high temperature of ≥45 °C during the vegetative and/or reproductive stages [41,43]. MCHS is used for a longer period of screening with mild high temperatures of 30–36 °C, ranging from several days to the entirety of the plant growth and developmental cycle [41,44]. The MCHS regime has been more widely applied than the DAHS regime to tomato plants for screening heat tolerance in terms of the reproductive parameters, including pollen germination and viability with flower and fruit characteristics in both greenhouses and open fields [20]. This is because the MCHS regime more closely resembles natural field conditions and/or it is difficult to maintain the DAHS regime in field conditions. Tomato plants exhibit different physiological responses to DAHS and MCHS [29]. The effects of HS on tomato cultivars vary significantly, depending on growth stages and experimental environment conditions including the temperature range, light intensity and quality, relative humidity and others; sometimes, different definitions of the same traits have caused variation in HS effects [41]. Therefore, it is indispensable to establish common screening methods and protocols to identify heat-tolerant tomato genotypes by considering the aforementioned climate conditions in a certain target area and/or expanded geological location.

3. Optimum Temperature Range and Heat Stress Threshold in Tomato Plants

Some studies have shown that a night temperature of 13 °C maintains a good fruit set, [45] while night temperatures ranging from 15–20 °C influence the increment of marketable yields of tomatoes [46]. However, the daily average temperature (DAT) is more crucial than the day or night temperature or the difference between day and night temperatures during the reproductive processes in tomato plants [20]. For example, the increments or differentials of day and night temperature did not show constant patterns of flower number and fruit weight, but a DAT of 29 °C considerably diminished the fruit number and weight, and the seed number per fruit in comparison with those in a DAT of 25 °C. Moreover, when the average night temperature was 19.2 °C, which is just below the upper critical point (20 °C), and the DAT was 26.8 °C led to a significant decrease in the fruit set of the tomato plants [30,45]. It has been demonstrated that a DAT of 25–30 °C is markedly optimal for the net assimilation rate [47]. DATs of 21–24 °C [48], 22–25 °C [49], and 22–26 °C [24] are optimal for fruit set and yield. An average temperature of approximately 21.3 °C, with an average day and night temperature of 27.3 and 15.1 °C [30], and 26.3 and 15.6 °C [50], are also beneficial for the fruit set and yield of tomato plants, respectively.

The threshold temperature of crops is defined as a value of DAT where a decline in crop growth begins and can be termed as the lower/base and upper threshold temperatures of plant development [4]. This has been determined via the environmental regulation of laboratory and field conditions. Lower and upper threshold temperatures in plant development represent points below and above those at which plant growth and development ceases, respectively [8]. The lower threshold temperatures vary depending on plant species, such as spinach (2 °C), pea (4.4 °C), pumpkin (13 °C), and tomato (15 °C) [51]. However, 0 °C is often considered to be a predicted lower threshold temperature for cool-season plants [52]. Notably, the upper threshold temperatures are critical for tropical and subtropical crops, which are important limiting factors for determining crop yields. The upper threshold temperatures also vary among plant species, such as wheat (26 °C), tomato (30 °C), corn (38 °C) and cotton (45 °C), and even among genotypes within the same species [8]. The determination of an exact upper threshold temperature is also difficult since the physiological responses of plants to different environmental stimuli, as well as to habitats, vary greatly [4,8]. For example, when the ambient temperature was over 35 °C as an upper threshold temperature, the seed germination rate, vegetative growth, flowering time, and fruit set and ripening were significantly inhibited in tomato plants [8]. On the other hand, an upper threshold temperature of only 30 °C can damage tomato plants at the period of seedling emergence [4,26], suggesting that the evaluation of threshold temperatures for tomato plants must be performed in each growing stage to apply these threshold temperatures for screening heat-tolerant genotypes.

4. The Response of Vegetative and Reproductive Growth to Heat Stress in Tomato Plants

4.1. Leaf Growth, Plant Height, and Stem Diameter

Many studies have been conducted to determine heat tolerance in tomato plants using diverse vegetative growth parameters, including leaf growth, PH, and stem diameter (SD) under HS. Leaf-growth parameters, such as the fresh and dry leaf weight and leaf area, have been assessed under different HS regimes. Abdelmageed et al. [53] reported that these parameters in three heat-tolerant tomato cultivars were less and smaller at 37/27 °C (day/night) and pre-heat shock than in 26/20 °C (day/night) with no pre-heat shock. In the other study, however, leaf number and area in three tomato cultivars were not significantly different between the control conditions (CK) of 26/20 °C (day/night) and heat stress conditions of 32/26 °C (day/night) [54]. Moreover, Zheng et al. [55] have investigated the leaf area of tomato plants under two HS conditions of 38/18 °C and 41/18 °C (day/night) and three relative humidity conditions (50%, 70%, and 90%). The leaf area was significantly reduced in two HT regimes with 50% relative humidity (RH), in comparison with a CK of 28/18 °C (day/night) in 50% RH, but it was similar and/or larger in two different HT regimes combined with higher RHs, suggesting an alteration of the HS response by RH. In our recently published study with 38 tomato accessions [56], the leaf length and width were not significantly affected under HT greenhouse conditions with CK and MCHS (19.7/35 °C and 20.2/38.8 °C of the average daily minimum/maximum temperatures, respectively).

The contrasting responses to HS among studies were also observed in PH and SD. In the HS regime of 36/28 °C (day/night) and a CK of 26/18 °C (day/night), there were no apparent differences in PH and SD between a tolerant and a susceptible genotype [39]. Zheng et al. [55] also assessed PH and SD in the aforementioned HS and CK conditions; the difference in PH under HS was remarkable, whereas that in SD was not significant when compared to CK. In our previous study, the PH and SD in most of the 38 tomato accessions increased regardless of fruit types [56]. Bhattarai et al. [57] have recently reported that the PH and SD among 18 cultivars were dramatically decreased in a constant HS regime maintaining 36/28 °C (day/night) in growth chambers, in comparison with those in greenhouses set to 26/20 °C (day/night). These contrasting results indicate that vegetative growth parameters cannot be general indicators for heat tolerance and may not be appropriate for indirect selection in tomato breeding programs for heat tolerance.

4.2. Pollen Development

In tomato plants, HS causes negative effects not only on pollen development and viability but also on ovule development, embryogenesis, and viability [20,58]. The fact that HT affects the fruit set or number more than flower number implies that HT has a greater effect on the process of fertilization. Indeed, reduced fertility is a common problem associated with HT during meiosis and fertilization periods in tomato plants [59]. When pollinated with pollens matured in HT, female plants grown in optimal temperature conditions did not produce fruits, whereas female plants grown in HT that were pollinated with pollens matured in optimal temperature conditions could bear fruits [28].

Favorable temperatures for pollen germination and pollen tube length are between 15 and 22 °C in vitro [60] and 25 °C in vivo [61]. Temperatures above 30 and 35 °C reduce the pollen germination rate and pollen tube growth [61,62]. Since the pollen development stage is very sensitive to HS and is critical for determining fruit set and yield in HS [9,41,63], most of the studies of HT tolerance in tomato plants have been focused on the pollens. HT significantly reduces pollen viability [8,28,37,64,65], germination [56,61] and number [37]. Frank et al. [66] reported that pollen viability is significantly diminished in flowers of three to seven days before the anthesis stage under DAHS (43–45 °C, 2 h), whereas the pollen germination rate and the number of pollen grains were not significantly reduced. The pollen's germinability was influenced by a higher DAHS (50 °C, 2 h) in flowers of 2, 7, and 9 days before anthesis in the tomato cultivar "MicroTom" [21]. In addition, the long period of MCHS (32/26 °C, day/night) for young tomato plants or 1–2 weeks before anthesis had a negative effect on pollen development [29]. Pressman et al. [67] showed that the total number of pollen grains, pollen germinability, and viability are noticeably decreased in MCHS (32/26 °C, day/night) in comparison with CK (28/22 °C, day/night). The number of pollen grains and the percentage of viable pollen grains are also lower in MCHS (32/26 °C, day/night) than those in CK (28/22 °C, day/night) [29].

Remarkably, the response of pollen traits is genotype-dependent. In our previous study, the pollen germination of 23 tomato accessions and the pollen tube length of 28 tomato accessions were significantly reduced under MCHS conditions among a total of 38 accessions [56]. Other studies also reported genotype-dependent HT tolerance in pollen germination [68] and viability [37]. Positive correlations were observed between fruit set and pollen viability [37] or pollen germination and tube length [56,68], although the correlation of the latter two was not significant. HT treatment applied to pollen donor plants before and during pollen release caused significant reductions in seed number and fruit set in comparison with HT treatment that was applied to the developing ovule after pollination [28]. These results suggest that the effects of HT are most significant during pollen maturation, rather than during pollen germination and tube growth or fertilization [69].

4.3. Ovule Development

HS also causes the abnormal development of female reproductive organs [20] or reduced female fertility [37] in tomato plants, although the female gametophyte is generally considered to have more heat tolerance than its male counterpart. It is noticeable that HS influences ovule development under open-field and greenhouse conditions, leading to the malfunction of both male and female reproductive organs at the same time [21,69]. Ovule development is significantly affected by MCHS. The fruit set, number and weight, and seediness were significantly reduced when DAT during the pollination inclined from 25–26 °C to 28–29 °C, even though pollens for pollination were collected under optimal temperatures of 26/22 °C (day/night) [20]. Similarly, Xu et al. [37] reported a reduction in female fertility and seediness in fruit under MCHS (32/26 °C, day/night) when hand-pollinating using pollens from both MCHS and optimal temperature samples (25/19 °C, day/night). In addition, the interactions of pollen and pistil are essential for pollen tube growth from the stigma to the ovules; therefore, a pistil exposed to HS would be dedicated to the considerable regulation of pollen performance in vivo [42,70–72]. These studies suggest the importance of the synergistic effect of male and female organs to achieve the resulting level of fertility under HS [73]. However, it is not likely that the evaluation of female fertility by screening a large number of germplasms is to be recommended, due to the labor-intensive evaluation process, including hand-pollination and counting seeds in fruits [63].

4.4. Flower and Fruit Development

Flowering and fruit set may be the most critical index in the evaluation of heat-tolerant tomato cultivars under HS [74]. Notably, a failure in normal pollen development and female fertility results in a drastic reduction in flower number, fruit weight, fruit set, and seediness under MCHS, in comparison with CK conditions [20,24,68]. For example, HS influenced floral abortion, which led to 80% of flower abscission and a resulting decrease in fruit set [9,75]. In addition, tomato plants exposed to MCHS (DAT of 34 °C/19 °C, day/night) caused 34% of flower abscission and decreased the fruit set by 71% [76]. Heat-tolerant genotypes have been successfully selected according to high fruit set and yield under HT in open fields and greenhouses or growth chambers (36/28 °C, day/night) [19,57], implying that these traits can be utilized for selecting heat-tolerant genotypes in tomato plants.

5. Physiological and Biochemical Responses of HS Tolerance in Tomato

HS directly influences the alteration of photosynthetic parameters, including the net photosynthetic rate (*Pn*), CO_2 assimilation, transpiration rate (*Tr*), stomata conductance (*Ci*), Photosystem II (PSII, *Fv/Fm*), and chlorophyll contents [33,77], which are closely related to delayed plant growth and development. The ability to adjust the accumulation of primary and secondary metabolites and proteins is also important for plants to display heat tolerance [5,78]; heat-tolerant tomato genotypes with a high fruit set and pollen viability might have this ability. In the plants, primary metabolites such as soluble sugars, glycine betaine, and proline accumulate in response to HS [8]. The production of these osmolytes under HS may increase the protein stability and membrane bilayer structure [79].

5.1. Photosynthesis

Photosynthetic apparatus under HS causes the severe malfunction of chloroplasts, which are dedicated to the generation of ATP and metabolites in plants [4,80]. Thus, good performance of the photosynthetic apparatus under HS would be seen in the capability of the plant to overcome stress conditions in response to HS [81]. In particular, since HS prohibits successful chlorophyll biosynthesis, chlorophyll content (chl*a* and chl*b*) could be utilized as a reliable evaluation index for identifying heat-tolerant plants [26]. In tomato plants, the chlorophyll *a*/*b* ratio decreased and the chlorophyll/carotenoid ratio increased in a heat-tolerant tomato cultivar under DAHS (45 °C, 2 h) compared to CK (25/20 °C, day/night), whereas *Pn*, the CO_2 assimilation rate, and PSII (*Fv/Fm*) were reduced in heat-susceptible cultivars [26]. In addition, PSII was sensitively stimulated by HS and *Fv/Fm*; the ratio representing the maximum quantum efficiency of PSII is often utilized to measure the normal or better performance of chloroplasts under HS [82].

5.2. Soluble Sugars

Soluble sugars are necessary for pollen viability and germination. An imbalance in the sugar metabolism caused by HS is associated with the failure of tomato plant fruit set [29]. When developing tomato anthers are continuously exposed to HT, the carbohydrate metabolism is altered, resulting in the reduction of the number of pollen grains per flower and pollen viability [67]. Heat-tolerant tomato genotypes have a higher carbohydrate concentration in pollen grains than susceptible ones under HS [21]. In addition, heat-tolerant tomato genotypes accumulate more soluble sugars in their leaves under HS than susceptible ones at the flowering and anthesis stages [39], possibly due to their better performance in maintaining carbohydrate synthesis under HS.

5.3. Proline

The development and fertility of pollens depend on local proline biosynthesis in mature pollen grains, as well as during the later microspore development stages [83]. Proline also functions as a molecular chaperone, regulating the protein structure and protecting cell damage in stress conditions [84,85]. In tomato plants, the disruption of proline transport to the anther may be a possible cause of the reduction in pollen viability [29]. Proline accumulation in pollens is affected more significantly in heat-susceptible tomato cultivars than in heat-tolerant ones [38]. The proline content in leaves can be a useful measure of stress in tomato plants [86]. Changes in proline content under HT, as well as the endogenous level of proline content, differ according to genotypes [87]. Seedlings of a heat-tolerant cultivar accumulate significantly less proline in leaves than those of a susceptible one in HT [36]. The proline content of a tolerant cultivar did not show significant change but that of a susceptible one showed a continuous increase during the HT treatment period [36].

To understand the relationship between proline content and HT tolerance in tomatoes and to test the possibility of using proline content in heat tolerance screening, we grew 43 tomato genotypes in greenhouses where the temperature set-point for ventilation was 28 °C and 40 °C for CK and MCHS, respectively. The proline content, pollen germination, pollen tube length, and fruit set and yield of 43 genotypes were investigated; correlation analyses were conducted according to Sherzod et al. [56]. In CK, the proline content in leaves was significantly correlated with pollen germination (r = 0.377 *) and fruit set (r = 0.415 **) (Table 1) but no significant correlation was observed between proline content and other traits in MCHS (Table 2). The significant correlation between pollen germination and fruit set in CK but not in MCHS may be due to differences in genotype-dependent proline accumulation in leaves [36], resulting in disrupted proline transport in HT [29]. The results indicate that the use of proline content in leaves for heat tolerance screening is still premature and further study is necessary.

Table 1. Correlation between biochemical and reproductive traits at control temperatures (28 °C).

	Proline Content	Pollen Germination	Pollen Tube Length	Fruit Yield	Fruit Set
Proline content	1				
Pollen germination	0.337 *	1			
Pollen tube length	0.139	0.488 **	1		
Fruit yield	0.009	−0.152	−0.113	1	
Fruit set	0.415 **	0.025	−0.207	0.127	1

* and ** indicate significant difference at $p < 0.05$ and $p < 0.01$ levels, respectively.

Table 2. Correlation between biochemical and reproductive traits at high temperatures (40 °C).

	Proline Content	Pollen Germination	Pollen Tube Length	Fruit Yield	Fruit Set
Proline content	1				
Pollen germination	0.288	1			
Pollen tube length	0.078	0.610 **	1		
Fruit yield	−0.175	0.075	−0.120	1	
Fruit set	−0.003	−0.037	−0.043	0.354 *	1

* and ** indicate significant difference at $p < 0.05$ and $p < 0.01$ levels, respectively.

5.4. Glycine Betaine

Glycine betaine is a compatible osmolyte and plays an important role in osmoregulation in plants. It is synthesized in both chloroplasts and cytoplasm, but only glycine betaine in chloroplasts is positively related to stress tolerance [88]. This implies that high glycine betaine content may not necessarily account for enhanced stress tolerance. In tomato plants, however, glycine betaine significantly increased in heat-stressed tomato plants, in comparison with non-stressed plants [89]. The exogenous application of glycine betaine to heat-stressed tomato plants enhanced seed germination, the expression of heat-shock genes, and the accumulation of heat-shock protein 70 [90] and fruit yield in open fields [91].

5.5. Secondary Metabolites

Secondary metabolites also play a critical role in pollen growth and germination, osmotic regulation, the scavenging of reactive oxygen species (ROS), and membrane fluidity, as well as the signaling pathways contributing to pollen viability and fruit set [92]. HT increased the accumulation of soluble phenolics and the activity of phenylalanine by ammonia-lyase, the principal enzyme in the biosynthesis of phenolic compounds, as an acclimated mechanism against HS [93]. A significant increase in total flavonoids including kaempferol was observed in tomato plants under HS conditions during the pollen development stage [94]. In addition, the accumulation of polyamines, such as spermidine and spermine, is required for pollen germination in tomato plants [62,95], while polyamine accumulation in transgenic tomato plants enhances HT tolerance [96]. However, no effect of HT on the level of polyamine was observed in a recent study of tomato plants [94].

5.6. Superoxide Dismutase

HS affects the overproduction of ROS, destroying essential cellular components and structural elements [97]. Superoxide dismutase (SOD) helps break down harmful oxygen molecules and is considered a defense enzyme against oxidative stress [98]. In tomato plants, DAHS (>40 °C) for three hours in both ambient and root zone temperatures caused a decrease in total specific SOD activity in two tomato genotypes, one of which has heat tolerance, but no significant difference in SOD activity was observed between the two cultivars [26]. In our recent study, however, HS with 38/30 °C (day/night) for three days resulted in a significant increase in SOD activity in two cultivars, and a significant difference between the two cultivars in terms of SOD activity was also observed [87]. The contrasting results may be due to the small number of genotypes used and the different environmental conditions. Further studies are needed, with a large number of genotypes in the same environment, to clarify the role of SOD activity in the heat tolerance of tomato plants.

6. Traits Related to Direct and Indirect Selection for Heat-Tolerant Genotypes in Tomato Plants

As discussed earlier, many physiological and biochemical traits are related to heat tolerance in tomato plants. However, most of the previous studies were based on a few genotypes and the results were genotype-dependent; therefore, these traits cannot be general predictors in screening or selecting heat-tolerant genotypes in the tomato. There have been a limited number of correlation studies between fruit set and/or yield as well as other traits in HS that use a large number of tomato genotypes having various degrees of heat tolerance. A correlation study is particularly important for indirect selection to identify heat-tolerant genotypes. In this section, we examine promising target traits for direct and indirect selection for heat-tolerant genotypes, based on correlation studies with a large number of genotypes. To our knowledge, no such correlation study was conducted in tomato plants among fruit traits and soluble sugars, glycine betaine, secondary metabolites, and SOD activity. The proline content did not show any significant correlation with traits related to heat tolerance in HS (Table 2). Therefore, this will not be discussed here.

6.1. Traits for Direct Selection for Heat Tolerance

Reproductive rather than vegetative growth is more vulnerable to HS in many crops including the tomato [69]. Vegetative traits were also not significantly correlated with fruit yield and fruit set in HT among 38 [56] and 13 tomato genotypes [37], respectively, despite fruit set and yield being considered as the main targeted traits for heat-tolerance screening in tomato plants. Therefore, most studies on heat tolerance in tomato plants have been focused on reproductive growth. HS significantly decreased the fruit set and the number of fruits per truss [31,37,56,99–101], which were significantly correlated with reduced fruit yield [56]. However, the effect of HS on flower number per truss is somewhat controversial; there were reports of a reduction [37,102,103], no change [20,28,29,104], and even an increase [56] in flower number under HS. The obvious decrease in fruit set or number but not in flower number under HT among tomato genotypes indicates that floral development may not be the key physiological trait that confers heat tolerance on tomato plants.

Fruit set and yield have been the main traits used for screening heat tolerance among tomato genotypes [63], although fruit drops after fruit set have also increased under HT [37,102]. Various tomato genotypes have been screened, based on fruit set and yield [24,63,68]. However, the fruit set is calculated from the ratio of the number of fruits divided by that of flowers and can be significantly affected by a reduction [37,102,103] or increase in the number of flowers [56]. Therefore, the fruit set in HS is inevitably affected by the changes in the number of flowers in HS, which can also affect the correlation between fruit set and other traits, resulting in a lower correlation with fruit yield than fruit number [56]. Therefore, breeders working with tomato genotypes who have increased the flower number per truss under HS should consider fruit number per truss instead of fruit set for the trait to achieve direct selection for heat tolerance.

Different fruit traits must be considered, depending on cultivars with different fruit sizes, in the screening of heat-tolerant genotypes since some traits associated with heat tolerance differ according to tomato fruit size [56]. For example, both the fruit number per truss and fruit set were significantly and positively correlated with fruit yield in cherry tomato genotypes (>50 g) but only the fruit number per truss was significantly correlated with fruit yield in large fruit types (<100 g) [56]. This is because an increase or decrease in flower number per truss does not significantly affect the fruit set in cherry tomato genotypes, which have much more flowers per truss than large fruit genotypes; however, the increase or decrease by one or two flowers in large fruit genotypes significantly affects the fruit set. In addition, in cherry tomato types, heat-tolerant genotypes can be pre-selected, simply by looking at the previous fruit yield data collected in optimum temperature conditions, because a significantly positive correlation was observed between fruit yields in CK and MCHS conditions that was not observed among large fruit genotypes [56].

6.2. *Traits for Indirect Selection for Heat Tolerance*

Vegetative growth parameters should be avoided regarding heat tolerance since recent studies with a large number of tomato accessions showed that these parameters were not significantly correlated with fruit set [37] and fruit yield [56] in HT, which are key indicators for heat tolerance. Besides this, there was no association between seedling and reproductive growth stages in HT [56]. Therefore, the selection of heat-tolerant tomato genotypes based on seedling performance and vegetative growth parameters in HT should be conducted very carefully, although selection in the early growth stages can facilitate the breeding process.

Pollen traits may be promising candidates for indirect selection when screening tomato accessions in HS. In particular, pollen viability and germination may be promising candidates for indirect selection for heat tolerance since many of the studies discussed above reported an association between pollen traits and fruit set or yield. However, careful consideration must be taken when applying pollen traits for the selection of heat-tolerant lines in breeding programs since the results of these studies were largely based on a few genotypes and showed genotype-dependent responses. Studies using a large number of genotypes showed somewhat contrasting results. Pollen germination was not significantly correlated with fruit set [68] and fruit yield [56], when using 11 and 38 genotypes, respectively, and both fruit set and yield when using 43 genotypes (Table 2) in HS. In another study with 13 tomato genotypes, significantly positive correlations were observed between pollen viability and fruit set and fruit number in tomato plants [37]. In addition, no significant correlation was observed between pollen tube length and fruit set or number [56,68]. These contrasting results might be due to different HS regimes and environmental conditions (pollens receiving heat stress in a Petri dish [68], and plants in pots [37] and in soil in greenhouses [56]). The procedures for pollen tests were also different, such as pollen viability [37] and germination [56,68]. A standardized procedure to screen for pollen tolerance to HS must be developed and applied in future studies and breeding programs.

Membrane thermostability may be a good candidate for indirect selection for heat tolerance and should generally be measured by ion or electrolyte leakage. In tomato plants, the level of ion leakage differed by tomato genotypes and significantly decreased under HS [37]. Heat-tolerant genotypes showed less electrolyte leakage than susceptible ones in tomato plants [26,36]. In a study using 13 tomato cultivars, fairly good correlations between ion leakage and fruit set ($r = -0.444$) or pollen viability ($r = -0.294$) were observed, although they were not significant [37]. In an experiment using a larger population (43 cultivars), a significantly negative correlation ($r = -0.9$) between electrolyte leakage and fruit yield was observed [57]. These correlation studies suggest the potential of ion or electrolyte leakage for heat tolerance screening in tomatoes.

The index of *Fv/Fm* has been utilized when screening large numbers of tomato genotypes in HS from seedlings to mature plant stages [82]. Specifically, *Fv/Fm* of 28 genotypes was measured under DAHS and MCHS conditions, following three different HS regimes, under conditions from a climate chamber to an open field. Tolerant genotypes that were selected by higher *Fv/Fm* in HS displayed a lower level of leaf heat injury, as well as a higher fruit yield [82]. Moreover, in a correlation experiment using four heat-tolerant and four susceptible cultivars, *Fv/Fm* was significantly correlated with plant dry weight under four days of HS ($r = 0.974$) in a climate chamber, which was significantly correlated with fruit dry weight ($r = 1.00$) and fruit set ($r = 0.984$) in field conditions [82]. The study may suggest that the parameters associated with photosynthetic apparatus can be applied to the rapid detection of heat-tolerant tomato plants during the early vegetative growth stage. However, correlation analyses were not performed with the same plant types or under the same environmental conditions; therefore, *Fv/Fm* is still far from sufficient as a predictor of HT tolerance. Further study is essential with a larger number of genotypes and trait investigations in the same plant type and environment.

7. Conclusions and Future Perspectives

According to a recent IPCC report, it is certainly a matter of major concern that climate change will be detrimental to food security. In particular, HS, an elevated temperature above a threshold level, can negatively influence plant behaviors during the entire growth and development cycle in both greenhouse and open-field conditions, thereby leading to the diminishing of the fruit set, quality, and yield of tomatoes. It is, therefore, essential to develop tomato cultivars that are tolerant of HS. Since the different organs and growth stages of tomato plants exhibit different sensitivities in response to HS, different HS regimes should be properly applied from seedlings to reproductive growth stages. In this review, we have discussed recent attempts in screening and breeding for heat tolerance in tomato plants under different HS regimes. In addition, we discussed the diverse selecting traits shown during the vegetative and reproductive growth stages and provided an association between the traits and the key biochemical factors, which can be effectively utilized to screen and/or select a large number of tomato genotypes under HS conditions. Finally, we proposed effective direct and indirect selection strategies to develop heat-tolerant tomato cultivars presenting promising candidate traits that are significantly correlated with fruit set and yield under HS.

Further studies are required with a large number of tomato genotypes to identify the key traits for indirect selection since there is still a lack of correlation studies for HS tolerance using a large population. In addition, further study is necessary to establish the effect of HS on the marketable yield of tomato plants. Seedling selection to accelerate the breeding process will apparently not be applied in the near future for the selection of heat-tolerant tomato genotypes due to the contrasting results of an association between the seedling and reproductive stages. To increase the efficiency of developing heat-tolerant tomato cultivars, it is also necessary to develop a marker-assisted selection system. Since the traits related to heat tolerance show quantitative inheritance, the development of molecular markers can be conducted with bi-parental quantitative trait loci mapping and genome-wide association studies, with plenty of sequence information being obtained from whole-genome sequencing, re-sequencing, and genotyping-by-sequencing. Emerging genome-editing techniques, including the CRISPR/Cas 9 system, can be adopted in the future to introduce or neutralize beneficial or deleterious genes, respectively, in elite tomato lines.

Author Contributions: Conceptualization, E.-Y.Y. and W.-B.C.; writing—original draft preparation, K.L. and W.-B.C.; writing—review and editing, K.L., S.N.R., H.-B.J., M.-C.C., O.-J.L., S.-G.K., E.-Y.Y. and W.-B.C.; supervision, E.-Y.Y. and W.-B.C. All authors have read and agreed to the published version of the manuscript.

Funding: This study was supported by a grant (Project No: PJ01669601 "A study on the high-temperature tolerance mechanism of hot pepper and tomato using an evaluation population") from the National Institute of Horticultural and Herbal Science, Rural Development Administration.

Data Availability Statement: The datasets presented in this study are available upon request to the corresponding author.

Conflicts of Interest: The authors declare no conflict of interest.

References

1. Hedhly, A.; Hormaza, J.I.; Herrero, M. Global warming and sexual plant reproduction. *Trends Plant Sci.* **2009**, *14*, 30–36. [CrossRef] [PubMed]
2. Zandalinas, S.I.; Fritschi, F.B.; Mittler, R. Global warming, climate change, and environmental pollution: Recipe for a multifactorial stress combination disaster. *Trends Plant Sci.* **2021**, *26*, 588–599. [CrossRef] [PubMed]
3. Jones, P.D.; New, M.; Parker, D.E.; Martin, S.; Rigor, I.G. Surface air temperature and its changes over the past 150 years. *Rev. Geophys.* **1999**, *37*, 173–199. [CrossRef]
4. Golam, F.; Prodhan, Z.H.; Nezhadahmadi, A.; Rahman, M. Heat tolerance in tomato. *Life Sci. J.* **2012**, *9*, 1936–1950.
5. Bita, C.; Gerats, T. Plant tolerance to high temperature in a changing environment: Scientific fundamentals and production of heat stress-tolerant crops. *Front. Plant Sci.* **2013**, *4*, 273. [CrossRef]
6. Battisti, D.S.; Naylor, R.L. Historical warnings of future food insecurity with unprecedented seasonal heat. *Science* **2009**, *323*, 240–244. [CrossRef]
7. Bhutia, K.; Khanna, V.; Meetei, T.; Bhutia, N. Effects of climate change on growth and development of chilli. *Agrotechnology* **2018**, *7*, 2. [CrossRef]
8. Wahid, A.; Gelani, S.; Ashraf, M.; Foolad, M.R. Heat tolerance in plants: An overview. *Environ. Exp. Bot.* **2007**, *61*, 199–223. [CrossRef]
9. Alsamir, M.; Mahmood, T.; Trethowan, R.; Ahmad, N. An overview of heat stress in tomato (*Solanum lycopersicum* L.). *Saudi J. Biol. Sci.* **2021**, *28*, 1654–1663. [CrossRef]
10. Blum, A. *Plant Breeding for Stress Environments*; CRC Press Inc.: Boca Raton, FL, USA, 1988; p. 223.
11. Pék, Z.; Helyes, L. The effect of daily temperature on truss flowering rate of tomato. *J. Sci. Food Agric.* **2004**, *84*, 1671–1674. [CrossRef]
12. Willcox, J.K.; Catignani, G.L.; Lazarus, S. Tomatoes and Cardiovascular Health. *Crit. Rev. Food Sci. Nutr.* **2003**, *43*, 1–18. [CrossRef]
13. Foolad, M.R. Genome mapping and molecular breeding of tomato. *Int. J. Plant Genom.* **2007**, *2007*, 64358. [CrossRef]
14. Kimura, S.; Sinha, N. Tomato (*Solanum lycopersicum*): A model fruit-bearing crop. *Cold Spring Harb. Protoc.* **2008**, *2008*, pdb-emo105. [CrossRef]
15. Vinson, J.A.; Hao, Y.; Su, X.; Zubik, L. Phenol antioxidant quantity and quality in foods: Vegetables. *J. Agric. Food Chem.* **1998**, *46*, 3630–3634. [CrossRef]
16. Minhthy, L.; Steven, J. Lycopene: Chemical and biological properties. *Food Technol.* **1999**, *53*, 38–45.
17. Van Ploeg, D.; Heuvelink, E. Influence of sub-optimal temperature on tomato growth and yield: A review. *J. Hortic. Sci. Biotechnol.* **2005**, *80*, 652–659. [CrossRef]
18. De Koning, A. The effect of different day/night temperature regimes on growth, development and yield of glasshouse tomatoes. *J. Hortic. Sci.* **1988**, *63*, 465–471. [CrossRef]
19. Ro, S.; Chea, L.; Ngoun, S.; Stewart, Z.P.; Roeurn, S.; Theam, P.; Lim, S.; Sor, R.; Kosal, M.; Roeun, M. Response of tomato genotypes under different high temperatures in field and greenhouse conditions. *Plants* **2021**, *10*, 449. [CrossRef]
20. Peet, M.M.; Willits, D.; Gardner, R. Response of ovule development and post-pollen production processes in male-sterile tomatoes to chronic, sub-acute high temperature stress. *J. Exp. Bot.* **1997**, *48*, 101–111. [CrossRef]
21. Firon, N.; Shaked, R.; Peet, M.; Pharr, D.; Zamski, E.; Rosenfeld, K.; Althan, L.; Pressman, E. Pollen grains of heat tolerant tomato cultivars retain higher carbohydrate concentration under heat stress conditions. *Sci. Hortic.* **2006**, *109*, 212–217. [CrossRef]

22. Silva, R.; Kumar, L.; Shabani, F.; Picanço, M. Assessing the impact of global warming on worldwide open field tomato cultivation through CSIRO-Mk3· 0 global climate model. *J. Agric. Sci.* **2017**, *155*, 407–420. [CrossRef]
23. Jones, B.J. *Tomato Plant Culture: In The Field, Greenhouse, and Home Garden*, 2nd ed.; CRC Press: Boca Raton, FL, USA, 2008; Volume 136.
24. Sato, S.; Peet, M.; Thomas, J. Physiological factors limit fruit set of tomato (*Lycopersicon esculentum* Mill.) under chronic, mild heat stress. *Plant Cell Environ.* **2000**, *23*, 719–726. [CrossRef]
25. Hedhly, A.; Hormaza, J.; Herrero, M. Flower emasculation accelerates ovule degeneration and reduces fruit set in sweet cherry. *Sci. Hortic.* **2009**, *119*, 455–457. [CrossRef]
26. Camejo, D.; Rodríguez, P.; Morales, M.A.; Dell'Amico, J.M.; Torrecillas, A.; Alarcón, J.J. High temperature effects on photosynthetic activity of two tomato cultivars with different heat susceptibility. *J. Plant Physiol.* **2005**, *162*, 281–289. [CrossRef]
27. Willits, D.; Peet, M. The effect of night temperature on greenhouse grown tomato yields in warm climates. *Agric. Forest Meteorol.* **1998**, *92*, 191–202. [CrossRef]
28. Peet, M.; Sato, S.; Gardner, R. Comparing heat stress effects on male-fertile and male-sterile tomatoes. *Plant Cell Environ.* **1998**, *21*, 225–231. [CrossRef]
29. Sato, S.; Kamiyama, M.; Iwata, T.; Makita, N.; Furukawa, H.; Ikeda, H. Moderate increase of mean daily temperature adversely affects fruit set of *Lycopersicon esculentum* by disrupting specific physiological processes in male reproductive development. *Ann. Bot.* **2006**, *97*, 731–738. [CrossRef]
30. Ansary, S. Breeding Tomato (*Lycopersicon esculentum* Mill.) Tolerant to High Temperature Stress. Ph.D. Thesis, Bidhan Chandra Krishi Viswavidyalaya, West Bengal, India, 2006.
31. Stevens, M.A. Genetic potential for overcoming physiological limitations on adaptability, yield, and quality in the tomato. *Hort-Sci.* **1978**, *13*, 673–678.
32. Qu, G.-Q.; Liu, X.; Zhang, Y.-L.; Yao, D.; Ma, Q.-M.; Yang, M.-Y.; Zhu, W.-H.; Yu, S.; Luo, Y.-B. Evidence for programmed cell death and activation of specific caspase-like enzymes in the tomato fruit heat stress response. *Planta* **2009**, *229*, 1269–1279. [CrossRef]
33. Zhou, R.; Yu, X.; Kjær, K.H.; Rosenqvist, E.; Ottosen, C.-O.; Wu, Z. Screening and validation of tomato genotypes under heat stress using *Fv/Fm* to reveal the physiological mechanism of heat tolerance. *Environ. Exp. Bot.* **2015**, *118*, 1–11. [CrossRef]
34. Zhou, R.; Wu, Z.; Wang, X.; Rosenqvist, E.; Wang, Y.; Zhao, T.; Ottosen, C.-O. Evaluation of temperature stress tolerance in cultivated and wild tomatoes using photosynthesis and chlorophyll fluorescence. *Hortic. Environ. Biotechnol.* **2018**, *59*, 499–509. [CrossRef]
35. Sun, M.; Jiang, F.; Zhang, C.; Shen, M.; Wu, Z. A new comprehensive evaluation system for thermo-tolerance in tomato at different growth stage. *J. Agric. Sci. Technol. B* **2016**, *6*, 152–168.
36. Rajametov, S.N.; Yang, E.Y.; Jeong, H.B.; Cho, M.C.; Chae, S.Y.; Paudel, N. Heat treatment in two tomato cultivars: A study of the effect on physiological and growth recovery. *Horticulturae* **2021**, *7*, 119. [CrossRef]
37. Xu, J.; Wolters-Arts, M.; Mariani, C.; Huber, H.; Rieu, I. Heat stress affects vegetative and reproductive performance and trait correlations in tomato (*Solanum lycopersicum*). *Euphytica* **2017**, *213*, 156. [CrossRef]
38. Din, J.U.; Khan, S.U.; Khan, A.; Qayyum, A.; Abbasi, K.S.; Jenks, M.A. Evaluation of potential morpho-physiological and biochemical indicators in selecting heat-tolerant tomato (*Solanum lycopersicum* Mill.) genotypes. *Hortic. Environ. Biotechnol.* **2015**, *56*, 769–776. [CrossRef]
39. Zhou, R.; Kjaer, K.; Rosenqvist, E.; Yu, X.; Wu, Z.; Ottosen, C.O. Physiological response to heat stress during seedling and anthesis stage in tomato genotypes differing in heat tolerance. *J. Agron. Crop Sci.* **2017**, *203*, 68–80. [CrossRef]
40. Driedonks, N.; Rieu, I.; Vriezen, W.H. Breeding for plant heat tolerance at vegetative and reproductive stages. *Plant Reprod.* **2016**, *29*, 67–79. [CrossRef]
41. Mesihovic, A.; Iannacone, R.; Firon, N.; Fragkostefanakis, S. Heat stress regimes for the investigation of pollen thermotolerance in crop plants. *Plant Reprod.* **2016**, *29*, 93–105. [CrossRef]
42. Yeh, C.-H.; Kaplinsky, N.J.; Hu, C.; Charng, Y.-y. Some like it hot, some like it warm: Phenotyping to explore thermotolerance diversity. *Plant Sci.* **2012**, *195*, 10–23. [CrossRef]
43. Shaheen, M.R.; Ayyub, C.M.; Amjad, M.; Waraich, E.A. Morpho-physiological evaluation of tomato genotypes under high temperature stress conditions. *J. Sci. Food Agric.* **2016**, *96*, 2698–2704. [CrossRef]
44. Paupière, M.J.; van Haperen, P.; Rieu, I.; Visser, R.G.; Tikunov, Y.M.; Bovy, A.G. Screening for pollen tolerance to high temperatures in tomato. *Euphytica* **2017**, *213*, 130. [CrossRef]
45. Went, F. The climatic control of flowering and fruit set. *Am. Nat.* **1950**, *84*, 161–170. [CrossRef]
46. Charles, W.; Harris, R. Tomato fruit-set at high and low temperatures. *Can. J. Plant Sci.* **1972**, *52*, 497–506. [CrossRef]
47. RA, K.N. Growth of tomato plants in different oxygen concentrations. *Photosynthetica* **1980**, *14*, 326–336.
48. Geisenberg, C.; Stewart, K. Field crop management. In *The Tomato Crop, a Scientific basis for Improvement*; Antherton, J.G., Rudich, J., Eds.; Chapman and Hall: New York, NY, USA, 1986; pp. 511–557.
49. Peet, M.M.; Bartholemew, M. Effect of night temperature on pollen characteristics, growth, and fruit set in tomato. *J. Am. Soc. Hortic. Sci.* **1996**, *121*, 514–519. [CrossRef]
50. Dhankar, S.; Dhankhar, B.; Sharma, N. Correlation and path analysis in tomato under normal and high temperature conditions. *Haryana J. Hortic. Sci.* **2001**, *30*, 89–91.

51. Choudhary, M.; Patel, V.; Siddiqui, M.W.; Mahdl, S.S.; Verma, R. *Climate Dynamics in Horticultural Science, Volume One: The Principles and Applications*; CRC Press: Florida, FL, USA, 2015.
52. Miller, P.; Lanier, W.; Brandt, S. *Using Growing Degree Days to Predict Plant Stages*; Ag/Extension Communications Coordinator, Communications Services; Montana State University-Bozeman: Bozeman, MO, USA, 2001; Volume 59717, pp. 994–2721.
53. Abdelmageed, A.H.A.; Gruda, N. Influence of heat shock pretreatment on growth and development of tomatoes under controlled heat stress conditions. *J. Appl. Bot. Food Qual.* **2007**, *81*, 26–28.
54. Zhou, R.; Yu, X.; Ottosen, C.-O.; Rosenqvist, E.; Zhao, L.; Wang, Y.; Yu, W.; Zhao, T.; Wu, Z. Drought stress had a predominant effect over heat stress on three tomato cultivars subjected to combined stress. *BMC Plant Biol.* **2017**, *17*, 24. [CrossRef]
55. Zheng, Y.; Yang, Z.; Xu, C.; Wang, L.; Huang, H.; Yang, S. The interactive effects of daytime high temperature and humidity on growth and endogenous hormone concentration of tomato seedlings. *HortScience* **2020**, *55*, 1575–1583. [CrossRef]
56. Sherzod, R.; Yang, E.Y.; Cho, M.C.; Chae, S.Y.; Chae, W.B. Physiological traits associated with high temperature tolerance differ by fruit types and sizes in tomato (*Solanum lycopersicum* L.). *Hortic. Environ. Biotechnol.* **2020**, *61*, 837–847. [CrossRef]
57. Bhattarai, S.; Harvey, J.T.; Djidonou, D.; Leskovar, D.I. Exploring morpho-physiological variation for heat stress tolerance in tomato. *Plants* **2021**, *10*, 347. [CrossRef]
58. Foolad, M.R. Breeding for abiotic stress tolerances in tomato. In *Abiotic Stresses: Plant Resistance through Breeding and Molecular Approaches*; Ashraf, M., Harris, P.J.C., Eds.; The Haworth Press: New York, NY, USA, 2005; pp. 613–684.
59. Giorno, F.; Wolters-Arts, M.; Mariani, C.; Rieu, I. Ensuring reproduction at high temperatures: The heat stress response during anther and pollen development. *Plants* **2013**, *2*, 489–506. [CrossRef]
60. Kakani, V.; Reddy, K.; Koti, S.; Wallace, T.; Prasad, P.; Reddy, V.; Zhao, D. Differences in in vitro pollen germination and pollen tube growth of cotton cultivars in response to high temperature. *Ann. Bot.* **2005**, *96*, 59–67. [CrossRef]
61. Karapanos, I.; Akoumianakis, K.; Olympios, C.; Passam, H.C. Tomato pollen respiration in relation to in vitro germination and pollen tube growth under favourable and stress-inducing temperatures. *Sex Plant Reprod.* **2010**, *23*, 219–224. [CrossRef]
62. Song, J.; Nada, K.; Tachibana, S. Suppression of S-adenosylmethionine decarboxylase activity is a major cause for high-temperature inhibition of pollen germination and tube growth in tomato (*Lycopersicon esculentum* Mill.). *Plant Cell Physiol.* **2002**, *43*, 619–627. [CrossRef]
63. Ayenan, M.A.T.; Danquah, A.; Hanson, P.; Ampomah-Dwamena, C.; Sodedji, F.A.K.; Asante, I.K.; Danquah, E.Y. Accelerating breeding for heat tolerance in tomato (*Solanum lycopersicum* L.): An integrated approach. *Agronomy* **2019**, *9*, 720. [CrossRef]
64. Berry, S.; Uddin, M.R. Effect of high temperature on fruit set in tomato cultivars and selected germplasm. *Hortscience* **1988**, *23*, 606–608.
65. Dane, F.; Hunter, A.G.; Chambliss, O.L. Fruit set, pollen fertility, and combining ability of selected tomato genotypes under high-temperature field conditions. *J. Am. Soc. Hortic. Sci.* **1991**, *116*, 906–910. [CrossRef]
66. Frank, G.; Pressman, E.; Ophir, R.; Althan, L.; Shaked, R.; Freedman, M.; Shen, S.; Firon, N. Transcriptional profiling of maturing tomato (*Solanum lycopersicum* L.) microspores reveals the involvement of heat shock proteins, ROS scavengers, hormones, and sugars in the heat stress response. *J. Exp. Bot.* **2009**, *60*, 3891–3908. [CrossRef]
67. Pressman, E.; Peet, M.M.; Pharr, D.M. The effect of heat stress on tomato pollen characteristics is associated with changes in carbohydrate concentration in the developing anthers. *Ann. Bot.* **2002**, *90*, 631–636. [CrossRef]
68. Abdul-Baki, A.A.; Stommel, J.R. Pollen viability and fruit set of tomato genotypes under optimum and high-temperature regimes. *HortScience* **1995**, *30*, 115–117. [CrossRef]
69. Zinn, K.E.; Tunc-Ozdemir, M.; Harper, J.F. Temperature stress and plant sexual reproduction: Uncovering the weakest links. *J. Exp. Bot.* **2010**, *61*, 1959–1968. [CrossRef]
70. Herrero, M.; Hormaza, J. Pistil strategies controlling pollen tube growth. *Sex Plant Reprod.* **1996**, *9*, 343–347. [CrossRef]
71. Herrero, M.; Arbeloa, A. Influence of the pistil on pollen tube kinetics in peach (*Prunus persica*). *Am. J. Bot.* **1989**, *76*, 1441–1447. [CrossRef]
72. Lord, E.M.; Russell, S.D. The mechanisms of pollination and fertilization in plants. *Annu. Rev. Cell Dev. Biol.* **2002**, *18*, 81–105. [CrossRef]
73. Snider, J.L.; Oosterhuis, D.M. How does timing, duration, and severity of heat stress influence pollen-pistil interactions in angiosperms? *Plant Signal. Behav.* **2011**, *6*, 930–933. [CrossRef]
74. Hanson, P.M.; Chen, J.-t.; Kuo, G. Gene action and heritability of high-temperature fruit set in tomato line CL5915. *HortScience* **2002**, *37*, 172–175. [CrossRef]
75. Ruan, Y.-L.; Jin, Y.; Yang, Y.-J.; Li, G.-J.; Boyer, J.S. Sugar input, metabolism, and signaling mediated by invertase: Roles in development, yield potential, and response to drought and heat. *Mol. Plant* **2010**, *3*, 942–955. [CrossRef]
76. Hazra, P.; Ansary, S.H.; Dutta, A.K.; Balacheva, E.; Atanassova, B. Breeding tomato tolerant to high temperature stress. *Acta Hortic.* **2009**, *830*, 241–248. [CrossRef]
77. Wise, R.; Olson, A.; Schrader, S.; Sharkey, T. Electron transport is the functional limitation of photosynthesis in field-grown Pima cotton plants at high temperature. *Plant Cell Environ.* **2004**, *27*, 717–724. [CrossRef]
78. Bokszczanin, K.L.; Fragkostefanakis, S.; Bostan, H.; Bovy, A.; Chaturvedi, P.; Chiusano, M.L.; Firon, N.; Iannacone, R.; Jegadeesan, S.; Klaczynskid, K. Perspectives on deciphering mechanisms underlying plant heat stress response and thermotolerance. *Front. Plant Sci.* **2013**, *4*, 315. [CrossRef] [PubMed]

79. Sung, D.-Y.; Kaplan, F.; Lee, K.-J.; Guy, C.L. Acquired tolerance to temperature extremes. *Trends Plant Sci.* **2003**, *8*, 179–187. [CrossRef]
80. Lee, K.; Kang, H. Roles of organellar RNA-binding proteins in plant growth, development, and abiotic stress responses. *Int. J. Mol. Sci.* **2020**, *21*, 4548. [CrossRef]
81. Driedonks, N.; Wolters-Arts, M.; Huber, H.; de Boer, G.-J.; Vriezen, W.; Mariani, C.; Rieu, I. Exploring the natural variation for reproductive thermotolerance in wild tomato species. *Euphytica* **2018**, *214*, 67. [CrossRef]
82. Poudyal, D.; Rosenqvist, E.; Ottosen, C.-O. Phenotyping from lab to field–tomato lines screened for heat stress using *Fv/Fm* maintain high fruit yield during thermal stress in the field. *Funct. Plant Biol.* **2018**, *46*, 44–55. [CrossRef]
83. Mattioli, R.; Biancucci, M.; El Shall, A.; Mosca, L.; Costantino, P.; Funck, D.; Trovato, M. Proline synthesis in developing microspores is required for pollen development and fertility. *BMC Plant Biol.* **2018**, *18*, 356. [CrossRef]
84. Nathalie, V.; Christian, H. Proline accumulation in plants: A review. *Amino Acids* **2008**, *35*, 753–759.
85. Szabados, L.; Savouré, A. Proline: A multifunctional amino acid. *Trends Plant Sci.* **2010**, *15*, 89–97. [CrossRef]
86. Claussen, W. Proline as a measure of stress in tomato plants. *Plant Sci.* **2005**, *168*, 241–248. [CrossRef]
87. Zhou, R.; Kong, L.; Yu, X.; Ottosen, C.-O.; Zhao, T.; Jiang, F.; Wu, Z. Oxidative damage and antioxidant mechanism in tomatoes responding to drought and heat stress. *Acta Physiol. Plant.* **2019**, *41*, 20. [CrossRef]
88. Chen, T.H.; Murata, N. Enhancement of tolerance of abiotic stress by metabolic engineering of betaines and other compatible solutes. *Curr. Opin. Plant Biol.* **2002**, *5*, 250–257. [CrossRef]
89. Rivero, R.M.; Mestre, T.C.; Mittler, R.; Rubio, F.; Garcia-Sanchez, F.; Martinez, V. The combined effect of salinity and heat reveals a specific physiological, biochemical and molecular response in tomato plants. *Plant Cell Environ.* **2014**, *37*, 1059–1073. [CrossRef]
90. Li, S.; Li, F.; Wang, J.; Zhang, W.; Meng, Q.; Chen, T.H.; Murata, N.; Yang, X. Glycinebetaine enhances the tolerance of tomato plants to high temperature during germination of seeds and growth of seedlings. *Plant Cell Environ.* **2011**, *34*, 1931–1943. [CrossRef]
91. Mäkelä, P.; Jokinen, K.; Kontturi, M.; Peltonen-Sainio, P.; Pehu, E.; Somersalo, S. Foliar application of glycinebetaine—a novel product from sugar beet—as an approach to increase tomato yield. *Ind. Crops Prod.* **1998**, *7*, 139–148. [CrossRef]
92. Paupière, M.J.; Van Heusden, A.W.; Bovy, A.G. The metabolic basis of pollen thermo-tolerance: Perspectives for breeding. *Metabolites* **2014**, *4*, 889–920. [CrossRef]
93. Rivero, R.M.; Ruiz, J.M.; Garcıa, P.C.; Lopez-Lefebre, L.R.; Sánchez, E.; Romero, L. Resistance to cold and heat stress: Accumulation of phenolic compounds in tomato and watermelon plants. *Plant Sci.* **2001**, *160*, 315–321. [CrossRef]
94. Paupière, M.J.; Müller, F.; Li, H.; Rieu, I.; Tikunov, Y.M.; Visser, R.G.; Bovy, A.G. Untargeted metabolomic analysis of tomato pollen development and heat stress response. *Plant Reprod.* **2017**, *30*, 81–94. [CrossRef]
95. Song, J.; Nada, K.; Tachibana, S. Ameliorative effect of polyamines on the high temperature inhibition of in vitro pollen germination in tomato (*Lycopersicon esculentum* Mill.). *Sci. Hortic.* **1999**, *80*, 203–212. [CrossRef]
96. Cheng, L.; Zou, Y.; Ding, S.; Zhang, J.; Yu, X.; Cao, J.; Lu, G. Polyamine accumulation in transgenic tomato enhances the tolerance to high temperature stress. *J. Integr. Plant Biol.* **2009**, *51*, 489–499. [CrossRef]
97. Mittler, R. Oxidative stress, antioxidants and stress tolerance. *Trends Plant Sci.* **2002**, *7*, 405–410. [CrossRef]
98. Gupta, N.; Agarwal, S.; Agarwal, V.; Nathawat, N.; Gupta, S.; Singh, G. Effect of short-term heat stress on growth, physiology and antioxidative defence system in wheat seedlings. *Acta Physiol. Plant.* **2013**, *35*, 1837–1842. [CrossRef]
99. El Ahmadi, A.B.; Stevens, M.A. Reproductive responses of heat-tolerant tomatoes to high temperature. *J. Am. Soc. Hortic. Sci.* **1979**, *104*, 686–691.
100. Bar-Tsur, A.; Rudich, J.; Bravdo, B. High temperature effects on CO_2 gas exchange in heat-tolerant and sensitive tomatoes. *J. Am. Soc. Hortic.* **1985**, *110*, 582–586.
101. Lohar, D.; Peat, W. Floral characteristics of heat-tolerant and heat-sensitive tomato (*Lycopersicon esculentum* Mill.) cultivars at high temperature. *Sci. Hortic.* **1998**, *73*, 53–60. [CrossRef]
102. Hanna, H.Y. Response of six tomato genotypes under summer and spring conditions in Louisiana. *HortScience* **1982**, *17*, 758–759.
103. Adams, S.; Cockshull, K.; Cave, C. Effect of temperature on the growth and development of tomato fruits. *Ann. Bot.* **2001**, *88*, 869–877. [CrossRef]
104. Sato, S.; Peet, M.M.; Gardner, R.G. Altered flower retention and developmental patterns in nine tomato cultivars under elevated temperature. *Sci. Hortic.* **2004**, *101*, 95–101. [CrossRef]

Review

'Breathing Out' under Heat Stress—Respiratory Control of Crop Yield under High Temperature

Nitin Sharma [1], Meenakshi Thakur [2], Pavithra Suryakumar [3], Purbali Mukherjee [3], Ali Raza [4,*], Channapatna S. Prakash [5] and Anjali Anand [3,*]

1 Department of Basic Sciences, College of Forestry, Dr. Yashwant Singh Parmar University of Horticulture and Forestry, Nauni, Solan 173230, India; nitinphysio@yspuniversity.ac.in
2 Department of Basic Sciences, College of Horticulture and Forestry, Dr. Yashwant Singh Parmar University of Horticulture and Forestry, Neri, Hamirpur 177001, India; thakurmeenakshi94@gmail.com
3 Division of Plant Physiology, ICAR-Indian Agricultural Research Institute, New Delhi 110012, India; pavi796@gmail.com (P.S.); purbalim2@gmail.com (P.M.)
4 Key Laboratory of Ministry of Education for Genetics, Breeding and Multiple Utilization of Crops, Center of Legume Crop Genetics and Systems Biology/College of Agriculture, Oil Crops Research Institute, Fujian Agriculture and Forestry University (FAFU), Fuzhou 350002, China
5 College of Arts & Science, Tuskegee University, Tuskegee, AL 36088, USA; cprakash@tuskegee.edu
* Correspondence: alirazamughal143@gmail.com (A.R.); anjaliiari@gmail.com or anjalianand@iari.res.in (A.A.)

Abstract: Respiration and photosynthesis are indispensable plant metabolic processes that are affected by elevated temperatures leading to disruption of the carbon economy of the plants. Increasing global temperatures impose yield penalties in major staple crops that are attributed to increased respiratory carbon loss, through higher maintenance respiration resulting in a shortage of non-structural carbohydrates and an increase in metabolic processes like protein turnover and maintenance of ion concentration gradients. At a cellular level, warmer temperatures lead to mitochondrial swelling as well as downregulation of respiration by increasing the adenosine triphosphate:adenosine diphosphate (ATP:ADP) ratio, the abscisic acid-mediated reduction in ATP transfer to the cytosol, and the disturbance in a concentration gradient of tricarboxylic acid (TCA) cycle intermediates, as well as increasing lipid peroxidation in mitochondrial membranes and cytochrome c release to trigger programmed cell death. In this review, we discuss the mechanistic insight into the heat stress-induced mitochondrial dysfunction that controls dark respiration in plants. Furthermore, the role of hormones in regulating the network of processes that are involved in retrograde signaling is highlighted. We also propose different strategies to reduce carbon loss under high temperature, e.g., selecting genotypes with low respiration rates and using genome editing tools to target the carbon-consuming pathways by replacing, relocating, or rescheduling the metabolic activities.

Keywords: acclimation; alternative oxidase; heat stress; mitochondria; maintenance; respiration

Citation: Sharma, N.; Thakur, M.; Suryakumar, P.; Mukherjee, P.; Raza, A.; Prakash, C.S.; Anand, A. 'Breathing Out' under Heat Stress—Respiratory Control of Crop Yield under High Temperature. *Agronomy* **2022**, *12*, 806. https://doi.org/10.3390/agronomy12040806

Academic Editor: Daniel Mullan

Received: 16 February 2022
Accepted: 24 March 2022
Published: 27 March 2022

1. Introduction

The rising temperature is an intrinsic component of global climate change that controls the carbon fluxes in all the crops. High temperature affects the major plant physiological processes, such as photosynthesis and respiration; therefore, it becomes important to estimate the plant carbon dioxide (CO_2) balance that finally decides the crop productivity [1–3].Through these two pathways, the terrestrial ecosystems exchange about 120 Gt of carbon per year with the atmosphere [4]. A rough estimate states that half of the CO_2 assimilated annually through photosynthesis is released back to the atmosphere by plant respiration [5–7], and merely 15–25% of the fixed carbon finally translates into yield [8,9]. The projected elevation in temperature beyond 2.0 °C by the end of the decade [10] may increase the magnitude of carbon loss exponentially in the physiological temperature range of 0 to 38 °C [11], which will further exacerbate in a species-and environment-dependent manner at higher temperatures between 48 and 60 °C [12–15].

The carbon lost through the 'breathing out' processes in plants can occur via two mechanisms, namely photorespiration and dark/mitochondrial respiration. These processes release CO_2, but dark respiration occurs regardless of light in the plant cells [16,17]. Biochemically, dark respiration is an enzymatically regulated, multistep, amphibolic process that produces ATP by the oxidation of glucose formed during photosynthesis. Glucose is initially broken into pyruvate during glycolysis, which is oxidized to form acetyl-CoA, releasing a molecule of CO_2. The acetyl-CoA then enters the tricarboxylic acid (TCA) cycle, where it is oxidized to CO_2 and also produces reductants (nicotinamide adenine dinucleotide: NADH; dihydroflavine-adenine dinucleotide: $FADH_2$) that pass through the mitochondrial electron transport chain (ETC). The oxidation of the reductants produces a proton gradient across the inner membrane of the mitochondria that drives the synthesis of ATP. High temperatures impact dark respiration in plants with an exponential increase [18], which can become detrimental due to irreversible damage to the enzymatic machinery [15]. Climate change prediction models have speculated a 3–20% decline in the yield of major crops like wheat, rice, maize, and soybean with every 1 °C increase in the global mean temperatures [19,20], which makes it pertinent to relate this loss to the waste of carbon due to respiration. The contribution of dark respiration in limiting the productivity of crops under elevated temperatures has not been extensively reviewed, in comparison to photorespiration. Therefore, our present review discusses the heat-induced alterations in dark respiration in plants and proposes strategies to reduce the carbon loss under the inevitable reality of a changing climate.

2. Respiratory Carbon Loss-A Constraint to Crop Yield

Respiration, rather than photosynthesis, may be the primary contributor to yield losses in a high temperature climate [11]. Low respiration rates are generally correlated with high crop yields [21,22]. Walker et al. [23] reported that photorespiration decreased soybean and wheat yields by 36% and 20%, respectively, in the United States. In another study, a 10–12% and 17–35% decrease in the yields of wheat and rice, respectively, was reported due to high temperatures [24]. The yield loss in wheat and rice due to high night temperature (HNT) is mainly ascribed to higher dark respiration, which increases the consumption of photoassimilates, thereby resulting in the reduction of non-structural carbohydrates (NSCs) in stem tissues [25,26]. Glaubitz et al. [27] reported that increasing night temperature from 25 °C to 35 °C resulted in increased leaf respiratory carbon losses in grapevines, as reflected by the decrease in NSCs of 0.025 and 0.041 mg g^{-1}dry weight, respectively. Such losses are consistent with metabolite profiling studies in wheat and rice, which revealed an increase in TCA cycle intermediates in leaves exposed to HNT, supporting increased respiration in the photosynthesizing tissue [25,28]. Xu et al. [29] suggest that increased dark respiration restrains source availability under the combined stress of high day and night temperatures, leading to a considerably more severe yield penalty due to carbon loss.

3. Heat-Induced Changes in the Proportion of Maintenance Respiration

Dark respiration (R_d) is typically partitioned into two functional components, i.e., growth respiration (R_g) and maintenance respiration (R_m), which are impacted upon by environmental stresses [9,30,31]. Figure 1 illustrates the differences in these components under elevated temperatures. Growth respiration is a dominant component of respiration in younger tissues, while the latter contributes majorly to the older tissues [32]. Growth respiration is defined as the amount of photoassimilates respired to provide energy for the synthesis of additional biomass [33]. It also provides a carbon skeleton and reductants to facilitate nutrient uptake/assimilation followed by biosynthesis of cellular components to drive the growth of tissues. Thus, the relationship between the growth rates of a species and temperature is actually a measure of the rate of the growth respiration component [34]. A recent analysis of 101 evergreen species growing in different biomes (boreal to tropical) showed that respiration increased with an increase in growth temperatures in accordance with previous studies [35,36]. Leaf form accounted for the response ratio of R_g to warming, as species with needle-like leaves had a significantly higher response (25 ± 9%) than broad-leaved ones [36].

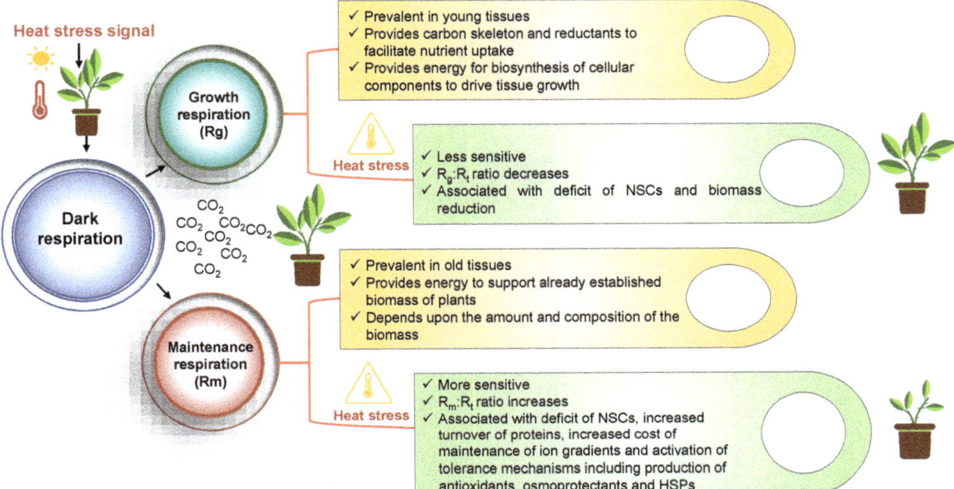

Figure 1. Growth respiration and maintenance respiration under elevated environmental temperature. HSPs: heat shock proteins; NSCs: non-structural carbohydrates; R_g: growth respiration; R_m: maintenance respiration; R_t: total respiration.

On the other hand, maintenance respiration comprises the respiratory processes that help in supporting the already established biomass of the plant [33]. It depends upon the amount and composition of the biomass, as both these factors undergo change depending on the environment and developmental stage of the plant. Although the role of both the components is integral to the life cycle of the plants, their estimation can only be done by employing physiological models [32,37–39]. The higher temperature responsiveness of R_m over R_g in mature tissues was concluded from various studies, e.g., Marigolds when exposed to a 10 °C increase in temperature resulted in a 43% to 55% increase in the proportion of maintenance respiration to total respiration (R_t) [40]. Additionally, a significant reduction in ATP content and total biomass was observed in rice plants subjected to 10 °C higher temperature at the reproductive stage than the ambient temperature (28 °C), thereby suggesting that energy produced by respiration under high temperature conditions was mainly attributed to maintenance respiration rather than growth respiration [32]. Mathematically, maintenance respiration is expressed as the product of maintenance respiration coefficient and plant size. The Q_{10} value (proportional increase in rate of respiration with a 10 °C rise in temperature) of the maintenance respiration coefficient varies between 1.35 and 3.0 depending upon the species, developmental stage, and environmental conditions as shown in the data compiled from various studies (Table 1). The sensitivity of Q_{10} to temperature indicates that the response of respiration to temperature cannot be represented by one value.

Table 1. Q_{10} values for maintenance respiration coefficient in various crops.

Crop	Experimental Temperature	Q_{10} Value	Reference
Marigold (*Tagetes patula*)	20 °C (Control) 30 °C (Elevated)	1.35–1.55	[40]
Barley (*Hordeum vulgare*)	15 °C (Control) 28 °C (Elevated)	3.00	[41]
Subterranean clover (*Trifolium subterraneum*)	10 °C (Control) 35 °C (Elevated)	1.85	[42]
Japanese knotweed (*Reynoutria japonica*)	15 °C (Control) 25 °C (Elevated)	1.90	[43]
Wheat (*Triticum aestivum*)	15 °C (Control) 20 °C (Elevated)	1.80	[44]
	10 °C (Control-Night temperature) 21 °C (Elevated-Night temperature)	1.97	[45]

4. Substrate Availability for Respiration under High Temperature

The considerable variation observed in the Q_{10}–temperature relationship is influenced by the supply of the respiratory substrate and the respiration capacity [4,46]. Environmental variables that affect the biosynthesis of the substrates [18,46] or increase the metabolism of energy consuming processes like turnover of proteins and maintenance of ion gradients [47], make Q_{10} values highly dynamic in response to temperature. Additional energy costs are incurred by mechanisms imparting heat tolerance in the crops, e.g., upregulation of the antioxidant defense system to counteract the upsurge in the level of reactive oxygen species (ROS), synthesis of osmoprotectants, and accretion of heat shock proteins (HSPs). The need for respiratory substrate in the plants is mainly met from the non-structural carbohydrates [25–27,48] and the protein turnover [11,32]. Studies on the effect of elevated night temperatures have shown that the high rate of nighttime respiration exerted pressure on the supply of NSCs, which subsequently reduced the biomass and yield of rice [25,26]. The concentration of sugars has been positively correlated with the rate of dark respiration in *Pinus* [49], *Quercus rubra* [50], and *Spinacia oleracea* [51]. The light control of carbohydrate synthesis affected the rate of dark respiration in *Geum urbanum* plants grown under 75% shade as it declined due to limited photosynthate supply, but Q_{10} declined only when the leaves experienced near darkness for long periods. It was concluded that intense shade for a prolonged period would cause a reduction in both respiration and Q_{10} due to adenylate restriction on respiration in addition to the substrate availability [4].

5. Regulation of Respiratory Flux at High Temperature

Adenylates (in particular the ratio of ATP to ADP and the concentration of ADP per se), are likely the most important in regulating respiratory flux at warm temperatures [52]. Adenylate control would indicate that the respiratory capacity at warmer temperatures exceeded the level required for cell processes to proceed [18], which in turn would lead to elevated ATP:ADP ratios or low ADP concentration, causing downregulation of respiration [53]. The increased leakiness of membranes at high temperatures could further contribute to substrate limitation because concentration gradients of TCA cycle intermediates are more difficult to maintain when mitochondrial membranes are excessively fluid [18].

6. Positive Correlation between Protein Turnover Cost and Respiratory Cost at High Temperature

Nitrogen (N) utilization processes, including nitrate reduction and ammonium assimilation, are thought to have high respiratory costs [54]. In fact, the estimates of construction respiration are greatly influenced by the form of N source, e.g., nitrate or ammonium [55]. The protein turnover rate increases with temperature, suggesting that the protein turnover cost is a major component of the N-utilization cost and dominates during maintenance respiration. Hachiya et al. [56] studied the protein turnover cost in *Petunia x hybrida* petals grown at three different temperatures (20, 25,and 35 °C) during the development of the petals. Most petals are non-photosynthetic; therefore, ATP and reducing equivalents are supplied mainly from the respiratory pathway. The integrated protein turnover cost on dry weight basis was similar between 20 and 25 °C but increased by more than four times at 35 °C, suggesting that the high temperature enhanced the cost of protein turnover, thereby increasing the total cost of N-utilization along with respiration in the petals.

7. Diurnal Dynamics of Respiration

The diurnal or diel cycle of plant growth interacts with the respiratory metabolism, which can be directly linked with the availability of respiratory metabolites regulating the process at different times of the day [57]. The photosynthate synthesized during the day supports carbon supply for the entire plant during the day, which is reduced to critical levels by the end of the night [58]. The strong coupling between carbon fixation through photosynthesis and loss due to respiration [59] indicates the diurnal fluctuation in rates of dark respiration as a result of changes in the concentration of various metabolites supporting the respiratory process [60]. In this case, the supply of sugars is stabilized over the day–night cycle, and the diel variation in respiration may be explained by changes in the availability of amino acids, proteins, organic acids, and/or lipids. These metabolites may drive respiration by supplying intermediates to the TCA cycle, reductants for ATP synthesis via oxidative phosphorylation, and carbon skeletons required for biosynthesis or nitrogen assimilation into amino acids [61].

Metabolomic studies have shown that warmer day (30 °C) and night (28 °C) temperatures lead to the accumulation of amino acids derived from shikimate pathways, such as phenylalanine, tyrosine, tryptophan, aspartic acid, lysine, proline, and γ-amino butyrate (GABA), in thermo-sensitive rice cultivars (DR2 and M202) but not in intermediate (IR64 and IRRI123) and temperature-tolerant cultivars (IR72 and Taipei 309) [28]. Similarly, in wheat, high night temperatures showed a prevalence of fumarate and alanine without any significant change in the level of glutamine, glutamate, and GABA [25]. The accumulation of TCA intermediates like malate and fumarate during the day, and citrate, aconitate, and succinate during the night [62,63], reiterates the circadian control of the TCA pathway, which is a hub for the process of respiration and can be markedly influenced by an increase in temperature [64]. Rashid et al. [64] assessed the influence of growth temperature and the diel cycle on the concentrations of metabolites involved in the respiratory network of rice. They raised the plants under 25 °C:20 °C, 30 °C:25 °C, and 40 °C:35 °C day:night cycles and measured the dark respiration and changes in metabolites at five time points spanning a single 24 h period and observed that shikimate pathway-derived aromatic amino acids were the only metabolites to interact in response to both the growth temperature and the day:night cycle. Cook et al. [65] reported increased concentrations of α-ketoglutarate, fumarate, malate, and citrate in *Arabidopsis* leaves when cooled from 20 °C to 4 °C. All these studies suggest that there are distinct respiratory metabolite adjustments to temperature and the diel cycle.Further, detailed experiments on the interaction of the diel cycle and temperature will generate a better understanding of the metabolites controlling dark respiration in plants. Therefore, the instantaneous measurement of respiration rates at a single point during the day can overlook the differential response prevalent during an extended period.

8. Thermal Acclimation of the Respiration Response in Plants under Heat Stress

Short periods of high temperatures show an exponential increase in dark respiration [66,67], whereas prolonged exposure can result in thermal acclimation of the respiration response to lessen the impact of continued carbon loss due to increasing temperatures [31,68]. Under thermal acclimation, the tissues that develop under the new temperature show a better homeostatic response to respiration than the ones formed before the acclimation temperature [31,66,69]. There are two types of thermal acclimation responses [18] that occur across plant types and biomes (Figure 2):(i) Type I acclimation, where warm acclimated leaves show lower short-term sensitivity to temperature, and the regulation by the existing respiratory enzymes causes a reduction in Q_{10} [46].(ii) Type II acclimation, which involves a change in the respiratory capacity due to change in the concentration of the respiratory enzymes or mitochondrial proteins, resulting in lower respiration across the temperature range and no change in Q_{10}. Type I acclimation is less efficient and occurs in leaves that mature prior to the temperature change. In contrast, Type IIis common in leaves that are formed later under higher temperatures with a high degree of homeostasis. The advantage of Type II acclimation is that it allows the plant to make both the physiological and developmental adjustments in the size and density of mitochondria [70], whereas Type I, which solely influences the physiological plasticity to temperature. Based on this fact, it was found that boreal evergreen tree species, which grow under changing temperatures, are more efficient in acclimation during their lifetime than deciduous species that seasonally shed their leaves [71]. A recent meta-analysis by Crous et al. [36] highlighted the differential respiration response across various biogeographical regions and leaf forms and found that the leaves of gymnosperms showed a 30–40% reduction in respiration rates at a common temperature of 25 °C compared to broadleaved evergreens at >10 °C warming.

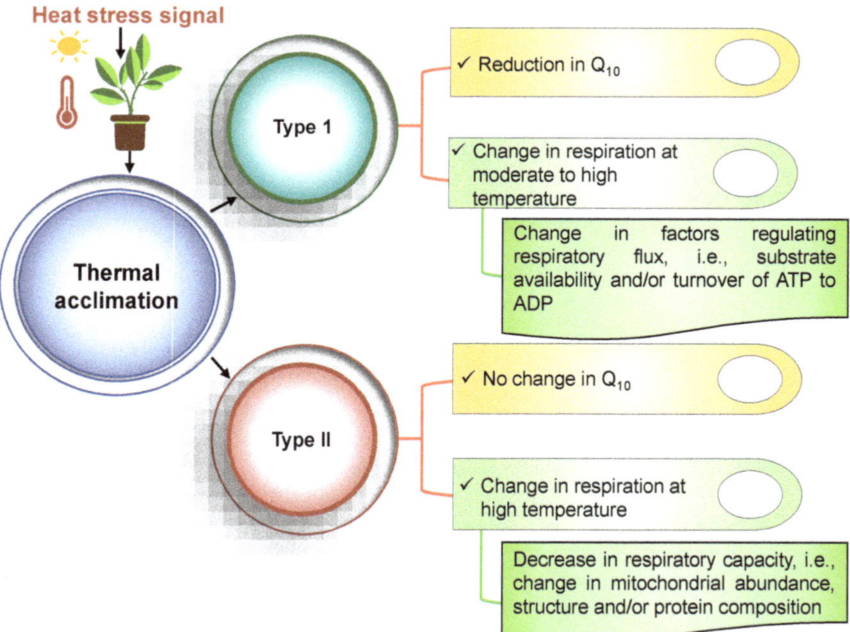

Figure 2. Types of thermal acclimation in plants in response to heat stress.

The dynamicity in the size, number, and signaling responses of mitochondria can cause a collective outcome during the acclimation response to meet the demand for metabolic energy, carbon skeleton, and reductants [72], and it is controlled by a network of genes [73]. The inability to acclimate is the consequence of mitochondrial disorganization under high temperatures that increase the leakiness of mitochondrial membrane and lipid peroxidation [74] along with the disruption of the TCA cycle, mitochondrial NADH pool, and ATP synthesis [75].

9. Mitochondrial Physiology under High Temperature

Temperature stress exerts a thermodynamic influence on the subcellular structures and intracellular macromolecules in the plant cells [76]. Since mitochondria maintain the energy requirements of the cells, they become the primary targets for structural and functional changes under stress [77], as illustrated in Figure 3. The phospholipid, cardiolipin (CL), is an important constituent of the inner mitochondrial membrane and contributes approximately 10% toward the total lipid content of the mitochondria [78]. The loss-of-function mutants of cardiolipin synthase (*cls*), involved in the synthesis of CL, confirmed its role in morphogenesis of mitochondria during heat stress in *Arabidopsis* via stabilizing the protein complex of mitochondrial fission factor DYNAMIN-RELATED PROTEIN 3 [79]. Additionally, CL is rich in polyunsaturated fatty acids and more vulnerable to lipid peroxidation [80] by the excess ROS produced during high temperature stress. The damaged CL increases the pore formation capacity of the membrane, resulting in the dephosphorylation of the mobile electron carrier cytochrome c and its release from the inner membrane towards the cytosol [81,82]. This reduces the cytochrome c activity and ATP synthesis and triggers the programmed cell death (PCD) response under stress [83,84]. A significant association has also been explained between the release of cytochrome c and Ca^{2+} dynamics during heat stress [85]. Complex I, II, and III are known to be the important sites for the production of ROS in the mitochondrial respiratory chain [86,87]. Increased ROS production is positively correlated to hyperpolarization of the mitochondrial membrane in cultured wheat cells under heat treatment. The depolarization of the membrane using the protonophore CCCP (carbonyl cyanide *m*-chlorophenylhydrazone) inhibited ROS production and oxidative phosphorylation [88]. High temperature-induced ROS production increases the cytosolic concentration of calcium that eventually finds its entry into the mitochondria and other organelles [89]. Amongst the various channels and transporters, the Ca^{2+}, voltage-dependent anion channels (VDAC) in the outer mitochondrial membrane, and the mitochondrial calcium uniporter complex (MCUC) in the inner mitochondrial membrane are involved in Ca^{2+} influx into the mitochondria [90]. The influx of Ca^{2+} through VDAC is free, while MCUC are pore-forming proteins that regulate the entry of Ca^{2+} into the mitochondria. Though transient changes in Ca^{2+} levels are detected in the mitochondria under a stressed environment, knowledge of the Ca^{2+} sensors still remains obscure.

During the stress response, ROS generation in the mitochondria communicates the signal to the nucleus through the mitochondrial retrograde signaling pathway. Retrograde signaling operates in the organelles like mitochondria and chloroplast when the organelles signal to the nucleus about its dysfunction in order to activate certain genes to carry the adaptive response [91]. The nuclear-encoded upregulation of alternative oxidase (*AOX1*) is the most prominent gene involved in the mitochondrial retrograde signaling pathway. It is a cyanide insensitive terminal oxidase in ETC that, along with alternative NADH dehydrogenases, does not generate a proton motive force that is required to produce ATP [92]. The impairment of the cytochrome pathway during stress makes the re-routing of electrons through the alternative respiratory pathway necessary as it reduces the accumulation of ROS [93,94].

In addition to the secondary messengers and metabolites discussed above, mitochondrial biogenesis and function are also controlled by plant hormones like abscisic acid (ABA), auxin (AUX), cytokinin (CK), jasmonic acid (JA), and salicylic (SA). Other hormones like brassinosteroid (BS), ethylene (ET), and gibberellic acid (GA) play a minor role in the signaling network [95].

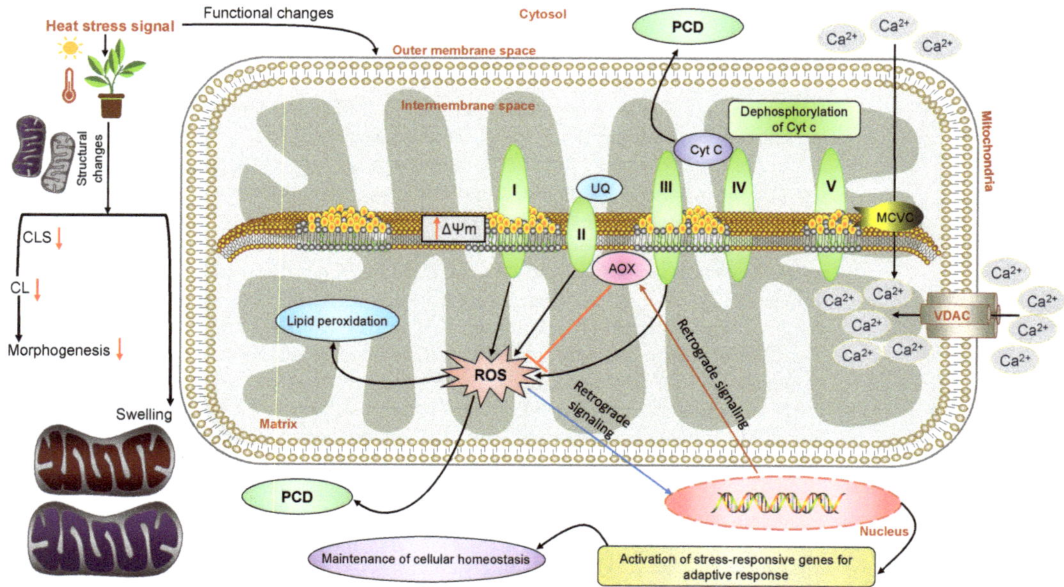

Figure 3. Structural and functional changes in mitochondria under heat stress. Cardiolipin (CL) involved in mitochondrial morphogenesis is reduced due to the low activity of cardiolipin synthase (CLS) under heat stress. The damaged CL results in depolarization of Cyt c ultimately leading to its release from the inner mitochondrial membrane into the cytosol and triggering PCD. ROS are generated at complexes I, II, and III. Under heat stress, overproduction of ROS in the inner mitochondrial membrane causes lipid peroxidation of phospholipids. ROS overproduction increases the cytosolic Ca^{2+} and influx into mitochondria *via* voltage-dependent anion channels (VDAC) in the outer mitochondrial membrane and the mitochondrial calcium uniporter complex (MCUC) in the inner mitochondrial membrane. ROS can communicate signals to the nucleus through retrograde signaling to activate genes for an adaptive response to maintain cellular homeostasis. *AOX1* is also upregulated *via* retrograde signaling, which ultimately inhibits ROS production and helps in maintaining cellular homeostasis.

10. Hormonal Regulation of Respiratory Metabolism under High Temperature

Heat stress alters the hormonal biosynthesis, stability, compartmentalization, and homeostasis within the plants [96]. The accumulation of hormones like ABA, ET, SA, CK, and JA may directly interact with mitochondrial functions in plants [97–100]. High concentration of SA interacts with mitochondrial ETC complexes I and III, whereas lower concentrations are observed as uncoupling agents [101,102]. SA oxidizes the ubiquinone (UQ) pool by altering the kinetics of dehydrogenases [103,104] and blocking the electron transport between succinate and UQ. Further, SA can also directly bind to the subunit of α-ketoglutarate dehydrogenase E2 (α-KGDH), an important enzyme of the TCA cycle, and act upstream to affect ETC during pathogen resistance to tobacco mosaic virus [105]. However, its role in abiotic stress tolerance has not been elucidated so far. Cytokinins like 6-benzylaminopurine, 6-(Δ^2-isopentenylamino) purine, and 6-furfuryl aminopurine target the mitochondrial respiration by restricting the electron transport from NADH to the cytochrome system in the stems of pea and hypocotyls of mung bean [106]. Cytokinin-like effects exhibited by N-(2-chloropyridyl)-N′-phenylurea inhibited the oxidation of malate, succinate, and NADH by the intact mitochondria of pea [107]. The respiratory control by AUXs was supported by previous studies, where decreased AUX levels and transport inhibited the functioning of mitochondrial respiratory chain complexes [108]. The mechanistic link between AUX signaling and perturbation in mitochondria was inferred by employing the potent inducers of UDP–glucosyl transfer as encoded by the gene *UGT74E2*, which evoked a common response in mitochondrial dysfunction and inhibition of auxin-related transcription in the meristematic tissues during stress response [109].

ABA can hinder ATP/ADP exchange by the mitochondrial adenine nucleotide translocators (ANTs) to cause a reduction in ATP transfer to the cytosol or renewal of ADP to the mitochondria [95]. Consequently, the reduced availability of ADP results in the activation of ROS production in the mitochondria, with detrimental consequences, as discussed earlier [110]. The transcription factor abscisic acid insensitive 4 (ABI 4) has been reported to be a repressor of the mitochondrial *AOX1* gene of *A. thaliana*. However, *AOX* expression was stimulated by the application of ABA; therefore, the repressor effect of ABI4 on AOX1a is likely to be a part of the complex regulatory circuit [92]. Moreover, altered expression of genes like ABA hypersensitive germination 11 (*AHG 11)* involved in the editing of NADH dehydrogenase subunit 4 (*NAD4*) [111], slow growth 2 (*SLO2)* [112] involved in editing three complex I genes, ABA overly sensitive 6 (*ABO6*) [113] involved in the splicing of complex I genes and lovastatin insensitive 1 (*LOI 1)* involved in the RNA editing of cytochrome c maturation [114], were associated with altered ABA responses.

The information regarding the regulation of mitochondrial function by ET during stress is lacking. Nevertheless, the increase in AOX activity has been simultaneously related to ET biosynthesis during ripening in climacteric fruits, and its blockage leads to the inhibition of respiratory increase [115]. The implication of crosstalk between ET and AOX during stress response may help in deciphering a connection that may probably exist with ROS inhibition during retrograde signaling. Brassinolide also induces increased AOX activity in tobacco by directly affecting the promoter of the *AOX* gene [116]. Evidence linking the network of pathways that are impacted by the high temperature-induced mitochondrial dysfunction needs to be strengthened to understand the checkpoints that finally determine the respiratory control of productivity.

11. Strategies to Reduce Carbon Loss

Based on the literature, we propose the following strategies that can help in reducing the loss at the cellular and plant level.

11.1. Selection of Genotypes with Low Rates of Respiration under High Night Temperature

Studies relating to the increase in carbon loss due to high night temperatures [11,117–119] emphasize the need to screen genotypes that maintain normal respiration rates under different environmental regimes. Rice plants grown under high minimum temperatures have generated data to show that cultivars like Nagina 22 [119], Abhishek, SahbhagiDhan, and Bakal [120], with insignificant changes in post-flowering respiration, showed a marginal reduction in grain yield [117]. The biodiversity existing for this trait in the wild or related crop species can be used for introgression in high yielding cultivars of various crops with the target of generating a positive carbon balance under future climate change scenarios.

11.2. Genome Editing to Target the Metabolic Processes Consuming Carbon

The genetic improvement of crops by using genome editing approaches like knockout, replacement base editing, and regulation of expression of desirable/undesirable genes can effectively target the metabolic processes that lead to futile carbon loss in crops. The few pathways that can be replaced, relocated, or rescheduled through this approach have been discussed (Figure 4).

11.2.1. Substitution of the Lignin Biosynthesis Pathway

The lignin biosynthesis pathway involving phenylalanine ammonia lyase (PAL) can be overridden by substitution with tyrosine ammonia lyase (TAL) as it provides a gain due to the formation of two NADPH per *p*-coumarate [121] and can potentially decrease growth respiration [32].

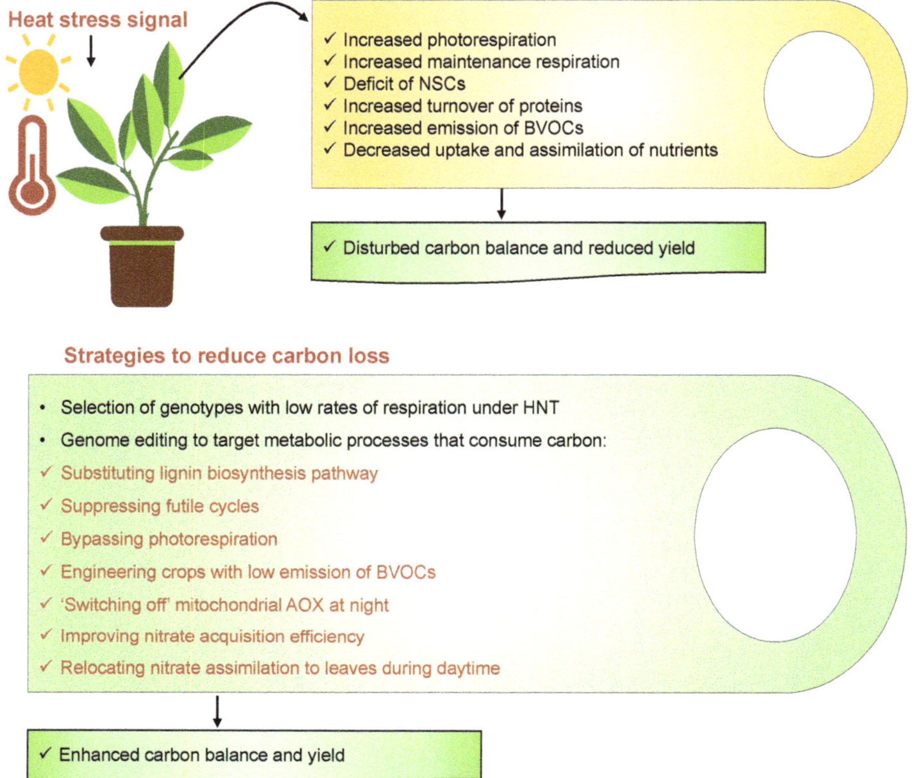

Figure 4. Respiratory carbon loss in plants and strategies to enhance yield under high temperature.

11.2.2. Suppression of Futile Cycles

A substantial proportion of the ATP generated during respiration is consumed by certain pathways that can be called 'futile' cycles. For example, the simultaneous synthesis and degradation of starch in leaves during the day [122], the simultaneous synthesis and degradation of sucrose [123], and cycling between fructose 6-phosphate and fructose 1,6-bisphosphate [124] are futile cycles. The suppression of these futile cycles will decrease the respiratory costs without exerting collateral damage on the metabolic machinery [32].

11.2.3. Designing Carbon Conserving Photorespiration

Photorespiratory bypass to eliminate the loss of CO_2 can be designed by incorporating synthetic routes through metabolic engineering. The reduction of glycolate to glycolaldehyde is a promising approach as it can assimilate 2-phosphoglycolate into the Calvin cycle without the loss of carbon. Screening the germplasm for highly stable and substrate-specific enzymes, such as acetyl-CoA synthetase and propionyl-CoA reductase, would help in favoring the reduction process over oxidation and generating a carbon-conserving pathway [125].

11.2.4. Engineering for Low Emission of Biogenic Volatile Organic Compounds

Plants release a considerable fraction of the assimilated carbon as biogenic volatile organic compounds (BVOCs). Temperature within the range of 20–40 °C has a strong influence on the activity of enzymes involved in the biosynthesis of BVOCs like isoperenes, monoterpenes, acetaldehyde, and (E)-2-hexenal. Though BVOCs impart thermal tolerance at high temperature [126,127], it happens at the cost of 10% of fixed carbon loss. Engineering cultivars with reduced emissions of BVOCs can act as a promising strategy to save carbon under high temperatures.

11.2.5. "Switching Off" Mitochondrial AOX at Night

The AOX pathway continues to remain operative at night, accounting for 10–50% of the total respiratory rate, resulting in a reduced ATP yield per unit of carbon oxidized [128]. Among the different isoforms of AOX identified so far, one is constitutive, whereas the rest are stress inducible [129]. Engineering the constitutive AOX with a light-specific promoter can lower the alternative pathway rate at night and raise it again during the day without compromising the carbon loss [32].

11.2.6. Improving Nitrate Acquisition and Relocating Nitrate Assimilation

Plants take up nitrogen mainly in the form of nitrates from the soil in an energy-intensive process [130], which is further reduced in the roots and shoots [131]. The cost of NO_3^- acquisition can be minimized by identification and elimination of NO_3^- leaks that take place via the nitrate excretion transporter (NAXT1) or alternatively, increasing the flux density of an optimized NO_3^- transporter on root hair cells [32]. Further, the cost of nitrate reduction is 1.72 kg glucose C respired per kg nitrate N reduced to ammonia [132]. If the entire or most of the nitrate assimilation during the daytime takes place in the leaves, then the excess of NADPH and ATP produced during light reaction can be exploited under high light. This would reduce the additional cost of sucrose respiration in the roots, which is required for the generation of the carbon skeleton [32].

12. Conclusions and Future Outlooks

The significant upsurge in respiration rate under climate warming rather than an antagonistic change in photosynthetic rate disrupts the carbon economy of the plant, resulting in a yield penalty. The mechanism responsible for this yield penalty is increased utilization of non-structural carbohydrates to carry out maintenance respiration to support increased turnover of proteins, maintenance of ion gradients, and activation of energetically expensive heat tolerance mechanisms, thereby creating an overall deficit of carbohydrates partitioned towards growth respiration, eventually reducing the total dry matter production. At a cellular level, warmer temperatures lead to mitochondrial swelling as well as downregulation of respiration by increasing the ATP:ADP ratio, the ABA-mediated reduction in ATP transfer to the cytosol, and the disturbance in a concentration gradient of TCA cycle intermediates, as well as increasing lipid peroxidation in mitochondrial membranes and enough cytochrome c release to trigger programmed cell death. In plants, distinct respiratory metabolic adjustments are available in response to high temperatures and the diel cycle. Plants show thermal acclimation of the respiration response to lessen the impact of carbon loss due to increasing temperatures. Genome editing approaches to reduce unnecessary carbon loss and to increase the energy utilization efficiency of processes are ways to escalate positive carbon balance. This can be addressed by replacing, relocating, or rescheduling the metabolic pathways like substituting the lignin biosynthesis pathway, suppressing futile cycles that decrease the respiratory costs, bypassing photorespiration via metabolic engineering, engineering cultivars with reduced emission of BVOCs and a low alternative pathway rate at night, minimizing the cost of NO_3^- acquisition, and relocating NO_3^- assimilation from roots and shoots to leaves during the daytime. Thus, cutting respiratory losses and increasing photosynthesis are the most effective solutions to beat the heat in the presently warming world for and sustain crop productivity in the long run.

Author Contributions: A.A., N.S. and M.T. conceptualized and prepared an outline. N.S., M.T., P.S. and P.M. performed the literature search and contributed to the original draft of the review. A.R. prepared the illustrations. A.A., C.S.P. and A.R. critically reviewed, edited, and finalized the draft. All authors have read and agreed to the published version of the manuscript.

Funding: This research received no external funding.

Data Availability Statement: Not applicable.

Acknowledgments: We are grateful to many scientists and colleagues for scientific discussions, which enabled the development of this up-to-date comprehensive review. We apologize to colleagues whose relevant work could not be cited due to space limitations.

Conflicts of Interest: The authors declare no conflict of interest.

References

1. Hüve, K.; Bichele, I.; Ivanova, H.; Keerberg, O.; Pärnik, T.; Rasulov, B.; Tobias, M.; Niinemets, U. Temperature responses of dark respiration in relation to leaf sugar concentration. *Physiol. Plant.* **2012**, *144*, 320–334. [CrossRef] [PubMed]
2. Heskel, M.A.; O'sullivan, O.S.; Reich, P.B.; Tjoelker, M.G.; Weerasinghe, L.K.; Penillard, A.; Atkin, O.K. Convergence in the temperature response of leaf respiration across biomes and plant functional types. *Proc. Natl. Acad. Sci. USA* **2016**, *113*, 3832–3837. [CrossRef] [PubMed]
3. Sadok, W.; Jagadish, S.K. The hidden costs of nighttime warming on yields. *Trends Plant Sci.* **2020**, *25*, 644–651. [CrossRef] [PubMed]
4. Schlesinger, W.H. *Biogeochemistry: An Analysis of Global Change*, 2nd ed.; Academic Press: San Diego, CA, USA, 1997.
5. Amthor, J.S. Terrestrial higher-plant response to increasing atmospheric [CO_2] in relation to the global carbon cycle. *Glob. Chang. Biol.* **1995**, *1*, 243–274 . [CrossRef]
6. Waring, R.H.; Landsberg, J.J.; Williams, M. Net primary production of forests: A constant fraction of gross primary production? *Tree Physiol.* **1998**, *18*, 129–134. [CrossRef]
7. Gifford, R.M. The global carbon cycle: A viewpoint on the missing sink. *Aust. J. Plant Physiol.* **1994**, *21*, 1–15. [CrossRef]
8. Gifford, R.M.; Thorne, J.H.; Hitz, W.D.; Giaquinta, R.T. Crop productivity and photoassimilate partitioning. *Science* **1984**, *225*, 801–808. [CrossRef]
9. Amthor, J.S. Evolution and applicability of a whole plant respiration model. *J. Theor. Biol.* **1986**, *122*, 473–490 . [CrossRef]
10. Masson-Delmotte, V.; Zhai, P.; Pirani, A.; Connors, S.L.; Pean, C.; Berger, S.; Caud, N.; Chen, Y.; Goldfarb, L.; Gomis, M.I.; et al. Summary for Policymakers. In *Climate Change 2021: The Physical Science Basis. Contribution of Working Group I to the Sixth Assessment Report of the Intergovernmental Panel on Climate Change. 2022*; Cambridge University Press: Cambridge, UK, 2021.
11. Li, G.; Chen, T.; Feng, B.; Peng, S.; Tao, L.; Fu, G. Respiration, rather than photosynthesis, determines rice yield loss under moderate high-temperature conditions. *Front. Plant Sci.* **2021**, *12*, 1287. [CrossRef]
12. Tjoelker, M.G.; Oleksyn, J.; Reich, P.B. Modelling respiration of vegetation: Evidence for a general temperature-dependent Q_{10}. *Glob. Chang. Biol.* **2001**, *7*, 223–230. [CrossRef]
13. Heskel, M.A.; Bitterman, D.; Atkin, O.K.; Turnbull, M.H.; Griffin, K.L. Seasonality of foliar respiration in two dominant plant species from the Arctic tundra: Response to long-term warming and short-term temperature variability. *Funct. Plant Biol.* **2014**, *41*, 287–300. [CrossRef] [PubMed]
14. Weerasinghe, L.K.; Creek, D.; Crous, K.Y.; Xiang, S.; Liddell, M.J.; Turnbull, M.H.; Atkin, O.K. Canopy position affects the relationships between leaf respiration and associated traits in a tropical rainforest in Far North Queensland. *Tree Physiol.* **2014**, *34*, 564–584. [CrossRef] [PubMed]
15. O'Sullivan, O.S.; Heskel, M.A.; Reich, P.B.; Tjoelker, M.G.; Weerasinghe, L.K.; Penillard, A.; Zhu, L.L.; Egerton, J.J.G.; Bloomfield, K.J.; Creek, D.; et al. Thermal limits of leaf metabolism across biomes. *Glob. Chang. Biol.* **2017**, *23*, 209–223. [CrossRef] [PubMed]
16. Kromer, S. Respiration during photosynthesis. *Annu. Rev. Plant Physiol. Plant Mol. Biol.* **1995**, *46*, 45–70. [CrossRef]
17. Dusenge, M.E.; Duarte, A.G.; Way, D.A. Plant carbon metabolism and climate change: Elevated CO_2 and temperature impacts on photosynthesis, photorespiration and respiration. *New Phytol.* **2018**, *221*, 32–49. [CrossRef]
18. Atkin, O.K.; Tjoelker, M.G. Thermal acclimation and the dynamic response of plant respiration to temperature. *Trends Plant Sci.* **2003**, *8*, 343–351. [CrossRef]
19. Zhao, C.; Liu, B.; Piao, S.; Wang, X.; Lobell, D.B.; Huang, Y.; Huang, M.; Yao, Y.; Bassu, S.; Ciais, P.; et al. Temperature increase reduces global yields of major crops in four independent estimates. *Proc. Natl. Acad. Sci. USA* **2017**, *114*, 9326–9331. [CrossRef]
20. Cavanagh, A.P.; South, P.F.; Bernacchi, C.J.; Ort, D.R. Alternative pathway to photorespiration protects growth and productivity at elevated temperatures in a model crop. *Plant Biotechnol. J.* 2021. [CrossRef]
21. Nunes-Nesi, A.; Carrari, F.; Lytovchenko, A.; Smith, A.M.; Loureiro, M.E.; Ratcliffe, R.G.; Sweetlove, L.J.; Fernie, A.R. Enhanced photosynthetic performance and growth as a consequence of decreasing mitochondrial malate dehydrogenase activity in transgenic tomato plants. *Plant Physiol.* **2005**, *137*, 611–622. [CrossRef]
22. Hauben, M.; Haesendonckx, B.; Standaert, E.; Van Der Kelen, K.; Azmi, A.; Akpo, H.; Van Breusegem, F.; Guisez, Y.; Bots, M.; Lambert, B.; et al. Energy use efficiency is characterized by an epigenetic component that can be directed through artificial selection to increase yield. *Proc. Natl. Acad. Sci. USA* **2009**, *106*, 20109–20114. [CrossRef]
23. Walker, B.J.; VanLoocke, A.; Bernacchi, C.J.; Ort, D.R. The costs of photorespiration to food production now and in the future. *Annu. Rev. Plant Biol.* **2016**, *67*, 107–129. [CrossRef] [PubMed]

24. Cai, C.; Yin, X.; He, S.; Jiang, W.; Si, C.; Struik, P.C.; Luo, W.; Li, G.; Xie, Y.; Xiong, Y.; et al. Responses of wheat and rice to factorial combinations of ambient and elevated CO_2 and temperature in FACE experiments. *Glob. Chang. Biol.* **2016**, *22*, 856–874. [CrossRef] [PubMed]
25. Impa, S.M.; Raju, B.; Hein, N.T.; Sandhu, J.; Prasad, P.V.V.; Walia, H.; Jagadish, S.K. High night temperature effects on wheat and rice: Current status and way forward. *Plant Cell Environ.* **2021**, *44*, 2049–2065. [CrossRef] [PubMed]
26. Xu, J.; Misra, G.; Sreenivasulu, N.; Henry, A. What happens at night? Physiological mechanisms related to maintaining grain yield under high night temperature in rice. *Plant Cell Environ.* **2021**, *44*, 2245–2261. [CrossRef] [PubMed]
27. Tombesi, S.; Cincera, I.; Frioni, T.; Ughini, V.; Gatti, M.; Palliotti, A.; Poni, S. Relationship among night temperature, carbohydrate translocation and inhibition of grapevine leaf photosynthesis. *Environ. Exp. Bot.* **2019**, *157*, 293–298 . [CrossRef]
28. Glaubitz, U.; Erban, A.; Kopka, J.; Hincha, D.K.; Zuther, E. High night temperature strongly impacts TCA cycle, amino acid and polyamine biosynthetic pathways in rice in a sensitivity-dependent manner. *J. Exp. Bot.* **2015**, *66*, 6385–6397. [CrossRef]
29. Xu, J.; Henry, A.; Sreenivasulu, N. Rice yield formation under high day and night temperatures-A prerequisite to ensure future food security. *Plant Cell Environ.* **2020**, *43*, 1595–1608. [CrossRef]
30. Lambers, H. Respiration in intact plants and tissues: Its regulation and dependence on environmental factors, metabolism and invaded organisms. In *Higher Plant Cell Respiration*; Douce, R., Day, D.A., Eds.; (Encyclopedia of Plant Physiology), New Series; Springer: Berlin, Germany, 1985; Volume 18.
31. Slot, M.; Kitajima, K. General patterns of acclimation of leaf respiration to elevated temperatures across biomes and plant types. *Oecologia* **2015**, *177*, 885–900. [CrossRef]
32. Amthor, J.S.; Bar-Even, A.; Hanson, A.D.; Millar, A.H.; Stitt, M.; Sweetlove, L.J.; Tyerman, S.D. Engineering strategies to boost crop productivity by cutting respiratory carbon loss. *Plant Cell* **2019**, *31*, 297–314. [CrossRef]
33. Amthor, J.S. The McCree-de Wit-Penning de Vries-Thornley respiration paradigms: 30 years later. *Ann. Bot.* **2000**, *86*, 1–20 . [CrossRef]
34. Adu-Bredu, S.; Yokota, T.; Hagihara, A. Temperature effect on maintenance and growth respiration coefficients of young, field-grown hinoki cypress (*Chamaecyparisobtusa*). *Ecol. Res.* **1997**, *12*, 357–362. [CrossRef]
35. Atkin, O.K.; Bloomfield, K.J.; Reich, P.B.; Tjoelker, M.G.; Asner, G.P.; Bonal, D.; Bönisch, G.; Bradford, M.G.; Cernusak, L.; Cosio, E.; et al. Global variability in leaf respiration in relation to climate, plant functional types and leaf traits. *New Phytol.* **2015**, *206*, 614–636. [CrossRef] [PubMed]
36. Crous, K.Y.; Uddling, J.; De Kauwe, M.G. Temperature responses of photosynthesis and respiration in evergreen trees from boreal to tropical latitudes. *New Phytol.* **2022**, *234*, 353–374. [CrossRef]
37. Thornley, J.H.M. Respiration, growth and maintenance in plants. *Nature* **1970**, *227*, 304–305. [CrossRef] [PubMed]
38. McCree, K.J. Maintenance requirements of white clover at high and low growth rates. *Crop Sci.* **1982**, *22*, 345–351. [CrossRef]
39. Thornley, J.H.M.; Johnson, I.R. Plant and Crop Modeling. Clarendon Press Oxford: Oxford, UK, 1990.
40. Van Iersel, M.W. Respiratory Q_{10} of marigold (*Tagetes patula*) in response to long-term temperature differences and its relationship to growth and maintenance respiration. *Physiol. Plant.* **2006**, *128*, 289–301. [CrossRef]
41. Winzeler, H.; Hunt, L.A.; Mahon, J.D. Ontogenetic changes in respiration and photosynthesis in a uniculm barley. *Crop Sci.* **1976**, *16*, 786–790. [CrossRef]
42. McCree, K.J.; Silsbury, J.H. Growth and maintenance requirements of subterranean clover. *Crop Sci.* **1978**, *18*, 13–18. [CrossRef]
43. Mariko, S.; Koizumi, H. Respiration for maintenance and growth in *Reynoutria japonica* ecotypes from different altitudes on Mt Fuji. *Ecol. Res.* **1993**, *8*, 241–246. [CrossRef]
44. Gifford, R.M. Whole plant respiration and photosynthesis of wheat under increased CO_2 concentration and temperature: Long-term vs. short-term distinctions for modelling. *Glob. Chang. Biol.* **1995**, *1*, 385–396. [CrossRef]
45. Tan, K.; Zhou, G.; Ren, S. Response of leaf dark respiration of winter wheat to changes in CO_2 concentration and temperature. *Chin. Sci. Bull.* **2013**, *58*, 1795–1800. [CrossRef]
46. Atkin, O.K.; Bruhn, D.; Tjoelker, M.G. Response of plant respiration to changes in temperature: Mechanisms and consequences of variations in Q_{10} values and acclimation. In *Advances in Photosynthesis and Respiration, Plant Respiration*; Lambers, H., Ribas-Carbo, M., Eds.; Springer: Dordrecht, The Netherlands, 2005; Volume 18. [CrossRef]
47. Atkin, O.K.; Macherel, D. The crucial role of plant mitochondria in orchestrating drought tolerance. *Ann. Bot.* **2009**, *103*, 581–597. [CrossRef] [PubMed]
48. Thakur, M.; Sharma, P.; Anand, A.; Pandita, V.K.; Bhatia, A.; Pushkar, S. Raffinose and hexose sugar content during germination are related to infrared thermal fingerprints of primed onion (*Allium cepa* L.) seeds. *Front. Plant Sci.* **2020**, *11*, 579037. [CrossRef] [PubMed]
49. Ögren, E. Maintenance respiration correlates with sugar but not nitrogen concentration in dormant plants. *Physiol. Plant.* **2000**, *108*, 295–299. [CrossRef]
50. Whitehead, D.; Griffin, K.L.; Turnbull, M.H.; Tissue, D.T.; Engel, V.C.; Brown, K.J.; Schuster, W.S.; Walcroft, A.S. Response of total night-time respiration to differences in total daily photosynthesis for leaves in a *Quercus rubra* L. canopy: Implications for modelling canopy CO_2 exchange. *Glob. Chang. Biol.* **2004**, *10*, 925–938. [CrossRef]
51. Noguchi, K.; Terashima, I. Different regulation of leaf respiration between *Spinacia oleracea*, a sun species, and *Alocasia odora*, a shade species. *Physiol. Plant.* **1997**, *101*, 1–7. [CrossRef]

52. Hoefnagel, M.H. Interdependence between chloroplasts and mitochondria in the light and the dark. *Biochim. Biophys. ActaBioenerg.* **1998**, *1366*, 235–255. [CrossRef]
53. Douce, R.; Neuburger, M. The uniqueness of plant mitochondria. *Annu. Rev. Plant Biol.* **1980**, *40*, 371–414. [CrossRef]
54. Zerihun, A.; Mckenzie, B.A.; Morton, J.D. Photosynthate costs associated with the utilization of different nitrogen-forms: Influence on the carbon balance of plants and shoot-root biomass partitioning. *New Phytol.* **1998**, *138*, 1–11. [CrossRef]
55. Williams, K.; Percival, F.; Merino, J.; Mooney, H.A. Estimation of tissue construction cost from heat of combustion and organic nitrogen content. *Plant Cell Environ.* **1987**, *10*, 725–734. [CrossRef]
56. Hachiya, T.; Terashima, I.; Noguchi, K. Increase in respiratory cost at high growth temperature is attributed to high protein turnover cost in *Petunia x hybrida* petals. *Plant Cell Environ.* **2007**, *30*, 1269–1283. [CrossRef] [PubMed]
57. Lee, C.P.; Eubel, H.; Millar, A.H. Diurnal changes in mitochondrial function reveal daily optimization of light and dark respiratory metabolism in Arabidopsis. *Mol. Cell. Proteom.* **2010**, *9*, 2125–2139. [CrossRef] [PubMed]
58. Blasing, O.E.; Gibon, Y.; Günther, M.; Höhne, M.; Morcuende, R.; Osuna, D.; Thimm, O.; Usadel, B.; Scheible, W.R.; Stitt, M. Sugars and circadian regulation make major contributions to the global regulation of diurnal gene expression in *Arabidopsis*. *Plant Cell* **2005**, *17*, 3257–3281. [CrossRef] [PubMed]
59. Shameer, S.; Ratcliffe, R.G.; Sweetlove, L.J. Leaf energy balance requires mitochondrial respiration and export of chloroplast NADPH in the light. *Plant Physiol.* **2019**, *180*, 1947–1961 . [CrossRef]
60. Gibon, Y.; Pyl, E.-T.; Sulpice, R.; Lunn, J.E.; Hohne, M.; Gunther, M.; Stitt, M. Adjustment of growth, starch turnover, protein content and central metabolism to a decrease of the carbon supply when Arabidopsis is grown in very short photoperiods. *Plant Cell Environ.* **2009**, *32*, 859–874. [CrossRef]
61. O'Leary, B.M.; Lee, C.P.; Atkin, O.K.; Cheng, R.; Brown, T.B.; Millar, A.H. Variation in leaf respiration rates at night correlates with carbohydrate and amino acid supply. *Plant Physiol.* **2017**, *174*, 2261–2273. [CrossRef]
62. Gibon, Y.; Usadel, B.; Blaesing, O.E.; Kamlage, B.; Hoehne, M.; Trethewey, R.; Stitt, M. Integration of metabolite with transcript and enzyme activity profiling during diurnal cycles in Arabidopsis rosettes. *Genome Biol.* **2006**, *7*, R76. [CrossRef]
63. Watanabe, C.; Sato, S.; Yanagisawa, S.; Uesono, Y.; Terashima, I.; Noguchi, K. Effects of elevated CO_2 on levels of primary metabolites and transcripts of genes encoding respiratory enzymes and their diurnal patterns in *Arabidopsis thaliana*: Possible relationships with respiratory rates. *Plant Cell Physiol.* **2014**, *55*, 341–357. [CrossRef]
64. Rashid, F.A.A.; Crisp, P.A.; Zhang, Y.; Berkowitz, O.; Pogson, B.J.; Day, D.A.; Masle, J.; Dewar, R.C.; Whelan, J.; Atkin, O.K. Molecular and physiological responses during thermal acclimation of leaf photosynthesis and respiration in rice. *Plant Cell Environ.* **2020**, *43*, 594–610. [CrossRef]
65. Cook, D.; Fowler, S.; Fiehn, O.; Thomashow, M. A prominent role for the CBF cold response pathway in configuring the low-temperature metabolome of Arabidopsis. *Proc. Natl. Acad. Sci.USA* **2004**, *101*, 15243–15248. [CrossRef]
66. Atkin, O.K.; Bruhn, D.; Hurry, V.M.; Tjoelker, M. The hot and the cold: Unraveling the variable response of plant respiration to temperature. *Funct. Plant Biol.* **2005**, *32*, 87–105. [CrossRef] [PubMed]
67. Reich, P.B.; Sendall, K.M.; Stefanski, A.; Wei, X.; Rich, R.L.; Montgomery, R.A. Boreal and temperate trees show strong acclimation of respiration to warming. *Nature* **2016**, *531*, 633–636. [CrossRef] [PubMed]
68. Smith, N.G.; Dukes, J.S. Short-term acclimation to warmer temperatures accelerates leaf carbon exchange processes across plant types. *Glob. Chang. Biol.* **2017**, *23*, 4840–4853. [CrossRef] [PubMed]
69. Smith, N.G.; Dukes, J.S. Plant respiration and photosynthesis in global-scale models: Incorporating acclimation to temperature and CO_2. *Glob. Chang. Biol.* **2013**, *19*, 45–63. [CrossRef]
70. Armstrong, A.F.; Logan, D.C.; Atkin, O.K. On the developmental dependence of leaf respiration: Responses to short-and long-term changes in growth temperature. *Am. J. Bot.* **2006**, *93*, 1633–1639. [CrossRef]
71. Tjoelker, M.G.; Oleksyn, J.; Reich, P.B. Acclimation of respiration to temperature and CO_2 in seedlings of boreal tree species in relation to plant size and relative growth rate. *Glob. Chang. Biol.* **1999**, *5*, 679–691. [CrossRef]
72. Rurek, M. Plant mitochondria under a variety of temperature stress conditions. *Mitochondrion* **2014**, *19*, 289–294. [CrossRef]
73. Logan, D.C. Mitochondrial fusion, division and positioning in plants. *Biochem. Soc. Trans.* **2010**, *38*, 789–795. [CrossRef]
74. Taylor, N.L.; Heazlewood, J.L.; Day, D.A.; Millar, A.H. Differential impact of environmental stresses on the pea mitochondrial proteome. *Mol. Cell. Proteom.* **2005**, *4*, 1122–1133. [CrossRef]
75. Hu, W.H.; Xiao, Y.A.; Zeng, J.J.; Hu, X.H. Photosynthesis, respiration and antioxidant enzymes in pepper leaves under drought and heat stresses. *Biol. Plant* **2010**, *54*, 761–765. [CrossRef]
76. Ruelland, E.; Zachowski, A. How plants sense temperature. *Environ. Exp. Bot.* **2010**, *69*, 225–232. [CrossRef]
77. Crawford, T.; Lehotai, N.; Strand, A. The role of retrograde signals during plant stress responses. *J. Exp. Bot.* **2018**, *69*, 2783–2795. [CrossRef] [PubMed]
78. Osman, C.; Voelker, D.R.; Langer, T. Making heads or tails of phospholipids in mitochondria. *J. Cell Biol.* **2011**, *192*, 7–16. [CrossRef]
79. Pan, R.H.; Jones, A.D.; Hu, J.P. Cardiolipin-mediated mitochondrial dynamics and stress response in Arabidopsis. *Plant Cell* **2014**, *26*, 391–409. [CrossRef] [PubMed]
80. Welchen, E.; Gonzalez, D.H. Cytochrome c, a hub linking energy, redox, stress and signaling pathways in mitochondria and other cell compartments. *Physiol. Plant.* **2016**, *157*, 310–321 101111/ppl12449. [CrossRef]

81. Zancani, M.; Casolo, V.; Petrussa, E.; Peresson, C.; Patui, S.; Bertolini, A.; De Col, V.; Braidot, E.; Boscutti, F.; Vianello, A. The permeability transition in plant mitochondria: The missing link. *Front. Plant Sci.* **2015**, *6*, 1120. [CrossRef]
82. Petereit, J.; Katayama, K.; Lorenz, C.; Ewert, L.; Schertl, P.; Kitsche, A.; Wada, H.; Frentzen, M.; Braun, H.-P.; Eubel, H. Cardiolipin supports respiratory enzymes in plants in different ways. *Front. Plant Sci.* **2017**, *8*, 72 . [CrossRef]
83. Van Aken, O.; Van Breusegem, F. Licensed to Kill: Mitochondria, Chloroplasts, and Cell Death. *Trends Plant Sci.* **2015**, *20*, 754–766. [CrossRef] [PubMed]
84. Pedrotti, L.; Weiste, C.; Nagele, T.; Wolf, E.; Lorenzin, F.; Dietrich, K.; Mair, A.; Weckwerth, W.; Teige, M.; Baena-Gonzalez, E.; et al. Snf1-RELATED KINASE1-controlled C/S_1- bZIP signaling activates alternative mitochondrial metabolic pathways to ensure plant survival in extended darkness. *Plant Cell* **2018**, *30*, 495–509. [CrossRef]
85. Dourmap, C.; Roque, S.; Morin, A.; Caubriere, D.; Kerdiles, M.; Beguin, K.; Perdoux, R.; Reynoud, N.; Bourdet, L.; Audebert, P.-A.; et al. Stress signalling dynamics of the mitochondrial electron transport chain and oxidative phosphorylation system in higher plants. *Ann. Bot.* **2020**, *125*, 721–736. [CrossRef]
86. Murphy, M.P. How mitochondria produce reactive oxygen species. *Biochem. J.* **2009**, *417*, 1–13. [CrossRef] [PubMed]
87. Jardim-Messeder, D.; Caverzan, A.; Rauber, R.; De souza Ferreira, E.; Margis-pinheiro, M.; Galina, A. Succinate dehydrogenase (mitochondrial complex II) is a source of reactive oxygen species in plants and regulates development and stress responses. *New Phytol.* **2015**, *208*, 776–789. [CrossRef] [PubMed]
88. Fedyaeva, A.V.; Stepanov, A.V.; Lyubushkina, I.V.; Pobezhimova, T.P.; Rikhvanov, E.G. Heat shock induces production of reactive oxygen species and increases inner mitochondrial membrane potential in winter wheat cells. *Biochemistry* **2014**, *79*, 1202–1210. [CrossRef] [PubMed]
89. Rikhvanov, E.G.; Fedoseeva, I.V.; Pyatrikas, D.V.; Borovskii, G.B.; Voinikov, V.K. Role of mitochondria in the operation of calcium signaling system in heat-stressed plants. *Russ. J. Plant Physiol.* **2014**, *61*, 141–153. [CrossRef]
90. Vothknecht, U.C.; Szabo, I. Channels and transporters for inorganic ions in plant mitochondria: Prediction and facts. *Mitochondrion* **2020**, *53*, 224–233. [CrossRef]
91. Jazwinski, S.M. The retrograde response: When mitochondrial quality control is not enough. *Biochim. Biophys. Acta-Mol. Cell Res.* **2013**, *1833*, 400–409. [CrossRef]
92. Giraud, E.; Van Aken, O.; Ho, L.H.; Whelan, J. The transcription factor ABI4 is a regulator of mitochondrial retrograde expression of ALTERNATIVE OXIDASE1a. *Plant Physiol.* **2009**, *150*, 1286–1296. [CrossRef]
93. Parsons, H.L.; Yip, J.Y.H.; Vanlerberghe, G.C. Increased respiratory restriction during phosphate-limited growth in transgenic tobacco cells lacking alternative oxidase. *Plant Physiol.* **1999**, *121*, 1309–1320. [CrossRef]
94. Cvetkovska, M.; Vanlerberghe, G.C. Alternative oxidase modulates leaf mitochondrial concentrations of superoxide and nitric oxide. *New Phytol.* **2012**, *195*, 32–39. [CrossRef]
95. Berkowitz, O.; Clercq, I.D.; Breusegem, F.V.; Whelan, J. Interaction between hormonal and mitochondrial signalling during growth, development and in plant defense responses. *Plant Cell Environ.* **2016**, *39*, 1127–1139. [CrossRef]
96. Maestri, E.; Klueva, N.; Perrotta, C.; Gulli, M.; Nguyen, H.T.; Marmiroli, N. Molecular genetics of heat tolerance and heat shock proteins in cereals. *Plant Mol. Biol.* **2002**, *48*, 667–681. [CrossRef] [PubMed]
97. Koornneef, A.; Pieterse, C. Cross talk in defense signaling. *Plant Physiol.* **2008**, *146*, 839–844. [CrossRef] [PubMed]
98. Wang, H.; Liang, X.; Huang, J.; Zhang, D.; Lu, H.; Liu, Z.; Bi, Y. Involvement of ethylene and hydrogen peroxide in induction of alternative respiratory pathway in salt-treated Arabidopsis calluses. *Plant Cell Physiol.* **2010**, *51*, 1754–1765. [CrossRef] [PubMed]
99. Verslues, P.E. ABA and cytokinins: Challenge and opportunity for plant stress research. *Plant Mol. Biol.* **2016**, *91*, 629–640. [CrossRef]
100. Zhu, J.K. Abiotic stress signaling and responses in plants. *Cell* **2016**, *167*, 313–324. [CrossRef]
101. Norman, C.; Howell, K.A.; Millar, A.H.; Whelan, J.M.; Day, D.A. Salicylic acid is an uncoupler and inhibitor of mitochondrial electron transport. *Plant Physiol.* **2004**, *134*, 492–501. [CrossRef]
102. Czarnocka, W.; Karpinski, S. Friend or foe? Reactive oxygen species production, scavenging and signaling in plant response to environmental stresses. *Free Radic. Biol. Med.* **2018**, *122*, 4–20. [CrossRef]
103. Van den Bergen, C.W.; Wagner, A.M.; Krab, K.; Moore, A.L. The relationship between electron flux and the redox poise of the quinone pool in plant mitochondria: Interplay between quinol-oxidizing and quinone reducing pathways. *Eur. J. Biochem.* **1994**, *226*, 1071–1078. [CrossRef]
104. Millar, A.H.; Atkin, O.K.; Lambers, H.; Wiskich, J.T.; Day, D.A. A critique of the use of inhibitors to estimate partitioning of electrons between mitochondrial respiratory pathways in plants. *Physiol. Plant.* **1995**, *95*, 523–532. [CrossRef]
105. Liao, Y.; Tian, M.; Zhang, H.; Li, X.; Wang, Y.; Xia, X.; Zhou, J.; Zhou, Y.; Yu, J.; Shi, K.; et al. Salicylic acid binding of mitochondrial alpha-ketoglutarate dehydrogenase E_2 affects mitochondrial oxidative phosphorylation and electron transport chain components and plays a role in basal defense against tobacco mosaic virus in tomato. *New Phytol.* **2015**, *205*, 1296–1307. [CrossRef]
106. Miller, C.O. Cytokinin modification of mitochondrial function. *Plant Physiol.* **1982**, *69*, 1274–1277. [CrossRef] [PubMed]
107. Alberto, D.; Couee, I.; Sulmon, C.; Gouesbet, G. Root-level exposure reveals multiple physiological toxicity of triazine xenobiotics in *Arabidopsis thaliana*. *J. Plant Physiol.* **2017**, *212*, 105–114. [CrossRef] [PubMed]
108. Zhang, S.; Zhang, D.; Yang, C. AtFtsH4 perturbs the mitochondrial respiratory chain complexes and auxin homeostasis in Arabidopsis. *Plant Signal. Behav.* **2014**, *9*, e29709. [CrossRef] [PubMed]

109. Kerchev, P.I.; Clercq, I.D.; Denecker, J.; Muhlenbock, P.; Kumpf, R.; Nguyen, L.; Audenaert, D.; Dejonghe, W.; Breusegem, F.V. Mitochondrial perturbation negatively affects auxin signaling. *Mol. Plant.* **2014**, *7*, 1138–1150. [CrossRef] [PubMed]
110. Turrens, J.F. Mitochondrial formation of reactive oxygen species. *J. Physiol.* **2003**, *552*, 335–344. [CrossRef] [PubMed]
111. Murayama, M.; Hayashi, S.; Nishimura, N.; Ishide, M.; Kobayashi, K.; Yagi, Y.; Asami, T.; Nakamura, T.; Shinozaki, K.; Hirayama, T. Isolation of Arabidopsis ahg11, a weak ABA hypersensitive mutant defective in nad4 RNA editing. *J. Exp. Bot.* **2012**, *63*, 5301–5310. [CrossRef]
112. Zhu, Q.; Dugardeyn, J.; Zhang, C.; Muhlenbock, P.; Eastmond, P.J.; Valcke, R.; De Coninck, B.; Oden, S.; Karampelias, M.; Cammue, B.P.A.; et al. The *Arabidopsis thaliana* RNA editing factor SLO2, which affects the mitochondrial electron transport chain, participates in multiple stress and hormone responses. *Mol Plant.* **2014**, *7*, 290–310. [CrossRef]
113. He, J.; Duan, Y.; Hua, D.; Fan, G.; Wang, L.; Liu, Y.; Chen, Z.; Han, L.; Qu, L.-J.; Gong, Z. DEXH box RNA helicase-mediated mitochondrial reactive oxygen species production in Arabidopsis mediates crosstalk between abscisic acid and auxin signaling. *Plant Cell* **2012**, *24*, 1815–1833. [CrossRef]
114. Sechet, J.; Roux, C.; Plessis, A.; Effroy, D.; Frey, A.; Perreau, F.; Biniek, C.; Krieger-Liszkar, A.; Macherel, D.; North, H.N.; et al. The ABA-deficiency suppressor locus HAS2 encodes the PPR protein LOI_1/MEF_{11} involved in mitochondrial RNA editing. *Mol. Plant.* **2015**, *8*, 644–656. [CrossRef]
115. Watkins, C.B. Advances in the use of 1-MCP. In *Advances in Postharvest Fruit and Vegetable Technology*; CRC Press: Boca Raton, FL, USA, 2015; pp. 117–145. [CrossRef]
116. Deng, X.G.; Zhu, T.; Zhang, D.W.; Lin, H.H. The alternative respiratory pathway is involved in brassinosteroid-induced environmental stress tolerance in *Nicotiana benthamiana*. *J. Exp. Bot.* **2015**, *66*, 6219–6232. [CrossRef]
117. Sharma, N.; Yadav, A.; Khetarpal, S.; Anand, A.; Sathee, L.; Kumar, R.R.; Singh, B.; Soora, N.K.; Pushkar, S. High day-night transition temperature alters nocturnal starch metabolism in rice (*Oryza sativa* L.). *Acta Physiol. Plant* **2017**, *39*, 74. [CrossRef]
118. Shi, W.; Yin, X.; Struik, P.C.; Solis, C.; Xie, F.; Schmidt, R.C.; Huang, M.; Zou, Y.; Ye, C.; Jagadish, S.K. High day-and night-time temperatures affect grain growth dynamics in contrasting rice genotypes. *J. Exp. Bot.* **2017**, *68*, 5233–5245. [CrossRef] [PubMed]
119. Bahuguna, R.N.; Solis, C.A.; Shi, W.; Jagadish, K.S. Post-flowering night respiration and altered sink activity account for high night temperature-induced grain yield and quality loss in rice (*Oryza sativa* L.). *Physiol.Plant.* **2017**, *159*, 59–73. [CrossRef] [PubMed]
120. Sharma, N.; Yadav, A.; Anand, A.; Khetarpal, S.; Kumar, D.; Trivedi, S.M. Adverse effect of increase in minimum temperature during early grain filling period on grain growth and quality in *indica* rice (*Oryza sativa*) cultivars. *Indian J. Agric. Sci.* **2017**, *87*, 883–888.
121. Barros, J.; Serrani-Yarce, J.C.; Chen, F.; Baxter, D.; Venables, B.J.; Dixon, R.A. Role of bifunctional ammonia-lyase in grass cell wall biosynthesis. *Nat.Plants* **2016**, *2*, 16050. [CrossRef] [PubMed]
122. Fernandez, O.; Ishihara, H.; George, G.M.; Mengin, V.; Flis, A.; Sumner, D.; Arrivault, S.; Feil, R.; Lunn, J.E.; Zeeman, S.C.; et al. Leaf starch turnover occurs inlong days and in falling light at the end of the day. *Plant Physiol.* **2017**, *174*, 2199–2212. [CrossRef]
123. Alonso, A.P.; Vigeolas, H.; Raymond, P.; Rolin, D.; Dieuaide-Noubhani, M. A new substrate cycle in plants. Evidence fora high glucose-phosphate-to-glucose turnover from in vivo steady-state and pulse-labeling experiments with [^{13}C] glucose and [^{14}C] glucose. *Plant Physiol.* **2005**, *138*, 2220–2232 . [CrossRef]
124. Hill, S.A.; Ap Rees, T. Fluxes of carbohydrate metabolism in ripening bananas. *Planta* **1994**, *192*, 52–60. [CrossRef]
125. Trudeau, D.L.; Edlich-Muthm, C.; Zarzycki, J.; Scheffen, M.; Goldsmith, M.; Khersonsky, O.; Avizemer, Z.; Fleishman, S.J.; Cotton, C.A.R.; Erb, T.J.; et al. Design and in vitro realization of carbon-conserving photorespiration. *Proc. Natl. Acad. Sci. USA* **2018**, *115*, E11455–E11464. [CrossRef]
126. Sharkey, T.D.; Singsaas, E.L. Why plants emit isoprene. *Nature* **1995**, *374*, 769. [CrossRef]
127. Loreto, F.; Forster, A.; Durr, M.; Csiky, O.; Seufert, G. On the monoterpene emission under heat stress and on the increased thermotolerance of leaves of *Quercus ilex* L. fumigated with selected monoterpenes. *Plant Cell Environ.* **1998**, *21*, 101–107. [CrossRef]
128. Cheah, M.H.; Millar, A.H.; Myers, R.C.; Day, D.A.; Roth, J.; Hillier, W.; Badger, M.R. Online oxygen kinetic isotope effects using membrane inlet mass spectrometry can differentiate between oxidases for mechanistic studies and calculation of their contributions to oxygen consumption in whole tissues. *Anal. Chem.* **2014**, *86*, 5171–5178. [CrossRef] [PubMed]
129. Millar, A.H.; Whelan, J.; Soole, K.L.; Day, D.A. Organization and regulation of mitochondrial respiration in plants. *Annu. Rev. Plant Biol.* **2011**, *62*, 79–104. [CrossRef] [PubMed]
130. Xu, G.; Fan, X.; Miller, A.J. Plant nitrogen assimilation and use efficiency. *Annu. Rev. Plant Biol.* **2012**, *63*, 153–182. [CrossRef] [PubMed]
131. Masclaux-Daubresse, C.; Daniel-Vedele, F.; Dechorgnat, J.; Chardon, F.; Gaufichon, L.; Suzuki, A. Nitrogen uptake, assimilation and remobilization in plants: Challenges for sustainable and productive agriculture. *Ann. Bot.* **2010**, *105*, 1141–1157. [CrossRef]
132. Cannell, M.G.R.; Thornley, J.H.M. Modeling the components of plant respiration: Some guiding principles. *Ann. Bot.* **2000**, *85*, 45–54. [CrossRef]

MDPI

Article

Agronomic and Physiological Indices for Reproductive Stage Heat Stress Tolerance in Green Super Rice

Syed Adeel Zafar [1,2,†], Muhammad Hamza Arif [1,3,†], Muhammad Uzair [1,*], Umer Rashid [3], Muhammad Kashif Naeem [1], Obaid Ur Rehman [1], Nazia Rehman [1], Imdad Ullah Zaid [1], Muhammad Shahbaz Farooq [1], Nageen Zahra [1], Bilal Saleem [1], Jianlong Xu [4], Zhikang Li [4], Jauhar Ali [5], Ghulam Muhammad Ali [1], Seung Hwan Yang [6,*] and Muhammad Ramzan Khan [1,*]

1 National Institute for Genomics and Advanced Biotechnology, National Agricultural Research Center (NARC), Park Road, Islamabad 45500, Pakistan
2 Department of Botany and Plant Science, University of California, Riverside, CA 92507, USA
3 Department of Biochemistry and Biotechnology, University of Gujrat, Gujrat 50700, Pakistan
4 Institute of Crop Science, Chinese Academy of Agricultural Sciences, Beijing 100081, China
5 International Rice Research Institute, Los Baños 4031, Laguna, Philippines
6 Department of Biotechnology, Chonnam National University, Yeosu 59626, Korea
* Correspondence: uzairbreeder@gmail.com (M.U.); ymichigan@jnu.ac.kr (S.H.Y.); mrkhan@parc.gov.pk (M.R.K.)
† These authors contributed equally to this work.

Abstract: Optimum growing temperature is necessary for maximum yield-potential in any crop. The global atmospheric temperature is changing more rapidly and irregularly every year. High temperature at the flowering/reproductive stage in rice causes partial to complete pollen sterility, resulting in significant reduction in grain yield. Green Super Rice (GSR) is an effort to develop an elite rice type that can withstand multiple environmental stresses and maintain yield in different agro-ecological zones. The current study was performed to assess the effect of heat stress on agronomic and physiological attributes of GSR at flowering stage. Twenty-two GSR lines and four local checks were evaluated under normal and heat-stress conditions for different agro-physiological parameters, including plant height (PH), tillers per plant (TPP), grain yield per plant (GY), straw yield per plant (SY), harvest index (HI), 1000-grain weight (GW), grain length (GL), cell membrane stability (CMS), normalized difference vegetative index (NDVI), and pollen fertility percentage (PFP). Genotypes showed high significant variations for all the studied parameters except NDVI. Association and principal component analysis (PCA) explained the genetic diversity of the genotypes, and relationship between the particular parameters and grain yield. We found that GY, along with other agronomic traits, such as TPP, SY, HI, and CMS, were greatly affected by heat stress in most of the genotypes, while PH, GW, GL, PFP, and NDVI were affected only in a few genotypes. Outperforming NGSR-16 and NGSR-18 in heat stress could be utilized as a parent for the development of heat-tolerant rice. Moreover, these findings will be helpful in the prevention and management of heat stress in rice.

Keywords: green super rice; heat stress; pollen fertility; association; grain yield

Citation: Zafar, S.A.; Arif, M.H.; Uzair, M.; Rashid, U.; Naeem, M.K.; Rehman, O.U.; Rehman, N.; Zaid, I.U.; Farooq, M.S.; Zahra, N.; et al. Agronomic and Physiological Indices for Reproductive Stage Heat Stress Tolerance in Green Super Rice. *Agronomy* **2022**, *12*, 1907. https://doi.org/10.3390/agronomy12081907

Academic Editors: Dilip R. Panthee, Channapatna S. Prakash, Xiling Zou, Daojie Wang and Ali Raza

Received: 26 May 2022
Accepted: 12 August 2022
Published: 14 August 2022

Publisher's Note: MDPI stays neutral with regard to jurisdictional claims in published maps and institutional affiliations.

1. Introduction

Rice (*Oryza sativa* L.) is considered as one of the most essential food crops around the world, especially Asia, Latin America, and Africa [1,2]. It constitutes nearly 20% of overall calorie intake worldwide [2], with up to 80% of calorie ingestion in Asia [3]. Global rice production is direly needed to increase at a growth rate of 1.0–1.2% and grain yield must increase by 0.6–0.9% annually to feed the rapidly growing population, comprising a projected increase of nearly 2 billion people up to the 2050s [4,5]. Agricultural crops are more prone to abiotic stresses due to irregular and unsteady climatic changes [6–9]. Atmospheric temperature is one of the most critical variables determining the seasonal growth and geographic cultivation and distribution of crops [10–12]. An increase in the global mean surface temperature of 0.85 °C was observed between 1880 to 2012 and future projections forecast a 3.0–5.0 °C increase by the end of this century [13] and 2.0 to 4.0 °C until 2050 in Southeast Asia, specifically [14]. Relative to the period from 1900 to 2000, the climatic observations through various models have projected a high (more than 90%) probability of temperature-increase during crop growing season in the tropical and subtropical regions of Asia, such as China, by the year 2100 [15,16].

According to projections, it is expected that environmental fluctuations, especially high temperature stress, may cause a 41% yield decline by end of this century [17,18]. High temperature stress destructively impacts the rice metabolism in all growth phases [19–21]. Rice seedlings are very sensitive to the critical high temperature of 35 °C [22]. Further elevation beyond the critical high temperature can be destructive and may lead to plant death at respective growth stages [18,23]. The frequent occurrence of extreme climatic events, such as high temperature, leads to adverse impacts on rice growth and development.

Flowering in rice is one of the most critical phases in the context of high temperature stress because it could reduce the grain yield due to pollen sterility, poor grain-filling, low grain weight accumulation, and undermined seed setting [24]. Rice is sensitive to heat stress and the threshold temperature for rice at the anthesis and flowering stages are 33.7 °C and 35 °C, respectively [25,26]. High temperature stress also impacts the physiological processes of rice, such as chlorophyll contents, photosynthesis, respiration, and RuBP carboxylase activities [27]. High temperature stress above 38 °C inhibits the spikelet formation associated with the decomposition and synthesis of cytokines [28], and spikelet differentiation aggravates spikelet degeneration and reduces the overall number of spikelets through peroxide accumulation, which destructs the cell division and construction [29–31]. Additionally, the incidence of high temperature stress inhibits the anther filling and panicle initiation phase, which may lead to a decrease in pollen activities inducted by the impeded development of pollen mother cells and abnormalities in the decomposition of the tapetum [32,33]. Recently, studies have shown that high temperature stress could cause spikelet sterility due to a reduction in pollen vitality, vigor, and viability, and also due to the inhibition of anther dehiscence [34,35] and obstruction of pollen tube germination [36,37]. High temperature stress could also cause the insufficient accumulation of nutrients in pollens, which may lead to a reduction in pollen activities, sugar transport, accumulation of peroxides, and carbon metabolism [38,39]. High temperature stress at the flowering stage also effects the stigma vitality and pollen tube elongation [37].

Self-adaptability in the rice plant and responses to high temperature stress greatly depends on several factors, such as the intensity and duration of high temperature stress, growth period, plant size and age, and rice genotypes. To address this challenge, natural variations in rice germplasm under drought stress could be utilized to evaluate the associated traits of stress-tolerant genotypes [40,41]. Target breeding programs could be exploited as an important and essential genetic breeding resource to stimulate the genetic variations through hybridization. Aiming at this, studies have been initiated focusing on green super rice (GSR), an elite and highly water- and nutrient-use efficient rice type [42–44]. Based on the information on cloned green genes and loci, large-scale cross- and backcross-breeding was conducted to generate IL populations and lines abundant in green traits with wild rice, core germplasm, and specific local varieties as the donors [45]. The GSR lines were developed by combining genes from different native and non-native sources and required fewer fertilizers, pesticides, and irrigations. These lines also have greater stress tolerance without compromising the high yield and quality [46].

This study was conducted with the aim of investigating and evaluating the mechanisms of high temperature stress-tolerance through the identification of high temperature-responsive morpho-physiological traits of different GSR lines along with local rice cultivars. GSR lines showed several genotypic differences on high temperature stress tolerance; however, the physiological and biochemical heat-tolerance mechanisms are rarely considered. Other major aim of the study was the investigation of mechanisms that how high temperature incidence on flowering impacts the growth, yield, and quality of rice. The acquired research knowledge will be the basis for sustainable GSR production systems and the breeding of novel rice genotypes in order to optimize the net grain yield and nutritional quality, ultimately moving towards human health by decreasing poverty in densely-populated rice regions.

2. Materials and Methods

2.1. Experimental Site and Design

Healthy seeds of 22 green super rice (GSR) varieties and four local controls (IR-6, Kisan Basmati, Kashmir Basmati, and NIAB-B-2016) were obtained from CAAS, China, and NIGAB, Pakistan, respectively. Previously, Kashmir Basmati was reported as a heat-tolerant genotype, while the NIAB-B-2016 as a susceptible variant [23]. All four controls are normally high-yielding rice varieties grown in Pakistan. The seeds were sown at National Institute for Genomics and Advanced Biotechnology (NIGAB), National Agricultural Research Center (33.684° N and 73.048° E), Islamabad, Pakistan. During the rice growing season of 2020, the seeds of selected lines were placed in trays containing 128 wells filled with a mixture of soil, sand, and peat moss, all containing essential nutrients. The transplantation of the 35-days-old seedlings was carried out on 20 July 2020. For this purpose, randomized complete block design (split plot) with three replicates was followed. Two sets of 26 genotypes were transplanted in the field (one for control and one for heat). Each plot consisted of five rows with 10 plants each. Row-to-row and plant-to-plant distance were kept at 30 cm [4]. Recommended agronomic practices were followed.

At the flowering stage (pre-anthesis), a tunnel was prepared to cover the plants with a polythene sheet and high temperature was maintained (40–45 °C) to apply heat stress from 10:00 a.m. to 03:00 p.m. After 03:00 pm, polythene sheets were removed daily to reduce the temperature (25–35 °C). Morpho-physiological parameters were recorded at maturity from the central five randomly-selected plants in order to remove the border effect [47]. After heat exposure, a pollen fertility test was performed for all genotypes to screen out heat-susceptible genotypes. All fresh leaf samples were collected and stored at −80 °C for further analysis.

2.2. Cell Membrane Thermostability

In the last week of the heat stress, the flag leaf samples from control and stressed plants were collected in pre-labeled 20 mL glass tubes. The leaf segments were treated with distilled water used for conductivity measurements. Cell membrane thermostability was measured by following the method of [48] and following formula:

$$\text{CMS\%} = \frac{1-\left(\frac{\text{T1}}{\text{T2}}\right)}{1-\left(\frac{\text{C1}}{\text{C2}}\right)} \times 100$$

where T and C refer to stressed and control plants, respectively. T1 and T2 are electrode conductance measurements before and after autoclave, while C1 and C2 are electrode conductivity measurements before and after autoclave.

2.3. Pollen Fertility Test

A pollen fertility test was performed for all 26 genotypes following the method of [4] using light microscope (Nikon digital sight DS-Fi2). Spikelets were collected from the heat-treated plants on the following day in the morning before anthesis and preserved in formaldehyde solution (FAA). FAA was prepared with 1:1:18 ratio of formaldehyde, acetic acid, and ethanol, respectively. Anthers were placed on slide and pollens were extracted by crushing the anthers with the help of forceps. Then 50 µL of 1% potassium iodide (I_2-KI) solution was added to the slide and covered with cover slip for visualization under compound microscope. Pollens that stained black with a circular shape were counted as fertile, while the irregular red-orange pollens were considered sterile [31]. At the end, pollen fertility percentage (PFP) was calculated as followed:

$$PFP = \frac{\text{No. of fertile pollens}}{\text{Total No. of pollens}} \times 100$$

2.4. NDVI (Normalized Difference Vegetation Index)

NDVI is a spectral calculation of the density of the green vegetations on a specific area. For NDVI, data was recorded from the three replicates, where a GreenSeeker™ Handheld Optical Sensor Unit (NTech Industries, Inc., Ukiah, CA, USA) was used, kept one meter above the plants during measurement [49].

2.5. Agronomic Parameters and Heat Susceptiblity Index

At the maturity stage, following agronomics parameters; plant height (PH, cm), number of tillers per plant (TPP), grain yield per plant (GY, g), straw yield per plant (SY, g), harvest index (HI) calculated by GY/total biomass, 1000-grain weight (GW, g), grain length (GL, mm), and normalized difference vegetative index (NDVI) were recorded. The heat susceptibility index (HSI) was measured for grain yield by using the formula:

$$1 - \left(\frac{Y}{Y_p}\right)/D,$$

where Y is the grain yield under heat stress genotypes and Y_p is the grain yield of genotypes under normal conditions. D is the stress intensity, which is measured by the formula:

$$\left(1 - \frac{X}{X_p}\right),$$

where X is the mean of the grain yield of heat stress genotypes and X_p is the mean of grain yield of genotypes under normal conditions [24].

2.6. Statistical Analysis

Recorded morpho-physiological data were analyzed with the help of Excel 2019. Analysis of variance and heritability were calculated in R. Packages "corrplot", "Ggally" and "factoextra" were used for correlation and principal component analysis.

3. Results

3.1. Assessment of Genetic Diversity Using Principal Component Analysis

Principal component analysis (PCA) was performed to study the genetic differences among the genotypes, and trait–genotype biplots were constructed from data recorded under control and heat stress (Figure 1). Under control conditions, PC1 and PC2 captured the 31.7% and 26.8% of the total variations (Figure 1A). Our results explained that GY, HI, GW, and NDVI showed opposite response to GL, PH, TPP, and SY. The GSR lines were more conserved because they were clustered near the origin, while the check varieties Kisan Basmati, Kashmir Basmati, NIAB-B-2016, and IR-6 showed more genetic variability because they spread far away from the center of origin (Figure 1A). Similarly, under heat stress, PC1 and PC2 showed 34.8% and 21.7% variations, respectively (Figure 1B). PFP, NDVI, HI, and GY were in opposite direction to the rest of the studied parameters (TPP, PH, SY, GL, GW, and HSI). In contrast to control conditions, both the GSR and check varieties fall away from the origin, which suggests that genotypes were more responsive to heat stress compared to the control (Figure 1B). Importantly, NGSR-3, NGSR-13, Kashmir Basmati and IR-6 were found near the apex of the biplot under heat stress, suggesting that these genotypes have the distinct genetic potential of the best tolerance to heat stress compared to others (Figure 1B).

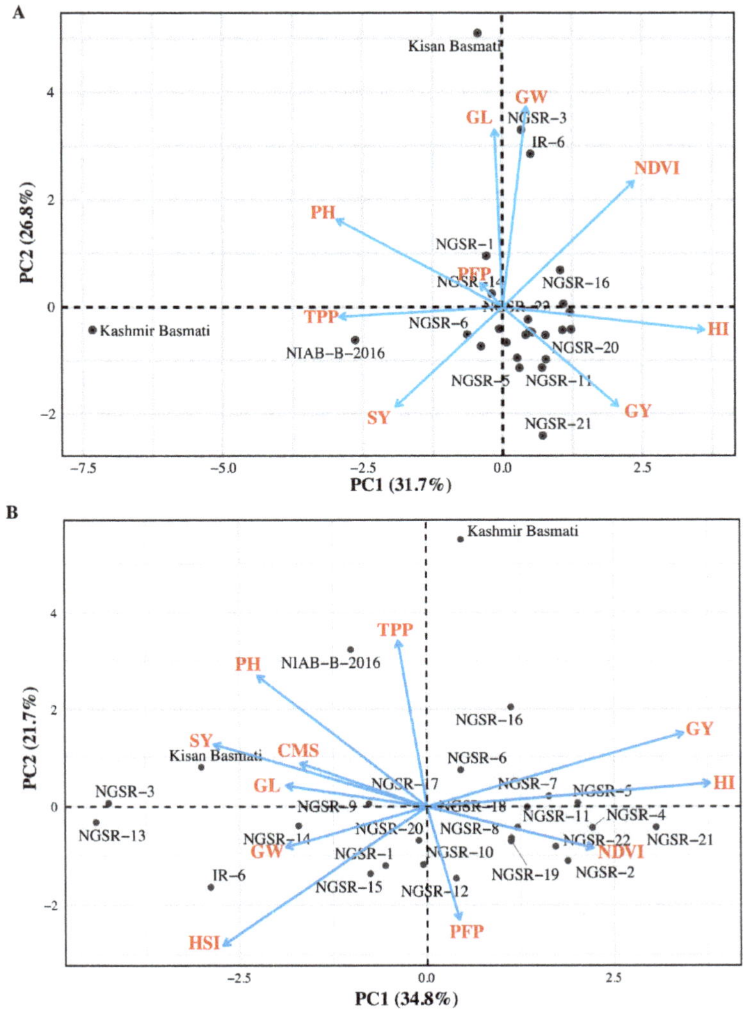

Figure 1. PCA under control (**A**) and heat stress (**B**) for 22 GSR lines and four controls and studied parameters. PH = Plant height (cm), TPP = Tillers per plant, GY = Grain yield per plant (g), SY = Straw yield per plant (g), HI = Harvest index, GW = 1000-Grain weight (g), GL = Grain length (mm), NDVI = Normalized difference vegetative index, PFP = Pollen fertility percentage, CMS = Cell membrane stability, and HIS = Heat susceptibility index.

3.2. *Analysis of Variance Showed Significant Variation in Green Super Rice*

To study the significant differences between the genotypes and heat treatment, the morpho-physiological data were subjected to analysis of variance (ANOVA). Results showed highly significant variations ($p < 0.001$) among the genotypes for all the studied parameters except NDVI (Table 1). Heat stress also showed a highly significant effect on the studied parameters except GW and NDVI. The interactions of genotypes × treatment were also highly significant ($p < 0.001$) for PH, SY, HI, and PFP. GY varied significantly ($p < 0.05$), while TPP, GW, GL, and NDVI were non-significant (Table 1). It is important to note that ANOVA showed significant effects of heat stress on most traits; however, heat stress affected PH, GW, GL, PFP, and NDVI in only a few genotypes and did not show a considerable effect overall (Figures 2–5).

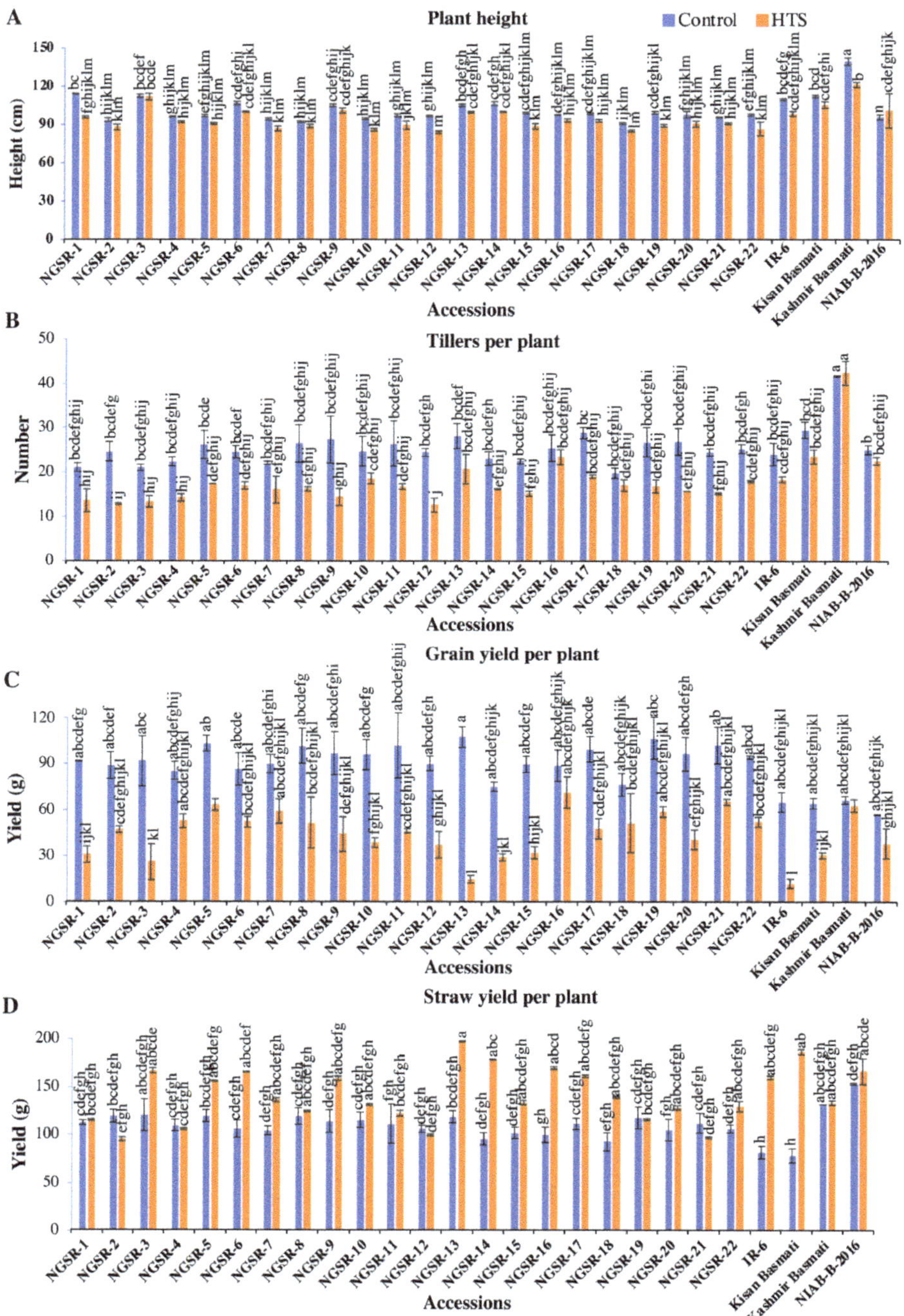

Figure 2. Evaluation of 22 GSR lines and four control variants under control and heat stress. Data for plant height (**A**), tillers per plant (**B**), grain yield per plant (**C**), and straw yield per plant (**D**) were recorded under both conditions. Tuckey's test was used for statistical differences. Different letters above the column varied significantly at $p < 0.05$.

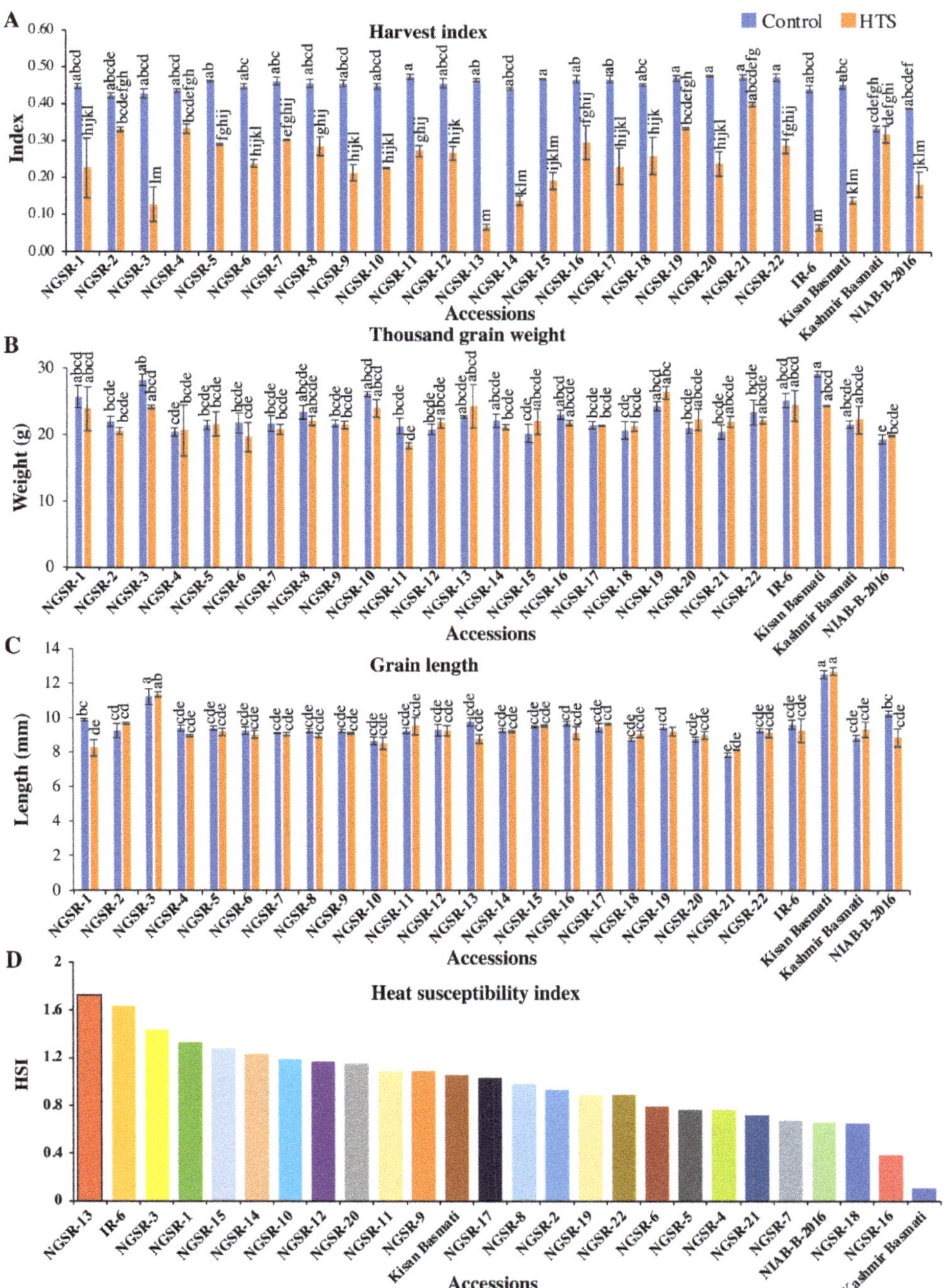

Figure 3. Evaluation of 22 GSR lines and four control types under control and heat stress. Data for harvest index per plant (**A**), thousand grain weight (**B**), and grain length (**C**) were recorded under both conditions. Tuckey's test was used for statistical differences. Different letters above the column varied significantly at $p < 0.05$. (**D**) Distribution of heat susceptibility index for grain yield.

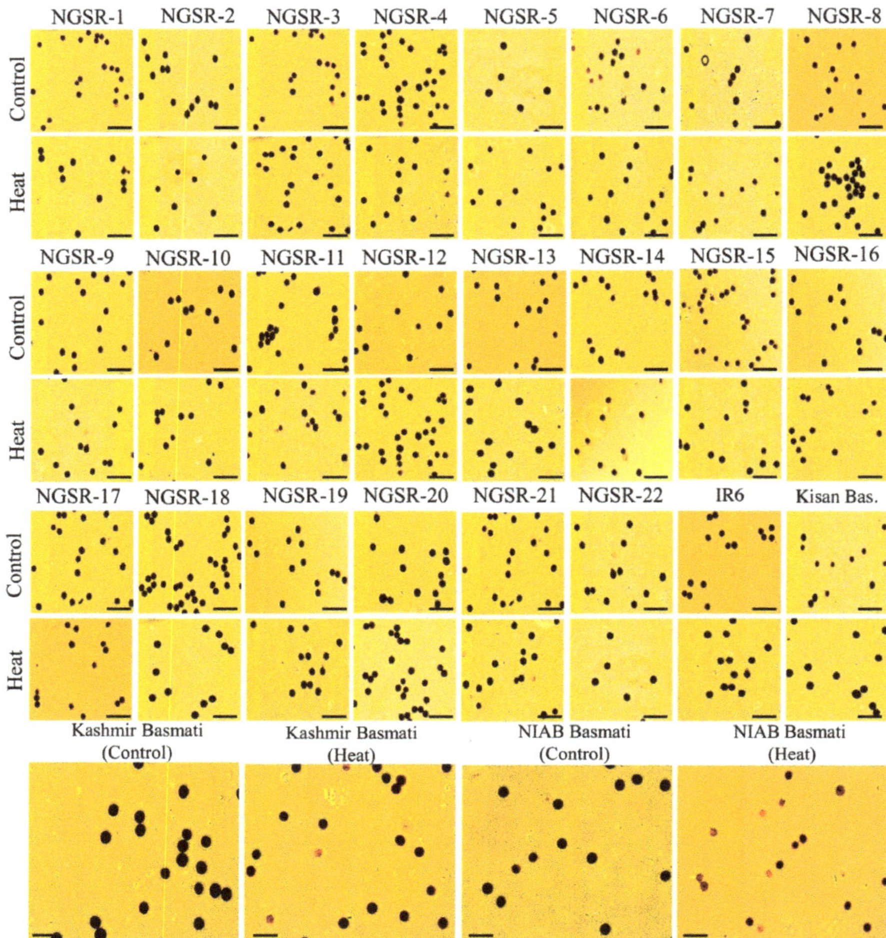

Figure 4. Pollen fertility test in 22 GSR lines and four control variants of rice. Viable pollens were stained deep brown color, while the sterile pollens were slightly or not stained (bars = 100 μm).

Table 1. Mean square values of morpho-physiological parameters.

SOV	G	Rep	T	G × T	H^2
DF	25	1	1	25	
PH	358.2 ***	1.4 ns	828.9 ***	76.7 ***	96.43
TPP	90.0 ***	15.9 ns	1155.6 ***	10.0 ns	92.60
GY	458 ***	461 ns	45,179 ***	318 *	64.85
SY	797 ***	920 ns	34,194 ***	1158 ***	63.36
HI	0.007 ***	0.007 ns	1.1028 ***	0.0078 ***	88.31
GW	18.26 ***	15.69 *	0.122 ns	4.122 ns	81.80
GL	3.09 ***	0.08 ns	1.1839 **	0.223 ns	95.71
NDVI	0.003 ns	0.000047 ns	0.006 ns	0.003 ns	35.83
PFP	110.0 ***	80.1 *	467.0 ***	107.4 ***	85.18

* $p < 0.05$, ** $p < 0.01$, *** $p < 0.001$, and ns = non-significant. SOV = Source of variation, DF = Degree of freedom, G = Genotype, Rep = Replication, T = Treatment, G × T = Genotype × Treatment interactions, H^2 = Heritability, PH = Plant height (cm), TPP = Tillers per plant, GY = Grain yield per plant (g), SY = Straw yield per plant (g), HI = Harvest index, GW = 1000-Grain weight (g), GL = Grain length (mm), NDVI = Normalized difference vegetative index, PFP = Pollen fertility percentage.

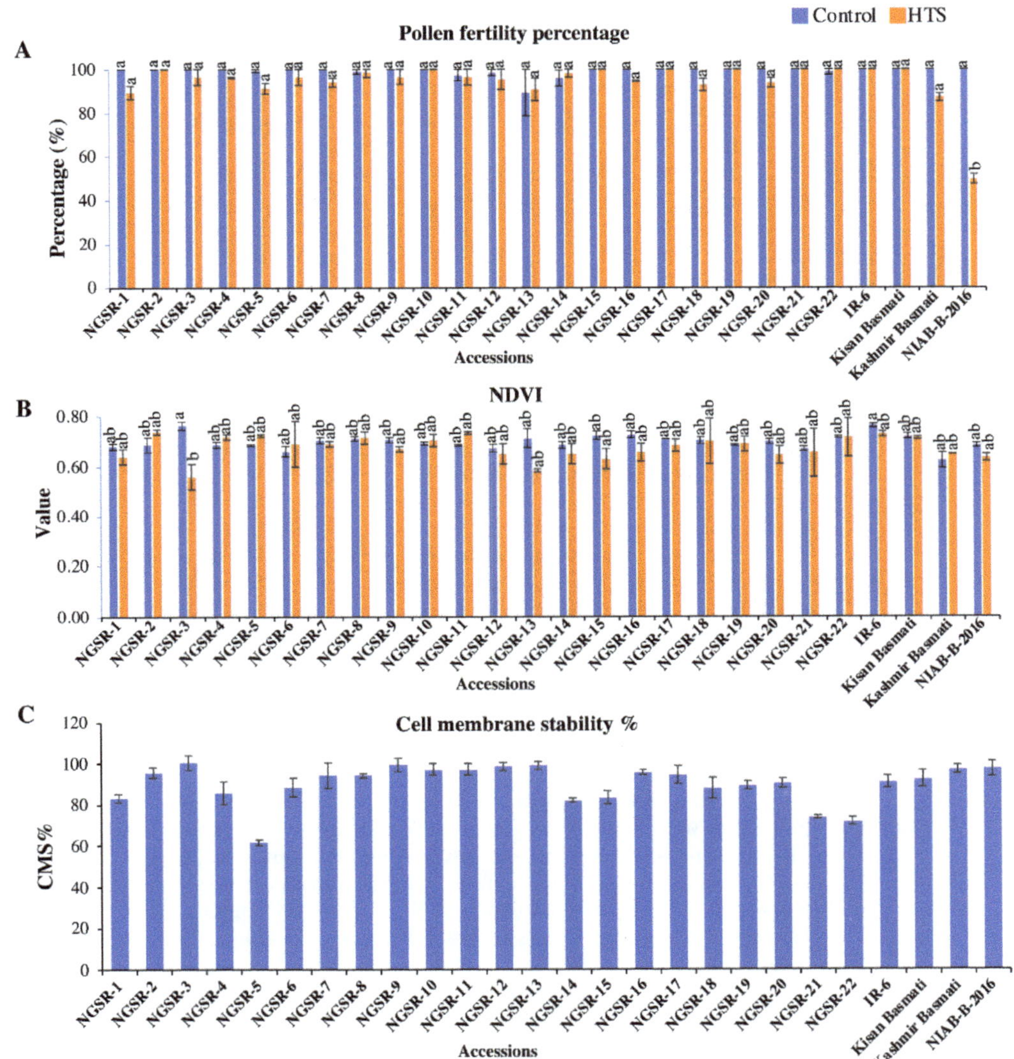

Figure 5. Effect of heat stress on pollen fertility (**A**), NDVI (normalized difference vegetative index, **B**), and cell membrane stability (CMS, **C**). Data was mean ± SE of three replicates. Tuckey's test was used for statistical differences. Different letters above the column varied significantly at $p < 0.05$.

3.3. Mean Performance of Green Super Rice under Heat Stress

Heat stress significantly reduced the GY and affected other yield-contributing agronomic traits, namely TPP, SY, HI, and CMS, in most of the studied GSR and local genotypes (Figures 2–5). However, the traits PH, GW, GL, PFP, and NDVI were not affected in most of the studied genotypes and probably do not play a role in grain yield variations (Figures 2–5). Regarding the PH, we did not observe a very strong effect of heat stress on the tested genotypes (Figure 2A). Although it decreased overall, PH was considerably decreased in NGSR-1, NGSR-19, IR-6, and Kashmir Basmati under heat stress (Figure 2A). TPP and GY were decreased significantly under heat stress in most of the genotypes, except for Kashmir Basmati (Figure 2B,C). Overall reduction in the TPP was more obvious in

GSR lines as compared to controls and Kashmir Basmati was the genotype with maximum (42.33) TPP (Figure 2B). For the GY, we observed a 5.68–86.52% decrease under heat stress. Interestingly, NGSR-16 and Kashmir Basmati showed no or little reduction in GY under heat stress and proved to be relatively heat-tolerant genotypes. Of all the genotypes, type IR-6 (11.83 g) was the worst performer and showed the maximum reduction (81.86%) under heat stress.

SY is another important agronomic trait that determines the overall plant biomass. We observed a sharp increase in SY in most genotypes under heat stress compared to the normal conditions (Figure 2D). The GSR lines NGSR-13 (197.17 g), NGSR-14 (178.5 g), and NGSR-16 (169.67 g) were the highest SY producers and showed a 39.84%, 46.47%, and 41.42% increase in SY, respectively (Figure 2D). The increase in SY suggest that plants have increased their vegetative growth and reduced their reproductive growth, which could be an avoidance mechanism.

HI is an important indicator of genotype performance, and a significant reduction in the HI was observed in all the studied genotypes except Kashmir Basmati (Figure 3A). It reduced from 0.40 to 0.07 but an overall less reduction was observed in GSR lines as compared to non-GSR checks (Figure 3A). The GSR lines NGSR-21 (0.40), NGSR-19 (0.34), NGSR-4 (0.33), and NGSR-2 (0.33) were the best performers. Highest reduction was observed in our high-temperature sensitive checks IR-6 (0.07), followed by Kisan Basmati (0.14) and NIAB-B-2016 (0.18).

Previously, drought- and heat-susceptibility indices have been widely used for the identification of tolerant genotypes, and genotypes with low values are considered tolerant ones [4,24,50]. Based on the heat susceptibility index (HSI), the GSR lines NGSR-16 and NGSR-18 showed the minimum HSI values 0.38 and 0.65, respectively, showing their maximum heat tolerance level. While the control variant Kashmir Basmati showed the minimum (0.11) HSI (Figure 3D). In contrast, the highest HSI was observed for NGSR-13 (1.72) and IR-6 (1.63), indicating the least heat tolerance in these genotypes. Overall, GSR lines performed well as compared to studied controls.

Pollen fertility has been used as an important indicator for the identification of heat-tolerant genotypes because it influences the seed setting and, ultimately, the grain yield. However, we observed the significant effect of heat stress on pollen fertility in only NIAB-B-2016, while all the remaining genotypes showed a higher fertility under heat stress. Most of the GSR genotypes were completely fertile under heat stress, and maximum sterility was observed in NGSR-1 (10.66%) and NGSR-13 (9.07%) (Figures 4 and 5A). These results suggest that these genotypes used some tolerance mechanism to avoid the deleterious effect of heat stress on PFP, and there were some other factors responsible for the reduction in grain yield under heat stress.

3.4. Normalized Difference Vegetative Index (NDVI)

To monitor the health of a plant, remote sensing is widely used, and data collected from NDVI can be utilized in the identification of tolerance. In this study, we observed that there was no significant reduction in GSR lines as well as controls. The genotypes NGSR-2, NGSR-11, and the control variants IR-6 and Kissan Basmati showed the maximum (>0.7) value of NDVI. Minimum NDVI was observed in NGSR-3 and NGSR-13, while the NIAB-B-2016 showed the minimum NDVI among the control variants (Figure 5B). These results indicate that the GSR lines have the ability to maintain the growth rate under heat stress.

3.5. Effect of Heat Stress on Cell Membrane Stability (CMS%)

CMS is used to assess the stress tolerance ability of plant cells under abiotic stresses. We observed that most of the GSR lines showed higher CMS% than the controls except NGSR-5, NGSR-14, NGSR-15, NGSR-21, and NGSR-22. Highest CMS% was observed in NGSR-3, followed by NGSR-9, NGSR-13, and NGSR-16, while Kashmir Basmati showed the maximum CMS% of the control variants (Figure 5C).

3.6. Association of Grain Yield and Other Parameters under Heat Stress

Knowledge of association among the yield, yield-related, and other agronomic parameters is very important because it provide the basic information regarding the selection of certain parameters, which can be utilized as marker of grain yield improvement. In control conditions (Figure 6A), PH showed positive highly significant association with TPP (r = 0.62 ***) and a negative but significant association with GY (r = −0.40 *) and HI (r = −0.65 ***). TPP negatively associated with HI (r = −0.50 *) and NDVI (r = −0.43 *). GY showed highly positive association (r = 0.65 ***) with HI. SY showed a negative but significant association with NDVI (r = −0.42 *) and HI (r = −0.49 *). There was a highly significant association between GL and GW (r = 0.65 ***).

Similarly, under heat stress, PH showed a positive association with TPP (r = 0.63 ***), SY (r = 0.51 **), and GL (r = 0.40 *). PH also showed a negative but significant association with HI (r = −0.39 *) and NDVI (r = −0.44 *, Figure 6B). GY showed a highly significant association with HI (r = 0.89 ***) and a negative but highly significant association with HSI (r = −0.88 ***). HI and SY also showed a negative but significant association (r = −0.74 ***). Similarly, HI showed a negative association with HSI (r = −0.72 ***).

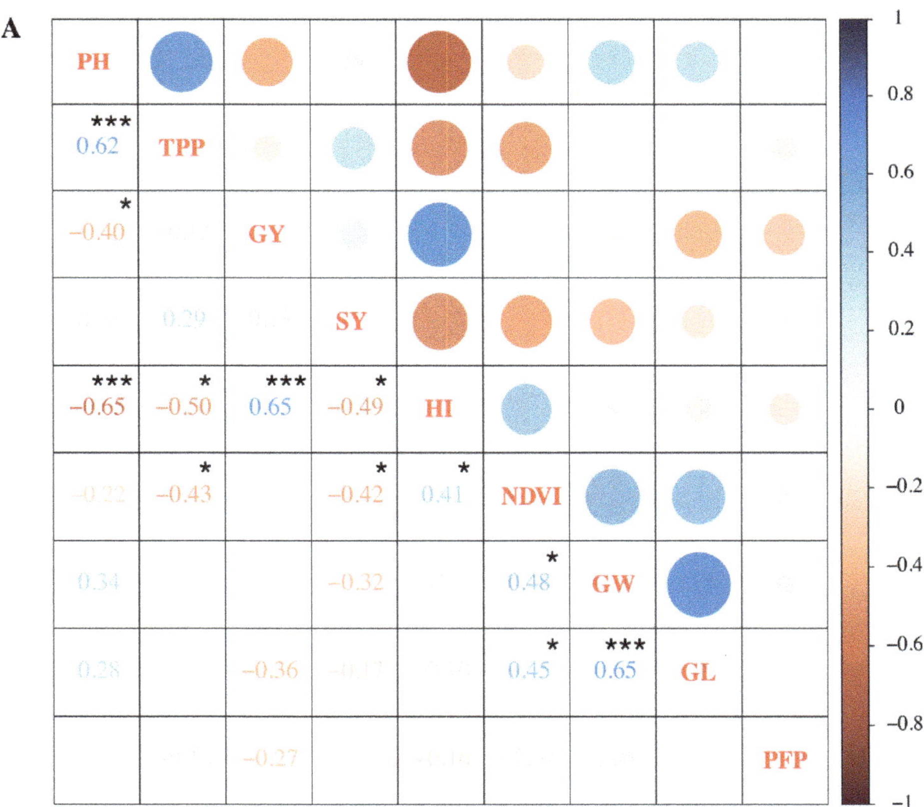

Figure 6. *Cont.*

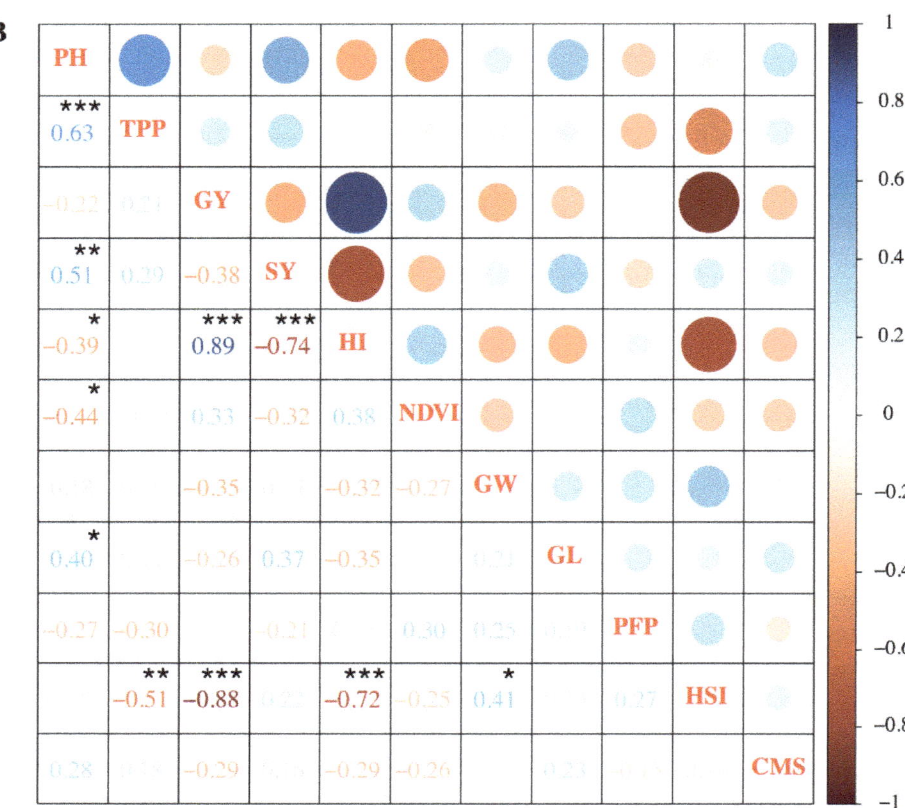

Figure 6. Association studies between agronomic parameters under control (**A**) and heat stress (**B**). PH = Plant height (cm), TPP = Tillers per plant, GY = Grain yield per plant (g), SY = Straw yield per plant (g), HI = Harvest index, GW = 1000-Grain weight (g), GL = Grain length (mm), NDVI = Normalized difference vegetative index, PFP = Pollen fertility percentage, his = Heat susceptibility index, and CMS = Cell membrane stability. * $p < 0.05$, ** $p < 0.01$, and *** $p < 0.001$.

4. Discussion

Rice is a very important cereal crop for majority of the world's population [51]. Previously, it had been reported that the origin of rice varieties was not related to the degree of heat tolerance [52]. In general, the different growth stages of rice behave differently towards heat stress, but the flowering stage is a particularly sensitive stage [53]. Previous studies showed that rice production was optimum at 32–36 °C and a reduction in yield was observed at higher temperatures beyond that level [32,53,54]. The global climate is changing rapidly, and during the 21st century the expected increase in the earth's temperature will be 2 to 4.5 °C [55]. Climate is intrinsically connected with agriculture and an increase in temperature will significantly reduce crop production [11,56]. It is reported that with every 1 °C increase in temperature, rice production will decrease by 2.6% [57]. The population of the world is also increasing day by day and is expected to reach 9 billion by 2050. Keeping in mind the increasing temperature and population and its demand, green super rice (GSR) was developed through the utilization of the world's best germplasm material (Figure 7). GSR has the potential to cope different environmental stresses and maintain an overall grain yield [44]. Furthermore, to our knowledge, GSR lines have never been evaluated for heat stress.

Figure 7. Salient features of GSR and basmati lines. (**A**) Comparison of GSR and basmati lines under control conditions. (**B**) Panicle comparison of basmati and GSR lines. (**C**,**D**) Basmati lines under control and heat stress conditions. (**E**) Seed comparison and (**F**) the effect of disease on GSR lines. Bars (**A**,**C**,**F** = 10 cm; **B**,**D** = 1 cm; **E** = 1 mm).

In this study, twenty-two GSR lines, along with four local Pakistani varieties (controls), were studied under normal and heat stress conditions for grain yield and morpho-physiological parameters. Several morpho-physiological parameters collectively contribute to the grain yield [5,24,58]. In this study, we observed a significant reduction in TPP, SY, HI, CMS, and, ultimately, the GY, for most of the genotypes under heat stress. However, certain traits, including PH, GW, GL, PFP, and NDVI, were less affected under heat stress and probably contributed towards overall heat tolerance. In general, the GSR lines were less affected as compared to local varieties for several traits, showing their potential for breeding heat-tolerant rice cultivars. This was possibly due to GSR having more photosynthates than the control varieties because they might absorb more resources or nutrients in a short period of time [59,60].

Heat stress at the flowering or anthesis stage caused pollen sterility, which may lead to the failure of fertilization and ultimately reduction in yield [11]. Pollen sterility, in most cases, is a major reason for reduced grain production in rice under heat stress [24]. Surprisingly, we did not observe a considerable loss of PFP under heat stress, showing that reduced pollen fertility is not a reason for reduced GY (Figure 4). Compatible with this finding, we observed an increase in SY under heat stress in most of the genotypes, suggesting that plants increased their vegetative growth and slowed down their reproductive growth, which is an important avoidance mechanism for heat tolerance. Heat stress causes cell injury, which leads to the leakage of ions in the susceptible genotypes [61]. The genotypes which show better cell membrane stability under heat stress are generally considered to be heat tolerant [24,62]. However, we did not observe a very straightforward trend of CMS with GY, suggesting that it may not be very reliable to screen tolerant genotypes only based on CMS. In addition, prolonged multi-generational heat stress at the flowering stage may cause the accumulation of mutations that enable the plants to become acclimated under heat stress conditions [63]. In the current study, we found NGSR-16 to be the most heat-tolerant GSR line as it showed a minimum or no reduction in the GY, TPP, HI, TGW, and CMS compared to rest of the GSR lines. Among the local variants and overall, Kashmir Basmati outperformed in all the traits with the least reduction in GY under heat stress compared to the control, suggesting a better source of breeding heat-tolerant cultivars. Kashmir Basmati has previously been identified as a heat-tolerant cultivar in a separate study in Pakistan [23]. It is interesting to note that Kashmir Basmati was originally bred in Pakistan as a cold-tolerant rice cultivar for the Kashmir region, where it tolerates cold stress by producing heat shock proteins [64]. This suggests that there could be a common mechanism for cold- and heat-tolerance, possibly involving heat shock proteins, which needs to be further studied to understand the biochemical mechanism of heat tolerance in GSR.

Stress breeding programs depend on the reliable selection indices to screen good germplasm. Thus, it is important to evaluate the association between the final yield and other agronomic indices. We used correlation and PCA analyses to study the association of GY with other traits. Results showed that the GY had a significant positive association with HI but a negative correlation with GW and GL, suggesting that increasing the GW and GL may decrease the final GY probably via decreasing the total number of grains. Thus, an optimum GW and GL would be required for an optimum GY, which involves a sophisticated balance of source–sink translocation.

5. Conclusions

This study was conducted to evaluate the heat tolerance potential of 22 elite GSR lines in comparison with local varieties as controls in Pakistan. We observed a significant variation among the studied germplasm for heat tolerance and identified several heat-tolerant GSR lines that could be used as a genetic resource for breeding programs. We found that GY, along with other morpho-physiological traits, such as TPP, SY, HI, and CMS, were seriously affected by heat stress in most of the genotypes, while PH, GW, GL, PFP, and NDVI were affected only in a few genotypes. Future studies are required to explore the underlying molecular and physiological mechanisms of heat tolerance in the selected tolerant genotypes.

Author Contributions: Conceptualization, S.A.Z., U.R., M.K.N. and M.R.K.; Data curation, S.A.Z., M.H.A., M.K.N., I.U.Z. and B.S.; Formal analysis, M.U., M.K.N., O.U.R., N.R., M.S.F. and N.Z.; Funding acquisition, G.M.A., S.H.Y. and M.R.K.; Investigation, S.A.Z. and N.Z.; Methodology, I.U.Z.; Resources, N.R., J.X., Z.L., J.A., G.M.A. and S.H.Y.; Software, M.H.A. and N.Z.; Supervision, U.R. and M.R.K.; Visualization, M.U. and B.S.; Writing—original draft, M.U. and O.U.R.; Writing—review and editing, S.A.Z., M.U., M.S.F., J.X., Z.L., J.A., S.H.Y. and M.R.K. All authors have read and agreed to the published version of the manuscript.

Funding: This study was conducted from the funds of Ministry of National Food Security and Research of Pakistan released for the PSDP-project (743) "Green Super Rice in Pakistan". The publication of the present work is supported by the National Research Foundation of Korea (NRF), grant funded by the Korean government (MSIT) (NRF-2021R1F1A1055482).

Institutional Review Board Statement: Not applicable.

Informed Consent Statement: Not applicable.

Data Availability Statement: All the relevant data are within the paper and its supporting information file.

Acknowledgments: We appreciate the efforts of PSDP and Ministry of National Food Security and Research (MNFS&R) for providing resources for this study. We are also thankful to Institute of Crop Science, Beijing, China, for providing the GSR lines.

Conflicts of Interest: The authors declare no conflict of interest.

Compliance with Ethical Standards: This manuscript does not contain any studies with human/animals performed by any of the authors.

References

1. Zafar, S.A.; Hameed, A.; Nawaz, M.A.; Ma, W.; Noor, M.A.; Hussain, M.; Mehboobur, R. Mechanisms and molecular approaches for heat tolerance in rice (*Oryza sativa* L.) under climate change scenario. *J. Integr. Agric.* **2018**, *17*, 726–738. [CrossRef]
2. Fukagawa, N.K.; Ziska, L.H. Rice: Importance for global nutrition. *J. Nutr. Sci. Vitaminol.* **2019**, *65*, S2–S3. [CrossRef] [PubMed]
3. Mahajan, G.; Sekhon, N.; Singh, N.; Kaur, R.; Sidhu, A. Yield and nitrogen-use efficiency of aromatic rice cultivars in response to nitrogen fertilizer. *J. New Seeds* **2010**, *11*, 356–368. [CrossRef]
4. Ahmad, H.; Zafar, S.; Naeem, M.; Shokat, S.; Inam, S.; Rehman, M.A.u.; Naveed, S.A.; Xu, J.; Li, Z.; Ali, G.; et al. Impact of Pre-Anthesis Drought Stress on Physiology, Yield-Related Traits, and Drought-Responsive Genes in Green Super Rice. *Front. Genet.* **2022**, *13*, 832542. [CrossRef]
5. Uzair, M.; Patil, S.B.; Zhang, H.; Kumar, A.; Mkumbwa, H.; Zafar, S.A.; Chun, Y.; Fang, J.; Zhao, J.; Khan, M.R. Screening Direct Seeding-Related Traits by Using an Improved Mesocotyl Elongation Assay and Association between Seedling and Maturity Traits in Rice. *Agronomy* **2022**, *12*, 975. [CrossRef]
6. Raza, A.; Tabassum, J.; Kudapa, H.; Varshney, R.K. Can omics deliver temperature resilient ready-to-grow crops? *Crit. Rev. Biotechnol.* **2021**, *41*, 1209–1232. [CrossRef] [PubMed]
7. Sun, F.; Yu, H.; Qu, J.; Cao, Y.; Ding, L.; Feng, W.; Khalid, M.H.B.; Li, W.; Fu, F. Maize ZmBES1/BZR1-5 Decreases ABA Sensitivity and Confers Tolerance to Osmotic Stress in Transgenic Arabidopsis. *Int. J. Mol. Sci.* **2020**, *21*, 996. [CrossRef]
8. Lu, F.; Li, W.; Peng, Y.; Cao, Y.; Qu, J.; Sun, F.; Yang, Q.; Lu, Y.; Zhang, X.; Zheng, L.; et al. ZmPP2C26 Alternative Splicing Variants Negatively Regulate Drought Tolerance in Maize. *Front. Plant Sci.* **2022**, *13*, 851531. [CrossRef]
9. Feng, W.; Liu, Y.; Cao, Y.; Zhao, Y.; Zhang, H.; Sun, F.; Yang, Q.; Li, W.; Lu, Y.; Zhang, X.; et al. Maize ZmBES1/BZR1-3 and -9 Transcription Factors Negatively Regulate Drought Tolerance in Transgenic Arabidopsis. *Int. J. Mol. Sci.* **2022**, *23*, 6025. [CrossRef]
10. Li, B.; Gao, K.; Ren, H.; Tang, W. Molecular mechanisms governing plant responses to high temperatures. *J. Integr. Plant Biol.* **2018**, *60*, 757–779. [CrossRef] [PubMed]
11. Waqas, M.A.; Wang, X.; Zafar, S.A.; Noor, M.A.; Hussain, H.A.; Azher Nawaz, M.; Farooq, M. Thermal Stresses in Maize: Effects and Management Strategies. *Plants* **2021**, *10*, 293. [CrossRef] [PubMed]
12. Raza, A. Metabolomics: A systems biology approach for enhancing heat stress tolerance in plants. *Plant Cell Rep.* **2020**, *41*, 741–763. [CrossRef] [PubMed]
13. Tollefson, J. How hot will Earth get by 2100? *Nature* **2020**, *580*, 443–446. [CrossRef] [PubMed]
14. Stocker, T. *Climate Change 2013: The Physical Science Basis: Working Group I Contribution to the Fifth Assessment Report of the Intergovernmental Panel on Climate Change*; Cambridge University Press: Cambridge, UK, 2014.
15. Sun, Q.; Miao, C.; Hanel, M.; Borthwick, A.G.; Duan, Q.; Ji, D.; Li, H. Global heat stress on health, wildfires, and agricultural crops under different levels of climate warming. *Environ. Int.* **2019**, *128*, 125–136. [CrossRef]
16. Shuangyi, L.; Zhihong, J.; Jieming, C.; Gang, T.; Shuyu, W. Response of temperature-related rice disaster to different warming levels under RCP8. 5 emission scenario in major rice production region of China. *Front. Clim.* **2022**, *3*, 736459. [CrossRef]
17. Shah, F.; Huang, J.; Cui, K.; Nie, L.; Shah, T.; Chen, C.; Wang, K. Impact of high-temperature stress on rice plant and its traits related to tolerance. *J. Agric. Sci.* **2011**, *149*, 545–556. [CrossRef]
18. Aryan, S.; Gulab, G.; Habibi, N.; Kakar, K.; Sadat, M.I.; Zahid, T.; Rashid, R.A. Phenological and physiological responses of hybrid rice under different high-temperature at seedling stage. *Bull. Natl. Res. Cent.* **2022**, *46*, 45. [CrossRef]
19. Chunlin, S.; Zongqiang, L.; Min Jiang, Y.S.; Yingxue, L.; Shouli, X.; Yang, L.; Shenbin, Y.; Gengkang, Y. An quantitative analysis of high temperature effects during meiosis stage on rice grain number per panicle. *Chin. J. Rice Sci.* **2017**, *31*, 658.
20. Hasanuzzaman, M.; Nahar, K.; Alam, M.; Roychowdhury, R.; Fujita, M. Physiological, biochemical, and molecular mechanisms of heat stress tolerance in plants. *Int. J. Mol. Sci.* **2013**, *14*, 9643–9684. [CrossRef]
21. Liu, W.; Yin, T.; Zhao, Y.; Wang, X.; Wang, K.; Shen, Y.; Ding, Y.; Tang, S. Effects of high temperature on rice grain development and quality formation based on proteomics comparative analysis under field warming. *Front. Plant Sci.* **2021**, *12*, 746180. [CrossRef]
22. Nagai, T.; Makino, A. Differences between rice and wheat in temperature responses of photosynthesis and plant growth. *Plant Cell Physiol.* **2009**, *50*, 744–755. [CrossRef] [PubMed]
23. Zafar, S.A.; Hameed, A.; Khan, A.S.; Ashraf, M. Heat shock induced morpho-physiological response in indica rice (*Oryza sativa* L.) at early seedling stage. *Pak. J. Bot* **2017**, *49*, 453–463.
24. Zafar, S.A.; Hameed, A.; Ashraf, M.; Khan, A.S.; Li, X.; Siddique, K.H. Agronomic, physiological and molecular characterisation of rice mutants revealed the key role of reactive oxygen species and catalase in high-temperature stress tolerance. *Funct. Plant Biol.* **2020**, *47*, 440–453. [CrossRef]

25. Jagadish, S.K.; Craufurd, P.Q.; Wheeler, T. High temperature stress and spikelet fertility in rice (*Oryza sativa* L.). *J. Exp. Bot.* **2007**, *58*, 1627–1635. [CrossRef] [PubMed]
26. Satake, T.; Yoshida, S. High temperature-induced sterility in indica rices at flowering. *Jpn. J. Crop Sci.* **1978**, *47*, 6–17. [CrossRef]
27. Mathur, S.; Agrawal, D.; Jajoo, A. Photosynthesis: Response to high temperature stress. *J. Photochem. Photobiol. B Biol.* **2014**, *137*, 116–126. [CrossRef]
28. Wu, C.; Cui, K.; Wang, W.; Li, Q.; Fahad, S.; Hu, Q.; Huang, J.; Nie, L.; Mohapatra, P.K.; Peng, S. Heat-induced cytokinin transportation and degradation are associated with reduced panicle cytokinin expression and fewer spikelets per panicle in rice. *Front. Plant Sci.* **2017**, *8*, 371. [CrossRef]
29. Ya-liang, W.; Yu-ping, Z.; Yan-hua, Z.; Hui, W.; Jing, X.; Hui-zhe, C.; Yi-kai, Z.; De-feng, Z. Effect of high temperature stress on rice spikelet differentiation and degeneration during panicle initiation stage. *Chin. J. Agrometeorol.* **2015**, *36*, 724.
30. Fu, G.; Zhang, C.; Yang, X.; Yang, Y.; Chen, T.; Zhao, X.; Fu, W.; Feng, B.; Zhang, X.; Tao, L. Action mechanism by which SA alleviates high temperature-induced inhibition to spikelet differentiation. *Chin. J. Rice Sci.* **2015**, *29*, 637–647.
31. Zafar, S.A.; Patil, S.B.; Uzair, M.; Fang, J.; Zhao, J.; Guo, T.; Yuan, S.; Uzair, M.; Luo, Q.; Shi, J. DEGENERATED PANICLE AND PARTIAL STERILITY 1 (DPS1) encodes a cystathionine β-synthase domain containing protein required for anther cuticle and panicle development in rice. *New Phytol.* **2020**, *225*, 356–375. [CrossRef]
32. Xu, Y.; Chu, C.; Yao, S. The impact of high-temperature stress on rice: Challenges and solutions. *Crop J.* **2021**, *9*, 963–976. [CrossRef]
33. Deng, Y.; Tian, X.; Wu, C.; Matsui, T.; Xiao, B. Early signs of heat stress-induced abnormal development of anther in rice. *Zhongguo Shengtai Nongye Xuebao Chin. J. Eco-Agric.* **2010**, *18*, 377–383. [CrossRef]
34. Zhao, L.; Zhu, Z.; Tian, X.; Kobayasi, K.; Hasegawa, T.; Zhang, Y.; Chen, Z.; Wang, C.; Matsui, T. Inheritance analysis of anther dehiscence as a trait for the heat tolerance at flowering in japonica hybrid rice (*Oryza sativa* L.). *Euphytica* **2016**, *211*, 311–320. [CrossRef]
35. Hu, Q.; Wang, W.; Lu, Q.; Huang, J.; Peng, S.; Cui, K. Abnormal anther development leads to lower spikelet fertility in rice (*Oryza sativa* L.) under high temperature during the panicle initiation stage. *BMC Plant Biol.* **2021**, *21*, 428. [CrossRef] [PubMed]
36. Wang, Y.; Wang, L.; Zhou, J.; Hu, S.; Chen, H.; Xiang, J.; Zhang, Y.; Zeng, Y.; Shi, Q.; Zhu, D. Research progress on heat stress of rice at flowering stage. *Rice Sci.* **2019**, *26*, 1–10. [CrossRef]
37. Zhang, C.; Li, G.; Chen, T.; Feng, B.; Fu, W.; Yan, J.; Islam, M.R.; Jin, Q.; Tao, L.; Fu, G. Heat stress induces spikelet sterility in rice at anthesis through inhibition of pollen tube elongation interfering with auxin homeostasis in pollinated pistils. *Rice* **2018**, *11*, 14. [CrossRef] [PubMed]
38. Santiago, J.P.; Sharkey, T.D. Pollen development at high temperature and role of carbon and nitrogen metabolites. *Plant Cell Environ.* **2019**, *42*, 2759–2775. [CrossRef] [PubMed]
39. Rieu, I.; Twell, D.; Firon, N. Pollen development at high temperature: From acclimation to collapse. *Plant Physiol.* **2017**, *173*, 1967–1976. [CrossRef] [PubMed]
40. Panda, D.; Mishra, S.S.; Behera, P.K. Drought tolerance in rice: Focus on recent mechanisms and approaches. *Rice Sci.* **2021**, *28*, 119–132. [CrossRef]
41. Xiong, H.; Yu, J.; Miao, J.; Li, J.; Zhang, H.; Wang, X.; Liu, P.; Zhao, Y.; Jiang, C.; Yin, Z. Natural variation in *OsLG3* increases drought tolerance in rice by inducing ROS scavenging. *Plant Physiol.* **2018**, *178*, 451–467. [CrossRef]
42. Ali, J.; Anumalla, M.; Murugaiyan, V.; Li, Z. Green super rice (GSR) traits: Breeding and genetics for multiple biotic and abiotic stress tolerance in rice. In *Rice Improvement*; Springer: Berlin/Heidelberg, Germany, 2021; pp. 59–97.
43. Zaid, I.U.; Zahra, N.; Habib, M.; Naeem, M.K.; Asghar, U.; Uzair, M.; Latif, A.; Rehman, A.; Ali, G.M.; Khan, M.R. Estimation of genetic variances and stability components of yield-related traits of Green Super Rice at multi-environmental conditions of Pakistan. *Agronomy* **2022**, *12*, 1157. [CrossRef]
44. Jewel, Z.A.; Ali, J.; Pang, Y.; Mahender, A.; Acero, B.; Hernandez, J.; Xu, J.; Li, Z. Developing green super rice varieties with high nutrient use efficiency by phenotypic selection under varied nutrient conditions. *Crop J.* **2019**, *7*, 368–377. [CrossRef]
45. Yu, S.; Ali, J.; Zhang, C.; Li, Z.; Zhang, Q. Genomic Breeding of Green Super Rice Varieties and Their Deployment in Asia and Africa. *Appl. Genet.* **2020**, *133*, 1427–1442. [CrossRef] [PubMed]
46. Zhang, Q. Strategies for developing green super rice. *Proc. Natl. Acad. Sci. USA* **2007**, *104*, 16402–16409. [CrossRef]
47. Chaturvedi, A.K.; Bahuguna, R.N.; Shah, D.; Pal, M.; Jagadish, S. High temperature stress during flowering and grain filling offsets beneficial impact of elevated CO_2 on assimilate partitioning and sink-strength in rice. *Sci. Rep.* **2017**, *7*, 8227. [CrossRef] [PubMed]
48. Tripathy, J.; Zhang, J.; Robin, S.; Nguyen, T.T.; Nguyen, H. QTLs for cell-membrane stability mapped in rice (*Oryza sativa* L.) under drought stress. *Theor. Appl. Genet.* **2000**, *100*, 1197–1202. [CrossRef]
49. Govaerts, B.; Verhulst, N. *The Normalized Difference Vegetation Index (NDVI) Greenseeker (TM) Handheld Sensor: Toward the Integrated Evaluation of Crop Management Part A: Concepts and Case Studies*; CIMMYT: Mexico City, Mexico, 2010.
50. Shanmugavadivel, P.; Sv, A.M.; Prakash, C.; Ramkumar, M.; Tiwari, R.; Mohapatra, T.; Singh, N.K. High resolution mapping of QTLs for heat tolerance in rice using a 5K SNP array. *Rice* **2017**, *10*, 28.
51. Zafar, S.A.; Uzair, M.; Khan, M.R.; Patil, S.B.; Fang, J.; Zhao, J.; Singla-Pareek, S.L.; Pareek, A.; Li, X. DPS1 regulates cuticle development and leaf senescence in rice. *Food Energy Secur.* **2021**, *10*, e273. [CrossRef]
52. Shi, W.; Ishimaru, T.; Gannaban, R.; Oane, W.; Jagadish, S. Popular rice (*Oryza sativa* L.) cultivars show contrasting responses to heat stress at gametogenesis and anthesis. *Crop Sci.* **2015**, *55*, 589–596. [CrossRef]

53. Cheabu, S.; Moung-Ngam, P.; Arikit, S.; Vanavichit, A.; Malumpong, C. Effects of heat stress at vegetative and reproductive stages on spikelet fertility. *Rice Sci.* **2018**, *25*, 218–226. [CrossRef]
54. Karwa, S.; Bahuguna, R.N.; Chaturvedi, A.K.; Maurya, S.; Arya, S.S.; Chinnusamy, V.; Pal, M. Phenotyping and characterization of heat stress tolerance at reproductive stage in rice (*Oryza sativa* L.). *Acta Physiol. Plant.* **2020**, *42*, 29. [CrossRef]
55. Haider, S.; Raza, A.; Iqbal, J.; Shaukat, M.; Mahmood, T. Analyzing the regulatory role of heat shock transcription factors in plant heat stress tolerance: A brief appraisal. *Mol. Biol. Rep.* **2022**, *49*, 5771–5785. [CrossRef] [PubMed]
56. Farooq, M.S.; Uzair, M.; Raza, A.; Habib, M.; Xu, Y.; Yousuf, M.; Yang, S.H.; Khan, M.R. Uncovering the Research Gaps to Alleviate the Negative Impacts of Climate Change on Food Security: A Review. *Front. Plant Sci.* **2022**, *13*, 927535. [CrossRef]
57. Easterling, W.E.; Aggarwal, P.K.; Batima, P.; Brander, K.M.; Erda, L.; Howden, S.M.; Kirilenko, A.; Morton, J.; Soussana, J.-F.; Schmidhuber, J. Food, fibre and forest products. *Clim. Chang.* **2007**, *2007*, 273–313.
58. Uzair, M.; Ali, M.; Fiaz, S.; Attia, K.; Khan, N.; Al-Doss, A.A.; Khan, M.R.; Ali, Z. The Characterization of Wheat Genotypes for Salinity Tolerance Using Morpho-Physiological Indices under Hydroponic Conditions. *Saudi J. Biol. Sci.* **2022**, *29*, 103299. [CrossRef] [PubMed]
59. Cheng, W.; Sakai, H.; Yagi, K.; Hasegawa, T. Interactions of elevated [CO_2] and night temperature on rice growth and yield. *Agric. For. Meteorol.* **2009**, *149*, 51–58. [CrossRef]
60. Farooq, J.; Khaliq, I.; Ali, M.A.; Kashif, M.; Rehman, A.U.; Naveed, M.; Ali, Q.; Nazeer, W.; Farooq, A. Inheritance pattern of yield attributes in spring wheat at grain filling stage under different temperature regimes. *Aust. J. Crop Sci.* **2011**, *5*, 1745–1753.
61. Ilík, P.; Špundová, M.; Šicner, M.; Melkovičová, H.; Kučerová, Z.; Krchňák, P.; Fürst, T.; Večeřová, K.; Panzarová, K.; Benediktyová, Z. Estimating heat tolerance of plants by ion leakage: A new method based on gradual heating. *New Phytol.* **2018**, *218*, 1278–1287. [CrossRef] [PubMed]
62. Rehman, S.U.; Bilal, M.; Rana, R.M.; Tahir, M.N.; Shah, M.K.N.; Ayalew, H.; Yan, G. Cell membrane stability and chlorophyll content variation in wheat (*Triticum aestivum*) genotypes under conditions of heat and drought. *Crop Pasture Sci.* **2016**, *67*, 712–718. [CrossRef]
63. Lu, Z.; Cui, J.; Wang, L.; Teng, N.; Zhang, S.; Lam, H.-M.; Zhu, Y.; Xiao, S.; Ke, W.; Lin, J. Genome-wide DNA mutations in Arabidopsis plants after multigenerational exposure to high temperatures. *Genome Biol.* **2021**, *22*, 160. [CrossRef]
64. Iqbal, N.; Farooq, S.; Arshad, R.; Hameed, A. Differential accumulation of high and low molecular weight heat shock proteins in Basmati rice (*Oryza sativa* L.) cultivars. *Genet. Resour. Crop Evol.* **2010**, *57*, 65–70. [CrossRef]

 agronomy

Article

Adaptability Mechanisms of Japonica Rice Based on the Comparative Temperature Conditions of Harbin and Qiqihar, Heilongjiang Province of Northeast China

Muhammad Shahbaz Farooq [1], Amatus Gyilbag [1], Ahmad Latif Virk [2] and Yinlong Xu [1,*]

1 Institute of Environment and Sustainable Development in Agriculture, Chinese Academy of Agricultural Sciences, Beijing 100081, China; mshahbazfarooq786@gmail.com (M.S.F.); amatusmike@yahoo.com (A.G.)
2 College of Agronomy and Biotechnology, China Agricultural University, Beijing 100193, China; virk3813b64@yahoo.com
* Correspondence: xuyinlong@caas.cn

Abstract: Japonica rice has been considerably impacted from climate change, mainly regarding temperature variations. Adjusting the crop management practices based on the assessment of adaptability mechanisms to take full advantage of climate resources during the growing season is an important technique for japonica rice adaptation to climate changed conditions. Research based on the adaptability mechanisms of japonica rice to temperature and other environmental variables has theoretical and practical significance to constitute a theoretical foundation for sustainable japonica rice production system. A contrived study was arranged with method of replacing time with space having four different japonica cultivars namely Longdao-18, Longdao-21, Longjing-21, and Suijing-18, and carried out in Harbin and Qiqihar during the years 2017–2019 to confer with the adaptability mechanisms in terms of growth, yield and quality. The formation of the grain-filling material for superior and inferior grains was mainly in the middle phase which shared nearly 60% of whole grain-filling process. Maximum yield was noticed in Longdao-18 at Harbin and Qiqihar which was 9500 and 13,250 kg/ha, respectively. The yield contributing components fertile tillers, number of grains per panicle, and 1000-grain weight were higher at Qiqihar; therefore, there was more potential to get higher yield. The data for grain-filling components demonstrated that the filling intensity and duration at Qiqihar was contributive to increase the grain yield, whereas the limiting agents to limit yield at Harbin were the dry weights of inferior grains. The varietal differences in duration and time of day of anthesis were small. Across all cultivars and both study sites, nearly 85% of the variation of the maximum time of anthesis could be justified with mean atmospheric temperature especially mean minimum temperature. Mean onset of anthesis was earliest in Longdao-21 at Harbin, whereas it was latest in Longdao-18 at Qiqihar. The maximum time to end anthesis and the longest duration of anthesis were taken by Longdao-18, i.e., 9.0 hasr and 4.2 h, respectively. Chalkiness and brown rice percentages were elevated at Qiqihar showing Harbin produced good quality rice. This study investigated the adaptability mechanisms of japonica rice under varying temperature conditions to distinguish the stress tolerance features for future sustainability and profitability in NEC. It was concluded that there is an adaptive value for anthesis especially regarding T_{min} and, moreover, earlier transplantation may produce tall plants. The results demonstrated that high temperature at the onset of anthesis at the start of the day enhanced the escape from high temperature later during the day. Early transplantation is recommended in NEC because earlier anthesis during humid days rendered for potential escape from high ambient temperature later during that day. Temperature influenced japonica rice significantly and coherently, whereas the influence of growing season precipitation was not significant. Daily mean sunshine influenced the japonica rice significantly, but the impact was less spatially coherent. The results foregrounded the response of the japonica rice to external driving factors focusing climate, but ignored socioeconomic suggesting emphasis on both driving factors to target future research and render important insights into how japonica rice can adapt in mid-high-latitude regions.

Citation: Shahbaz Farooq, M.; Gyilbag, A.; Virk, A.L.; Xu, Y. Adaptability Mechanisms of Japonica Rice Based on the Comparative Temperature Conditions of Harbin and Qiqihar, Heilongjiang Province of Northeast China. *Agronomy* **2021**, *12*, 2367. https://doi.org/10.3390/agronomy11112367

Academic Editor: Arnd Jürgen Kuhn

Received: 6 October 2021
Accepted: 16 November 2021
Published: 22 November 2021

Keywords: japonica rice; adaptability mechanisms; grain-filling; anthesis; grain yield; Northeast China

1. Introduction

Global mean surface temperatures are expected to be higher from the present by 1–3 °C at the end of year 2100 [1]. China's climate has become drier and warmer compared to the 20th century [2]. Northeast China (NEC), one of the major rice producing regions in China, experienced the most obvious warming since last century [3], but the most evident warming has been observed since the 1980s with an annual mean temperature rise of 1.0–2.5 °C. In NEC, reduction in precipitation was seen during summer as the mean rainfall has been decreasing since 1965 [3], whereas increase in temperature has been observed in winter [4]. In NEC, the temperature was higher during 1920–1930, after three decades, it started to decrease, and thereafter again during the 1970s–1980s, it started to become higher [5]. For NEC, the average rise in daily minimum temperature was more obvious than the daily maximum temperature which noticeably narrowed the diurnal temperature range [6]. There is vulnerability to semi-arid areas in NEC because of periodic drought stress as most of the lakes are even disappearing because of declining precipitation and ground water levels.

Production of cereals and majorly rice is one of the major characteristics of food security and grown in over 100 countries around the globe, fulfilling the dietary requirements of millions of people, and considered as an extremely thermosensitive cereal [7,8]. Heat stress events are expected to become frequent, and intensely impact crop growth and grain yield [9–11]. In recent decades, the global temperature has increased due to activities of continuously increasing global population such as deforestation, spread of industrial setups, and enhanced emissions of greenhouse gases (GHGs) [12,13]. Extreme climatic events are adverse for crop growth and development, such as heat stress produce impacts on net yield [14]. High temperature stress on reproductive growth stage rice has become a global issue. Therefore, researching the mechanisms of impacts of climatic variability during different rice growth stages and tolerance against this variability to minimize the losses have become interest among global scientists.

Cereals share 27% of total cultivable area in China where rice is the major crop, sharing 35% of the total food demand nationwide [10,15]. NEC harvests 20% of the China's marketable food grain where rice shares highest quantity [16,17]. Rice is considered as a highly climate-sensitive cereal, and NEC has been observed as one of the most susceptible regions to climate change [10]. Several studies have shown an increase in mean surface temperature with an average warming trend of 0.38–0.65 °C per decade during last five decades [18] which favored the cultivation of seasonal flooded rice. Seasonally grown flooded rice in NEC has brought significant changes in recent decades as it is a major source of methane emissions [19], as over 10% of global methane emissions are being released in the atmosphere due to rice cultivation [20]. Consequently, the dynamic changes in the rice statistics and relationship with climatic variabilities in NEC along with other causes of GHGs emissions are of great importance for eco-efficient japonica rice sustainability [21,22].

In NEC particularly in Heilongjiang Province, rice cultivation has been motivated among local communities by many features such as balance in market prices and climatic variabilities [23–25]. Over the last three decades (1980–2010), rice production in Heilongjiang Province has been increased from 3 to 13% of total national rice production, mostly owing to the speedy growth of rice cultivating areas in NEC [15]. Many studies have done the investigations on variation of rice production due to the impacts of climatic variabilities in NEC—though up till now the outcomes are still confusing with none of the sound adjustive measures—by assessing the adaptability mechanisms regionally [26–28]. Ref. [27] revealed that net grain yield is reduced due to the effect of climate warming, but research conducted in South China and NEC unveiled a boost in rice grain yield at high-latitude regions [26,28].

Rice grain yield is comprised of two major fundamentals: rice yield and planting area [29]. Previous studies uncovered that a nearly 92% increase (about 4.23 mha) in single rice cropping regions in China has occurred in NEC between 1949 and 2013 [23]. Only native yield analyses cannot reflect the natural resource management and food security issues behind higher production of rice perfectly [15,21,30]. Moreover, the primary association among climate variability with japonica rice growth and development, adaptability mechanisms of japonica rice, and production have received fewer attention in high latitudes of China.

The japonica rice growth has been severely affected due to high temperature above the normal range in areas where the temperature has surpassed the optimum range (28/22 °C). It has been reported that rice yield decreased by 7–8% with an increase of each 1 °C temperature at the maximum daytime/minimum night time from 28/21 to 34/27 °C, respectively [31,32]. Moreover, rice production was greatly impacted due to variation in internal climate with an increase in the interannual climate predicted to be highly variable under frequent temperature stress events during the reproductive growth stages [33]. Therefore, this prediction rejects the hypothesis of expected benefits of estimated rise in atmospheric CO_2 on rice plant growth [34].

Among all critical growth stages, booting and flowering are comparatively more sensitive to temperature stresses [35,36]. During early stages of booting, the plant is occupied with low panicles, often at or below flood water level, and is safer due to plant tissues. However, cells undergoing the meiosis have been noticed with damages of cold temperature stress [37,38] during microspore release from tetrads [39]. Sensitive stage of booting starts approximately 7 and 15 d between panicles' initiation and the end of panicle initiation, respectively [40,41]. The upper part of the plant and the spikelets exposed and emerged during the flowering phase are more vulnerable to temperature stress [38,42], which may cause failure or damage of the pollens [38,43,44]. Climatic variability greatly affects the grain yield due to impacts on grain-filling. There are several explanations for poor grain-filling and low grain weight of the superior and inferior spikelets such as low enzyme activity in the conversion of sucrose to starch [11,45–47], hormonal imbalance [11,45], and assimilating transportation barriers [46,48]. It has been revealed that at the early grain-filling stage, the concentrations of soluble carbohydrates in the inferior spikelets are higher than those in the superior spikelets, suggesting that assimilating the supply is not the main reason for poor spikelet grain-filling among inferior grains [47].

Warming stress at flowering and grain-filling stages can reduce the net grain yield through spikelet sterility and shortening the duration of the grain-filling phase [49,50]. The growing degree days (GDD) for a specific cultivar for flowering are almost the same when grown under varying temperature conditions within the temperature ranges of optimum and base temperatures. Growth of superior and inferior grains was faster at higher temperatures but with a reduced grain-filling period [51]. There is an inverse correlation of the length of daily average temperature with the ripening period; therefore, the temperature below or above the optimum range will reduce the grain-filling period. Poor grain-filling decreases the grain weight as a result of rice plant exposure to frequent and continuous high temperature stress during the grain-filling stage [50]. Meanwhile, higher temperature stress during the grain-filling stage enhances the demand for more assimilations avoiding the production of chalky grains [52]. Higher temperature also impacts the developmental and cellular processes leading towards poor grain quality [53,54]. Drought prevalence during grain-filling adversely impacts the grain weight of superior and inferior grains and also reduces the grain quality [55]. Considering the declining water resources in NEC, the future research studies must be focusing on a genotype selection tool in future breeding varietal development programs for screening of drought tolerant japonica rice cultivars with considerations of the adaptability mechanisms of specific cultivars during the grain-filling period for efficient grain-filling duration and rate.

The research gap in NEC is calling the researchers' focus to address climate change impacts on japonica rice growth and yield, thereby suggesting the possible concrete adjustive measures for sustainable japonica rice production systems in NEC. Climatic variabilities have already been exacerbated under climate change, e.g., temperature stress including high and low, humidity, drought, soil salinity, and submergence [8]. Higher temperature stress can greatly damage rice yield by two principles: firstly, high maximum temperature stress combined with higher humidity causing spikelet sterility and reduced quality of grains [54]. Secondly, through higher night-time temperature stress which usually reduces the process of assimilates accumulation. Thus, if response mechanisms could have been investigated at regional and local scales of NEC, then it could possibly help in development of improved rice germplasm with better resistance against specific climatic stress.

Past research in NEC has not focused on the japonica rice adaptation to climate change in NEC. Limited literature is available to apprehend the adaptability mechanisms of the japonica rice cultivars under varying temperature conditions of NEC. Majorly, previous studies have ignored to comprehend the transitions in eco-physiology of japonica rice cultivars to temperature variations. Furthermore, a lack in understanding of the self-adaptability of japonica rice for its necessary threatened the adaptation which was possible with suitable outside interventions. A lack of evaluation of adaptability mechanisms and thereby possible adjustive measures reduced the adaptation process of japonica rice in NEC. To evaluate the sound possible adjustive measures against environmental variabilities in NEC, it is necessary to analyze the adaptability mechanisms of japonica rice cultivars to different temperature conditions. Comparative assessment of japonica rice adaptability mechanisms under climatic variations at regional and local scales of NEC is necessary to overcome the main research gap of past studies. Rice originally is a semiaquatic phylogenetic plant with unique features of susceptibility and self-adaptability against climatic variability [56] which help to possibly adjust the rice production system. Therefore, there are considerable risks to japonica rice system sustainability branching from climatic variability, but addressing the adaptability mechanisms at local scales in NEC and then delivering necessary adjustive strategies can produce a sustainable and wide range of japonica rice production system under varying climatic conditions to encourage the regional sustainability of japonica rice in NEC [57]. Therefore, this study hypothesized that deep investigations of adaptability mechanisms among short- and long-duration japonica rice cultivars under varying temperature conditions pave the way for better adaptation with possible adjustive measures in management practices. To have concrete estimations of the adaptability mechanisms of japonica rice to different temperature changed conditions, this study was designed with the following objectives: (1) providing deep insights into the adaptability mechanisms of japonica rice to climatic driving factors at different growth phases; (2) identifying and evaluating possible potential adjustive measures in management practices to adapt and sustain japonica rice production.

2. Material and Methods

2.1. Description of Study Area

This research was conducted in one of the three provinces of NEC, i.e., Heilongjiang located between 121°13′–135°05′ E longitude and 43°22′–53°24′ N latitude. The northernmost province of China has a territory of 454,000 km^2 and population of 38.18 million with a continental monsoon climate. Annual temperature in Heilongjiang Province ranges between −4 and 4 °C. Winter is long and frigid, whereas summer is short and cool. Annual rainfall averages 500–600 mm, where 70% is received in summer. Its topography is dominated by a few mountain ranges which accounts for 59% of the total area. The interior of the province is relatively flat with low altitude. After a year of land reclamation, Heilongjiang Province has become one of the most important bases of agricultural products like rice.

From Heilongjiang Province, two regions were selected for this research, i.e., Harbin, the capital city of Heilongjiang Province and the other was Qiqihar. Harbin city is situated between 45°25′–45°30′ N latitude and 126°20–126°25′ E longitude. The north of Harbin is occupied by low hilly areas and mountains, whereas the general terrain is high towards southeast side and low towards northwest side. The mean annual temperature is 3.2 °C and the mean annual frost-free season is 130 days. The annual precipitation ranges between 400 to 600 mm. Winter at Harbin is dry with freezing cold where the 24-h mean temperature in January is −17.6 °C. Spring and autumn are constituted by brief transition phases with continuously varying wind direction. Summers can be become hot with a July average temperature of 23.1 °C. Qiqihar is situated between 47°21′15.65″ N latitude and 123°55′5.47″ E longitude and it is the second largest city in the Heilongjiang, located in the west-central part of Heilongjiang Province. It has a cold, monsoon-influenced, and humid continental climate with long, bitterly cold, but dry winter where 24-h average temperature in January is −18.6 °C, but the annual mean temperature is 3.9 °C. Spring and autumn are mild with short and frequent transitions. Summers are usually warm and humid, where the 24-h average temperature in July is 23.2 °C, and the average annual precipitation is 415 mm, most of which comes in summer.

2.2. Study Plan and Data Source

The study was conducted during the rice growing seasons of 2017, 2018, and 2019. The study was conducted in a randomized complete block design (RCBD) with three replications. The seed rate to raise the nursery was selected @35 kg per hectare for all selected cultivars. Transplantation was done manually, taking 2–3 seedlings at 15 × 15 spacing. Weeding was done thrice during the growing season, the first 15 days after transplanting (DAT), the second 30 DAT, and the third 45 DAT. Pendimethalin herbicide was also sprayed at 8 DAT with optimum moisture condition. Net plot size was 6 × 3 m. Fertilizer management was done based on the local recommendations, i.e., nitrogen (N), phosphorus (P), and potassium (K) were amended at a recommended rate of 90-60-60 kg ha^{-1}, respectively. The sources for N application were synthetic Urea fertilizer (46% N) and compost made of poultry manure (1% N), whereas the source of P was synthetic diammonium phosphate (DAP). The supply of N and P from compost was calculated. All the remaining P and K were applied as basal dose of synthetic sources. All compost was applied as the basal dose, and the remaining required N was applied in three equal splits of synthetic urea as the basal dose, at active tillering and panicle initiation. The study was conducted in RCBD design as there were two factors involved: The first was different cultivars and second was different temperature sites. The first factor involved four different cultivars chosen based on the local adoptability in NEC—namely, Longdao-18, Longdao-21, Longjing-21, and Suijing-18—whereas the second factor included different temperature sites of Heilongjiang where all 4 cultivars were grown randomly, thereby recommending the possible adjust-control measures as an adaptation process. The earlier two cultivars were late-maturing and the late two were early-maturing. Japonica rice growth duration is a basic and critical variable of crop production and shifts in growth duration either shortening or lengthening are beneficial for the implementation of sustainable japonica rice system. Different growth duration cultivars were selected to compare the overall performance of short- and long-duration japonica rice in terms of adaptability mechanisms, and to identify the essential characteristics of short- and long-duration japonica rice cultivated under varying climatic conditions of NEC. Currently, it is well known that further changes in growth durations of japonica rice would change the flexibilities of crop rotation and ultimately intensify the crop systems under wide-scale farming set-ups.

2.2.1. Crop Data

Specific leaf area (SLA) was calculated where one sided area of a fresh leaf was divided by its dry weight. For SLA, leaf area was calculated by the manual destructive method. SLA was measured at four growth stages, i.e., tillering, booting, heading, and maturity.

Crop growth rate (CGR) was also calculated by recording the dry biomass at the above-mentioned four growth stages selected for SLA. Hunt, in 1978, gave the following formula for the calculation of CGR:

$$CGR = \frac{W_2 - W_1}{t_2 - t_1}$$

where W_2 and W_1 are the dry biomass weights at the two respective growth stages and the difference of t_2 and t_1 is the time difference between the two respective growth stages.

Plant height was recorded at different growth stages by randomly selecting the 20 plant samples at each growth stage and the maturity average was taken. The number of productive tillers was counted by randomly selecting the 1-m^2 area in each plot. For calculation of spike weight, spike length, and number of grains per panicle, 20 panicles of primary tillers were taken randomly from each plot and then the average was taken for each of these three parameters. To estimate the 1000-grain weight, 1000 grains were randomly weighed by taking five samples from each plot, and then the average was taken. The final grain yield was calculated after threshing the crop which was done at 14% grain moisture level. The record for time taken by a specific growth stage, namely phenological data record, was also noted for sowing, transplanting, tillering, booting, heading, grain-filling, and maturity. To have a record for dry weight accumulation and grain-filling rate at grain-filling stage, each plot was labeled with 200 panicles and the date of the labeling day was 0 days (d). Samples were taken at 1, 4, 8, 12, 16, 20, 26, 32, 38, and 44 d after labeling. A total of 10 spikes were taken each time, and separation and counting of superior and inferior grains was done. Grains were counted and separated through basic ideas about superior and inferior grains, i.e., grains of the three primary branches directly at the top were the superior ones, whereas the grains of the three branches at the bottom of the panicle were the inferior ones. After separation, superior and inferior grains were separately dried to have dry weight accumulation and grain-filling rate record for each plot. The dry weight accumulation was measured in mg $grain^{-1}$, whereas the grain-filling rate was calculated in mg $grain^{-1}$ day^{-1}. Using Richard's growth equation with reference to the formula given by [58], the grain-filling rate was calculated:

$$G = kW/N(1 - (\frac{W}{A})^N)$$

where W is the grain weight (mg), A is the final grain weight (mg), t is the time in days (d) after anthesis, and B, k, and N are the constants/coefficients calculated after regression (data not given in results).

For calculation of time of day of anthesis (TOA) and duration of anthesis, a square of 1 m^2 area was selected. Every square was named as the sub-plot and was observed every day during the entire flowering period every 30 min or less, from sunrise until the termination of anthesis on the last spikelets about midday or early afternoon. Onset of anthesis is defined as the time of day when at least 5 panicles in the observational sub-plot started anthesis of at least one opened spikelet visible per panicle. The maximum of anthesis is when all panicles of the sub-population of panicles attained anthesis of at least one spikelet opened on every panicle, and the end of anthesis is when all the panicles in sub-plot terminated the anthesis as shown by stamens' droopiness, change of color of stamens, and spikelet closure. The TOA was expressed as hours after sunrise (hasr) and duration of anthesis was noted in hours (h).

2.2.2. Meteorological Data

Average daily weather parameters were recorded during each growing season of japonica rice at both study sites through manual installation of the automated and computed weather station in experimental plots which was interlinked with the main weather station of respective study site. Variation in weather parameters was calculated every five minutes. The meteorological data record during the growth season for both sites was made for average, maximum, and minimum atmospheric temperature (°C), soil temperature (°C) at different depths, relative humidity (%), daily precipitation (mm), CO_2 concentration (ppm), and daily radiation accumulation (MJ/m^2).

2.2.3. Statistical Analysis

For analysis of variance (ANOVA), Tukey's HSD test at 0.05 probability level was used for the comparison of the differences among cultivars' means. Duncan's multiple range test (DMRT) was also used to measure the specific differences among treatment means. One-way ANOVA (as the study design was RCBD) was run through Tukey's HSD test and it provided the differences in treatment means, but it did not provide any information regarding which means are different. Thus, DMRT was used to have clear differentiation between pairs of means. For the statistical analysis of data, "Statistix-8.1" software was used, whereas to draw the figures and graphs, "SigmaPlot-14.0" and "Microsoft Excel-2016" were used.

3. Results

3.1. Yield Components Data

The data for yield components, the importance, and magnitude of factors recorded during all three study years are given in Tables 1 and 2. Average aggregated data for yield components represented that the mean values were evidentiarily varying among cultivars within a site as well as between study sites. Discoursing the plant height, maximum values were ascertained in Longdao-18 at both Harbin and Qiqihar study regions viz. 105.6 cm and 113.5 cm, respectively, during rice growth season in 2018. Comparatively, the mean values regarding plant height at both study sites showed less values in 2017 and 2019 (Table 1). Suijing-18 followed the same trend and stood second after longdao-18 at Harbin, but at Qiqihar, Longdao-21 showed higher mean values after Longdao-18 (Table 1). Mean values for spike length were highest in Longdao-21 during all three study years with the highest value of 22.6 cm in 2018 followed by Longdao-18 at Harbin, whereas at Qiqihar, mean maximum spike length was noticed in Longdao-18 during all study years with highest value of 21.3 cm in 2018 followed by Longjing-21 as presented in Table 1. This trend was the same in 2017, but the average values were less as compared to 2018 (Table 1). Moreover, cultivars showed increased or decreased values in 2019 for spike length as compared to the previous two years because of positive or negative correlated effects of internal growth make-up of cultivars and prevailing environmental conditions. The number of productive tillers is important in ciphering the overall yield compared to the total number of tillers. Productive tillers were counted per hill for all cultivars and the highest numbers were seen in Suijing-18 at Harbin with mean values of 17 and 15, respectively, in 2018 and 2019, whereas a similar trend was seen at Qiqihar with mean values of 13 and 12, respectively in 2018 and 2019 (Table 1). All cultivars showed a decreasing trend for productive tillers at Qiqihar in 2019, but Longjing-21 comparatively showed increasing values as shown in Table 1. Mean values regarding grains per panicle were highest in Longdao-18 at Harbin with the highest values of 161 and 151 in 2018 and 2017, respectively, but the trends varied with other cultivars; for example, Longjing-21 showed higher number of grains per panicle at Qiqihar region, but at Harbin it produced a smaller number of grains. The mean values for net grain yield were highest for all cultivars in 2018 than in 2017 and 2019 at Harbin where the maximum grain yield was observed in Longdao-18 in 2018, which was 9500 kg/ha. At Qiqihar, the same trend was seen, where Longdao-18 produced 13,250 kg/ha (Table 1). The increasing trend in the net grain yield among all cultivars at Harbin in 2018 compared to 2017 and 2019 can be justified undergoing that the yield components' values were greater in respective cultivars. Comparatively, all cultivars negated the same trends at both study regions for net grain yield as the net grain yield was highest in 2018 and decreased in 2019 except for Longjing-18, wherein the yield increased in 2019 as speechified in Table 1.

Table 1. Impacts of different environmental conditions on yield and yield components of four different cultivars at Harbin and Qiqihar during rice growing seasons of 2017, 2018, and 2019 (a, b, c, d = DMRT test values to differentiate the treatment means in different traits for different cultivars).

Site	Cultivar	Year	Plant Height (cm) mv ± sd * a **	Prod. Tillers/Hill mv ± sd	Spike Weight (g) mv ± sd	Grains/Panicle mv ± sd	1000-Grain Weight (g) mv ± sd	Spike Length (cm) mv ± sd	Grain Yield (kg/ha) mv ± sd
Harbin	Longdao-18	2017	103.1 ± 3.1 a	13 ± 2 b	57.4 ± 3.7 a	151 ± 15 a	23.5 ± 0.4 c	21.5 ± 0.7 a	9367 ± 369 a
		2018	105.6 ± 3.5 a	14 ± 2 b	59.1 ± 9.4 a	161 ± 15 a	25.3 ± 0.7 b	20.8 ± 0.7 b	9500 ± 400 a
		2019	100.5 ± 4.0 a	13 ± 1 b	55.9 ± 5.5 a	146 ± 11 a	22.5 ± 0.4 c	21.6 ± 1.6 a	7705 ± 297 a
	Longdao-21	2017	93.4 ± 4.7 c	12 ± 1 b	60.8 ± 7.3 a	144 ± 23 a	23.7 ± 1.1 c	20.0 ± 1.5 b	9017 ± 283 a
		2018	95.8 ± 4.2 c	14 ± 1 bc	63.4 ± 4.2 a	151 ± 23 a	26.0 ± 0.6 ab	22.6 ± 1.5 a	9166 ± 289 a
		2019	95.5 ± 1.6 a	12 ± 2 bc	58.6 ± 9.8 a	144 ± 13 a	22.7 ± 1.1 c	22.4 ± 0.5 a	7167 ± 133 a
	Longjing-21	2017	95.1 ± 1.8 bc	11 ± 1 c	46.9 ± 5.4 b	106 ±14 b	26.9 ± 0.8 a	16.3 ± 0.7 c	7050 ± 296 c
		2018	96.7 ± 2.5 bc	12 ± 1 c	49.9 ± 5.7 b	146 ± 15 a	27.0 ± 0.6 a	17.4 ± 0.7 c	7166 ± 305 c
		2019	96.9 ± 2.9 a	11 ± 2 c	42.0 ± 6.2 b	101 ± 18 b	25.9 ± 0.8 a	16.2 ± 0.5 c	6962 ± 117 a
	Suijing-18	2017	98.1 ± 1.6 b	15 ± 1 a	53.1 ± 4.9 ab	141 ± 7 a	25.2 ± 0.4 b	17.4 ± 0.6 c	8200 ± 176 b
		2018	100.0 ± 1.3 b	17 ± 1 a	56.1 ± 7.0 ab	149 ± 8 a	26.1 ± 0.7 ab	18.5 ± 0.6 c	8333 ± 208 b
		2019	97.2 ± 2.5 a	15 ± 1 a	49.6 ± 1.8 ab	136 ± 13 a	24.1 ± 0.2 b	18.6 ± 0.9 b	7309 ± 98 a
Qiqihar	Longdao-18	2017	111.7 ± 0.4 a	11 ± 1 a	34.2 ± 9.1 b	154 ± 4 a	24.1 ± 0.6 a	20.5 ± 1.0 a	12,267 ± 453 a
		2018	113.5 ± 0.5 a	12 ± 1 a	36.9 ± 12.3 b	158 ± 4.0 a	25.2 ± 0.3 a	21.3 ± 1.0 a	13,267 ± 351 a
		2019	93.9 ± 2.9 a	11 ± 1 a	30.6 ± 8.1 c	155 ± 9 a	22.2 ± 1.0 a	19.6 ± 0.4 ab	7350 ± 300 c
	Longdao-21	2017	99.2 ± 0.6 b	11 ± 1 ab	40.3 ± 5.1 ab	153 ± 7 a	24.6 ± 0.1 a	19.8 ± 0.8 ab	11,133 ± 305 a
		2018	100.7 ± 0.7 b	12 ± 2 a	43.5 ± 7.9 ab	150 ± 7 ab	25.7 ± 0.2 a	20.1 ± 0.8 a	13,133 ± 350 a
		2019	83.1 ± 1.9 c	12 ± 1 a	38.4 ± 7.8 b	144 ± 11 b	22.6 ± 2.9 a	21.0 ± 0.7 a	9217 ± 75 b
	Longjing-21	2017	94.6 ± 0.5 c	10 ± 1 ab	51.8 ± 3.7 a	133 ± 4 b	24.2 ± 0.3 a	19.0 ± 1.4 ab	9733 ± 208 b
		2018	96.5 ± 0.5 c	11 ± 1 a	53.1 ± 11.8 a	132 ± 3 c	25.3 ± 0.4 a	20.9 ± 1.4 a	10,500 ± 225 b
		2019	86.7 ± 0.5 bc	13 ± 2 a	58.4 ± 6.8 a	138 ± 7 bc	22.9 ± 2.1 a	18.0 ± 0.9 ab	11,383 ± 120 a
	Suijing-18	2017	94.1 ± 1.7 c	10 ± 1 b	35.8 ± 7.9 b	142 ± 9 ab	24.0 ± 0.2 a	18.1 ± 0.3 b	9634 ± 57 b
		2018	94.5 ± 1.2 d	13 ± 1 a	37.2 ± 11.7 b	143 ± 8 bc	25.1 ± 0.3 a	19.2 ± 0.3 a	9700 ± 100 c
		2019	92.4 ± 2.2 ab	12 ± 2 a	30.2 ± 4.3 c	134 ± 13 c	18.5 ± 2.0 b	17.6 ± 1.2 b	7048 ± 150 d

* mean values ± standard deviation, ** DMRT test values to differentiate the groups of treatment means.

Table 2. Impacts of different environmental conditions prevailed at Harbin and Qiqihar on yield and yield components traits of four different cultivars.

Region		Year	SL * (cm)	PT/P **	G/P ***	SS ⁰ (%)	1000-GW ¹ (g)	GY ² (kg/ha)
Harbin	MV	2017	20.07	11.00	121.00	0.93	21.53	9367.73
		2018	21.17	13.00	124.00	0.95	22.09	9500.00
		2019	18.6	14.00	118.00	0.90	20.10	7313.09
	CV (%)	2017	11.18	6.80	11.79	2.53	4.32	7.52
		2018	11.37	7.01	12.13	2.75	4.59	7.83
		2019	10.81	6.23	11.94	2.61	4.11	7.08
Qiqihar	MV	2017	18.84	12.00	139.00	0.86	24.90	10,528.75
		2018	19.16	13.00	137.00	1.11	25.10	10,978.12
		2019	18.19	11.00	145.00	1.19	21.2	9217.00
	CV (%)	2017	17.39	12.89	9.23	7.64	5.78	5.28
		2018	17.62	13.03	9.52	8.12	6.21	5.73
		2019	16.28	13.08	8.98	7.92	5.89	5.51

* spike length, ** productive tillers per plant, *** grains per panicle, ⁰ seed set, ¹ 1000-grain weight, ² grain yield.

Here the subject matter is exploring why the values for yield and yield components were higher in 2018. The most possible explanation is the larger differences in prevailing environmental conditions during the respective study years which suited well with the requirements of the respective cultivars. Overall values regarding yield contributing parameters were highest in Longdao-18 comparative with other cultivars, but maximum 100-grain weight was recorded in Longjing-21 with a value of 27.0 g in 2018. Unlike Harbin, the interaction among all cultivars at Qiqihar was non-significant as they reported almost the same 1000-grain weight during 2017 and 2018, where the highest value was recorded in Longdao-21 with a value of 25.7 g in 2018, as shown in Table 1. Considering the spike weight, all cultivars at Harbin had higher values in 2018 than 2017 and 2019, where the highest value was noticed in Longdao-21 which was 63.4 g, whereas Longjing-21 had a higher spike weight at Qiqihar having value of 58.4 g in 2019. Based on the recorded findings, insights into the adaptability mechanisms in terms of yield and yield components bristle the suitability of average prevailed environmental conditions with respective growth phases of all cultivars. Therefore, conclusively the higher yield was recorded in 2018, followed by 2017, and minimum values were observed in 2019. Table 2 represents the mean value of all genotypes regarding the variations in yield and yield components' traits to understand the importance and magnitude of the environmental variables, and their impacts on traits. Table 2 presents the impacts of different environmental conditions prevailed at Harbin and Qiqihar on yield and yield components traits of four different cultivars.

3.2. Crop Growth Rate (CGR) and Specific Leaf Area (SLA)

One of the main purposes of this study was to approximate how different climatic conditions influence the growth of rice at different stages and to suggest the most suitable and possible managemental adjustments. Leaf area is the informant to evaluate how the crop is growing and what the final yield would be.

SLA determines the canopy growth and expansion by effecting the total leaf area per plant. Moreover, it also determines the canopy light interception and light use efficiency (LUE). It is a crucial invariable for plant growth estimation because it ascertains how much new leaf area to deploy for each unit of biomass production. The variation in SLA for all four cultivars at both study regions during three growing years is shown in Figure 1. A progressive increase in SLA was observed up to the end of the grand growth stage and measured at four growing stages viz. tillering, booting, heading, and maturity. All cultivars had larger differences in SLA when comparing the two study sites. SLA increased gradually till the grain-filling stage. After that, SLA started to decrease for both sites. At Harbin, the maximum SLA was recorded in Longdao-18 during all study seasons with a maximum value of 32.88 m^2 kg^{-1} in 2018. In the same way, at Qiqihar Longdao-18 had the highest SLA in 2018 with a value of 30.95 m^2 kg^{-1}, whereas the minimum was seen in Longjing-21 with a value of 19.50 m^2 kg^{-1}. At booting, Longjing-21 and Longdao-21 had almost the same SLA as shown in Figure 1. Comparing the mean SLA trend at Harbin and Qiqihar, the former showed better values for SLA than the latter. Overall, mean SLA was highest during 2018 followed by 2017, and the minimum was seen in 2019 (Figure 1) at both study sites. Overall, the values for SLA were higher at Harbin during all study years than Qiqihar. Thus, between study sites, there were no significant differences in SLA among all cultivars, but within a site the difference was significant.

CGR can be outlined as the per unit area dry matter accumulation. More specifically, it can be defined as a measure of mass increase in crop biomass per unit area per unit time. CGR exhibited a very similar trend as with SLA which gradually increased. Then after grain-filling as crop advanced towards maturity, CGR started to decrease. Maximum CGR at Harbin was observed in Longdao-21 in 2018 which was 26.94 g m^{-2} day^{-1}, but in 2017 and 2019, the highest values for CGR were noticed in Longjing-21 with values of 21.66 and 21.02 g m^{-2} day^{-1}, respectively. Mean values of CGR from tillering to maturity was similarly higher in Longdao-18, and Suijing-18 at Harbin. Due to the variation in prevailing environmental components at respective growth stages, the cultivars had larger differences in CGR values as shown in Figure 2. Though the highest values of CGR were highest in Longjing-21 at Harbin, a decreasing trend was fragmentally rapid compared to other cultivars, as in maturity the CGR value was comparatively less than 10 g m^{-2} day^{-1}, whereas for other cultivars it was higher than 10 g m^{-2} day^{-1} as presented in Figure 2. At Qiqihar, the trend was different, maximum CGR unlike Harbin was recorded in Longdao-18 with values of 29.39 and 24.23 g m^{-2} day^{-1} in 2018 and 2019, respectively due to substantial and fluent growth along the whole crop growth period. But at maturity the CGR values for Longdao-18 were less as in maturity Longdao-21 had higher CGR values (Figure 2). Overall, the mean CGR values were higher in Longdao-18 at Qiqihar as for Harbin. Minimum values for CGR were observed in Longjing-21 during 2018 as shown in Figure 2. The decreasing trend for Suijing-18 was rapid compared to other cultivars.

At Qiqihar, in 2018, CGR was recorded at four growth stages, whereas during the other two years it was recorded at three growth stages. The mean values for CGR during 2019 were higher among all cultivars in Harbin than in Qiqihar, but during the other two years (2017 and 2018) the mean values were higher at Qiqihar than at Harbin (Figure 2). All cultivars had significant differences among their CGR values within a site, whereas between two study sites the comparative trend was highly significant. Water and temperature are considered as the most influential factors impacting the growth of crop. Their impact and relevance of results against different environmental components are given in discussion part of the paper.

3.3. *Variation in Time of Day of Anthesis and Duration*

Time of day of anthesis was recorded in 2018 and 2019 and was explicated as hours after sunrise (hasr) whereas duration was recorded during same years elucidated as hours (h) (Table 3) because length of day and time of solar noon changed between environments. Mean onset of anthesis was earliest in Longdao-21 during 2018 and 2019 study years at Harbin with values of 5.7 hasr and 4.8 hasr, respectively, whereas mean onset was latest in Longdao-18 (Table 3). Longjing-21 and Suijing-18 were intermediate. Time to attain maximum anthesis was minimum in Longdao-21 in 2019 with value of 5.9 hasr. Maximum time to end anthesis was taken by Longdao-18 which was 9.0 hasr in 2019. Differences between two study environments were highly significant for time of day of anthesis values and standard error mean was small because of higher number daily recordings throughout the flowering phase. In Qiqihar, the mean onset of anthesis was earliest in Longdao-18 and latest in Longjing-21 with values of 5.0 and 5.4 hasr in 2018, whereas in 2019 it was earliest in Longjing-21 (Table 3). The mean maximum anthesis time was earliest in Longdao-18, and latest in Longjing-21 in 2018, but in 2019 it was latest in Suijing-18 and Longdao-21 with a value of 7.7 hasr. Longdao-21 took more time than other cultivars to reach the mean end of anthesis.

The differences in time of day of onset of anthesis were endured to maximum anthesis (when all spikelets on that day were open) and end of flowering on that day (when all spikelets had closed again). Thus, the degree of variation in the duration of anthesis was less than the time of day of anthesis between 2.9 h for Suijing-18 to 4.2 h for Longdao-21 at Harbin, whereas it was 2.9 in Harbin and 3.8 h in Qiqihar (Table 3). Genotypic variations in time of day of anthesis and duration of anthesis were modest and did not have any consistency between environments.

Within a given study site, there was no significant effect of environment on the time of day of anthesis due to small difference in the variability of environmental conditions. Across cultivars and environments, nonetheless, variable factors of time of day of anthesis were correlated with all observed components of environment except solar radiation. The probable prognosticator variable was the daily minimum temperature along with mean higher temperature. Consequently, low values of minimum air temperature were linked with delayed onset and end of anthesis. Environment components related to environmental humidity such as relative humidity or potential evapotranspiration were typically associated with anthesis variables describing dry atmospheric conditions delayed the onset and end of anthesis. Day length had a positive association with anthesis variables, but solar radiation did not show any significant effect. Possibly, the apparent day length upshots were caused by strong associations among environmental components and day length effects as climatic variables are not independent because linked to season. Probably, the mean low minimum temperature delayed the mean onset of anthesis, and it can be hypothecated that anthesis in reality occurred at a different time of day but at the same ambient prevailed temperature. This hypothesis can be proven false as all cultivars began anthesis at almost the same temperature at Harbin, but this temperature was significantly lower at Qiqihar in 2019. Among all cultivars, anthesis began significantly at low temperatures in Qiqihar than in Harbin. Consequently, the detained anthesis under low daily minimum temperatures made anthesis to begin under warmer conditions but this impact did not bring any isothermal pattern of onset of anthesis across the study sites. Therefore, there is no single defined value of critical temperature for onset of anthesis. Table 4 presents variations in environmental variables prevailed during anthesis at both study sites.

3.4. Grain-Filling Data

Grain weight accumulation and the grain-filling rate for all four selected cultivars at Harbin were recorded during the 2018 and 2019 growing years and are presented in Figures 3 and 4, respectively, whereas at Qiqihar they are showcased in Figures 5 and 6, respectively. Grain weight accumulation for superior grains showed a typical S-shaped trend line with high grain-filling rates, whereas the dry weights of inferior grains, though increased throughout the grain-filling period, had very low-filling rates. The record for grain-filling components was comprised of 44 days for Harbin, but for Qiqihar because of varying environmental conditions, it was fragmentally short in 2018. At Harbin, dry weight accumulation for superior and inferior grains was utmost in Longdao-21 with values of 25.40 and 23.81 mg grain^{-1} in 2018 and 2019, respectively, Longdao-18 had lesser values for dry weight accumulation which were 23.08 and 23.09 as shown in Figures 3 and 4. The dry weight accumulation for inferior grains increased at extremely high rates, and as the end of grain-filling approached, the dry weight of inferior grains in Longdao-18 became almost the same as of superior grains. Superior grains among all cultivars accumulated higher dry weights in 2018 than 2019. Moreover, the dry weight accumulation among inferior grains of all cultivars was less in 2019 due to low filling-rate. Therefore, the dry weights of inferior grains during 2019 were less comparative to 2018 (Figures 3 and 4).

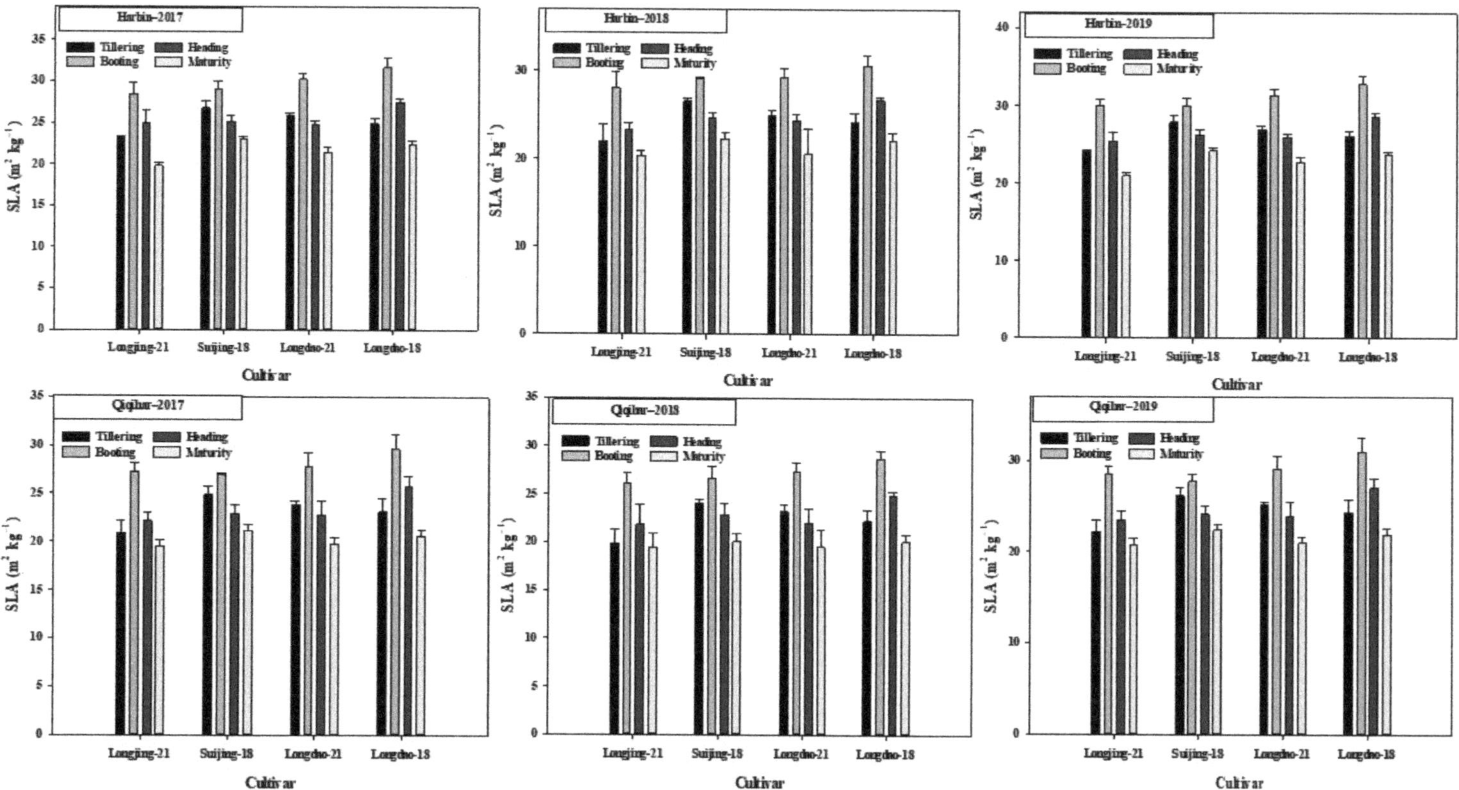

Figure 1. Impact of variation in environmental components on specific leaf area (SLA) (m^2 kg^{-1}) of all four cultivars during 2017, 2018, and 2019 at Harbin and Qiqihar.

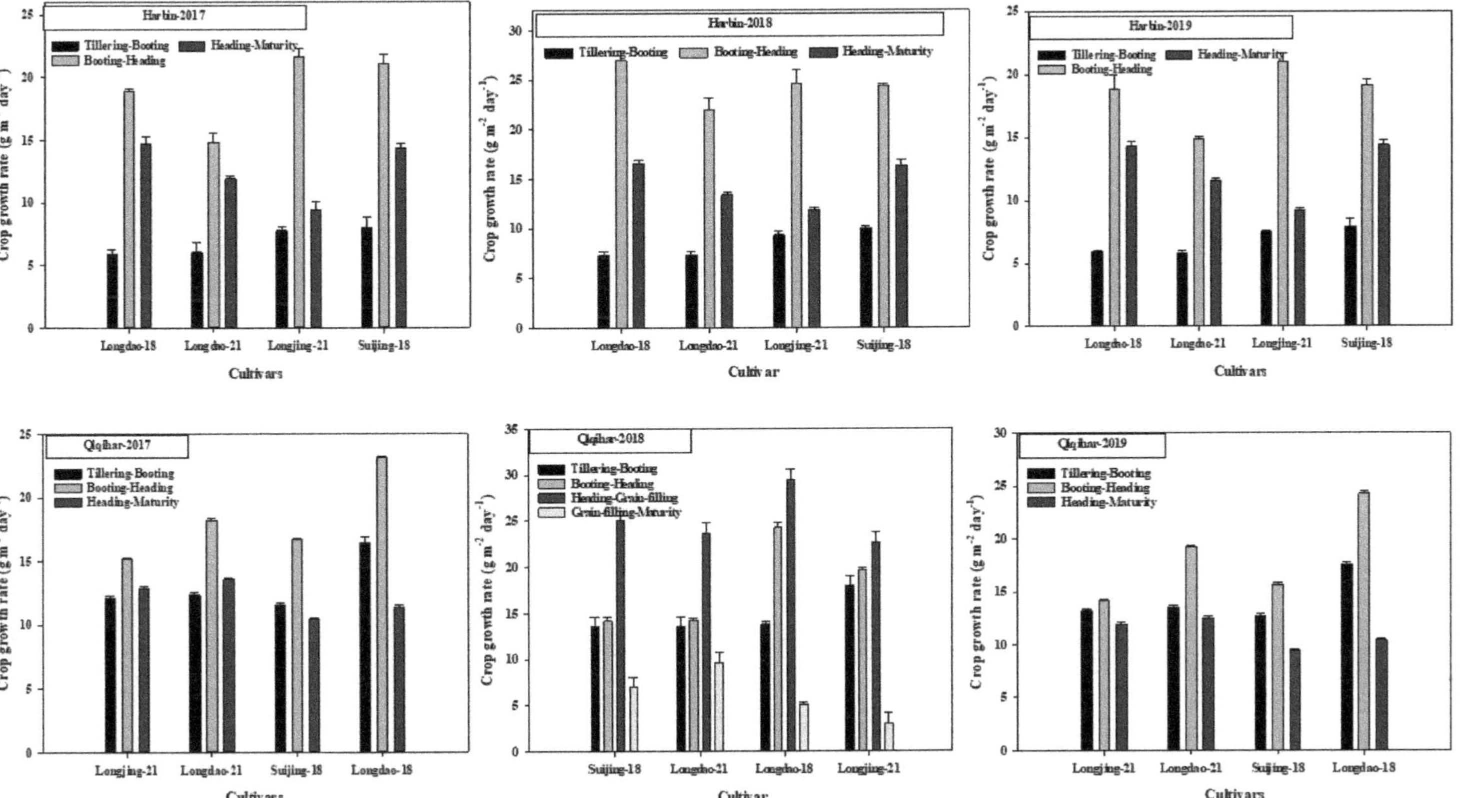

Figure 2. Impact of variation in environmental components on crop growth rate (CGR) (g m^2 day^{-1}) of all four cultivars during 2017, 2018, and 2019 at Harbin and Qiqihar.

Table 3. Impact of environmental variability on time of day of anthesis (hours after sunrise, hasr) and duration of anthesis (time from onset to end of anthesis (h) for 4 cultivars in 2 environments). (a, b, c, d = DMRT test values to differentiate the treatment means in different traits for different cultivars).

Site	Cultivar	Time of Day of Anthesis (hasr) Mean Onset mv ± sd * a **		Mean Max mv ± sd		Mean End mv ± sd		Duration (h) Mean mv ± sd	
		2018	2019	2018	2019	2018	2019	2018	2019
Harbin	Longdao-18	5.9 ± 0.1 a	5.3 ± 0.1 a	6.6 ± 0.1 a	6.3 ± 0.1 a	9.0 ± 0.1 a	9.0 ± 0.1 a	3.1 ± 0.1 ab	3.7 ± 0.2 ab
	Longdao-21	5.7 ± 0.1 a	4.8 ± 0.1 c	6.7 ± 0.2 a	5.9 ± 0.1 b	9.0 ± 0.1 a	8.7 ± 0.1 b	3.1 ± 0.1 a	4.2 ± 0.4 a
	Longjing-21	5.8 ± 0.1 a	5.1 ± 0.1 b	6.6 ± 0.3 a	6.2 ± 0.1 a	9.0 ± 0.1 a	8.7 ± 0.1 b	3.1 ± 0.1 ab	3.9 ± 0.1 ab
	Suijing-18	5.8 ± 0.1 a	5.1 ± 0.1 b	6.7 ± 0.1 a	6.3 ± 0.1 a	8.8 ± 0.1 a	8.7 ± 0.1 b	2.9 ± 0.1 b	3.6 ± 0.2 b
Qiqihar	Longdao-18	5.0 ± 0.1 b	6.5 ± 0.2 a	6.1 ± 0.2 b	7.4 ± 0.1 c	8.7 ± 0.1 b	8.9 ± 0.1 b	3.6 ± 0.1 ab	2.4 ± 0.2 ab
	Longdao-21	5.1 ± 0.2 b	6.4 ± 0.1 a	6.1 ± 0.1 b	7.7 ± 0.1 a	8.9 ± 0.1 a	9.1 ± 0.1 a	3.8 ± 0.1 a	2.7 ± 0.1 a
	Longjing-21	5.4 ± 0.1 a	6.2 ± 0.1 a	6.7 ± 0.1 a	7.5 ± 0.1 b	8.7 ± 0.1 b	9.1 ± 0.1 a	3.3 ± 0.1 b	2.9 ± 0.1 a
	Suijing-18	5.1 ± 0.1 b	6.4 ± 0.1 a	6.6 ± 0.3 a	7.7 ± 0.1 a	8.7 ± 0.1 ab	9.1 ± 0.1 a	3.7 ± 0.1 ab	2.7 ± 0.1 ab

* mean values ± standard deviation, ** DMRT test values to differentiate the groups of treatment means

Table 4. Environmental variables prevailed at anthesis in 2018 and 2019 at Harbin and Qiqihar.

Cultivars	Region	Year	T_{avg} (°C)	T_{max} (°C)	T_{min} (°C)	CO_2 (ppm)	Rad. Accum. (MJ/m^2)	RH (%)	Soil Temp. (5 cm) (°C)	Soil Temp. (10 cm) (°C)
Suijing-18	Harbin	2018	23.58	28.80	18.08	396.63	12.48	87.61	23.80	23.35
		2019	24.97	27.13	16.65	384.11	11.93	82.13	21.42	20.17
	Qiqihar	2018	23.26	28.36	18.89	374.36	10.53	85.59	23.99	22.53
		2019	23.57	28.14	18.51	388.28	12.91	84.37	23.60	22.48
Longjing-21	Harbin	2018	23.12	28.31	17.65	408.87	12.07	85.78	21.87	22.51
		2019	24.51	27.82	17.31	374.58	12.08	81.43	21.75	20.04
	Qiqihar	2018	22.73	27.78	18.63	399.35	11.06	83.42	23.47	21.97
		2019	23.10	27.79	17.74	402.18	13.31	82.94	23.41	21.93
Londao-21	Harbin	2018	22.91	27.98	18.39	407.64	12.83	84.39	24.78	21.98
		2019	24.18	26.79	16.23	386.75	12.32	81.95	20.93	20.21
	Qiqihar	2018	23.56	28.49	17.95	396.63	10.82	84.38	24.06	22.17
		2019	22.92	27.86	17.97	406.36	13.11	83.11	22.90	22.01
Longdao-18	Harbin	2018	23.01	28.23	17.91	402.53	12.30	86.75	22.47	23.63
		2019	24.40	26.92	17.10	392.76	12.58	80.24	22.01	19.63
	Qiqihar	2018	23.13	27.92	19.05	401.65	10.35	82.88	23.28	22.31
		2019	23.08	28.03	17.18	407.13	12.76	81.42	23.14	22.37

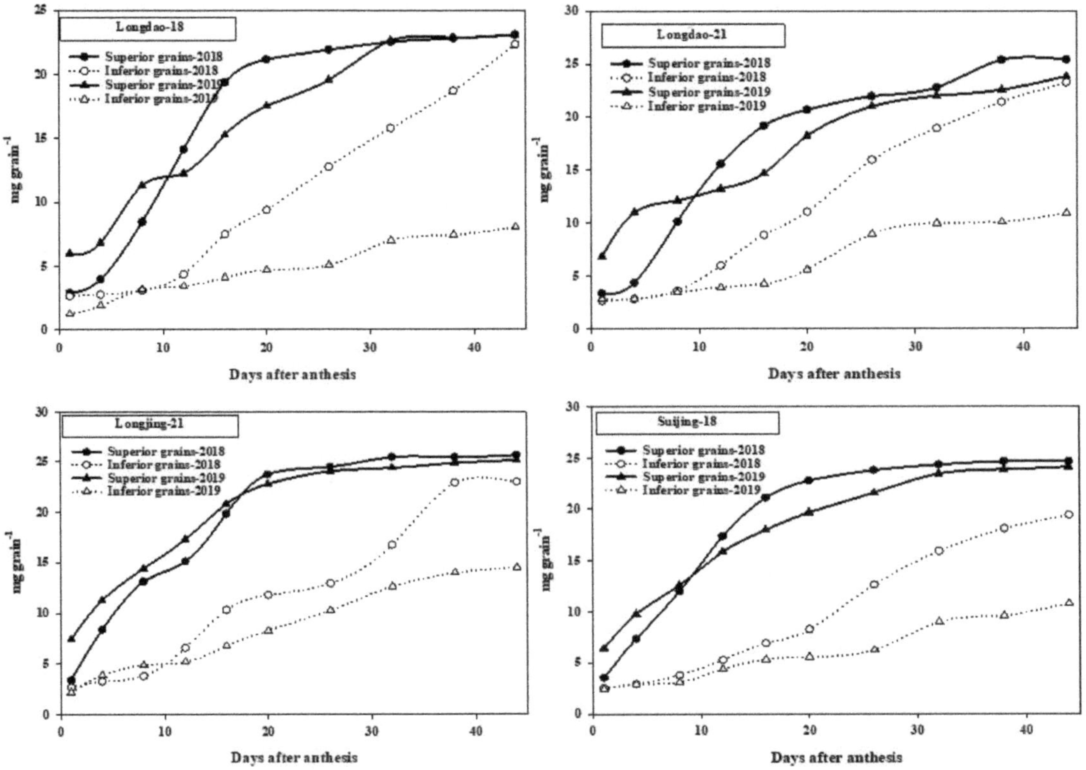

Figure 3. Impact of variation in environmental components on grain weight accumulation (mg grain^{-1}) of all four cultivars during 2018, and 2019 at Harbin.

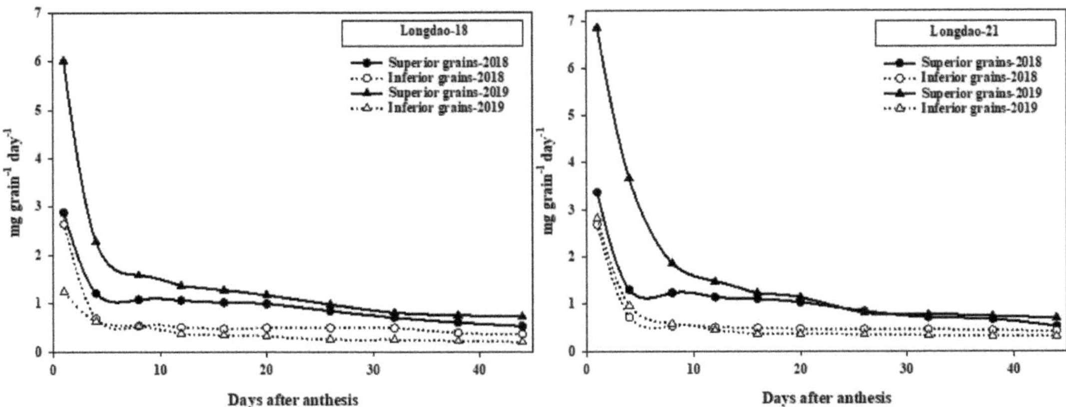

Figure 4. *Cont.*

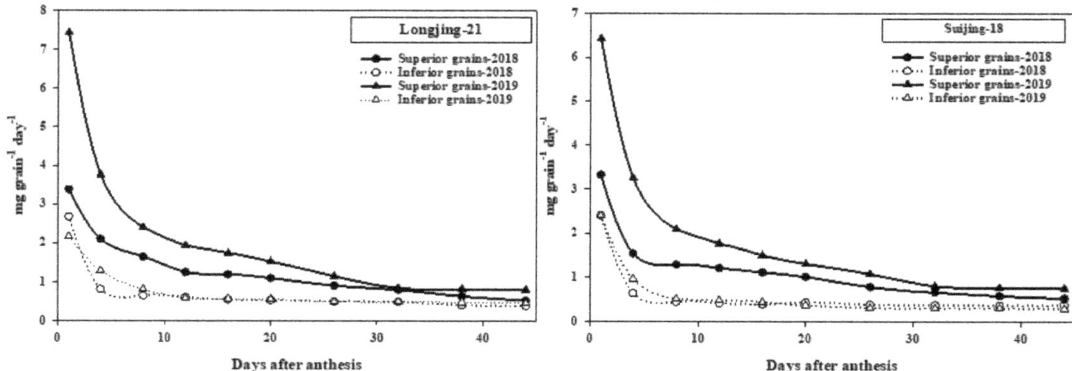

Figure 4. Impact of variation in environmental components on grain-filling rate (mg grain^{-1} day^{-1}) of all four cultivars during 2018 and 2019 at Harbin.

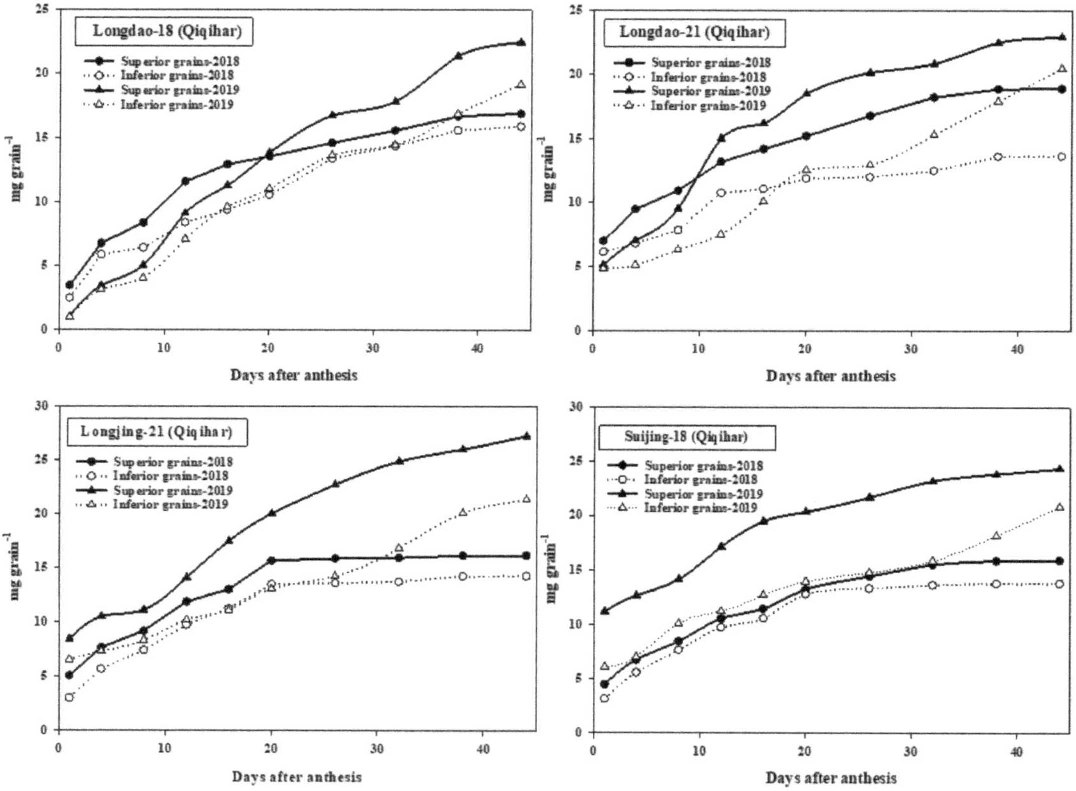

Figure 5. Impact of variation in environmental components on grain weight accumulation (mg grain^{-1}) of all four cultivars during 2018 and 2019 at Qiqihar.

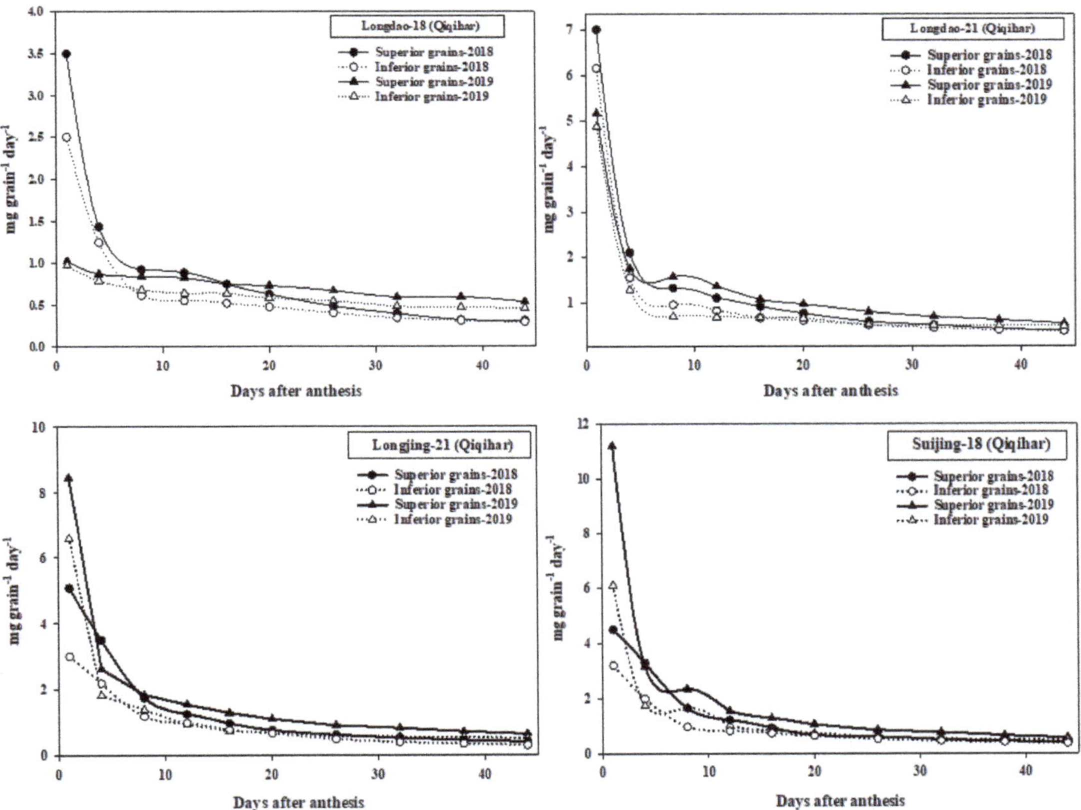

Figure 6. Impact of variation in environmental components on grain-filling rate (mg grain^{-1} day^{-1}) of all four cultivars during 2018 and 2019 at Qiqihar.

Comparing the grain weight accumulation of cultivars for both study sites, the cultivars showed low dry weight accumulation at Qiqihar during 2018 as the highest value for superior grains was recorded in Longdao-21 which was 18.91 mg grain^{-1}, whereas in 2019, the values were higher in Qiqihar and the highest was observed in Longjing-21 with value of 27.31 mg grain^{-1} (Figures 5 and 6). The grain weight accumulation among inferior grains for all cultivars at Qiqihar was low comparative to Harbin. The highest dry weight accumulation for inferior grains was seen in Longdao-18 with value of 15.91 mg grain^{-1} in 2018 whereas in 2019 it was highest in Longjing-21 with value of 21.37 mg grain^{-1} as the filling rate was higher among inferior grains in 2019 as shown in Figures 5 and 6. The varying trend for dry weight accumulation and grain-filling rate was different in 2018 and 2019 as the mean values for grain weight accumulation for all cultivars were less in 2018. Moreover, the mean values of dry weight accumulation for inferior grain among all cultivars were exceedingly high in 2019 compared to the dry weights of superior grains among all cultivars in 2018 (Figures 5 and 6). The environmental variables prevailed during the grain-filling stage at Harbin and Qiqihar for 2018 and 2019, as given in Table 5.

The grain-filling rate for superior and inferior grains at Harbin for all cultivars was comparatively high up till harvest and showed uttered loop-shaped trend lines among all cultivars during both seasons of 2018 and 2019 (Figures 3 and 4). The filling-rate trend line for all cultivars at Qiqihar did not have typical loop-shape expression between superior and inferior grains. Qiqihar had higher filling-rates for superior as well as inferior grains for Longdao-18 and Longdao-21 in 2019, whereas in Longjing-21 and Suijing-18 it was higher for superior grains but for inferior grains, and the filling-rate was almost the same among all cultivars during both study years. Comparing the two study seasons, the grain-filling rate for superior grains was higher in 2019 but for the inferior grain it was higher in 2018 except for Longjing-21, where filling rate was nearly the same during both study years for inferior grains (Figure 3).

At Qiqihar, the low-filling rate for inferior grains could not be associated with temperature differences between superior and inferior grains, as the T_{max} and T_{min} during both study years were comparatively unvarying, more significantly in 2018. Low-filling rates among inferior grains at Qiqihar directed slow grain weight accumulation, thus slow and incomplete filling of the inferior grains resulted in a continuous increase of grain weight up till harvest. Having interaction comparison for grain-filling between study years, it was noticed that grain-weight accumulation for superior and inferior grains was significantly higher during 2018 and the grain-filling rate for superior grains was 2.5 times advanced than for inferior grains. During 2018, 25 days after anthesis, it was noticed that the filling-rate became almost the same for inferior grains as for the superior grains (Figures 3 and 4). Therefore, the environmental variants fluctuated during the grain-filling growth phase, and both study years brought variations in grain-filling rate and ultimately the grain weight accumulation among cultivars at both sites.

Among all environmental variables, temperature is considered as one of the main variants affecting grain-filling phase; therefore, the fluctuations in daily mean temperature most probably are the causative component in bringing changes in the filling-rate. Temperature suitability at the grain-filling stage at Harbin had strongly favored the reason behind higher grain weight accumulation at Harbin than Qiqihar as the mean daily temperature during the grain-filling growth phase was more suitable at Harbin than Qiqihar. The mean growing temperature necessarily required for healthy grain-filling in japonica rice is 20–27 °C, and the average temperature at start of grain-filling at Harbin was more feasible than in Qiqihar. Transplantation of nursery was done on different dates at Harbin and Qiqihar, which caused an obvious time difference in attaining the peak of grain-filling curve for superior as well as inferior grains, demonstrating that the difference in environmental variables had varying degree of influence on each cultivar. Based on the total growth period of all cultivars, it was observed that the completion of the grain-filling phase for inferior grains between Longdao-18 and Longjing-21 varied by 7 and 4 days, respectively, and Qiqihar was earlier than Harbin in 2018. The grain-filling among inferior grains in Longdao-18 varied by 5 days between Harbin and Qiqihar, where grain-filling was completed early at Harbin than at Qiqihar in 2018. It has been foreshadowed that the different environmental components at two different study sites had impacted differently on two early-maturing varieties of the second accumulative temperate zone and first accumulative temperate zone, and had a great impact on dry weight accumulation of the grain filling of the Suijing-18 and Longjing-21, whereas the dry weights and grain-filling rate were less influenced in Longdao-21 and Longdao-18 during both study years.

Overall, the grain-filling growth phase majorly consisted of three sub-phases, viz. starting sub-phase, middle sub-phase, and later sub-phase, to have better consideration regarding the impact of environmental components on respective stage of grain-filling. Based on this division, it was observed that during the starting, middle, and late sub-phases of grain-filling at Harbin, the contribution rates were 39.43%, 61.54%, and 29.80%, respectively, in 2018, whereas, in 2019 it varied at rates of 37.23%, 59.22%, and 33.51%, respectively. Concludingly, the grain-filling for superior and inferior grains among all cultivars under each study sites during both study years were mainly constituted in the middle sub-phase of grain-filling growth phase, which accounted for almost 60% of the whole grain-filling.

3.5. Quality Assessment

The quality of japonica rice was observed for two study years in 2018 and 2019 and is presented in Table 6. The observed data demonstrated that the chalkiness degree and brown rice percentage were higher in Qiqihar than Harbin among all cultivars during both study years as given in Table 5. At Harbin, higher protein contents were observed in Longjing-21 during both study years 2018 and 2019 with values of 9.34% and 9.15%, respectively, whereas at Qiqihar, Longjing-21 had the highest protein contents in 2018, but in 2019 they were high in Suijing-18 with a value of 7.70% (Table 6). Amylose contents were highest in Longdao-18 during both study seasons at Harbin, but at Qiqihar, a different trend was observed, as in 2018, they were the highest in Longdao-18, but in 2019, Longdao-21 had comparatively high amylose contents. The length to width ratio of japonica rice grains among all cultivars was also higher at Harbin with very little difference. Overall, the mean fine rice percentages were advanced during both seasons at Harbin than Qiqihar. Generally, it can be concluded that the quality of japonica rice was better at Harbin relative to Qiqihar, but not with big differences as shown in Table 6. Moreover, the variation in time of phenological phases during the three study years at Harbin and Qiqihar is presented in Table 7.

3.6. The Relationship between Climatic Variables and Japonica Rice Growth and Yield

The correlative analysis between environmental variables and rice growth and yield denoted that temperature is the major and significant component in impacting the rice growth over remaining variables. Therefore, climatic changes in Heilongjiang Province majorly referred to the changes in temperature (T_{min} and T_{max}). There was no significant correlation observed between rice growth and precipitation, mentioning that rainfall had not been the main controlling variable to rice yield due to well conditional irrigation facilities, though precipitation during anthesis impacted the flowering at Qiqihar. Therefore, based on the observed results, it is suggested that temperature-based indices over all climatic variables such as GDD and meteorological standard index should be applied in future studies covering NEC to observe the overall relationship analysis. In this study, cold stress events during sensitive growth periods caused chilling injuries which suggests necessarily incorporating chilling injury indices and diurnal variations of the temperature in future climatic-rice studies in NEC, as past studies denoted only one temperature component (T_{min}, T_{max}, or T_{avg}) was considered to observe the temperature variation impacts on rice yield in NEC. The approved methods to evaluate the impacts in past studies were national standard indices, meteorological standards indices, or cumulative temperature indices that can only consider one temperature variable, strongly ignoring the diurnal variations of the temperature. Therefore, the results of this study suggested utilizing a GDD method in evaluation of temperature impacts on critical growth phases and interannual shifts in japonica rice yield in NEC as this method considers different threshold levels.

Both high and low temperature stresses at sensitive growth stages cause injuries to japonica rice. Boosting the high temperature tolerance in rice during sensitive growth stages may prove vital under varying and warming climates. This study provided the evidence that how tolerance comprises several components of escape to high temperature stress: firstly, initiation of panicle emergence, time of spikelet openings against the occurrence of temperature stress during a day, and self-adaptability and absolute tolerance under high temperature stress. The variability of climatic components especially high temperature and impacts on growth at Harbin and Qiqihar provided essential basis for evaluation of impacts of warming on rate of spikelet anthesis. Generally, flowering in both indica and japonica rice varieties occurs over a five-day period, but in Harbin and Qiqihar continued to a 7-d period depending on the cultivars and growing conditions where maximum spikelet anthesis reached around 8 to 9 h. Although the cultivars were the same at both sites, it is worth understanding that the cultivars flowered earlier during the day at Harbin than Qiqihar with more than 95% spikelets by nearly 8.5 h. This observance provided a useful and potential escape mechanism that should be introduced in breeding programs. The daily average temperature and monthly mean precipitation at Harbin and Qiqihar are presented in Figures 7 and 8, respectively.

Table 5. Environmental variables prevailed during grain-filling growth stage in 2018 and 2019 at Harbin and Qiqihar.

Cultivars	Region	Year	T_{avg} (°C)	T_{max} (°C)	T_{min} (°C)	CO_2 (ppm)	RH (%)	Soil Temp. (5 cm) (°C)	Soil Temp. (10 cm) (°C)	Rad. Accum. (MJ/m^2)
Suijing-18	Harbin	2018	20.1	26.1	15.0	407.6	82.8	21.7	20.2	17.6
		2019	19.2	25.4	14.2	386.2	80.1	22.2	20.7	17.9
	Qiqihar	2018	18.3	24.8	12.6	416.1	80.9	19.7	18.3	16.7
		2019	19.1	25.4	13.5	409.7	78.1	21.0	20.3	16.9
Longjing-21	Harbin	2018	20.3	26.0	15.9	402.5	82.7	21.8	20.5	17.7
		2019	21.1	27.8	16.3	385.5	80.4	22.7	21.4	17.9
	Qiqihar	2018	18.7	24.4	13.6	401.5	81.4	19.9	18.7	16.6
		2019	18.9	25.7	14.4	403.9	79.1	20.4	19.3	16.3
Londao-21	Harbin	2018	19.9	25.8	14.0	398.3	83.3	21.2	19.4	16.8
		2019	20.8	26.1	15.1	396.4	81.9	21.7	20.2	16.0
	Qiqihar	2018	16.9	22.5	11.5	402.5	77.3	18.6	17.1	16.8
		2019	17.5	23.6	13.1	438.3	79.1	20.0	18.6	17.1
Longdao-18	Harbin	2018	20.1	25.2	14.9	379.9	83.3	21.4	20.6	17.0
		2019	21.5	26.2	16.3	396.8	81.4	22.1	19.6	17.5
	Qiqihar	2018	17.5	23.9	12.5	401.3	79.8	19.2	17.2	16.5
		2019	18.1	24.3	14.8	417.2	80.6	21.0	19.7	17.1

Table 6. Impact of variation in environmental variables on quality of japonica rice cultivars during study years of 2018 and 2019 at Harbin and Qiqihar (BR = brown rice; FR = fine rice; L-W = length–width; GL = grain length; GW = grain width; a, b, c, d = DMRT test values to differentiate the treatment means in different traits for different cultivars).

Region	Cultivar	Year	Protein (%) mv ± sd * a **	Amylose (%) mv ± sd	BR (%) mv ± sd	FR (%) mv ± sd	L-W Ratio mv ± sd	GL (mm) mv ± sd	GW (mm) mv ± sd	Chalkiness mv ± sd
Harbin	Longdao-18	2018	7.92 ± 0.5 c	18.91 ± 0.1 a	77.63 ± 2.7 a	67.90 ± 2.9 a	2.01 ± 0.07 a	5.12 ± 0.07 a	2.51 ± 0.1 b	1.01 ± 0.07 bc
		2019	7.21 ± 0.6 c	17.90 ± 0.3 a	75.71 ± 2.1 a	68.71 ± 2.2 a	2.20 ± 0.01 a	5.73 ± 0.05 ab	2.71 ± 0.01 b	1.20 ± 0.12 b
	Longdao-21	2018	7.83 ± 0.7 c	18.35 ± 0.5 ab	76.24 ± 5.6 a	65.97 ± 4.2 a	2.02 ± 0.01 a	4.84 ± 0.03 b	2.32 ± 0.02 c	0.52 ± 0.18 c
		2019	7.10 ± 0.2 c	17.53 ± 0.3 ab	74.42 ± 5.2 a	66.53 ± 5.1 a	2.11 ± 0.01 b	5.51 ± 0.02 a	2.62 ± 0.03 c	0.71 ± 0.11 c
	Longjing-21	2018	9.34 ± 0.5 a	16.82 ± 0.8 c	78.95 ± 4.1 a	68.71 ± 3.8 a	1.83 ± 0.03 b	4.62 ± 0.09 c	2.72 ± 0.03 a	2.09 ± 0.4 a
		2019	9.15 ± 0.5 a	15.81 ± 0.7 c	76.12 ± 4.7 a	69.03 ± 3.1 a	1.51 ± 0.01 c	4.41 ± 0.04 b	2.91 ± 0.02 a	2.31 ± 0.10 a
	Suijing-21	2018	8.76 ± 0.6 b	17.77 ± 0.5 b	77.67 ± 3.3 a	68.36 ± 2.0 a	1.93 ± 0.03 b	4.82 ± 0.08 b	2.71 ± 0.04 a	1.10 ± 0.61 bc
		2019	8.37 ± 0.1 b	17.23 ± 0.3 b	75.36 ± 3.4 a	68.75 ± 2.8 a	2.01 ± 0.1 b	5.13 ± 0.09 ab	2.83 ± 0.02 a	1.21 ± 0.21 b

Table 6. *Cont.*

Region	Cultivar	Year	Protein (%) mv ± sd * a **	Amylose (%) mv ± sd	BR (%) mv ± sd	FR (%) mv ± sd	L-W Ratio mv ± sd	GL (mm) mv ± sd	GW (mm) mv ± sd	Chalkiness mv ± sd
Qiqihar	Longdao-18	2018	6.85 ± 0.5 c	17.92 ± 0.1 a	79.62 ± 2.1 a	65.78 ± 2.5 a	1.91 ± 0.02 a	4.30 ± 0.25 ab	2.31 ± 0.02 b	0.81 ± 0.21 c
		2019	6.31 ± 0.5 c	16.30 ± 0.4 ab	78.31 ± 2.8 a	64.71 ± 2.7 a	1.10 ± 0.02 b	4.93 ± 0.28 a	2.90 ± 0.01 a	1.21 ± 0.03 b
	Longdao-21	2018	6.67 ± 0.33 c	17.57 ± 0.4 ab	78.13 ± 6.1 a	63.34 ± 3.1 a	1.92 ± 0.04 a	4.31 ± 0.1 ab	2.31 ± 0.01 b	0.82 ± 0.28 c
		2019	6.23 ± 0.3 c	16.65 ± 0.7 a	77.54 ± 6.1 a	62.39 ± 3.1 a	1.82 ± 0.04 a	4.80 ± 0.11 a	2.63 ± 0.01 b	0.91 ± 0.09 b
	Longjing-21	2018	7.81 ± 0.53 a	15.51 ± 0.9 c	79.49 ± 4.3 a	66.11 ± 3.9 a	1.80 ± 0.02 ab	4.30 ± 0.06 ab	2.53 ± 0.07 a	2.31 ± 0.43 a
		2019	7.33 ± 0.3 b	14.83 ± 0.2 c	78.87 ± 4.3 a	65.10 ± 3.9 a	1.71 ± 0.02 ab	4.32 ± 0.09 b	2.60 ± 0.08 b	2.91 ± 0.07 a
	Suijing-21	2018	7.23 ± 0.6 b	16.82 ± 0.3 b	78.96 ± 4.3 a	65.32 ± 2.1 a	1.73 ± 0.03 b	4.43 ± 0.09 a	2.43 ± 0.06 a	1.21 ± 0.41 b
		2019	7.70 ± 0.6 a	14.10 ± 0.6 c	77.63 ± 4.3 a	64.39 ± 2.1 a	1.93 ± 0.03 a	4.41 ± 0.10 b	2.52 ± 0.09 b	1.53 ± 0.11 b

* mean values ± standard deviation, ** DMRT test values to differentiate the groups of treatment means

Table 7. Impact of variation in environmental variables on time of phenological phases of japonica rice cultivars during study years of 2017, 2018, and 2019 at Harbin and Qiqihar.

Site	Cultivar	Year	Sowing	Transplanting	Emergence	Tillering	Booting	Heading	Grain-Filling	Maturity
Harbin	Longdao-21	2017	4/18	5/24	6/04	6/20	7/27	8/10	8/16	9/18
		2018	4/17	5/18	5/30	6/11	7/20	7/28	8/03	9/20
		2019	4/17	5/16	5/28	6/13	7/25	8/01	8/04	9/23
	Longdao-18	2017	4/18	5/24	6/04	6/19	7/22	8/06	8/12	9/17
		2018	4/17	5/18	5/30	6/11	7/18	7/23	7/31	9/20
		2019	4/17	5/16	5/28	6/13	7/24	7/31	8/06	9/25
	Longjing-21	2017	4/18	5/24	6/04	6/18	7/17	8/04	8/08	9/14
		2018	4/17	5/18	5/30	6/11	7/12	7/22	7/28	9/15
		2019	4/17	5/16	5/28	6/13	7/19	7/24	7/29	9/20
	Suijing-18	2017	4/18	5/24	6/04	6/18	7/16	8/04	8/12	9/15
		2018	4/17	5/18	5/30	6/11	7/15	7/23	8/01	9/15
		2019	4/17	5/16	5/28	6/13	7/18	7/23	7/28	9/20
Qiqihar	Longdao-21	2017	4/17	5/27	6/05	6/27	8/2	8/19	8/24	9/29
		2018	4/17	5/31	6/09	6/21	7/23	8/05	8/12	10/03
		2019	4/17	5/29	6/07	6/21	7/23	7/29	8/05	9/25
	Longdao-18	2017	4/17	5/27	6/05	6/27	7/30	8/19	8/24	9/26
		2018	4/17	5/31	6/09	6/21	7/23	8/05	8/12	10/03
		2019	4/17	5/29	6/07	6/21	7/23	7/29	8/05	9/25
	Longjing-21	2017	4/17	5/27	6/05	6/23	7/29	8/13	8/20	9/22
		2018	4/17	5/31	6/09	6/21	7/19	7/29	8/07	9/27
		2019	4/17	5/29	6/07	6/21	7/19	7/23	7/29	9/17
	Suijing-18	2017	4/17	5/27	6/05	6/23	7/29	8/13	8/20	9/24
		2018	4/17	5/31	6/09	6/21	7/19	7/29	8/07	9/27
		2019	4/17	5/29	6/07	6/21	7/19	7/23	7/29	9/17

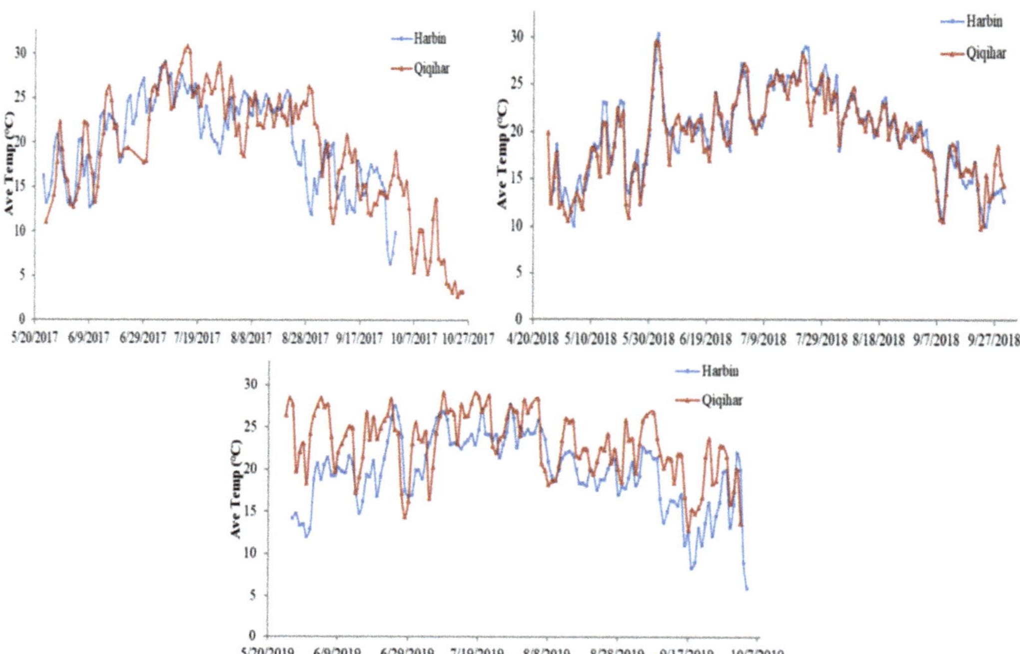

Figure 7. Daily average temperature conditions at Harbin and Qiqihar during rice growing seasons of 2017, 2018, and 2019.

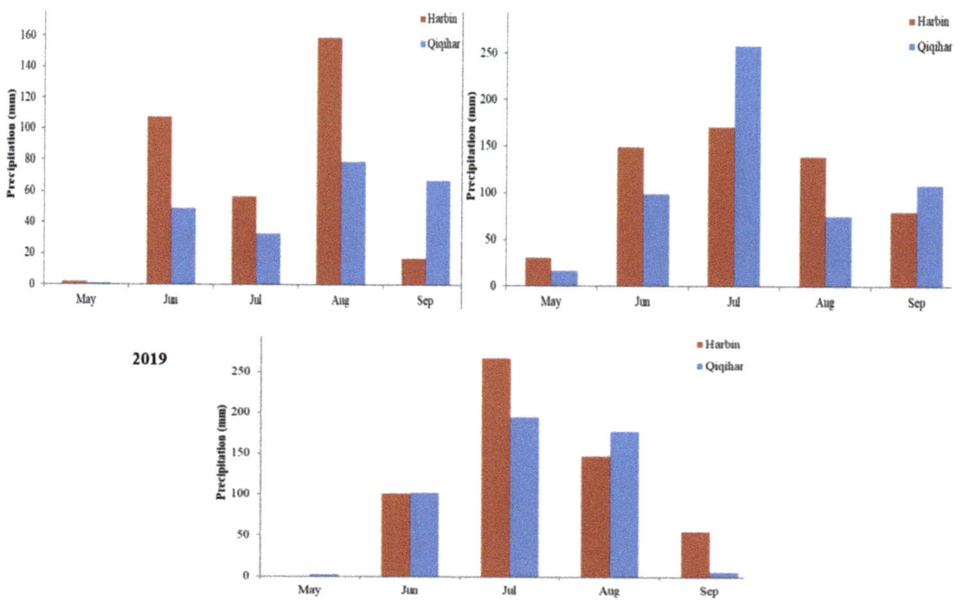

Figure 8. Monthly average precipitation at Harbin and Qiqihar during rice growing seasons of 2017, 2018, and 2019.

At Harbin, the peak anthesis occurred nearly 25–40 min with mean high temperature of 28.4 °C during a day, which presumptively indicated a thermal response mechanism of general rate of development towards spikelet temperature exposure whose optimum temperature might be apparently different. Additionally, the rate of spikelet opening during the day increased proportionally with temperature at both sites, as after 3 d anthesis was observed at a peak at a temperature of 30.8 at Harbin and 30.3 °C at Qiqihar. However, when temperature prevailed above these values, this caused the spikelet opening to reduce by 23% at Harbin when temperature range was above 31 °C and by 36% in Qiqihar when temperature was above than 32 °C. By contrast, warming temperature stress delayed or reduced the spikelet opening during the day at both sites. Although warming did not impact the anthesis so adversely, the number of superior grains during grain-filling was strongly reduced which meant that spikelet sterility in japonica rice at both sites increased with increased days (d) of exposure warming. Spikelet sterility in japonica rice is associated with a smaller number of germinated pollens or low number of overall viable pollens on the stigma; therefore, the warming stress caused acute changes in anther dehiscence before and during anthesis. Thus, it is concluded that temperature had a significant interaction with the rate and duration of spikelet opening.

A significant positive correlation was noticed between temperature variation from transplantation to maturity and japonica rice yield, declaring that the decline in heat during the growing period generally caused the decline in rice yield due to the injuries caused by chilling as seen in Qiqihar during later growth stages. Sterility among panicles due to exposure to cold days showed the strong positive correlation with a number of cold days during anthesis period, inferring that the characterized low temperature events in July caused the reduction in yield due to cold injuries. Precipitation periods at both study sites did not show the same anomaly during both study years.

Generally, it was debated that positive change in rice yield per unit area in NEC happened due to inclusion of improved non-climatic factors, but the correlation between climatic variables and rice yield in this study also demonstrated that rice yield was not happening only due to improvements in technology, but change share could be attributed to the suitable change in climatic components, particularly shifts in temperature. Cold injuries during anthesis caused by a delayed type of chilling due to heavy rainfall during the growing season had significant impacts on japonica rice yield which could decrease with each 1 °C of temperature increase. Frequent prevalence of cold temperature periods during sensitive growth stages such as anthesis and grain-filling in July and August caused serious concerns to the rice yield. Generally, it was observed that every 1 °C rise in temperature anomaly during the early growth period from panicle initiation to booting and late growth period from heading to flowering caused reduction in rice yield. Injuries caused by precipitation and cold periods cannot be ignored as in this study two type of cold injuries during later growth stages in July and August were observed: the sterile type and delayed type. Though shifts in temperature during previous decades favored rice yield and rice land-use, if climate continues to change with the anomaly since the 1970s, it would cause serious threats to rice yield through T_{min} and T_{max} stresses.

Although global warming has had great attributions with increases in temperature in NEC across the last four decades, the extent of variation in temperature (T_{avg}, T_{min}, and T_{max}) indicated variations in the three-year study from transplantation to maturity. Moreover, this study noticed heavy rainfalls once or twice during growing seasons, but historical study trends showed a decline in precipitation, inferring this decline in precipitation may cause threats to rice yield. Therefore, the current study necessarily denoted a few major threats that are needed to be addressed in Heilongjiang Province; firstly, relatively unaccented rise in temperature during sensitive growth phases of japonica rice may threaten the rice growth and yield. Secondly, increase in precipitation during critical growth phases and decline in precipitation during whole growing season may call for serious concerns. Thirdly, no distinctive decrease in cold injuries whether sterile-type or delayed-type chilling injuries during sensitive growth periods may threaten the overall japonica rice productivity.

4. Discussion

4.1. Yield and Yield Components

Plant height is one of the main components contributing to overall biological production. Comparing the plant height of both sites, it was determined that rice plants were taller at Qiqihar than Harbin. The temperature requirement for enhanced aerial growth of japonica rice is 18–33 °C after transplanting. Therefore, plant height usually increases till the heading stage approaches, where the plant ceases its vegetative growth. Plants on both sites increased their aerial growth till the heading stage and showed higher plant height values but if comparing the interaction, it was showing the values were marginally higher at Qiqihar though interaction was non-significant. These results are in accordance with findings reported by [51] who concluded that the enhancement in plant height was steeper under high temperature than normal ambient conditions. During early weeks, plant height increased slowly, but later, it increased more steeply as the ambient temperature was high. Rice grain yield in any given environment is usually determined by yield components (panicle length, productive tillers, and grains per panicle) developed at different phenophases. It was determined that the cultivars grown in a specific environment, the grain yield is impacted by the respective prevailed environmental conditions plant experienced at different growth stages. Rice production systems along an altitude gradient, for example in Heilongjiang Province, have been traditionally graded into three types of altitudes, i.e., low-, mid-, and high-altitude environments. Cultivars specifically chosen according to a region's environment were bred for those environments and well adapted to those areas based on local cropping calendar aiming higher yields. Due to climatic variabilities, there is an executed relationship between cultivars' adaptation and the respective growing environment conditions, since environmental conditions would keep on varying significantly every year, e.g., temperature, intensity, and frequency of precipitation, intensity, and the accumulation of solar radiation may become more intense or mild [8,59]. Thereby, fluctuating environmental conditions may bring in new combinations such as lower or higher temperature, which may cause new combination with pest existence along the altitude (Weerakoon et al. 2008). Moreover, high temperature at anthesis may bring in new combinations of fertility of spikelets or appearance of new pests across the gradient depending on availability of water [60,61]. Thus, the variations in yield and yield components observed at both sites revealed the possible existence of new combinations that supported the increase in yield values or harmed the overall grain yield. Therefore, based on the adaptability mechanisms of japonica rice in terms of yield and yield components, possible adjustive measures are necessarily suggested to optimize the yield loss through adjustments in agronomic practices for example shifts in planting dates for nursery, changes in dates for transplantation or changes in methods and types of external inputs which may lead towards significant shifts in japonica rice production and duration across altitude gradient for its sustainability [62].

Other logical justification for yield variation was growing cultivars not adapted to a specific environment, different from the ones it was adapted for, which increased the risk of whole crop failure or may be risk in production loss and vice versa. The results suggested that yield sustainability in such cases among different environments could be attained with shifts in agronomic management practices through possible adjustments where yield target could be achieved by having plentiful crop production under selectively favorable high-yielding climatic conditions [63,64]. Our results are also in line with Lu et al. (2008) who reported that the changes in yield components and grain yield in different cultivars within a region and among multiple selected regions can be justified by possibility of non-adaptability of a cultivar to a specific environment or may be temperature and precipitation changes on a specific growth stage [65]. The variations in yield and yield components are also supported by other reports which found that cold as well as heat stress can cause spikelet sterility and can disturb the pathways for source-sink in japonica rice [62].

The findings of this study revealed that only temperature does not impact the grain yield for all genotypes among different study sites, rather than shifts that happened due to the combined effects of other environmental components prevailing during the different growth and developmental stages. The results of this study uncovered how different environments acted upon the individual yield component at a respective growth stage, e.g., panicle length enhancement, 1000-grain weight, productive tillers, etc. Based on the findings, it was observed that variation in total number of productive tillers brought changes in overall grain yield where the increased number of productive tillers per hill with fertile spikelets per panicle supported the yield increase. These results are in consistence with [66] who found that an increase in the total number of tillers and reducing the unfertile tillers per hill does not have more positive impact on the yield. However, productive tillers with a high number of fertile spikelets impacted the yield positively, so having a more productive tiller number with a high number of fertile spikelets is most important among yield components to increase the grain yield across different environments and different planting dates [67]. The results of yield components are also supported by research which found that tillers per hill had little influence on the net grain yield, but productive tillers had great impact, as the fertility of tillers was found to be the environment-dependent trait [68,69]. It was observed that grains per panicle could be regarded as the ultimate sink potential, but had less environment dependency and showed more dependence on genetic control [69,70], though an indirect influence of temperature on panicle length was noticed [70]. It was concluded that the number of total filled spikelets is a clearly temperature-dependent trait and influence can only be reduced by avoiding prejudicial environmental conditions.

4.2. Variation in Time of Day of Anthesis (Hasr)

Under a continuously changing global climate, extreme cold or hot stress events are likely to be more frequent in the future depending on the regions where rice will be subjected to untoward abiotic stresses. Therefore, this study suggested the need to improve the resistance against climatic stresses in japonica rice genotypes at reproductive stages, especially during anthesis to get adapted under highly dynamic climatic conditions [71]. Moreover, the results depicted that intensifying the absolute stress tolerance in japonica rice could make it possible to carry out the important physiological processes (such as pollen germination, pollination, anther dehiscence, fertilization) to have a higher rate of spikelet fertility under stressful conditions [72]. The cultivars at Qiqihar took a longer time for anthesis and had longer duration of daily anthesis, which favored higher spikelet fertility which is also reported by [60] who found that anthesis under varying environments might feasibly determine the fertility of spikelets. Temperature prevalence at the study sites was more in the optimum range during the anthesis and preceding events at Harbin than Qiqihar, and less intensity and frequency of precipitation positively influenced the anthesis. Similar results were reported by [60] who indicated that cold responsiveness among cultivars might cause the infertility of the spikelets. Generally, anther dehiscence may affect the number of pollen grains on the stigma [60]. However, the reason behind anther dehiscence at both sites was that anthers still dehisced under stress due to spikelet flowering and poor swelling of pollen grains, which might cause in losing their viability, resulting in unfertilized pollen as reported by [41]. Strong variations were seen regarding onset and end of anthesis between study sites, whereas the duration of anthesis showed less variations. Across two study sites, atmospheric T_{min} averaged over the 7 days preceding any respective anthesis event was the one of the major causes behind almost all variations as observed by [73], whereas higher temperature impacted spikelets to open earlier in the morning, but no significant influence of solar radiation observed on anthesis duration. There was no environmental influence on anthesis time within a site because of insufficient environmental variabilities, but the effects were caused by other factors such as irrigational management practices, fertilizer amendments etc. The results are supported by [73] who observed that reduction in day length by 1 h (or application of a dark treatment before anthesis time) could delay or advance the onset of flowering. Ref. [74] proposed that rice spikelet sterility is influenced by thermal stress majorly at two critical periods, one during microscopic stage at meiosis and the second two weeks later during anthesis when pollination is about to start. The first phase is usually affected by cold or chilling stress but rarely by heat.

4.3. Japonica Rice Quality of Superior and Inferior Grains

Temperature variation more specifically high temperature influence the quality of rice if prevailed during grain-filling phase [75–77]. The rate and extent of grain-filling of japonica rice depends on the arrangement and position of grains in the spikelet and panicle. Mainly the superior grains are positioned at the primary branches that increase their weight due to higher translocation rate. Meanwhile, inferior grains are located at the secondary branches with low and slow translocation rates which make them unsuitable for human consumption [78,79]. The same case has been noticed in the current study where the maximum and average grain-filling rate and grain quality were significantly affected by temperature variations at both experimental locations. The length–width ratio of superior grains at Harbin was powerfully but negatively correlated with the maximum temperature. Our results are in consistence with [80] who observed that higher temperature increased the grain-filling rate but it crumbled the grain weight and quality. Maximum grain weight was also negatively correlated with length-width ratio of superior grains at Harbin site, whereas at Qiqihar the fine rice percentage was positively strongly correlated with the occurrence of maximum temperature. The remaining quality parameters including whole milled rice, chalkiness degree, and length–width ratio were negatively correlated with the prevalence of maximum temperature. However, amylose contents in superior grains had no acquaintance with maximum temperature at Qiqihar, but amylose contents in superior grains were highly positively correlated with the maximum grain weight. These results are in agreement with [81,82], who found that the retention of endosperm starch has been controlled by genetic make-up and environmental factors during progressive plant development. It was observed that the variations in ambient temperature could enhance the apparent amylose contents and bring adjustments in primary structure of starch granules such as crystalline structure and granular shape, thus bringing major changes in the quality of storage starches. Distinctiveness in overall amylose contents depended chiefly on specific rice cultivars, however, it was suggested that such fluctuations in cold weather conditions had role to widened amylose contents in the same cultivar as also reported by [81,82].

It was interpreted that for inferior grains, protein and amylose contents had a strong negative correlation with initial growth phase. In contrast, the same quality indicators were found with strong positive significant relationship with maximum temperature at Harbin. Similarly, the length to width ratio was also negatively correlated with the beingness of maximum temperature for inferior grains at Harbin. Similar result was reported by [81,82] who indicated that amylose contents in rice endosperm were reported to be determined by the ambient temperature at an early development stage (5–15 days after anthesis at 25.8 °C). If temperature variation continues to prevail even during night, then grain will be occupying higher degree of chalkiness [83]; therefore, chalkiness degree was positively correlated with the existence of maximum temperature at the same site in inferior grains. Ref. [84] also described that the induction of heat stress during grain-filling stage among different cultivars crumbled the overall grain quality and grain yield by 53–83%. Perceptibly, amylose contents with the maximum grain weight and protein contents with average grain-filling rate were significantly correlated among inferior grains at Harbin and Qiqihar, respectively. Amylose and protein contents were negatively correlated with the number of days of filling for inferior grains as proved by [47]. Similar to our findings [85], who reported that owing to high temperatures during the ripening phase, abnormal morphology and coloration occur in rice, probably due to decreased enzymatic activity in grain-filling, respiratory ingestion of assimilation products, and decreased sink activity.

4.4. Grain-Filling of Superior and Inferior Grains

Based on the findings of the current study and limited knowledge, the grain-filling growth phase was positively correlated with the environmental conditions at Harbin and negatively correlated at Qiqihar because stressful environmental conditions that prevailed during the grain-filling period were fractionally imprudent. Dry weight accumulation for superior as well as inferior grains was elevated in Longdao-21 followed by Longjing-21 and Suijing-18. Longdao-18 had attenuated values for dry weight accumulation as the rate of grain growth was faster and therefore, grain-filling period was shorter as higher temperatures approached [51,86]. Our results are consistent with findings of [49] who reported that the high temperature at flowering and grain-filling reduced grain yield through spikelet sterility and shortened grain-filling period. For a specific cultivar, the GDD necessary for flowering were comparatively assonant at different growing temperatures within the temperature approximate range between base and optimum temperatures. The findings also confirmed that grain weight accumulation for inferior grains was relatively low at Qiqihar probably due to the anticipated fluctuations in temperature at the grain-filling phase. In contrast, dry weight accumulation at Harbin was comparatively higher. These results are in line with [51] who reported that high temperature encouraged the ripening of grains and shortened the grain-filling phase. Ref. [87] explicated that the high temperature negatively impacted the rate of photosynthesis through diminution in root activities. It has also been observed previously that a high temperature at the flowering and grain-filling phase reduced net grain yield by enhancing spikelet sterility and shortening the grain-filling period [49,88]. The findings of this study illustrated where the prevalence of environmental conditions, i.e., solar radiation and temperature were in optimal range during flowering and grain-filling, the grain-filling rate, and duration potentiated there. It has been observed previously that duration of grain-filling phase directly depends on optimal solar radiation and temperature which determines the final grain yield [89].

4.5. Prevailed Environmental Components and Different Growth Phases

The possible forthcoming menace to japonica rice production and quality is changes in climate that will impact rice morphology, growth, physiology, biology, and ultimately, causing serious food security threats [50]. The relationship between environmental variables and growth phases of japonica rice excavated major alterations at both experimental sites. During early crop stages, average temperature, sunshine hours, solar radiation, and relative humidity had little impact on the initial growth cycle (from transplanting to booting). But these factors exerted negative correlation with the initial specific growth period of the japonica rice. However, in current study, the second half growth period of the rice plant is mostly affected by the prevailing environmental factors at both sites. Soil temperature at different depths, average sunshine hours, and daily radiation had strong negative correlation from booting to maturity at Harbin, whereas relative humidity was positively correlated with the later crop stages at Harbin. In contrast, average sunshine hours significantly bestowed ($r = 0.958^*$) from booting to maturity at Qiqihar along with daily radiation accumulation that was statistically insignificant. These findings are consistent with [90] who demonstrated that fluctuations in day and night temperatures and other environmental components impacted growth, yield and yield contributing components, and quality due to higher temperature stress and also affected physiological processes. Japonica grain quality became poorer when either higher day or night temperatures were applied to the panicle or the whole plant. The logical reason behind a decrease in the grain quality due to high night temperature was not because of the deficiency of carbohydrates in the leaves and the culms, as exposing the vegetative parts of the plant to this temperature did not reduce the quality of rice grain [91].

4.6. Impacts of Environmental Factors on Specific Leaf Area and Crop Growth Rate

One of the main measurements to note the crop photosynthesis is leaf area measurement. At different growth stages, it was aimed to brief the changing relationship between crop growth and leaf area development among different japonica rice cultivars grown under contrasting environments. The results of current study are supported by [92] who concluded that the temperature-dependent processes in leaf area development such as appearance and elongation of the leaf responded positively to high temperatures at different growth stages. However, as higher temperature continued to prevail till the sensitive growth stage such as flowering, the biomass production was reduced because of combined effects of other environmental components for example radiation interception by the plant and its absorption efficiency. Leaf area was higher at Harbin than Qiqihar and the decreasing trend after heading at Qiqihar was steeper. Temperature stress, either cold or heat, impacted the vegetative as well as reproductive growth stages and brought changes in a specific growth phase. It has been reported previously that temperature variation caused a decrease in leaf area and total dry matter accumulation [93,94]. The results of this study are in agreement with findings of [95] who demonstrated that leaf area development and maturity of crop strongly depended on temperature fluctuations, and variations in altitudes. Crop duration strongly influenced by changes in temperature and altitude, and seasonal mean temperature varied due to the altitudinal temperature gradient by 7 °C per km at 60% air humidity.

Increased leaf area was probably due to constant relative humidity, and inversions in day and night temperatures at a specific study site. Rendering that plant growth is compelled by photosynthetic carbon fixation during the daytime [96], higher growth rates could probably occur under higher day temperature, since maximum assimilation rates for japonica rice were in the range of 30–35 °C regardless of the growing temperature [94]. In addition, the process of respiration increased under higher night temperatures, which devoured a large quantity of daily available assimilates, therefore limiting the biomass accumulation [97]. In contrast, under semi-arid environments, leaf area development [98] and stomatal conductance [99] of japonica rice were observed to be strongly positively correlated with night temperature.

It was noticed that shifts in day and night temperatures solely did not significantly impact the crop growth rate and total dry matter, but had a significant effect on the zoning between plant organs and leaf area development. Leaf area development and total plant dry matter were higher under high night temperature which supports our findings of increased in leaf area development under high night temperature in duration of constant relative humidity. Under field conditions, relative humidity in the night was usually closer to 100% and considerably declined during the day. In temperature-controlled or greenhouses or growth chambers, diurnal relative humidity often showed less fluctuations, and even though the absolute quantity of water in air remains constant, relative humidity proportionally decreased with increasing temperature. Therefore, findings of this study indicated that leaf area development responded to temperature applicable only to field conditions may not be applicable to controlled conditions. Figure 9 represents the conceptual conclusions of the study conducted at Harbin and Qiqihar of NEC.

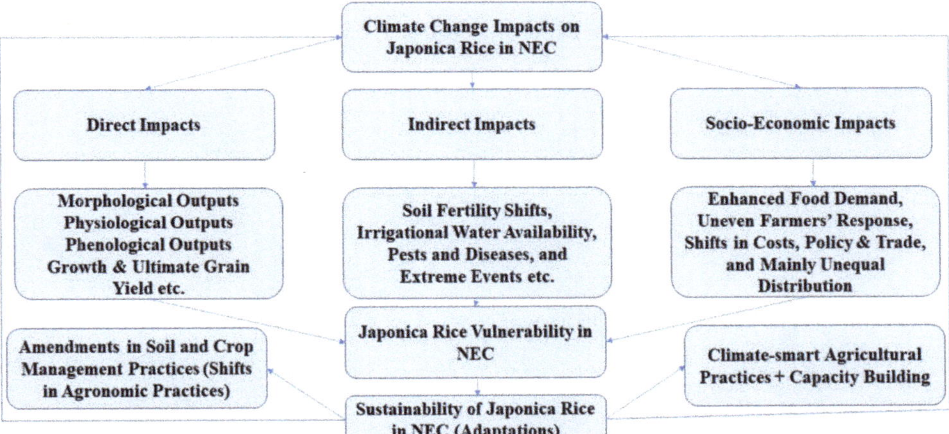

Figure 9. Conceptual conclusive remarks for future sustainability of japonica rice in NEC based on 3-year experiment (2017–2019).

Relative humidity not only influenced the plant growth response to shifts in temperature, but also had a strong direct impact on crop growth rate and leaf area. High humidity during the day light period combined with low humidity during the night dark period resulted in higher crop growth rate than in other possible combinations of low and high, day and night relative humidity, but in general a positive impact of higher relative humidity on crop growth rate has been reported [100]. SLA was not only highly affected by water availability but also by relative humidity across both sites. Although a strong positive correlation was noticed between SLA and shifts in night temperature, it was retracted that the relationship between changes in day and night temperature rather than night temperature itself authorized SLA. Findings of this study are supported by previous study who showed that low soil temperatures especially in rootzone decreased SLA [101], whereas another study showed a decrease in SLA under high night temperature [102]. Contrastingly, another study reported a strong positive correlation between SLA and temperature where an increase in SLA was seen with increase in temperature amplitude especially high day time temperature [103]. Moreover, it has been reasoned that SLA started to decrease when leaf expansion was more affected by variations in environmental factors rather than photosynthesis. Under warm and humid days, SLA started to decrease, and any decrease in SLA might be owing to low area development rate during the night, triggered by low temperature, low water availability. and relative humidity, whereas, based on the collected results, Figure 10 represents the clear two-dimensional visualization of the environment-by-trait table encoded as a grid of colored cells to understand the similarities between different environments and traits.

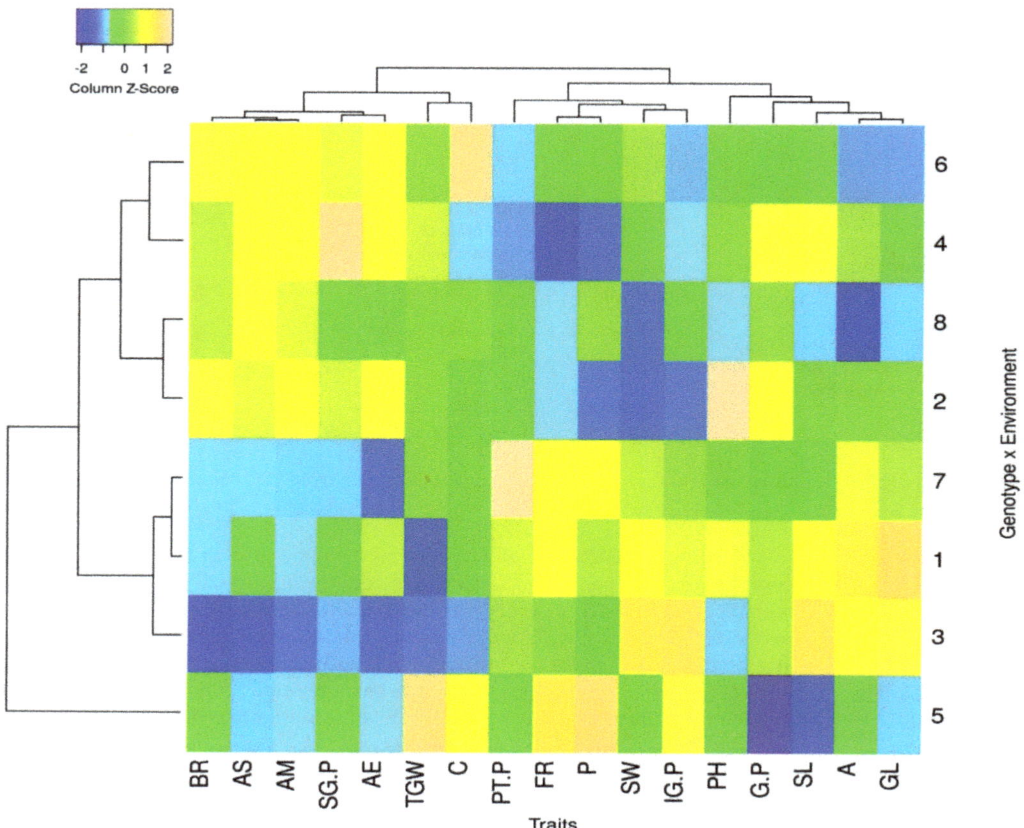

Figure 10. Two-dimensional visualization of environment-by-trait table to differentiate the similarities between the environments and traits (PT.P: productive tillers/plant, G.P: grains/panicle, TGW: 1000-grain weight (g), PH: plant height (cm), SW: spike weight (g), SL: spike length (cm), IG.P: inferior grains/panicle, SG.P: superior grains/panicle, GY: grain yield, BR: brown rice (%), FR: fine rice (%), GL: grain length (mm), AS: anthesis start (hasr), AM: anthesis mid (hasr), AE: anthesis end (hasr), C: chalkiness, A: amylose (%), P: protein (%), 1, 2: Longdao-18 at Harbin and Qiqihar, respectively, 3, 4: Longdao-21 at Harbin and Qiqihar, respectively, 5, 6: Longjing-21 at Harbin and Qiqihar, respectively, 7, 8: Suijing-18 at Harbin and Qiqihar, respectively).

5. Conclusions

This research provides indications of strong impacts of variations in weather, notably the effects on critical growth phases such as the grain-filling stage and time of day of anthesis in japonica rice. The current study evidenced the correlative potent impacts of varying environmental conditions on different growth stages of japonica rice and ultimately effects on grain-filling of superior and inferior grains, anthesis, yield and yield components, and quality of rice. This research provides an adaptive value, especially for scenarios of global warming, where heat induces a great spikelet sterility expectedly to be a major constraint to high net grain yield. Taking as another factor for escape from climatic stresses, the collective and aggregated duration of flowering at the panicle stage and quantification of population was also done, denoting a risk diffusing mechanism.

Adaptability mechanism of japonica rice observed on anthesis depicts those prevailing high temperatures on the start of the day at earlier anthesis intensified the exceedance of escape from even the higher temperature stress later during that day. Humid environmental conditions on earlier anthesis made the rice plant capable of potentially escaping from higher ambient temperature late during the day. The observed variations in the phenology of japonica rice rendered that those cultivars transplanted earlier produced a higher net yield, and provided the positive correlation between yield and transplantation, i.e., the earlier the transplantation, the taller plants will be and higher the net grain yield will be. Moreover, undergoing the correlation between transplantation dates and net grain yield, cultivars with earlier transplantation dates escaped well from the high precipitation and low temperature stresses during later growth stages such as anthesis. Short duration cultivars are recommended in Heilongjiang specifically to avoid the low temperature stress periods on later growth stages majorly on anthesis and grain-filling. Models that predict the temperature-based panicle sterility in rice are necessarily needed in future research focusing NEC to abstract the temporal malleability of the anthesis process along with precise simulation of spikelet temperature during the critical growth phases of japonica rice.

Different trends for air and soil temperature, sunshine, and precipitation impacted the phenological variables and ultimately had impacts on the growth and production of early and late maturing cultivars. Since the phenological variables of rice are mainly controlled by climatic components and management practices, better adaptation through shifts in management practices should be encouraged which is majorly controlled by farmers. Using NEC's data of weather variables and rice production in current study, the positive and negative correlative responses of japonica rice to environmental variables were empirically identified. Adverse impacts of abnormal weather may invite the changes in soil fertility at a specific growth stage, therefore motivation for incorporation of management measures based on climate smart agriculture is necessary to avoid the worsened impacts on production. The abnormalities in temperature may lead to a shortage of inputs (such as labor), impacting the rice production. Thus, in summary both direct and indirect impacts of climatic variabilities on japonica rice yield cannot be ignored.

Aiming to sustain the future japonica rice production, awareness of climate-smart agriculture and optimized use of inputs is necessary. Strengthening the technological programs to offset the negative impacts of climate variabilities is indispensable. Pre- and post-disaster measures taken by relevant local authorities are necessary by rationalizing the optimized japonica rice farming. It is also proposed that more advanced statistical techniques for deep studies integrated with mechanized approaches should be explored for deeper investigations of impacts of climatic variables on different growth stages. Additionally, based on the observed results, it is suggested that temperature-based indices over all climatic variables such as GDD, meteorological standard indices, etc. should be applied in future climate-rice studies in NEC to observe the relationship analysis. Cold periods during critical growth phases caused chilling injuries and yield decline which suggested to necessarily have chilling injury indices in future research, but previous research in NEC denoted that only one temperature component (T_{min}, T_{max}, or T_{avg}) was considered to observe the temperature variation impacts on rice yield due to traditional evaluation methods. The most commonly and approved methods to evaluate the impacts in past studies were national standard indices, meteorological standards indices, or cumulative temperature indices etc. that can only consider one temperature variable ignoring the diurnal variations of the temperature. Therefore, the results of this study suggested the need of have modern GDD methods in evaluation of temperature impacts on critical growth phases and to have interannual shifts in japonica rice yield in NEC because these methods consider different threshold levels of the environmental variables such as temperature.

Author Contributions: Data collection and data analysis of the research was done by first and second authors M.S.F. and A.G., respectively. The third author A.L.V. helped in the write up process of the article, whereas the mentor of the research group and corresponding author of this article Y.X. did the financial arrangements for the conduction of the field trials. Moreover, he helped for necessary revisions in the draft. All authors have read and agreed to the published version of the manuscript.

Funding: This study was financially supported by the Climate Change Adaptation Lab of Institute of Environment and Sustainable Development in Agriculture (IEDA), Chinese Academy of Agricultural Sciences (CAAS).

Institutional Review Board Statement: Not applicable.

Informed Consent Statement: Not applicable.

Acknowledgments: The authors would like to express their gratitude for technical support by the field staff and workers at the research sites of Heilongjiang Academy of Agricultural Sciences (HAAS) at Harbin and Qiqihar.

Conflicts of Interest: The authors declare no conflict of interest.

References

1. Kaufmann, R.K.; Zhou, L.; Tucker, C.J.; Slayback, D.; Shabanov, N.V.; Myneni, R.B. Ipcc Ar5. *Report* **2002**, *107*, 1535. [CrossRef]
2. Piao, S.; Ciais, P.; Huang, Y.; Shen, Z.; Peng, S.; Li, J.; Fang, J. The impacts of climate change on water resources and agriculture in China. *Nature* **2010**, *467*, 43–51. [CrossRef] [PubMed]
3. Liu, Y.; Wang, D.; Gao, J.; Deng, W. Land Use/Cover Changes, the Environment and Water Resources in Northeast China. *Environ. Manage* **2005**, *36*, 691–701. [CrossRef] [PubMed]
4. Samol, P.; Umponstira, C.; Klomjek, P.; Thongsanit, P. Responses of Rice Yield and Grain Quality to High Temperature in Open-Top Chamber to Predict Impact of Future Global Warming in Thailand. *Aust. J. Crop. Sci.* **2015**, *9*, 886–894.
5. Masutomi, Y.; Takahashi, K.; Harasawa, H.; Matsuoka, Y. Impact Assessment of Climate Change on Rice Production in Asia in Comprehensive Consideration of Process/Parameter Uncertainty in General Circulation Models. *Agric. Ecosyst. Environ.* **2009**, *131*, 281–291. [CrossRef]
6. Asseng, S.; Ewert, F.; Martre, P.; Rötter, R.P.; Lobell, D.B.; Cammarano, D.; Kimball, B.A.; Ottman, M.J.; Wall, G.W.; White, J.W.; et al. Rising Temperatures Reduce Global Wheat Production. *Nat. Clim. Chang.* **2015**, *5*, 143–147. [CrossRef]
7. Paterson, R.R.M.; Lima, N. Further Mycotoxin Effects from Climate Change. *Food Res. Int.* **2011**, *44*, 2555–2566. [CrossRef]
8. Wassmann, R.; Jagadish, S.V.K.; Heuer, S.; Ismail, A.; Redona, E.; Serraj, R.; Singh, R.K.; Howell, G.; Pathak, H.; Sumfleth, K. *Chapter 2 Climate Change Affecting Rice Production. The Physiological and Agronomic Basis for Possible Adaptation Strategies*, 1st ed.; Elsevier Inc.: Amsterdam, The Netherlands, 2009; Volume 101. [CrossRef]
9. Pramanick, B.; Brahmachari, K.; Kar, S.; Mahapatra, B.S. Can Foliar Application of Seaweed Sap Improve the Quality of Rice Grown under Rice–Potato–Greengram Crop Sequence with Better Efficiency of the System? *J. Appl. Phycol.* **2020**, *32*, 3377–3386. [CrossRef]
10. Li, Z.; Tang, H.; Yang, P.; Wu, W.; Chen, Z.; Zhou, Q.; Zhang, L.; Zou, J. Spatio-Temporal Responses of Cropland Phenophases to Climate Change in Northeast China. *J. Geogr. Sci.* **2012**, *22*, 29–45. [CrossRef]
11. Wang, E.; Martre, P.; Zhao, Z.; Ewert, F.; Maiorano, A.; Rötter, R.P.; Kimball, B.A.; Ottman, M.J.; Wall, G.W.; White, J.W.; et al. The Uncertainty of Crop Yield Projections Is Reduced by Improved Temperature Response Functions. *Nat. Plants* **2017**, *3*, 1–13. [CrossRef]
12. Andronova, N.G.; Schlesinger, M.E. Causes of Global Temperature Changes during the 19th and 20th Centuries. *Geophys. Res. Lett.* **2000**, *27*, 2137–2140. [CrossRef]
13. Crowley, T.J. Causes of Climate Change over the Past 1000 Years. *Science* **2000**, *289*, 270–277. [CrossRef] [PubMed]
14. Kang, S.; Eltahir, E.A.B. North China Plain Threatened by Deadly Heatwaves Due to Climate Change and Irrigation. *Nat. Commun.* **2018**, *9*, 1–9. [CrossRef] [PubMed]
15. Xia, T.; Wu, W.B.; Zhou, Q.B.; Yu, Q.Y.; Verburg, P.H.; Yang, P.; Lu, Z.J.; Tang, H.J. Spatio-Temporal Changes in the Rice Planting Area and Their Relationship to Climate Change in Northeast China: A Model-Based Analysis. *J. Integr. Agric.* **2014**, *13*, 1575–1585. [CrossRef]
16. Peng, S.; Tang, Q.; Zou, Y. Current Status and Challenges of Rice Production in China. *Plant Prod. Sci.* **2009**, *12*, 3–8. [CrossRef]
17. Tong, C.; Hall, C.A.S.; Wang, H. Land Use Change in Rice, Wheat and Maize Production in China (1961–1998). *Agric. Ecosyst. Environ.* **2003**, *95*, 523–536. [CrossRef]
18. Kedra, M. Regional Response to Global Warming: Water Temperature Trends in Semi-Natural Mountain River Systems. *Water* **2020**, *12*, 283. [CrossRef]
19. Muller, A.; Jawtusch, J.; Gattinger, A. Mitigating Greenhouse Gases in Agriculture; Brot für die Welt: 2011; Volume 88. Available online: https://orgprints.org/id/eprint/19989/1/gatti.pdf (accessed on 2 October 2021).
20. Reay, D.; Sabine, C.; Smith, P.; Hymus, G. *Intergovernmental Panel on Climate Change*; Fourth Assessment Report Geneva, Switzerland: Inter-Gov-Ernmental Panel on Climate Change; Cambridge University Press: Cambridge, UK, 2007. [CrossRef]
21. Yu, Q.; Wu, W.; You, L.; Zhu, T.; van Vliet, J.; Verburg, P.H.; Liu, Z.; Li, Z.; Yang, P.; Zhou, Q.; et al. Assessing the Harvested Area Gap in China. *Agric. Syst.* **2017**, *153*, 212–220. [CrossRef]
22. Sakamoto, T.; Van Nguyen, N.; Ohno, H.; Ishitsuka, N.; Yokozawa, M. Spatio-Temporal Distribution of Rice Phenology and Cropping Systems in the Mekong Delta with Special Reference to the Seasonal Water Flow of the Mekong and Bassac Rivers. *Remote Sens. Environ.* **2006**, *100*, 1–16. [CrossRef]

23. Li, T.; Hasegawa, T.; Yin, X.; Zhu, Y.; Boote, K.; Adam, M.; Bregaglio, S.; Buis, S.; Confalonieri, R.; Fumoto, T.; et al. Uncertainties in Predicting Rice Yield by Current Crop Models under a Wide Range of Climatic Conditions. *Glob. Chang. Biol.* **2015**, *21*, 1328–1341. [CrossRef]
24. Dong, J.; Xiao, X.; Zhang, G.; Menarguez, M.A.; Choi, C.Y.; Qin, Y.; Luo, P.; Zhang, Y.; Moore, B. Northward Expansion of Paddy Rice in Northeastern Asia during 2000–2014. In *Geophysical Research Letters*; Blackwell Publishing Ltd.: Hoboken, NJ, USA, 2016; pp. 3754–3761. [CrossRef]
25. Kontgis, C.; Schneider, A.; Ozdogan, M. Mapping Rice Paddy Extent and Intensification in the Vietnamese Mekong River Delta with Dense Time Stacks of Landsat Data. *Remote Sens. Environ.* **2015**, *169*, 255–269. [CrossRef]
26. Lobell, D.B.; Schlenker, W.; Costa-Roberts, J. Climate Trends and Global Crop Production since 1980. *Science* **2011**, *333*, 616–620. [CrossRef] [PubMed]
27. Liu, L.; Wang, E.; Zhu, Y.; Tang, L. Contrasting Effects of Warming and Autonomous Breeding on Single-Rice Productivity in China. *Agric. Ecosyst. Environ.* **2012**, *149*, 20–29. [CrossRef]
28. Tao, F.; Zhang, Z.; Shi, W.; Liu, Y.; Xiao, D.; Zhang, S.; Zhu, Z.; Wang, M.; Liu, F. Single Rice Growth Period Was Prolonged by Cultivars Shifts, but Yield Was Damaged by Climate Change during 1981-2009 in China, and Late Rice Was Just Opposite. *Glob. Chang. Biol.* **2013**, *19*, 3200–3209. [CrossRef] [PubMed]
29. Iizumi, T.; Ramankutty, N. How Do Weather and Climate Influence Cropping Area and Intensity? In *Global Food Security*; Elsevier B.V.: Amsterdam, The Netherlands, 2015; pp. 46–50. [CrossRef]
30. Gao, J.; Liu, Y. Climate Warming and Land Use Change in Heilongjiang Province, Northeast China. *Appl. Geogr.* **2011**, *31*, 476–482. [CrossRef]
31. Baker, J.T.; Allen, L.H.J. 1 Temperature Treatments as Daytime J Nighttime Bulb Air Temperature. *J. Agric. Meteorol.* **1993**, *48*, 575–582. [CrossRef]
32. Rasul, G.; Chaudhry, Q.Z.; Mahmood, A.; Hyder, K.W. Effect of Temperature Rise on Crop Growth & Productivity. *Pakistan J. Meteorol.* **2011**, *8*, 53–62.
33. Sheehy, J.E.; Mitchell, P.L.; Ferrer, A.B. Decline in Rice Grain Yields with Temperature: Models and Correlations Can Give Different Estimates. *Field Crop. Res.* **2006**, *98*, 151–156. [CrossRef]
34. Krishnan, P.; Swain, D.K.; Chandra Bhaskar, B.; Nayak, S.K.; Dash, R.N. Impact of Elevated CO2 and Temperature on Rice Yield and Methods of Adaptation as Evaluated by Crop Simulation Studies. *Agric. Ecosyst. Environ.* **2007**, *122*, 233–242. [CrossRef]
35. De Souza, N.M.; Marschalek, R.; Sangoi, L.; Weber, F.S. Spikelet Sterility in Rice Genotypes Affected by Temperature at Microsporogenesis. *Rev. Bras. Eng. Agric. Ambient.* **2017**, *21*, 817–821. [CrossRef]
36. Counce, P.A.; Keisling, T.C.; Mitchell, A.J. A Uniform, Objective, and Adaptive System for Expressing Rice Development. *Crop Sci.* **2000**, *40*, 436–443. [CrossRef]
37. Samejima, H.; Kikuta, M.; Katura, K.; Menge, D.; Gichuhi, E.; Wainaina, C.; Kimani, J.; Inukai, Y.; Yamauchi, A.; Makihara, D. A Method for Evaluating Cold Tolerance in Rice during Reproductive Growth Stages under Natural Low-Temperature Conditions in Tropical Highlands in Kenya. *Plant Prod. Sci.* **2020**, *23*, 466–476. [CrossRef]
38. Farrell, T.C.; Fox, K.M.; Williams, R.L.; Fukai, S. Genotypic Variation for Cold Tolerance during Reproductive Development in Rice: Screening with Cold Air and Cold Water. *Field Crop. Res.* **2006**, *98*, 178–194. [CrossRef]
39. Mamun, E.; Alfred, S.; Cantrill, L.; Overall, R.; Sutton, B. Effects of Chilling on Male Gametophyte Development in Rice. *Cell Biol. Int.* **2006**, *30*, 583–591. [CrossRef] [PubMed]
40. Peterson, M.; Lin, S.; Jones, D.; Rutger, J. Cool Night Temperatures Cause Sterility in Rice. *Calif. Agric.* **1974**, *28*, 12–14.
41. Zeng, Y.; Zhang, Y.; Xiang, J.; Uphoff, N.T.; Pan, X.; Zhu, D. Effects of Low Temperature Stress on Spikelet-Related Parameters during Anthesis in Indica–Japonica Hybrid Rice. *Front. Plant Sci.* **2017**, *8*, 1350. [CrossRef] [PubMed]
42. Van Oort, P.A.J.; Saito, K.; Zwart, S.J.; Shrestha, S. A Simple Model for Simulating Heat Induced Sterility in Rice as a Function of Flowering Time and Transpirational Cooling. *Field Crop. Res.* **2014**, *156*, 303–312. [CrossRef]
43. Wassmann, R.; Jagadish, S.V.K.; Sumfleth, K.; Pathak, H.; Howell, G.; Ismail, A.; Heuer, S. Regional Vulnerability of Climate Change Impacts on Asian Rice Production and Scope for Adaptation. *Adv. Agron.* **2009**, *102*, 91–133.
44. Maruyama, D.; Hamamura, Y.; Takeuchi, H.; Susaki, D.; Nishimaki, M.; Kurihara, D.; Kasahara, R.D.; Higashiyama, T. Independent Control by Each Female Gamete Prevents the Attraction of Multiple Pollen Tubes. *Dev. Cell* **2013**, *25*, 317–323. [CrossRef]
45. Tang, L.; Zhu, Y.; Hannaway, D.; Meng, Y.; Liu, L.; Chen, L.; Cao, W. RiceGrow: A Rice Growth and Productivity Model. *NJAS Wageningen J. Life Sci.* **2009**, *57*, 83–92. [CrossRef]
46. Wang, F.; Peng, S.B. Yield Potential and Nitrogen Use Efficiency of China's Super Rice. *J. Integr. Agric.* **2017**, *16*, 1000–1008. [CrossRef]
47. Yang, J.; Zhang, J. Grain-Filling Problem in "super" Rice. *J. Exp. Bot.* **2010**, *61*, 1–5. [CrossRef] [PubMed]
48. Zhou, Q.; Ju, C.X.; Wang, Z.Q.; Zhang, H.; Liu, L.J.; Yang, J.C.; Zhang, J.H. Grain Yield and Water Use Efficiency of Super Rice under Soil Water Deficit and Alternate Wetting and Drying Irrigation. *J. Integr. Agric.* **2017**, *16*, 1028–1043. [CrossRef]
49. Tian, X.H.; Matsui, T.; Li, S.H.; Lin, J.C. High Temperature Stress on Rice Anthesis: Research Progress and Prospects. *Chin. J. Appl. Ecol.* **2007**, *18*, 2632–2636.
50. Krishnan, P.; Ramakrishnan, B.; Reddy, K.R.; Reddy, V.R. *High-Temperature Effects on Rice Growth, Yield, and Grain Quality*, 1st ed.; Elsevier Inc.: Amsterdam, The Netherlands, 2011; Volume 111. [CrossRef]

51. Oh-e, I.; Saitoh, K.; Kuroda, T. Effects of High Temperature on Growth, Yield and Dry-Matter Production of Rice Grown in the Paddy Field. *Plant Prod. Sci.* **2007**, *10*, 412–422. [CrossRef]
52. Kobata, T.; Uemuki, N. High Temperatures during the Grain-Filling Period Do Not Reduce the Potential Grain Dry Matter Increase of Rice. *Agron. J.* **2004**, *96*, 406. [CrossRef]
53. Fábián, A.; Sáfrán, E.; Szabó-Eitel, G.; Barnabás, B.; Jäger, K. Stigma Functionality and Fertility Are Reduced by Heat and Drought Co-Stress in Wheat. *Front. Plant Sci.* **2019**, *10*, 244. [CrossRef]
54. Barnabás, B.; Jäger, K.; Fehér, A. The Effect of Drought and Heat Stress on Reproductive Processes in Cereals. *Plant. Cell Environ.* **2007**, *31*, 11–38. [CrossRef]
55. Alghabari, F.; Ihsan, M.Z. Effects of Drought Stress on Growth, Grain Filling Duration, Yield and Quality Attributes of Barley (*Hordeum vulgare* L.). *Bangladesh J. Bot.* **2018**, *47*, 421–428. [CrossRef]
56. Mohanty, S.; Wassmann, R.; Nelson, A.; Moya, P.; Jagadish, S. Rice and Climate Change: Significance for Food Security and Vulnerability. *IRRI Discuss. Pap. Ser.* **2013**, *49*, 1–14.
57. Challinor, A.J.; Wheeler, T.R.; Craufurd, P.Q.; Ferro, C.A.T.; Stephenson, D.B. Adaptation of Crops to Climate Change through Genotypic Responses to Mean and Extreme Temperatures. *Agric. Ecosyst. Environ.* **2007**, *119*, 190–204. [CrossRef]
58. Zhu, Q.; Cao, X.; Luo, Y. Growth analysis on the process of grain filling in rice. *Acta Agron. Sin.* **1988**, *14*, 182–193.
59. Meehl, G.A.; Stocker, T.F.; Collins, W.D.; Friedlingstein, P.; Gaye, T.; Gregory, J.M.; Kitoh, A.; Knutti, R.; Murphy, J.M.; Noda, A.; et al. *Global Climate Projections*; Cambridge University Press: Cambridge, UK, 2007.
60. Rang, Z.W.; Jagadish, S.V.K.; Zhou, Q.M.; Craufurd, P.Q.; Heuer, S. Effect of High Temperature and Water Stress on Pollen Germination and Spikelet Fertility in Rice. *Environ. Exp. Bot.* **2010**, *70*, 58–65. [CrossRef]
61. Kocmánková, E.; Trnka, M.; Juroch, J.; Dubrovský, M.; Semerádová, D.; Možný, M.; Žalud, Z. Impact of Climate Change on the Occurrence and Activity of Harmful Organisms. *Plant Protect. Sci.* **2009**, *45*, S48. [CrossRef]
62. Shrestha, S.; Asch, F.; Dingkuhn, M.; Becker, M. Cropping Calendar Options for Rice–Wheat Production Systems at High-Altitudes. *Field Crop. Res.* **2011**, *121*, 158–167. [CrossRef]
63. Acuña, T.L.B.; Lafitte, H.R.; Wade, L.J. Genotype × Environment Interactions for Grain Yield of Upland Rice Backcross Lines in Diverse Hydrological Environments. *Field Crop. Res.* **2008**, *108*, 117–125. [CrossRef]
64. Peng, S.; Bouman, B.; Visperas, R.M.; Castañeda, A.; Nie, L.; Park, H.K. Comparison between Aerobic and Flooded Rice in the Tropics: Agronomic Performance in an Eight-Season Experiment. *Field Crop. Res.* **2006**, *2–3*, 252–259. [CrossRef]
65. Bajracharya, J.; Rana, R.B.; Gauchan, D.; Sthapit, B.R.; Jarvis, D.I.; Witcombe, J.R. Rice Landrace Diversity in Nepal. Socio-Economic and Ecological Factors Determining Rice Landrace Diversity in Three Agro-Ecozones of Nepal Based on Farm Surveys. *Genet. Resour. Crop Evol.* **2010**, *57*, 1013–1022. [CrossRef]
66. Ao, H.; Peng, S.; Zou, Y.; Tang, Q.; Visperas, R.M. Reduction of Unproductive Tillers Did Not Increase the Grain Yield of Irrigated Rice. *Field Crop. Res.* **2010**, *116*, 108–115. [CrossRef]
67. Moradpour, S.; Koohi, R.; Babaei, M.; Khorshidi, M.G. Effect of Planting Date and Planting Density on Rice Yield and Growth Analysis (Fajr Variety). *Int. J. Agric. Crop Sci.* **2013**, *5*, 267.
68. Zhu, J.; Zhou, Y.; Liu, Y.; Wang, Z.; Tang, Z.; Yi, C.; Tang, S.; Gu, M.; Liang, G. Fine Mapping of a Major QTL Controlling Panicle Number in Rice. *Mol. Breed.* **2011**, *27*, 171–180. [CrossRef]
69. Akinwale, M.G.; Gregorio, G.; Nwilene, F.; Akinyele, B.O.; Ogunbayo, S.A.; Odiyi, A.C. Heritability and Correlation Coefficient Analysis for Yield and Its Components in Rice (*Oryza sativa* L.). *Afr. J. Plant Sci.* **2011**, *5*, 207–212.
70. Kovi, M.R.; Bai, X.; Mao, D.; Xing, Y. Impact of Seasonal Changes on Spikelets per Panicle, Panicle Length and Plant Height in Rice (*Oryza sativa* L.). *Euphytica* **2011**, *179*, 319–331. [CrossRef]
71. Matsui, T.; Omasa, K.; Horie, T. Comparison between Anthers of Two Rice (*Oryza saliva* L.) Cultivars with Tolerance to High Temperatures at Flowering or Susceptibility. *Plant Prod. Sci.* **2001**, *4*, 36–40. [CrossRef]
72. Jagadish, S.V.K.; Craufurd, P.Q.; Wheeler, T.R. Phenotyping Parents of Mapping Populations of Rice for Heat Tolerance during Anthesis. *Crop Sci.* **2008**, *48*, 1140. [CrossRef]
73. Kobayasi, K.; Matsui, T.; Yoshimoto, M.; Hasegawa, T. Effects of Temperature, Solar Radiation, and Vapor-Pressure Deficit on Flower Opening Time in Rice. *Plant Prod. Sci.* **2010**, *13*, 21–28. [CrossRef]
74. Jagadish, S.; Craufurd, P.; Wheeler, T. High Temperature Stress and Spikelet Fertility in Rice (*Oryza sativa* L.). *J. Exp. Bot.* **2007**, *58*, 1627–1635. [CrossRef]
75. Cooper, N.T.W.; Siebenmorgen, T.J.; Counce, P.A. Effects of Nighttime Temperature During Kernel Development on Rice Physicochemical Properties. *Cereal Chem. J.* **2008**, *85*, 276–282. [CrossRef]
76. Lisle, A.J.; Martin, M.; Fitzgerald, M.A. Chalky and Translucent Rice Grains Differ in Starch Composition and Structure and Cooking Properties. *Cereal Chem. J.* **2000**, *77*, 627–632. [CrossRef]
77. Ambardekar, A.A.; Siebenmorgen, T.J.; Counce, P.A.; Lanning, S.B.; Mauromoustakos, A. Impact of Field-Scale Nighttime Air Temperatures during Kernel Development on Rice Milling Quality. *Field Crop. Res.* **2011**, *122*, 179–185. [CrossRef]
78. Yang, J.; Zhang, J. Grain Filling of Cereals under Soil Drying. *New Phytol.* **2006**, *169*, 223–236. [CrossRef] [PubMed]
79. Yang, J.; Zhang, J.; Wang, Z.; Liu, K.; Wang, P. Post-Anthesis Development of Inferior and Superior Spikelets in Rice in Relation to Abscisic Acid and Ethylene. *J. Exp. Bot.* **2006**, *57*, 149–160. [CrossRef] [PubMed]

80. Lin, C.J.; Li, C.Y.; Lin, S.K.; Yang, F.H.; Huang, J.J.; Liu, Y.H.; Lur, H.S. Influence of High Temperature during Grain Filling on the Accumulation of Storage Proteins and Grain Quality in Rice (*Oryza sativa* L.). *J. Agric. Food Chem.* **2010**, *58*, 10545–10552. [CrossRef] [PubMed]
81. Asaoka, M.; Okuno, K.; Fuwa, H. Effect of Environmental Temperature at the Milky Stage on Amylose Content and Fine Structure of Amylopectin of Waxy and Nonwaxy Endosperm Starches of Rice (*Oryza sativa* L.). *Agric. Biol. Chem.* **1985**, *49*, 373–379. [CrossRef]
82. Ahmed, N.; Maekawa, M.; Tetlow, I.J. Effects of Low Temperature on Grain Filling, Amylose Content, and Activity of Starch Biosynthesis Enzymes in Endosperm of Basmati Rice. *Aust. J. Agric. Res.* **2008**, *59*, 599. [CrossRef]
83. Counce, P.A.; Bryant, R.J.; Bergman, C.J.; Bautista, R.C.; Wang, Y.-J.; Siebenmorgen, T.J.; Moldenhauer, K.A.K.; Meullenet, J.-F.C. Rice Milling Quality, Grain Dimensions, and Starch Branching as Affected by High Night Temperatures. *Cereal Chem. J.* **2005**, *82*, 645–648. [CrossRef]
84. Ali, F.; Waters, D.L.E.; Ovenden, B.; Bundock, P.; Raymond, C.A.; Rose, T.J. Australian Rice Varieties Vary in Grain Yield Response to Heat Stress during Reproductive and Grain Filling Stages. *J. Agron. Crop Sci.* **2019**, *205*, 179–187. [CrossRef]
85. Tsukaguchi, T.; Iida, Y. Effects of Assimilate Supply and High Temperature during Grain-Filling Period on the Occurrence of Various Types of Chalky Kernels in Rice Plants (*Oryza sativa* L.). *Plant Prod. Sci.* **2008**, *11*, 203–210. [CrossRef]
86. Yoshida, S.; Hara, T.; Hara, T. Effects of Air Temperature and Light on Grain Filling of an Indica and a Japonica Rice (*Oryza sativa* L.) under Controlled Environmental Conditions. *Soil Sci. Plant Nutr.* **1977**, *23*, 93–107. [CrossRef]
87. Shah, F.; Huang, J.; Cui, K.; Nie, L.; Shah, T.; Chen, C.; Wang, K. Impact of High-Temperature Stress on Rice Plant and Its Traits Related to Tolerance. *J. Agric. Sci.* **2011**, *149*, 545–556. [CrossRef]
88. Xie, X.; Li, B.; Li, Y.; Shen, S. High Temperature Harm at Flowering in Yangtze River Basin in Recent 55 Years. *Jiangsu J. Agric. Sci.* **2009**, *25*, 28–32.
89. Yang, L.; Wang, Y.; Kobayashi, K.; Zhu, J.; Huang, J.; Yang, H.; Wang, Y.; Dong, G.; Liu, G.; Han, Y.; et al. Seasonal Changes in the Effects of Free-Air CO_2 Enrichment (FACE) on Growth, Morphology and Physiology of Rice Root at Three Levels of Nitrogen Fertilization. *Glob. Chang. Biol.* **2008**, *14*, 1844–1853. [CrossRef]
90. Yang, Z.; Zhang, Z.; Zhang, T.; Fahad, S.; Cui, K.; Nie, L.; Peng, S.; Huang, J. The Effect of Season-Long Temperature Increases on Rice Cultivars Grown in the Central and Southern Regions of China. *Front. Plant Sci.* **2017**, *8*, 1908. [CrossRef] [PubMed]
91. Morita, S.; Shiratsuchi, H.; Takahashi, J.I.; Fujita, K. Effect of High Temperature on Grain Ripening in Rice Plants—Analysis of the Effects of High Night and High Day Temperatures Applied to the Panicle and Other Parts of the Plant. *Jpn. J. Crop Sci.* **2004**, *73*, 77–83. [CrossRef]
92. Tardieu, F. Plant Response to Environmental Conditions: Assessing Potential Production, Water Demand, and Negative Effects of Water Deficit. *Front. Physiol.* **2013**, *4*, 17. [CrossRef] [PubMed]
93. Aghaee, A.; Moradi, F.; Zare-Maivan, H.; Zarinkamar, F.; Irandoost, H.P.; Sharifi, P. Physiological responses of two rice (*Oryza sativa* L.) genotypes to chilling stress at seedling stage. *African J. Biotechnol.* **2011**, *10*, 7617–7621.
94. Nagai, T.; Makino, A. Differences Between Rice and Wheat in Temperature Responses of Photosynthesis and Plant Growth. *Plant Cell Physiol.* **2009**, *50*, 744–755. [CrossRef] [PubMed]
95. Pitman, A.J.; Narisma, G.T.; Pielke, R.A.; Holbrook, N.J. Impact of Land Cover Change on the Climate of Southwest Western Australia. *J. Geophys. Res.* **2004**, *109*, D18109. [CrossRef]
96. Graf, A.; Schlereth, A.; Stitt, M.; Smith, A.M. Circadian Control of Carbohydrate Availability for Growth in Arabidopsis Plants at Night. *Proc. Natl. Acad. Sci. USA* **2010**, *107*, 9458–9463. [CrossRef]
97. Peraudeau, S.; Roques, S.O.; Quiñones, C.; Fabre, D.; Van Rie, J.; Ouwerkerk, P.B.F.; Jagadish, K.S.V.; Dingkuhn, M.; Lafarge, T. Increase in Night Temperature in Rice Enhances Respiration Rate without Significant Impact on Biomass Accumulation. *Field Crop. Res.* **2015**, *171*, 67–78. [CrossRef]
98. Stuerz, S.; Sow, A.; Muller, B.; Manneh, B.; Asch, F. Leaf Area Development in Response to Meristem Temperature and Irrigation System in Lowland Rice. *Field Crop. Res.* **2014**, *163*, 74–80. [CrossRef]
99. Stuerz, S.; Sow, A.; Muller, B.; Manneh, B.; Asch, F. Canopy Microclimate and Gas-Exchange in Response to Irrigation System in Lowland Rice in the Sahel. *Field Crop. Res.* **2014**, *163*, 64–73. [CrossRef]
100. Hirai, G.I.; Okmura, T.; Takeuchi, S.; Tanaka, O.; Chujo, H. Studies on the Effect of the Relative Humidity of the Atmosphere on the Growth and Physiology of Rice Plants: Effects of Relative Humidity during the Light and Dark Periods on the Growth. *Plant Prod. Sci.* **2000**, *3*, 129–133. [CrossRef]
101. Kuwagata, T.; Ishikawa-Sakurai, J.; Hayashi, H.; Nagasuga, K.; Fukushi, K.; Ahamed, A.; Takasugi, K.; Katsuhara, M.; Murai-Hatano, M. Influence of Low Air Humidity and Low Root Temperature on Water Uptake, Growth and Aquaporin Expression in Rice Plants. *Plant Cell Physiol.* **2012**, *53*, 1418–1431. [CrossRef] [PubMed]
102. Peraudeau, S.; Lafarge, T.; Roques, S.; Quiñones, C.O.; Clement-Vidal, A.; Ouwerkerk, P.B.F.; Van Rie, J.; Fabre, D.; Jagadish, K.S.V.; Dingkuhn, M. Effect of Carbohydrates and Night Temperature on Night Respiration in Rice. *J. Exp. Bot.* **2015**, *66*, 3931–3944. [CrossRef] [PubMed]
103. Sunoj, V.S.J.; Shroyer, K.J.; Jagadish, S.V.K.; Prasad, P.V.V. Diurnal Temperature Amplitude Alters Physiological and Growth Response of Maize (*Zea mays* L.) during the Vegetative Stage. *Environ. Exp. Bot.* **2016**, *130*, 113–121. [CrossRef]

Article

Leaf Pigments, Surface Wax and Spectral Vegetation Indices for Heat Stress Resistance in Pea

Endale Geta Tafesse [1], Thomas D. Warkentin [1], Steve Shirtliffe [1], Scott Noble [2] and Rosalind Bueckert [1,*]

[1] Department of Plant Sciences, College of Agriculture and Bio-Resources, University of Saskatchewan, Saskatoon, SK S7N 5A8, Canada; endale.tafesse@usask.ca (E.G.T.); tom.warkentin@usask.ca (T.D.W.); steve.shirtliffe@usask.ca (S.S.)

[2] Department of Mechanical Engineering, College of Engineering, University of Saskatchewan, Saskatoon, SK S7N 5A9, Canada; scott.noble@usask.ca

* Correspondence: rosalind.bueckert@usask.ca; Tel.: +1-306-966-8826

Abstract: Pea is a grain legume crop commonly grown in semi-arid temperate regions. Pea is susceptible to heat stress that affects development and reduces yield. Leaf pigments and surface wax in a crop canopy make the primary interaction with the environment and can impact plant response to environmental stress. Vegetation indices can be used to indirectly assess canopy performance in regard to pigment, biomass, and water content to indicate overall plant stress. Our objectives were to investigate the contribution of leaf pigments and surface wax to heat avoidance in pea canopies, and their associations with spectral vegetation indices. Canopies represented by 24 pea cultivars varying in leaf traits were tested in field trials across six environments with three stress levels in western Canada. Compared with the control non-stress environments, heat stress reduced leaf lamina and petiole chlorophyll a, chlorophyll b, and carotenoid concentrations by 18–35%, and increased leaf lamina chlorophyll a/b ratio, anthocyanin and wax concentrations by 24–28%. Generally, greater leaf pigment and wax concentrations were associated with cooler canopy temperature and high heat tolerance index (HTI) values. Upright cultivars had higher HTI values, whereas the lowest HTI was associated with normal leafed vining cultivars. Vegetation indices, including photochemical reflectance index (PRI), green normalized vegetation index (GNDVI), normalized pigment chlorophyll ratio index (NPCI), and water band index (WBI), had strong correlations with HTI and with heat avoidance traits. This study highlights the contribution of pigments and wax as heat avoidance traits in crop canopies, and the potential application of spectral measurements for selecting genotypes with more heat resistant vegetation.

Keywords: pea; heat-stress; wax; lamina; petiole; canopy type; chlorophyll; anthocyanin; vegetation indices

Citation: Tafesse, E.G.; Warkentin, T.D.; Shirtliffe, S.; Noble, S.; Bueckert, R. Leaf Pigments, Surface Wax and Spectral Vegetation Indices for Heat Stress Resistance in Pea. *Agronomy* **2022**, *12*, 739. https://doi.org/10.3390/agronomy12030739

Academic Editors: Channapatna S. Prakash, Ali Raza, Xiling Zou and Daojie Wang

Received: 17 February 2022
Accepted: 16 March 2022
Published: 19 March 2022

Publisher's Note: MDPI stays neutral with regard to jurisdictional claims in published maps and institutional affiliations.

1. Introduction

Pea (*Pisum sativum* L.) is a pulse crop that is widely grown in temperate regions for its nutritious seed and soil fertility benefits [1]. Unfortunately, pea is susceptible to heat stress, which causes impaired photosynthesis, accelerated senescence, and abortion of reproductive organs including flowers and pods, all culminating in reduced yield [2–4]. Due to weather alterations such as increased air temperature and severe drought caused by climate change, crop production is becoming increasingly challenging in many parts of the world [5]. For example, in the province of Saskatchewan, Canada, the worlds' leading producer and exporter of pea, the 2021 cropping season was the most heat and drought stressed in the last several decades, causing about 37% reduction in pea seed yield compared to the average of the previous five years (https://agriculture.canada.ca (accessed on 25 January 2022)). Pea heat stress arises in spring and summer-grown crops on days when air temperature exceeds a threshold of 28 °C, and when heat shock occurs from temperature > 34 °C for several hours during sensitive stages [3,6]. Although the extent of heat sensitivity varies with phenology, heat stress can impede crop performance at any developmental stage [2,4].

To cope with heat and other sub-optimal environmental factors, plants have developed various amendments to their morpho-anatomical form and physiological and biochemical functions, as avoidance or tolerance strategies [7,8]. These strategies can be broadly categorized into long-term alterations to morphological architecture and phenological patterns, or short-term heat aversion mechanisms such as through transpirational cooling and reflection of radiation overload on plant canopies [9–11]. For example, spectral reflectance in the ultraviolet (UV) and infrared regions makes plants avoid or minimize radiation and heat load [10,12,13]. Such reflectance of excess heat can be affected by the amount and composition of epicuticular waxes [14]. Vegetation indices (VI), derived from spectral data, are useful proxies to qualitatively or quantitatively estimate traits associated with growth, biomass, pigment composition, and water content in a single leaf and at the canopy level in plant populations [15,16]. A recent study on wheat revealed the use of spectral data in predicting leaf epicuticular wax concentration [17].

Epicuticular wax making an outermost layer over plant surfaces protects the plant from extreme weather variables and contributes to the plant's survival under stressful environments [18]. In pea, epicuticular wax reduces residual transpiration, minimizing water loss to help maintain tissue water status under drought stress [19]. Likewise, pigments may be involved in heat tolerance through heat dissipation and protection of essential plant processes [9,10]. Recently, Arafa et al. [20] reported pea seed priming with carrot extracts rich in carotenoids enhanced the plant's biochemical functions, and contributed to greater yield and stress tolerance. Stay-green, a trait characterized by delayed plant senescence, contributes to improved yield under both drought and heat stress conditions [21].

Although selection for thicker leaf epicuticular wax as a drought tolerance trait has resulted in improved cultivars in several crops [19,22–24], its contribution to heat tolerance is usually overlooked. Similarly, leaf pigments and their association with heat tolerance or avoidance have not been sufficiently addressed. We hypothesized that increased leaf pigments and wax concentrations would contribute to pea heat stress avoidance, and a substantial range of concentration of these biochemical compounds would be distributed across diverse pea germplasm. Based on the association with leaf wax and pigments, vegetation indices may serve as proxies to indirectly determine plant's resistance to heat stress and, therefore, amenable to high throughput field phenotyping. Our specific objectives were: (1) to investigate heat stress effects on pea canopies varying in leaf pigment and wax concentrations, (2) to examine the contributions of leaf pigments and wax concentrations in heat avoidance in a diverse range of pea cultivars, and, (3) to determine how spectral vegetation indices associate with leaf pigments and wax concentrations.

2. Materials and Methods

2.1. Pea Germplasm and Growth Conditions

Twenty-four diverse pea cultivars, adapted to western Canada and described in Table 1, were tested in field trials for three years (2014–2016) at two locations, Rosthern (52°66′ N, 106°33′ W; Orthic Black Chernozem) and Saskatoon (52°12′ N, 106°63′ W; Dark Brown Chernozem), in Saskatchewan, Canada. The study consisted of six trial sets (environments): Rosthern 2014 (RL14), Saskatoon 2014 (SL14), Rosthern 2015 (RL15), Saskatoon 2015 (SL15), Saskatoon 2016 (SL16) and Saskatoon 2016 with a normal seeding date (SN16). All trial sets except SN16 were intentionally late seeded late by 20 to 30 days from the regular seeding date. Late seeding delayed reproduction and flowering duration into mid-July to early-August, where daytime maximum air temperatures rose to 27–35 °C for several days, imposing heat stress on pea.

Table 1. Canopy type, description, and name of 24 pea cultivars evaluated for heat resistance traits at Rosthern and Saskatoon, Canada, in 2014–16.

Canopy Type [a]	Description	Cultivars [b]
1-n-u-dg	normal leaf, upright habit, dark-green canopy	MPG87, MFR043, TMP 15213
2-n-v-bg	normal leaf, vining habit, bright-green canopy	Naparnyk, TMP 15116, TMP 15181, Torsdag, 40-10
3-n-v-dg	normal leaf, vining habit, dark-green canopy	Mini, Rally, Superscout
4-sl-u-bg	semi-leafless, upright habit, bright-green canopy	Kaspa, CDC Sage, Aragorn, Eclipse
5-sl-u-dg	semi-leafless, upright habit, dark-green canopy	03H107P-04HO2026, 03H267-04HO2006, CDC Golden, CDC Vienna, CDC Meadow, Delta
6-sl-v-dg	semileafless, vining habit, bright-green canopy	TMP 15179, TMP 15206
7-n-u-bg	normal leaf, upright habit, bright-green canopy	TMP 15202

[a] n, normal leaf; sl, semi-leafless; u, upright habit; v, vining habit; bg, bright-green color; dg, dark-green color.; [b] CDC, Crop Development Center; TMP, temporary accession designation.

The six environments were grouped into three environmental stress levels (ESL) based on temperature, vapor pressure deficit and precipitation: control (RL14 and SN16), intermediate (SL14 and SL16), and stress (RL15 and SL15). For every trial set of the six environments, a randomized complete block design was used with four replications. A standard plant breeding plot size, 1.37 m width $\times$ 3.66 m length with three raw seeding, was used. Errors associated with edge effects were minimized by bordering the plots by other pea plots. Plants were grown under best management practices recommended for pea production in western Canada. As details of the crop husbandry, including the type and active ingredients of herbicides used for weed control and fertilization, were described by Tafesse et al. [4], here we provide a brief summary of the practices. To control weeds, herbicides were applied in fall several months before seeding and during the trial seasons. The trials were seeded into cereal stubble, no fertilizer was applied, but seeds were inoculated with rhizobia for atmospheric nitrogen fixation. At maturity, a few days before harvesting, a desiccant was applied to facilitate uniform drying and the plots were combine harvested. Weather data for each of the six trial sets (environments) was presented by Tafesse et al. [4]. Generally, environments RL14 and SN16 had relatively cooler air temperature and sufficient rainfall, and were designated as control ESL. Environment SL14 and SL16 had relatively hotter conditions with sufficient rainfall and were considered as intermediate ESL, and environments RL15 and SL15 had the hottest and driest conditions compared to the other four environments and were designated as stress ESL.

2.2. Leaf Sample Collection and Area Determination

Pea plants have compound leaves consisting of stipules, leaflets, petioles, rachis, and tendrils [25]. The pea cultivars used in this study were normal and semi-leafless leaf types. Normal leafed cultivars had a wide flat leaf lamina surface from stipules, leaflets, and relatively short petioles and rachis. Semi-leafless cultivars had stipules, a longer petiole and rachis, with more tendrils, but no leaflets. Unless otherwise stated, we refer to the flat leaf surface as 'lamina', and the petiole, rachis plus tendrils as 'petiole'. Fully expanded young leaf samples from the second or third main stem node, counting down from the apical tip, were sampled for chlorophyll, carotenoid, anthocyanin and wax measurements. Three leaf samples were collected from each plot and placed in plastic bags in an ice box cooler to avoid evaporative loss prior to transport to the laboratory. The leaf samples were collected twice during the pea growing season, at early flowering, and the full seed stages. Leaf samples were clipped at the main stem node, separated into lamina and petiole components, and then lamina and petiole scanned separately for each plot using winRHIZO (Regent Instruments Inc, Quebec City, Canada) to determine their respective projected surface areas (cm^2). For each plot and time of collection, one leaf was used for chlorophyll and carotenoid extraction in acetone, another for anthocyanin in acidified ethanol, and the third leaf was used for wax extraction in chloroform.

2.3. Chlorophyll and Carotenoid Determination

Chlorophyll and carotenoid measurements were performed according to Lichtenthaler [26]. A 1.22 cm^2 stipule disc, and the entire petiole, using the same scanned tissue above, were each placed in 10 mL glass tubes with a tight cap, 3.5 mL of 100% acetone was added, and samples incubated for 6 h at room temperature for complete pigment extraction. Samples were then vortex mixed and centrifuged for 5 min at 5000 rpm. The supernatant was used for absorbance (A) measurement using an Agilent 8453 diode array spectrophotometer with 1.6 $\pm$ 0.5 nm resolution, equipped with Chem Station software for UV-visible spectroscopy (Agilent Technologies, Santa Clara, CA, USA), at wavelengths 470, 645, 662 and 710 nm. Concentrations of chlorophyll a, chlorophyll b, total chlorophyll, and total carotenoid were determined in $\mu g\ cm^{-2}$ using the following equations:

$$\text{Chlorophyll a } (\mu g\ mL^{-1}) = (11.24(A_{662} - A_{710})) - (2.04(A_{645} - A_{710})) \quad (1)$$

$$\text{Chlorophyll b } (\mu g\ mL^{-1}) = (20.13(A_{645} - A_{710})) - (4.19(A_{662} - A_{710})) \quad (2)$$

$$\text{Total Chlorophyll } (\mu g\ mL^{-1}) = (7.05(A_{662} - A_{710})) + (18.09(A_{645} - A_{710})) \quad (3)$$

$$\text{Total carotenoid } (\mu g\ mL^{-1}) = (1000(A_{470} - A_{710})) - (1.90 \text{ Chlorophyll a}) - (63.14 \text{ Chlorophyll b}) \quad (4)$$

In order to present the data in $\mu g\ cm^{-2}$, the results from the above equations were multiplied by the extraction volume (3.5 mL), and then divided by the sample (lamina or petiole) projected area (cm^2).

2.4. Anthocyanin Determination

A 1.33 cm^2 disc cut from stipule, and the entire petiole were each used per leaf, and anthocyanin extraction was done according to Abdel-Aal and Hucl [27]. Samples were placed in 5 mL glass tubes with 3 mL of acidified ethanol (85:15, V/V) ratio using 95% ethanol and 1.428N HCl) at pH 1, tubes were capped and then incubated overnight at room temperature for extraction. Samples were vortex mixed and centrifuged at 5000 rpm for 5 min. The supernatant solution was measured with spectrophotometer and absorbance was read at 535 and 663 nm, wavelength peaks for absorbance of anthocyanin and chlorophyll a, respectively. Total anthocyanin concentration in $\mu g\ cm^{-2}$ was calculated according to Murray and Hackett [28] as:

$$\text{Anthocyanin } (\mu g\ ml^{-1}) = A_{535} - 0.24(A_{663}) \quad (5)$$

2.5. Bulk Wax Determination

Leaf lamina and petiole wax were extracted and quantified according to the methods used on pea [19]. Details of the method used for bulk wax extraction and quantification is exactly as presented by Tafesse et al. [29]. The method can be summarized as follows: first, bulk was extracted from the leaf surfaces by dipping the leaf tissue (lamina or petiole) samples into 10 mL chloroform for 15 s in 100 mL glass tubes. To evaporate the chloroform, the tubes were then placed in a water bath at 70 °C for 30 min. Then 5 mL reagent (acidic potassium dichromate) was added to each tube containing the wax, and boiled at 100 °C for half an hour. After cooling, 5 mL distilled water was added to each tube, vortexed, and spectral absorbance was measured at 590 nm with a spectrophotometer. Finally, wax concentrations were calculated from the spectral data using a standard curve equation that was developed from a linear ($R^2 > 0.98$) relationship of known concentrations of beeswax [29].

2.6. Spectral Reflectance and Vegetation Indices

Spectral reflectance measurements on stipules were taken on three to five occasions per plot for each of the six environments during the crop reproductive phase using a portable spectroradiometer (Model PSR-1100F, Spectral Evolution Inc., Lawrence, MA, USA). This instrument enabled hyperspectral readings with a range of 320–1126 nm and

1.6 nm sampling interval, and a total of 512 discrete narrow bands. A 1-m fiber-optic cable with industry-standard interface with the instrument, controlled by a PSR-1100 Pistol Grip, enabled us to specifically capture reflectance from stipules for spectral measurements. A stipule of a fully expanded leaf at the second or third node counting from the tip of the pea main stem, fully exposed to the sun, was measured on sunny and usually hot days around solar noon (between 11:00 and 14:00 h) from the same direction, avoiding shadows, cloud, and any other interference we could control. Before measurements, reflectance was taken on a white plate that provided maximum reflection, and leaf reflectance was measured by holding the fiber sensor within 3 cm from the stipule surface approximately within a viewing angle of 80–90°. The reference reflectance was repeatedly taken every 15 min (equivalent to once every 12 plots) to adjust for the changing irradiance from the sun, and more frequently if clouds stopped measurements.

Vegetation and pigment indices, including normalized difference vegetation index (NDVI), green normalized difference vegetation index (GNDVI), photochemical reflectance index (PRI), normalized pigment chlorophyll ratio index (NPCI), and water band index (WBI), were each calculated according to Rouse et al. [30], Gitelson et al. [31], Gamon et al. [32], Peñuelas et al. [33], and Peñuelas et al. [34], respectively, as follows:

$$NDVI = (R_{nir} - R_r) \div (R_{nir} + R_r) \tag{6}$$

$$GNDVI = (R_{nir} - R_g) \div (R_{nir} + R_g) \tag{7}$$

$$PRI = (R_{531} - R_{570}) \div (R_{531} + R_{570}) \tag{8}$$

$$NPCI = (R_{531} - R_{570}) \div (R_{531} + R_{570}) \tag{9}$$

$$WBI = R_{900} \div R_{970} \tag{10}$$

where; R, reflectance; nir, near infrared band (bandwidth 760–860, center band 820 nm), r, red band (bandwidth 650–700 nm, center band 675 nm); g, green (bandwidth 530–580, center band 555 nm). The center bands were rounded to the nearest whole number (for example 530.5 nm was 531 nm). For vegetation indices calculated from two or more single bands such as WBI, the nearest whole number band was used as the center band.

2.7. Canopy Temperature

Canopy temperature was measured five to eight times per plot in each environment during late vegetative and reproductive growth using a handheld infrared (IR) thermometer (Model 6110.4ZL, Everest Interscience Inc., Tucson, AZ, USA). Measurements were taken within 3 h centred on solar noon when pea transpiration was at its maximum rate, assuming no drought stress-related closure, with the sun unobstructed by cloud, and when there was low wind pressure. The infrared thermometer was held for six seconds approximately 30 cm above the canopy at 15° field of view pointing down for a wider canopy view. During the six seconds of viewing the canopy, the thermometer averaged over a range of measurements and stabilized to the mean value that was used as one data point to represent a plot reading. The reading did not include any ground or soil surface, only green vegetation, and predominantly upright vegetation and the upper half of the canopy.

2.8. Heat Tolerance Index

Heat tolerance index (HTI), a concept that indicates the extent of yield reduction due to heat stress compared to the potential yield under control condition, was determined according to Fernandez [35] and used to separate cultivars yield response into heat sensitive and heat tolerant. The seed yield was obtained by small plot combine harvest of individual plots at maturity.

$$HTI = \frac{(Yield_c)\,(Yield_h)}{(Yield_{c.ave})^2} \tag{11}$$

where $Yield_c$ is seed yield for each cultivar in a replication under non-heat stress (control conditions), $Yield_h$ is seed yield for each cultivar in a replication under heat stress, $Yield_{c.ave}$ is the grand mean seed yield from all control plots of all replications per environment under non-heat stress conditions. When HTI is close to zero or zero, crops do not yield under heat and are heat sensitive. When HTI is high (>1), then the cultivar would be deemed heat tolerant for yield compared to the grand mean yield under the control conditions.

2.9. Data Analysis

Univariate analysis of the variables chlorophyll a, chlorophyll b, chlorophyll a/b, total carotenoid, total anthocyanin, and total wax concentrations from lamina and petiole, NDVI, GNDVI, PRI, NPCI and WBI were computed by using the mixed procedure of SAS, Version 9.4, SAS Institute. Before undertaking the analysis variance (ANOVA), normal distribution of residuals and homogeneity of variances, the two major ANOVA assumptions, were checked according to Shapiro-Wilk and Levene and tests, respectively [36,37] and these assumptions were met for each variable. Then ANOVA with the least square difference (LSD) test ($p < 0.05$) was performed on each variable. The designation of environments into three stress levels, "control", "intermediate" and "stress", was based on the intensity of stress at each environment as described by Tafesse et al. [4]. For growth habit, leaf type, and canopy color, seven combinations of canopy groups, designated as "type" were used for testing the effect of canopy type. Thus, the three main treatment factors were environmental stress that we simply referred to as 'environmental stress level (ESL)', canopy 'type', and 'cultivar'. The effects of ESL, type, cultivar, ESL by cultivar, and ESL by type nested in cultivar interactions were treated as fixed effects, and block nested in environment was treated as a random effect. Whenever the interaction term was significant, a separate analysis was performed for each of the three ESLs and the results of the 'control' and 'stress' levels are shown while the result of the 'intermediate' that generally lay between the two ESLs is omitted in figures to save space. Pearson correlations test were performed among the variables of canopy temperature, pigments, wax, and vegetation indices, and significance was declared at $p < 0.05$ for combined data using four environments, control and stress.

3. Results

3.1. Treatment Effects on Pigment, Wax and Vegetation Indices

The main treatment effects of ESL, canopy type and cultivar significantly affected all pigment and wax traits. The one exception was petiole anthocyanin which was not significantly affected by cultivar. Vegetation indices including PRI, GNDVI, WBI, and NPCI were significantly affected by all main effect treatment factors. The interaction of ESL by type impacted all traits except petiole anthocyanin and the majority of the vegetation indices. Only PRI, WBI and NPCI were affected by the interaction of ESL by type. The ESL by cultivar interaction affected lamina chlorophyll a, lamina and petiole chlorophyll b, lamina carotenoid, lamina and petiole wax concentration, GNDVI, WBI and NPCI, but not other traits (Table 2).

For ESL, compared to the control, stress decreased mean lamina chlorophyll a and chlorophyll b concentrations by 22.8, and 34.9%, respectively. In contrast, stress increased the corresponding chlorophyll a/b ratio, anthocyanin, and bulk wax concentrations by 23.9, 24.5, and 28.4%, respectively (Table 2). The increased chlorophyll a/b ratio under stress resulted from a greater reduction in chlorophyll b concentration than in chlorophyll a (Table 2; Figure 1A,B).

Table 2. Univariate analysis of variance (ANOVA) results showing significance of cultivar, environmental stress level (ESL), canopy type (T) main effects and their interactions on leaf pigments, wax, and vegetation indices of 24 pea cultivars grown across six environments at Rosthern and Saskatoon, 2014–16. Means of traits are presented for the three stress levels of control, intermediate, and stress. The control ESL are 2014 late seeding date at Rosthern and 2016 normal seeding date at Saskatoon; intermediate ESL are 2014 and 2016 late seeding date at Saskatoon; and stress ESL are 2015 late seeding date at Rosthern and Saskatoon, Canada.

Variable	Tissue	ESL	T	Cultivar	ESL * T(C)	ESL * C	ESL Treatment Means		
							Control	Intermediate	Stress
Chlorophyll a (μg cm^{-2})	Lamina	***	***	***	**	**	33.5 a	28.5 b	25.9 c
	Petiole	*	***	***	*	NS	20.9 a	18.5 ab	16.9 b
Chlorophyll b (μg cm^{-2})	Lamina	***	***	***	**	**	11.0 a	8.3 b	7.2 c
	Petiole	*	***	***	**	*	9.2 a	6.7 b	6.0 b
Chlorophyll a/b ratio	Lamina	***	***	**	***	NS	3.1 c	3.57 b	3.9 a
	Petiole	**	***	***	**	NS	2.5 b	2.93 a	2.6 ab
Carotenoids (μg cm^{-2})	Lamina	**	***	***	*	*	8.4 a	7.83 a	6.79 b
	Petiole	***	***	***	**	NS	5.8 a	4.98 b	4.7 c
Anthocyanins (μg cm^{-2})	Lamina	*	***	***	***	NS	1.06 b	1.10 b	1.32 a
	Petiole	*	***	NS	NS	NS	1.34 ab	1.29 b	1.37 a
Bulk wax (μg cm^{-2})	Lamina	**	***	***	***	**	23.6 b	27.34 a	30.3 a
	Petiole	**	***	***	**	*	41.6 b	42.45 b	53.4 a
NDVI	Lamina	**	NS	*	NS	NS	0.81 a	0.77 b	0.76 b
PRI	Lamina	**	***	***	*	NS	−0.01 a	−0.02 b	−0.02 b
GNDVI	Lamina	**	***	***	NS	*	0.64 a	0.61 b	0.60 b
WBI	Lamina	***	***	***	*	*	1.10 a	1.08 b	1.08 b
NPCI	Lamina	**	***	***	***	*	0.29 c	0.33 b	0.36 a

*, **, and *** indicates significant at 0.05, 0.01, and 0.001 probability levels, respectively. NS, not significantly different at 0.05 probability level. For each variable within each raw, means labeled by same letter are not significantly different at 0.05 probability level. For each stress level, N = 192 from 24 cultivars, two environments and four replications. GNDVI, green normalized difference vegetation index; NDVI, normalized difference vegetation index; NPCI, normalized pigment chlorophyll ratio index.; PRI, photochemical reflectance index; WBI, water band index.

As reproduction proceeded from early flowering to pod filling, the leaf lamina chlorophyll a, chlorophyll b, carotenoid, and wax concentrations increased by 20, 18, 5, and 39%, respectively, and the corresponding anthocyanin concentration decreased by 20%. Similarly, petiole chlorophyll a, chlorophyll b, carotenoid, and wax concentrations increased by 28, 10, 6, and 53%, respectively (Figure 2). Generally, leaf lamina had >32% greater chlorophyll and carotenoid concentrations than those found in the petiole. On the other hand, higher anthocyanin and wax concentrations were found in the petioles compared with the leaf lamina (Table 2). Under both control and stress conditions, cultivars with dark green canopies, including Superscout, Rally, MPG87, Mini, and CDC Vienna, were associated with greater lamina chlorophyll and carotenoid concentrations. In contrast, the bright green cultivars including TMP15116, Naparnyk, CDC Sage, and Torsdag had less (<32 μg cm^{-2}) lamina chlorophyll and carotenoid concentrations (Figure 1A,B,D).

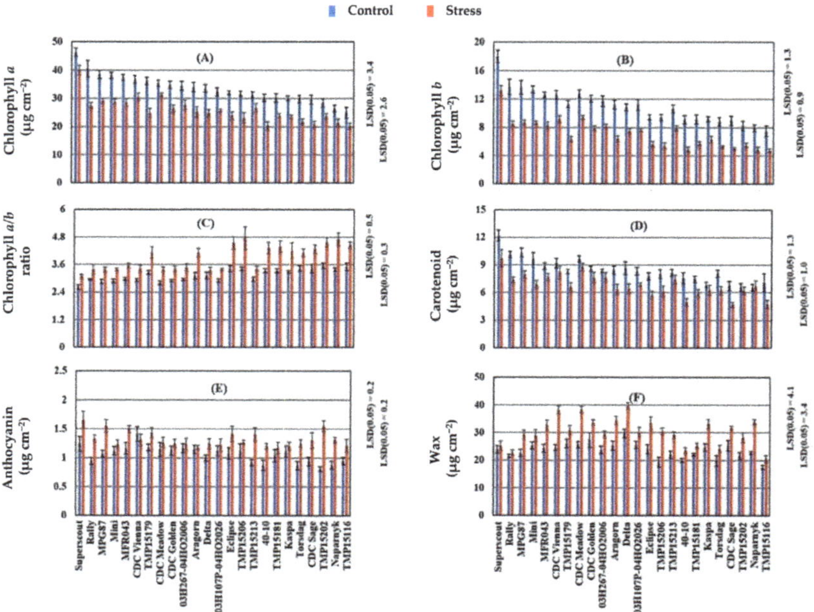

Figure 1. Mean lamina pigment and wax concentrations of 24 pea cultivars grown in control (blue) and stress (red) conditions, with (**A**) chlorophyll a, (**B**) chlorophyll b, (**C**) chlorophyll a/b ratio, (**D**) carotenoid, (**E**) anthocyanin and (**F**) wax concentrations (μg cm^{-2}). Each bar represents the mean values, and error bars are the standard error of the mean. For each bar, N = 16 (two environments × four replications × two growth stage samples) for each of control or stress condition. The LSD values for each of the stress levels is shown in the figure. The control conditions are 2014 late seeding date at Rosthern and 2016 normal seeding date at Saskatoon; and the stress conditions are 2015 late seeding date at Rosthern and Saskatoon, Canada.

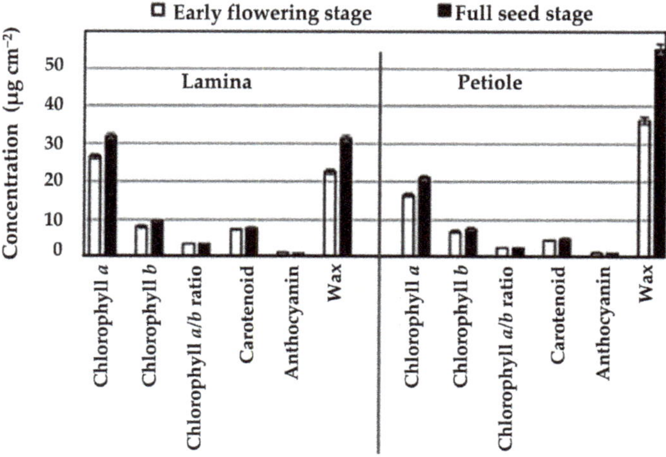

Figure 2. Chlorophyll a, b, a/b, carotenoid, anthocyanin and wax concentrations (μg cm^{-2} of tissue) in leaf lamina and petiole at early flowering and flower termination stages of pea grown in six field environments in Saskatchewan, Canada (2014 to 2016). Each bar is the mean value averaged over 24 cultivars, six environments and four replications per environment, i.e., N = 576 for each variable.

Generally, for chlorophyll and carotenoid concentrations under both control and heat stress, the normal leafed vining cultivars with dark green canopies had the greatest chlorophyll a, chlorophyll b, and carotenoid concentrations; but the least chlorophyll and carotenoid concentrations were found in normal leafed vining cultivars with bright green color (Figure 3A,B,D). Under control and heat stress, bright green cultivars had a higher chlorophyll a/b ratio than dark green cultivars regardless of the growth habit and leaf type (Figure 3C). Under control conditions, normal leafed vining cultivars with bright green canopies had lower anthocyanin concentration than all other types; but under heat stress this type had a relatively greater anthocyanin concentration than other types (Figure 3E). For leaf wax, under control conditions, semi-leafless cultivars had the same wax concentration regardless of their canopy habit and color. The lowest wax concentrations were found in normal leafed vining cultivars with bright green canopies (Figure 3F). Under heat stress, upright semi-leafless cultivars with dark green canopies had the greatest wax concentrations, and normal leafed vining cultivars had the lowest wax concentrations (Figure 3F).

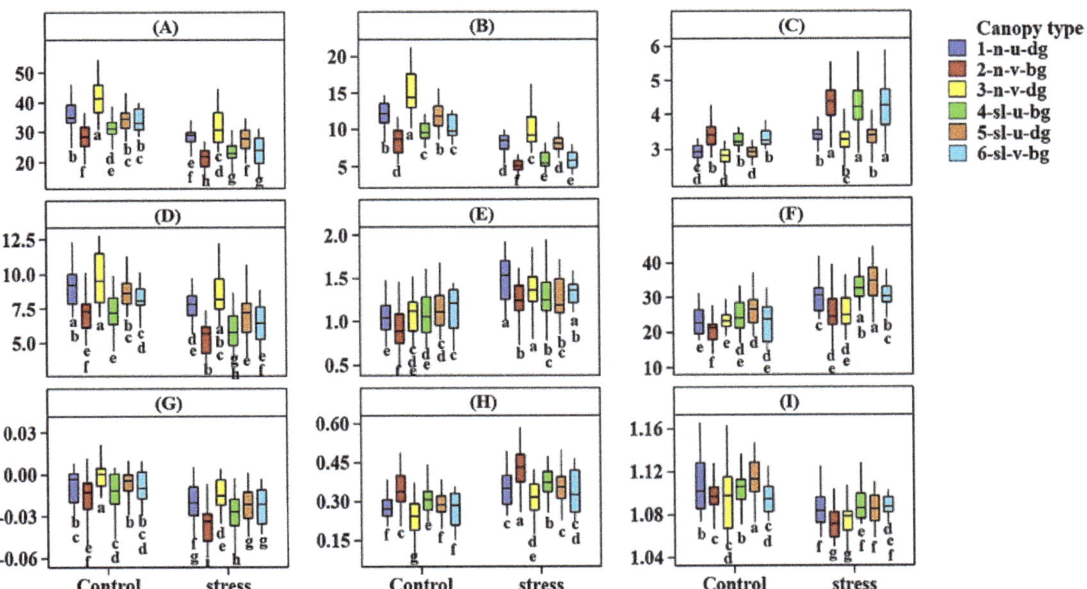

Figure 3. Box plot showing the interaction effects of heat stress and canopy type on: chlorophyll a (**A**), chlorophyll b (**B**), chlorophyll a/b ratio (**C**), carotenoid (**D**), anthocyanin (**E**), wax (**F**) concentrations (μg cm^{-2}); and vegetation indices: PRI (**G**), NPCI (**H**), and WBI (**I**) measured on pea leaf stipules under control and stress environments at Rosthern and Saskatoon, Canada, 2014–16. The size of box represents 50% of the middle data, the line in the middle of the box is the median, and the whiskers represent the range of the data. Boxes labeled with same letters within trait are not significantly different at $p < 0.05$. N = 24 for type 1, 40 for type 2, 24 for type 3, 32 for type 4, 48 for type 5, and 16 for type 6. The control conditions are 2014 late seeding date at Rosthern and 2016 normal seeding date at Saskatoon; and the stress conditions are 2015 late seeding date at Rosthern and Saskatoon, Canada.; Canopy type legend: n, sl, u, v, bg, and dg represents normal leaf, semi-leafless; upright habit, vining habit, bright-green color, and dark-green color, respectively; and vegetation indices: PRI, photochemical reflectance index; NPCI, normalized pigment chlorophyll ratio index; WBI, water band index.

3.2. Response of Vegetation Indices

The control environment had greater NDVI and GNDVI values than the stress and intermediate environments. Although the values of most vegetation indices were in the 'normal' range for healthy vegetation, heat stress and control treatments differed (Table 2). For PRI, under both control and heat stress, dark green cultivars had greater PRI than bright green cultivars regardless of leaf type and canopy habit (Figure 3G). Under both control and stress conditions, normal leafed vining cultivars with bright green canopies had greater NCPI, suggesting more stress than all other canopy combinations (Figure 3H). For WBI, for the control, semi-leafless upright cultivars with dark green canopies had a greater WBI than vining cultivars regardless of leaf type and canopy color. Furthermore, under heat stress, semi-leafless upright cultivars with dark green canopies had the greatest WBI, inferring a high leaf water content compared to all other canopy types. For WBI and heat stress, the cultivar ranking matched with the cultivar ranking for wax concentration. Water band index is associated with leaf water content, so the greater WBI value in upright and semi-leafless cultivars implied that these cultivars maintained greater leaf water content under heat stress.

3.3. Heat Tolerance Index

Pea cultivars significantly varied in HTI, calculated from the relative seed yield of cultivars under control and heat stress, with values ranging from 0.35 to 1.25 (Figure 4A). Generally, upright cultivars with dark green canopies had a greater HTI, with the smallest HTI in normal leafed vining cultivars with dark green canopies (Figure 4B). Cultivars with greater (>1) HTI included CDC Meadow, TMP 15 2013, CDC Golden, Naparynk and TMP 15181 (Figure 4A). Heat tolerance index was negatively correlated with canopy temperature (Figure 5C), positively correlated with lamina wax concentration and WBI (Figure 5E,F).

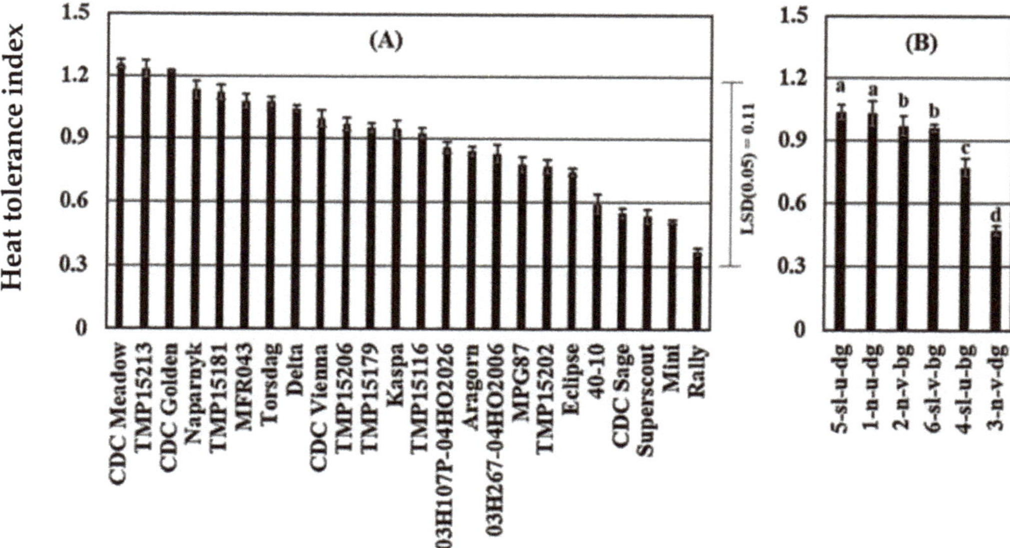

Figure 4. Heat tolerance index by cultivars (**A**), and canopy type (**B**) of 24 pea cultivars grown in four field environments (control and stress) in western Canada. N = 8 for each cultivar in panel A; and N = 12 for type 1, 20 for type 2, 12 for type 3, 16 for type 4, 24 for type 5, and 8 for canopy type 6 in panel B. Error bars are standard errors of the mean. In panel B, canopy types labeled with same letters are not significantly different at $p < 0.05$. Legend for canopy type: n, normal leaf; sl, semi-leafless; u, upright habit; v, vining habit; bg, bright-green color; dg, dark-green color.

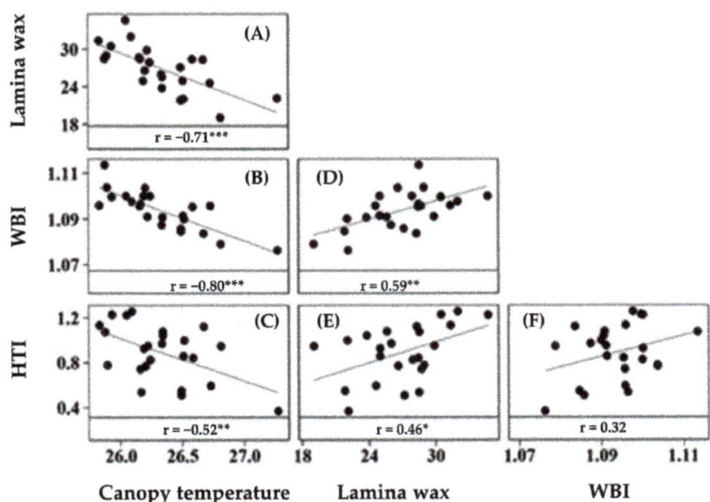

Figure 5. Matrix plot showing correlation between canopy temperature (CT) and lamina wax (**A**), CT and WBI (**B**), CT and HTI (**C**), Lamina wax and WBI (**D**), Lamina wax and HTI (**E**), and WBI and HTI (**F**) of 24 pea cultivars grown in four field environments (control and stress) at Rosthern and Saskatoon, western Canada, in 2014–16. Each symbol is a cultivar averaged over 16 observations (four environments by four replications). *, **, and *** indicates significant at 0.05, 0.01, and 0.001 probability levels, respectively. HTI, heat tolerance index; WBI, water band index.

3.4. Phenotypic Correlation among Pigments, Wax, Vegetation Indices, Canopy Temperature, and Heat Tolerance Index

Leaf spectral reflectance is mainly affected by pigment and wax compositions and concentrations. Our result showed that most vegetation indices had significant correlation with pigment and wax concentrations, canopy temperature and HTI (Table 3 and Figure 5). The PRI and GNDVI had similar trends, correlating positively with chlorophyll, carotenoid, and anthocyanin concentrations, and negatively with chlorophyll a/b ratio. In contrast, NPCI had the opposite pattern to PRI and GNDVI. Correlations among lamina wax concentration, WBI, canopy temperature, and HTI were of specific interest and illustrated in Figure 5. Water band index correlated positively with lamina wax concentration, and negatively with canopy temperature. Finally, Heat tolerance index was negatively correlated with canopy temperature (Figure 5C), and positively correlated with lamina wax concentration and WBI (Figure 5E,F).

Table 3. Pearson correlation test showing associations among canopy temperature, total chlorophyll, chlorophyll a/b ratio, carotenoid, anthocyanin, wax, photochemical reflectance index (PRI), green normalized vegetation index (GNDVI), water band index (WBI), and normalized pigment chlorophyll ratio index (NPCI) of 24 pea cultivars grown under field conditions under control (upper right diagonal) and stress (lower left diagonal) environments averaged over two environmental levels and four replications. The control conditions are 2014 late seeding date at Rosthern and 2016 normal seeding date at Saskatoon; and the stress conditions are 2015 late seeding date at Rosthern and Saskatoon, Canada.

Variable	Canopy Temperature	Total Chlorophyll	Chlorophyll a/b Ratio	Carotenoid	Antho-cyanin	Wax	PRI	GNDVI	WBI	NPCI
Canopy temperature		0.23	−0.03	0.29	−0.31	−0.46 *	0.06	0.07	−0.55 *	−0.07
Total chlorophyll	−0.16		−0.86 ***	0.95 ***	0.61 **	0.37	0.74 ***	0.64 **	0.12	−0.77 ***
Chlorophyll a/b ratio	0.19	−0.80 ***		−0.85 ***	−0.60 **	−0.46 *	−0.61 **	−0.47 *	−0.22	0.55 **
Carotenoid	−0.18	0.92 ***	−0.76 ***		0.53 **	0.23	0.68 **	0.62 *	−0.01	−0.66 **
Anthocyanin	−0.03	0.50 *	−0.20	0.47 *		0.45 *	0.44 *	0.37	0.25	−0.43 *
Wax	−0.72 ***	0.15	−0.18	0.26	−0.07		0.34	0.17	0.50 *	−0.31
PRI	−0.21	0.81 ***	−0.73 ***	0.72 ***	0.39	0.09		0.67 **	0.21	−0.81 ***
GNDVI	−0.13	0.78 ***	−0.65 **	0.64 **	0.55 **	−0.05	0.74 ***		0.40 *	−0.79 ***
WBI	−0.67 **	0.12	−0.05	0.13	0.01	0.52 **	0.24	0.15		−0.31
NPCI	0.10	−0.74 ***	0.61 **	−0.64 **	−0.48 *	−0.11	−0.75 ***	−0.70 **	−0.16	

*, **, and *** indicates significant level at the 0.05, 0.01, and 0.001 probability levels, respectively.

4. Discussion

4.1. Leaf Pigment Concentrations as Heat Resistance Traits

We found that heat stress and the significant cultivar by environment interaction lowered chlorophyll a and chlorophyll b concentrations in leaf lamina and petiole, with the reduction more pronounced in heat sensitive cultivars. Photosyntheric pigments are prone to heat and other environmental stresses. Recently, Giordano et al. [38] reviewed the reduction of photosynthetic pigments in response to heat stress, and such reduction led to the reduction of photosynthetic activities related to photosystem II. Cultivars that were able to maintain chlorophyll concentration under heat stress had greater HTI values, and therefore greater heat resistance, implying that chlorophyll concentration was likely linked to plant heat response. As chlorophyll is an integral component in light absorption and transfer; chlorophyll loss or degradation leads to reduced photosynthesis and coupled oxidative damage which consequently reduces growth and yield [7,11]. Under heat and excess radiation stress, chlorophyll loss arises either due to limited biosynthesis caused by enzyme malfunctioning [39], or due to rapid degradation caused by heat and radiation damage. Chlorophyll loss also occurs naturally in senescing plants, and stress induces tissue senescence [40].

Interestingly, chlorophyll a/b ratio increased under heat stress in both leaf lamina and petiole in our research, likely due to rapid chlorophyll b degradation compared to that of chlorophyll a, suggesting a differential susceptibility in light-harvesting chlorophyll a/b-binding protein complex [41]. Although chlorophyll a/b ratio changes were associated with plant heat response, the literature is inconsistent in how chlorophyll a/b ratio changes with stress in crops. Feng et al. [42] found decreased chlorophyll a/b ratio was associated with heat tolerance in wheat, but Cui et al. [43] reported the opposite on a cool season perennial grass tall fescue (*Festuca arundinacea*). While the optimal range of chlorophyll a/b ratio needs further study, we noted that pea cultivars with either high (>4.0), or low (<2.5) chlorophyll a/b ratio had low heat tolerance indices (Figures 1C and 5A), suggesting damage at the antenna complex or reaction center, respectively, as reported by Feng et al. [42]. We found, generally, that upright cultivars with dark green canopies had low chlorophyll a/b ratio, and greater HTI than vining cultivars with bright green canopies (Figures 2C and 4B), inferring that upright canopies were less stressed.

Leaf lamina carotenoid concentration had a similar pattern as chlorophyll concentration and decreased due to the heat stress (Figures 1C and 3D). In published research, carotenoid biosynthesis and accumulation were influenced by multiple factors including light and temperature stresses [44]. Although heat stress resulted in a decreased concentration of carotenoid, there was a significant difference among the pea cultivars. Cultivars better able to maintain relatively stable carotenoid concentration under heat stress had greater HTI (Figures 1D and 4A), implying that greater or maintained carotenoid concentration reduced heat damage on pea seed yield. Carotenoids are antenna pigments and have direct influence on photosynthesis, their two major roles being light harvesting during photosynthesis, and minimizing photo-oxidative damage of chlorophyll molecules by dissipating excess energy in the form of heat [10,45] by the Mehler-ascorbate-peroxidase cycle at Photosystem I [46].

Anthocyanin concentration increased with heat-stress (Table 2; Figure 1E), a pattern opposite to chlorophyll and carotenoid concentrations. Anthocyanin production is enhanced in response to most environmental stresses including cold, heat, drought, and light [47]. However, stressful environments also trigger formation of reactive oxygen species and free radicals [48]. To protect plants from the damaging effects of reactive oxygen species, high levels of anti-oxidants are required, and anthocyanins fulfill such a protective role [47]. Anthocyanins protect chloroplasts by reducing incident light, and they have an anti-oxidant role through scavenging reactive oxygen species [49]. Unlike chlorophyll and carotenoid, anthocyanin concentration was greater in petiole than lamina, and anthocyanin concentration declined in leaf lamina during reproduction, indicating anthocyanin biosynthesis was growth-stage dependent and younger leaves produced more. Anthocyanins also function like sunscreen for leaves, where anthocyanins form a layer and damaging radiation does not penetrate internal sensitive tissue. In addition to heat and UV protection, increased anthocyanin concentration under heat stress is associated with enhanced water uptake and decreased transpiration [7].

4.2. Wax as Heat Resistance Trait

While the roles of epicuticular wax as a drought tolerance trait have been extensively reported in cereals and brassica crops [19,22,23,50], a heat avoidance role for wax has rarely been addressed. We found significant variation, ranging from 23 to 53 $\mu g\ cm^{-2}$, in lamina and petiole bulk wax concentrations under heat and control conditions. Wax composition and concentration shows variation within and across crop species [19,22,23]. Our results showed that compared to the control, heat-stress resulted in a 28% increase in total leaf wax concentration. Moreover, during reproduction, from early flowering to full seed stage, wax concentration increased by >45% in heat and control environments (Figure 2). Part of this wax increase can be due to a reduction in leaf expansion during the season as crops experience diminished water supply, and part of this increase is likely due to increased induction of leaf wax biosynthesis. Overall, our results indicate that genetic factors (cultivar), plant age and heat stress jointly contributed to effects on leaf wax biosynthesis. In addition to heat stress, various stresses such as drought, cold, salinity, and mechanical damage have each contributed to increased wax load in crops [19,23].

For heat avoidance, epicuticular wax has two major roles. First, guarding leaves and stems from radiation and heat loads by reflecting ultraviolet, visible and infrared wavelengths. In a pilot study in which extra wax was applied to pea leaf surfaces under field conditions, we recorded radiation reflectance in the visible and near-infrared region and found reflectance here was positively associated with wax concentration [51]. Second, by minimizing water loss through reduced stomatal and residual (i.e., non-stomatal) transpiration, several groups associated epicuticular wax with improved drought tolerance [22,52,53]. Drought and heat stress usually occur together, and drought stress aggravates plant heat stress. Heat stress can be moderated if the plant is able to maintain and conserve sufficient water in leaves and tissues for transpirational cooling while minimizing non-transpirational losses. Our results showed that greater wax concentration was generally associated with a cooler canopy temperature, and a higher heat tolerance index (Figure 5A,E).

We discovered that upright canopies have an advantage in stress, an important finding in pea where leaf type determines the fate of upright crops to stay upright or lodge and suffer high temperatures early in vegetative growth. Upright canopies have also been linked to lower canopy temperatures versus lodged canopies in wheat [54]. Our pea cultivars with upright growth habits and semi-leafless leaf type, both stress hardy traits, were also associated with higher wax concentration under heat stress (Figure 3F). Wax accumulation was positively associated with WBI in both control and heat stress conditions; WBI is a proxy for leaf water content, indicating that leaf surface wax minimized water loss (Table 3). Thus, leaf wax indirectly functioned as a heat tolerance trait because sufficient water supply was able to moderate heat stress by 2 °C [51]. Similarly, Camarillo-Castillo et al. [17] reported the importance of leaf epicuticular wax in enhancing light reflecting both in the visible and near infrared regions, which likely contribute to the dissipation of heat and excess energy [52]. Generally, glaucousness or waxy leaves were associated with high water potential that contributed to cooler canopy [19]. We concluded that greater lamina and petiole wax concentrations minimized heat stress by guarding pea from excess radiation and heat, and they also helped maintain leaf water content by lowering residual transpiration.

4.3. Spectral Reflectance Association with Heat Resistance Traits

Recent advancements in large scale and more accurate phenotyping techniques largely rely on remotely sensed data. These measurements focus on leaf and canopy traits including vegetation area, pigments, canopy temperature and plant water status; all these being associated with a crops's overall physiological status [34,55]. For example, indices derived from reflectance in the visible and near infrared regions such as NDVI and its derivatives indicate vigor and biomass, vegetation greenness, photosynthesis efficiency, and rate of senescence [56,57]. In soybean, Dhanapal et al. [58] demonstrated useful correlations between leaf and canopy measured pigments and canopy measured VIs that are applicable for high throughput field phenotyping. A dark green canopy index was able to distinguish dark green genotypes from regular soybean genotypes, and showed several steps in nitrogen metabolism and transport, photosynthesis and senescence across a range of germplasm and environments [59]. Sexton et al. [60] reported that photosynthetic capacity of plants can be effectively determined non-destructively from hyperspectral reflectance in the short-wave infrared regions. More recently, Camarillo-Castillo et al. [17] showed the application of spectral data to assess epicuticular wax concentration in wheat leaves and the benefits of such information to indirectly select stress resistant wheat cultivars. Their study also showed the ability of spectral measurement to effectively predict leaf epicuticular wax concentration. Several single nucleotide polymorphism markers and candidate gene associated vegetation indices including NDVI, PRI, NPCI, and WBI were reported from recent studies conducted on pea [29,61]. These studies clearly indicated the importance of spectral data and vegetation indices in detecting plant stress responses and the associated genetic factors controlling such responses. We found that both genotype and environment had significant effect on pea pigment, and also on wax concentrations. Alteration of pigment and wax concentrations under various environments suggest direct involvement in avoiding or tolerating stress. Reflectance in the visible wavelengths (400–700 nm) are influenced mainly by leaf chlorophyll, carotenoid and anthocyanin concentrations and compositions [15,32,34].

Heat stress degrades photosynthetic pigments, and hampers the photosynthesis processes at different levels, and such effects can be indirectly traced from spectral reflectance. Our results demonstrated a positive correlation between GNDVI and chlorophyll concentration (Table 3). Vegetation indices derived from reflectance in the near infrared region including WBI are proxies for tissue water status [34]. We found WBI was negatively correlated with canopy temperature (Figure 5B), and positively correlated with wax concentration (Figure 5D). Another group of VIs are those derived from the reflectance in the visible spectral region including PRI and NPCI, proxies for pigment concentration and function, and photosynthesis [32,33]. Significant positive correlation was observed between PRI and chlorophyll concentration, and NPCI was associated with limited pigment and high stress. Such strong and consistent association of VIs with pigment, wax, canopy temperature and other stress related traits indicate the potential of the VIs specifically

GNDVI, PRI, NPCI and WBI, as measurement proxies in heat stress studies for pea and other crops.

5. Conclusions

Our results on pea demonstrated several novel findings. Firstly, heat stress reduced chlorophyll a, chlorophyll b, and carotenoid concentrations, but increased wax and anthocyanin concentrations, and chlorophyll a/b ratio in leaves. Generally, leaf pigments (chlorophyll, carotenoid, and anthocyanin) both from petiole and lamina were positively correlated with heat tolerance index and contributed to lower canopy temperature. Secondly, surface wax contributed to heat resistance presumably by reflecting excess radiation and heat from the plant canopy; and by minimizing water loss through reduced stomatal and residual (i.e., non-stomatal) transpiration. Thirdly, cultivars with the semi-leafless leaf type, upright habit, and dark-green canopies were associated with high (>1) HTI under the heat stress environments, inferring that these traits conferred heat resistance and upright, dark green canopies result in more stress resistant crops. Finally, vegetation indices including GNDVI, PRI, NPCI, and WBI measured from stipules showed consistent relationships with pigment and wax concentrations and other heat tolerance traits, suggesting these indices can be useful proxies in future heat stress studies, and for high throughput phenotyping for heat stress resistance.

Author Contributions: Conceptualization, R.B. and T.D.W.; methodology, E.G.T. and R.B.; software, E.G.T. and R.B.; validation, R.B., T.D.W., S.S. and S.N.; formal analysis, E.G.T.; investigation, E.G.T. and R.B.; resources, R.B. and T.D.W.; data curation, E.G.T.; writing—original draft preparation, E.G.T.; writing—review and editing, E.G.T., R.B., T.D.W., S.S. and S.N.; visualization, E.G.T., T.D.W. and R.B.; supervision, R.B. and T.D.W.; project administration, R.B.; funding acquisition, R.B. All authors have read and agreed to the published version of the manuscript.

Funding: This work was supported by the Saskatchewan Agriculture Development Fund, Saskatchewan Pulse Crop Development Board, and Western Grains Research Foundation.

Institutional Review Board Statement: Not applicable.

Informed Consent Statement: Not applicable.

Data Availability Statement: This manuscript includes the essential data used either as tables or figures in the Section 3.

Acknowledgments: The authors thank B. Louie, J. Denis, Z. Wang, S. Ryu, and R. Xiang for their assistance in field measurements, and the Crop Development Centre, University of Saskatchewan Pulse Crop Breeding staff for seeding the trials and plot management.

Conflicts of Interest: The authors declare that they have no conflict of interest.

References

1. Dahl, W.J.; Foster, L.M.; Tyler, R.T. Review of the health benefits of peas (*Pisum sativum* L.). *Br. J. Nutr.* **2012**, *108*. [CrossRef] [PubMed]
2. Guilioni, L.; Wery, J.; Tardieu, F. Heat Stress-induced Abortion of Buds and Flowers in Pea: Is Sensitivity Linked to Organ Age or to Relations between Reproductive Organs? *Ann. Bot.* **1997**, *80*, 159–168. [CrossRef]
3. Bueckert, R.A.; Wagenhoffer, S.; Hnatowich, G.; Warkentin, T.D. Effect of heat and precipitation on pea yield and reproductive performance in the field. *Can. J. Plant Sci.* **2015**, *95*, 629–639. [CrossRef]
4. Tafesse, E.G.; Warkentin, T.D.; Bueckert, R.A. Canopy architecture and leaf type as traits of heat resistance in pea. *Field Crops Res.* **2019**, *241*. [CrossRef]
5. IPCC Summary for policymakers. Managing the Risks of Extreme Events and Disasters to Advance Climate Change Adaptation. In Proceedings of the International Symposium on Adaptation of Vegetables and other Food Crops in Temperature and Water Stress, Geneva, Switzerland, 2–3 May 2012; Core Writing Team, Pachauri, R.K.L.A.M., Eds.; IPCC: Geneva, Switzerland, 2014; Volume 9781107025, p. 151.
6. Sadras, V.O.; Lake, L.; Chenu, K.; McMurray, L.S.; Leonforte, A. Water and thermal regimes for field pea in Australia and their implications for breeding. *Crop Pasture Sci.* **2012**, *63*, 33–44. [CrossRef]
7. Wahid, A.; Gelani, S.; Ashraf, M.; Foolad, M.R. Heat tolerance in plants: An overview. *Environ. Exp. Bot.* **2007**, *61*, 199–223. [CrossRef]

8. Dirks, I.; Raviv, B.; Shelef, O.; Hill, A.; Eppel, A.; Aidoo, M.K.; Hoefgen, B.; Rapaport, T.; Gil, H.; Geta, E.; et al. Green roofs: What can we learn from desert plants? *Isr. J. Ecol. Evol.* **2016**, *62*, 58–67. [CrossRef]
9. Powles, S.B. Photoinhibition of Photosynthesis Induced by Visible Light. *Annu. Rev. Plant Physiol.* **1984**. [CrossRef]
10. Havaux, M. Increased Thermal Deactivation of Excited Pigments in Pea Leaves Subjected to Photoinhibitory Treatments. *Plant Physiol.* **1989**, *89*, 286–292. [CrossRef] [PubMed]
11. Hasanuzzaman, M.; Nahar, K.; Alam, M.M.; Roychowdhury, R.; Fujita, M. Physiological, biochemical, and molecular mechanisms of heat stress tolerance in plants. *Int. J. Mol. Sci.* **2013**, *14*, 9643–9684. [CrossRef] [PubMed]
12. Gamon, J.A.; Surfus, J.S. Assessing leaf pigment content and activity with a reflectometer. *New Phytol.* **1999**, *143*, 105–117. [CrossRef]
13. Holmes, M.G.; Keiller, D.R. Effects of pubescence and waxes on the reflectance of leaves in the ultraviolet and photosynthetic wavebands: A comparison of a range of species. *Plant Cell Environ.* **2002**, *25*, 85–93. [CrossRef]
14. Munaiz, E.D.; Townsend, P.A.; Havey, M.J. Reflectance spectroscopy for non-destructive measurement and genetic analysis of amounts and types of epicuticular waxes on onion leaves. *Molecules* **2020**, *25*, 3454. [CrossRef] [PubMed]
15. Hatfield, J.L.; Gitelson, A.A.; Schepers, J.S.; Walthall, C.L. Application of spectral remote sensing for agronomic decisions. *Agron. J.* **2008**, *100*, S-117–S-131. [CrossRef]
16. Xue, J.; Su, B. Significant remote sensing vegetation indices: A review of developments and applications. *J. Sens.* **2017**, *2017*, 1353691. [CrossRef]
17. Camarillo-Castillo, F.; Huggins, T.D.; Mondal, S.; Reynolds, M.P.; Tilley, M.; Hays, D.B. High-resolution spectral information enables phenotyping of leaf epicuticular wax in wheat. *Plant Methods* **2021**, *17*, 1–17. [CrossRef]
18. Xue, D.; Zhang, X.; Lu, X.; Chen, G.; Chen, Z.H. Molecular and evolutionary mechanisms of cuticular wax for plant drought tolerance. *Front. Plant Sci.* **2017**, *8*, 1–12. [CrossRef] [PubMed]
19. Sánchez, F.J.; Manzanares, M.; De Andrés, E.F.; Tenorio, J.L.; Ayerbe, L. Residual transpiration rate, epicuticular wax load and leaf colour of pea plants in drought conditions. Influence on harvest index and canopy temperature. *Eur. J. Agron.* **2001**, *15*, 57–70. [CrossRef]
20. Arafa, S.A.; Attia, K.A.; Niedbała, G.; Piekutowska, M.; Alamery, S.; Abdelaal, K.; Alateeq, T.K.; Ali, M.A.M.; Elkelish, A.; Attallah, S.Y. Seed Priming boost adaptation in pea plants under drought stress. *Plants* **2021**, *10*, 2201. [CrossRef]
21. Borrell, A.K.; van Oosterom, E.J.; Mullet, J.E.; George-Jaeggli, B.; Jordan, D.R.; Klein, P.E.; Hammer, G.L. Stay-green alleles individually enhance grain yield in sorghum under drought by modifying canopy development and water uptake patterns. *New Phytol.* **2014**, *203*, 817–830. [CrossRef]
22. Ebercon, A.; Blum, A.; Jordan, W.R. A Rapid Colorimetric Method for Epicuticular Wax Contest of Sorghum Leaves. *Crop Sci.* **1977**, *17*, 179–180. [CrossRef]
23. Shepherd, T.; Griffiths, D. The effects of stress on plant cuticular\rwaxes. *New Phytol.* **2006**, *171*, 469–499. [CrossRef] [PubMed]
24. Guo, J.; Xu, W.; Yu, X.; Shen, H.; Li, H.; Cheng, D.; Liu, A.; Liu, J.; Liu, C.; Zhao, S.; et al. Cuticular Wax Accumulation Is Associated with Drought Tolerance in Wheat Near-Isogenic Lines. *Front. Plant Sci.* **2016**, *7*, 1–10. [CrossRef] [PubMed]
25. Tattersall, A.D.; Turner, L.; Knox, M.R.; Ambrose, M.J.; Ellis, T.H.N.; Hofer, J.M.I. The mutant crispa reveals multiple roles for Phantastica in pea compound leaf development. *Plant Cell* **2005**, *17*, 1046–1060. [CrossRef] [PubMed]
26. Lichtenthaler, H.K. Chlorophylls and Carotenoids: Pigments of Photosynthetic Biomembranes. *Methods Enzymol.* **1987**, *148*, 350–382. [CrossRef]
27. Abdel-Aal, E.S.M.; Hucl, P. A rapid method for quantifying total anthocyanins in blue aleurone and purple pericarp wheats. *Cereal Chem.* **1999**, *76*, 350–354. [CrossRef]
28. Murray, J.R.; Hackett, W.P. Dihydroflavonol Reductase-Activity in Relation to Differential Anthocyanin Accumulation in Juvenile and Mature Phase *Hedera-Helix* L. *Plant Physiol.* **1991**, *97*, 343–351. [CrossRef] [PubMed]
29. Tafesse, E.G.; Gali, K.K.; Reddy Lachagari, V.B.; Bueckert, R.; Warkentin, T.D. Genome-wide association mapping for heat and drought adaptive traits in pea. *Genes* **2021**, *12*, 1897. [CrossRef] [PubMed]
30. Rouse, J.W.; Haas, R.H.; Schell, J.A.; Deering, D.W. *Monitoring Vegetation Systems in the Great Plains with ERTS*; NASA: Washington, DC, USA, 1974; Volume I.
31. Gitelson, A.A.; Kaufman, Y.J.; Merzlyak, M.N. Use of a green channel in remote sensing of global vegetation from EOS- MODIS. *Remote Sens. Environ.* **1996**, *58*, 289–298. [CrossRef]
32. Gamon, J.A.; Serrano, L.; Surfus, J.S. The photochemical reflectance index: An optical indicator of photosynthetic radiation use efficiency across species, functional types, and nutrient levels. *Oecologia* **1997**, *112*, 492–501. [CrossRef]
33. Peñuelas, J.; Gamon, J.A.; Fredeen, A.L.; Merino, J.; Field, C.B. Reflectance indices associated with physiological changes in nitrogen- and water-limited sunflower leaves. *Remote Sens. Environ.* **1994**, *48*, 135–146. [CrossRef]
34. Penuelas, J.; Filella, I.; Biel, C.; Serrano, L.; Save, R. The reflectance at the 950-970 nm region as an indicator of plant water status. *Int. J. Remote Sens.* **1993**, *14*, 1887–1905. [CrossRef]
35. Fernandez, G.J. Effective Selection Criteria for Assessing Plant Stress Tolerance. In Proceedings of the International Symposium on Adaptation of Vegetables and other Food Crops in Temperature and Water Stress, Taiwan, 13–18 August 1992; pp. 257–270.
36. SS, S.; MB, W. An analysis of variance test for normality (Complete samples). *Biometrika* **1965**, *52*, 591–611.
37. Levene, H. Robust tests for equality of variances. In *Contributions to Probability and Statistics: Essays in Honor of Harold Hotelling*; Stanford University Press: Redwood City, CA, USA, 1960; pp. 278–292.

38. Giordano, M.; Petropoulos, S.A.; Rouphael, Y. Response and defence mechanisms of vegetable crops against drought, heat and salinity stress. *Agriculture* **2021**, *11*, 463. [CrossRef]
39. Dutta, S.; Mohanty, S.; Tripathy, B.C. Role of Temperature Stress on Chloroplast Biogenesis and Protein Import in Pea. *Plant Physiol.* **2009**, *150*, 1050–1061. [CrossRef] [PubMed]
40. Maunders, M.J.; Brown, S.B.; Woolhouse, H.W. The appearance of chlorophyll derivatives in senescing tissue. *Phytochemistry* **1983**, *22*, 2443–2446. [CrossRef]
41. Plumley, G.; Schmidt, G.W. Light-Harvesting Chlorophyll alb Complexes: Lnterdependent Pigment Synt hesis and Protein Assembly. *Plant Cell* **1995**, *7*, 689–704. [CrossRef] [PubMed]
42. Feng, B.; Liu, P.; Li, G.; Dong, S.T.; Wang, F.H.; Kong, L.A.; Zhang, J.W. Effect of Heat Stress on the Photosynthetic Characteristics in Flag Leaves at the Grain-Filling Stage of Different Heat-Resistant Winter Wheat Varieties. *J. Agron. Crop Sci.* **2014**, *200*, 143–155. [CrossRef]
43. Cui, L.; Li, J.; Fan, Y.; Xu, S.; Zhang, Z. High temperature effects on photosynthesis, PSII functionality and antioxidant activity of two Festuca arundinacea cultivars with different heat susceptibility. *Bot. Stud.* **2006**, *47*, 61–69.
44. Othman, R.; Mohd Zaifuddin, F.A.; Hassan, N.M. Carotenoid biosynthesis regulatory mechanisms in plants. *J. Oleo Sci.* **2014**, *63*, 753–760. [CrossRef] [PubMed]
45. Demmig-Adams, B.; Adams, W.W. The role of xanthophyll cycle carotenoids in the protection of photosynthesis. *Trends Plant Sci.* **1996**, *1*, 21–26. [CrossRef]
46. Heldt, H.-W.; Piechulla, B. *Plant Biochemistry*, 5th ed.; Academic press: London, UK, 2021.
47. Chalker-Scott, L. Environmental significance of anthocyanins in plant stress responses. *Photochem. Photobiol.* **1999**, *70*, 1–9. [CrossRef]
48. Tripathy, B.C.; Oelmüller, R. Reactive oxygen species generation and signaling in plants. *Plant Signal. Behav.* **2012**, *7*, 1621–1633. [CrossRef]
49. Steyn, W.J.; Wand, S.J.E.; Holcroft, D.M.; Jacobs, G. Anthocyanins in vegetative tissues: A proposed unified function in photoprotection. *New Phytol.* **2002**, *155*, 349–361. [CrossRef]
50. Willick, I.R.; Lahlali, R.; Vijayan, P.; Muir, D.; Karunakaran, C.; Tanino, K.K. Wheat flag leaf epicuticular wax morphology and composition in response to moderate drought stress are revealed by SEM, FTIR-ATR and synchrotron X-ray spectroscopy. *Physiol. Plant* **2018**. [CrossRef]
51. Tafesse, E.G. Heat Stress Resistance in Pea (*Pisum sativum* L.) Based on Canopy and Leaf Traits. Ph.D. Thesis, University of Saskatchewan, Saskatoon, SK, Canada, 2018; pp. 1–166.
52. Jordan, W.R.; Shouse, P.J.; Blum, A.; Miller, F.R.; Monk, R.L. Environmental Physiology of Sorghum. II. Epicuticular Wax Load and Cuticular Transpiration1. *Crop Sci.* **1984**, *24*, 1168. [CrossRef]
53. Jenks, M.A.; Rich, P.J.; Peters, P.J.; Axtell, J.D.; Ashworth, E.N. Epicuticular Wax Morphology of Bloomless (bm) Mutants in Sorghum bicolor. *Int. J. Plant Sci.* **1992**. [CrossRef]
54. Acreche, M.M.; Slafer, G.A. Lodging yield penalties as affected by breeding in Mediterranean wheats. *Field Crop. Res.* **2011**. [CrossRef]
55. Peñuelas, J.; Munné-Bosch, S.; Llusià, J.; Filella, I. Leaf reflectance and photo- and antioxidant protection in field-grown summer-stressed Phillyrea angustifolia. Optical signals of oxidative stress? *New Phytol.* **2004**. [CrossRef]
56. Babar, M.A.; Reynolds, M.P.; Van Ginkel, M.; Klatt, A.R.; Raun, W.R.; Stone, M.L. Spectral reflectance indices as a potential indirect selection criteria for wheat yield under irrigation. *Crop Sci.* **2006**. [CrossRef]
57. Lake, L.; Sadras, V.O. Screening chickpea for adaptation to water stress: Associations between yield and crop growth rate. *Eur. J. Agron.* **2016**. [CrossRef]
58. Dhanapal, A.P.; Ray, J.D.; Singh, S.K.; Hoyos-Villegas, V.; Smith, J.R.; Purcell, L.C.; Fritschi, F.B. Genome-wide association mapping of soybean chlorophyll traits based on canopy spectral reflectance and leaf extracts. *BMC Plant Biol.* **2016**, *16*. [CrossRef] [PubMed]
59. Kaler, A.S.; Abdel-Haleem, H.; Fritschi, F.B.; Gillman, J.D.; Ray, J.D.; Smith, J.R.; Purcell, L.C. Genome-Wide Association Mapping of Dark Green Color Index using a Diverse Panel of Soybean Accessions. *Sci. Rep.* **2020**, *10*, 1–11. [CrossRef] [PubMed]
60. Sexton, T.; Sankaran, S.; Cousins, A.B. Predicting photosynthetic capacity in tobacco using shortwave infrared spectral reflectance. *J. Exp. Bot.* **2021**, *72*, 4373–4383. [CrossRef] [PubMed]
61. Tafesse, E.G.; Gali, K.K.; Reddy Lachagari, V.B.; Bueckert, R.; Warkentin, T.D. Genome-wide association mapping for heat stress responsive traits in field pea. *Int. J. Mol. Sci.* **2020**, *21*, 2043. [CrossRef] [PubMed]

Article

Low Temperature Effect on Different Varieties of *Corchorus capsularis* and *Corchorus olitorius* at Seedling Stage

Susmita Dey [1], Ashok Biswas [1,*], Siqi Huang [1], Defang Li [1,*], Liangliang Liu [2], Yong Deng [1], Aiping Xiao [1], Ziggiju Mesenbet Birhanie [1], Jiangjiang Zhang [1], Jianjun Li [1] and Youcai Gong [1]

1 Annual Bast Fiber Breeding Laboratory, Institute of Bast Fiber Crops, Chinese Academy of Agricultural Sciences, Changsha 410205, China; susmita.ag4sau@gmail.com (S.D.); huangsiqi@caas.cn (S.H.); dengyong@caas.cn (Y.D.); xiaoaiping@caas.cn (A.X.); zegje23@gmail.com (Z.M.B.); 82101179212@caas.cn (J.Z.); lijianjun@caas.cn (J.L.); gongyoucai@caas.cn (Y.G.)

2 Department of Chemistry, Institute of Bast Fiber Crops, Chinese Academy of Agricultural Sciences, Changsha 410205, China; liuliangliang@caas.cn

* Correspondence: ashok.ag1sau@gmail.com (A.B.); lidefang@caas.cn (D.L.); Tel.: +86-13873129468 (A.B.); +88-01723-167332 (D.L.)

Abstract: To address the demand for natural fibers, developing new varieties that are resistant to abiotic stress is necessary. The present study was designed to investigate the physiological and biochemical traits of three varieties of *C. capularis* (Y49, Y38, and Y1) and four varieties *C. olitorius* (T8, W57, M33, M18) under low temperature to identify the cold-tolerant varieties and elucidate the mechanisms involved in enhancing cold tolerance. Research findings revealed that the varieties Y49 and M33 exhibited the highest chlorophyll and carotenoid content. Biochemical profiles revealed that varieties Y49 and M33 were found to be able to withstand low-temperature stress by accumulating different enzymatic and non-enzymatic antioxidants, such as superoxide dismutase (SOD), peroxidase (POD), catalase (CAT), ascorbate peroxidase (APx), glutathione (GSH), and phenolics, which participated in reducing the content of malondialdehyde (MDA) and hydrogen peroxide (H_2O_2) caused by low temperature. Osmolytes compounds, such as total soluble sugar, significantly increased in Y49 and M33; and proline content decreased in all varieties except Y49 and M33 after low-temperature exposure. The rise in these osmolytes molecules can be a defense mechanism for the jute's osmotic readjustment to reduce the oxidative damage induced by low temperature. Furthermore, PCA and hierarchical cluster analysis distinguished the seven varieties into three separate groups. Results confirmed that group I (Y49 and M33 varieties) were low-temperature tolerant, group II (M18, W57) were intermediate, whereas III groups (Y38, T8, and Y1) were low temperature susceptible. PCA also explained 88.36% of the variance of raw data and clearly distinguished three groups that are similar to the cluster heat map. The study thus confirmed the tolerance of selected varieties that might be an efficient adaptation strategy and utilized them for establishing breeding programs for cold tolerance.

Keywords: antioxidant activities; low-temperature stress; seedling stage; physiological and biochemical response

Citation: Dey, S.; Biswas, A.; Huang, S.; Li, D.; Liu, L.; Deng, Y.; Xiao, A.; Birhanie, Z.M.; Zhang, J.; Li, J.; et al. Low Temperature Effect on Different Varieties of *Corchorus capsularis* and *Corchorus olitorius* at Seedling Stage. *Agronomy* **2021**, *11*, 2547. https://doi.org/10.3390/agronomy11122547

Academic Editors: Channapatna S. Prakash, Ali Raza, Xiling Zou and Daojie Wang

Received: 11 November 2021
Accepted: 13 December 2021
Published: 15 December 2021

1. Introduction

Low temperature is a significant abiotic factor in China limiting plant dispersion on land, impeding plant growth and crop productivity, yield, and quality, and limiting the geographical area suited for cultivating a specific plant species [1–3]. Crops are exposed to periods of intense cold in many parts of the world [4], and tropical plants are more susceptible to chilling than plants growing in cold climates [5].

Jute is an annual herb in the Malvaceae family, with two commercial species, *C. capsularis* and *C. olitorious* [6]. Jute is produced in twenty countries, although Bangladesh, India, and China account for 85% of global production [7,8]. Jute fiber is naturally occurring, soft, shiny, longest, strongest, and most recyclable derived from stem bark [8]. Recently, jute has gained popularity as herbal medicine, renewable biofuel, and paper fiber [8–10]. Thus, the global demand for jute is increasing [11]. As global environmental awareness rises, more people are actively purchasing environmentally friendly products. While jute is a natural fiber, they have many composite-like properties like rigidity and flexibility. Jute would be a unique source to supply the global demand for eco-friendly fibers.

It provides luxuriant growth in a warm and humid climate with temperatures between 24 °C and 37 °C for optimum fiber yield, and the growth rate gradually slows down when temperatures fall below this range [12]. It has been reported that jute is a short-day fiber plant. The majority of biological responses that occur in jute are temperature dependent. Any early planting provides premature flowering and reduces plant growth and yield of fiber due to thermal sensitivity. It was reported that some varieties can be planted early with the absence of premature flowering in appropriate sowing time [12]. In China's subtropics, where cold weather is unpredictable, extending the growing season (early planting and late harvesting) is crucial. Intensive cropping during late March and early April could make jute more profitable in China's subtropics and warm temperate zones, as well as in jute growing countries. It was proved that appropriate sowing and harvesting could allow facilities to fit the crop in three cropped patterns [12]. If an intensive cropping pattern could be established in which jute is produced from late February to early March, it is believed that jute could be more profitably cultivated in Asia's jute producing countries [13]. For this purpose, new varieties that are tolerant to low temperatures will be required; therefore, new varieties should be developed to endure various biotic and abiotic stresses.

However, there is no proof of physiological or biochemical investigation for jute's cold tolerance mechanism. In a previous study carried out only on the molecular level, where DNA fingerprinting randomly amplified polymorphic DNA (RAPD) and automated amplified fragment length, polymorphism (AFLP) was used to detect or distinguish between cold-tolerant and cold-sensitive jute species was assessed [13]. Findings indicated that eight primer combinations distinguished the two cold-sensitive and four cold-tolerant jute populations using 93 polymorphic fragments. Understanding low-temperature adaptation is crucial to developing cold-tolerant crops. It has been reported that extreme temperature can cause changes in numerous physiological, biochemical, molecular, and metabolic processes, including membrane fluidity, enzyme activity, and homeostatic metabolism, which can impact agriculture [14] by overproducing reactive oxygen species (ROS), such as superoxide anions (O_2^-) and hydrogen peroxide (H_2O_2). Interestingly, plants have antioxidant systems with various enzymatic and non-enzymatic components to protect them from the injury caused by reactive oxygen species ROS [15]. Plants have evolved many antioxidant systems and osmolytes to cope with stress. These include superoxide dismutase (SOD), peroxidase (POD), catalase (CAT), glutathione (GSH), and proline, all of which contribute to scavenging H_2O_2 with different mechanisms under stressful conditions and avoid oxidative damage [16–18]. Understanding low-temperature adaption processes are crucial for the development of cold-tolerant crops. This research report summarizes the physiological responses of a representative sample of jute varieties to cold stress, as ascertained by quantification of photosynthetic parameters, ROS-mediated damage, antioxidant accumulation, and osmolyte accumulation, all of which differentiate the sensitive and tolerance jute phenomes. The main purpose of the present study was to evaluate low temperature physiological, biochemical, and antioxidant defense responses to verify the tolerance level of seven selected *C. capsularis* and *C. olitorius* varieties from earlier experiments.

2. Materials and Methods

2.1. Plant Materials, Growth Conditions, and Cold Stress Treatment

Seeds of *Corchorus capularis* (Y49, Y38, and Y1) and *Corchorus olitorius* (T8, W57, M33, M18) varieties were collected from the Institute of Bast Fiber Crops (IBFC), Chinese Academy of Agricultural Sciences (CAAS) Changsha, Hunan (Table 1). These varieties were previously screened out from large populations by studying low-temperature stress based on their germination rate, survival rate, visual scoring under cold stress and physiological, and biochemical parameters. The selected jute varieties were distinguished as tolerant (Y49 and M33), intermediate (W57 and M18), and sensitive (Y38, T8 and Y1) to low temperature and used in this present study. Seeds of all varieties were carefully rinsed with sterile deionized water after 10 min of surface sterilization with 10% NaClO. Then seeds were placed in a 19 × 14.5 × 6 cm germination box with three layers of sterile filter paper. The box was placed in a brightly illuminated incubator set at 25 °C with daily water top ups for germination. After three days of sprouting, seedlings were moved to quarter-strength Hoagland nutritional solution containing 5.79 mmol L^{-1} calcium and placed in culture pots (40 × 20 cm) $(NO_3)_2$, 8.02 μmmol L^{-1} KNO_3, 1.35 mmol L^{-1} $NH_4H_2PO_4$, 4.17 mmol L^{-1} $MgSO_4$, 8.90 μmol L^{-1} $MnSO_4$, 48.3 μmol L^{-1} H_3BO_3, 0.94 μmol L^{-1} $ZnSO_4$, 0.20 μmol L^{-1} $CuSO_4$, 0.015 μmol L^{-1} $(NH_4)_2MoO_4$, and 72.6 μmol L^{-1} Fe-EDTA for subsequent growth [19]. Seedlings were grown in a culture room with a day/night temperature regime of 28/16 °C, a photoperiod of 16 h/8 h (light/dark), relative humidity of 60%, and a light intensity of 300 μmol $m^{-2}s^{-1}$. Every other day, the nutrition solution was replenished.

Table 1. Origin of 7 varieties of *C. capsularis* and *C. olitorius*.

	Species	Variety	Origin (Province/Country)
Y49	*C. capsularis*	Huangma 971	Hunan
Y38	*C. capsularis*	Miandianyuanguo	Myanmar
Y1	*C. capsularis*	Longxihongpi	Longxi county, Guangdong
T8	*C. olitorius*	T8	Zhejiang
W57	*C. olitorius*	W57	Zhejiang
M33	*C. olitorius*	Funong 5	Fujian
M18	*C. olitorius*	Maliyengshengchangguo	Mali

Five-week-old morphologically uniform seedlings were selected for treatment and transferred to another chamber for low-temperature stress. The treatment chamber's temperature was set at 5 °C to simulate the low-temperature condition, and the plant maintained an optimal condition (28 °C) that was regarded as control. After 24 h of low-temperature stress, three to four fully expanded leaves were collected. Each sample contained a minimum of three individual plants of similar varieties mixed to form one sample. After collection, each sample tube was immediately steeped in liquid nitrogen and stored at −80 °C until analysis.

2.2. Determination of Photosynthetic Pigment Contents

About 0.1 g fresh leaf samples were homogenized with 10 mL (4.5:4.5:1) mixed solution of ethanol, acetone, and distilled water until the green-colored leaf sample turned white. After that, absorbance readings were recorded at 645, 663, and 470 nm, and concentration (mg/g FW) of chlorophyll *a* (Chl *a*), Chlorophyll *b* (Chl *b*), total chlorophyll (Chl t), and carotenoid were calculated by the formula [20].

$$\text{Chlorophyll } a \text{ (mg/g leaf fresh weight)} = [12.7\,(OD_{663}) - 2.69\,(OD_{645})] \times V/1000 \times W \quad (1)$$

$$\text{Chlorophyll } b \text{ (mg/g leaf fresh weight)} = [22.9(OD_{645}) - 4.68\,(OD_{663})] \times V/1000 \times W \quad (2)$$

$$\text{Total chlorophyll (mg/g leaf fresh weight)} = [20.2\,(OD_{645}) + 8.02\,(OD_{663})] \times V/1000 \times W \quad (3)$$

$$\text{Carotenoid (mg/g leaf fresh weight)} = [OD470 + (0.114 * OD_{663}) - (0.638 * OD_{645})] \times V/1000 \times W \quad (4)$$

According to Sairam et al., the chlorophyll stability index (CSI) was developed [21]. It is calculated as follows: CSI = (Total Chl under stress/Total Chl under control) × 100.

2.3. Determination of Osmolyte Contents

The proline concentrations were determined using a slightly modified method reported by Bates, 1973 [22]. The proline content of fresh leaf samples (0.1 g) was assayed using aqueous 3% sulfosalicylic acid (5 mL) in a water bath for 10 min. Following this, the mixture was allowed to cool to ambient temperature, and then the resulting extract was filtered using filter paper. The supernatant from the second step was then combined with ninhydrin and acetic acid to make a 2.0 mL solution. The combination was then maintained in a boiling water bath for 30 min, and the reactions were terminated in an ice bath. After adding 5 mL of toluene, the mixture was left in the dark for 5 h. The absorbance of colored toluene at 520 nm was measured using the UV/Vis spectrophotometer (UV 2700, Shimadzu, Japan), with toluene serving as a blank. The amount of proline was tested using a standard curve constructed with L-proline common solution.

The amount of total soluble sugar (TSS) was determined using the anthrone method proposed by Yemn and Willis [23]. About 0.1 g fresh samples were placed in 10 mL cuvette, and after adding 10 mL distilled water, samples were heated at 100 °C for 1 h and then filtered into 25 mL volumetric flasks. The volumetric flask was filled up to mark by distilled water. Following that, 0.5 mL extracts, 0.5 mL mixed reagents (1 g anthrone + 50 mL ethyl acetate), 5 mL H_2SO_4 (98%), and 1.5 mL distilled water were added. After heating the mixture to 100 °C for a minute, the 630 nm absorbance was measured. Sucrose solution was used as a standard sample. The concentration of soluble sugar was measured using glucose as a standard solution.

2.4. Determination of Oxidative Damage and Enzymatic Antioxidant Activities

To determine the degree of oxidative damage and different antioxidant activity, 0.2 g fresh leaf was extracted with 5% thiobarbituric acid (TBA) dissolved in 5% trichloroacetic acid (TCA) for MDA content detection and homogenized in 0.2 M phosphate buffers (pH 7.0~7.5) for SOD, POD, APx, and CAT activity detection and GSH content. The assessed activities of MDA, H_2O_2, GSH, SOD, POD, CAT, and APx using the assay test kits were purchased from Nanjing Jian Cheng Bioengineering Institute in Nanjing, China. Protein content was determined using the Bradford protein colorimetry method with bovine serum albumin (BSA) as a protein standard [24]. Briefly, in a 50 mL 95% ethanol solution, 100 mg Coomassie Brilliant Blue G-250 was dissolved (C_2H_5OH). Following that, 100 mL of 85% phosphoric acid (H_3PO_4) was carefully added while stirring and then diluted with distilled water to a total volume of 1 L. The solution was filtered and maintained at a temperature of 4 °C. For the measurements, 20 μL extract and 200 μL Bradford solution were combined and incubated for 5 min before determining the absorbance at 595 nm using a UV/Vis spectrophotometer (UV 2700, Shimadzu, Japan).

2.5. Determination of Non-Enzymatic Antioxidant Compounds

The leaf sample (0.2 g) was homogenized in 10 mL of 80% ethanol to determine the total phenolic and flavonoids. The ethanolic extract was then centrifuged at 12,000× *g* for 20 min at 4 °C, and the supernatants were collected in the flask. The supernatant was then utilized to determine the total phenolic and flavonoid content.

The total flavonoid content (TFC) in the sample was determined using a modified aluminum chloride assay reported [25] with rutin as standard. Briefly, 1 mL of the extract was placed in a 10 mL volumetric tube. To begin, each volumetric flask was filled with 2 mL of 0.1 M $AlCl_3$ and incubated for 5 min. Then, 3 mL of 1 M CH_3COOK was added; the volume was topped up 10 mL with 80% ethanol and thoroughly mixed. At 420 nm, the absorbance was measured against a blank using a UV/Vis spectrophotometer (UV 2700, Shimadzu, Japan). The results were expressed in mg rutin equivalents per gram dry weight.

The concentration of total polyphenol content (TPC) was determined using the Folin–Ciocalteu colorimetric technique with minor modifications [26]. Briefly, 1.5 mL distilled water was added to 0.5 mL extracted plant samples in a test tube. After adding 0.2 mL FC reagent, the mixture was gently oscillated and maintained at room temperature for 4 min. Following that, each sample was vortexed with 0.8 mL of newly prepared 10% (*w*/*w*) Na_2CO_3. The mixtures were left for 1 h in a dark room condition to ensure a good reaction. The absorbance of the solution was determined using a UV/Vis spectrophotometer (UV

2700, Shimadzu, Japan) at 765 nm compared to the reagent blank. To estimate all of the determinations, three biological replications were carried out. The total phenolic content was reported in milligrams of gallic acid standard equivalent (mg) dry weight (mg of GAE/g DW).

3. Statistical Analysis

The data were examined using the one-way analysis of variance (ANOVA) using SPSS 16. The effect of treatment of each variety was determined compared to the control, and the statistical differences between control and treatment were performed using a least significant difference (LSD) test when $p < 0.05$. The correlation coefficient was determined using Pearson's correlation coefficient. All data were transformed to stress tolerance indices prior to Pearson's correlation, principal component analysis (PCA), and cluster analysis. The stress tolerance index was described as the observed value of a target trait when subjected to a particular level of stress divided by its mean value under control. To investigate the relationship between varieties and cluster features, principal component analysis (PCA) and cluster heat map analysis were performed using the Origin software and an online ClustVis application, respectively.

4. Results

4.1. Effect of Cold Stress on Photosynthetic Pigment Contents

Low temperature impacts chlorophyll level in plant species based on their cold tolerance [27,28]. To determine the effect of cold stress on photosynthesis, we measured chlorophyll *a* and *b* concentrations and total chlorophyll in jute plant leaves. The results of our experiment indicated that the production of photosynthetic pigments (Chl *a*, Chl *b*, and Total Chl) varied significantly between selected varieties, with variation patterns being comparable across all varieties (Table 2). It was observed that the total chlorophyll concentration of leaves declined with cold stress compared to their control. Greater decrease was observed in the varieties Y38 and T8, whereas a lower reduction rate was observed in Y49 and M33 varieties with the highest chlorophyll stability index (CSI) as 97.96% and 90.97%, respectively. Our findings suggested that these varieties were less harmed at 5 °C and more resistant to chilling injury than others, implying that they possessed a greater cold tolerance. In the case of the carotenoid content, a similar trend was observed as chlorophyll; except for Y49, all varieties showed declined trend compared to control. Meanwhile, the Y38 and T8 recorded significantly decreased by 18.46% and 19.62%, respectively, relative to control, and an increase and slight decrease was observed in Y49 and M33, respectively. This result suggests that Y49 and M33 have more low-temperature tolerance potential, whereas Y38 and T8 demonstrate a low degree of tolerance.

4.2. Oxidative Stress Evaluation

Lipid peroxidation, reflected by MDA content, usually accompanies ROS accumulation. Our experiment examined the effects of MDA content, and it showed that concentrations were considerably greater than the control. The highest increase was observed in Y38 (91.26%) and Y1 (65.81%), followed by T8 (48.82%). Whereas low level of MDA content was observed in Y49 (17.25%) and M33 (11.50%) varieties compared to optimal temperature (Figure 1a). The above result indicated that Y49 and M33 varieties suffered the least and cell membrane experienced little damage under cold injury, whereas Y1, Y38, and T8 varieties might have suffered severe and irreversible damage. The result of this study demonstrated that H_2O_2 content significantly increased under cold treatment conditions compared to control based on mean comparisons (Figure 1b). However, the lowest increase was noted in the varieties Y49 (11.64%) and M33 (17.25%). Whereas the Y38, T8, and M18 varieties showed a greater increase of H_2O_2 by 30.70~46.61% under low-temperature treatment than in the optimum temperature.

Table 2. Changes of chlorophyll content compared to control, chlorophyll stability index (CSI), and carotenoid in different varieties under exposure to cold stress.

Genotypes	Chlorophyll									Carotenoid	
	Chl *a*			Chl *b*			Total Chl				
	Control	Treatment	CSI%	Control	Treatment	CSI%	Control	Treatment	CSI%	Control	Treatment
Y49	1.77 ± 0.01^{a}	1.73 ± 0.0^{a}	97.81	0.61 ± 0.0^{a}	0.60 ± 0.01^{b}	98.42	2.39 ± 0.0^{a}	2.34 ± 0.0^{b}	97.97	0.127 ± 0.0^{b}	0.129 ± 0.0^{a}
Y1	1.93 ± 0.02^{a}	1.23 ± 0.01^{b}	64.14	0.50 ± 0.04^{a}	0.77 ± 0.02^{b}	152.87	2.43 ± 0.01^{a}	2.00 ± 0.03^{b}	82.44	0.13 ± 0.0^{a}	0.12 ± 0.0^{b}
Y38	1.94 ± 0.0^{a}	1.10 ± 0.02^{b}	56.85	0.44 ± 0.02^{a}	0.31 ± 0.02^{b}	69.18	2.39 ± 0.01^{a}	1.41 ± 0.01^{b}	59.17	0.12 ± 0.01^{a}	0.10 ± 0.0^{b}
M33	1.82 ± 0.0^{b}	1.85 ± 0.0^{a}	101.67	0.55 ± 0.04^{a}	0.31 ± 0.01^{b}	55.62	2.37 ± 0.0^{a}	2.16 ± 0.02^{b}	90.97	0.12 ± 0.0^{a}	0.10 ± 0.0^{b}
T8	1.75 ± 0.03^{a}	1.10 ± 0.02^{b}	63.23	0.46 ± 0.0^{a}	0.44 ± 0.0^{b}	94.93	2.20 ± 0.01^{a}	1.54 ± 0.02^{b}	69.88	0.11 ± 0.0^{a}	0.09 ± 0.0^{b}
W57	1.51 ± 0.0^{a}	1.21 ± 0.0^{b}	80.43	0.32 ± 0.0^{a}	0.25 ± 0.0^{b}	78.95	1.83 ± 0.01^{a}	1.46 ± 0.0^{b}	80.17	0.10 ± 0.0^{a}	0.09 ± 0.0^{a}
M18	1.61 ± 0.02^{a}	1.24 ± 0.03^{b}	77.03	0.53 ± 0.0^{a}	0.48 ± 0.01^{a}	110.51	2.09 ± 0.0^{a}	1.77 ± 0.03^{b}	84.71	0.109 ± 0.0^{a}	0.105 ± 0.0^{b}

Significant difference at $p < 0.05$ probability level using LSD test. Values in the table represent means and standard errors. Different lowercases indicate difference significant at 0.05 level.

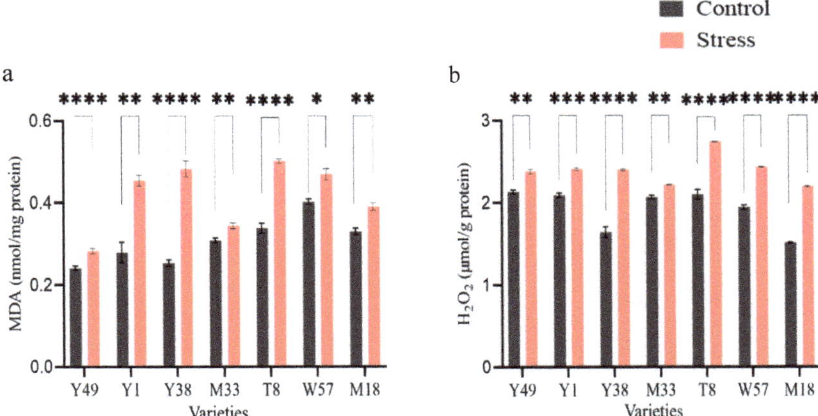

Figure 1. Influence of cold stress on leaf antioxidant and enzymatic antioxidants in jute varieties. Plant under control (28 °C) and stress (5 °C) showed a response in: (**a**) Lipid peroxidation MDA, (**b**) Hydrogen peroxide (H_2O_2). Different significant level marked with **** $p < 0.0001$, *** $p \leq 0.001$, ** $p \leq 0.01$, * $p \leq 0.05$, and ns mean non-significant.

4.3. Effect of Cold Stress on Osmolytes Contents

The present investigation has shown that proline content is higher in leaves of stress treatment than under room temperature when exposed to 5 °C (Figure 2a). The most elevated proline was recorded in Y49, followed by M33 with 67.84% and 60.0% increased rate on cold stress compared to control condition, whereas the lowest increase was recorded in T8 (7.89%) and Y38 (11.39%)

In comparison to the control group after 24 h at 5 °C, our results revealed that total soluble sugar content of the leaves in most of the varieties demonstrated decline except Y49 and M33 (significantly increased by 16.77~24.99% and reached in highest in low-temperature treatment compared to other) (Figure 2b). Conversely, the highest rate of decrease was observed in the varieties T8 (30.07%) and Y38 (27.81%) compared to other varieties in stress treatment.

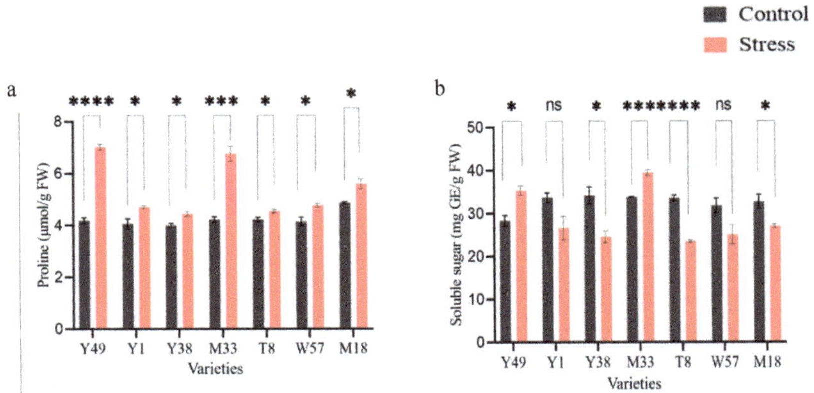

Figure 2. Response of jute varieties to low temperature treatments, with control (28 °C) and stress (5 °C) showing differential response in osmolyte content. (**a**) Proline content, (**b**) Soluble sugar content. Different significant level marked with **** $p < 0.0001$, *** $p \leq 0.001$, * $p \leq 0.05$, and ns mean non-significant.

4.4. Enzymatic Antioxidant Activities (SOD, POD, CAT, and APx)

Low-temperature treatment caused a significant decrease in SOD activity in all varieties except Y49 and M33 in cold-induced leaves shown in Figure 3a. The activity of SOD during low-temperature stress was dramatically reduced in most of the varieties compared with control. In contrast, the Y49 and M33 had an increasing trend, which only increased by 5.16% and 14.13%, respectively.

Our experiment revealed that POD activity significantly decreased in Y38, T8 followed by W57 cultivar compared to control (Figure 3b). Conversely, POD activity was increased higher in both Y49 and M33 cultivars. There was a substantial difference in CAT activity between all varieties between control and stress conditions in our study. CAT activity decreased significantly in most of the varieties compared to the control (Figure 3c). There was a much lower level of CAT activity in Y1, Y38, and T8 after low temperature stresses Y49 and M33 than the control, whereas Y49 and M33 increased the CAT activity by 50.87% and 25.80%, respectively.

Cold treatment also had a significant effect on APx activity in the treatment leaves; however, cold treated leaves of the Y49, Y33, M33, T8, and M18 varieties exhibited an increasing trend in APx activity, whereas T8 and W57 exhibited a decreasing trend (Figure 3d). It was observed that Y49 and M33 enhanced by 47.5% and 84.79%, respectively compared with those of the non-stress condition. Increased APX activity in low temperature-treated leaves of Y49 and M33 may indicate that Apx interferes with cold signal transduction.

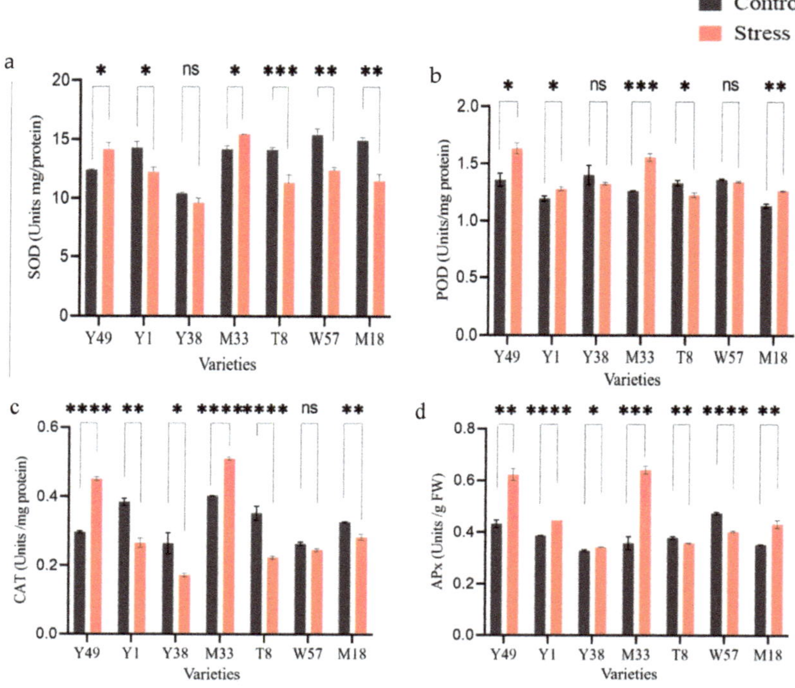

Figure 3. Influence of cold stress on leaf antioxidant and enzymatic antioxidants in jute varieties. Plants under control (28 °C) and stress (5 °C) showed a response in: (**a**) Superoxidase dismutase (SOD) activity, (**b**) Peroxidase (POD) activity (**c**) Catalase (CAT) activity and (**d**) APx activity. Different significant level marked with **** $p < 0.0001$, *** $p \leq 0.001$, ** $p \leq 0.01$, * $p \leq 0.05$, and ns mean non-significant.

4.5. Non-Enzymatic Antioxidant Activities (TFC, TPC, and GSH)

Our findings showed that the controlled and low temperature-treated leaves accumulated phenolic compounds differently during cold stress. Determining phenolic contents in response to stress can help research the cultivation tolerance mechanism and crop loss minimization. When exposed to stress, phenolic compounds serve as antioxidants and activate the cell's enzyme system [29]. In Figure 4a, it was noticed that most of the varieties showed a decreasing trend in low-temperature treatment, whereas the Y49 and M33 varieties showed significantly accelerated by 17.71% and 32.94%, respectively, and most reduced level was observed in T8 (12.08%) and Y1 (11.75%) compared to non-stressed conditions.

Like TPC, total flavonoid content (TFC) was enormously accelerated in Y49 (23.96%) and M33 (4.94%); whereas a high reduction rate was recorded in Y1, T38, T8, and W57 in stress conditions compared to control (Figure 4b). In the case of GSH, after 24 h low temperature stress, GSH contents were significantly increased in varieties Y49 (78.66%), M33 (45.65%), and M18 (4.86%) (Figure 4c). In contrast, the rest of the varieties demonstrated a decrease from 10.82~47.99%. The higher GSH level and homeostasis enhanced the antioxidant and glyoxalase systems' activity to alleviate cold-induced damage in the cold tolerant.

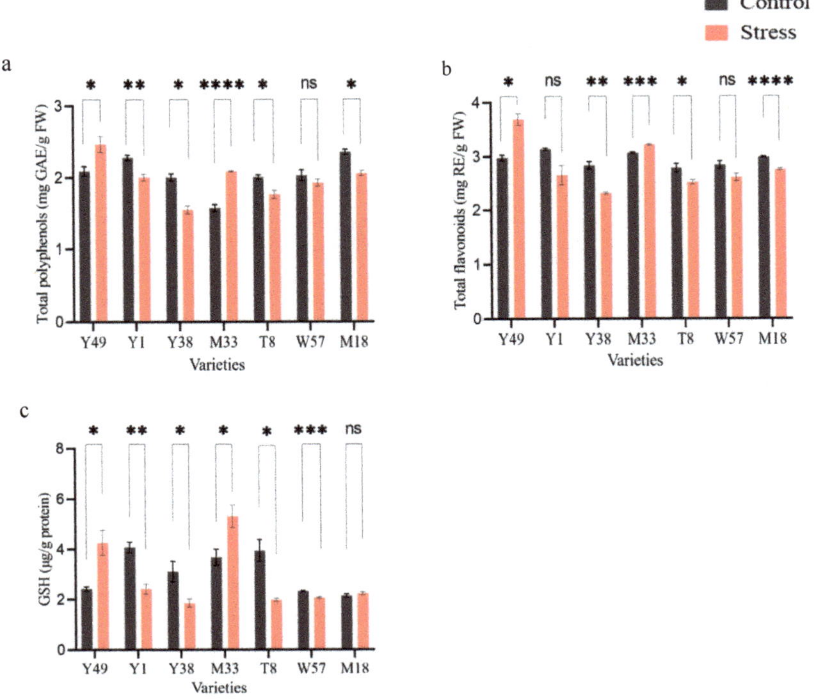

Figure 4. Influence of cold stress on enzymatic and non-enzymatic antioxidants in jute varieties. Plants under control (28 °C) and stress (5 °C) showed a response in: (**a**) Total polyphenol, (**b**) Total flavonoid, and (**c**) GSH activity. Different significant level marked with **** $p < 0.0001$, *** $p \leq 0.001$, ** $p \leq 0.01$, * $p \leq 0.05$, and ns mean non-significant.

4.6. Principle Component Analysis (PCA)

PCA is a multivariate exploratory technique used to reduce the multidimensionality of the facts and provide a two-dimensional map that explains the determined variance. In the PCA analysis, eigenvalues greater than 1 were regarded as significant and loading plots allowed for easy visualization of biochemical parameters and varieties of *C. capsularis* and

C. olitorius (Figure 5). Cumulative PCA biplot contributes 88.36% of the total variability of the studied parameters, while PC1 accounts for 78.98% and PC2 accounts for 9.38% of the original data in this study. The angle of the trait vectors reflected the correlation of variables. A lower angle between distinct factors pointing in the same direction suggested a strong correlation between the respective varieties' classification criteria. The broad spectrum distribution of measured parameters in this biplot showed the differential correlation (positively and negatively to different PC groups) with each other. SOD, APx, proline, soluble sugar, total polyphenols (TPC), and POD were clustered on the right upper side of the biplot with positive loading, indicating that these parameters exhibited a significant degree of positive correlation among themselves. Total flavonoids, CAT, GSH, total chlorophyll, and carotenoid were all located on the right lower side of the biplot, indicating a positive association between these measures. MDA and H_2O_2 were detected on the left upper section, indicating that these parameters had a strong negative and substantial association with one another.

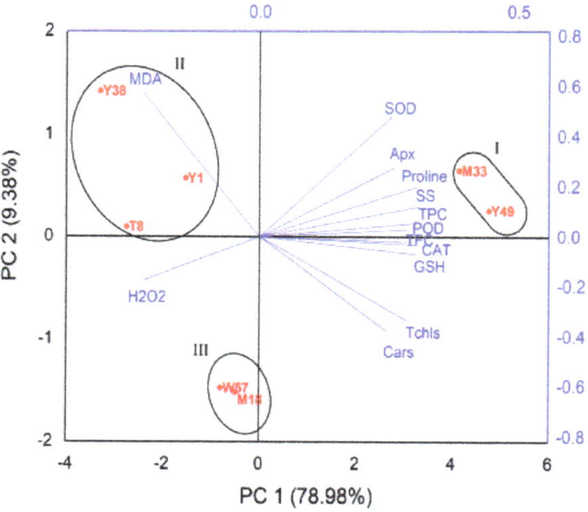

Figure 5. Biplot for the first two principal components (PC1 and PC2) was shown using the principal component analysis (PCA) with all measured parameters and 7 varieties.

Under the low-temperature stress, the varieties Y49 and M33 were clustered together on the right side of the biplot with positive loading, and these two varieties were considered for low-temperature tolerance potential. The varieties Y38, T8, and Y1 were clustered together in the direction of MDA and H_2O_2 to the upper area of the biplot, and relatively poor performance of different enzymatic and non-enzymatic antioxidants activities indicated increased susceptibility to low-temperature stress. Whereas the varieties W57 and M18 shifted to the lower portion and were classified as intermediate to cold stress.

4.7. Cluster Heat Map Analysis

Further, cluster heat map approach to hierarchical cluster analysis using the average linkage via Ward's method of agglomeration. According to the heatmap, seven jute varieties were divided into three main groups based on their varieties' potential (Figure 6) consistent with PCA. The distribution pattern showed that cluster-I contained Y49 and M33 had the highest mean STI (1.26) values based on physiological and biochemical parameters. Therefore, the varieties in Cluster-I can be considered as low temperature-tolerant potential varieties. On the other hand, three varieties viz. Y1, T8, and Y38 were clustered in Cluster-II represented that sensitive group with the lowest mean STI (0.93). The rest of two varieties (W57 and M18) were found in Cluster-III with moderate mean STI (0.98) value. Comparing the score plots to the cluster heat map revealed that the PCA score plots corresponded to the HCA scores, suggesting tolerant, intermediate, and sensitive varieties.

Figure 6. Cluster heat map results were obtained based on biochemical parameters in low-temperature stress conditions.

4.8. Genotypic Variation under Cold Stress

In our study, radar plot examination of STI values revealed that the jute varieties responded differently to cold treatment (Figure 7). The varieties Y49 and M33 had the highest average STI values, while the variety T8 had the lowest for all parameters except MDA and H_2O_2 concentration (Table 3). Under cold stress, the Y1, Y38, and T8 varieties collected the highest STI of MDA and H_2O_2 content, whereas M33 and Y49 varieties accumulated the lowest STI of MDA and H_2O_2 under the same conditions. Additionally, Y49 and M33 varieties maintained the higher STI value when exposed to cold stress conditions.

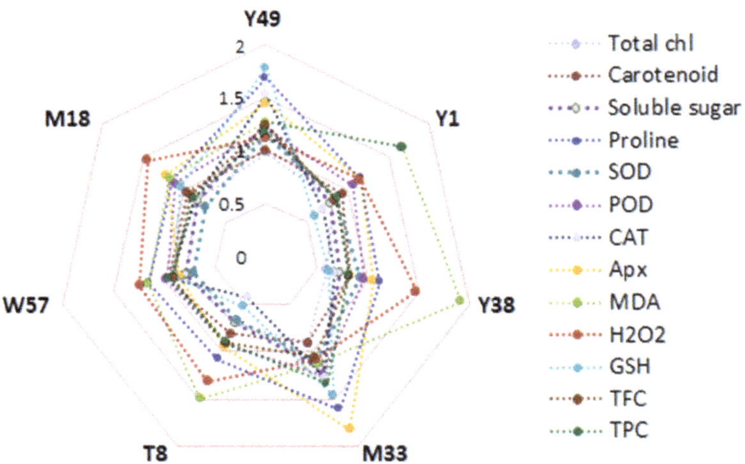

Figure 7. Radar plot represents the varietal variations in biochemical traits at low-temperature stress.

Table 3. Stress tolerance index (STI) with their mean.

Genotypes	Total chl	Carotenoid	Soluble Sugar	Proline	SOD	POD	CAT	APx	MDA	H_2O_2	GSH	TFC	TPC	Mean STI
Y49	0.98 ± 0.0 ^{a}	1.00 ± 0.01 ^{a}	1.25 ± 0.04 ^{a}	1.68 ± 0.06 ^{a}	1.14 ± 0.04 ^{a}	1.21 ± 0.07 ^{a}	1.52 ± 0.01 ^{a}	1.44 ± 0.07 ^{b}	1.17 ± 0.02 ^{c}	1.12 ± 0.01 ^{c}	1.79 ± 0.27 ^{a}	1.24 ± 0.05 ^{a}	1.18 ± 0.03 ^{b}	1.29
Y1	0.66 ± 0.01 ^{c}	0.94 ± 0.01 ^{bc}	0.79 ± 0.08 ^{b}	1.17 ± 0.06 ^{b}	0.86 ± 0.0 ^{bc}	1.07 ± 0.03 ^{bc}	0.69 ± 0.05 ^{d}	1.16 ± 0.0 ^{cd}	1.66 ± 0.17 ^{ab}	1.15 ± 0.02 ^{c}	0.59 ± 0.03 ^{cd}	0.85.05 ^{c}	0.88 ± 0.02 ^{cd}	0.96
Y38	0.59 ± 0.0 ^{e}	0.82 ± 0.02 ^{d}	0.72 ± 0.05 ^{b}	1.11 ± 0.04 ^{b}	0.92 ± 0.03 ^{b}	0.95 ± 0.06 ^{cd}	0.67 ± 0.05 ^{d}	1.05 ± 0.0 ^{cde}	1.91 ± 0.10 ^{a}	1.47 ± 0.06 ^{a}	0.61 ± 0.06 ^{cd}	0.82 ± 0.02 ^{c}	0.77 ± 0.02 ^{d}	0.96
M33	0.91 ± 0.03 ^{b}	0.99 ± 0.01 ^{a}	1.17 ± 0.02 ^{a}	1.60 ± 0.02 ^{a}	1.10 ± 0.02 ^{a}	1.24 ± 0.02 ^{a}	1.27 ± 0.01 ^{b}	1.82 ± 0.17 ^{a}	1.12 ± 0.01 ^{c}	1.07 ± 0.01 ^{c}	1.46 ± 0.11 ^{a}	1.05 ± 0.0 ^{b}	1.33 ± 0.04 ^{a}	1.24
T8	0.70 ± 0.01 ^{d}	0.80 ± 0.0 ^{d}	0.70 ± 0.0 ^{b}	1.08 ± 0.03 ^{b}	0.80 ± 0.05 ^{c}	0.92 ± 0.03 ^{d}	0.64 ± 0.0 ^{d}	0.95 ± 0.01 ^{de}	1.49 ± 0.06 ^{b}	1.31 ± 0.03 ^{b}	0.52 ± 0.08 ^{d}	0.91 ± 0.03 ^{c}	0.88 ± 0.04 ^{cd}	0.90
W57	0.80 ± 0.0 ^{c}	0.9 ± 0.01 ^{ab}	0.79 ± 0.05 ^{b}	1.16 ± 0.06 ^{b}	0.80 ± 0.02 ^{c}	0.99 ± 0.01 ^{bcd}	0.94 ± 0.02 ^{c}	0.85 ± 0.0 ^{e}	1.17 ± 0.05 ^{c}	1.25 ± 0.01 ^{b}	0.89 ± 0.02 ^{bc}	0.92 ± 0.03 ^{c}	0.95 ± 0.05 ^{c}	0.96
M18	0.85 ± 0.01 ^{c}	0.96 ± 0.01 ^{b}	0.83 ± 0.04 ^{b}	1.15 ± 0.03 ^{b}	0.77 ± 0.05 ^{c}	1.11 ± 0.01 ^{ab}	0.87 ± 0.02 ^{c}	1.23 ± 0.03 ^{bc}	1.18 ± 0.03 ^{c}	1.45 ± 0.0 ^{a}	1.05 ± 0.05 ^{b}	0.92 ± 0 ^{c}	0.90 ± 0.01 ^{c}	1.02

Significant difference at $p < 0.05$ probability level using LSD test. Values in the table represent of means and standard errors. Different lowercases indicate difference significant at 0.05 level.

4.9. Correlation of Various Biochemical Parameters

Pearson's correlation analysis was used to investigate the association between the physiological and biochemical characteristics of *C. capsularis* and *C. olitorius* under conditions of cold stress (Figure 8). Although most physiological and biochemical markers tested correlated significantly, some indices were more tightly related than others. Different physiological traits like photosynthetic pigment content (total chlorophyll) were observed to be positively correlated with different enzymatic and non-enzymatic antioxidants but negatively associated with reactive oxygen species (like H_2O_2 and MDA contents). At the same time, a strong positive significant positive correlation was observed between MDA and H_2O_2. It was observed that, in most cases, MDA and H_2O_2 made a high negative correlation with other enzymatic and non-enzymatic antioxidants. This high negative correlation suggests lipid peroxidation caused by cold stress was the main reason for decreasing non-enzymatic antioxidant activity and the occurrence of severe damage to jute varieties. In comparison, other enzymatic antioxidants such as SOD, POD, CAT, APx, and GSH exhibit a substantial positive connection with non-enzymatic antioxidants and physiological indices. Pearson correlation coefficient data supported the cluster analysis conclusion.

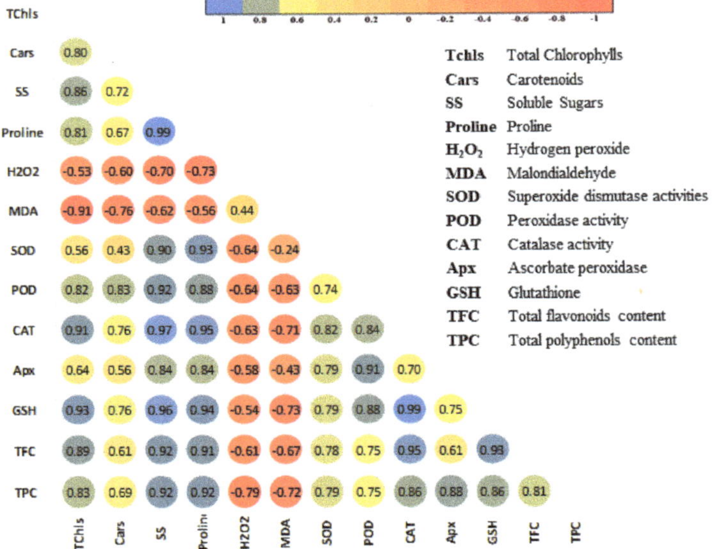

Figure 8. Pearson's correlation analysis of different physiological and biochemical traits under low-temperature stress.

5. Discussion

Like other tropical and subtropical plants, jute seedlings could be susceptible to chilling temperatures, and the injuries are results of their susceptibility. This circumstance shows dramatic reductions in the rates of many physiological characteristics when they are under chilling stress. Chlorophyll is a necessary and vital biomolecule in photosynthesis, serving as an absorber of light and a converter of light energy [30]. When plants are subjected to low-temperature stress, chlorophyll biosynthesis is impaired, resulting in a decrease in light harvesting [31]. It has been reported that plants with a high tolerance for cold keep a constant chlorophyll content, while plants with low cold tolerance experience a decrease in chlorophyll content [28]. The present study reported that except M33, all varieties significantly reduced the chlorophyll content. This may be due to a cold-induced increase in the activity of the chlorophyll degrading enzyme, chlorophyllase reported by Noreen, 2009 [32]. It was also observed that varieties Y49 showed stable or less decreased, whereas Y38 and T8 had a high level of changes compared to control. This indicates that leaf chlorophyll content was better protected in Y49 and M33 varieties, probably because of the high antioxidant enzyme activities.

Our study observed that varieties Y49 and M33 showed a less declining trend at chilling than other varieties compared to their control condition. In our research, nearly stable and small changes in chlorophyll and carotenoid contents in Y49 and M33 are consistent with *B. oleracea varacephalais* a plant with good cold tolerance reported by Atici et al. [33]. Like this, high chlorophyll stability index (CSI) under chilling stress in Y49 and M33 varieties are considered superior among the varieties studied. Due to their genetic heterogeneity and the difference in the method of defense, the examined varieties in our study behaved differently than at the same low temperature.

In response to cold stress, total chlorophyll, stable, or low decrease was noticed in Y49 (1.17%) and M33 (14.08%) while Y38 and T8 recorded the highest change as decreased. The lower fall in carotenoid concentration in response to cold stress compared to in chlorophyll *a* or *b* content may reflect the activity of xanthophyll cycle carotenoids in releasing thermal energy and protecting PSII reaction centers. It has been suggested that a decrease in chlorophyll biosynthesis in plants exposed to cold temperatures is partially due to the reduction of 5-aminolevulinic acid biosynthesis [34]. Reduced photosynthetic pigments diminish light absorption, and the greater decrease in carotenoid content than in chlorophyll *a* or *b* content in Y49 and M33 may be related to the release of heat energy and protection of PSII reaction centers by xanthophyll cycle carotenoids [35,36]. Numerous studies have established that low-temperature stress impairs photosynthesis, as evidenced by decreases in photosynthetic rate and pigment concentrations [37]

Proline is a well-established suitable osmolyte required to maintain osmotic balance and stabilize cellular structures in plants under a variety of abiotic conditions [38]. Plants protect their tissues from low-temperature damage by accumulating proline in the cells leading to a better osmotic adjustment by eliminating stress-induced excess H^+ and protecting enzymes from denaturation [39]. It was believed that more significant proline accumulation during stress conditions might account for a portion of plants' increased tolerance to cold stress conditions by mitigating the ROS-induced oxidative damage [40]. In this experiment, results showed that low-temperature stress significantly increased proline content where a high increasing trend was observed in Y49 and M33 by 67.84% and 60%, indicating that these two varieties improve the cold tolerance by scavenging ROS produced under stress condition. Whereas lower increase was observed in Y38 (11.39%) and T8 (7.89%), which reflected that these two have less radical scavenger activity. Moreover, in support of the outcomes of the present study, several other scholars reported that tolerant varieties accumulated a higher proline content than sensitive varieties during the low-temperature treatment period in seedling of sugarcane [41] and grafted watermelon [42].

It has been proved that soluble sugars play a critical role in the process of cold tolerance. Soluble sugars protect plant cells from cold stress-induced damage in a variety of ways, including as osmoprotectants, nutrition, and by reacting with the lipid bilayer [43]. In our study, varieties Y49 and M33 recorded high soluble sugar accumulation, where other varieties displayed a decreasing trend compared to the control. The highest reduction was displayed in Y1 (31.84%), followed by T8 and Y38 with 30.94% and 32.84 % respectively. Thus, more accumulated varieties attained low-temperature tolerance by increasing membrane cryostability. Increased membrane cryostability is required for freezing tolerance since membrane instability is the main source of plant injury [44]. Considerable research indicated that cold-tolerant varieties accumulated higher soluble sugar, especially sucrose, which decreased significantly in the susceptible varieties [45,46].

It has been reported that cold stress increased uncontrolled ROS production, resulting in lipid peroxidation, protein degradation, DNA degradation, and mutation. Lastly, it affected cellular metabolism and physiology, impairing the plant's membrane stability. Normally, the breakdown of unsaturated fatty acids produced MDA as the primary product in biological membranes. Likewise, greater H_2O_2 accumulation in many cell compartments, including chloroplasts, mitochondria, and apoplastic space, relates to oxidative damage in plants under cold stress [47]. Both MDA and H_2O_2 are useful indicators for detecting and monitoring oxidative stress in plants [48,49]. In the current study's treatment groups, lipid peroxidation was enhanced and triggered cell-produced MDA accumulation. It has been shown that cold stress increases lipid peroxidation and H_2O_2 concentrations, but lesser accumulation observed in tolerant varieties suggests protection against oxidative damage through a better regulating mechanism to control the synthesis of more MDA and H_2O_2 [47]. Our research showed that at low-temperature stress, MDA levels and H_2O_2 increased more pronouncedly in the Y1, Y38, and T8 varieties than other varieties compared to control, indicating that the damage caused by cold injury in these varieties was more severe than others. Among the tested varieties, less augmented MDA levels and H_2O_2 were noticed in Y49 and M33 varieties, indicating a greater ROS scavenging mechanism effectiveness and higher tolerance to cold stress. The current research findings were similar to those previously reported on sugarcane seedlings; tolerant seedlings showed the lower MDA content while susceptible seedlings demonstrated higher MDA levels [41]. It also noted that MDA content was lower in the treatment groups while proline content was more elevated, and other radical scavenging enzymes were observed [50]. The enzymatic antioxidant system may prevent the degradation of polyunsaturated fatty acids. Thus, increased proline under cold stress helps adjust osmotic levels by reducing the MDA content and improving the cell membrane.

Due to the increased electron leakage to molecular oxygen, unfavorable conditions promote the generation of reactive oxygen species (ROS) such as H_2O_2 (hydrogen peroxide), $O_2{}^-$ (superoxide), and OH_- (hydroxyl) radicals (Arora et al., 2002). An increase in ROS accumulation under abiotic stress parallels increased lipid peroxidation. To mitigate H_2O_2 induced oxidative damage and lipid peroxidation caused by the accumulation of MDA, plants increase their defense mechanisms against ROS by enhancing the ability of variety components both enzymatic and non-enzymatic to detoxify ROS [51]. In our study, a significant increase of SOD, CAT, POD, and APx were observed in the varieties of Y49 and T8 compared to other varieties and control conditions. It has been discovered that increasing enzyme activity results in a decrease in MDA and H_2O_2 concentrations below those found in control plants [48]. SOD and CAT serve as the initial line of defense for plants' antioxidative machinery. They prevent the production of more hazardous reactive oxygen species (ROS) and play a vital role in intracellular H_2O_2 signaling [52]. In the initial stage, SOD catalyzes the dismutation of $O_2{}^-$ to H_2O_2 and O_2 molecules. Hydrogen peroxide is less toxic than superoxide radicals. On the other hand, POD enzymes catalyze the conversion of H_2O_2 to H_2O and O_2. Then, H_2O_2 is detoxified by APx, POD, and CAT in different organelles and antioxidant cycles [51]. CAT is an antioxidant enzyme with a high capacity for rapidly H_2O_2 scavenging and is more engaged in H_2O_2 detoxification (removes H_2O_2 by breaking it down to create H_2O and oxygen and oxidizes H^+ donors via peroxide consumption), which is required for cold stress tolerance [53,54]. In the current investigation, varieties (Y49 and M33) with a relatively high CAT activity accumulated less H_2O_2 and MDA than varieties with a relatively low CAT activity and vice versa. In comparison, following 24 h of low-temperature treatment at 5 °C, the activity of CAT increased up to 2.3-fold in cold-treated leaves compared to control leaves. Similar results were also obtained from different studies; Zhang et al. stated that cold-tolerant banana varieties demonstrated a significant increase in CAT activity under cold stress [55] and winter-type wheat revealed much higher CAT activity than the spring type soybean plants when exposed to cold [56].

GSH can be produced in both the cytosol and the chloroplast of the plant's leaves, which is the primary component of plants' non-enzymatic antioxidant system. ROS detoxification in the chloroplast is known to be primarily carried out by the ascorbate–GSH cycle due to its high reductive potential and electron donor properties; GSH could scavenge H_2O_2, or react non-enzymatically with $^1O_2$, $O_2{}^{\bullet-}$, and $^{\bullet}OH$ and protects the various biomolecules by forming adducts (glutathiolated) or reducing them in the presence of ROS or organic free radicals [57]. However, GSH is a strong antioxidant in its own right; its key role is to renew another hydrophilic antioxidant, ascorbic acid, mainly through the Asc-GSH cycle. The APx directly reduces H_2O_2 into H_2O and O_2, utilizing ascorbic acid (AA) as a reducing agent. Many researchers reported that increased APx activity can reduce ROS levels and promote resistance to oxidative stress, whereas reduced APx activity can decrease the cold tolerance of plants [58]. In our investigation, Y49 and M33 accumulated high GSH by 78.66% and 45.65%, whereas APx activity increased by 44.35% and 71.78%, respectively compared to control. This resulted in higher efficiency of the H_2O-H_2O and ascorbate-glutathione cycles. On the other hand, APx is more likely to be responsible for fine-tuning ROS in the signaling pathway, whereas CAT may be accountable for removing excess ROS during stress [59,60]. On the other hand, much reduced APx content was observed in Y38 and T8; as a result, these varieties cannot neutralize ROS under cold stress and showed susceptibility to cold stress. Phenols (flavonoids, polyphenols) are a large class of specialized metabolites found in plant tissue that exhibit antioxidant activity due to their structure (aromatic ring with $-OH$ or $-OCH_3$ substituents) [61]. They have a high capacity for electron or hydrogen atom donation due to their quick stabilization of generated phenol radicals. Additionally, by trapping lipid alkoxy radicals, they will directly capture $^1O_2$ and reduce lipid peroxidation [62]. This procedure is helpful for avoiding chilling harm and cell collapse during periods of cold stress [63]. According to our results (Figure 4a,b), low temperature significantly increases total flavonoid in Y49 (23.97%) and M33 (4.94%), and polyphenols content increased 17.72% and 32.95% for Y49 and M33, respectively. In contrast, other varieties showed decreasing trend compared to that of the control. For flavonoids highest decrease was recorded in Y1, Y38, and T8 (15.49%, 18.16%, 9.24%, respectively), whereas polyphenols in Y38 (22.62%) were followed by Y1 (11.75%), and (12.02%). It has been reported that increased phenolic levels have been shown to contribute to ROS detoxification, enhance phenolic compound accumulation in plant cell walls, and increase cell wall thickness, showing that these compounds have a role in stress tolerance at low temperatures [61]. The findings of this study were similar to previous publications where the synthesis of phenolic compounds in plant tissue was seen under abiotic stress [64–66].

Therefore, plants were resistant and adapted to low temperatures by alleviating oxidative stress caused by low temperature and thus protecting the photosynthetic system [67,68]. Current experimental results revealed that different antioxidant and non-enzymatic antioxidant activities at low-temperature stress had been changed; those are relevant to cold tolerance. It was noted that the synergistic interactions of the SOD-POD-CAT-APx system were found to be efficient in preventing oxidative damage in jute plants exposed to cold stresses. Overall, the results indicated that cold indices, PCA, and cluster heat map generated a wide range of variability and could be used as credible approaches for screening jute varieties and identifying tolerant varieties based on physiological and biochemical performance under cold stress. From both of PCA and heat map, it was observed that all varieties were placed in three separate groups. In cluster heat map analysis, Group I (Y49 and M33 varieties) was low-temperature tolerance (MSTI 1.26) with improved physiological and biochemical traits such as total chlorophyll, soluble sugar, proline, and different enzymatic and non-enzymatic antioxidants activities. Whereas Group-II (Y38, T8, and Y1) was low-temperature susceptible (MSTI 0.93) by displayed low level activates of different antioxidants and osmolytes, thus these groups were identified as low-temperature susceptible varieties. Groups-III (M18, W57) with MSTI 0.98 were moderately tolerant or susceptible due to intermediate physiological and biochemical activities. Many researchers revealed that cluster analysis could be a promising tool to screen the desirable varieties based on the similarity [69]. The study thus explained the cold-tolerant mechanisms in jute and verified the cold tolerance level of selected varieties.

6. Conclusions

The selection of stress-tolerant genotypes might be a promising approach to alleviate the detrimental effects of abiotic stress and cultivate natural fiber crop productivity in the cooler regions. In the present study, the results show that the cold-induced inhibition of growth was significantly ameliorated in Y49 and M33 varieties, as manifested by physiological indices, such as much better adaptation with much higher chlorophyll, proline, and soluble sugar contents; lower levels of ROS, and lipid peroxidation; higher enzymatic antioxidant activities, especially SOD, POD, CAT, APx; and higher levels of non-enzymatic activities like TFC, TPC, and GSH. On the contrary, Y38, T8, and Y1 varieties exhibited more sensitivity by low temperature by increasing cell membrane damage through a high level of MDA and H_2O_2. Furthermore, for PCA or cluster heat map, seven varieties created three different groups effectively, where Cluster I (Y49 and M33) indicated the low-temperature tolerant varieties. To summarize, our data demonstrate the role of several antioxidants and non-enzymatic activities in modifying physiological and biochemical responses associated with cold tolerance. Thus, the results would provide the theoretical guidance to the evaluation of jute varieties under cold stress, cold tolerance response mechanisms, and cropping adjustment on cooler regions.

Author Contributions: Conceptualization, S.D., D.L. and A.B.; Methodology, S.D., D.L. and A.B.; Supervision, D.L., S.H., Y.D. and A.B.; Data curation, S.D. and A.B.; Formal analysis, A.B., S.D. and Z.M.B.; Validation S.H., J.Z., Y.G. and L.L.; Investigation Y.D., A.X., J.L., Original draft writing, S.D. and A.B.; Review and editing, A.B. and D.L., Funding acquisition, D.L. All authors have read and agreed to the published version of the manuscript.

Funding: This research was supported by the Chinese Agriculture Technology Research System (CARS -16-E04) and Agricultural Science and Technology Innovation Program (ASTIP-IBFC03).

Institutional Review Board Statement: Not applicable.

Informed Consent Statement: Not applicable.

Data Availability Statement: On request, the corresponding author will provide the data used to support the findings of this study.

Acknowledgments: The authors would like to express their gratitude to the Chemistry Department of the Institute of Bast Fiber Crops for allowing them to utilize their facilities for this work.

Conflicts of Interest: The authors disclose that they have no conflict interests.

References

1. Raza, A.; Su, W.; Hussain, M.A.; Mehmood, S.S.; Zhang, X.; Cheng, Y.; Zou, X.; Lv, Y. Integrated Analysis of Metabolome and Transcriptome Reveals Insights for Cold Tolerance in Rapeseed (*Brassica napus* L.). *Front. Plant Sci.* **2021**, *12*, 1796. [CrossRef]
2. Szczerba, A.; Płażek, A.; Pastuszak, J.; Kopeć, P.; Hornyák, M.; Dubert, F. Effect of low temperature on germination, growth, and seed yield of four soybean (*Glycine max* L.) cultivars. *Agronomy* **2021**, *11*, 800. [CrossRef]
3. Yang, Q.S.; Gao, J.; He, W.D.; Dou, T.X.; Ding, L.J.; Wu, J.H.; Li, C.Y.; Peng, X.X.; Zhang, S.; Yi, G.J. Comparative transcriptomics analysis reveals difference of key gene expression between banana and plantain in response to cold stress. *BMC Genom.* **2015**, *16*, 446. [CrossRef]
4. Wang, W.; Chen, Q.; Hussain, S.; Mei, J.; Dong, H.; Peng, S.; Huang, J.; Cui, K.; Nie, L. Pre-sowing Seed Treatments in Direct-seeded Early Rice: Consequences for Emergence, Seedling Growth and Associated Metabolic Events under Chilling Stress. *Sci. Rep.* **2016**, *6*, 19637. [CrossRef]
5. Kumar, R. A Review Report: Low Temperature Stress for Crop Production. *Int. J. Pure Appl. Biosci.* **2018**, *6*, 575–598. [CrossRef]
6. Saleem, M.H.; Rehman, M.; Zahid, M.; Imran, M.; Xiang, W.; Liu, L. Morphological changes and antioxidative capacity of jute (*Corchorus capsularis*, Malvaceae) under different color light-emitting diodes. *Rev. Bras. Bot.* **2019**, *42*, 581–590. [CrossRef]
7. Kumari, K.; Singh, P.K.; Kumari, S.; Singh, K.M. Dynamics of Jute Export in India. *Int. J. Curr. Microbiol. Appl. Sci.* **2020**, *9*, 3769–3774. [CrossRef]
8. Islam, M.M. Biochemistry, Medicinal and Food values of Jute (*Corchorus capsularis* L. and *C. olitorius* L.) leaf: A Review. *Int. J. Enhanc. Res. Sci. Technol. Eng.* **2013**, *2*, 35–44.
9. Dansi, A.; Adjatin, A.; Adoukonou-Sagbadja, H.; Faladé, V.; Yedomonhan, H.; Odou, D.; Dossou, B. Traditional leafy vegetables and their use in the Benin Republic. *Genet. Resour. Crop Evol.* **2008**, *55*, 1239–1256. [CrossRef]

10. Zeghichi, S.; Kallithraka, S.; Simopoulos, A.P. Nutritional composition of molokhia (*Corchorus olitorius*) and stamnagathi (*Cichorium spinosum*). *World Rev. Nutr. Diet.* **2003**, *91*, 1–21. [CrossRef]
11. Yang, Z.; Lu, R.; Dai, Z.; Yan, A.; Tang, Q.; Cheng, C.; Xu, Y.; Yang, W.; Su, J. Salt-stress response mechanisms using de novo transcriptome sequencing of salt-tolerant and sensitive *Corchorus* spp. Genotypes. *Genes* **2017**, *8*, 226. [CrossRef]
12. Islam, M.; Ali, S. Agronomic Research Advances in Jute Crops of Bangladesh. *AASCIT J. Biol.* **2018**, *3*, 34–46.
13. Hossain, M.B.; Awal, A.; Rahman, M.A.; Haque, S.; Khan, H. Distinction between Cold-sensitive and -tolerant Jute by DNA Polymorphisms. *J. Biochem. Mol. Biol.* **2003**, *36*, 427–432. [CrossRef]
14. Bajwa, V.S.; Shukla, M.R.; Sherif, S.M.; Murch, S.J.; Saxena, P.K. Role of melatonin in alleviating cold stress in Arabidopsis thaliana. *J. Pineal Res.* **2014**, *56*, 238–245. [CrossRef]
15. Apel, K.; Hirt, H. Reactive oxygen species: Metabolism, oxidative stress, and signal transduction. *Annu. Rev. Plant Biol.* **2004**, *55*, 373–399. [CrossRef]
16. Chen, Y.E.; Cui, J.M.; Li, G.X.; Yuan, M.; Zhang, Z.W.; Yuan, S.; Zhang, H.Y. Effect of salicylic acid on the antioxidant system and photosystem II in wheat seedlings. *Biol. Plant.* **2016**, *60*, 139–147. [CrossRef]
17. Rentel, M.C.; Knight, M.R.; Kingdom, U. Oxidative Stress-Induced Calcium Signaling. *Plant Physiol.* **2004**, *135*, 1471–1479. [CrossRef]
18. Erdal, S.; Genisel, M.; Turk, H.; Dumlupinar, R.; Demir, Y. Modulation of alternative oxidase to enhance tolerance against cold stress of chickpea by chemical treatments. *J. Plant Physiol.* **2015**, *175*, 95–101. [CrossRef]
19. Hoagland, D.R.; Arnon, D.I. *Water-Culture Method Growing Plants without Soil;* California Agricultural Experiment Station: Berkely, CA, USA, 1950; Volume 347, pp. 29–31.
20. Lichtenthaler, H.K.; Wellburn, A.R. Determinations of total carotenoids and chlorophylls a and b of leaf extracts in different solvents. *Biochem. Soc. Trans.* **1983**, *11*, 591–592. [CrossRef]
21. Sairam, R.K.; Deshmukh, P.S.; Shukla, D.S. Tolerance of drought and temperature stress in relation to increased antioxidant enzyme activity in wheat. *J. Agron. Crop Sci.* **1997**, *178*, 171–178. [CrossRef]
22. Bates, L.S.; Waldren, R.P.; Teare, I.D. Rapid determination of free proline for water-stress studies. *Plant Soil* **1973**, *207*, 205–207. [CrossRef]
23. Yemm, E.W.; Willis, A. The estimaion of carbohydrates in plant extracts by anthrone. *Biochem. J.* **1954**, *57*, 197. [CrossRef]
24. Bradford, M.M. A Rapid and sensitive method for quantitation of microgram quantities of protein utilizing principle of protein-dye binding. *Anal. Biochem.* **1976**, *72*, 248–254. [CrossRef]
25. Biswas, A.; Dey, S.; Li, D.; Liu, Y.; Zhang, J.; Huang, S.; Pan, G.; Deng, Y. Comparison of Phytochemical Profile, Mineral Content, and in Vitro Antioxidant Activities of Corchorus capsularis and Corchorus olitorius Leaf Extracts from Different Populations. *J. Food Qual.* **2020**, *2020*, 2931097. [CrossRef]
26. Xu, B.; Chang, S.K.C. Effect of soaking, boiling, and steaming on total phenolic contentand antioxidant activities of cool season food legumes. *Food Chem.* **2008**, *110*, 1–13. [CrossRef]
27. Ensminger, I.; Busch, F.; Huner, N.P.A. Photostasis and cold acclimation: Sensing low temperature through photosynthesis. *Physiol. Plant.* **2006**, *126*, 28–44. [CrossRef]
28. Kalaji, H.M.; Jajoo, A.; Oukarroum, A.; Brestic, M.; Zivcak, M.; Samborska, I.A.; Cetner, M.D.; Łukasik, I.; Goltsev, V.; Ladle, R.J. Chlorophyll a fluorescence as a tool to monitor physiological status of plants under abiotic stress conditions. *Acta Physiol. Plant.* **2016**, *38*, 102. [CrossRef]
29. Zainol, M.K.; Abd-Hamid, A.; Yusof, S.; Muse, R. Antioxidative activity and total phenolic compounds of leaf, root and petiole of four accessions of *Centella asiatica* (L.) Urban. *Food Chem.* **2003**, *81*, 575–581. [CrossRef]
30. Xu, S.C.; Li, Y.P.; Hu, J.; Guan, Y.J.; Ma, W.G.; Zheng, Y.Y.; Zhu, S.J. Responses of antioxidant enzymes to chilling stress in tobacco seedlings. *Agric. Sci. China* **2010**, *9*, 1594–1601. [CrossRef]
31. Wu, J.; Lightner, J.; Warwick, N.; Browse, J. Low-temperature damage and subsequent recovery of fab1 mutant arabidopsis exposed to 2 °C. *Plant Physiol.* **1997**, *113*, 347–356. [CrossRef]
32. Noreen, Z.; Ashraf, M. Changes in antioxidant enzymes and some key metabolites in some genetically diverse cultivars of radish (*Raphanus sativus* L.). *Environ. Exp. Bot.* **2009**, *67*, 395–402. [CrossRef]
33. Atici, Ö.; Demir, Y.; Kocaçalişkan, İ. Effects of low temperature on winter wheat and cabbage leaves. *Biol. Plant.* **2003**, *46*, 603–606. [CrossRef]
34. Tewari, A.K.; Tripathy, B.C. Temperature-stress-induced impairment of chlorophyll biosynthetic reactions in cucumber and wheat. *Plant Physiol.* **1998**, *117*, 851–858. [CrossRef] [PubMed]
35. Demmig-Adams, B.; Gilmore, A.M.; Iii, W.W.A. In vivo functions of carotenoids in higher plants. *FASEB J.* **1996**, *10*, 403–412. [CrossRef]
36. Demmig-Adams, B.; Adams, W.W. Photoprotection in an ecological context: The remarkable complexity of thermal energy dissipation. *New Phytol.* **2006**, *172*, 11–21. [CrossRef] [PubMed]
37. Boese, S.R.; Huner, N.P.A. Effect of growth temperature and temperature shifts on spinach leaf morphology and photosynthesis. *Plant Physiol.* **1990**, *94*, 1830–1836. [CrossRef] [PubMed]
38. Kaur, G.; Asthir, B.J.B.P. Proline: A key player in plant abiotic stress tolerance. *Biol. Plant.* **2015**, *59*, 609–619. [CrossRef]
39. Huang, M.; Guo, Z. Responses of antioxidative system to chilling stress in two rice cultivars differing in sensitivity. *Biol. Plant.* **2005**, *49*, 81–84. [CrossRef]

40. Ben Rejeb, K.; Abdelly, C.; Savouré, A. How reactive oxygen species and proline face stress together. *Plant Physiol. Biochem.* **2014**, *80*, 278–284. [CrossRef]
41. Huang, X.; Chen, M.H.; Yang, L.T.; Li, Y.R.; Wu, J.M. Effects of Exogenous Abscisic Acid on Cell Membrane and Endogenous Hormone Contents in Leaves of Sugarcane Seedlings under Cold Stress. *Sugar Tech* **2015**, *17*, 59–64. [CrossRef]
42. Lu, K.; Sun, J.; Li, Q.; Li, X.; Jin, S. Effect of cold stress on growth, physiological characteristics, and calvin-cycle-related gene expression of grafted watermelon seedlings of different gourd rootstocks. *Horticulturae* **2021**, *7*, 391. [CrossRef]
43. Yuanyuan, M.; Yali, Z.; Jiang, L.; Hongbo, S. Roles of plant soluble sugars and their responses to plant cold stress. *Afr. J. Biotechnol.* **2009**, *8*, 2004–2010. [CrossRef]
44. Uemura, M.; Steponkus, P.L. A contrast of the plasma membrane lipid composition of oat and rye leaves in relation to freezing tolerance. *Plant Physiol.* **1994**, *104*, 479–496. [CrossRef] [PubMed]
45. De Freitas, G.M.; Thomas, J.; Liyanage, R.; Lay, J.O.; Basu, S.; Ramegowda, V.; do Amaral, M.N.; Benitez, L.C.; Braga, E.J.B.; Pereira, A. Cold tolerance response mechanisms revealed through comparative analysis of gene and protein expression in multiple rice genotypes. *PLoS ONE* **2019**, *14*, e0218019. [CrossRef]
46. Morsy, M.R.; Almutairi, A.M.; Gibbons, J.; Yun, S.J.; De Los Reyes, B.G. The OsLti6 genes encoding low-molecular-weight membrane proteins are differentially expressed in rice cultivars with contrasting sensitivity to low temperature. *Gene* **2005**, *344*, 171–180. [CrossRef] [PubMed]
47. Fahimirad, S.; Karimzadeh, G.; Ghanati, F. Cold-induced Changes of Antioxidant Enzymes Activity and Lipid Peroxidation in Two Canola (*Brassica napus* L.) Cultivars. *J. Plant Physiol. Breed.* **2013**, *3*, 1–11.
48. Khalofah, A.; Migdadi, H.; El-harty, E. Antioxidant enzymatic activities and growth response of quinoa (Chenopodium quinoa willd) to exogenous selenium application. *Plants* **2021**, *10*, 719. [CrossRef]
49. Wakeel, A.; Xu, M.; Gan, Y. Chromium-induced reactive oxygen species accumulation by altering the enzymatic antioxidant system and associated cytotoxic, genotoxic, ultrastructural, and photosynthetic changes in plants. *Int. J. Mol. Sci.* **2020**, *21*, 728. [CrossRef] [PubMed]
50. Posmyk, M.M.; Kontek, R.; Janas, K.M. Antioxidant enzymes activity and phenolic compounds content in red cabbage seedlings exposed to copper stress. *Ecotoxicol. Environ. Saf.* **2009**, *72*, 596–602. [CrossRef]
51. Mittler, R. Oxidative stress, antioxidants and stress tolerance. *Trends Plant Sci.* **2002**, *7*, 405–410. [CrossRef]
52. Skyba, M.; Petijová, L.; Košuth, J.; Koleva, D.P.; Ganeva, T.G.; Kapchina-Toteva, V.M.; Čellárová, E. Oxidative stress and antioxidant response in *Hypericum perforatum* L. plants subjected to low temperature treatment. *J. Plant Physiol.* **2012**, *169*, 955–964. [CrossRef]
53. Auh, C.K.; Scandalios, J.G. Spatial and temporal responses of the maize catalases to low temperature. *Physiol. Plant.* **1997**, *101*, 149–156. [CrossRef]
54. Sevengor, S.; Yasar, F.; Kusvuran, S.; Ellialtioglu, S. The effect of salt stress on growth, chlorophyll content, lipid peroxidation and antioxidative enzymes of pumpkin seedling. *Afr. J. Agric. Res.* **2011**, *6*, 4920–4924. [CrossRef]
55. Zhang, Q.; Zhang, J.Z.; Chow, W.S.; Sun, L.L.; Chen, J.W.; Chen, Y.J.; Peng, C.L. The influence of low temperature on photosynthesis and antioxidant enzymes in sensitive banana and tolerant plantain (*Musa* sp.) cultivars. *Photosynthetica* **2011**, *49*, 201–208. [CrossRef]
56. Balestrasse, K.B.; Tomaro, M.L.; Batlle, A.; Noriega, G.O. The role of 5-aminolevulinic acid in the response to cold stress in soybean plants. *Phytochemistry* **2010**, *71*, 2038–2045. [CrossRef] [PubMed]
57. Krasnovsky, A.A., Jr. Singlet molecular oxygen in photobiochemical systems: IR phosphorescence studies. *Membr. Cell Biol.* **1998**, *12*, 665–690.
58. Tatari, M.; Fotouhi Ghazvini, R.; Mousavi, A.; Babaei, G. Comparison of some physiological aspects of drought stress resistance in two ground cover genus. *J. Plant Nutr.* **2018**, *41*, 1215–1226. [CrossRef]
59. Gill, S.S.; Tuteja, N. Reactive oxygen species and antioxidant machinery in abiotic stress tolerance in crop plants. *Plant Physiol. Biochem.* **2010**, *48*, 909–930. [CrossRef]
60. Almela, L.; Fernández-López, J.A.; Roca, M.J. High-performance liquid chromatographic screening of chlorophyll derivatives produced during fruit storage. *J. Chromatogr. A* **2000**, *870*, 483–489. [CrossRef]
61. Šamec, D.; Karalija, E.; Šola, I.; Vujčić Bok, V.; Salopek-Sondi, B. The role of polyphenols in abiotic stress response: The influence of molecular structure. *Plants* **2021**, *10*, 118. [CrossRef]
62. Sharma, P.; Jha, A.B.; Dubey, R.S.; Pessarakli, M. Reactive Oxygen Species, Oxidative Damage, and Antioxidative Defense Mechanism in Plants under Stressful Conditions. *J. Bot.* **2012**, *2012*, 217037. [CrossRef]
63. Sarikamiş, G.; Çakir, A. Influence of low temperature on aliphatic and indole glucosinolates in broccoli (*Brassica oleracea* var. *italica* L.). *Acta Hortic.* **2016**, *1145*, 79–84. [CrossRef]
64. Dixon, R.A.; Achnine, L.; Kota, P.; Liu, C.J.; Reddy, M.S.S.; Wang, L. The phenylpropanoid pathway and plant defence—A genomics perspective. *Mol. Plant Pathol.* **2002**, *3*, 371–390. [CrossRef] [PubMed]
65. Weidner, S.; Karolak, M.; Karamać, M.; Kosińska, A.; Amarowicz, R. Phenolic compounds and properties of antioxidants in grapevine roots (*Vitis vinifera* L.) under drought stress followed by recovery. *Acta Soc. Bot. Pol.* **2009**, *78*, 97–103. [CrossRef]
66. Swigonska, S.; Amarowicz, R.; Król, A.; Mostek, A.; Badowiec, A.; Weidner, S. Influence of abiotic stress during soybean germination followed by recovery on the phenolic compounds of radicles and their antioxidant capacity. *Acta Soc. Bot. Pol.* **2014**, *83*, 209–218. [CrossRef]

67. Hakam, N.; Simon, J.P. Effect of low temperatures on the activity of oxygen-scavenging enzymes in two populations of the C4 grass *Echinochloa crus-galli*. *Physiol. Plant.* **1996**, *97*, 209–216. [CrossRef]
68. Gechev, T.; Willekens, H.; Van Montagu, M.; Inzé, D.; Van Camp, W.; Toneva, V.; Minkov, I. Different responses of tobacco antioxidant enzymes to light and chilling stress. *J. Plant Physiol.* **2003**, *160*, 509–515. [CrossRef]
69. Rajametov, S.N.; Lee, K.; Jeong, H.B.; Cho, M.C.; Nam, C.W.; Yang, E.Y. Physiological traits of thirty-five tomato accessions in response to low temperature. *Agriculture* **2021**, *11*, 792. [CrossRef]

MDPI
St. Alban-Anlage 66
4052 Basel
Switzerland
Tel. +41 61 683 77 34
Fax +41 61 302 89 18
www.mdpi.com

Agronomy Editorial Office
E-mail: agronomy@mdpi.com
www.mdpi.com/journal/agronomy

www.ingramcontent.com/pod-product-compliance
Lightning Source LLC
LaVergne TN
LVHW071610170726
843515LV00008B/2367

* 9 7 8 3 0 3 6 5 8 2 5 6 6 *